CRC Press
Computer Engineering Series

Series Editor
Udo W. Pooch
Texas A&M University

Algorithms and Data Structure in C++
Alan Parker, Georgia Institute of Technology

Computer System and Network Security
Gregory B. White, United States Air Force Academy
Eric A. Fisch, Texas A&M University
Udo W. Pooch, Texas A&M University

Digital Signal Processing Algorithms: Number Theory, Convolution, Fast Fourier Transforms and Applications
Hari Krishna Garg, National University of Singapore

Discrete Event Simulation: A Practical Approach
Udo W. Pooch, Texas A&M University
James A. Wall, Simulation Consultant

Distributed Simulation
John A. Hamilton, Jr., United States Military Academy
David A. Nash, United States Military Academy
Udo W. Pooch, Texas A&M University

Handbook of Software Engineering
Udo W. Pooch, Texas A&M University

Microprocessor-Based Parallel Architecture for Reliable Digital Signal Processing Systems
Alan D. George, Florida State University
Lois Wright Hawkes, Florida State University

Spicey Circuits: Elements of Computer-Aided Circuit Analysis
Rahul Chattergy, University of Hawaii

Telecommunications and Networking
Udo W. Pooch, Texas A&M University
Denis P. Machuel, Telecommunications Consultant
John T. McCahn, Networking Consultant

Hari Krishna Garg
Department of Electrical Engineering
National University of Singapore

DIGITAL SIGNAL PROCESSING ALGORITHMS

Number Theory, Convolution, Fast Fourier Transforms, and Applications

CRC Press
Boca Raton Boston London New York Washington, D.C.

Library of Congress Cataloging-in-Publication Data

Garg, Hari Krishna.
Digital signal processing algorithms : number theory, convolution, fast fourier transforms, and applications / Hari Krishna Garg.
p. cm.
Includes bibliographical references and index.
ISBN 0-8493-7178-3
1. Signal processing--Digital techniques--Mathematics.
2. Algorithms. I. Title.
TK5102.9.G37 1998
621.382'2'015118—dc21 97-52809
CIP

Direct all inquiries to CRC Press LLC, 2000 Corporate Blvd., N.W., Boca Raton, Florida 33431.

International Standard Book Number 0-8493-7178-3
Library of Congress Card Number 97-52809
Printed in the United States of America 1 2 3 4 5 6 7 8 9 0
Printed on acid-free paper

dedicated to

sarojini agarwal, *my mother*

who taught me 1, 2, 3, and much more

bal krishna, *my father*

who taught me $\cos^2 t + \sin^2 t = 1$ and more

Preface

This is a story of numbers, digital signal processing, computational algorithms, Fourier transforms, and coding theory, and some more. The author has been fascinated and intrigued by numbers for a long time. It is as exciting and spell-binding to play with numbers as it is gazing at the stars in the heavenly skies on a clear night. With the figures of great giants such as Gauss, Fermat, Euler, Ramanujam, and Newton pointing towards the stars to explain and unravel the innermost mysteries, myths, and secrets, I feel like a kid. I have walked through the galaxies of numbers holding and sometimes pulling on their hands.

This book is about fast digital signal processing algorithms, their intimate relationships to mathematical concepts, and application of number theory to designing them. It is meant to provide a text that covers all aspects of polynomial algebra as it pertains to the design of fast algorithms. A fast algorithm can be termed as a computational procedure for a task that requires fewer computations than the obvious method. In this regard, we are focused on the ingredients and recipes for the cooking of fast algorithms for the computational tasks that arise in digital signal processing. Three of the most popular tasks are analyzed with a view to describe fast computationally efficient algorithms for them. These tasks are cyclic convolution, acyclic convolution, and discrete Fourier transform.

A common thread present in the algorithms for cyclic and acyclic convolution is the Chinese remainder theorem, a technique described in ancient writings for breaking a large size problem into a number of independent smaller size problems. We have developed new versions of the theorem for certain number systems that arise in digital signal processing. A by-product of this exercise is the new insight into algebraic code construction for these unconventional number systems.

This book has provided us with challenges, fun, and satisfaction as we have continued our journey into digital signal processing algorithms, number theory, fast Fourier transforms, coding theory, and beyond.

An attempt has been made to make this book as self-contained as possible in terms of its coverage of the topics and their selection. The readers are expected to be familiar with the basics of error control coding for latter chapters.

We hope that the readers will share the joy of learning, discovery, and the journey into the universe of numbers and benefit from the modern day applications and advancements on the Planet Earth.

hari krishna garg, BTech, MEng, PhD, MBA
Singapore

Acknowledgments

The work on this book was started in the Fall of 1995 after HKG joined the Department of Electrical Engineering, National University of Singapore, Singapore. It has been a very rewarding and enriching experience to return to Asia in general and Singapore in particular after working in the U.S. for ten years.

Many individuals contributed to the writing of this book. All the graduate students who worked on ideas that are being included are to be thanked for their perseverance. Dr. Kuo-Yu Lin worked on parts of the material that are included here during his graduate studies. Mr Li Ming is the expert hand behind the figures in the book. There have been many more who contributed to or suffered because of this project. Thanks to you all.

The gurus who taught HKG through their teachings and writings are to be thanked. This includes teachers at IIT Delhi, Concordia University, Syracuse University, and many other institutions. I owe a debt of gratitude to all.

It has been a unique privilege to have known Professors Bradley J. Strait, Norman Balabanian, and Donald D. Weiner, all of Syracuse University. They are leaders in their respective areas of research, teaching, and human endeavor. Professor Balabanian was a guide during the early years of HKG's academic life at Syracuse, a city that is a die-hard fan of its university.

The affiliation that started with HKG spending one year of sabbatical leave at National University of Singapore has grown tremendously since Fall of 1995. HKG wishes to acknowledge Professor Ah C. Liew, Vice-Dean, External Relations, Faculty of Engineering, Professor Daniel S.H. Chan, Head, Department of Electrical Engineering, Associate Professor Swee P. Yeo, Deputy Head, Department of Electrical Engineering, and Professor Mook S. Leong, Head, Division of Communications and Microwave, from NUS. They are genuine leaders within the University in Singapore and the research community that knows no national boundaries. They have been extremely supportive of HKG's research and teaching activities at NUS. During this time, HKG has been affiliated with Associate Professors Chi C. Ko and Lawrence W.C. Wong on several signal processing and communications related projects. Dr. B. Farhang is to be thanked for reading several chapters of the first draft of the book.

We express our sincere thanks to Russ Hall, Acquiring Editor, Felicia Shapiro, Associate Editor, and the staff at CRC Press for making our journey a part of theirs and sticking by when turbulence was encountered.

The love and affection that we received from our friends and families are gratefully acknowledged. Thanks Kamal, Frederick, Ritu, Anuradha, Ravi, Arun, Asha Rani, Vivek, Nalini, Neetu, Raj, Shivam, Rajat, Baboo, Roopam, Rocky, Anjali, Anuj, Shivani, Chukku-Rukku, and Lori for being there for us.

The research work that is being presented is supported by the Academic Research Grant No. RP960699, awarded by the National University of Singapore.

Contents

List of Figures

List of Tables

Paper two

List of Abbreviations and Important Symbols

ABFT	Algorithm Based Fault Tolerance
ADD	Addition
AICE-CRT	American-Indian-Chinese Extension of CRT-P
BCH	Bose-Chaudhary-Hoquenghem
CNTT	Complex Number Theoretic Transform
CRT-I	Chinese Remainder Theorem for Integers
CRT-P	Chinese Remainder Theorem for Polynomials
CRT-PR	Chinese Remainder Theorem for Polynomials in Rings
DFT	Discrete Fourier Transform
FFT	Fast Fourier Transform
FIR	Finite Impulse Response
FNT	Fermat Number Transform
gcd	greatest common divisor (also known as highest common factor)
IDFT	Inverse DFT
iff	if and only if
lcm	least common multiple
l.d.	linearly dependent
lhs	left-hand side
l.i.	linearly independent
LI	Lagrange Interpolation
MNT	Mersenne Number Transform
MULT	Multiplication
NC	Non-Commutative
NTT	Number Theoretic Transform
RADD	Real ADD
rhs	right-hand side
RMULT	Real MULT
RP	Reconstruction Polynomials (in CRT-P)
RS	Reed Solomon
SEC	Single Error Correcting
SFC	Single Fault Correcting
SFC-DFD	Single Fault Correcting-Double Fault Detecting
VLSI	Very Large Scale Integration
WSLC	Word Sequence Length Constraint
(a, b)	*gcd* of integers a and b
$d(\underline{a}, \underline{b})$	Hamming distance between the vectors a and b
$wt(\underline{a})$	Hamming weight of $\underline{a}$
I	Integers
Z	Rational Numbers
Re	Real Numbers

CI	Complex Integers
CZ	Complex Rational Numbers
C	Complex Numbers
$CF(N)$	Cyclotomic Fields
$CCF(N)$	Complex Cyclotomic Fields
$GF(p)$	Finite Field, p a Prime Integer
$GF(q)$	Finite Field $q = p^m$
$Z(M)$	Finite Integer Ring
$CZ(M)$	Finite Complex Integer Ring
$Z(q, Q(\theta))$	Finite Integer Polynomial Ring
$CZ(q, Q(\theta))$	Finite Complex Integer Polynomial Ring

from times unknown to the times to be known

as t goes from $-\infty$ to $+\infty$, mathematically speaking

Chapter 1

Introduction

Design of fast, computationally efficient algorithms has been a major focus of research activity in digital signal processing. Such algorithms are based on the mathematical structure inherent in the computational task. In many instances, these algorithms also lead to suitable architectures for hardware implementation. The emphasis in most cases is on designing the algorithms and their respective implementations in a way so as to perform the required computations in the least amount of time. A more concrete measure of the performance of the algorithms is their computational complexity. Computational complexity of an algorithm is measured in terms of the number of arithmetic operations it requires to compute the quantities of interest. This measure will be adopted throughout this book. In this regard, parallel processing has also received much attention in the research community. Interestingly, the mathematical concepts that lead to fast algorithms also lead to parallel realizations.

It is clear that number theory plays a fundamental role in the design of fast algorithms for performing some of the most computationally intensive tasks arising in digital signal processing. Examples of such tasks include cyclic and acyclic convolution, correlation, discrete Fourier transformation, and solving a system of linear equations. Many algorithms are independent of the number systems over which the sequences are defined and depend only on the indices of the sequences. These indices are always integers. Such algorithms are valid for sequences over any number system. A flurry of research activity has resulted in numerous algorithms that exploit the number-theoretic properties of the discrete-time data and the indices of such data.

1.1 Outline

This book is a collection of mathematical ideas and number-theoretic concepts that are fundamental to the understanding of existing digital signal processing algorithms and vital to developing new algorithms for the future. The computational tasks that we study are as follows:

1. Cyclic convolution for one- and higher-dimensional sequences;

2. Acyclic convolution for one- and higher-dimensional sequences; and

3. Discrete Fourier transform (DFT) for one- and higher-dimensional sequences.

The author has found the topic of DFT algorithms, collectively known as the fast Fourier Transform (FFT) particularly challenging. FFTs have been traced back to the days of Gauss. Their modern day applications are far too varied and diverse. Therefore, one finds research papers and articles on FFTs in so many places that it becomes quite a task to cover it all. We have remained focused on the mathematical aspects of all the algorithms throughout the book. All the results that we feel are essential are proven in order to make this book self-contained.

Along other lines, coding theory has flourished as an art and science to protect numeric data from errors, thereby improving the reliability and overall performance of data processing systems. However, by and large, with certain exceptions, coding theory is developed and studied as it applies to digital communication systems. This is evident from a large number of texts on coding theory as it relates to the disciplines of information theory and communication theory. Exceptions to this are the applications of coding theory concepts to fault tolerance in computing systems.

We also present techniques for fault-tolerant processing of integer sequences. This is the other facet of digital signal processing techniques and is as fundamental as the computational algorithms. We present methods that are new and depart from the classical treatment of coding theory techniques that can be found in books dealing with the subject.

An attempt has been made to make the book useful and relevant to a large audience by covering all the different number systems that arise in digital signal processing systems. This includes the following number systems:

1. Integers, $\mathbb{I}$,

2. Rational numbers, Z,
3. Real numbers, Re,
4. Complex integers, CI,
5. Complex rationals, CZ,
6. Complex numbers, C,
7. Cyclotomic fields, $CF(N)$,
8. Complex cyclotomic fields, $CCF(N)$,
9. Finite field, $GF(p)$, p a prime integer,
10. Finite Field $GF(q)$, $q = p^m$,
11. Finite integer ring, $Z(M)$,
12. Finite complex integer ring, $CZ(M)$,
13. Finite integer polynomial ring, $Z(q, Q(\theta))$,
14. Finite complex integer polynomial ring, $CZ(q, Q(\theta))$.

Some of these number systems are used when multidimensional data sequences are converted to a number of one-dimensional data sequences. In such instances, the multidimensional sequences may be defined in simple number systems such as Z and/or C, but their equivalent one-dimensional sequences are defined in extension fields that can be $CF(N)$ and $CCF(N)$. The same statements hold for processing sequences defined in finite fields and integer rings.

An attempt is made to present techniques and results for readers interested in digital signal processing algorithms over (i) traditional number systems such as the rational and complex number systems, and (ii) nontraditional number systems such as the finite fields and rings. Readers interested only in (i) and (ii) are referred to as **Group A** and **Group B readers**, respectively. The material is organized in a way that the readers interested only in the complex, real, and rational number systems can skip the sections and/or chapters on finite fields and rings without any break in the continuity. Those readers who are interested in algorithm construction over finite fields and rings are advised to cover the entire book as many, if not all, results on infinite fields and rings are also useful in their context.

1.2 The Organization

Chapters 2 to 7 constitute Part I of this book. Part I is on mathematical fundamentals crucial to the design of fast digital signal processing algorithms.

Chapter 2 is on elementary number theory for integers as it relates to the computational problems that arise in digital signal processing. This chapter is a collection of results that are already available in the literature. The focus in this chapter is on mathematical properties of integers crucial to the understanding of computational problems such as cyclic and acyclic convolution of sequences, DFTs, and number-theoretic-transforms (NTTs). The definitions and other material related to groups, rings, and fields are presented. The Chinese remainder theorem for integers (CRT-I) is established. The CRT-I constitutes the heart of many digital signal processing algorithms and is used quite extensively for simplifying computations.

Chapter 3 is on the algebra of polynomials defined on a field. The Chinese remainder theorem for polynomials (CRT-P) is established. Lagrange interpolation (LI) is presented as a special case of CRT-P for computing convolution. Polynomial algebra over finite fields is also presented. Finally, the discussion on the order of an element is completed. Section 3.7 can be skipped by *Group A readers.*

In Chapter 4, theoretical aspects of convolution and DFT computation are discussed. The interrelationship between the DFT and convolution is established and basic computational techniques are outlined. The computation of convolution is cast as a bilinear form. A matrix exchange property that is used to obtain equivalent bilinear forms having different computational features is established.

In Chapter 5, a detailed analysis of cyclotomic polynomial factorization is undertaken. Their precise form is described for the rational, complex rational number systems, and cyclotomic extension fields. Applications of these ideas to the design of digital signal processing algorithms are also described. This chapter is a key to understanding one- and multidimensional cyclic convolution algorithms.

In Chapter 6, cyclotomic polynomial factorization is presented over the finite fields and their extensions. This chapter can be skipped by *Group A readers.*

Chapter 7 is on polynomial algebra and cyclotomic factorization in finite integer rings and their polynomial extensions. New forms of CRT-P, termed as CRT-P for rings or CRT-PR, are described. CRT-PR is an outgrowth of the author's research work in the past four years. It is our understanding that it leads to a complete solution to many of the computational problems in finite rings. NTTs are also cast as a special case of the general framework presented here. Results that are important to one as well as multidimensional digital signal processing algorithms are described. This chapter can be skipped by *Group A readers.*

Chapters 8 to 12 constitute Part II of this book. Part II is on fast algorithms for computing acyclic and cyclic convolution of one- and multidimensional data sequences with the exception of Chapter 12 that deals with fault-tolerant techniques for integer sequences.

Chapter 8 is on fast algorithms for computing acyclic (also known as aperiodic) convolution of data sequences. All possible number systems are considered. Many of the techniques depend only on the indices of the data sequences that are always integers. Such techniques are valid for sequences over any number system. Section 8.6 can be skipped by *Group A readers*.

Chapter 9 is on fast algorithms for computing one-dimensional cyclic convolution of data sequences. A detailed and thorough investigation is undertaken of the various matrices that arise in the bilinear form of such algorithms. Agarwal-Cooley algorithm is presented that converts a one-dimensional cyclic convolution to multidimensional cyclic convolution. Section 9.7 can be skipped by *Group A readers*.

Chapter 10 is on multidimensional cyclic convolution algorithms. A number of approaches and algorithms are presented. The structure of these algorithms depends on the lengths of the sequences and the number system over which they are defined. Section 10.6 can be skipped by *Group A readers*.

Chapter 11 takes an alternative approach to the design of one- and multidimensional algorithms over finite integer rings. It is rather unconventional. The main idea is to show that a large number of algorithms that are valid over Z and CZ are also valid over the finite integer rings and their extensions under certain non-restrictive conditions. This chapter is interesting and should be fun reading for both *Group A* and *B readers*.

Chapter 12 is on fault-tolerant techniques for integer sequences. It departs from the earlier chapters in Part II and goes into the algebraic construction of fault tolerant techniques. The mathematical tools required to construct these techniques remain the same ones required for constructing fast digital signal processing algorithms. Only their usage is different.

Chapters 13 and 14 constitute Part III of this book. Part III is on fast Fourier transform (FFT) algorithms. This part brings all the concepts together to describe some of the most widely used algorithms in digital signal processing systems.

Chapter 13 is on FFT algorithms for one-dimensional data sequences. There are numerous approaches and refinements that have been described in the literature for FFTs. We remain focused on their mathematical aspects.

Interestingly, most FFTs compute one-dimensional DFT by first converting it to a multidimensional DFT. A computationally significant aspect of FFTs is their interrelationship with cyclic convolution algorithms. Cyclic convolution can be computed using FFTs and now we see that DFT can be computed using cyclic convolutions.

Chapter 14 is on FFT algorithms for computing multidimensional DFT. Chapters 13 and 14 interlace due to the close relationship between the structure of FFT algorithms for one- and multidimensional DFTs.

A set of four research papers constitutes Part IV of this book. These papers are a result of author's research in the past two years. They deal with one- and multidimensional convolution, cyclotomic factorization, design of error control techniques, all for sequences defined in finite rings.

PART I

COMPUTATIONAL NUMBER THEORY

Scalar and Polynomial Algebra

Thoughts on Part I

The main purpose of this part is to establish the mathematical concepts, theorems, and results that are critical to our understanding of the various algorithms. As these results are developed, care is taken to cast them in a way that exposes their close connection to digital signal processing algorithms. In this regard, the contents of this section deal with the following topics:

1. Number systems that arise in digital signal processing systems and applications; these include infinite as well as finite number systems.

2. Polynomial extension fields and rings.

3. Number theoretic results pertaining to data sequences and their indices; these include the Chinese remainder theorem for integers.

4. Polynomial algebra over infinite and finite number systems; these include the Chinese remainder theorem for polynomials defined over fields (infinite and finite) and finite integer rings.

5. Concept of the primitive roots and their existence for different number systems.

6. Theoretical aspects of various convolution and DFT algorithms.

7. Cyclotomic polynomial factorization for the number systems that one is likely to encounter in the design of digital signal processing algorithms.

It is our intention to cover all aspects of mathematical and algebraic results that form the basis of digital signal processing algorithms' construction. This is done with the objective of making the book useful to novices as well as experts. The material is organized in a way that the readers interested only in the complex, real, and rational number systems can skip the sections and/or chapters on finite fields and rings without any break in the continuity. Those readers who are interested in algorithm construction over finite fields and rings are advised to cover all portions of the book as the results on infinite fields and rings are also useful in their context.

There are plenty of mathematical ideas, notations, and equations in the chapters that follow, so the author has decided to keep his thoughts free of them in this brief exposition.

Thoughts on Part I

Chapter 2

Computational Number Theory

There is a fundamental need to define a number system over which the computations are to be performed. In the present context, we are interested in the study of number systems based on integers, rational, real, and complex numbers, and the computational algorithms that employ the corresponding arithmetic. In this chapter, we define such number systems, the mathematical properties of integers, and the affiliated algebra of interest.

2.1 Groups, Rings, and Fields

Definition 2.1 Group. Consider a set G containing M elements and a single valued arithmetic operation $\square$ defined between any two elements of G. The set G is called a **group** if it satisfies the following properties:

A. Closure: For arbitrarily chosen $a, b \in G$, if $c = a \square b$, then $c \in G$.

B. Associative law: $(a \square b) \square c = a \square (b \square c)$ for all $a, b, c \in G$.

C. Identity: There exists an element $e \in G$ such that $a \square e = a$, for all $a \in G$.

D. Inverse: For every $a \in G$, there exists $b \in G$, such that $a \square b = e$.

Thus a group is a number system with one arithmetic operation. A group G that satisfies the commutative law ($a \,\square\, b = b \,\square\, a$) is called a **commutative** or an **abelian** group. The number of elements, M, in a group G is called the **order** of G. The arithmetic operation can be either addition (+) or multiplication (·).

Definition 2.2 Ring. A set R containing M elements is called a **ring** if it satisfies the following properties:

A. R is a commutative group under +. The additive identity (called zero) and the additive inverse of $a \in \mathsf{R}$ are denoted by 0 and $-a$, respectively.

B. Closure: For arbitrarily chosen $a, b \in \mathsf{R}$, if $c = a \cdot b$, then $c \in \mathsf{R}$.

C. Associative law: $(a \cdot b) \cdot c = a \cdot (b \cdot c)$ for all $a, b, c \in \mathsf{R}$.

D. Distributive law: $a \cdot (b + c) = a \cdot b + a \cdot c$ for all $a, b, c \in \mathsf{R}$.

A ring R that satisfies the commutative law under '·' is called a *commutative* ring. The number of elements, M, in a ring R is called the *order* of R. Clearly a ring is a number system with two arithmetic operations, namely '+' and '·'.

Definition 2.3 Field. A **field** F is defined as a commutative ring that contains a multiplicative identity and a multiplicative inverse for every non-zero element in F.

A field is a complete number system over which all the arithmetic operations are defined. The multiplicative identity (called one or unity) and the multiplicative inverse of $a \in \mathsf{F}$ are denoted by 1 and a^{-1}, respectively. The number of elements, M, in a field F is called the *order* of F.

Based on the above definitions, it is straightforward to verify that the sets of integers and complex integers are a commutative ring (hereafter called simply a ring), and the sets of rational, real, complex rational, and complex numbers are fields. The set of complex integers is defined as the set containing all numbers of the type $a + j\,b$, where $a, b \in \mathtt{I}$, the set of integers, and $j = \sqrt{-1}$. A similar definition holds for the set of complex rational numbers. The order in each case is infinite. This is summarized in Table 2.1. The last three entries in Table 2.1 will be the focus of attention later in this and subsequent chapters.

In this book, the focus is on algorithms for processing data sequences defined over these number systems. The ring of integers will be of crucial importance for two reasons: (i) integer data sequences arise in a number of signal processing applications, and (ii) the indices of data sequences defined over any number system are always defined over the integer ring.

Table 2.1: Properties of Some Sets of Numbers

Sets of Numbers	Ring	Field	Order
Integers, I	√		∞
Complex Integers, CI	√		∞
Rational Numbers, Z		√	∞
Complex Rationals, CZ		√	∞
Real Numbers, Re		√	∞
Complex Numbers, C		√	∞
Finite Integers, $Z(M)$, M: composite	√		M
Finite Complex Integers, $CZ(M)$, M: composite	√		M^2
Finite Fields, $GF(q)$, $q = p^m$, p: prime		√	q

2.2 Elements of Number Theory

Given two positive integers A and M, we may divide A by M and write

$$A = Q\,M + R, \;\; 0 \le R < M. \tag{2.2.1}$$

Here the integers Q and R are called the **quotient** and the **remainder**, respectively. Without any loss in generality, we assume throughout this book that all integers are non-negative in order to simplify our description and analysis. If $R = 0$, then M divides A or M is a factor of A. This is denoted by $M \mid A$. We also write (2.2.1) as

$$R \equiv A \bmod M, \tag{2.2.2}$$

or

$$R = \langle A \rangle_M, \tag{2.2.3}$$

and read it as "A is equivalent or congruent to R mod (for modulo) M." The integers M and R are also called modulus and residue, respectively. The mod operation satisfies certain basic properties listed in the following:

A. Given three integers, A_1, A_2, and M such that $\langle A_1 \rangle_M = \langle A_2 \rangle_M$, then $M \mid (A_1 - A_2)$,

B. $\langle A_1 + A_2 \rangle_M = \langle \langle A_1 \rangle_M + \langle A_2 \rangle_M \rangle_M$, and

C. $\langle A_1 \cdot A_2 \rangle_M = \langle \langle A_1 \rangle_M \cdot \langle A_2 \rangle_M \rangle_M$.

Definition 2.4 Greatest common divisor. Given two integers A and M, the largest integer D that divides both the integers A and M is called the **greatest common divisor** (*gcd*) and is denoted by

$$D = (A, M). \tag{2.2.4}$$

The greatest common divisor may also be denoted by $gcd(A, M)$ or simply (A, M). If $D = 1$, then the integers A and M have no factors in common and are said to be **relatively prime** or **coprime**. The algorithm for determining D is called the **Euclid's algorithm**. It consists in repeated application of (2.2.1) as follows. Since $D \mid A$ and $D \mid M$, we have $D \mid R$ from (2.2.1). In other words,

$$(A, M) = (M, R), \quad 0 \le R < M. \tag{2.2.5}$$

Applying (2.2.1) to M and R, we get

$$M = Q_1 R + R_1, \quad 0 \le R_1 < R. \tag{2.2.6}$$

Once again, $(M, R) = (R, R_1)$. We repeat this process and obtain a series of decreasing positive integers $R_2, R_3, \ldots$, such that

$$\begin{aligned} R &= Q_2 R_1 + R_2, \quad 0 \le R_2 < R_1 \\ R_1 &= Q_3 R_2 + R_3, \quad 0 \le R_3 < R_2 \\ &\cdots \\ R_{l-2} &= Q_l R_{l-1} + R_l, \quad 0 \le R_l < R_{l-1} \\ R_{l-1} &= Q_{l+1} R_l. \end{aligned} \tag{2.2.7}$$

Since $R > R_1 > R_2 > \ldots \ge 0$, this process must terminate, say at the $(l + 1)$-th step, $R_{l+1} = 0$. Equations (2.2.1), (2.2.5) to (2.2.7) imply that

$$D = (A, M) = (M, R_1) = (R_1, R_2) = \ldots = (R_{l-1}, R_l) = R_l.$$

A flowchart for computing (A, M) is shown in Figure 2.1. Also (2.2.1), (2.2.6), and (2.2.7) may be written as

$$\begin{aligned}
R &= A - Q\,M \\
R_1 &= M - Q_1\,R \\
R_2 &= R - Q_2\,R_1 \\
&\ldots \\
R_l &= R_{l-2} - Q_l\,R_{l-1}.
\end{aligned} \tag{2.2.8}$$

Using the first two expressions from above, we may express R_1 as a linear combination of A and M. Similarly, using the second and the third expressions, we may express R_2 as a linear combination of A and M, and so on. Finally, R_l ($= D$) can be written as

$$D = R_l = \alpha\,A + \beta\,M. \tag{2.2.9}$$

The fact that *gcd* of two integers can be expressed as their linear combination will be used frequently in solving systems of linear congruences. If A and M are coprime, then (2.2.9) becomes

$$1 = \alpha\,A + \beta\,M. \tag{2.2.10}$$

The definition of *gcd* and the above analysis can be extended to more than two integers, say $A_1, A_2, \ldots, A_N$. Once again,

$$D = (A_1, A_2, \ldots, A_N), \tag{2.2.11}$$

and we can express D as

$$D = \sum_{i=1}^{N} \alpha_i A_i\,. \tag{2.2.12}$$

Example 2.1 Consider the process of computing the *gcd* of $A = 56$ and $M = 20$. The various steps are

$$\begin{aligned}
56 &= 2 \cdot 20 + 16 \\
20 &= 1 \cdot 16 + 4 \\
16 &= 4 \cdot 4.
\end{aligned}$$

Thus, $Q = 2$, $R = 16$, $Q_1 = 1$, $R_1 = 4$, $Q_2 = 4$, $R_2 = 0$, $D = 4$. The process terminates at $l = 2$. To express D as a linear sum of A and M, we proceed as follows:

$$16 = 56 - 2 \cdot 20$$
$$4 = 20 - 1 \cdot 16 = 20 - (56 - 2 \cdot 20) = 3 \cdot 20 - 56.$$

Consider the linear congruence in an unknown X as

$$AX \equiv C \bmod M. \tag{2.2.13}$$

Equivalently, we wish to find two integers X and Y such that

$$A\,X + M\,Y = C. \tag{2.2.14}$$

Since $D \mid A$ and $D \mid M$, it divides the sum $A\ X + M\ Y$ for all X and Y. Therefore, for a solution to (2.2.13) to exist, a necessary condition is $D \mid C$. On the other hand, if $D \mid C$ $(C = C_1 D)$, then using (2.2.9), we may write

$$C = (C_1 \alpha)\, A + (C_1 \beta)\, M,$$

and the solution to (2.2.13) is obtained as $X = C_1 \alpha$. It may be verified easily that if $C_1 \alpha$ is a solution to (2.2.13), then so is $C_1 \alpha + (M/D) \cdot \theta$, $\theta = 0, \ldots, D - 1$. The D integers $C_1 \alpha + (M/D) \cdot \theta$, $\theta = 0, \ldots, D - 1$, are incongruent mod M. In summary, the congruence in (2.2.13) has a solution if and only if (*iff*) $D \mid C$, $D = (A, M)$. Also, if $D \mid C$, then there are D distinct or incongruent solutions. Thus, if $D = 1$, that is, the integers A and M are coprime, then a solution always exists and, moreover, it is unique. Also, if $D = 1$ and $C = 1$, then X is called the inverse of A and is denoted by A^{-1}. We emphasize that A^{-1} exists whenever $(A, M) = 1$ and is unique.

Example 2.2 Consider the congruence

$$8\,X \equiv 16 \bmod 20.$$

Since $D = (8, 20) = 4$ divides C, there exists a solution. Writing

$$4 = 20 - 2 \cdot 8$$

we get

$$16 = 4 \cdot 20 - 8 \cdot 8.$$

Thus, $X = -8 \bmod 20 = 12$ is a solution to the congruence. All the $D = 4$ incongruent solutions mod 20 are given by $12 + 5 \cdot \theta$, $\theta = 0, 1, 2, 3$, or 12, 17, 2, 7.

Definition 2.5 Prime numbers. An integer is called **prime** if it has only 1 and itself as its factors. An integer is called **composite** if it is not prime.

Prime integers are denoted by p in this work. The **fundamental theorem of arithmetic** states that a composite integer M can be uniquely factored as (**standard factorization** of M),

$$M = \prod_{i=1}^{t} p_i^{a_i} = \prod_{i=1}^{t} q_i , \tag{2.2.15}$$

where p_i is the i-th prime factor, a_i is a positive integer exponent, and $q_i = p_i^{a_i}$. For a prime p, $(A, p) = 1$ for $0 < A < p$ and therefore, a solution to the congruence

$$A\,X \equiv C \bmod p$$

will always exist and be unique, $0 < A, C < p$.

2.3 Integer Rings and Fields

Consider the set of integers, $\mathcal{G} = \{0, 1, 2, ..., M - 1\}$ and let the arithmetic operations of '+' and '·' between two elements of $\mathcal{G}$ be defined mod M, that is, if $C = A + B$, then

$$C = \langle A + B \rangle_M,$$

and if $C = A \cdot B$, then

$$C = \langle A \cdot B \rangle_M.$$

Using the properties of modulo operation, it can be verified that $\mathcal{G}$ forms a ring. In order to compute the multiplicative inverse of $A \in \mathcal{G}$, $A \neq 0$, we have to compute $X \in \mathcal{G}$ that satisfies

$$A\,X \equiv 1 \bmod M.$$

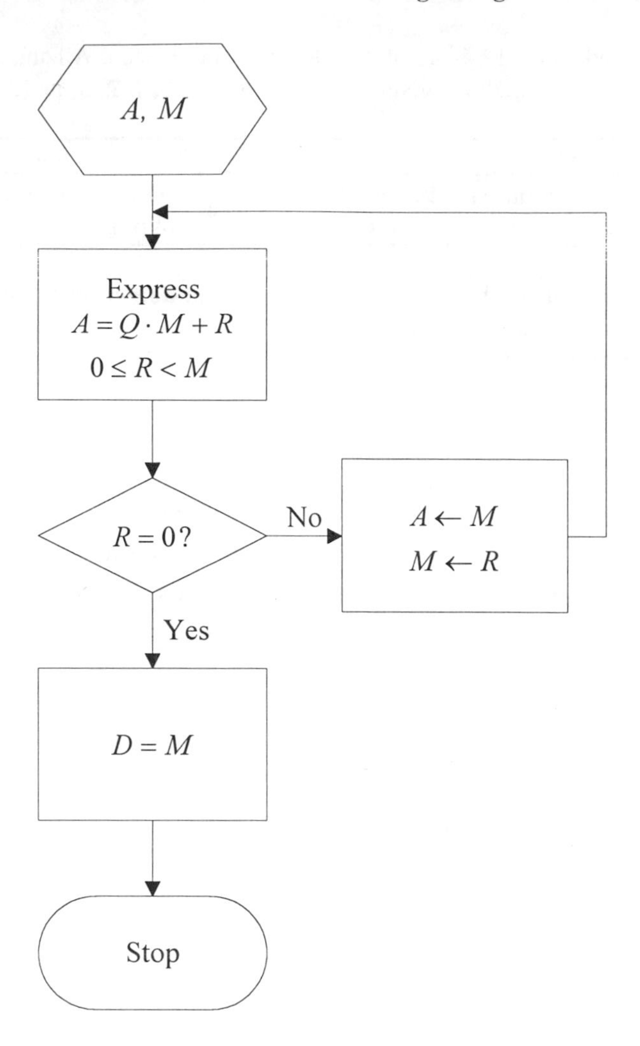

Figure 2.1 A Flowchart for Computing (*A*, *M*)

Following the analysis for the linear congruence in (2.2.13), we see that X exists and is unique *iff* $(A, M) = 1$. This condition is trivially satisfied for all $A \in G$, $A \neq 0$, if $M = p$, a prime integer. In this case, all non-zero elements G in will have a multiplicative inverse. However, for composite M, only those elements of G will have a multiplicative inverse that are coprime to M.

Based on this discussion, we conclude that the set $G = \{0, 1, ..., M - 1\}$ forms an integer ring, $Z(M)$, for composite M and a field, $GF(p)$, for $M = p$, a prime. The order of $Z(M)$ and $GF(p)$ is M and p, respectively, which are finite. Therefore, $Z(M)$ and $GF(p)$ are termed a finite integer ring and a finite integer field, respectively. This is summarized in Table 2.1. The notation $GF(q)$ stands for 'the **Galois field** of order q.' We reaffirm that the arithmetic operation of '+' and '·' are defined mod M for $Z(M)$ and mod p for $GF(p)$. In $Z(M)$, only those elements that are coprime with M have a multiplicative inverse.

2.4 Chinese Remainder Theorem for Integers

The Chinese remainder theorem in two versions (one for integers and the second for polynomials) constitutes the heart of the majority of the fast algorithms in digital signal processing. In this section, the **Chinese remainder theorem for integers** (CRT-I) is stated and proved. The second version will be presented in Chapter 3.

Consider the following extension of the linear congruence in (2.2.13) in an unknown X,

$$X \equiv x_i \bmod m_i, \quad i = 1, 2, ..., k. \tag{2.4.1}$$

The integers x_1, x_2, ..., x_k are known and the moduli m_1, m_2, ..., m_k are pairwise coprime integers, that is, $(m_i, m_j) = 1$, $i \neq j$; $i, j = 1, 2, ..., k$.

Theorem 2.1 Chinese remainder theorem for integers. There exists a unique solution $X \in Z(M)$, $M = \prod_{i=1}^{k} m_i$ to the k congruences in (2.4.1). It can be computed as

$$X \equiv \sum_{j=1}^{k} x_j T_j \left(\frac{M}{m_j} \right) \bmod M, \tag{2.4.2}$$

where the integers T_j are precomputed using the congruences

$$T_j \left(\frac{M}{m_j} \right) \equiv 1 \bmod m_j. \tag{2.4.3}$$

Proof: Since the moduli $m_1, m_2, ..., m_k$ are pairwise coprime and $M_j = (M/m_j) = \prod_{i=1, i \neq j}^{k} m_i$, we have $(M/m_j, m_j) = 1$. Therefore, the congruence in (2.4.3) can always be solved for T_j. The integers T_j are unique and have the property $(T_j, m_j) = 1, j = 1, 2, ..., k$. It is clear from the form of X in (2.4.2) that $X \in Z(M)$. Taking mod m_i on both sides of (2.4.2), we get

$$\begin{aligned} X \bmod m_i &\equiv \left(\sum_{j=1}^{k} x_j T_j \left(\frac{M}{m_j} \right) \right) \bmod m_i \\ &\equiv \left(\sum_{j=1}^{k} \left(x_j T_j \left(\frac{M}{m_j} \right) \right) \bmod m_i \right) \bmod m_i \\ &\equiv x_i T_i \left(\frac{M}{m_i} \right) \bmod m_i \\ &\equiv x_i \bmod m_i . \end{aligned}$$

The last step is based on (2.4.3). Thus, there is one solution to the k congruences in (2.4.1) as expressed in (2.4.2). Now we prove the uniqueness of the solution. Suppose there are two solutions to (2.4.1) in $Z(M)$, say X and Y. Without any loss in generality, we assume that $X > Y$. Also, since X and Y belong to $Z(M)$, they satisfy $0 \leq X, Y < M$. Substituting X and Y in (2.4.1), we get

$$X \equiv x_i \bmod m_i \tag{2.4.4a}$$

and

$$Y \equiv x_i \bmod m_i, \quad i = 1, 2, ..., k. \tag{2.4.4b}$$

Subtracting (2.4.4b) from (2.4.4a) leads to

$$X - Y \equiv 0 \bmod m_i, \quad i = 1, 2, ..., k.$$

This implies that $m_i \mid (X - Y), i = 1, 2, ..., k$. Since the moduli are pairwise relatively coprime, we have $(m_1 m_2 \cdots m_k) \mid (X - Y)$ or $M \mid (X - Y)$. This is not possible as $0 \leq X - Y < M$. Therefore, $X - Y = 0$ or $X = Y$, thereby implying that the solution is unique.

It is clear from (2.4.3) that $(T_i, m_i) = 1$ and one can employ the CRT-I to solve for an integer T, $0 < T < M$, such that

$$T \equiv T_i \bmod m_i, \quad i = 1, 2, ..., k. \tag{2.4.5}$$

Since$(T_i, m_i) = 1$, we have $(T, M) = 1$. One can alternatively express the CRT-I reconstruction in (2.4.2) as

$$X \equiv T \sum_{j=1}^{k} x_j \left(\frac{M}{m_j} \right) \bmod M. \tag{2.4.6}$$

The CRT-I forms the basis for representing an integer $X \in Z(M)$ as a residue vector $\underline{X} = [x_1\ x_2\ \ldots\ x_k]$, $x_i \in Z(m_i)$, where $x_i \equiv X \bmod m_i$. We express this relationship as

$$X \leftrightarrow \underline{X} = [x_1\ x_2\ \ldots\ x_k]. \tag{2.4.7}$$

Also, given two integers X, $Y \in Z(M)$ and their corresponding residue vectors $\underline{X} = [x_1\ x_2\ \ldots\ x_k]$, $\underline{Y} = [y_1\ y_2\ \ldots\ y_k]$, we have

$$X + Y \leftrightarrow \underline{X} + \underline{Y} = [x_1 + y_1\ \ x_2 + y_2\ \ \ldots\ \ x_k + y_k], \tag{2.4.8}$$

$$X \cdot Y = \underline{X} \cdot \underline{Y} = [x_1 \cdot y_1\ \ x_2 \cdot y_2\ \ \ldots\ \ x_k \cdot y_k]. \tag{2.4.9}$$

The above two equations must be interpreted carefully. The sum and the product $(X + Y)$ and $(X \cdot Y)$ in the left-hand side (*lhs*) are defined over $Z(M)$, that is, they are performed mod M, while the sum and the product in the i-th component in the right-hand side (*rhs*) are defined over $Z(m_i)$. This follows directly from the properties of the mod operation and the definitions of X and Y. Figure 2.2 shows the organization of CRT-I based computation in $Z(M)$.

The CRT-I is a fundamental method of partitioning a large ring $Z(M)$ into a number of smaller but independent rings $Z(m_i)$ over which the various computations may be performed in parallel. This is the direct sum property of the CRT-I and is expressed as

$$Z(M) = Z(m_1) \oplus Z(m_2) \oplus \cdots \oplus Z(m_k). \tag{2.4.10}$$

The CRT-I is used extensively to convert a one-dimensional data sequence into a two- or higher-dimensional data sequence by applying its inherent mapping properties to the index of the sequence. The data sequence itself may or may not be integer valued. Consider the following example.

EXAMPLE 2.3 For a data sequence of length $n = 15$, the index i goes from 0 to 14, that is, $i \in Z(n)$. Since $15 = 3 \cdot 5$ and $(3, 5) = 1$, let $n_1 = 3$ and $n_2 = 5$. We apply CRT-I to convert a one-dimensional data sequence into a two-dimensional data sequence having indices (i_1, i_2) such that $i_1 \equiv i \bmod n_1$ and $i_2 \equiv i \bmod n_2$. The complete mapping is given by

$0 \to (0\ 0)$ $1 \to (1\ 1)$ $2 \to (2\ 2)$ $3 \to (0\ 3)$ $4 \to (1\ 4)$
$5 \to (2\ 0)$ $6 \to (0\ 1)$ $7 \to (1\ 2)$ $8 \to (2\ 3)$ $9 \to (0\ 4)$
$10 \to (1\ 0)$ $11 \to (2\ 1)$ $12 \to (0\ 2)$ $13 \to (1\ 4)$ $14 \to (2\ 4)$.

In this case, $T_1 = 2$ and $T_2 = 2$ are obtained by solving the congruences $T_1 n_2 \equiv 1 \bmod n_1$ and $T_2 n_1 \equiv 1 \bmod n_2$, respectively. We can compute i by employing $i \equiv (T_1 n_2 i_1 + T_2 n_1 i_2) \bmod n$ or $i \equiv (10\, i_1 + 6\, i_2) \bmod 15$.

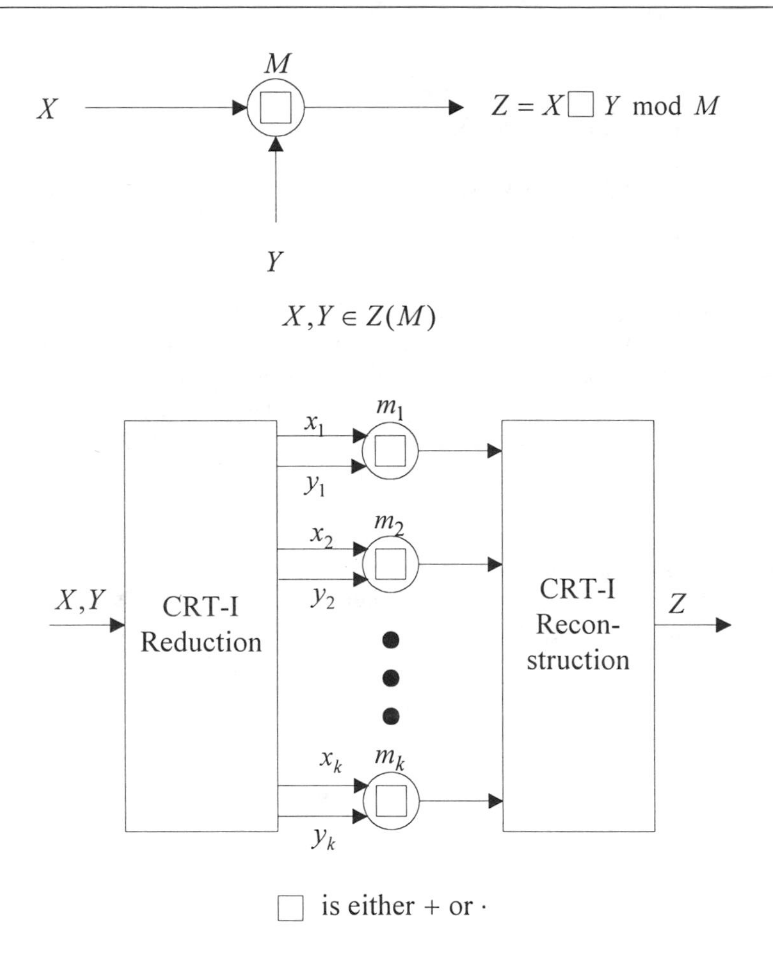

Figure 2.2 Organization of CRT-I Based Computation in *Z*(*M*)

Figure 2.3 shows the process of converting a one-dimensional sequence of length $n = n_1 \cdot n_2$, $(n_1, n_2) = 1$, into a two-dimensional sequence of length $n_1 \times n_2$, and vice versa. Figure 2.4 shows the overall configuration of the digital signal processing algorithm based on CRT-I.

In this book, we are primarily interested in the direct sum property of the CRT-I as stated in (2.4.10) and its applications to processing data sequences defined over various number systems summarized in Table 2.1.

2.5 Number Theory for Finite Integer Rings

In this section, we describe the number theoretic results for the finite integer ring $Z(M)$. These results will play a crucial role in the design of computationally efficient algorithms for digital signal processing. These results pertain to the concepts of **permutation, Euler's totient function,** and **order of an element.**

2.5.1 Permutation

Consider the ring $Z(M)$ and the set $\mathsf{S} = \{S_i, i = 0, 1, 2, ..., M-1\}$. The set S is a listing of all the elements of $Z(M)$ in a certain order. Let S_i be the ith element in S. For a given integer A such that $(A, M) = 1$, consider the set $\mathsf{P} = \{C_i, i = 0, 1, ..., M-1\}$ whose elements are defined as

$$C_i = A \cdot S_i \bmod M. \tag{2.5.1}$$

It is clear that $C_i \in Z(M)$. Since $(A, M) = 1$, we have $C_i \neq C_j$, $i \neq j$. This holds due to the fact that if $C_i = C_j$, $i \neq j$ then $(C_i - C_j) \bmod M \equiv 0$ or $A \cdot (S_i - S_j) \equiv 0 \bmod M$ or $M \mid A \cdot (S_i - S_j)$ or $M \mid (S_i - S_j)$. This is absurd as $-M < (S_i - S_j) < M$. Thus, $\mathsf{S} = \mathsf{P}$, the only difference being the different order in which the elements of $Z(M)$ are listed in them. We say that the ordering C_i is obtained by *permuting* the ordering S_i, $i = 0, 1, ..., M-1$. For example, the ordering 0, 3, 6, 1, 4, 7, 2, 5 is obtained by permuting the ordering 0, 1, 2, 3, 4, 5, 6, 7 with $A = 3$ and $M = 8$. Note that since $(A, M) = 1$, permutation is reversible. The original ordering can be obtained by applying the permutation with $A^{-1} \bmod M$ to the ordering obtained by permutation by A. In this example, $3^{-1} \bmod 8 = 3$.

2.5.2 Euler's Totient Function

Consider the set S of all those elements of $Z(M)$ that are coprime to M. The **Euler's totient function** of the integer M, denoted by $\phi(M)$, is defined as the cardinality of the set S. Let $\mathsf{S} = \{B_1, B_2, ..., B_{\phi(M)}\}$ and A be an arbitrary element

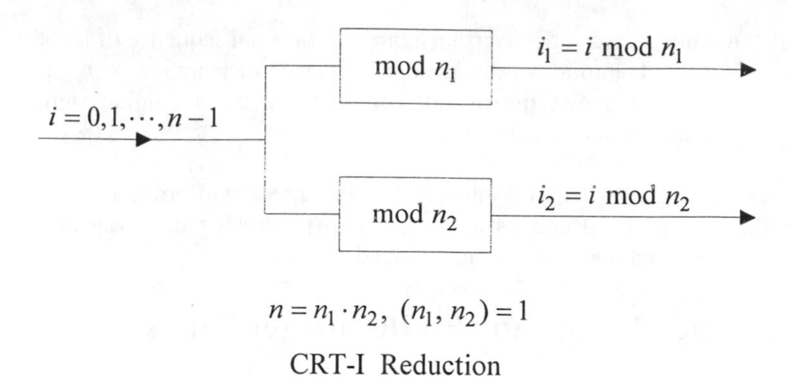

CRT-I Reduction

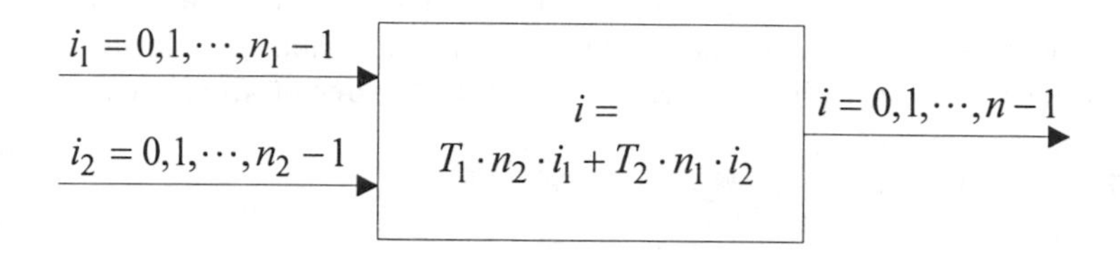

CRT-I Reconstruction

Figure 2.3 Process of Converting a One-dimensional Sequence into a Two-dimensional Sequence and Vice Versa

---

in S. Define the set P = $\{C_1, C_2, ..., C_{\phi(M)}\}$ whose elements are given as

$$C_i \equiv A \cdot B_i \bmod M,\quad i = 1, 2, ..., \phi(M). \tag{2.5.2}$$

Since $(A, M) = (B_i, M) = 1$, we have $(C_i, M) = 1$, $0 < C_i < M$. Therefore, $C_i \in$ S. Also $C_i \neq C_j$, $i \neq j$. This holds due to the fact that if $C_i = C_j$, $i \neq j$, then $(C_i - C_j) \bmod M \equiv 0$ which leads to $A \cdot (B_i - B_j) \equiv 0 \bmod M$ or $M \mid A \cdot (B_i - B_j)$ or $M \mid (B_i - B_j)$. This is absurd as $-M < (B_i - B_j) < M$. Consequently, P = S and the two sets differ only in the order in which the elements are listed in them. Once again, the ordering in P is obtained by permuting the ordering in S and vice versa. We may also write

$$\prod_{i=1}^{\phi(M)} C_i \equiv \prod_{i=1}^{\phi(M)} B_i \bmod M. \tag{2.5.3}$$

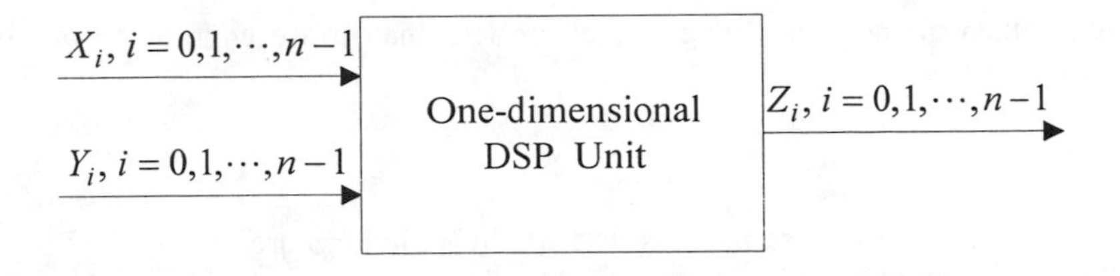

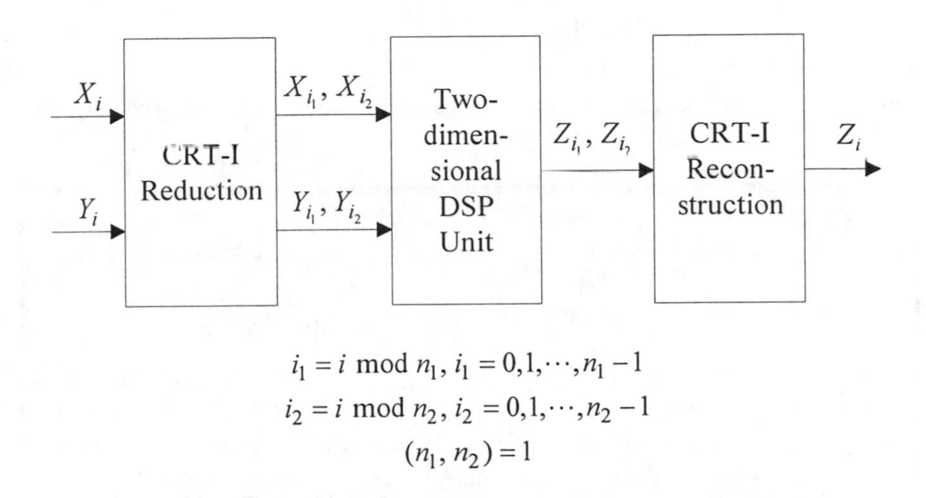

$$i_1 = i \bmod n_1,\ i_1 = 0,1,\cdots,n_1-1$$
$$i_2 = i \bmod n_2,\ i_2 = 0,1,\cdots,n_2-1$$
$$(n_1, n_2) = 1$$

Figure 2.4 Configuration of DSP Algorithms Based on CRT-I Mapping of Indices

---

Substituting for C_i from (2.5.2) in (2.5.3) and multiplying both sides by the multiplicative inverse of $\prod_{i=1}^{\phi(M)} B_i \bmod M$ (such an inverse exists and is unique as $(B_i, M) = 1$ from the definition of S), we get the statement of the well-known **Euler's theorem**.

Theorem 2.2 Euler's theorem. For given integers A and M such that $(A, M) = 1$,

$$A^{\phi(M)} \equiv 1 \bmod M. \qquad (2.5.4)$$

We now turn to the computation of $\phi(M)$ for a given value of M. We assume that the factorization of M in (2.2.15) is known. If $M = p$, a prime integer, then all non-zero integers that are less than p are coprime to p and, therefore,

$$\phi(p) = p - 1. \tag{2.5.5}$$

If $M = p^a$, then only those integers that are divisible by p are not coprime to p. All the integers that are less than p^a and have p as a factor can be expressed as $p \cdot \alpha$, $1 \le \alpha \le (p^{a-1} - 1)$. Therefore, there are $(p^a - 1) - (p^{a-1} - 1) = (p^a - p^{a-1})$ integers that are coprime to M and we get

$$\phi(p^a) = p^a - p^{a-1} = p^{a-1}(p-1). \tag{2.5.6}$$

For values of M containing more than one prime factor in its factorization, we first establish an intermediate result.

Theorem 2.3 If $(A, B) = 1$, then $\phi(A \cdot B) = \phi(A) \cdot \phi(B)$.
Proof: Using the CRT-I, an integer N less than $A \cdot B$ can be expressed as

$$N \equiv (T_1 B n_1 + T_2 A n_2) \bmod A \cdot B,\ 0 \le n_1 < A,\ 0 \le n_2 < B.$$

Since $(T_1, A) = (T_2, B) = 1$, $(N, A) = 1$ *iff* $(n_1, A) = 1$. Similarly, $(N, B) = 1$ *iff* $(n_2, B) = 1$. Therefore, $(n, A \cdot B) = 1$ *iff* $(n_1, A) = 1$ and $(n_2, B) = 1$. There are $\phi(A)$ values for n_1 and $\phi(B)$ values for n_2 that satisfy this constraint. As a result, there are $\phi(A) \cdot \phi(B)$ values of N that satisfy the constraint $(N, A \cdot B) = 1$.

Given the factorization of M in (2.2.15) and the value of Euler's totient function in (2.5.6), a direct application of Theorem 2.3 leads to

$$\phi(M) = \prod_{i=1}^{t} p_i^{a_i - 1}(p_i - 1). \tag{2.5.7}$$

For a given integer M and all its factors d_1, d_2, and so on, consider a partitioning of the set $S = \{1, 2, ..., M\}$ into subsets S_1, S_2,... such that S_i consists of all those elements of S that have *gcd* with M equal to d_i, that is $s_{ij} \in S_i$ if $(s_{ij}, M) = d_i$. As *gcd* of two integers is unique, each element of S belongs to one and only one partition. Since each element s_{ij} in S_i has *gcd* with M equal to d_i, we can alternatively say that the elements in S_i are such that $(s_{ij}/d_i, M/d_i) = 1$. Thus S_i has exactly $\phi(M/d_i)$ elements. Since the total number of elements in all the partitions is equal to the number of elements in S, we have

$$\sum_{d_i|M} \phi(M/d_i) = M \, . \tag{2.5.8}$$

Since the summation runs for all possible factors of M and, if d is a factor, then so is M/d, and we can also write

$$\sum_{d_i|M} \phi(M/d_i) = \sum_{d_i|M} \phi(d_i) = M \, . \tag{2.5.9}$$

Setting $M = p$ in Theorem 2.3, and noting that $\phi(p) = p - 1$, we get the special case of Euler's theorem known as the **Fermat's theorem**.

Theorem 2.4 Fermat's theorem. For integers A such that $0 < A < p$, p being prime,

$$A^{p-1} \equiv 1 \bmod p \, . \tag{2.5.10}$$

The Euler's totient function can be employed in a variety of ways. If $(A, M) = 1$, then the unique solution to the congruence $A\,X \equiv C \bmod M$ in (2.2.13) can simply be expressed as

$$X \equiv C \cdot A^{\phi(M)-1} \bmod M. \tag{2.5.11}$$

Thus the solution for the integers T_j in the CRT-I reconstruction in (2.4.3) can be obtained as

$$T_j \equiv \left(M/m_j\right)^{\phi(m_j)-1} . \tag{2.5.12}$$

2.5.3 Order of an Element

Consider those integers in $Z(M)$ that are coprime to M. There are $\phi(M)$ such integers. Let A be any one of them. Examine the sequence of integers obtained as $A^i \bmod M$, $i = 1, 2, \ldots$. Since there are only $\phi(M)$ possible values for such a sequence, it must repeat itself in a periodic manner after a while. Let $n + 1$ be the smallest value of i for which $A = A^{n+1} \bmod M$. In other words, n is the *smallest* value for which

$$A^n \equiv 1 \bmod M. \tag{2.5.13}$$

This value of n is called the **order** of A and is denoted by $\Theta(A)$. The sequence A^i mod M, $i = 1, 2, \ldots$ gives rise to distinct values for $i = 1, 2, \ldots, n$. A flowchart for computing the order of A in $Z(M)$ is shown in Figure 2.5.

Some basic properties of $\Theta(A)$ follow. Dividing $\phi(M)$ by n, we may write

$$\phi(M) = Q\,n + r, \quad 0 \le r < n.$$

Raising A to the power of both sides of this expression, using the Euler's theorem, and recalling that $A^n \equiv 1 \bmod M$, we get $A^r \equiv 1 \bmod M$, $0 \le r < n$. This is not possible as n, by definition, is the smallest value for which $A^n \equiv 1 \bmod M$. The only possible value for r, therefore, is $r = 0$. Thus $n \mid \phi(M)$. This result is summarized in the following theorem.

Theorem 2.5 The order of an integer in $Z(M)$ divides $\phi(M)$.

In a similar manner, we can show that if there exists an integer s such that $A^s \equiv 1 \bmod M$, then $n \mid s$.

Given an element of A of order n, consider integers of the type A^i mod M, $i = 1, 2, \ldots$. If $\Theta(A^i) = c$, then $A^{i \cdot c} \equiv 1 \bmod M$ and therefore $n \mid (i \cdot c)$. Also $(A^i)^n \equiv (A^n)^i \equiv 1 \bmod M$. Therefore, $c \mid n$. If $(i, n) = 1$, then the conditions $n \mid (i \cdot c)$ and $c \mid n$ imply that $c = n$. In general, let $(i, n) = d$. Then we may write $i = d \cdot e$, $(e, n) = 1$. The lowest value of c that satisfies the two conditions is $c = n/d$. In summary,

Theorem 2.6 If order of A is n, then order of A^i is n/d, $d = (i, n)$. If $(i, n) = 1$, then order of A is equal to order of A^i.

Consider two elements A_1 and A_2 having order n_1 and n_2 in $Z(M)$, respectively. Let these elements be such that $(n_1, n_2) = 1$. If an element is defined as $A = A_1 \cdot A_2$, then we have the following result for the order of A.

Theorem 2.7 If two elements A_1 and A_2 in $Z(M)$ have orders that are coprime, then the order of the product is the product of the orders.

Proof: Consider the element $A = A_1 \cdot A_2$. Since $A_1^{n_1} \equiv 1 \bmod M$ and $A_2^{n_2} \equiv 1 \bmod M$, we have $A_1^{n_1 n_2} \equiv 1 \bmod M$ and $A_2^{n_1 n_2} \equiv 1 \bmod M$. Therefore, $(A_1 A_2)^{n_1 n_2} \equiv 1 \bmod M$ or $A^{n_1 n_2} \equiv 1 \bmod M$. Let n be the order of the element A, that is, $A^n \equiv 1 \bmod M$. Since $A^{n_1 n_2} \equiv 1 \bmod M$, we have $n \mid (n_1 n_2)$. $A^n = 1$

also leads to $A^{nn_1} \equiv 1 \bmod M$ and $A^{nn_2} \equiv 1 \bmod M$. Replacing A by $A_1 \cdot A_2$ in these equations and realizing that $A_1^{n_1} \equiv 1 \bmod M$ and $A_2^{n_2} \equiv 1 \bmod M$, we get

$$A_2^{nn_1} \equiv 1 \bmod M$$

and

$$A_1^{nn_2} \equiv 1 \bmod M .$$

These equations imply that $n_1 \mid (n \cdot n_2)$ and $n_2 \mid (n \cdot n_1)$ or, equivalently, $n_1 \mid n$ and $n_2 \mid n$ as $(n_1, n_2) = 1$. Since $(n_1, n_2) = 1$, we have $(n_1 \cdot n_2) \mid n$. The two relations $n \mid (n_1 \cdot n_2)$ and $(n_1 \cdot n_2) \mid n$ lead to $n = n_1 \cdot n_2$.

The order of any integer $A \in Z(M)$, $(A, M) = 1$, divides $\phi(M)$. Those integers that have order exactly equal to the maximum value of $\phi(M)$ are called **primitive roots** or **elements** of $Z(M)$. If A is a primitive element in $Z(M)$, then $(A, M) = 1$. Also, A^i, $i = 1, 2, ..., \phi(M)$, gives rise to $\phi(M)$ distinct elements in $Z(M)$ that are coprime with M. Combining these two statements, we get the following theorem.

Theorem 2.8 If a ring $Z(M)$ contains a primitive element A ($\Theta(A) = \phi(M)$), then all those $\phi(M)$ elements in $Z(M)$ that are coprime with M can be expressed as A^i, $i = 1, 2, ..., \phi(M)$.

We now turn to establishing the existence of primitive roots for several important cases. We begin with $M = p$. For $M = p$, $Z(M)$ is same as the $GF(p)$. A primitive root of $GF(p)$ will have order equal to $p - 1$, if it exists. The element 1 has order equal to 1. It is trivially seen that there is no other element of order 1 in $GF(p)$. Let A be any other element of $GF(p)$ having order equal to n. Then the all the $\phi(n)$ elements of the type A^i, $(i, n) = 1$, have order equal to n. It will be shown in Chapter 3 that all elements of order n are included in the set $S = \{A^i, 0 < i < n, (i, n) = 1\}$ and there are no other elements of $GF(p)$ that have order equal to n. If true, then it would imply that either there is no element of order n or that there are exactly $\phi(n)$ elements of order n. In general, $GF(p)$ contains exactly $\phi(n)$ elements of order n for every divisor n of $p - 1$. Also, $Z(M)$ contains a primitive element for $M = p^a$ and $M = 2 \cdot p^a$ for $p > 2$. When M is a power of 2, then primitive roots exist for $M = 2$ and 4 only. We prove these and other related results in Chapter 3.

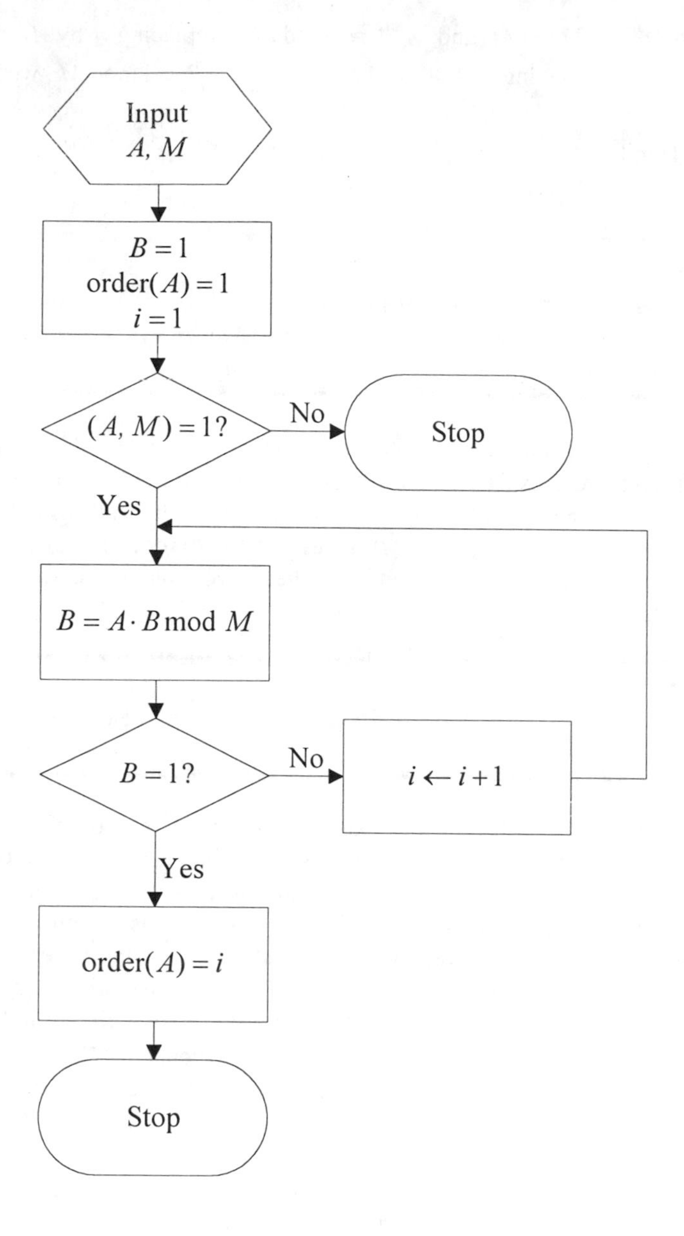

Figure 2.5 A Flowchart for Computing Order of A in $Z(M)$

Notes

We have used several excellent texts in number theory to put together this chapter. The books by McClellan and Rader [1979] and Nussbaumer [1981] present number theory as it relates to the field of digital signal processing. We have used Peterson and Weldon [1972], Lin and Costello [1983], and McWilliams and Sloane [1977] extensively to learn the algebra of finite fields and its applications to the various engineering disciplines.

It is our effort to present the various results in number theory in two distinct categories, first being the category of results that are of importance in the development and understanding of algorithms for digital signal processing of data sequences defined over the rational, real complex rational, and complex number systems. The second category consists of those results in number theory that are of importance in the development and understanding of algorithms for digital signal processing of data sequences defined over finite integer rings, fields, and their extensions. In all cases, the indices of data sequences are defined over the integer ring. We will make this distinction in the description of the algorithms as well. The objective in drawing such a distinction is that a part of the digital signal processing community may not be interested in the results pertaining to sequences defined over finite integer rings, fields, and their extensions. Similarly, a part of the digital signal processing community may not be interested in the results pertaining to sequences defined over the rational, real complex rational, and complex number systems. It is our intent to provide a complete coverage of all the results in a comprehensive manner.

Among other books on number theory, the books by Dickson [1920], MacDuffee [1940], Hua [1988], and Shapiro [1983] come to our mind as very valuable sources of information. The Chinese remainder theorem for integers can be found in any text on number theory including those mentioned here. Our presentation was particularly influenced by Blahut [1984], McClellan and Rader [1979], Nussbaumer [1981], Soderstrand et al. [1986], and Krishna [1994].

Bibliography

[2.1] J.H. McClellan and C.M. Rader, *Number Theory in Digital Signal Processing*, Prentice Hall, Inc., 1979.

[2.2] H.J. Nussbaumer, *Fast Fourier Transform and Convolution Algorithms*, Springer-Verlag, 1981.

[2.3] W.W. Peterson and E.J. Weldon, Jr., *Error Correcting Codes*, II Edition, MIT Press, 1972.

[2.4] S. Lin and D.J. Costello, Jr., *Error Control Coding: Fundamentals and Applications*, Prentice Hall, 1983.

[2.5] F.J. McWilliams and N.J.A. Sloane, *The Theory of Error Correcting Codes*, North Holland Publishing Co., 1977.

[2.6] L.E. Dickson, *History of the Theory of Numbers*, Vol. I-V, Carnegie Institution of Washington, 1920.

[2.7] C.C. MacDuffee, *An Introduction to Abstract Algebra*, John Wiley & Sons, Inc., 1940.

[2.8] L.-K., Hua, *Introduction to Number Theory*, Springer-Verlag, 1982.

[2.9] H.N. Shapiro, *Introduction to the Theory of Numbers*, John Wiley & Sons, Inc., 1983.

[2.10] R.E. Blahut, *Fast Algorithms for Digital Signal Processing*, Addison-Wesley Publishing Co., 1984.

[2.11] M.S. Soderstrand, W.K. Jenkins, G.A. Jullien, and F.J. Taylor, *Residue Number System Arithmetic: Modern Applications in Digital Signal Processing*, IEEE Press, 1986.

[2.12] H. Krishna, B. Krishna, K.-Y. Lin, and J.-D Sun, *Computational Number Theory and Digital Signal Processing*, CRC Press, 1994.

Problems

2.1 Draw a flowchart for computing α and β such that $D = \alpha A + \beta M$.

2.2 From first principles, construct the field containing two elements by identifying its elements and defining the operations of multiplication and addition. Such a field is denoted by $GF(2)$.

2.3 In a manner similar to Problem 2.2, construct the fields containing 3, 4, and 5 elements. They are denoted by $GF(3)$, $GF(4)$, and $GF(5)$, respectively.

2.4 Show that the operations of multiplication and addition in $GF(2)$, $GF(3)$, and $GF(5)$ are defined mod 2, 3, and 5 respectively, while this is not the case for $GF(4)$.

2.5 Consider the set containing elements of the type $a + j\, b$, where $a, b \in$ Re and j is defined as $j^2 = -j - 1$. Define the arithmetic operations for the elements in the set and show that it constitutes a field. Specifically, express the multiplicative inverse of the element $a + j\, b$ as the element $c + j\, d$. Is this the same as the set of complex numbers? Justify.

2.6 Using the Euclid's algorithm, compute the *gcd* of 63 and 65. Write (63, 65) as a linear sum of 63 and 65.

2.7 Compute 657^{209} mod 25.

2.8 Expressing (2.2.1) as $A = Q_0 M + R_0$, show that the repeated division for the Euclid's algorithm in (2.2.7) can be written as the continued fraction expansion,

$$\frac{A}{M} = Q_0 + \cfrac{1}{Q_1 + \cfrac{1}{Q_2 + \ddots + \cfrac{1}{Q_{l+1}}}}.$$

2.9 Given the continued fraction expansion in Problem 2.8, define a series of *convergents*

$$\frac{A_i}{M_i} = Q_0 + \cfrac{1}{Q_1 + \cfrac{1}{Q_2 + \ddots + \cfrac{1}{Q_i}}}.$$

Show that A_i and M_i satisfy the recurrence $A_i = Q_i\, A_{i-1} + A_{i-2}$ and $M_i = Q_i\, M_{i-1} + M_{i-2}$ for $i > 1$. Using mathematical induction, show that $M_i A_{i+1} - A_i M_{i+1} = (-1)^i$.

2.10 Find all possible values of X that satisfy the recurrence $65\, X \equiv 30 \bmod 105$, $X < 105$.

2.11 Consider the CRT-I for $m_1 = 7$, $m_2 = 10$, and $m_3 = 33$. What is the largest value of X that can be uniquely determined from the congruences in (2.4.1). Compute all the quantities required in the CRT-I reconstruction. Using these values, compute X having residues 4, 9, and 22 mod 7, 10, and 33, respectively.

2.12 Permute the sequence of integers {1, 5, 6, 2, 3, 7, 8, 4, 0} using the integer $A = 4$.

2.13 What integer should be used for permutation to get back the original ordering from the permuted sequence in Problem 2.12.

2.14 Verify the Euler's theorem for $A = 77$ and $M = 450$.

2.15 Compute the order of 7, 49, 11, 121, and 77 in $Z(450)$.

Chapter 3

Polynomial Algebra

In many digital signal processing applications, one is required to process data sequences defined over a given number system. This number system may either be a field or a ring of finite or infinite order. In this chapter, we present some fundamental results in the theory of polynomial algebra for polynomials defined over a field. We observe here that all the results valid for polynomials defined over infinite fields are also valid for polynomials defined over finite fields. The order of the field does not play a very critical role in the fundamental characterization of polynomial algebra. The important distinction to be drawn is between the polynomials defined over fields and rings. Many basic results in polynomial algebra over a field are not valid for polynomial algebra over a ring.

The exposition in this chapter is similar to the exposition in the Chapter 2.

DEFINITION 3.1 Given a field F or a ring R, a **polynomial** $A(u)$ is defined as

$$A(u) = A_0 + A_1 u + \ \dots + A_n u^n, \tag{3.0.1}$$

where u denotes an indeterminate quantity over F or R, and the coefficients A_0, A_1, ..., A_n belong to either F (for a polynomial defined over F) or R (for a polynomial defined over R).

The degree of $A(u)$, $deg(A(u))$, is the largest integer i for which $A_i \neq 0$. All the elements of F or R can be expressed as polynomials of degree 0 and are termed as scalars. The arithmetic operations of addition '+' and multiplication '·' between two polynomials are defined as

$$A(u)+B(u)=\sum_{l}\left(A_i+B_i\right)u^l \tag{3.0.2}$$

and

$$A(u)\cdot B(u)=\sum_{l}\left(\sum_{j}A_j\cdot B_{l-j}\right)u^l, \tag{3.0.3}$$

respectively. Here '+' and '·' in the *rhs* of (3.0.2) and (3.0.3) are defined over either F or R.

Given a data sequence of length $n + 1$ expressed as a vector of length $n + 1$, that is, $\underline{A} = [A_0 \;\; A_1 \;\cdots\; A_n]$, it can equivalently be represented as a polynomial $A(u)$ as in (3.0.1). This relationship is expressed as $\underline{A} \leftrightarrow A(u)$. The polynomial $A(u)$ is also known as the **generating function** of the sequence A_i, $i = 0, 1, \ldots, n$.

Two of the most widely occurring operations in digital signal processing are the computation of acyclic and cyclic convolution of two data sequences. Given two data sequences A_i, $i = 0, 1, \ldots, k - 1$, of length k and B_j, $j = 0, 1, \ldots, d - 1$, of length d, the **acyclic convolution** consists in the computation of the sequence C_l, $l = 0, 1, \ldots, n - 1$, $n = k + d - 1$, defined as

$$C_l = \sum_{\substack{l=i+j \\ 0\le i\le k-1 \\ 0\le j\le d-1}} A_i B_j, \quad l = 0,1,\ldots,n-1. \tag{3.0.4}$$

If $C(u)$ is the generating function of the sequence C_l, $l = 0, 1, \ldots, n - 1$, then comparing (3.0.4) with (3.0.3), we may write

$$C(u) = A(u) \cdot B(u), \tag{3.0.5}$$

where $A(u)$ and $B(u)$ are the generating functions of the sequences A_i, $i = 0, 1, \ldots, k - 1$, and B_j, $j = 0, 1, \ldots, d - 1$. The degree of the polynomials $A(u)$, $B(u)$, and $C(u)$ can be, at most, $k - 1$, $d - 1$, and $n - 1$, respectively; $n = k + d - 1$.

Given two data sequences A_i, $i = 0, 1, \ldots, n - 1$, of length n and B_j, $j = 0, 1, \ldots, n - 1$, of length n, the **cyclic convolution** consists in the computation of the sequence C_l, $l = 0, 1, \ldots, n - 1$, defined as

$$C_l = \sum_{\substack{l=(i+j)\bmod n \\ 0 \le i \le n-1 \\ 0 \le j \le n-1}} A_i B_j, \quad l = 0, 1, \ldots, n-1. \tag{3.0.6}$$

If $C(u)$ is the generating function of the sequence C_l, $l = 0, 1, \ldots, n-1$, then we may write

$$C(u) = A(u) \cdot B(u) \bmod (u^n - 1), \tag{3.0.7}$$

where $A(u)$ and $B(u)$ are the generating functions of the sequences A_i, $i = 0, 1, \ldots, n-1$, and B_j, $j = 0, 1, \ldots, n-1$. The degree of the polynomials $A(u)$, $B(u)$, and $C(u)$ can be, at most, $n-1$.

The fact that acyclic or cyclic convolution of two sequences can equivalently be expressed as a polynomial product has led to a flurry of research activity to derive computationally efficient algorithms for computing the product of two polynomials. The main idea is to exploit the mathematical properties of the number system over which the indices of the sequences (integers) and the sequences themselves are defined. In this regard, the focus of this book is on sequences defined over the infinite and finite fields and rings.

Before proceeding further with design of algorithms for computing the polynomial product, one has to understand the essential elements of the algebra of polynomials. These are established in the following.

Consider the set of all polynomials defined over a field F or a ring R. It can be easily shown that under the arithmetic operations of '+' and '·' between two polynomials as defined in (3.0.2) and (3.0.3), respectively, this set forms a ring. In the next section, we describe polynomial algebra for polynomials defined over a field F. Polynomial algebra for polynomials defined over a ring will be described in Chapter 7.

3.1 Algebra of Polynomials Over a Field

All the polynomials in this section are defined over a field F. Given two polynomials $A(u)$ and $B(u)$ of degree k and d, respectively, ($A_k \neq 0$, $B_d \neq 0$), if $C(u) = A(u) \cdot B(u)$, then $deg(C(u)) = k + d$ as $C_{k+d} = A_k \cdot B_d \neq 0$. This leads to the following theorem.

Theorem 3.1 If $C(u) = A(u) \cdot B(u)$, then $deg(C(u)) = deg(A(u)) + deg(B(u))$.

Given two polynomials $A(u)$ and $M(u)$, we may divide $A(u)$ by $M(u)$ and write

$$A(u) = Q(u)\, M(u) + R(u), \;\; 0 \le deg(R(u)) < deg(M(u)). \tag{3.1.1}$$

The polynomials $Q(u)$ and $R(u)$ are called the *quotient* and the *remainder*, respectively. They are unique as if they are non-unique; then there exist $Q_1(u)$, $Q_2(u)$ such that $Q_1(u) \neq Q_2(u)$, and $R_1(u)$, $R_2(u)$ such that $R_1(u) \neq R_2(u)$, $0 \le deg(R_1(u)), deg(R_2(u)) < deg(M(u))$ that satisfy

$$A(u) = Q_1(u)\, M(u) + R_1(u) = Q_2(u)\, M(u) + R_2(u)$$

or

$$[Q_1(u) - Q_2(u)]\, M(u) = R_2(u) - R_1(u).$$

This is not possible as the degree of the non-zero polynomial in *rhs* is less than $deg(M(u))$ while the degree of the non-zero polynomial in *lhs* is greater than $deg(M(u))$.

If $R(u) = 0$ in (3.1.1), then $M(u)$ divides $A(u)$ or $M(u)$ is a factor of $A(u)$. This is denoted by $M(u) \mid A(u)$. We also write (3.1.1) as

$$R(u) \equiv A(u) \bmod M(u) \tag{3.1.2}$$

or

$$R(u) = < A(u) >_{M(u)} \tag{3.1.3}$$

and read it as "$A(u)$ is congruent to $R(u)$ mod $M(u)$." The polynomials $M(u)$ and $R(u)$ are also called the modulus and the residue polynomials, respectively. The polynomial modulo operation satisfies certain basic properties listed in the following:

A Given three polynomials $A_1(u)$, $A_2(u)$, and $M(u)$ such that $< A_1(u) >_{M(u)} = < A_2(u) >_{M(u)}$, then $M(u) \mid (A_1(u) - A_2(u))$,

B $< A_1(u) + A_2(u) >_{M(u)} = << A_1(u) >_{M(u)} + < A_2(u) >_{M(u)} >_{M(u)}$,

C $< A_1(u) \cdot A_2(u) >_{M(u)} = < < A_1(u) >_{M(u)} \cdot < A_2(u) >_{M(u)} >_{M(u)}$.

Given two polynomials $A(u)$ and $M(u)$, the largest degree polynomial $D(u)$ that divides both $A(u)$ and $M(u)$ is called their *gcd* and is denoted by

$$D(u) = (A(u), M(u)). \tag{3.1.4}$$

If $D(u) \mid A(u)$, then $f \cdot D(u) \mid A(u)$ for all f in F, $f \neq 0$, and vice versa. Therefore, all the factor polynomials can be specified only within a scalar multiple of the field elements. Consequently, we will assume that the *gcd* polynomial is such that the coefficient of its highest degree is equal to 1. Polynomials that satisfy this property are called **monic**.

Definition 3.2 A polynomial $A(u)$ is called **monic** if the coefficient of its highest degree is equal to 1.

If $D(u) = 1$, then $A(u)$ and $M(u)$ have no factors in common and are said to be *relatively prime* or *coprime*. The algorithm for determining $D(u)$ is called the *Euclid's algorithm*. It consists in repeated application of (3.1.1) as follows. Since $D(u) \mid A(u)$ and $D(u) \mid M(u)$, we have $D(u) \mid R(u)$ from (3.1.1). In other words,

$$(A(u), M(u)) = (M(u), R(u)), \;\; 0 \leq deg(R(u)) < deg(M(u)). \tag{3.1.5}$$

Applying (3.1.1) to $M(u)$ and $R(u)$, we get

$$M(u) = Q_1(u)\, R(u) + R_1(u), \;\; 0 \leq deg(R_1(u)) < deg(R(u)). \tag{3.1.6}$$

Once again, $(M(u), R(u)) = (R(u), R_1(u))$. We repeat this process and obtain a series of polynomials $R_2(u)$, $R_3(u)$ and so on with decreasing degrees such that

$$\begin{aligned}
&R(u) = Q_2(u)\, R_1(u) + R_2(u), \;\; 0 \leq deg(R_2(u)) < deg(R_1(u)),\\
&R_1(u) = Q_3(u)\, R_2(u) + R_3(u), \;\; 0 \leq deg(R_3(u)) < deg(R_2(u)),\\
&\ldots.\\
&R_{l-2}(u) = Q_l(u)\, R_{l-1}(u) + R_l(u), \;\; 0 \leq deg(R_l(u)) < deg(R_{l-1}(u)),\\
&R_{l-1}(u) = Q_{l+1}(u)\, R_l(u).
\end{aligned} \tag{3.1.7}$$

Since $deg(R(u)) > deg(R(u)) > \ldots \geq 0$, this process must terminate, say at the $(l + 1)$-th step, $R_{l+1}(u) = 0$. Equations (3.1.1), (3.1.5) to (3.1.7) imply that

$$\begin{aligned}
D(u) &= (A(u), M(u)) = (M(u), R_1(u)) = (R_1(u), R_2(u))\\
&= \ldots = (R_{l-1}(u), R_l(u)) = R_l(u).
\end{aligned}$$

Also, (3.1.1), (3.1.6), and (3.1.7) may be written as

$$\begin{aligned}
&R(u) = A(u) - Q(u)\, M(u)\\
&R_1(u) = M(u) - Q_1(u)\, R(u)\\
&R_2(u) = R(u) - Q_2(u)\, R_1(u)\\
&\ldots\\
&R_l(u) = R_{l-2}(u) - Q_l(u)\, R_{l-1}(u).
\end{aligned} \tag{3.1.8}$$

Using the first two expressions from above, we may express $R_1(u)$ as a linear combination of $A(u)$ and $M(u)$. Similarly, using the second and the third expression, we may express $R_2(u)$ as a linear combination of $A(u)$ and $M(u)$ and so on. Finally, $R_l(u)$ (= $D(u)$) can be written as

$$D(u) = R_l(u) = \alpha(u)\, A(u) + \beta(u)\, M(u). \tag{3.1.9}$$

The fact that *gcd* of two polynomials can be expressed as their linear combination will be used frequently in solving polynomial congruences. If $A(u)$ and $M(u)$ are coprime, then (3.1.9) becomes

$$1 = \alpha(u)\, A(u) + \beta(u)\, M(u). \tag{3.1.10}$$

Consider the polynomial congruence in an unknown polynomial $X(u)$ as

$$A(u)\, X(u) \equiv C(u) \bmod M(u). \tag{3.1.11}$$

Equivalently, we wish to find two polynomials $X(u)$ and $Y(u)$ such that

$$A(u)\, X(u) + M(u)\, Y(u) = C(u). \tag{3.1.12}$$

Since $D(u) \mid A(u)$ and $D(u) \mid M(u)$, it divides the sum $A(u)\, X(u) + M(u)\, Y(u)$ for all $X(u)$ and $Y(u)$. Therefore, for a solution to (3.1.11) to exist, a necessary condition is $D(u) \mid C(u)$. On the other hand, if $D(u) \mid C(u)$ ($C(u) = C_1(u)\, D(u)$), then using (3.1.9), we may write

$$C(u) = [C_1(u)\, \alpha(u)]\, A(u) + [C_1(u)\, \beta(u)]\, M(u),$$

and the solution to (3.1.11) is obtained as $X(u) = C_1(u)\, \alpha(u)$. It may be verified that if $C_1(u)\, \alpha(u)$ is a solution to (3.1.11), then so is

$$C_1(u)\, \alpha(u) + [M(u)/D(u)] \cdot \theta(u),\ 0 \le deg(\theta(u)) < deg(D(u)).$$

The polynomials $C_1(u)\, \alpha(u) + [M(u)/D(u)] \cdot \theta(u)$, $0 \le deg(\theta(u)) < deg(D(u))$ are incongruent mod $M(u)$.

In summary, the congruence in (3.1.11) has a solution *iff* $D(u) \mid C(u)$, $D(u) = (A(u), M(u))$. Thus, if $D(u) = 1$, that is, the polynomials $A(u)$ and $M(u)$ are coprime, then a solution to (3.1.11) always exists and is unique. Also, if $D(u) = 1$ and $C(u) = 1$, then $X(u)$ is called the inverse of $A(u)$ and is denoted by $A^{-1}(u)$. We emphasize that $A^{-1}(u)$ exists whenever $(A(u), M(u)) = 1$ and is unique.

Definition 3.3 A polynomial $M(u)$ is called **irreducible** if it has only a scalar and itself as its factors. Once again, all irreducible polynomials will be assumed to be monic. A polynomial that is not irreducible is termed as **reducible**.

Given two polynomials $M_1(u)$ and $M_2(u)$ such that $M_1(u) \mid A(u)$, $M_2(u) \mid A(u)$, and $(M_1(u), M_2(u)) = 1$, then $[M_1(u) \cdot M_2(u)] \mid A(u)$. This can be proved using (3.1.1) and the discussion on the congruence in (3.1.11). Finally, every polynomial $A(u)$ of degree n over a field F can be uniquely factored as

$$A(u) = \sum_{i=0}^{n} A_i u^i = A_n \prod_{i=1}^{l} p_i^{a_i}(u) = A_n \prod_{i=1}^{l} q_i(u), \tag{3.1.13}$$

where $p_i(u)$ is the i-th irreducible polynomial and a_i is a positive exponent. This is known as the **unique factorization property of polynomials**. We emphasize that this factorization depends on the field F. A polynomial that is irreducible over one field may factor further over another. Two properties immediately follow from the factorization in (3.1.13). They are

$$deg(A(u)) = n = \sum_{i=1}^{l} \deg(q_i(u)) = \sum_{i=1}^{l} \alpha_i deg(p_i(u))$$

and

$$(q_i(u),\ q_j(u)) = 1,\ \ i, j = 1, 2, ..., l;\ i \neq j.$$

Example 3.1 This example is presented to reinforce the concept that the factorization of a given polynomial in terms of its irreducible factors depends on the field F. Consider the factorization of $u^{12} - 1$ over the fields C, Re, Z, and CZ. These are given by

$$u^{12} - 1 = \prod_{i=0}^{11} \left(u - e^{2\pi i j/12}\right) \quad \text{I}$$

$$u^{12} - 1 = (u-1)(u+1)(u^2+1)(u^2-u+1) \cdot (u^2+u+1)(u^2-\sqrt{3}u+1)(u^2+\sqrt{3}u+1) \quad \text{II}$$

$$u^{12} - 1 = (u-1)(u+1)(u^2+1)(u^2-u+1)(u^2+u+1)(u^4-u^2+1) \quad \text{III}$$

$$u^{12} - 1 = (u-1)(u+1)(u-j)(u+j)(u^2-u+1)$$

$$\cdot\left(u^2 + u + 1\right)\left(u^2 - ju - 1\right)\left(u^2 + ju - 1\right) \quad \text{IV}$$

It is easy to verify these factorizations. They are over the fields C, Re, Z, and CZ, respectively. Due to its fundamental importance in digital signal processing, the factorization of $u^n - 1$ over these and many other fields of interest will be the focus of our attention in Chapters 5, 6, and 7.

For an irreducible polynomial $P(u)$, $(A(u), P(u)) = 1$ for $0 \le deg(A(u)) < deg(P(u))$ and, therefore, a solution to the congruence

$$A(u)\, X(u) \equiv C(u) \bmod P(u)$$

will always exist and be unique, $0 \le deg(A(u)), deg(C(u)) < deg(P(u))$.

Definition 3.4 Formal Derivative. Given a polynomial $A(u) = \sum_{i=0}^{n} A_i u^i$ over a field F or a ring R, the **formal derivative** of $A(u)$ is defined as the polynomial $B(u) = \sum_{i=1}^{n} i \cdot A_i u^{i-1}$ and is denoted by $B(u) = A'(u)$.

This definition is consistent with the definition of the derivative in the usual sense and satisfies the various properties associated with derivatives, such as if $A(u) = X(u) \cdot Y(u)$, then $A'(u) = X'(u)\, Y(u) + X(u)\, Y'(u)$.

3.2 Roots of a Polynomial

Given $A(u)$ over a field F and a scalar $\alpha \in$ F, consider dividing $A(u)$ by $(u - \alpha)$. Using (3.1.1), we get

$$A(u) = Q(u)\,(u - \alpha) + R(u), \tag{3.2.1}$$

where $0 = deg(R(u)) < deg(u - \alpha) = 1$. Therefore, $R(u) = R$, $R \in$ F. Setting $u = \alpha$ in both sides of (3.2.1), we get

$$A(\alpha) = R. \tag{3.2.2}$$

The scalar α is said to be a **root** of the polynomial $A(u)$ if $A(\alpha) = 0$ or equivalently $R = 0$. From (3.2.1), we see that if α is a root of a polynomial $A(u)$, then $(u - \alpha) \mid A(u)$.

If α_1 is a root of $A(u)$, the (3.2.1) becomes

$$A(u) = Q(u)\,(u - \alpha_1). \tag{3.2.3}$$

Let α_2 be a second root of $A(u)$, $\alpha_2 \neq \alpha_1$. Dividing $Q(u)$ by $(u - \alpha_2)$, we may write

$$Q(u) = Q_1(u)\,(u - \alpha_2) + R_2. \tag{3.2.4}$$

Substituting (3.2.4) in (3.2.3), $A(u)$ can be expressed as

$$A(u) = Q_1(u)\,(u - \alpha_2)\,(u - \alpha_1) + R_2\,(u - \alpha_1). \tag{3.2.5}$$

Setting $u = \alpha_2$ gives

$$R_2\,(\alpha_2 - \alpha_1) = 0, \tag{3.2.6}$$

where $\alpha_2 - \alpha_1 \neq 0$ as $\alpha_2 \neq \alpha_1$. Over a field, the multiplicative inverse of $(\alpha_2 - \alpha_1)$ exists and (3.2.6) implies that $R_2 = 0$. From (3.2.5), we have

$$[(u - \alpha_1)\,(u - \alpha_2)] \mid A(u).$$

This leads to the following theorem.

Theorem 3.2 A polynomial $A(u)$ of degree n can have at most n roots over a field.

3.3 Polynomial Fields and Rings

Consider the set G of all polynomials of degrees up to $m - 1$, that is,

$$G = \left\{ A(u) = \sum_{i=0}^{m-1} A_i u^i,\ A_i \in F \right\}.$$

Let the arithmetic operations of '+' and '·' between two elements of G be defined mod $M(u)$, $deg(M(u)) = m$. If $C(u) = A(u) + B(u)$, then

$$C(u) = \langle A(u) + B(u) \rangle_{M(u)}$$

and if $C(u) = A(u) \cdot B(u)$, then

$$C(u) = \langle A(u) \cdot B(u) \rangle_{M(u)},\quad A(u), B(u) \in G.$$

Using the properties of modulo operation, it can be verified that $\mathcal{G}$ forms a ring.

In order to compute the multiplicative inverse of $A(u) \in \mathcal{G}$, $A(u) \neq 0$, we have to compute a polynomial $X(u) \in \mathcal{G}$ that satisfies

$$A(u)\, X(u) \equiv 1 \bmod M(u).$$

Based on the analysis of the congruence in (3.1.11), we see that $X(u)$ exists and is unique *iff* $(A(u), M(u)) = 1$. This condition is trivially satisfied for all $A(u) \in \mathcal{G}$ if $M(u) = P(u)$, an irreducible polynomial. However, for a reducible polynomial $M(u)$, only those elements of $\mathcal{G}$ will have a multiplicative inverse that are coprime to $M(u)$. Consequently, the set $\mathcal{G}$ constitutes a field if $M(u)$ is irreducible over F. Such a field is termed as an **extension field** of F. Similarly, the set $\mathcal{G}$ constitutes a ring if $M(u)$ is reducible. Once again, it is of importance to note that the irreducibilty of a given polynomial depends on the field F. For example, $u^2 - 5$ is irreducible over Z, the field of rational numbers, while it factors as $u^2 - 5 = \left(u - \sqrt{5}\right)\left(u + \sqrt{5}\right)$ over Re, the field of real numbers.

Let the field F be the set of real numbers Re. The set $\mathcal{G}$ defined as

$$\mathcal{G} = \{A_0 + j\,A_1;\, A_0, A_1 \in \mathsf{Re}\},$$

where the arithmetic operations of '+' and '·' between two elements of $\mathcal{G}$ are defined mod $j^2 + 1$ is the well-known field of complex numbers $\mathcal{C}$. Note that the indeterminate is changed from u to j in the representation of complex numbers. This change of indeterminate is useful in distinguishing polynomial representation of the elements of extension field from the remainder of our discussion on polynomial algebra. It is easily verified that $j^2 + 1$ is irreducible over Re. Also if field F is the set of rational numbers Z, then set $\mathcal{G}$ defined as $\mathcal{G} = \{A_0 + j\,A_1;\, A_0, A_1 \in \mathsf{Z}\}$, where the arithmetic operations of '+' and '·' between two elements of $\mathcal{G}$ are defined mod $j^2 + 1$ is the field of complex rational numbers $\mathcal{CZ}$.

We end this section by reiterating that an extension field is one whose elements are polynomials having coefficients over a field F. The arithmetic operations in the extension field are defined modulo a polynomial that is irreducible in the field F. The field F is also termed as the **ground field**. In many cases, the irreducible polynomial used for field extension may be pre-specified. Such is the case when $j^2 + 1$ is used to extend Re and Z to get the fields C and CZ, respectively. In other cases, one may choose it to simplify the various computations involved. In the remainder of the book, the term 'field' will be used without drawing any distinction between fields and their polynomial extensions, unless it is necessary. We shall return to these results one more time in Chapter 6 and study them in great depth in the context of finite fields and their extensions.

Example 3.2 This example is a continuation of Example 3.1. Consider the extension field obtained from the ground field Re having polynomials up to degree 1 as its elements with the arithmetic operations defined mod $j^2 + j + 1$. This is not the same as C. It is easily verified that $j^2 + j + 1$ is irreducible over Re. In such a field, we have $(5 + 2j) \cdot (3 + 4j) = 7 + 18j$. In this field, $u^{12} - 1$ factors as follows:

$$u^{12} - 1 = (u-1)(u+1)(u-j)(u+j)(u+1+j)(u-1-j) \cdot (u^2+1)(u^2+j)(u^2-1-j).$$

A systematic procedure to obtain these factorizations will be described in Chapter 5.

3.4 Chinese Remainder Theorem for Polynomials

As will be demonstrated in Chapters 4, 8, 9, and 10, the **Chinese remainder theorem for polynomials** (CRT-P) is extensively used in digital signal processing for performing computations involving polynomial products. All the polynomials in this section are defined over a field F.

Consider the following extension of the linear congruence in (3.1.11) in an unknown polynomial $X(u)$,

$$X(u) \equiv x_i(u) \bmod m_i(u), \quad i = 1, 2, \ldots, k. \tag{3.4.1}$$

The polynomials $x_i(u)$, $i = 1, 2, \ldots, k$ are known and the moduli $m_i(u)$, $i = 1, 2, \ldots, k$ are pairwise coprime polynomials, that is, $(m_i(u), m_j(u)) = 1$, $i \neq j$; $i, j = 1, 2, \ldots, k$.

Theorem 3.3 Chinese Remainder Theorem for Polynomials. There exists a unique solution $X(u)$, $deg(X(u)) < deg(M(u))$, $M(u) = \prod_{i=1}^{k} m_i(u)$ to the k congruences in (3.4.1) and can be computed as

$$X(u) \equiv \sum_{j=1}^{k} x_j(u) T_j(u) \left[\frac{M(u)}{m_j(u)} \right] \bmod M(u), \tag{3.4.2}$$

where the polynomials $T_j(u)$ are precomputed using the congruences:

$$T_j(u) \left[\frac{M(u)}{m_j(u)} \right] \equiv 1 \bmod m_j(u). \tag{3.4.3}$$

Proof: Since the moduli $m_i(u)$, $i = 1, 2, ..., k$ are pairwise coprime and $\frac{M(u)}{m_j(u)} = \prod_{\substack{i=1 \\ i \neq j}}^{k} m_i(u)$, we have $\left(M(u)/m_j(u), m_j(u) \right) = 1$. Therefore, the congruence in (3.4.3) can always be solved for $T_j(u)$. The polynomials $T_j(u)$, $deg(T_j(u)) < deg(m_j(u))$ are unique and have the property that

$$(T_j(u), m_j(u)) = 1, \; j = 1, 2, ..., k.$$

It is clear from the form of $X(u)$ in (3.4.2) that $deg(X(u)) < deg(M(u))$. Taking mod $m_i(u)$ on both sides of (3.4.2), we get

$$\begin{aligned} X(u) \bmod m_i(u) &\equiv \left\{ \sum_{j=1}^{k} x_j(u) T_j(u) \left[\frac{M(u)}{m_j(u)} \right] \right\} \bmod m_i(u) \\ &\equiv x_i(u) T_i(u) \left[\frac{M(u)}{m_i(u)} \right] \bmod m_i(u) \\ &\equiv x_i(u) \bmod m_i(u). \end{aligned}$$

The last step is based on (3.4.3). Thus there is one solution to the k congruences in (3.4.1) as expressed in (3.4.2). Now, we prove the uniqueness of the solution. Suppose there are two solutions to (3.4.1), say $X(u)$ and $Y(u)$, $0 \leq deg(X(u))$, $deg(Y(u)) < deg(M(u))$. Substituting $X(u)$ and $Y(u)$ in (3.4.1), we get

$$X(u) \equiv x_i(u) \bmod m_i(u), \tag{3.4.4}$$
$$Y(u) \equiv x_i(u) \bmod m_i(u), \; i = 1, 2, ..., k. \tag{3.4.5}$$

Subtracting (3.4.5) from (3.4.4) leads to

$$X(u) - Y(u) \equiv 0 \bmod m_i(u), \; i = 1, 2, ..., k.$$

This implies that

$$m_i(u) \mid (X(u) - Y(u)),\ i = 1, 2, ..., k.$$

Since the moduli are pairwise coprime, we have $\prod_{i=1}^{k} m_i(u) \mid (X(u) - Y(u))$ or $M(u) \mid (X(u) - Y(u))$. This is not possible as $0 \le deg(X(u) - Y(u)) < deg(M(u))$. Therefore, $X(u) - Y(u) = 0$ or $X(u) = Y(u)$. This implies that the solution is unique.

The process of obtaining $x_i(u)$ from a given polynomial $X(u)$ is termed as the **CRT-P reduction** and the process of obtaining $X(u)$ from its given residues is termed as the **CRT-P reconstruction**. The CRT-P forms the basis of representing a polynomial $X(u)$, $deg(X(u)) < deg(M(u))$, as a vector

$$\underline{X}(u) = [x_1(u) \cdots x_k(u)],\ deg(x_i(u)) < deg(m_i(u)),$$

where $x_i(u) \equiv X(u) \bmod m_i(u)$. We express this relationship as

$$X(u) \leftrightarrow \underline{X}(u) = [x_1(u)\ \ x_2(u)\ \ ...\ \ x_k(u)]. \tag{3.4.6}$$

Also, given two polynomials $X(u)$ and $Y(u)$, $0 \le deg(X(u))$, $deg(Y(u)) < deg(M(u))$, and their corresponding residue vectors $\underline{X}(u) = [x_1(u) \ ... \ x_k(u)]$ and $\underline{Y}(u) = [y_1(u) \ ... \ y_k(u)]$, we have

$$X(u) + Y(u) \leftrightarrow \underline{X}(u) + \underline{Y}(u)$$
$$= [x_1(u) + y_1(u)\ \ x_2(u) + y_2(u)\ \ ...\ \ x_k(u) + y_k(u)], \tag{3.4.7}$$

$$X(u) \cdot Y(u) \leftrightarrow \underline{X}(u) \cdot \underline{Y}(u)$$
$$= [x_1(u) \cdot y_1(u)\ \ x_2(u) \cdot y_2(u)\ \ ...\ \ x_k(u) \cdot y_k(u)]. \tag{3.4.8}$$

The sum $X(u) + Y(u)$ and the product $X(u) \cdot Y(u)$ in the *lhs* are defined mod $M(u)$ while the sum and the product in the i-th component in the *rhs* are defined mod $m_i(u)$. This follows directly from the properties of the mod operation and the definition of $X(u)$ and $Y(u)$.

The CRT-P is a fundamental method of partitioning a large degree polynomial $X(u)$ into a number of smaller degree but independent polynomials $x_i(u)$ which may be processed in parallel. This is the direct sum property of the CRT-P and is expressed as

$$\langle X(u) \rangle_{M(u)} = \langle X(u) \rangle_{m_1(u)} \oplus \langle X(u) \rangle_{m_2(u)} \oplus \cdots \oplus \langle X(u) \rangle_{m_k(u)}. \tag{3.4.9}$$

Once all the polynomials involved in the CRT-P reconstruction are computed, there can be several different methods in which the intermediate quantities are obtained before the final sum in the CRT-P reconstruction is performed. Some of these ways are:

Method 1. Compute the product $y_j(u) = T_j(u) \cdot x_j(u) \bmod m_j(u)$ and then form the product $X_j(u) = M_j(u) \cdot y_j(u) \bmod M(u)$, where $M_j(u) = \prod_{i=1}^{k} m_i(u) / m_j(u)$. In fact, $y_j(u)$ and $X_j(u)$ need not be formed mod $m_j(u)$ and $M(u)$, respectively. The reconstruction now expressed as

$$X(u) = \sum_{j=1}^{k} X_j(u) \bmod M(u) \tag{3.4.10}$$

takes into account these operations automatically.

Method 2. Compute the product $y_j(u) = M_j(u) \cdot x_j(u)$ and then form the product $X_j(u) = T_j(u) \cdot y_j(u) \bmod M(u)$. Once again, $X_j(u)$ need not be formed mod $M(u)$.

Method 3. Compute the product $B_j(u) = T_j(u) \cdot M_j(u)$ *a priori* and then compute $X_j(u) = B_j(u) \cdot x_j(u) \bmod M(u)$.

All these ways are equally valid and differ in the way the computations involved in the CRT-P reconstruction are organized. However, different organizations may have different features that may make them preferable over others in different computational environments.

The CRT-P as described here will be a starting point of our mathematical analysis for deriving fast algorithms for computing the cyclic and acyclic convolution of sequences defined over the various fields and rings.

Example 3.3 Consider the CRT-P corresponding to the factorization of $u^{12} - 1$ over $\mathbb{Z}$. From Example 3.1, we have

$$u^{12} - 1 = (u-1)(u+1)(u^2+1)(u^2-u+1)(u^2+u+1)(u^4-u^2+1).$$

Let $m_1(u) = u - 1$, $m_2(u) = u + 1$, $m_3(u) = u^2 + 1$, $m_4(u) = u^2 - u + 1$, $m_5(u) = u^2 + u + 1$, $m_6(u) = u^4 - u^2 + 1$. The polynomials $M_j(u)$ and $T_j(u)$ (obtained using the Euclid's algorithm) are given by

$$M_1(u) = 1 + u + u^2 + \ldots + u^{11}, \quad T_1(u) = 1/12,$$

$M_2(u) = -1 + u - u^2 + u^3 - u^4 + \ldots + u^{11}$, $T_2(u) = -1/12$,
$M_3(u) = -1 + u^2 - u^4 + u^6 - u^8 + u^{10}$, $T_3(u) = (-1/12) \cdot 2$,
$M_4(u) = -1 - u + u^3 + u^4 - u^6 - u^7 + u^9 + u^{10}$, $T_4(u) = (1/12) \cdot (u-2)$,
$M_5(u) = -1 + u - u^3 + u^4 - u^6 + u^7 - u^9 + u^{10}$, $T_5(u) = (1/12) \cdot (-u-2)$,
$M_6(u) = -1 - u^2 + u^6 + u^8$, $T_6(u) = (1/12) \cdot (2u^2 - 4)$.

3.5 CRT-P in the Matrix Form

Conceptually, data sequences are converted to polynomials for deriving computationally efficient algorithms. In real-time computations, the matrix-vector forms may attract more attention. In this section, we complete the analysis on the CRT-P by expressing it in matrix form.

Assume that a monic polynomial $M(u)$, defined over the field F, has k relatively prime factors, $m_1(u)$, $m_2(u)$, ..., $m_k(u)$. Every polynomial $X(u)$ with degree less than the degree of $M(u)$ can be represented by an unique vector $\underline{X}(u)$ having k components $x_1(u)$, $x_2(u)$, ..., $x_k(u)$, where $x_i(u)$, the i-th residue polynomial corresponding to $m_i(u)$, is given in (3.4.1). Let the coefficient vectors of $X(u)$ and $x_i(u)$ be $\underline{X}$ and $\underline{x}_i$, respectively. If polynomial degrees are $deg(X(u)) = n - 1$, $deg(M(u)) = d$, $d \geq n$, $deg(m_i(u)) = d_i$, and $d = d_1 + d_2 + \cdots + d_k$, then vector $\underline{X}$ has length n and $\underline{x}_i$ has length d_i. Based on (3.4.1), there exists a matrix $\mathbf{A}_i$ of dimensions $d_i \times n$ such that $\underline{X}$, $\underline{x}_i$, and $\mathbf{A}_i$ are related as

$$\underline{x}_i = \mathbf{A}_i \underline{X}, \quad i = 1, 2, \ldots, k. \tag{3.5.1}$$

The expression in (3.5.1) states the individual relation between $\underline{X}$ and $\underline{x}_i$ for $i = 1, 2, ..., k$. Combining all k equations as shown in (3.5.1) together, we obtain the following matrix equation:

$$\begin{bmatrix} \underline{x}_1 \\ \underline{x}_2 \\ \vdots \\ \underline{x}_k \end{bmatrix}_{d\times 1} = \begin{bmatrix} \mathbf{A}_1 \\ \mathbf{A}_2 \\ \vdots \\ \mathbf{A}_k \end{bmatrix}_{d\times n} \underline{X}\,. \tag{3.5.2}$$

Note that matrix $\mathbf{A}_i$ depends only on the monic factor polynomial $m_i(u)$, $i = 1, 2, ..., k$, and is computed *a priori*.

The CRT-P reconstruction of the polynomial $X(u)$ based on its residue polynomials has been shown in (3.4.2). The matrix form of the CRT-P reconstruction is

$$\underline{X} = \sum_{i=1}^{k} \mathbf{B}_i \underline{x}_i . \tag{3.5.3}$$

Each matrix $\mathbf{B}_i$ has dimensions $n \times d_i$. Substituting (3.5.2) into (3.5.3), we obtain

$$\underline{X} = \begin{bmatrix} \mathbf{B}_1 & \mathbf{B}_2 & \cdots & \mathbf{B}_k \end{bmatrix} \begin{bmatrix} \mathbf{A}_1 \\ \mathbf{A}_2 \\ \vdots \\ \mathbf{A}_k \end{bmatrix} \underline{X}. \tag{3.5.4}$$

Let

$$\mathbf{A}^t = \begin{bmatrix} \mathbf{A}_1^t & \mathbf{A}_2^t & \cdots & \mathbf{A}_k^t \end{bmatrix},$$

$$\mathbf{B} = \begin{bmatrix} \mathbf{B}_1 & \mathbf{B}_2 & \cdots & \mathbf{B}_k \end{bmatrix},$$

and $\mathbf{I}_{n \times n}$ be the identity matrix. The superscript t denotes matrix transposition. Then (3.5.4) implies that

$$\mathbf{B}\,\mathbf{A} = \mathbf{I}_{n \times n}. \tag{3.5.5}$$

The above matrix formulation is obtained if method 3 is employed for CRT-P reconstruction. Similarly, if the polynomial products $T_j(u) \cdot x_j(u)$ and $M_j(u) \cdot y_j(u)$ are represented as the matrix-vector products $\mathbf{T}_j\, \underline{x}_j$ and $\mathbf{M}_j\, \underline{y}_j$, respectively, then method 1 for CRT-P reconstruction leads to a further factorization of the $\mathbf{B}$ matrix as

$$\mathbf{B} = \begin{bmatrix} \mathbf{M}_1 & \mathbf{M}_2 & \cdots & \mathbf{M}_k \end{bmatrix} \begin{bmatrix} \mathbf{T}_1 & & & \\ & \mathbf{T}_2 & & \\ & & \ddots & \\ & & & \mathbf{T}_k \end{bmatrix}$$

$$= \begin{bmatrix} \mathbf{M}_1\mathbf{T}_1 \mid \mathbf{M}_2\mathbf{T}_2 \mid \cdots \mid \mathbf{M}_k\mathbf{T}_k \end{bmatrix}. \tag{3.5.6}$$

A similar factorization of the $\mathbf{B}$ matrix may also be obtained for method 2 for CRT-P reconstruction.

Example 3.4 Let us continue Example 3.3 and cast the various quantities in the matrix-vector form. Corresponding to the 6 factors of $u^{12}-1$, the various blocks in the matrices **A** and **B** are given by

$$\mathbf{A}=\begin{bmatrix}\mathbf{A}_1\\ \mathbf{A}_2\\ \mathbf{A}_3\\ \mathbf{A}_4\\ \mathbf{A}_5\\ \mathbf{A}_6\end{bmatrix}=\left[\begin{array}{cccccccccccc}
1 & 1 & 1 & 1 & 1 & 1 & 1 & 1 & 1 & 1 & 1 & 1\\ \hline
1 & -1 & 1 & -1 & 1 & -1 & 1 & -1 & 1 & -1 & 1 & -1\\ \hline
1 & 0 & -1 & 0 & 1 & 0 & -1 & 0 & 1 & 0 & -1 & 0\\
0 & 1 & 0 & -1 & 0 & 1 & 0 & -1 & 0 & 1 & 0 & -1\\ \hline
1 & 0 & -1 & -1 & 0 & 1 & 1 & 0 & -1 & -1 & 0 & 1\\
0 & 1 & 1 & 0 & -1 & -1 & 0 & 1 & 1 & 0 & -1 & -1\\ \hline
1 & 0 & -1 & 1 & 0 & -1 & 1 & 0 & -1 & 1 & 0 & -1\\
0 & 1 & -1 & 0 & 1 & -1 & 0 & 1 & -1 & 0 & 1 & -1\\ \hline
1 & 0 & 0 & 0 & -1 & 0 & -1 & 0 & 0 & 0 & 1 & 0\\
0 & 1 & 0 & 0 & 0 & -1 & 0 & -1 & 0 & 0 & 0 & 1\\
0 & 0 & 1 & 0 & 1 & 0 & 0 & 0 & -1 & 0 & -1 & 0\\
0 & 0 & 0 & 1 & 0 & 1 & 0 & 0 & 0 & -1 & 0 & -1
\end{array}\right],$$

$$\mathbf{B}=\begin{bmatrix}\mathbf{B}_1 & \mathbf{B}_2 & \mathbf{B}_3 & \mathbf{B}_4 & \mathbf{B}_5 & \mathbf{B}_6\end{bmatrix}$$

$$=(1/12)\left[\begin{array}{c|c|cc|cc|cc|cccc}
1 & 1 & 1 & 0 & 2 & 1 & 2 & -1 & 4 & 0 & 2 & 0\\
1 & -1 & 0 & 1 & 1 & 2 & -1 & 2 & 0 & 4 & 0 & 2\\
1 & 1 & -1 & 0 & -1 & 1 & -1 & -1 & 2 & 0 & 4 & 0\\
1 & -1 & 0 & -1 & -2 & -1 & 2 & -1 & 0 & 2 & 0 & 4\\
1 & 1 & 1 & 0 & -1 & -2 & -1 & 2 & -2 & 0 & 2 & 0\\
1 & -1 & 0 & 1 & 1 & -1 & -1 & -1 & 0 & -2 & 0 & 2\\
1 & 1 & -1 & 0 & 2 & 1 & 2 & -1 & -4 & 0 & -2 & 0\\
1 & -1 & 0 & -1 & 1 & 2 & -1 & 2 & 0 & -4 & 0 & -2\\
1 & 1 & 1 & 0 & -1 & 1 & -1 & -1 & -2 & 0 & -4 & 0\\
1 & -1 & 0 & 1 & -2 & -1 & 2 & -1 & 0 & -2 & 0 & -4\\
1 & 1 & -1 & 0 & -1 & -2 & -1 & 2 & 2 & 0 & -2 & 0\\
1 & -1 & 0 & -1 & 1 & -1 & -1 & -1 & 0 & 2 & 0 & -2
\end{array}\right].$$

The various blocks in the factorization of **B** in (3.5.6) are given by:

$$\mathbf{M}_1=\begin{bmatrix}1 & 1 & 1 & 1 & 1 & 1 & 1 & 1 & 1 & 1 & 1 & 1\end{bmatrix}^t,\ \mathbf{T}_1=(1/12)[1],$$

$$\mathbf{M}_2 = \begin{bmatrix} -1 & 1 & -1 & 1 & -1 & 1 & -1 & 1 & -1 & 1 & -1 & 1 \end{bmatrix}^t,$$

$\mathbf{T}_2 = (1/12)[-1]$,

$$\mathbf{M}_3 = \begin{bmatrix} -1 & 0 & 1 & 0 & -1 & 0 & 1 & 0 & -1 & 0 & 1 & 0 \\ 0 & -1 & 0 & 1 & 0 & -1 & 0 & 1 & 0 & -1 & 0 & 1 \end{bmatrix}^t,$$

$\mathbf{T}_3 = (1/12)[-2]$,

$$\mathbf{M}_4 = \begin{bmatrix} -1 & -1 & 0 & 1 & 1 & 0 & -1 & -1 & 0 & 1 & 1 & 0 \\ 0 & -1 & -1 & 0 & 1 & 1 & 0 & -1 & -1 & 0 & 1 & 1 \\ 1 & 0 & -1 & -1 & 0 & 1 & 1 & 0 & -1 & -1 & 0 & 1 \end{bmatrix}^t,$$

$$\mathbf{T}_4 = \left(\frac{1}{12}\right) \begin{bmatrix} -2 & 0 \\ 1 & -2 \\ 0 & 1 \end{bmatrix},$$

$$\mathbf{M}_5 = \begin{bmatrix} -1 & 1 & 0 & -1 & 1 & 0 & -1 & 1 & 0 & -1 & 1 & 0 \\ 0 & -1 & 1 & 0 & -1 & 1 & 0 & -1 & 1 & 0 & -1 & 1 \\ 1 & 0 & -1 & 1 & 0 & -1 & 1 & 0 & -1 & 1 & 0 & -1 \end{bmatrix}^t,$$

$$\mathbf{T}_5 = \left(\frac{1}{12}\right) \begin{bmatrix} -2 & 0 \\ -1 & -2 \\ 0 & -1 \end{bmatrix},$$

$$\mathbf{M}_6 = \begin{bmatrix} -1 & 0 & -1 & 0 & 0 & 0 & 1 & 0 & 1 & 0 & 0 & 0 \\ 0 & -1 & 0 & -1 & 0 & 0 & 0 & 1 & 0 & 1 & 0 & 0 \\ 0 & 0 & -1 & 0 & -1 & 0 & 0 & 0 & 1 & 0 & 1 & 0 \\ 0 & 0 & 0 & -1 & 0 & -1 & 0 & 0 & 0 & 1 & 0 & 1 \\ 1 & 0 & 0 & 0 & -1 & 0 & -1 & 0 & 0 & 0 & 1 & 0 \end{bmatrix}^t,$$

$$\mathbf{T}_6 = \left(\frac{1}{12}\right) \begin{bmatrix} -4 & 0 & 2 & 0 & 0 & 0 \\ 0 & -4 & 0 & 2 & 0 & 0 \\ 0 & 0 & -4 & 0 & 2 & 0 \\ 0 & 0 & 0 & -4 & 0 & 2 \end{bmatrix}^t.$$

The structure of all these matrices will be analyzed in Chapter 9, where close form expressions will be derived for them for the cases of interest in digital signal processing.

3.6 Lagrange Interpolation

In this section, we describe the **Lagrange interpolation (LI)** algorithm for evaluating polynomials and demonstrate it to be a special case of the CRT-P when all the polynomial factors are of degree one. Let $X(u)$ be a polynomial of degree $n-1$ and represented as $X(u)=\sum_{i=0}^{n-1} X_i u^i$. LI deals with the problem of interpolating $X(u)$ through n points. This suggests a method for determining the coefficients of $X(u)$ based on the values of $X(u)$ evaluated at n points A_i, $i = 1, 2, ..., n$; namely, $X(A_1), X(A_2), ..., X(A_n)$. Since

$$X(A_i) = X(u) \bmod (u - A_i),$$

LI is same as CRT-P when the modulo polynomial $M(u)$ is set equal to the product of n degree-one polynomials $(u - A_i)$. In other words,

$$m_i(u) = (u - A_i), \quad i = 1, 2, ..., n.$$

According to (3.4.2), $X(u)$ can be reconstructed using CRT-P or LI as follows:

$$X(u)=\sum_{i=1}^{n} X(A_i)\prod_{\substack{j=1\\ j\neq i}}^{n}\frac{(u-A_j)}{(A_i-A_j)}. \tag{3.6.1}$$

Recall that the CRT-P requires that all the factor polynomials $m_i(u) = (u - A_i)$ be pairwise relatively prime. Since A_i, $i = 1, 2, ..., n$ are elements of a field, the only constraint for a unique solution is $A_i \neq A_j$, $i \neq j$.

Alternatively, the LI algorithm can be represented as a **Vandermonde** system of linear equations:

$$\begin{bmatrix} 1 & A_1 & A_1^2 & \cdots & A_1^{n-1} \\ 1 & A_2 & A_2^2 & \cdots & A_2^{n-1} \\ \vdots & \vdots & \vdots & \ddots & \vdots \\ 1 & A_n & A_n^2 & \cdots & A_n^{n-1} \end{bmatrix} \begin{bmatrix} X_0 \\ X_1 \\ \vdots \\ X_{n-1} \end{bmatrix} = \begin{bmatrix} X(A_1) \\ X(A_2) \\ \vdots \\ X(A_n) \end{bmatrix}. \tag{3.6.2}$$

Let the Vandermonde matrix be denoted by **V**. The determinant of **V** is given by

$$det(\mathbf{V})=\prod_{i=1}^{n-1}\prod_{j=i+1}^{n}(A_j-A_i). \tag{3.6.3}$$

The condition for the existence of a unique solution to the Vandermonde system shown in (3.6.2) is $det(\mathbf{V}) \neq 0$ or $A_i \neq A_j$, $i \neq j$. Consequently, we may interpret LI as a special case of the CRT-P when all the polynomial factors are of degree one.

The Vandermonde system of linear equations can be solved by either the LI in (3.6.1) (which is a special case of CRT-P reconstruction) or the order-recursive algorithm for the Vandermonde systems.

3.7 Polynomial Algebra Over *GF*(*p*)

In this section, we return to our study of the finite field $GF(p)$ and establish certain results that are of importance in deriving fast algorithms for digital signal processing of data sequences defined over arbitrary fields and rings.

Let $A_1, A_2, ..., A_{p-1}$ be the $p-1$ non-zero elements of $GF(p)$ and let A denote any one of them. From Fermat's theorem (Theorem 2.4), A satisfies the relation $A^{p-1} \equiv 1 \bmod p$ or

$$A^{p-1} - 1 \equiv 0 \bmod p. \tag{3.7.1}$$

In other words, order of A is either $p-1$ or a factor of $p-1$. Therefore, A is a root of the polynomial $u^{p-1} - 1$ in $GF(p)$. Since $u^{p-1} - 1$ is a polynomial of degree $p-1$, it can only have $p-1$ roots. Thus, $u^{p-1} - 1$ has all the $p-1$ non-zero elements of $GF(p)$ as its roots and the unique factorization of $u^{p-1} - 1$ over $GF(p)$ is given by

$$u^{p-1} - 1 = \prod_{i=1}^{p-1} (u - A_i) \bmod p, \quad A_i \in GF(p), A_i \neq 0. \tag{3.7.2}$$

Theorem 3.4

$$\left(u^{\alpha} - 1, u^{\beta} - 1\right) = u^{(\alpha,\beta)} - 1. \tag{3.7.3}$$

Proof: Let $\alpha > \beta$. We divide $u^{\alpha} - 1$ by $u^{\beta} - 1$ to get

$$u^{\alpha} - 1 = \left(u^{\alpha-\beta} + u^{\alpha-2\beta} + \ldots + u^{\alpha-c\beta}\right)\left(u^{\beta} - 1\right) + u^{r} - 1 \tag{3.7.4}$$

corresponding to the expression

$$\alpha = c\,\beta + r,\ 0 \le r < \beta. \tag{3.7.5}$$

It is clear from (3.7.4) and (3.7.5) that

$$\left(u^{\alpha} - 1, u^{\beta} - 1\right) = \left(u^{\beta} - 1, u^{r} - 1\right), \tag{3.7.6}$$

where

$$(\alpha, \beta) = (\beta, r). \tag{3.7.7}$$

If $r = 0$, that is $\beta \mid \alpha$, then $\left(u^{\beta} - 1\right) \big| \left(u^{\alpha} - 1\right)$ and vice versa. This statement along with a recursive application of (3.7.6) and (3.7.7) proves the theorem.

A consequence of the above theorem is that for every factor d of $p - 1$, that is $d \mid (p - 1)$, $(u^d - 1) \mid (u^{p-1} - 1)$. Combining this with the factorization of $u^{p-1} - 1$ in (3.7.2), we get

$$u^{p-1} - 1 = \left(u^{d} - 1\right)C(u) = \prod_{i=1}^{p-1}\left(u - A_i\right),\ deg(C(u)) = p - 1 - d.$$

Since $deg(C(u)) = p - 1 - d$, it can have only $p - 1 - d$ roots in $GF(p)$. Therefore, there must be d elements among the $p - 1$ non-zero elements A_1, A_2, ..., A_{p-1} in $GF(p)$ that are roots of $u^d - 1$. Let these d elements be B_1, B_2, ..., B_d. This gives factorization of $u^d - 1$ for every $d \mid (p - 1)$ over $GF(p)$ as

$$u^{d} - 1 = \prod_{i=1}^{d}\left(u - B_i\right),\ B_i \in GF(p),\ B_i \ne 0. \tag{3.7.8}$$

The above factorization also implies that the element B_i has order equal to d or a factor of d. This is consistent with the Fermat's theorem. There are exactly d elements that have d or a factor of d as their order. Now consider the factorization of $p - 1$ as

$$p - 1 = r_1^{a_1} \cdot r_2^{a_2} \cdots r_l^{a_l} = q_1 \cdot q_2 \cdots q_l\,,$$

and let $d = 1$. Then, order of $A = 1$ is d. Let $d = r_1$. From (3.7.8), we see that there are r_1 distinct elements in $GF(p)$ having order r_1 or a factor of r_1. Since r_1 is prime, there must be exactly $r_1 - 1$ elements in $GF(p)$ of order r_1. Similarly, there are $r_1^{i} - r_1^{i-1}$ elements in $GF(p)$ of order r_1^{i}, $i = 1, 2, ..., a_1$. All of these

$r_1^i - r_1^{i-1}$ elements are roots of the polynomial $u^{r_1^i} - 1$ but not of $u^{r_1^{i-1}} - 1$. Let A_j denote an element in $GF(p)$ of order $q_j = r_j^{a_j}$, $j = 1, 2, ..., l$. A fundamental property of these elements is stated in the following.

Theorem 3.5 In $GF(p)$, there is an element of order $p - 1$.
Proof: Consider the element $A = A_1 \cdot A_2$. Since $A_1^{q_1} = 1$, $A_2^{q_2} = 1$, we have $A_1^{q_1 q_2} = 1$ and $A_2^{q_1 q_2} = 1$. Therefore, $(A_1 A_2)^{q_1 q_2} = 1$ or $A^{q_1 q_2} = 1$. Let n be the order of the element A, that is $A^n = 1$. Since $A^{q_1 q_2} = 1$, we have $n \mid (q_1 \cdot q_2)$. $A^n = 1$ also leads to $A^{nq_1} = 1$ and $A^{nq_2} = 1$. Replacing A by $A_1 \cdot A_2$ in these equations and realizing that $A_1^{q_1} = 1$ and $A_2^{q_2} = 1$, we get $A_2^n = 1$ and $A_1^n = 1$. These equations imply that $q_1 \mid n$ and $q_2 \mid n$. Since $(q_1, q_2) = 1$, we have $(q_1 \cdot q_2) \mid n$. The two relations $n \mid (q_1 \cdot q_2)$ and $(q_1 \cdot q_2) \mid n$ lead to $n = q_1 \cdot q_2$. Since the element $A_1 A_2$ has order $q_1 \cdot q_2$, by induction, the element $A = A_1 \cdot A_2 \cdots A_l$ has order $\prod_{i=1}^{l} r_i^{a_i} = p - 1$.

The above theorem will play a critical role in the design of fast algorithms for digital signal processing. It guarantees that there is a primitive element A in $GF(p)$. Since A has order $p - 1$, every non-zero element A_i in $GF(p)$ can be expressed as

$$A_i = A^i, \quad i = 1, 2, ..., p - 1.$$

The above analysis also leads to the following generalization of Theorem 3.5.

Theorem 3.6 In $GF(p)$, there are exactly $\phi(d)$ elements of order d for every divisor d of $p - 1$. If A is a primitive element of $GF(p)$, then A^i is also a primitive element of $GF(p)$ for all values of i that satisfy $(i, p - 1) = 1$.

Example 3.5 Consider $GF(37)$. The order of every non-zero element divides 37 − 1 = 36. Therefore, there are elements of order 1, 2, 3, 4, 6, 9, 12, 18, 36. In fact, there are exactly $\phi(1) = 1$, $\phi(2) = 1$, $\phi(3) = 2$, $\phi(4) = 2$, $\phi(6) = 2$, $\phi(9) = 6$, $\phi(12) = 4$, $\phi(18) = 6$, $\phi(36) = 12$ elements of order 1, 2, 3, 4, 6, 9, 12, 18, and 36, respectively. We identify 2 as a primitive element of order 36 in $GF(37)$. The other primitive elements of order 36 are given by 2^i, $(i, 36) = 1$. They are 2, $2^5 = 32$, $2^7 = 17$, $2^{11} = 13$, $2^{13} = 15$, $2^{17} = 18$, $2^{19} = 35$, $2^{23} = 5$, $2^{25} = 20$, $2^{29} = 24$, $2^{31} = 22$, $2^{35} = 19$. Similarly, for the element 2^i, $(i, 36) > 1$, the order is $36/(i, 36)$. The elements of various orders can now be enumerated as follows:

order = 1 : element is $2^{36} = 1$;
order = 2 : element is $2^{18} = 36$;
order = 3 : elements are $2^{12} = 26$, $2^{24} = 10$;
order = 4 : elements are $2^{9} = 31$, $2^{27} = 6$;
order = 6 : elements are $2^{6} = 27$, $2^{30} = 11$;
order = 9 : elements are $2^{4} = 16$, $2^{8} = 34$, $2^{16} = 9$, $2^{20} = 33$, $2^{28} = 12$, $2^{32} = 7$;
order = 12: elements are $2^{3} = 8$, $2^{15} = 23$, $2^{21} = 29$, $2^{33} = 14$;
order = 18: elements are $2^{2} = 4$, $2^{10} = 25$, $2^{14} = 30$, $2^{22} = 21$, $2^{26} = 3$, $2^{34} = 28$.

3.8 Order of an Element: Continued

We now complete our discussion on the order of an element in $Z(M)$ for $M = 2^a$, $M = p^a$, $M = 2 \cdot p^a$. For $M = 2, 4$, the situation is trivial with 1 being the primitive root in the former case and 3 being the primitive root in the latter. Since we consider only those elements for order that are relatively coprime to M, we only need to consider odd integers for $M = 2^a$, $a > 2$. Any odd integer can be expressed as $4k + I$, where $I = 1$ or -1. Let A be any such integer. Then

$$A^{2^{a-2}} \bmod 2^a \equiv (4k+I)^{2^{a-2}} \bmod 2^a \equiv \sum_{i=0}^{2^{a-2}} {}^{2^{a-2}}C_i (4k)^i I^{2^{a-2}-i} \bmod 2^a$$

$$\equiv I^{2^{a-2}} + \sum_{i=1}^{2^{a-2}} {}^{2^{a-2}}C_i (4k)^i I^{2^{a-2}-i} \bmod 2^a .$$

Using the definition of ${}^nC_i$, it is seen that the second term (the entire summation) in the above expression contains 2^a as a factor. Therefore,

$$A^{2^{a-2}} \bmod 2^a \equiv I^{2^{a-2}} \bmod 2^a \equiv 1 \bmod 2^a .$$

Here the last relation is obtained by realizing that $I = 1$ or -1 and $a > 2$. This relation implies that the order of every element in $Z(M)$, $M = 2^a$, divides 2^{a-2}. For $a = 3$, an element of order 2^{a-2} is given by $A = 3$.

Now we find one element A having order exactly 2^{a-2} by showing that $A^{2^{a-k}} \neq 1 \bmod 2^a$, $k > 2$, $a > 3$. Consider the element $A = 3 = 1 + 2 \cdot 1$. For $A = 3$, we have the following lemma:

Lemma 3.1 For $A = 3$ and $a > 3$,

$$A^{2^{a-3}} \equiv 1 + 2^{a-1} \bmod 2^a . \quad (3.8.1)$$

Proof: The proof is based on induction. The statement is true for $a = 4$. Let us say that it is true for all $a > 3$. Therefore, we may write

$$A^{2^{a-3}} \equiv 1 + 2^{a-1} \bmod 2^a$$

or

$$A^{2^{a-3}} = 1 + 2^{a-1} + b2^a.$$

Squaring both sides to get a similar relation for $a + 1$, we get

$$A^{2^{a-2}} = \left(1 + 2^{a-1} + b2^a\right)^2 = 1 + 2^{2a-2} + b^2 2^{2a} + 2^a + 2b2^a + 2b2^{2a-1}.$$

Noting that $2a - 2 > a + 1$ for $a > 3$, and taking mod 2^{a+1}, we get

$$A^{2^{a-2}} \equiv 1 + 2^a \bmod 2^{a+1}.$$

Therefore, the expression is valid for $a + 1$. By induction, it is valid for all $a >$ 3.

This lemma implies that $A^{2^{a-3}} \neq 1 \bmod 2^a$. Also $A^{2^{a-3}} \neq 1 \bmod 2^a$ further implies that $A^{2^{a-3}} \neq 1 \bmod 2^{a+k}$, $k > 0$ which leads to $A^{2^{a-k}} \neq 1 \bmod 2^a$, $k > 3$. Lemma 3.1 takes care of the case $k = 3$. In other words, $A^{2^{a-k}} \neq 1 \bmod 2^a$, $k > 2$, $a > 3$. Since order of A divides 2^{a-2}, it must, therefore, be exactly equal to 2^{a-2}.

Let us now consider the case $M = p^a$. In this case, we first show that there is an element of order $p - 1$ in $Z(p^a)$. Let A be a primitive element in $GF(p)$. Therefore, $A^{p-1} \equiv 1 \bmod p$, while $A^i \bmod p \neq 1$ for $i < p - 1$. Also, it is trivial that $(A, p) = 1$. Given A, we show that the integer $A^{p^{a-1}}$ has order $p - 1$ in $Z(p^a)$. Since $(A, p) = 1$, using the Euler's theorem, we have

$$\left(A^{p^{a-1}}\right)^{p-1} \equiv 1 \bmod p^a.$$

Therefore, order of $A^{p^{a-1}}$ divides $p - 1$. Let n be the order of $A^{p^{a-1}}$. Then

$$\left(A^{p^{a-1}}\right)^n \equiv 1 \bmod p^a.$$

Taking mod p on both sides, we get

$$A^{np^{a-1}} \equiv 1 \bmod p.$$

Therefore, n is also the order of $A^{p^{a-1}}$ in $GF(p)$. Since A has order $p - 1$ in $GF(p)$, and $(p - 1, p) = 1$,

$$\text{order of } A \text{ in } GF(p) = \text{order of } A^{p^{a-1}} \text{ in } GF(p).$$

Consequently, $n = p - 1$.

Now we show that there is an element of order p^{a-1} in $Z(p^a)$. Consider the integer $A = 1 + p$. For A in this form, we have the following lemma.

Lemma 3.2 For the integer $A = 1 + p$ in $Z(p^a)$,

$$A^{p^{a-1}} \equiv 1 \bmod p^a.$$

Proof: Replacing A by $1 + p$ in the above expression, we have

$$A^{p^{a-1}} \equiv (1+p)^{p^{a-1}} \bmod p^a \equiv \sum_{k=0}^{p^{a-1}} {}^{p^{a-1}}C_k p^k \equiv 1 + \sum_{k=1}^{p^{a-1}} {}^{p^{a-1}}C_k p^k \bmod p^a.$$

All the terms in the summation contain p^a as a factor. The statement of the lemma follows.

Lemma 3.2 implies that the order of $A = 1 + p$ is a factor of p^{a-1}. The case for $a = 1$ is trivial. We now show that for none of the integers $n = p^i$, $i = 2, 3, \ldots, a - 2$, the expression $A^n \equiv 1 \bmod p^a$ is satisfied. Let us begin with $i = a - 2$.

Lemma 3.3 For $A = 1 + p$ and $a > 1$, the following congruence is valid:

$$A^{p^{a-2}} \equiv 1 + p^{a-1} \bmod p^a.$$

Proof: Once again, we prove the statement by induction. By inspection, the congruence is valid for $a = 2$. Let us say it is true for any a. Thus

$$A^{p^{a-2}} \equiv 1 + p^{a-1} \bmod p^a$$

or

$$A^{p^{a-2}} = 1 + p^{a-1} + bp^a.$$

To obtain a similar expression for $a + 1$, raise both sides of the above expression to the power p to get

$$A^{p^{a-1}} = \left(1 + p^{a-1} + bp^a\right)^p = \sum_{k=0}^{p} {}^pC_k \left(p^{a-1} + bp^a\right)^{p-k}$$

$$= 1 + p\left(p^{a-1} + bp^a\right) + \sum_{k=1}^{p-2} {}^pC_k \left(p^{a-1} + bp^a\right)^{p-k} + \left(p^{a-1} + bp^a\right)^p.$$

Since p is prime, ${}^pC_k$, $k = 1, 2, ..., p - 2$ has p as a factor. Therefore, all the terms in the summation have $p^{2(a-1)+1}$ as a factor and are divisible by p^{a+1}. Similarly, the last term has $p^{p(a-1)}$ as a factor and, therefore, is divisible by p^{a+1}. Taking mod p^{a+1} on both sides, we get

$$A^{p^{a-1}} \equiv 1 + p^a \bmod p^{a+1}.$$

Therefore, the expression is valid for $a + 1$. By induction, it is valid for all $a >$ 1.

This lemma implies that $A^{p^{a-2}} \neq 1 \bmod p^a$. Also, $A^{p^{a-2}} \neq 1 \bmod p^a$ further implies that

$$A^{p^{a-2}} \neq 1 \bmod p^{a+k}, \quad k > 0$$

which leads to

$$A^{p^{a-k}} \neq 1 \bmod p^a, \quad k > 2.$$

Lemma 3.3 takes care of the case $k = 2$. Thus, for $A = 1 + p$, $A^{p^{a-k}} \neq 1 \bmod p^a$, $k > 1$, $a > 3$. Since order of A divides p^{a-1}, it must, therefore, be exactly equal to p^{a-1}.

Let E and F be two integers in $Z(p^a)$ having order p^{a-1} and $(p - 1)$, respectively. In the above discussion, we have found one value for both of these integers given by

$$E = 1 + p$$

and

$$F = A^{p^{a-1}},$$

where A is a primitive element in $GF(p)$. Define an element F as

$$G = E \cdot F. \tag{3.8.2}$$

Since $(p, p - 1) = 1$, the orders of E and F are mutually prime. Applying Theorem 2.7, we get the following theorem:

Theorem 3.7 The order of the element G in (3.8.2) is equal to $p^{a-1}(p-1)$ and therefore G is a primitive element in $Z(p^a)$.

Finally, we turn to the case $M = 2 \cdot p^a$. The following theorem identifies the primitive element in this case.

Theorem 3.8 If G is a primitive element in $Z(p^a)$, then a primitive element in $Z(2 \cdot p^a)$ is given by

$$H = G + p^a. \tag{3.8.3}$$

Proof: As G is an element in $Z(p^a)$, $G < p^a$, and $H < 2 \cdot p^a$. It is easily verified that

$$H^{p^{a-1}(p-1)} \bmod p^a \equiv \left(G + p^a\right)^{p^{a-1}(p-1)} \bmod p^a \equiv G^{p^{a-1}(p-1)} \bmod p^a \equiv 1. \tag{3.8.4}$$

Also E is even (one such value is $1 + p$) which implies that G is even and H is odd. Thus,

$$H^{p^{a-1}(p-1)} \bmod 2 \equiv 1. \tag{3.8.5}$$

Using the CRT-I on the modulo relations in (3.8.4) and (3.8.5), we get

$$H^{p^{a-1}(p-1)} \bmod 2p^a \equiv 1. \tag{3.8.6}$$

Also $H^n \bmod 2 \cdot p^a$ cannot equal 1 for $n < p^{a-1}(p-1)$ as it would imply

$$H^n \bmod p^a \equiv G^n \bmod p^a \equiv 1 \bmod p^a, \quad n < p^{a-1}(p-1),$$

thereby violating the fact that G is an element of order $p^{a-1}(p-1)$ in $Z(p^a)$.

We summarize our discussion on the primitive elements by way of the following observations:

A. Primitive elements exist in $Z(M)$ only for values of M of the type p^a, $2 \cdot p^a$.

B. A primitive element in $Z(p^a)$ can be expressed in the form

$$G = E \cdot F,$$

where E is an element of order p^{a-1} in $Z(p^a)$, and

$$F = A^{p^{a-1}},$$

A being a primitive element in $GF(p)$. A possible choice for E is $E = 1 + p$.

C. A primitive element in $Z(2 \cdot p^a)$ has the form

$$H = G + p^a,$$

G being a primitive element in $Z(p^a)$.

D. For $M = 2^a$, $a > 2$, the maximum value for the period of any element is given by 2^{a-2}. The integer 3 is one element whose order is 2^{a-2}. For $M = 4$, 3 has order equal to 2; $M = 2$ is a trivial case.

These results are summarized in Table 3.1.

We end this chapter by stating that we have established all these results in the present and the previous chapter with the exception of Observation **A** that is left as an exercise.

Table 3.1 Summary of Primitive Roots in Finite Integer Rings and Fields

Ring or Field	Primitive Elements
$Z(2)$	1
$Z(4)$	3
$Z(2^a)$, $a > 2$	3; order $= 2^{a-2} < \phi(2^a)$. No element of order $\phi(2^a)$ exists.
$GF(p)$	A, $A^{p-1} = 1 \bmod p$ $A^i \neq 1 \bmod p,\ i < p-1$
$Z(p^a)$	$G = (1+p) \cdot A^{p^{a-1}}$
$Z(2 \cdot p^a)$	$G + p^a$

No primitive element exists in any other finite integer ring.

Notes

This chapter is a potpourri of results relevant to computing the convolution (cyclic and acyclic) and discrete Fourier transform of data sequences defined over the rational, real, complex rational, and complex number systems in a computationally efficient manner. Further results that are useful only to digital signal processing of data sequences defined over finite fields and rings will be presented in later chapters.

Polynomial algebra over a field is well established and we have selected those results that are useful in the further development and understanding of algorithms described in subsequent chapters. We have used the books by McClellan and Rader [1979], Nussbaumer [1981], and Blahut [1984] to enhance and select the topics in number theory that are relevant to digital signal processing. Specifically, the proof of the Chinese remainder theorem for polynomials and part of discussion on primitive roots in finite integer rings follow Blahut [1984] quite closely. The order recursive algorithm for solving a Vandermonde system of linear equations mentioned in Section 3.6 can be found in Golub and Van Loan [1983]. The book by D.M. Burton [1994] explains many of the results in an easy-to-follow manner.

We have chosen to limit our attention to only those results in number theory and polynomial algebra that are of relevance in the design of digital signal processing algorithms. In our approach, the *quadratic residue does not* play an important role. Consequently, we do not elaborate on it much. Finally, we have proven every result in order to make this chapter completely self contained with perhaps the exception of the unique factorization property of polynomials over a field. This has been our philosophy throughout this book.

Bibliography

[3.1] R.E. Blahut, *Fast Algorithms for Digital Signal Processing*, Addison Wesley Publishing Co., 1984.

[3.2] J.H. McClellan and C.M. Rader, *Number Theory in Digital Signal Processing*, Prentice Hall, Inc., 1979.

[3.3] H.J. Nussbaumer, *Fast Fourier Transforms and Convolution Algorithms*, Springer-Verlag, 1981.

[3.4] G.H. Golub and C.F. Van Loan, *Matrix Computations*, The John Hopkins University Press, 1983.

[3.5] D.M. Burton, *Elementary Number Theory*, Wm. C. Brown Publishers, 1994.

Problems

3.1 Let the sequences 1, 1, 1, 1, 1, 1 and 1, −1, 1, −1, 1, −1 be defined over $\mathbb{Z}$. Express them as polynomials.

3.2 Compute the cyclic and acyclic convolution of the sequences in Problem 3.1.

3.3 Compute the convolutions in Problem 3.2 as appropriate polynomial products.

3.4 Given $A(u) = 1 + u + u^2 + u^4 + u^6$ and $M(u) = 1 - u + u^2$ defined over $\mathbb{Z}$, compute $A(u)$ mod $M(u)$. Is the residue polynomial monic?

3.5 Repeat Problem 3.4 if the given polynomials are defined over $GF(3)$.

3.6 Compute the *gcd* of the polynomials $A(u) = 1 + u^2 + u^3 + u^6 + u^8 + u^9$ and $M(u) = -1 + u^4$ over $\mathbb{Z}$ and express it as a linear sum of $A(u)$ and $M(u)$.

3.7 Consider the following congruence over $\mathbb{Z}$, $[1 + u + u^2 + u^3 + u^4 + u^5] \cdot M(u) \equiv 1 + u + 2u^2 + u^3 + u^4 \bmod (1 + u^2 + u^4)$.

(a) Does a solution to the congruence exist? If yes, then find a solution.
(b) Is the solution unique? If no, then express the general form of the solution.

3.8 Solve the congruence $A(u)\ X(u) \equiv 1 \bmod M(u)$ over Z, $A(u) = 1 + u + u^2$, $M(u) = 1 - u + u^2$.

3.9 Can the congruence in Problem 3.8 be solved over $GF(5)$? If yes, then compute the solution.

3.10 Based on the Euclid's algorithm for polynomials, one can obtain similar continued fraction expansion for polynomials as in Problem 2.8. Establish the results in Problems 2.8 and 2.9 for polynomials.

3.11 Given the polynomials $m_1(u) = 1 + u$, $m_2(u) = 1 + u^2$, $m_3(u) = 1 - u + u^2$, $m_4(u) = 1 + u^3 + u^6$, compute the various polynomials required for the CRT-P reconstruction.

3.12 This is a continuation of Problem 3.11. Given $X(u) = 1 + u + u^2 + u^5 + u^8 + u^9$, compute its residues and then verify the CRT-P by constructing $X(u)$ from these residues.

3.13 Can we obtain an extension field of Re by using the polynomial $j^2 - j + 1$? If yes, then compute $(2 + 3\,j)\ (4 + j)$ in such a field.

3.14 Verify the factorization $u^7 - 1 = \prod_{i=1}^{7}\left(u - j^i\right)$ over the extension field of Z obtained by using the irreducible polynomial $1 + j + j^2 + j^3 + j^4 + j^5 + j^6$.
Note: One quick way is to set $u = j^i$ in the given factorization and show that j^i is a root of $u^7 - 1$ in the extension field.

3.15 Verify the factorization given in Examples 3.1 and 3.2.

3.16 By enumeration, find primitive roots in $GF(2)$, $GF(3)$, $GF(7)$, $GF(11)$, and $GF(127)$.

3.17 If they exist, find primitive roots in $Z(16)$, $Z(27)$, $Z(343)$, and $Z(242)$. Also state the order of the primitive roots in each case. Express as many elements as a power of the primitive element in $Z(27)$ as possible.

3.18 Consider $M = m_1 \cdot m_2$, $(m_1, m_2) = 1$. Show that the order of every element in $Z(M)$ divides the *least common multiple* (*lcm*) of $\phi(m_1)$ and $\phi(m_2)$ and that

$$lcm(\phi(m_1), \phi(m_2)) < \phi(m_1) \cdot \phi(m_2)$$

for all values of M except $M = 2, 4, p^a, 2 \cdot p^a$.
Note: This establishes observation **A** made towards the end of the chapter.

Chapter 4

Theoretical Aspects of the Discrete Fourier Transform and Convolution

In this chapter, we study the theoretical aspects of digital signal processing algorithms for the computation of discrete Fourier transform (DFT) and convolutions. The various polynomial algebra results are cast in the matrix-vector form and certain lower bounds are derived for the computational complexity of the convolution algorithms. It is shown that in certain special cases of interest, the representation of the algorithm can be suitably altered without effecting the final results. This alteration can lead to computational simplifications and a different organization of the algorithm. All the quantities in this chapter are defined over a field F. We begin our description by defining and analyzing the DFT.

4.1 The Discrete Fourier Transform

Definition 4.1 Discrete Fourier transform (DFT). Given a length n data sequence A_l, $l = 0, 1, \ldots, n-1$, or equivalently a degree $n-1$ polynomial $A(u)$, the **discrete Fourier transform (DFT)** in a field F is defined as the sequence $\boldsymbol{A}_i$, $i = 0, 1, \ldots, n-1$, given by

$$\boldsymbol{A}_i = \sum_{l=0}^{n-1} A_l \alpha^{li}, \quad i = 0, 1, ..., n-1, \tag{4.1.1}$$

where α is an element of order n in F; that is, $\alpha^n = 1$ in F and $\alpha^i \neq 1$, $i < n$. Representing the DFT as a matrix-vector product, we have

$$\begin{bmatrix} \boldsymbol{A}_0 \\ \boldsymbol{A}_1 \\ \vdots \\ \boldsymbol{A}_{n-1} \end{bmatrix} = \begin{bmatrix} 1 & 1 & 1 & \cdots & 1 & 1 \\ 1 & \alpha & \alpha^2 & \cdots & \alpha^{n-2} & \alpha^{n-1} \\ 1 & \alpha^2 & \alpha^4 & \cdots & \alpha^{2(n-2)} & \alpha^{2(n-1)} \\ \vdots & \vdots & \vdots & \ddots & \vdots & \vdots \\ 1 & \alpha^{n-2} & \alpha^{(n-2)2} & \cdots & \alpha^{(n-2)(n-2)} & \alpha^{(n-2)(n-1)} \\ 1 & \alpha^{n-1} & \alpha^{(n-1)2} & \cdots & \alpha^{(n-1)(n-2)} & \alpha^{(n-1)(n-1)} \end{bmatrix} \begin{bmatrix} A_0 \\ A_1 \\ \vdots \\ A_{n-1} \end{bmatrix} \tag{4.1.2}$$

or

$$\underline{\boldsymbol{A}} = \Lambda \, \underline{A}. \tag{4.1.3}$$

Definition 4.2 Inverse discrete Fourier transform. Given the DFT $\boldsymbol{A}_i$, $i = 0, 1, ..., n-1$, of a sequence A_l, $l = 0, 1, ..., n-1$, the original data sequence, A_l, $l = 0, 1, ..., n-1$, is said to constitute the **inverse discrete Fourier transform (IDFT)** of the DFT sequence $\boldsymbol{A}_i$, $i = 0, 1, ..., n-1$.

Theorem 4.1 Given the DFT $\boldsymbol{A}_i$, $i = 0, 1, ..., n-1$ of a sequence A_l, $l = 0, 1, ..., n-1$, the IDFT can be computed as

$$A_l = (1/n)\sum_{i=0}^{n-1} \boldsymbol{A}_i \alpha^{-il}, \quad l = 0, 1, ..., n-1, \tag{4.1.4}$$

which can alternately be represented as the matrix-vector product

$$\underline{A} = \Lambda^{-1} \underline{\boldsymbol{A}}$$

or

$$\begin{bmatrix} A_0 \\ A_1 \\ \vdots \\ A_{n-1} \end{bmatrix} = \left(\frac{1}{n}\right) \begin{bmatrix} 1 & 1 & 1 & \cdots & 1 & 1 \\ 1 & \alpha^{-1} & \alpha^{-2} & \cdots & \alpha^{-(n-2)} & \alpha^{-(n-1)} \\ 1 & \alpha^{-2} & \alpha^{-4} & \cdots & \alpha^{-2(n-2)} & \alpha^{-2(n-1)} \\ \vdots & \vdots & \vdots & \ddots & \vdots & \vdots \\ 1 & \alpha^{-(n-2)} & \alpha^{-(n-2)2} & \cdots & \alpha^{-(n-2)(n-2)} & \alpha^{-(n-2)(n-1)} \\ 1 & \alpha^{-(n-1)} & \alpha^{-(n-1)2} & \cdots & \alpha^{-(n-1)(n-2)} & \alpha^{-(n-1)(n-1)} \end{bmatrix} \begin{bmatrix} \boldsymbol{A}_0 \\ \boldsymbol{A}_1 \\ \vdots \\ \boldsymbol{A}_{n-1} \end{bmatrix}. \tag{4.1.5}$$

Proof: In other words, if the (i, l)-th element of the DFT matrix Λ is α^{il}, then the (l, i)-th element of the IDFT matrix Λ^{-1} is $(1/n)\,\alpha^{-li}$. We establish this result by showing that for the forms of Λ and Λ^{-1} matrices in (4.1.2) and (4.1.5), the product $[(\Lambda)\,(\Lambda^{-1})] = \mathbf{I}_{n\times n}$, an identity matrix. In this case,

$$
\begin{aligned}
&(i, k)\text{-th element of } [(\Lambda)(\Lambda^{-1})] \\
&= \sum_{l=0}^{n-1} \big((i,l)\text{th element of } \Lambda\big)\cdot\big((l,k)\text{th element of } \Lambda^{-1}\big) \\
&= \sum_{l=0}^{n-1} \alpha^{il}(1/n)\alpha^{-lk} = (1/n)\sum_{l=0}^{n-1}\alpha^{l(i-k)} \qquad (4.1.6)
\end{aligned}
$$

For $i = k$, $i - k = 0$ and all the terms in the above sum become equal to 1. Therefore, the (i, k)-th element is 1 for $i = k$. For $i \neq k$, $i - k \neq 0$ and the sum corresponds to a geometric series with the first term = 1, number of terms equal to n, and the common ratio = $\alpha^{(i-k)}$. This leads to

$$\sum_{l=0}^{n-1}\alpha^{l(i-k)} = \left(1-\alpha^{n(i-k)}\right)\Big/\left(1-\alpha^{(i-k)}\right) = 0\,. \qquad (4.1.7)$$

The last equality holds due to the fact that α is an element of order n in F and, therefore, $\alpha^n = 1$ and $\alpha^{(i-k)} \neq 1$, $0 < i, k < n$. We recognize that if α is an element of order n in F, then so is α^{-1}. Consequently, the (i, k)-th element of the matrix product $[(\Lambda)\,(\Lambda^{-1})]$ is 1 for $i = k$ and is 0 for $i \neq k$ which implies that $[(\Lambda)(\Lambda^{-1})] = \mathbf{I}_{n\times n}$, thereby proving the statement of the theorem.

The factor of $(1/n)$ in the IDFT matrix can also be distributed as $1/\sqrt{n}$ in the DFT matrix and $1/\sqrt{n}$ in the IDFT matrix. A comparison of (4.1.1) and (4.1.4) reveals an interesting aspect of the DFT and IDFT computation. The IDFT computation is identical to the DFT computation with the α, the nth root of unity in F replaced by α^{-1}. We recognize that if α is an nth root of unity in the field F, then so is α^{-1}. Therefore, the computation of DFT and IDFT can be performed in an identical manner.

A major part of this book deals with the DFT computation defined in C, the field of complex numbers. In C, the n-th root of unity (denoted by α thus far) is denoted by ω and is given by

$$\omega = e^{-2\pi j/n},\; j = \sqrt{-1}\,. \qquad (4.1.8)$$

In other fields, the n-th root of unity may take different forms. For example, in the extension field of $\mathbf{Z}$ obtained by using the irreducible polynomial $1 + j + j^2 + j^3 + j^4 + j^5 + j^6$ (Problem 3.14), the 7-th root of unity is given by j which can be used to define a DFT of length 7 in this field. This aspect will be discussed thoroughly in Chapter 5.

Definition 4.3 Fast Fourier transform (FFT). The fast algorithms for the computation of DFT are collectively known as the *fast Fourier transform* (FFT).

The DFT and FFT in one and higher dimensions will be studied in great depths in Chapters 13 and 14.

4.2 Basic Formulation of Convolution

In the context of convolution, the focus of this book is on computational algorithms for the following tasks:

Task A. Acyclic convolution. Given two data sequences A_i, $i = 0, 1, \ldots, k-1$, of length k and B_j, $j = 0, 1, \ldots, d-1$, of length d, the *acyclic convolution* consists in the computation of the sequence C_l, $l = 0, 1, \ldots, n-1$, $n = k + d - 1$, defined as

$$C_l = \sum_{\substack{l=i+j \\ 0 \le i \le k-1 \\ 0 \le j \le d-1}} A_i B_j, \quad l = 0,1,\ldots,n-1. \tag{4.2.1}$$

Using the generating functions of the input and output sequences, we may write

$$C(u) = A(u) \cdot B(u). \tag{4.2.2}$$

Casting acyclic convolution in (4.2.1) in the matrix-vector form, we have

$$\begin{bmatrix} C_0 \\ C_1 \\ \vdots \\ C_{n-1} \end{bmatrix}_{n\times 1} = \begin{bmatrix} A_0 & 0 & & \cdots & 0 \\ A_1 & A_0 & & & 0 \\ \vdots & A_1 & & & \vdots \\ A_{k-1} & \vdots & & A_0 & \\ 0 & \ddots & & A_1 & A_0 \\ \vdots & & & & A_1 \\ & & & A_{k-1} & \vdots \\ 0 & & \cdots & 0 & A_{k-1} \end{bmatrix}_{n\times d} \begin{bmatrix} B_0 \\ B_1 \\ \vdots \\ B_{d-1} \end{bmatrix}_{d\times 1}. \tag{4.2.3}$$

or

$$\underline{C} = \mathbf{X}\,\underline{B}. \tag{4.2.4}$$

Task B. Cyclic convolution. Given two data sequences, A_i, $i = 0, 1, ..., n-1$, of length n and B_j, $j = 0, 1, ..., n-1$, of length n, *cyclic convolution* consists in the computation of the sequence C_l, $l = 0, 1, ..., n-1$, defined as

$$C_l = \sum_{\substack{l=(i+j)\bmod n \\ 0\le i\le n-1 \\ 0\le j\le n-1}} A_i B_j,\; l = 0,1,\ldots,n-1. \tag{4.2.5}$$

Using the generating functions of the input and output sequences, we may write

$$C(u) = A(u) \cdot B(u) \bmod (u^n - 1). \tag{4.2.6}$$

Casting cyclic convolution in (4.2.5) in the matrix-vector form, we have

$$\begin{bmatrix} C_0 \\ C_1 \\ \vdots \\ C_{n-1} \end{bmatrix}_{n\times 1} = \begin{bmatrix} A_0 & A_{n-1} & \cdots & A_1 \\ A_1 & A_0 & \cdots & A_2 \\ \vdots & \vdots & \ddots & \vdots \\ A_{n-1} & A_{n-2} & \cdots & A_0 \end{bmatrix}_{n\times n} \begin{bmatrix} B_0 \\ B_1 \\ \vdots \\ B_{n-1} \end{bmatrix}_{n\times 1} \tag{4.2.7}$$

or

$$\underline{C} = \mathbf{Y}\,\underline{B}. \tag{4.2.8}$$

Task C. Acyclic correlation. Given two data sequences A_i, $i = 0, 1, ..., k-1$, of length k and B_j, $j = 0, 1, ..., d-1$, of length d, *acyclic correlation* consists in the computation of the sequence C_l, $l = -(k-1), -(k-2), ..., d-1$, (length $n = k + d - 1$), defined as

$$C_l = \sum_{\substack{j=l+i \\ 0\le i\le k-1 \\ 0\le j\le d-1}} A_i B_j,\; l = -(k-1), -(k-2), \ldots, d-1\,. \tag{4.2.9}$$

The indices can also be adjusted in a way that the above sum is defined for $l = 0, 1, ..., n-1$. Casting correlation in (4.2.9) in the matrix-vector form, we have

$$\begin{bmatrix} C_{-(k-1)} \\ C_{-(k-2)} \\ \vdots \\ C_{d-1} \end{bmatrix}_{n\times 1} = \begin{bmatrix} A_{k-1} & 0 & & \cdots & 0 \\ A_{k-2} & A_{k-1} & & & 0 \\ \vdots & A_{k-2} & & & \vdots \\ A_0 & \vdots & & A_{k-1} & \\ 0 & \ddots & & A_{k-2} & A_{k-1} \\ \vdots & & & & A_{k-2} \\ & & & A_0 & \vdots \\ 0 & & \cdots & 0 & A_0 \end{bmatrix}_{n\times d} \begin{bmatrix} B_0 \\ B_1 \\ \vdots \\ B_{d-1} \end{bmatrix}_{d\times 1} \tag{4.2.10}$$

or

$$\underline{C} = \mathbf{X}' \, \underline{B}. \tag{4.2.11}$$

Comparing (4.2.10) with (4.2.3), we observe that the computation in (4.2.10) is the same as the one in (4.2.3) except that one of the sequences (the A_i sequence in the present formulation) is reversed. Therefore, any algorithm for the computation of acyclic correlation can be employed for the computation of acyclic convolution and vice versa. Therefore, we will focus only on the computation of acyclic convolution as expressed in (4.2.2) or (4.2.3).

Task D. Cyclic correlation. Given two data sequences, A_i, $i = 0, 1, ..., n-1$, of length n and B_j, $j = 0, 1, ..., n-1$, of length n, *cyclic correlation* consists in the computation of the sequence C_l, $l = 0, 1, ..., n-1$, defined as

$$C_l = \sum_{\substack{j=(l+i)\bmod n \\ 0\le i\le n-1 \\ 0\le j\le n-1}} A_i B_j, \; l = 0,1,\ldots,n-1\,. \tag{4.2.12}$$

Casting correlation in (4.2.12) in the matrix-vector form, we have

$$\begin{bmatrix} C_0 \\ C_1 \\ \vdots \\ C_{n-1} \end{bmatrix}_{n\times 1} = \begin{bmatrix} A_0 & A_1 & \cdots & A_{n-1} \\ A_{n-1} & A_0 & \cdots & A_{n-2} \\ \vdots & \vdots & \ddots & \vdots \\ A_1 & A_2 & \cdots & A_0 \end{bmatrix}_{n\times n} \begin{bmatrix} B_0 \\ B_1 \\ \vdots \\ B_{n-1} \end{bmatrix}_{n\times 1} \tag{4.2.13}$$

or

$$\underline{C} = \mathbf{Y}' \, \underline{B}. \tag{4.2.14}$$

Comparing (4.2.13) with (4.2.7), we observe that the computation in (4.2.13) is the same as the one in (4.2.7) except that one of the sequences (the A_i sequence in this formulation as well) is reversed. Therefore, any algorithm for the computation of cyclic correlation can be employed for the computation of cyclic convolution and vice versa. Therefore, we will focus only on the computation of cyclic convolution as expressed in (4.2.5) or (4.2.6).

4.3 Bounds on the Multiplicative Complexity

In this section, we present results relevant to establishing the lower bounds associated with the computation of acyclic and cyclic convolution expressed as matrix-vector products in (4.2.3) and (4.2.7). These results are based on the concept of rank of a matrix whose elements are defined over fields extended by indeterminate quantities. A complete discussion of this and other related concepts is far more complex and general to be of significance to our discussion on the lower bounds. The lower bounds on the multiplicative complexity of convolution algorithms constitute a very special case in the theory of computational complexity.

In the domain of arithmetic complexity, algebraic problems such as function evaluation are analyzed to determine the number of arithmetic operations required by an algorithm. Let F be the given field and $u_1, u_2, ..., u_r$ be indeterminates over F. The extension of F, denoted by $\mathsf{F}[u_1, u_2, ..., u_r]$, is the smallest commutative ring R such that R contains $\mathsf{F} \cup \{u_1, u_2, ..., u_r\}$. The model of computation that is employed is the *straight line program model*, wherein a computation consists of a sequence of instructions of the type

$$f_i \leftarrow g_i \square h_i$$

where $\square$ is one of the operations '+', '–' or '×;' f_i is a variable not appearing in any previous step; and g_i and h_i are either indeterminates, elements of F or variable names appearing on the left of the arrow at a previous step. An element of F appearing in computation is called a *constant*. A computation computes E, a set of expressions in $\mathsf{F}[u_1, u_2, ..., u_r]$, with respect to F if, for each expression e in E, there is some variable f in the computation such that the value of $f = e$.

Definition 4.4 The **multiplicative complexity** of an expression e is defined as the minimum number of instructions of the specific type $f_i \leftarrow g_i \times h_i$ required to compute e, where instructions involving multiplications (MULTs) by constants are not included.

4.3.1 Multiplicative Complexity of Bilinear Forms

Here we consider the problem of computing a system φ of k bilinear forms that can be formulated in terms of computing the product of a $(k \times s)$ matrix $\mathbf{X}$ and a column vector $\underline{Y}$. Thus, the problem may be represented as

$$\underline{\varphi} = \mathbf{X}\,\underline{Y}. \tag{4.3.1}$$

The elements of the matrix **X** are linear forms of the type $\sum_{i=1}^{r} a_i x_i$, $a_i \in \mathsf{F}$, in the indeterminates $x_1, x_2, ..., x_r$ and $\underline{Y}$ is a column vector $(y_1\, y_2 \cdots y_s)'$.

It is well known that without division, bilinear forms can be computed as linear combinations of products of pairs of linear forms in the indeterminates. An algorithm to compute φ is an expression of the form

$$\mathbf{C}\,[\mathbf{A}\,\underline{X} \otimes \mathbf{B}\,\underline{Y}], \tag{4.3.2}$$

where **A**, **B**, and **C** are matrices over F having dimensions $(n \times r)$, $(n \times s)$, and $(k \times n)$, respectively; $\underline{X}$ is the column vector $(x_1\; x_2\; \cdots\; x_r)'$; and $\otimes$ denotes a component-by-component MULT of vectors. The algorithm $\mathbf{C}\,[\mathbf{A}\,\underline{X} \otimes \mathbf{B}\,\underline{Y}]$ for computing $\mathbf{X}\,\underline{Y}$ is said to be **non-commutative** (NC). Since the straight-line program model of evaluating $\mathbf{C}\,[\mathbf{A}\,\underline{X} \otimes \mathbf{B}\,\underline{Y}]$ requires n MULTs, the multiplicative complexity of the algorithm is n. We now state various results that establish lower bounds for the multiplicative complexity of the algorithms used to compute the systems of bilinear forms.

Definition 4.5 Row rank. Let $\mathsf{F}^m[x_1, x_2, \cdots, x_r]$ be the m-dimensional vector space with components from $\mathsf{F}[x_1, x_2, \cdots, x_r]$ and F^m be the m-dimensional vector space with components from F. A set of vectors $\{\underline{V}_1, ..., \underline{V}_a\}$ from $\mathsf{F}^m[x_1, x_2, \cdots, x_r]$ is *linearly independent* modulo F^m, if for $C_1, C_2, \cdots, C_a$ in F, $\sum_{i=1}^{a} C_i \underline{V}_i$ in F^m implies that all the C_i are zero. The **row rank** of a $(k \times s)$ matrix **X** modulo F^s (referred to as row rank of **X** in the sequel) is the number of linearly independent rows of **X** modulo F^s. The **column rank** of **X** modulo F^k is defined analogously.

Theorem 4.2 A row-oriented lower bound on multiplications. Let $\mathbf{X}\,\underline{Y}$ be a system of bilinear forms over F. If the row rank of **X** is a (over F), then any computation of $\mathbf{X}\,\underline{Y}$ requires at least a MULTs.

Theorem 4.3 A column-oriented lower bound on multiplications. Let $\mathbf{X}\,\underline{Y}$ be a system of bilinear forms over F. If the column rank of **X** is b (over F), then any computation of $\mathbf{X}\,\underline{Y}$ requires at least b MULTs.

Theorem 4.4 A row- and column-oriented bound on multiplications. Let $\mathbf{X}\,\underline{Y}$ be a system of bilinear forms over F. If $\mathbf{X}$ has a sub-matrix $\mathbf{W}$ with a rows and b columns such that for any vectors $\underline{U}$ and $\underline{V}$ in F^a and F^b, respectively, $\underline{U}^t\,\mathbf{W}\,\underline{V}$ is an element of F *iff* either $\underline{U} = \underline{0}$ or $\underline{V} = \underline{0}$, then any computation of $\mathbf{X}\,\underline{Y}$ requires at least $(a + b - 1)$ MULTs.

It is easily observed that the computation of convolutions in (4.2.3) and (4.2.7) constitute a bilinear form. Therefore, results associated with the bilinear forms are directly applicable to (4.2.3) and (4.2.7). In subsequent chapters, we will cast the overall algorithms for acyclic and cyclic convolution as a bilinear form in (4.3.2). Using the observation that the structure of the matrix $\mathbf{X}$ for acyclic convolution in (4.2.3) is such that there is exactly one row vector for which an indeterminate appears in a given position, we see that the row rank of the matrix $\mathbf{X}$ is $n = k + d - 1$. Based on Theorem 4.2, we conclude that any algorithm for the acyclic convolution computation in (4.2.3) will require at least $k + d - 1$ MULTs. It is interesting to note here that the column rank of the $\mathbf{X}$ matrix in (4.2.3) is d and if we were to apply Theorem 4.3 to $\mathbf{X}$, then the minimum number of MULTs will turn out to be d, a bound which is meaningless in presence of the lower bound of $k + d - 1$ MULTs. In this case, Theorem 4.4 may also be used to obtain the lower bound of $k + d - 1$ MULTs. We will demonstrate in Section 4.4.1 that the LI provides us with an algorithm that requires exactly $k + d - 1$ MULTs, thereby meeting this lower bound with equality. Similarly, the structure of the matrix $\mathbf{Y}$ for cyclic convolution in (4.2.7) is such that all the indeterminates in the first position of any row (or the first position of every column) are distinct. Therefore, row rank of $\mathbf{Y}$ = column rank of $\mathbf{Y} = n$. Based on Theorem 4.2 (or Theorem 4.3), we conclude that any algorithm for the cyclic convolution computation in (4.2.7) will require at least n MULTs. A DFT based algorithm for computing the cyclic convolution meets the lower bound of n MULTs with equality.

4.3.2 Dual of a Bilinear Form

Definition 4.6 Dual of a bilinear form. Let $\underline{Z}^t$ be the column vector $(z_1\ z_2\ \cdots\ z_k)^t$ and $\mathbf{C}\,[\mathbf{A}\,\underline{X} \otimes \mathbf{B}\,\underline{Y}]$ be a computation of φ. The P-dual of the computation is the computation $\mathbf{A}^t\,[\mathbf{C}^t\,\underline{Z} \otimes \mathbf{B}\,\underline{Y}]$; the R-dual of the computation is the computation $\mathbf{B}^t\,[\mathbf{A}\,\underline{X} \otimes \mathbf{C}^t\,\underline{Z}]$.

The procedure to obtain the P-dual is as follows. Given the bilinear form $\underline{\varphi} = \mathbf{X}\,\underline{Y}$, consider the trilinear form

$$\underline{Z}^t\,\underline{\varphi} = \underline{Z}^t\,\mathbf{X}\,\underline{Y} \qquad (4.3.3)$$

and express it as

$$\underline{Z}^t\, \underline{\varphi} = \underline{Z}^t\, \mathbf{X}\, \underline{Y} = \underline{X}^t\, \mathbf{Z}\, \underline{Y}. \tag{4.3.4}$$

The P-dual of the bilinear form $\underline{\varphi} = \mathbf{X}\, \underline{Y}$ is the bilinear form $\underline{\kappa} = \mathbf{Z}\, \underline{Y}$. In this connection, the following theorem is also of interest.

Theorem 4.5 There is a computation for the system of expressions represented by $\mathbf{C}\, [\mathbf{A}\, \underline{X} \otimes \mathbf{B}\, \underline{Y}]$ having n MULTs *iff* there is a computation having n MULTs for its P-dual $\mathbf{A}^t\, [\mathbf{C}^t\, \underline{Z} \otimes \mathbf{B}\, \underline{Y}]$, its R-dual $\mathbf{B}^t\, [\mathbf{A}\, \underline{X} \otimes \mathbf{C}^t\, \underline{Z}]$; the vector reversed system of expressions $\mathbf{C}\, [\mathbf{B}\, \underline{Y} \otimes \mathbf{A}\, \underline{X}]$, and the vector reversed systems of expressions for the P-dual and R-dual represented by $\mathbf{A}^t\, [\mathbf{B}\, \underline{Y} \otimes \mathbf{C}^t\, \underline{Z}]$ and $\mathbf{B}^t\, [\mathbf{C}^t\, \underline{Z} \otimes \mathbf{A}\, \underline{X}]$, respectively.

We now establish the following result.

Theorem 4.6 The multiplicative complexity of the computation

$$\phi(u) = Z(u) \cdot Y(u) \bmod P(u) \tag{4.3.5}$$

over the field F is $2n - 1$ for a polynomial $P(u)$ that is irreducible over F, where $deg(Z(u)) = deg(Y(u)) = n - 1$ and $deg(P(u)) = n$.
Proof: Let the irreducible polynomial $P(u)$ be given by

$$P(u) = -\sum_{i=0}^{n-1} p_{i,1} u^i + u^n. \tag{4.3.6}$$

Here we have assumed that $P(u)$ is monic without any loss in generality. Given (4.3.6), for any computation defined mod $P(u)$, we may write

$$u^n \equiv \sum_{i=0}^{n-1} p_{i,1} u^i \bmod P(u). \tag{4.3.7}$$

A recursive application of (4.3.7) leads to

$$u^{n+j} \equiv \sum_{i=0}^{n-1} p_{i,j+1} u^i \bmod P(u),\ j = 1, 2, \ldots \tag{4.3.8}$$

Based on (4.3.7) and (4.3.8), the computation in (4.3.5) may be expressed as

$$\phi(u)=\sum_{i=0}^{n-1}\phi_i u^i=(z_0y_0)u+(z_0y_1+z_1y_0)u^2+\ldots+\left(\sum_{l=0}^{n-1}z_ly_{n-1-l}\right)u^{n-1}$$
$$+\left(\sum_{l=1}^{n-1}z_ly_{n-l}\right)\sum_{i=0}^{n-1}p_{i,1}u^i+\left(\sum_{l=2}^{n-1}z_ly_{n+1-l}\right)\sum_{i=0}^{n-1}p_{i,2}u^i+\ldots$$
$$+\left(\sum_{l=j}^{n-1}z_ly_{n-1+j-l}\right)\sum_{i=0}^{n-1}p_{i,j}u^i+\ldots+(z_{n-1}y_{n-1})\sum_{i=0}^{n-1}p_{i,n-1}u^i. \tag{4.3.9}$$

Expressed as a bilinear form, (4.3.9) becomes

$$\underline{\phi}=\left\{\begin{bmatrix} z_0 & 0 & \cdots & 0 & 0\\ z_1 & z_0 & 0 & \ddots & 0\\ z_2 & z_1 & z_0 & 0 & \vdots\\ \vdots & \ddots & \ddots & \ddots & 0\\ z_{n-1} & \cdots & z_2 & z_1 & z_0\end{bmatrix}+\begin{bmatrix} 0 & z_{n-1}p_{0,1} & z_{n-2}p_{0,1} & \cdots & z_1p_{0,1}\\ 0 & z_{n-1}p_{1,1} & z_{n-2}p_{1,1} & \cdots & z_1p_{1,1}\\ 0 & z_{n-1}p_{2,1} & z_{n-2}p_{2,1} & \cdots & z_1p_{2,1}\\ \vdots & \vdots & \vdots & \cdots & \vdots\\ 0 & z_{n-1}p_{n-1,1} & z_{n-2}p_{n-1,1} & \cdots & z_1p_{n-1,1}\end{bmatrix}\right.$$
$$+\begin{bmatrix} 0 & 0 & z_{n-1}p_{0,2} & \cdots & z_2p_{0,2}\\ 0 & 0 & z_{n-1}p_{1,2} & \cdots & z_2p_{1,2}\\ 0 & 0 & z_{n-1}p_{2,2} & \cdots & z_2p_{2,2}\\ \vdots & \vdots & \vdots & \cdots & \vdots\\ 0 & 0 & z_{n-1}p_{n-1,2} & \cdots & z_2p_{n-1,2}\end{bmatrix}$$
$$\left.+\ldots+\begin{bmatrix} 0 & 0 & \cdots & 0 & z_{n-1}p_{0,n-1}\\ 0 & 0 & \cdots & 0 & z_{n-1}p_{1,n-1}\\ 0 & 0 & \cdots & 0 & z_{n-1}p_{2,n-1}\\ \vdots & \vdots & \vdots & \cdots & \vdots\\ 0 & 0 & \cdots & 0 & z_{n-1}p_{n-1,n-1}\end{bmatrix}\right\}\begin{bmatrix} y_0\\ y_1\\ \vdots\\ y_{n-1}\end{bmatrix}. \tag{4.3.10}$$

Based on the above expression, consider the bilinear form

$$\underline{K} = \begin{bmatrix} x_0 & x_1 & x_2 & \cdots & x_{n-1} \\ x_1 & x_2 & \cdot^{\cdot^{\cdot}} & \cdot^{\cdot^{\cdot}} & \sum_{i=0}^{n-1} x_i p_{i,1} \\ x_2 & \cdot^{\cdot^{\cdot}} & \cdot^{\cdot^{\cdot}} & \sum_{i=0}^{n-1} x_i p_{i,1} & \cdot^{\cdot^{\cdot}} \\ \vdots & x_{n-1} & \cdot^{\cdot^{\cdot}} & \cdot^{\cdot^{\cdot}} & \sum_{i=0}^{n-1} x_i p_{i,n-2} \\ x_{n-1} & \sum_{i=0}^{n-1} x_i p_{i,1} & \cdot^{\cdot^{\cdot}} & \sum_{i=0}^{n-1} x_i p_{i,n-2} & \sum_{i=0}^{n-1} x_i p_{i,n-1} \end{bmatrix} \begin{bmatrix} y_0 \\ y_1 \\ \vdots \\ y_{n-1} \end{bmatrix}. \tag{4.3.11}$$

or

$$\underline{K} = \mathbf{X}\, \underline{Y}. \tag{4.3.12}$$

Using the procedure described to obtain the dual form, it can be verified that the bilinear form in (4.3.11) is the P-dual of the bilinear form in (4.3.10).

Based on Theorem 4.5, we can say that the bilinear forms in (4.3.10) and (4.3.11) have the same multiplicative complexity. We will employ Theorem 4.4 in order to analyze the P-dual form for a lower bound on its multiplicative complexity. Setting $\mathbf{W} = \mathbf{X}$, $a = n$, $b = n$, in the statement of Theorem 4.4, we have

$$\begin{aligned} \underline{U}^t\, \mathbf{X}\, \underline{V} &= [u_0\ u_1\ \cdots\ u_{n-1}]\, \mathbf{X}\, [v_0\ v_1\ \cdots\ v_{n-1}]^t \\ &= (u_0 v_0) x_0 + (u_0 v_1 + u_1 v_0) x_1 + \ldots + \left(\sum_{l=0}^{n-1} u_l v_{n-1-l} \right) x_{n-1} \\ &+ \left(\sum_{l=1}^{n-1} u_l v_{n-l} \right) \sum_{i=0}^{n-1} p_{i,1} x_i + \left(\sum_{l=2}^{n-1} u_l v_{n+1-l} \right) \sum_{i=0}^{n-1} p_{i,2} x_i + \ldots \\ &+ \left(\sum_{l=j}^{n-1} u_l v_{n-1+j-l} \right) \sum_{i=0}^{n-1} p_{i,j} x_i + \ldots + (u_{n-1} v_{n-1}) \sum_{i=0}^{n-1} p_{i,n-1} x_i\,. \end{aligned} \tag{4.3.13}$$

For $\underline{U}^t\, \mathbf{X}\, \underline{V}$ to be an element in F, we must have coefficient of x_0 = coefficient of $x_1 = \ldots =$ coefficient of $x_{n-1} = 0$. It is seen from (4.3.13) that

coeff. of x_0

$$= u_0 v_0 + \left(\sum_{l=1}^{n-1} u_l v_{n-l} \right) p_{0,1} + \left(\sum_{l=2}^{n-1} u_l v_{n+1-l} \right) p_{0,2} + \ldots + (u_{n-1} v_{n-1}) p_{0,n-1}\,.$$

In general,

$$\text{coeff. of } x_j = \left(\sum\nolimits_{l=0}^{j} u_l v_{j-l}\right) + \left(\sum\nolimits_{l=1}^{n-1} u_l v_{n-l}\right) p_{j,1} + \ldots + \left(u_{n-1} v_{n-1}\right) p_{j,n-1},$$

and so on. From these expressions for x_0, x_1, ..., x_{n-1}, we observe that the coefficient of x_j is the coefficient of u^j in the computation $U(u)Y(u) \bmod P(u)$ or

$$X(u) = \sum_{i=0}^{n-1} x_i u^i = \left(\sum_{i=0}^{n-1} u_i u^i\right)\left(\sum_{i=0}^{n-1} v_i u^i\right) \bmod P(u) = U(u)V(u) \bmod P(u). \tag{4.3.14}$$

Consequently, the condition that

$$\text{coefficient of } x_0 = \text{coefficient of } x_1 = \ldots = \text{coefficient of } x_{n-1} = 0$$

for $\underline{U}^t \mathbf{X}\, \underline{V}$ to be an element in F, simplifies to

$$U(u)\, V(u) \bmod P(u) \equiv 0 \bmod P(u).$$

Since $P(u)$ is given to be an irreducible polynomial of degree n over F, and $deg(U(u)) = deg(V(u)) = n - 1$, this condition is satisfied *iff* either $U(u) = 0$ ($\underline{U} = \underline{0}$) or $V(u) = 0$ ($\underline{V} = \underline{0}$) or both $= 0$. Returning back to the statement of Theorem 4.4, we see that the computation of the bilinear form in (4.3.10) and, therefore, the computation of the bilinear form in (4.3.11) requires at least $n + n - 1 = 2n - 1$ MULTs.

This theorem can be extended to the case when $P(u)$ is a polynomial of degree n and is of the form $P(u) = (Q(u))^s$, $Q(u)$ being an irreducible polynomial over F. This leads to the following theorem.

Theorem 4.7 The multiplicative complexity of the computation

$$\phi(u) = Z(u) \cdot Y(u) \bmod P(u) \tag{4.3.15}$$

over the field F is $2n - 1$ for a polynomial $P(u)$ that is power of an irreducible polynomial $Q(u)$ over F, that is $P(u) = (Q(u))^s$, where $deg(Z(u)) = deg(Y(u)) = n - 1$ and $deg(P(u)) = n$.

4.4 Basic Formulation of Convolution Algorithms

4.4.1 The Acyclic Convolution Algorithm

If F is sufficiently large, then a CRT-P (or LI) based algorithm for computing acyclic convolution of length n can be described as follows. Choose any n elements of F, say $f_1, f_2, \ldots, f_n$, and let $P(u) = \prod_{i=1}^{n}(u - f_i)$. The computation of acyclic convolution $C(u) = A(u) \cdot B(u)$, $deg(A(u)) = k - 1$, $deg(B(u)) = d - 1$, $deg(C(u)) = n - 1$, $n = k + d - 1$, in (4.2.2) is performed in three steps:

Step 1. Evaluate the polynomials $A(u)$ and $B(u)$ at $f_1, f_2, \ldots, f_n$ to get

$$A_i = A(u) \bmod (u - f_i) = A(f_i),$$
$$B_i = B(u) \bmod (u - f_i) = B(f_i), \quad i = 1, 2, \ldots, n.$$

Step 2. Evaluate the product $C(u) \equiv A(u)\, B(u) \bmod (u - f_i)$ using

$$C_i \equiv C(u) \bmod (u - f_i) = C(f_i) = A(u)\, B(u) \bmod (u - f_i)$$
$$= A(f_i) \cdot B(f_i), \quad i = 1, 2, \ldots, n.$$

Step 3. Solve the Vandermonde system of linear equations or use the CRT-P reconstruction in (3.6.1) to obtain the polynomial $C(u)$ from C_i, $i = 1, 2, \ldots, n$.

This algorithm for computing the acyclic convolution is also known as the *Cook-Toom algorithm*. Steps 1 and 3 require only additions and MULTs by the elements of the field. In the present framework, the MULTs by field elements (termed constants) are not counted towards multiplicative complexity (even though such MULTs may be non-trivial in nature). Therefore, Steps 1 and 3 require no MULTs. Step 2 requires n MULTs. Thus, the overall LI based acyclic convolution algorithm requires exactly n MULTs, thereby meeting the lower bound on multiplicative complexity. The overall algorithm can be cast in the matrix form as $\mathbf{C}[\mathbf{A}\ \underline{A} \otimes \mathbf{B}\ \underline{B}]$.

Even though the above algorithm for computation of acyclic convolution meets the lower bound in terms of multiplicative complexity, it may suffer from several drawbacks. It is valid only over those fields that contain at least n distinct elements. In Chapter 8, we will show that the algorithm under discussion can be slightly modified, thereby requiring F to contain $n - 1$ distinct elements in place of n. Second, Steps 1 and 3 may require additions and MULTs that are non-trivial and make the algorithm unattractive from the point of view of overall realization. In such cases, it may be desirable to have an algorithm that has a slightly higher multiplicative complexity in Step 2 but rather simple operations in Steps 1 and 3.

In situations where the LI based algorithm becomes impractical, we may employ the CRT-P in Theorem 3.3 for computing acyclic convolution in the following manner. Choose a polynomial $P(u)$ having degree n such that $P(u) = \prod_{i=1}^{a} p_i(u)$, where each of the polynomials $p_i(u)$ is either irreducible or power of an irreducible polynomial over F, $deg(p_i(u)) = n_i$, $n = n_1 + n_2 + ... + n_a$. The computation of the acyclic convolution $C(u) = A(u) \cdot B(u)$, $deg(A(u)) = k - 1$, $deg(B(u)) = d - 1$, $deg(C(u)) = n$, $n = k + d - 1$, in (4.2.2) is performed in three steps:

Step 1. Evaluate the polynomials $A(u)$ and $B(u)$ mod $p_i(u)$ to get

$$a_i(u) \equiv A(u) \bmod p_i(u),$$
$$b_i(u) \equiv B(u) \bmod p_i(u), \quad i = 1, 2, ..., a.$$

Step 2. Evaluate the product $C(u) \equiv A(u)\, B(u) \bmod p_i(u)$ using

$$c_i(u) \equiv C(u) \bmod p_i(u) \equiv a_i(u)\, b_i(u) \bmod p_i(u), \quad i = 1, 2, ..., a.$$

Step 3. Use the CRT-P reconstruction in (3.4.2) to obtain the polynomial $C(u)$ from $c_i(u)$, $i = 1, 2, ..., a$.

The CRT-P can be further employed to obtain the algorithm for small degree polynomial MULT in Step 2. Once again, Steps 1 and 3 require additions and MULTs only by the elements of the field and, therefore, these operations are not counted towards multiplicative complexity. Thus, the multiplicative complexity of the CRT-P based acyclic convolution algorithm is determined by the multiplicative complexity of Step 2. As per theorems 4.6 and 4.7, the computation $c_i(u) = a_i(u)\, b_i(u) \bmod p_i(u)$ requires at least $2n_i - 1$ MULTs. Therefore, the overall algorithm requires at least $\sum_{i=1}^{a} (2n_i - 1) = 2n - a$ MULTs. From the point of view of this lower bound, it will be *desirable* to choose the modulo polynomial $P(u)$ to have as many irreducible factors as possible. This analysis leads to the following theorem.

Theorem 4.7 The CRT-P based algorithm for the computation of $A(u)\, B(u) \bmod P(u)$, $deg(P(u)) = n$ requires at least $2n - a$ MULTs, where a = number of distinct factors of the modulo polynomial $P(u)$ over F.

Once again, the overall algorithm can be cast in the matrix form as $\mathbf{C}\,[\mathbf{A}\,\underline{A} \otimes \mathbf{B}\,\underline{B}]$.

4.4.2 The Cyclic Convolution Algorithm

Consider the computation of cyclic convolution as expressed in (4.2.6). In this case, the modulo polynomial is fixed as $u^n - 1$. If the given field F contains a n-th root of unity, then the factorization of $u^n - 1$ is given by

$$u^n - 1 = \prod_{i=1}^{n} \left(u - \alpha^i\right), \tag{4.4.1}$$

where α represents a n-th root of unity in F. For example, if $\mathsf{F} = C$, then the n-th root of unity is given by

$$\alpha = \omega = e^{-2\pi j/n},\, j = \sqrt{-1}\,. \tag{4.4.2}$$

A CRT-P (or LI) based algorithm for computing cyclic convolution of length n can be described as follows. The computation of the cyclic convolution

$$C(u) = A(u) \cdot B(u) \bmod u^n - 1,\; deg(A(u)) = deg(B(u)) = deg(C(u)) = n - 1,$$

in (4.2.6) is performed in three steps:

Step 1. Evaluate the polynomials $A(u)$ and $B(u)$ at α^i to get

$$\boldsymbol{A}_i = A(u) \bmod (u - \alpha^i) = A(\alpha^i),$$
$$\boldsymbol{B}_i = B(u) \bmod (u - \alpha^i) = B(\alpha^i),\;\; i = 0, 1, ..., n - 1.$$

or

$$\boldsymbol{A}_i = \sum_{l=0}^{n-1} A_l \alpha^{il} \tag{4.4.3}$$

and

$$\boldsymbol{B}_i = \sum_{l=0}^{n-1} B_l \alpha^{il}\;,\;\; i = 0, 1, .., n - 1. \tag{4.4.4}$$

Step 2. Evaluate the product $C(u) = A(u)\, B(u) \bmod (u - \alpha^i)$ using

$$\boldsymbol{C}_i = C(\alpha^i) = C(u) \bmod (u - \alpha^i)$$
$$= A(u)\, B(u) \bmod (u - \alpha^i) = \boldsymbol{A}_i \cdot \boldsymbol{B}_i\,,\;\; i = 0, 1, ..., n - 1.$$

Step 3. Use the CRT-P reconstruction in (3.6.1) to obtain the polynomial $C(u)$ from C_l, $l = 0, 1, ..., n - 1$.

Steps 1 and 3 require only additions and MULTs by the elements of the field. Therefore, Steps 1 and 3 require no MULTs in the present framework. Step 2 requires n MULTs. Thus, the overall CRT-P based cyclic convolution algorithm requires exactly n MULTs, thereby meeting the lower bound on multiplicative complexity. The overall algorithm can also be cast in the matrix form as $\mathbf{C}[\mathbf{A}\,\underline{A} \otimes \mathbf{B}\,\underline{B}]$.

A quick comparison of the above algorithm for the computation of the cyclic convolution with the definitions of DFT and IDFT reveals that it can be also be described as follows.

Step 1. Compute DFT of the input data sequences A_l and B_l, $l = 0, 1, ..., n-1$. Let the DFTs be denoted by $\boldsymbol{A}_i$ and $\boldsymbol{B}_i$, $i = 0, 1, ..., n-1$.

Step 2. Compute DFT of the output sequence by taking the point-by-point product of the DFTs of the input sequences, that is,

$$\boldsymbol{C}_i = \boldsymbol{A}_i \cdot \boldsymbol{B}_i, \quad i = 0, 1, ..., n-1.$$

Step 3. Compute IDFT of the sequence $\boldsymbol{C}_i$, $i = 0, 1, ..., n-1$, to obtain the output sequence C_l, $l = 0, 1, ..., n-1$.

The various matrices in the bilinear form $\mathbf{C}\,[\mathbf{A}\,\underline{A} \otimes \mathbf{B}\,\underline{B}]$ are seen to be

$$\mathbf{A} = \mathbf{B} = \begin{bmatrix} 1 & 1 & 1 & \cdots & 1 & 1 \\ 1 & \alpha & \alpha^2 & \cdots & \alpha^{n-2} & \alpha^{n-1} \\ 1 & \alpha^2 & \alpha^4 & \cdots & \alpha^{2(n-2)} & \alpha^{2(n-1)} \\ \vdots & \vdots & \vdots & \ddots & \vdots & \vdots \\ 1 & \alpha^{n-2} & \alpha^{(n-2)2} & \cdots & \alpha^{(n-2)(n-2)} & \alpha^{(n-2)(n-1)} \\ 1 & \alpha^{n-1} & \alpha^{(n-1)2} & \cdots & \alpha^{(n-1)(n-2)} & \alpha^{(n-1)(n-1)} \end{bmatrix},$$

and

$$\mathbf{C} = \left(\frac{1}{n}\right) \begin{bmatrix} 1 & 1 & 1 & \cdots & 1 & 1 \\ 1 & \alpha^{-1} & \alpha^{-2} & \cdots & \alpha^{-(n-2)} & \alpha^{-(n-1)} \\ 1 & \alpha^{-2} & \alpha^{-4} & \cdots & \alpha^{-2(n-2)} & \alpha^{-2(n-1)} \\ \vdots & \vdots & \vdots & \ddots & \vdots & \vdots \\ 1 & \alpha^{-(n-2)} & \alpha^{-(n-2)2} & \cdots & \alpha^{-(n-2)(n-2)} & \alpha^{-(n-2)(n-1)} \\ 1 & \alpha^{-(n-1)} & \alpha^{-(n-1)2} & \cdots & \alpha^{-(n-1)(n-2)} & \alpha^{-(n-1)(n-1)} \end{bmatrix}.$$

The DFT approach for computing cyclic convolution can also be treated as a special case of LI or the Cook-Toom algorithm when all the interpolating points are expressed as consecutive powers of α, the n-th root of unity. Due to its fundamental importance in digital signal processing, we will analyze the matrix structures associated with this algorithm in great detail in Chapters 13 and 14.

4.4.3 Converting Acyclic to a Cyclic Convolution

A length n acyclic convolution $C(u) = A(u) \cdot B(u)$, $deg(A(u)) = k - 1$, $deg(B(u)) = d - 1$, $deg(C(u)) = n - 1$, $n = k + d - 1$, can also be defined as a cyclic convolution of length $\geq n$. Let us say that we wish to convert it to a cyclic convolution of length equal to n. This is accomplished by appending $d - 1$ and $k - 1$ zeros to $A(u)$ and $B(u)$ polynomials, respectively, thereby converting them to degree $n - 1$ polynomials. The original acyclic convolution is converted to a cyclic convolution by recasting it as the computation $C(u) = A(u) \cdot B(u)$ mod $(u^n - 1)$. The three-step algorithm for such a cyclic convolution based acyclic convolution can be summarized as follows.

Step 1. Compute DFT of the $A(u)$ and $B(u)$ polynomials after appending $d - 1$ and $k - 1$ zeros to them.

Step 2. Compute DFT of the output polynomial by taking the point-by-point of the DFTs of $A(u)$ and $B(u)$.

Step 3. Compute IDFT of the DFT of the output polynomial obtained in Step 2.

In Step 1 of the algorithm advantage may be taken by noting that the last $d - 1$ and $k - 1$ coefficients of the input polynomials $A(u)$ and $B(u)$ are always zero in such a framework. Once again, the overall algorithm requires only $n = k + d - 1$ MULTs in Step 2.

4.4.4 Cyclic Convolution Beyond DFT

In situations where n-th root of unity does not exist in the field F over which cyclic convolution is defined, $u^n - 1$ does not factor into n degree 1 polynomials. As a result, the DFT approach is no longer valid in such a case. Examples of such fields are Z, the field of rational numbers, Re, the field of real numbers, and CZ, the field of complex rational numbers. Exceptions are Z for $n = 2$, Re for $n = 2$, and CZ for $n = 2, 4$. Several distinct approaches can be adopted in such a case. One such approach is to factorize $u^n - 1$ over F in terms of factors having degree greater than 1. Let us say that this factorization is given by

$$u^n - 1 = \prod_{i=1}^{s} p_i(u). \tag{4.4.5}$$

Based on this factorization, a CRT-P based algorithm for cyclic convolution can be described in three steps as follows.

Step 1. CRT-P reduction. Compute the residue polynomials $A_i(u)$ and $B_i(u)$, $i = 1, \ldots, s$, as given by

$$a_i(u) \equiv A(u) \bmod p_i(u)$$
$$b_i(u) \equiv B(u) \bmod p_i(u), \quad i = 1, 2, \ldots, s.$$

Step 2. Small degree polynomial MULT. Compute the residue $c_i(u) \bmod p_i(u)$ as

$$c_i(u) \equiv a_i(u) \cdot b_i(u) \bmod p_i(u), \quad i = 1, 2, \ldots, s.$$

Step 3. CRT-P reconstruction. Employ the CRT-P in (3.4.2) to obtain $C(u)$ from its residues $c_i(u) \bmod p_i(u)$, $i = 1, 2, \ldots, s$.

Several observations are in order with regard to the algorithm outlined above. The CRT-P can be further employed to obtain the algorithm for small degree polynomial MULT in Step 2. Applying Theorem 4.7 to this formulation, we see that the minimum number of MULTs required for the overall algorithm (only Step 2 requires MULT in the present context) is $2n - s$. Therefore, one is motivated to factorize the polynomial $u^n - 1$ into as many factors as possible in order to maximize s. This will be accomplished by *factorizing* $u^n - 1$ *in terms of its irreducible factors* which is the focus of the next chapter. Finally, this algorithm can also be cast in the bilinear form $\mathbf{C}\,[\mathbf{A}\,\underline{A} \otimes \mathbf{B}\,\underline{B}]$.

4.5 Matrix Exchange Property

In many instances, one faces a computation of the type

$$C(u) \equiv A(u)B(u) \bmod (u^n - \lambda), \tag{4.5.1}$$

where $deg(C(u)) = deg(A(u)) = deg(B(u)) = n - 1$, and λ is an indeterminate or an element of the field F over which the computation is defined. Casting (4.5.1) as a matrix-vector product, we have

$$\begin{bmatrix} C_0 \\ C_1 \\ \vdots \\ C_{n-1} \end{bmatrix} = \begin{bmatrix} A_0 & \lambda A_{n-1} & \cdots & \lambda A_2 & \lambda A_1 \\ A_1 & A_0 & \lambda A_{n-1} & \ddots & \lambda A_2 \\ \vdots & A_1 & A_0 & \ddots & \vdots \\ A_{n-2} & \ddots & \ddots & \ddots & \lambda A_{n-1} \\ A_{n-1} & A_{n-2} & \cdots & A_1 & A_0 \end{bmatrix} \begin{bmatrix} B_0 \\ B_1 \\ \vdots \\ B_{n-1} \end{bmatrix} \tag{4.5.2}$$

or

$$\underline{C} = \mathbf{X}\, \underline{B}. \tag{4.5.3}$$

It is clear from this formulation that $\lambda = 1$ gives the cyclic convolution. Another interesting case is $\lambda = -1$ which gives rise to the computation $A(u)B(u)$ mod $(u^n + 1)$. This is called **skew circular convolution** in the digital signal processing literature.

The bilinear algorithm for the computation in (4.5.2) can be cast as $\mathbf{C}\,[\mathbf{A}\,\underline{A} \otimes \mathbf{B}\,\underline{B}]$. If the i-th row of $\mathbf{A}$ is $\underline{A}_i^t$, then i-th element in the matrix-vector product $\mathbf{A}\,\underline{A}$ is given by $\underline{A}_i^t \cdot \underline{A}$. Based on this representation, define a diagonal matrix Ω having its i-th diagonal element $\Omega_i = \underline{A}_i^t \cdot \underline{A}$. Thus, the bilinear algorithm $\mathbf{C}\,[\mathbf{A}\,\underline{A} \otimes \mathbf{B}\,\underline{B}]$ can also be expressed as

$$\underline{C} = \mathbf{C}\, \Omega\, \mathbf{B}\, \underline{B}. \tag{4.5.4}$$

A comparison of (4.5.3) with (4.5.4) reveals that the bilinear algorithm can be considered an appropriate way to obtain the factorization of the coefficient $\mathbf{X}$ matrix as

$$\mathbf{X} = \mathbf{C}\, \Omega\, \mathbf{B}. \tag{4.5.5}$$

Let $\mathbf{J}$ be the $n \times n$ dimensional contra-identity matrix having 1 on the main anti-diagonal and zeros elsewhere, that is, if J_{ij} is the (i, j)-th element of $\mathbf{J}$ then J_{ij} is 1 for $i + j = n + 1$ and zero otherwise. It is easily seen that if $\underline{D}_i^t$ is the i-th row of an $n \times n$ matrix $\mathbf{D}$, then pre-MULT of $\mathbf{D}$ by $\mathbf{J}$ (= $\mathbf{J}\,\mathbf{D}$) is a matrix $\mathbf{E}$ whose i-th row is given by $\underline{D}_{n+1-i}^t$, $i = 1, 2, \ldots, n$. Similarly, if $\underline{S}_i$ is the i-th column of an $n \times n$ matrix $\mathbf{D}$, then post-MULT of $\mathbf{D}$ by $\mathbf{J}$ (= $\mathbf{D}\,\mathbf{J}$) is a matrix $\mathbf{F}$ whose i-th column is given by $\underline{S}_{n+1-i}$, $i = 1, 2, \ldots, n$. Applying these two properties to the coefficient matrix $\mathbf{X}$ in (4.5.2), we obtain the following result:

$$\mathbf{X}^t = \mathbf{J}\, \mathbf{X}\, \mathbf{J}. \tag{4.5.6}$$

Taking transpose of (4.5.6), we get

$$\mathbf{X} = \mathbf{J}\,\mathbf{X}'\,\mathbf{J}. \qquad (4.5.7)$$

Taking transpose of **X** in (4.5.5) and substituting in (4.5.7), we get

$$\mathbf{X} = (\mathbf{J}\,\mathbf{B}')\,\Omega\,(\mathbf{C}'\,\mathbf{J}). \qquad (4.5.8)$$

The above identities lead to the following theorem.

Theorem 4.8 Matrix Exchange Property. If $\mathbf{X} = \mathbf{C}\,\Omega\,\mathbf{B}$ is a factorization of the coefficient matrix **X** in (4.5.2), then $\mathbf{X} = (\mathbf{J}\,\mathbf{B}')\,\Omega\,(\mathbf{C}'\,\mathbf{J})$ is also a factorization of **X**. More importantly, if $\mathbf{C}\,[\mathbf{A}\,\underline{A} \otimes \mathbf{B}\,\underline{B}]$ is a bilinear algorithm for the computation in (4.5.2) based on $\mathbf{X} = \mathbf{C}\,\Omega\,\mathbf{B}$, then $(\mathbf{J}\,\mathbf{B}')\,[\mathbf{A}\,\underline{A} \otimes (\mathbf{C}'\,\mathbf{J})\,\underline{B}]$ is also a bilinear algorithm for the computation in (4.5.2) based on the factorization $\mathbf{X} = (\mathbf{J}\,\mathbf{B}')\,\Omega\,(\mathbf{C}'\,\mathbf{J})$.

As described, it is trivial to obtain the matrices $(\mathbf{J}\,\mathbf{B}')$ and $(\mathbf{C}'\,\mathbf{J})$ once the matrices **B** and **C** are known. This theorem can be employed to exchange the various matrices involved in the bilinear algorithm in a way so as to put the overall algorithm in as convenient a form as possible. For example, in those applications where (i) the matrices **A** and **B** have simple elements such as 0, 1, −1 while the elements of the **C** matrix are not so simple; and (ii) one of the data sequences, say $\underline{B}$, is fixed, it may be highly desirable to pre-compute and store $(\mathbf{C}'\,\mathbf{J})\,\underline{B}$ and employ the bilinear algorithm expressed as $(\mathbf{J}\,\mathbf{B}')\,[\mathbf{A}\,\underline{A} \otimes (\mathbf{C}'\,\mathbf{J})\,\underline{B}]$. Even otherwise, it gives us additional freedom in the organization of the various computations involved in the bilinear algorithm.

Notes

The usefulness of bilinear forms for digital signal processing and other related areas came to the author's attention while studying the connection between algorithms for digital signal processing, systems of bilinear forms, and binary linear codes. The important references in this context are Lempel and Winograd [1977], Brockett and Dobkin [1973], and Krishna [1987]. Our exposition of the multiplicative complexity of bilinear forms closely follows the more general approach in Aho, Hopcroft, and Ullman [1974], Hopcroft and Musinki [1973], and Lempel and Winograd [1977]. One person who has made significant contributions to the study of computational aspects of convolution and discrete Fourier transform is S. Winograd. Some of his work as it relates to the material presented in this chapter can be found in Winograd [1977, 1978, 1980 (a), 1980 (b)].

We would like to reaffirm our approach in presenting the various topics. We focused on the computational aspects of the algorithms for the computation of convolutions and discrete Fourier transform. In this regard, the theoretical analysis of various bound on computational complexity provides merely a glimpse into the construction of actual algorithms. Consequently, we have chosen to present the general theorems on the bounds on multiplicative complexity without actually including their proof. It does not hinder understanding the derivations and, ultimately, the use of the algorithms.

Bibliography

[4.1] A. Lempel and S. Winograd, "A New Approach to Error-correcting Codes," *IEEE Transactions on Information Theory*, Vol. IT-23, no. 4, pp. 503-508, July 1977.

[4.2] R.W. Brockett and D. Dobkin, "On the Optimal Evaluation of a Set of Bilinear Forms," *Proceedings Fifth Annual ACM Symposium on Theory of Computing*, pp. 88-95, 1973.

[4.3] H. Krishna, *Computational Complexity of Bilinear Forms: Algebraic Coding Theory and Applications to Digital Communication Systems*, Lecture Notes in Control and Information Sciences, Vol. 94, Springer-Verlag, 1987.

[4.4] A.V. Aho, J.E. Hopcroft, and J.D. Ullman, *The Design and Analysis of Computer Algorithms*, Addison-Wesley Publishing Co., 1974.

[4.5] J. Hopcroft and J. Musinki, "Duality Applied to the Complexity of Matrix Multiplication and Other Bilinear Forms," *SIAM Journal on Computing*, Vol. 2, pp. 159-173, Sept. 1973.

[4.6] S. Winograd, "Some Bilinear Forms Whose Multiplicative Complexity Depends on the Field of Constants," *Mathematical Systems Theory*, Vol. 10, no. 2, pp. 169-180, 1977.

[4.7] S. Winograd, "On computing the Discrete Fourier Transform," *Math. Comp.*, Vol. 32, no. 141, pp. 175-199, Jan. 1978.

[4.8] S. Winograd, *Arithmetic Complexity of Computations*, SIAM Publications, Philadelphia, PA, 1980 (a).

[4.9] S. Winograd, "On Multiplication of Polynomials modulo a Polynomial," *SIAM Journal on Computing*, Vol. 9, no. 2, pp. 225-229, May 1980 (b).

Problems

4.1 Consider the complex MULT $e + jf = (a + jb)(c + jd)$, where $a, b, c, d, e, f \in \mathsf{Z}$, the field of rational numbers. Express it as a bilinear form defined over Z. Show that it requires at least 3 MULTs in Z. Derive an algorithm that meets this bound. Express the algorithm in the form $\mathbf{C}\,[\mathbf{A}\,\underline{A} \otimes \mathbf{B}\,\underline{B}]$. Compare it to the straightforward method in terms of number of MULTs and additions.

4.2 Consider the problem of multiplying a $(m \times n)$ matrix with a $(n \times k)$ matrix. Express it as a bilinear form and derive a lower bound for the multiplicative complexity of the computation. For the case $m = n = k = 2$, there is an algorithm that meets the lower bound of 7 MULTs. Can you derive such an algorithm (known as the Strassen's algorithm)?

4.3 Derive an algorithm for a length 3 cyclic convolution valid over Z. Express this algorithm in the form $\mathbf{C}\,[\mathbf{A}\,\underline{A} \otimes \mathbf{B}\,\underline{B}]$. Now alter the matrices suitably to get an algorithm for a length 3 cyclic convolution valid over $GF(2)$, $GF(5)$, and $Z(25)$.

4.4 Derive an algorithm for a length 4 cyclic convolution over C. Determine its multiplicative and additive complexity.

4.5 Derive an algorithm for a length 4 cyclic convolution over Z. Determine its multiplicative and additive complexity. Which algorithms would you prefer for processing data sequences defined over Z, Re and C? Give reasons for your choices.

4.6 Cast the algorithms derived in problems 4.4 and 4.5 in the form $\mathbf{C}\,[\mathbf{A}\,\underline{A} \otimes \mathbf{B}\,\underline{B}]$.

4.7 Write a 6-point acyclic convolution over Z by expressing it as the computation $A(u)\,B(u) \bmod P(u)$, $P(u) = u^2\,(u + 1)\,(u - 1)\,(u^2 + 1)$. Is this algorithm valid over $GF(2)$? Justify your answer.

4.8 Apply the matrix exchange property to obtain another bilinear form for cyclic convolution of length 4 over Z.

4.9 Given a polynomial $X(u)$ of degree 5, consider the residue computation $x_1(u) \equiv X(u) \bmod u^2 + u + 1$. How many additions are required in this computation?

4.10 Given a polynomial $X(u)$ of degree 5, consider the residue computation $x_2(u) \equiv X(u) \bmod u^2 - u + 1$. How many additions are required in this computation? If the residue computations for $x_1(u)$ (Problem 4.9) and $x_2(u)$ were to be done together, determine the additive complexity for the joint computation.

4.11 Derive an algorithm for the cyclic computation of length 6 over Z. Determine its additive and multiplicative complexity.

4.12 Derive the algorithm for the acyclic convolution of length 8 over C obtained by selecting the modulo polynomial $P(u)$ as $P(u) = (u + 1)(u - 1)(u + j)(u - j)(u + 1 - j)(u + 1 + j)(u - 1 - j)(u - 1 + j)$.

4.13 Show that the form of CRT-P reconstruction for the LI or the Cook-Toom algorithm is same as the IDFT when all the interpolating points are consecutive powers of α, the n-th root of unity.

Chapter 5

Cyclotomic Polynomial Factorization and Associated Fields

Among all the ingredients that go into the construction of good algorithms for digital signal processing, the key ones are number theory, polynomial algebra, the Chinese remainder theorem, and field theory. Aspects of number theory that pertain to the digital signal processing of data sequences defined over the infinite fields of rational numbers, Z, real numbers, Re, complex rational numbers, CZ, and complex numbers, C, were described in Chapter 2. Chapter 3 dealt with the polynomial algebra and the CRT-P for data sequences defined over an arbitrary field. Based on these ideas, the basic formulation of the digital signal processing algorithms for computing cyclic and acyclic convolution was presented in Chapter 4.

One theme that emerges from all of this analysis is that the CRT and the polynomial factorization play key roles in breaking down a large size computation into a number of smaller size computations that can be performed in parallel. The polynomials that appear repeatedly in these algorithms are the polynomial $u^n - 1$ and its factors over the field under study. In this chapter, we pursue the factorization of the $u^n - 1$ over the infinite fields Z, Re, CZ, C, CF, and CCF. Chapter 6 deals with the factorization of $u^n - 1$ over the finite fields $GF(p)$ and $GF(q)$, $q = p^m$. Chapter 7 is on the cyclotomic polynomial factorization of $u^n - 1$ over finite integer rings and their polynomial extensions. Readers interested only in the design of digital signal processing algorithms for sequences defined over infinite fields may leave out Chapters 6 and 7 without any loss in continuity. Similarly, readers interested only in the design of digital signal processing algorithms for sequences defined over finite fields may directly jump to Chapter 6 without any loss in continuity. Chapter 6 is crucial for those readers who are interested in the design of digital signal processing algorithms for sequences defined over finite integer rings. It is our firm belief that finite fields are fundamental to the understanding of digital signal processing algorithm design for sequences defined over finite integer rings.

Two more number systems have been added to the list of infinite fields, one denoted by CF and the second denoted by CCF. They will be termed as the cyclotomic field and its complex extension, respectively. A more general form of these fields was first introduced in Section 3.3 as the extension field of a ground field F. Example 3.2 is on a CF. Similarly, Problem 3.14 is on polynomial factorization over CF. We will formalize CF and CCF in this chapter. One wonders where the cyclotomic and complex cyclotomic fields might occur in the domain of digital signal processing. We will show in subsequent chapters that their importance lies in simplifying computation of two and higher dimensional cyclic convolutions and DFTs.

It is important to note that for any algorithm that employs CRT-P, it is advantageous to use as many factor polynomials as possible for a given modulo polynomial. Therefore, we always seek the factorization of a given modulo polynomial ($u^n - 1$ in this chapter) in terms of its irreducible factors in the field over which the computation is defined. This will be the case in all situations throughout this book.

5.1 Cyclotomic Polynomial Factorization Over Complex and Real Numbers

The cyclotomic factorization of the $u^n - 1$ over the infinite fields C and Re is the simplest. Over C, an n-th root of unity always exists and is given by

$$\omega = e^{-2\pi j/n}. \tag{5.1.1}$$

Based on this root, the factorization of $u^n - 1$ over C is given by

$$u^n - 1 = \prod_{i=0}^{n-1} \left(u - \omega^i\right). \tag{5.1.2}$$

For $n = 2$, the factorization consists of only rational entries,

$$u^2 - 1 = (u - 1)(u + 1). \tag{5.1.3}$$

Similarly, for $n = 4$, the factorization consists of simple complex rational entries,

$$u^4 - 1 = (u - 1)(u + 1)(u - j)(u + j). \tag{5.1.4}$$

Over **Re**, the polynomial factors of $u^n - 1$ have degree 2 at most. This is based on the property that if a polynomial has z (a complex number) and z^* (its conjugate) as its roots, then it has real-valued coefficients. It is straightforward to use this property to get the factorization of $u^n - 1$ over **Re**. Such a factorization is given in the following.

***n* even**

$$u^n - 1 = (u-1)(u+1)\prod_{i=1}^{n/2-1}\left(u^2 - 2\cos\theta_i u + 1\right), \tag{5.1.5}$$

***n* odd**

$$u^n - 1 = (u-1)\prod_{i=1}^{(n-1)/2}\left(u^2 - 2\cos\theta_i u + 1\right), \tag{5.1.6}$$

where

$$\cos\theta_i = \frac{\left(\omega^i + \omega^{-i}\right)}{2} = \cos\left(\frac{2\pi i}{n}\right). \tag{5.1.7}$$

5.2 Cyclotomic Polynomial Factorization Over Rational Numbers

The cyclotomic factorization of the $u^n - 1$ over the field of rational numbers, **Z**, is of fundamental importance in digital signal processing as it involves exact arithmetic with integer coefficients. For $n = 2, 3, 4$, it is easily verified that

$$\begin{aligned} u^2 - 1 &= (u - 1)\cdot(u + 1) \\ u^3 - 1 &= (u - 1)\cdot(u^2 + u + 1) \\ u^4 - 1 &= (u - 1)\cdot(u + 1)\cdot(u^2 + 1). \end{aligned}$$

In general, the cyclotomic factorization of the $u^n - 1$ over **Z** begins with the factorization of $u^n - 1$ over **C** and combines the various degree one factors over

C in a systematic way such that the resulting polynomials have rational coefficients. Based on the n-th root of unity given by $\omega = e^{-2\pi j/n}$, the factorization of $u^n - 1$ over C is given by

$$u^n - 1 = \prod_{i=1}^{n} \left(u - \omega^i\right). \tag{5.2.1}$$

For a given integer n and all its factors d_1, d_2, and so on, consider a partitioning of the set $\Gamma_n = \{1, 2, ..., n\}$ into subsets $\Gamma_{n,d_1}, \Gamma_{n,d_2}, ...$ such that Γ_{n,d_k} consists of all those elements of Γ_n that have the *gcd* with n equal to d_k, that is $s_{ik} \in \Gamma_{n,d_k}$ if $(s_{ik}, n) = d_k$. Since *gcd* of two integers is unique, each element of Γ_n belongs to one and only one partition. The factorization of $u^n - 1$ in (5.2.1) can alternately be written as

$$u^n - 1 = \prod_{d_k} \left(\prod_{s_{ik} \in \Gamma_{n,d_k}} \left(u - \omega^{s_{ik}}\right) \right). \tag{5.2.2}$$

Consider the inner product separately and define

$$C_{n/d_k}(u) = \prod_{s_{ik} \in \Gamma_{n,d_k}} \left(u - \omega^{s_{ik}}\right). \tag{5.2.3}$$

Since an element s_{ik} in Γ_{n,d_k} has *gcd* with n equal to d_k, we can alternatively say that the elements in Γ_{n,d_k} are such that $(s_{ik}/ d_k, n / d_k) = 1$. Thus, Γ_{n,d_k} has exactly $\phi(n / d_k)$ elements. Using the property that $(s_{ik}/ d_k, n / d_k) = 1$, we can write $s_{ik} = d_k \cdot r_{ik}$, $(r_{ik}, n/ d_k) = 1$ and $r_{ik} < (n/ d_k)$. Corresponding to $s_{ik} \in \Gamma_{n,d_k}$, let $r_{ik} \in \Psi_{n,d_k}$. We reaffirm that Ψ_{n,d_k} contains all those integers that are less than n / d_k and coprime to it. Thus Ψ_{n,d_k} also has exactly $\phi(n / d_k)$ elements. Replacing $s_{ik} = r_{ik} \cdot d_k$ in (5.2.3) leads to

$$C_{n/d_k}(u) = \prod_{r_{ik} \in \Psi_{n,d_k}} \left(u - \omega^{d_k r_{ik}}\right).$$

As ω is an n-th root of unity, ω^{d_k} is an (n/ d_k)th roots of unity and the product is on all those $\phi(n/ d_k)$ values of r_{ik} that are coprime to n/ d_k. If d_k is a factor of n, then so is n/ d_k. Therefore, instead of n/ d_k, we can alternately write the above expression in terms of d_k as follows:

$$C_{d_k}(u) = \prod_{(r,d_k)=1} \left(u - \omega_{d_k}^r\right), \tag{5.2.4}$$

where ω_{d_k} is an d_k-th root of unity. Removing the subscript k in the above expression, we can write

$$C_d(u) = \prod_{(r,d)=1} \left(u - \omega_d^r\right). \tag{5.2.5}$$

It is clear that $deg(C_d(u)) = \phi(d)$. Using (5.2.5) in (5.2.2), a factorization of $u^n - 1$ is obtained having the form

$$u^n - 1 = \prod_{d|n} C_d(u). \tag{5.2.6}$$

The expressions (5.2.5) and (5.2.6) provide a complete factorization of $u^n - 1$ over Z.

> **Definition 5.1 Cyclotomic Polynomials.** The **cyclotomic polynomials** over the field of rational numbers, Z, are defined as in (5.2.5) with the factorization of $u^n - 1$ over Z given as in (5.2.6).

Let us apply (5.2.5) and (5.2.6) to obtain the factors of $u^n - 1$ for a given n. First, the factors of n are listed. For a factor equal to d, we list all those $\phi(d)$ integers that are less than d and coprime to it. The d-th root of unity is given by $\omega_d = e^{-2\pi j/d}$. Then $C_d(u)$ is obtained using (5.2.5). The expression in (5.2.6) can also be used to determine $C_d(u)$ in an order recursive manner. For $n = 1$, it is trivial. Thus $C_1(u) = u - 1$. For $n = 2$, we have

$$u^2 - 1 = C_1(u)\, C_2(u).$$

Since $C_1(u)$ is already known, $C_2(u)$ can be determined from long division. Also since $u^2 - 1$ and $C_1(u)$ have rational (even better they have integer) coefficients, $C_2(u)$ also has integer coefficients. Similarly

$$u^3 - 1 = C_1(u)\, C_3(u)$$

and $C_3(u)$ can be determined as

$$C_3(u) = u^2 + u + 1.$$

At the nth stage, $C_1(u)$, $C_2(u)$, ..., $C_{n-1}(u)$ have been determined each having integer coefficients. We write (5.2.6) as

$$u^n - 1 = \left\{ \prod_{\substack{d|n \\ d<n}} C_d(u) \right\} C_n(u) \tag{5.2.7}$$

and carry out long division to obtain $C_d(u)$ having integer coefficients. This analysis leads to the following theorem.

Theorem 5.1 All the cyclotomic polynomials have integer coefficients.

Let us consider some special cases. For $n = p$, a prime number, the only divisors are 1 and p. In this case (5.2.7) becomes

$$u^p - 1 = C_1(u)\, C_p(u) = (u - 1)\, C_p(u) \tag{5.2.8}$$

or

$$C_p(u) = (u^p - 1)/(u - 1) = 1 + u + u^2 + \ldots + u^{p-1}. \tag{5.2.9}$$

For $n = p^a$, the only divisors are $1, p, p^2, \ldots, p^{a-1}, p^a$. For $a = 2$, (5.2.7) gives

$$u^{p^2} - 1 = C_1(u) C_p(u) C_{p^2}(u).$$

Since $C_1(u)\, C_p(u) = u^p - 1$ from (5.2.8), we get

$$C_{p^2}(u) = \left(u^{p^a} - 1\right) / \left(u^p - 1\right) = \sum_{i=0}^{p-1} u^{ip}. \tag{5.2.10}$$

Similarly, for $a = 3$,

$$C_{p^3}(u) = \left(u^{p^3} - 1\right) / \left(u^{p^2} - 1\right) = \sum_{i=0}^{p-1} u^{ip^2} \tag{5.2.11}$$

and, in general,

$$\prod_{i=1}^{a-1} C_{p^i}(u) = u^{p^{a-1}} - 1, \tag{5.2.12}$$

$$C_{p^a}(u) = \left(u^{p^a} - 1\right) / \left(u^{p^{a-1}} - 1\right) = \sum_{i=0}^{p-1} u^{ip^{a-1}} . \tag{5.2.13}$$

For $p = 2$, (5.2.13) becomes

$$C_{2^a}(u) = \left(u^{2^a} - 1\right) / \left(u^{2^{a-1}} - 1\right) = 1 + u^{2^{a-1}} . \tag{5.2.14}$$

We now turn towards establishing the general form of the cyclotomic polynomials for a given n (or equivalently its standard factorization in (2.2.15)).

Theorem 5.2 Let $n = m \cdot p^a$, where $(m, p) = 1$. Then

$$C_n(u) = C_m\left(u^{p^a}\right) / C_m\left(u^{p^{a-1}}\right). \tag{5.2.15}$$

Proof: The proof is based on the identity

$$u^h - A^h = \prod_{i=0}^{h-1} \left(u - A\omega_h^i\right). \tag{5.2.16}$$

Recall from Theorem 2.3 that any integer r such that $(r, m \cdot p^a) = 1$ can be expressed as $r = b \cdot m + c \cdot p^a$, where $(b, p^a) = 1$ and $(c, m) = 1$. Therefore, the cyclotomic polynomial $C_n(u)$ is given by

$$C_n(u) = \prod_{(r,n)=1} \left(u - \omega_n^r\right) = \prod_{(c,m)=1} \prod_{(b,p^a)=1} \left(u - \omega_n^{bm+cp^a}\right). \tag{5.2.17}$$

Since $b < p^a$ and $(b, p^a) = 1$, b takes values in the set $\Delta = \{0, 1, 2, ..., p^a - 1\} - \{p \cdot 0, p \cdot 1, ..., p \cdot (p^{a-1} - 1)\}$. Thus (5.2.17) becomes

$$C_n(u) = \prod_{(c,m)=1} \prod_{b=0}^{p^a-1} \left(u - \omega_n^{bm+cp^a}\right) \Bigg/ \prod_{(c,m)=1} \prod_{b=0}^{p^{a-1}-1} \left(u - \omega_n^{bpm+cp^a}\right)$$

$$= \prod_{(c,m)=1} \prod_{b=0}^{p^a-1} \left(u - \omega_n^{cp^a} \cdot \omega_n^{bm}\right) \Bigg/ \prod_{(c,m)=1} \prod_{b=0}^{p^{a-1}-1} \left(u - \omega_n^{cp^a} \cdot \omega_n^{bpm}\right). \tag{5.2.18}$$

From the definition of the root of unity, we have

$$\omega_n^{cp^a} = \omega_m^c,\ \omega_n^m = \omega_{p^a},\ \omega_n^{pm} = \omega_{p^{a-1}}.$$

Employing above relations and (5.2.16) with $h = p^a$, $A = \omega_m^c$ for the inner product in the numerator and $h = p^{a-1}$, $A = \omega_m^c$ for the inner product in the denominator of (5.2.18), we get

$$C_n(u) = \prod_{(c,m)=1} \left(u^{p^a} - \omega_m^{cp^a}\right) \Big/ \prod_{(c,m)=1} \left(u^{p^{a-1}} - \omega_m^{cp^{a-1}}\right). \qquad (5.2.19)$$

Since $(c, m) = 1$ and $(m, p^a) = 1$, $p^a \cdot c \bmod m$ defines a permutation on those integers c that are coprime to m. Thus, $p^a \cdot c \bmod m$ takes values in the same set as c and (5.2.19) simplifies to

$$C_n(u) = \prod_{(c,m)=1} \left(u^{p^a} - \omega_m^c\right) \Big/ \prod_{(c,m)=1} \left(u^{p^{a-1}} - \omega_m^c\right) = C_m\left(u^{p^a}\right) \Big/ C_m\left(u^{p^{a-1}}\right).$$

This theorem along with (5.2.13), (5.2.14), and the standard factorization of n in (2.2.15) can used to obtain the cyclotomic polynomial $C_n(u)$ for all values of n. Another result that can be established by using the factorization of the integer m and the above theorem is as follows.

Lemma 5.1 For any integer m and prime p,

$$C_{mp^a}(u) = C_{mp^k}\left(u^{p^{a-k}}\right),\ k = 1, 2, \ldots, a. \qquad (5.2.20)$$

Proof: Left as an exercise.

Some special cases of interest are presented in the following.

Case 1. $n = 2 \cdot m$, m odd, $m > 1$. We apply the above theorem with $p = 2$, $a = 1$. The only value that b takes in (5.2.17) is $b = 1$. Thus (5.2.17) simplifies to

$$C_n(u) = \prod_{(c,m)=1} \left(u - \omega_n^{m+2c}\right) = \prod_{(c,m)=1} \left(u - \omega_n^m \omega_n^{2c}\right).$$

Simplifying this using $\omega_n^m = -1$, $\omega_n^{2c} = \omega_m^c$, we get

$$C_n(u) = \prod_{(c,m)=1} \left(u + \omega_m^c\right) = \prod_{(c,m)=1} \left(-u - \omega_m^c\right) = C_m(-u). \tag{5.2.21}$$

Case 2. $n = p \cdot q$, the product of two distinct odd primes. We apply the above theorem with $m = q$, $a = 1$ and get

$$C_n(u) = C_q(u^p) / C_q(u).$$

Substituting for $C_q(u)$ from (5.2.9), we get

$$C_n(u) = [(u^{pq} - 1)\,(u - 1)] / [(u^p - 1)\,(u^q - 1)]. \tag{5.2.22}$$

Dividing $u^{pq} - 1$ by $u^p - 1$ and expanding $(1 - u^q)^{-1}$ as $1 + u^q + u^{2q} + \ldots$ and so on, we get

$$C_n(u) = \left[\sum_{i=0}^{q-1} \sum_{k=0,\ldots} u^{ip+kq}\right](1 - u), \tag{5.2.23}$$

$deg(C_n(u)) = (p - 1)\,(q - 1)$. It is easy to verify that $i\,p\ + k\,q$ take on distinct values for $i = 0, 1, \ldots, q - 1$. Therefore, the polynomial in the summation in $[\cdot]$ of (5.2.23) has coefficients that are either 0 or 1. Consequently, $C_n(u)$ has coefficients that are 0, 1, or -1.

Case 3. $n = 2 \cdot p \cdot q$. Using Case 1 result, $C_{2pq}(u) = C_{pq}(-u)$. Using (5.2.23), we see that $C_{2pq}(u)$ also has coefficients that are 0, 1, or -1.

Case 4. $n = 2^c \cdot q^b$. Using Theorem 5.2 with $m = 2^c$, we can write

$$C_n(u) = C_{2^c}\left(u^{q^b}\right) / C_{2^c}\left(u^{q^{b-1}}\right) = \left(1 + u^{2^{c-1} q^b}\right) / \left(1 + u^{2^{c-1} q^{b-1}}\right). \tag{5.2.24}$$

Alternatively, we can write

$$C_{2^c q^b}(u) = C_{2q}(w) = C_q(-w), \tag{5.2.25}$$

where

$$w = u^{2^{c-1} q^{b-1}}. \tag{5.2.26}$$

Case 5. $n = p^a \cdot q^b$. We apply Theorem 5.2 with $m = q^b$ to get

$$C_n(u) = C_{q^b}\left(u^{p^a}\right) \Big/ C_{q^b}\left(u^{p^{a-1}}\right).$$

Using (5.2.13) to substitute for $C_{q^b}(u)$, we get

$$C_{p^a q^b}(u) = \frac{\left\{\left(u^{p^a q^b} - 1\right)\left(u^{p^{a-1} q^{b-1}} - 1\right)\right\}}{\left\{\left(u^{p^a q^{b-1}} - 1\right)\left(u^{p^{a-1} q^b} - 1\right)\right\}} \tag{5.2.27}$$

or

$$C_{p^a q^b}(u) = C_{pq}(w), \tag{5.2.28}$$

where

$$w = u^{p^{a-1} q^{b-1}}. \tag{5.2.29}$$

Case 6. $n = 2^c \cdot p^a \cdot q^b$. We apply Theorem 5.2 with $m = 2^c \cdot q^b$ to get

$$C_n(u) = C_{2^c q^b}\left(u^{p^a}\right) \Big/ C_{2^c q^b}\left(u^{p^{a-1}}\right).$$

Using (5.2.25) to substitute for $C_{2^c q^b}(u)$, we get

$$C_{2^c p^a q^b}(u) = \frac{\left\{\left(u^{2^{c-1} p^a q^b} + 1\right)\left(u^{2^{c-1} p^{a-1} q^{b-1}} + 1\right)\right\}}{\left\{\left(u^{2^{c-1} p^a q^{b-1}} + 1\right)\left(u^{2^{c-1} p^{a-1} q^b} + 1\right)\right\}} \tag{5.2.30}$$

or

$$C_{2^c p^a q^b}(u) = C_{2pq}(w) = C_{pq}(-w), \tag{5.2.31}$$

where

$$w = u^{2^{c-1} p^{a-1} q^{b-1}}. \tag{5.2.32}$$

Based on the above analysis and (5.2.13), (5.2.14), and (5.2.20) to (5.2.32), we get the following theorem.

> **Theorem 5.3** The *coefficients* of $C_n(u)$ take values that are 0, 1, or -1 if n has at most two distinct odd prime numbers in its standard factorization.

The smallest value of n that contains the product of 3 distinct odd primes is $n = 3 \cdot 5 \cdot 7 = 105$. Therefore, for $n < 105$, all the $C_n(u)$ have coefficients that take values that are 0, 1, or -1.

Case 7. $n = p \cdot q \cdot r$, the product of 3 distinct odd primes. In this case,

$$C_n(u) = \frac{\{(u^{pqr}-1)(u^p-1)(u^q-1)(u^r-1)\}}{\{(u^{pq}-1)(u^{qr}-1)(u^{rp}-1)(u-1)\}}. \tag{5.2.33}$$

If n has only odd prime numbers in its factorization, then

$$n = \prod_i p_i^{a_i}, \quad p_i > 2, \tag{5.2.34}$$

and the cyclotomic polynomials satisfy the relation,

$$C_n(u) = C_{\prod_i p_i}(w), \tag{5.2.35}$$

where

$$w = u^\eta, \tag{5.2.36}$$

$$\eta = \prod_i p_i^{a_i - 1}. \tag{5.2.37}$$

Similarly, given factorization of n as

$$n = 2^c \prod_i p_i^{a_i}, \tag{5.2.38}$$

cyclotomic polynomials satisfy the relation

$$C_n(u) = C_{2\prod_i p_i}(w) = C_{\prod_i p_i}(-w), \tag{5.2.39}$$

where

$$w = u^\eta, \tag{5.2.40}$$

$$\eta = 2^{c-1} \prod_i p_i^{a_i - 1}. \tag{5.2.41}$$

If a cyclotomic polynomial $C_n(u)$, $n \neq 1$ has ω as a root, then it also has ω^{-1} as a root. Therefore, all cyclotomic polynomial $C_n(u)$, except for $n = 1$, are symmetric in nature. $C_1(u)$ is anti-symmetric in nature. A degree t polynomial $A(u) = \sum_{i=0}^{t} A_i u^i$ is called **symmetric** or **anti-symmetric** if it satisfies the property $A_i = A_{t-i}$ or $A_i = -A_{t-i}$, respectively. Even though the coefficients of $C_n(u)$ are not guaranteed to take values that are 0, 1, or −1, the symmetry property of $C_n(u)$ polynomials can be used to show that if n is the product of three distinct primes p, q, and r, $r < q < p$, then all coefficients of $C_n(u)$ lie in the range $-(r-1)$ to $(r-1)$.

An important aspect of the cyclotomic polynomials is their irreducibility over the field of rational numbers, Z. The following theorem makes exactly that statement.

Theorem 5.4 The cyclotomic polynomial are *irreducible* over Z.

The proof is omitted here. A key observation as far as this particular aspect of cyclotomic polynomial is concerned is that though cyclotomic polynomials are irreducible over Z, they may factor further over some other extension of Z. A summary of the results in this section is presented in Table 5.1.

5.3 Cyclotomic Fields and Cyclotomic Polynomial Factorization

As described in Section 3.3, an extension field can be obtained from the ground field by defining its elements as polynomials of degree up to $m - 1$ over the ground field. The arithmetic operations of '+' and '·' are performed modulo a degree m polynomial which is irreducible over the ground field. In the present context, let us specialize these statements in the following manner.

Definition 5.2 Cyclotomic field. Let the ground field be Z and the irreducible polynomial be $C_N(v)$, the N-th cyclotomic polynomial. The resulting extension field is termed as **cyclotomic field** having polynomials of degree up to $\phi(N) - 1$ in the indeterminate v as its elements. A cyclotomic field is denoted by CF in general. Also, CF(N) is used to denote the CF obtained when the irreducible polynomial is $C_N(u)$.

In this section, we are interested in cyclotomic fields CF(N) and factorization of $u^n - 1$ in them.

Table 5.1 On Creating Cyclotomic Polynomials

$n = 1$	$C_n(u) = u - 1$
$n = p$; p: prime	$C_n(u) = 1 + u + u^2 + \ldots + u^{p-1}$
$n = p \cdot q$ p, q: odd prime	$C_n(u) = [(u^{pq} - 1)(u - 1)]/[(u^p - 1)(u^q - 1)]$ or $C_n(u) = \left[\sum_{i=0}^{q-1} \sum_{k=0,\ldots} u^{ip+kq}\right](1 - u)$
$n = p \cdot q \cdot r$ p, q, r: odd prime	$C_n(u) = \dfrac{\{(u^{pqr} - 1)(u^p - 1)(u^q - 1)(u^r - 1)\}}{\{(u^{pq} - 1)(u^{qr} - 1)(u^{rp} - 1)(u - 1)\}}$
$n = \prod_i p_i^{a_i}, p_i > 2$	$C_n(u) = C_{\prod_i p_i}(u^\eta)$, $\eta = \prod_i p_i^{a_i - 1}$
$n = 2^c \prod_i p_i^{a_i}, p_i > 2$	$C_n(u) = C_{\prod_i p_i}(-u^\eta)$, $\eta = 2^{c-1} \prod_i p_i^{a_i - 1}$
$n = \prod_i p_i, p_i > 2$	Use $C_{mp}(u) = C_m(u^p)/C_m(u)$ for $(m, p) = 1$ recursively.

Example 5.1 Complex rational numbers. Consider the extension field Z extended by $C_4(v)$. Since $C_4(v) = v^2 + 1$, this is the field CF(4) and its elements are of the type $A + v\,B$, $A, B \in$ Z, with the multiplication between two elements defined modulo $v^2 + 1$ as $(A + v\,B)\,(C + v\,D) = (A\,C - B\,D) + v\,(A\,D + B\,C)$. This is immediately recognized as complex multiplication in the usual sense. Thus, CF(4) is same as the field of complex rational numbers and, in this case, the indeterminate v is replaced by the indeterminate j.

5.3.1 Cyclotomic Polynomial Factorization Over Cyclotomic Fields

The factorization of $u^n - 1$ over CF(N) begins with its factorization over Z. Since CF(N) is an extension of Z, the factorization of $u^n - 1$ over Z in terms of cyclotomic polynomials is also valid over CF(N). However, many of the cyclotomic polynomials that are irreducible in Z may factor further in CF. The fundamental question we face is whether a given cyclotomic polynomial $C_n(u)$ factorizes further over CF(N). In those situations where the answer is yes, the structure of the resulting factors must also be determined. For example, the smallest value of i for which $C_M(v)$ divides $v^N - 1$ is $i = N$. Therefore, in $\mathsf{CF}(N)$, v is an N-th root of unity and we can write

$$u^N - 1 = \prod_{i=1}^{N} \left(u - v^i\right). \tag{5.3.1}$$

The validity of the above factorization in $\mathsf{CF}(N)$ can be verified by setting $u = v^i$ and verifying that v^i is a root of $u^N - 1$ in $\mathsf{CF}(N)$, $i = 1, 2, ..., N$. If N is odd, then $-v$ is a primitive root of order $2N$ in $\mathsf{CF}(N)$, and we have

$$u^{2N} - 1 = \prod_{i=1}^{2N} \left(u - (-v)^i\right). \tag{5.3.2}$$

Similarly, in $\mathsf{CF}(N)$, $v^{(N/d)}$ is a d-th root of unity and we can write

$$u^d - 1 = \prod_{i=1}^{d} \left(u - \left(v^{N/d}\right)^i\right), \tag{5.3.3}$$

and if d is odd, then $-v^{(N/d)}$ is a $2d$-th root of unity giving the factorization

$$u^{2d} - 1 = \prod_{i=1}^{2d} \left(u - \left(-v^{N/d}\right)^i\right). \tag{5.3.4}$$

In general, the key to answer this question comes from the application of Dedekind's "reciprocity theorem" of Galois theory. The statement of this fundamental theorem is as follows.

Theorem 5.5 Dedekind's reciprocity theorem. Let $f(u)$ and $g(u)$ be irreducible polynomials with coefficients in F. Let $f(u)$ decompose into irreducible factors, $f(u) = f_1(u) \cdot f_2(u) \cdots f_\mu(u)$ over $\mathsf{F}(b)$, where b is a root of $g(u)$. Similarly, let $g(u)$ decompose as $g(u) = g_1(u) \cdot g_2(u) \cdots g_\gamma(u)$ over $\mathsf{F}(a)$, where a is a root of $f(u)$. Then $\mu = \gamma$ and the $g_i(u)$ can be reordered in such a way that the ratio of $deg(f_k(u))$ and $deg(g_k(u))$ is the same for all $k = 1, 2, ..., \mu$.

We apply this theorem with $f(u) = C_n(u)$, $g(u) = C_N(u)$, $\mathsf{F} = \mathsf{Z}$, $b = \omega_N$, the N-th root of unity, and $a = \omega_n$, the n-th root of unity. Then $\mathsf{F}(a) = \mathsf{CF}(n)$ and $\mathsf{F}(b) = \mathsf{CF}(N)$. There are three cases to consider (a) $\mu = 1$, (b) $1 < \mu < deg(C_n(u))$

$= \phi(n)$, and (c) $\mu = deg(C_n(u)) = \phi(n)$. For $\mu = 1$, $C_n(u)$ is irreducible over $\mathsf{CF}(N)$ and $C_N(u)$ is irreducible over $\mathsf{CF}(n)$. For $\mu = deg(C_n(u)) = \phi(n)$, $C_n(u)$ factorizes as product of degree 1 polynomials over $\mathsf{CF}(N)$ and, in this case, $\mathsf{CF}(N)$ is a splitting field for $C_n(u)$.

Definition 5.3 Splitting field. Let $f(u)$ be a polynomial with coefficients in a field F. A **splitting field** for $f(u)$ is a field G which contains F and which has the property that $f(u)$ can be written as a product of polynomials of first degree with coefficients in G.

If $d = d_1 \cdot d_2 \cdots d_a$ is a decomposition into pairwise coprime factors, then ω_d (d-th root of unity) can be uniquely represented as

$$\omega_d = \omega_1\, \omega_2 \cdots \omega_a, \tag{5.3.5}$$

corresponding to the decomposition of the residue class ring mod d into direct sum of residue class rings mod d_i based on the CRT-I (refer to (2.4.10)). This leads to

$$\mathsf{CF}(d) = \mathsf{CF}(d_1)\, \mathsf{CF}(d_2) \cdots \mathsf{CF}(d_a), \tag{5.3.6}$$

that is, $\mathsf{CF}(d)$ is the compositum of the fields $\mathsf{CF}(d_i)$. It can alternately be argued that the only roots of unity in $\mathsf{CF}(d)$ are of the form $\pm\omega_d^i$. Based on (5.3.5) and (5.3.6), we see that the only roots of unity in common between $\mathsf{CF}(n)$ and $\mathsf{CF}(N)$ are of the form $\pm\omega_d^i$, $d = (n, N)$.

Consequently, for $\mu = 1$, $(n, N) = 1$, which implies that $C_n(u)$ is irreducible over $\mathsf{CF}(N)$ and $C_N(u)$ is irreducible over $\mathsf{CF}(n)$ as $\mathsf{CF}(N) \cap \mathsf{CF}(n) = \mathsf{Z}$ for $(n, N) = 1$. Similarly, for $(n, N) = n$, $\mathsf{CF}(n) \subset \mathsf{CF}(N)$ and $\mu = deg(C_n(u)) = \phi(n)$. In this case, $C_n(u)$ factorizes as product of degree 1 polynomials over $\mathsf{CF}(N)$ and $\mathsf{CF}(N)$ is a splitting field for $C_n(u)$. Analogous statements hold for $(n, N) = N$.

In general, if $(n, N) = d$, then $\mathsf{CF}(N) \cap \mathsf{CF}(n) = \mathsf{CF}(d)$ and $\mu = \phi(d)$. Thus each of $C_n(u)$ and $C_N(u)$ will factor into $\phi(d)$ factors in the fields $\mathsf{CF}(N)$ and $\mathsf{CF}(n)$, respectively. To obtain these factorizations, we proceed as follows.

Case 1. $n = d \cdot e$, $(d, e) = 1$, $(e, N) = 1$. In this case, an integer $i < n$, $(i, n) = 1$ can be expressed as $i = (d\,k + e\,l) \bmod n$, $k < e$, $l < d$, $(k, e) = 1$, and $(l, d) = 1$. Clearly, l takes values that are coprime to d. We partition the values of i for which $(i, n) = 1$ into $\phi(d)$ sets, one for each of the $\phi(d)$ possible values of l. Let the values for l be l_1, l_2, and so on. Thus, $C_d(u)$ can be expressed as

$$C_n(u) = \prod_{(k,e)=1}\left(u - \omega_n^{dk+el_1}\right)\cdot \prod_{(k,e)=1}\left(u - \omega_n^{dk+el_2}\right)\cdots$$

$$= \prod_{(k,e)=1}\left(u - \omega_n^{el_1}\omega_n^{dk}\right)\cdot \prod_{(k,e)=1}\left(u - \omega_n^{el_2}\omega_n^{dk}\right)\cdots. \tag{5.3.7}$$

It is clear that $\omega_n^e = \omega_d = v^{N/d}$, where the last equality follows from the fact that in $CF(N)$ v is an N-th root of unity (refer to (5.3.3) above). Similarly, $\omega_n^d = \omega_e$ and (5.3.7) simplifies to

$$C_n(u) = \prod_{(k,e)=1}\left(u - v^{(N/d)l_1}\omega_e^k\right)\cdot \prod_{(k,e)=1}\left(u - v^{(N/d)l_2}\omega_e^k\right)\cdots \tag{5.3.8}$$

$$= \left\{\left(v^{(N/d)l_1}\right)^{\phi(e)} \prod_{(k,e)=1}\left(v^{-(N/d)l_1}u - \omega_e^k\right)\right\}$$

$$\cdot \left\{\left(v^{(N/d)l_2}\right)^{\phi(e)} \prod_{(k,e)=1}\left(v^{-(N/d)l_2}u - \omega_e^k\right)\right\}\cdots \tag{5.3.9}$$

$$= C_{l_1,n}(u)C_{l_2,n}(u)\cdots, \tag{5.3.10}$$

where

$$C_{l_i,n}(u) = \left(v^{(N/d)l_i}\right)^{\phi(e)} \prod_{(k,e)=1}\left(v^{-(N/d)l_i}u - \omega_e^k\right)$$

$$= \left(v^{(N/d)l_i}\right)^{\phi(e)} C_e\left(v^{-(N/d)l_i}u\right), \quad i = 1, 2, \ldots . \tag{5.3.11}$$

The last equality is based on the fact that ω_e is an e-th root of unity.

Case 2. $n = d_1 \cdot e$, $d \mid d_1$, and d_1 is the smallest integer such that $(d_1, e) = 1$. Let the factorization of d and d_1 be given by

$$d = \prod_i p_i^{a_i}$$

and

$$d_1 = \prod_i p_i^{b_i}, \; b_i \geq a_i > 0.$$

Using Lemma 5.1 in an recursive manner, we get

$$C_{d_1 e}(u) = C_{de}\left(u^{\prod p_i^{b_i - a_i}}\right). \tag{5.3.12}$$

Now Case 1 can be used to obtain the $\phi(d)$ factors.

It is very important to observe that the coefficients of the factors of $C_n(u)$ are elements of $CF(N)$. Therefore, they are polynomials in v of degree up to $\phi(N) - 1$. However, a quick glance at (5.3.11) reveals the important property that *the coefficients of the factors of* $C_n(u)$ *in* $CF(N)$ *are polynomials of the type* v^l, *that is, they are polynomials in v with only one non-zero coefficient.*

This completes a major part of our analysis for cyclotomic factorization. In the following, we present several cases of cyclotomic factorization of importance to digital signal processing and highlight their interesting features. We begin with the case $N = 4$.

5.3.2 Complex Cyclotomic Polynomials

For $N = 4$, the cyclotomic field is $CF(4)$ and is the same as the field of complex rational numbers, CZ. In such a field, the indeterminate is represented by j instead of v. Following this convention, it is clear that $C_4(j) = j^2 + 1$. Only three values are possible for $d = (n, N)$; they are 1, 2, and 4. For $d = 1$ or 2, $C_n(u)$ is irreducible over $CF(4)$. Therefore, only those cyclotomic polynomials $C_n(u)$ factorize further for which $4 \mid n$. Following the analysis done previously, we also have two cases in this instance.

Case 1. $n = 4 \cdot e$, $(4, e) = 1$. In this case, $d = 4$, $\phi(d) = 2$, and the possible values for l_i are 1 and 3. Setting these values in (5.3.11) and replacing v by j, we get

$$C_{1,n}(u) = j^{\phi(e)} C_e(-ju) \tag{5.3.13}$$

and

$$C_{3,n}(u) = (-j)^{\phi(e)} C_e(ju). \tag{5.3.14}$$

Case 2. $n = 4^m \cdot e$, $(2, e) = 1$, $m > 1$. In this case, we follow Case 2 and (5.3.12) to get

$$C_{1,n}(u) = j^{\phi(e)} C_e\left(-ju^{2^{m-2}}\right) \tag{5.3.15}$$

and

$$C_{3,n}(u) = (-j)^{\phi(e)} C_e\left(ju^{2^{m-2}}\right). \tag{5.3.16}$$

The polynomials $C_{1,n}(u)$ and $C_{2,n}(u)$ are termed the **complex cyclotomic polynomials**. Since the coefficients of $C_e(u)$ are integers, the coefficients of $C_{1,n}(u)$ and $C_{2,n}(u)$ are complex conjugates of each other and are complex integers. In addition, their values alternate as either integers or purely imaginary integers starting with the coefficient of u^0 which is 1 or -1. Summarizing this analysis, we have the following lemma.

Lemma 5.2 The *coefficients* of the complex cyclotomic polynomials take complex integer values. These values alternate as either integers or purely imaginary integers. If n has at most two odd prime integers in its factorization, then the coefficients of the complex cyclotomic polynomials take values in the set $\{0, 1, -1, j, -j\}$.

5.3.3 n: a Power of 2

In this case, let $n = 2^b$ and $N = 2^a$. If $a = b$, that is, $n = N$, then the factorization in (5.3.1) holds. Similarly, if $b < a$, that is $n \mid N$, then the factorization in (5.3.3) holds. In both cases, CF(N) is the splitting field of $C_n(u)$. Let us consider the case $b > a$. In this case, $d = (2^a, 2^b) = 2^a$ and $C_n(u)$ factorizes further over CF(N) and has $\phi(2^a) = 2^{a-1}$ factors. Such a factorization is given by

$$C_n(u) = u^{2^{b-1}} + 1 = \prod_{k=0}^{2^{b-1}-1}\left(u^{2^{b-a}} - v^{2k+1}\right). \tag{5.3.17}$$

Example 5.2 If n is a power of 2, that is, $n = 2^a$, $a > 1$, then $C_n(u)$ always factorizes over CZ and the factors are given by

$$C_n(u) = u^{n/2} + 1 = C_{1,n}(u)C_{3,n}(u) = \left(u^{n/4} - j\right)\left(u^{n/4} + j\right). \tag{5.3.18}$$

Example 5.3 Consider the factorization of $C_{20}(u)$ over CZ. Since $4|20$, $C_{20}(u)$ factorizes over CZ. Setting $d = 4$, $e = 5$ in the analysis for the complex cyclotomic polynomials, we get

$$C_{1,20}(u) = (j)^{\phi(5)} C_5(-ju) = u^4 + ju^3 - u^2 - ju + 1,$$
$$C_{3,20}(u) = (-j)^{\phi(5)} C_5(ju) = u^4 - ju^3 - u^2 + ju + 1.$$

Example 5.4 Consider the factorization of $C_{15}(u)$ over $\mathsf{CF}(5)$ generated by the cyclotomic polynomial $C_5(v) = 1 + v + v^2 + v^3 + v^4$. In this case, $(15, 5) = 5$ and, therefore, $C_{15}(u)$ will have $\phi(5) = 4$ factors over $\mathsf{CF}(5)$. Four possible values for l_i are 1, 2, 3, and 4. Setting $n = 15$, $N = 5$, $d = 5$, $e = 3$, and $\phi(e) = 2$ in (5.3.11), we get

$$C_{l_i,15}(u) = v^{2l_i} C_3\left(v^{-l_i} u\right)$$

or

$$C_{1,n}(u) = u^2 + vu + v^2,$$
$$C_{2,n}(u) = u^2 + v^2u + v^4,$$
$$C_{3,n}(u) = u^2 + v^3u + v,$$
$$C_{4,n}(u) = u^2 + v^4u + v^3.$$

Example 5.5 Consider the factorization of $C_{15}(u)$ over $\mathsf{CF}(3)$ generated by the cyclotomic polynomial $C_3(v) = 1 + v + v^2$. In this case, $(15, 3) = 3$ and, therefore, $C_{15}(u)$ will have $\phi(3) = 2$ factors over $\mathsf{CF}(3)$. Two possible values for l_i are 1 and 2. Setting $n = 15$, $N = 3$, $d = 3$, $e = 5$, and $\phi(e) = 4$ in (5.3.11), we get

$$C_{l_i,n}(u) = v^{4l_i} C_5\left(v^{-l_i} u\right)$$

or

$$C_{1,n}(u) = u^4 + vu^3 + v^2u^2 + u + v,$$
$$C_{2,n}(u) = u^4 + v^2u^3 + vu^2 + u + v^2.$$

Example 5.6 Consider the factorization of $u^{60} - 1$ over CZ. Such a factorization begins the factorization of $u^{60} - 1$ over Z. The factors of 60 are 1, 2, 3, 4, 5, 6, 10, 12, 15, 20, 30, and 60. Therefore, the irreducible factors of $u^{60} - 1$ over Z are $C_1(u)$, $C_2(u)$, $C_3(u)$, $C_4(u)$, $C_5(u)$, $C_6(u)$, $C_{10}(u)$, $C_{12}(u)$, $C_{15}(u)$, $C_{20}(u)$, $C_{30}(u)$, and $C_{60}(u)$; the degrees being $\phi(1) = 1$, $\phi(2) = 1$, $\phi(3) = 2$, $\phi(4) = 2$, $\phi(5) = 4$, $\phi(6) = 3$, $\phi(10) = 4$, $\phi(12) = 4$, $\phi(15) = 8$, $\phi(20) = 8$, $\phi(30) = 8$, and $\phi(60) = 16$, respectively. Since $4|4$, $4|12$, $4|20$, and $4|60$, $C_4(u)$, $C_{12}(u)$, $C_{20}(u)$, and $C_{60}(u)$ factorize further over CZ. Setting $d = 4$ in each case and $e =$

1, e = 3, e = 5, and e = 15 in the analysis for the complex cyclotomic polynomials, we get

$$C_{1,4}(u) = u - j,$$
$$C_{3,4}(u) = u + j,$$
$$C_{1,12}(u) = (j)^{\phi(3)}C_3(-ju) = u^2 + ju - 1,$$
$$C_{3,12}(u) = (-j)^{\phi(3)}C_3(ju) = u^2 - ju - 1,$$
$$C_{1,20}(u) = (j)^{\phi(5)}C_5(-ju) = u^4 + ju^3 - u^2 - ju + 1,$$
$$C_{3,20}(u) = (-j)^{\phi(5)}C_5(ju) = u^4 - ju^3 - u^2 + ju + 1,$$
$$C_{1,60}(u) = (j)^{\phi(15)}C_{15}(-ju) = u^8 - ju^7 - ju^5 - u^4 + ju^3 + ju + 1,$$
$$C_{1,60}(u) = (-j)^{\phi(15)}C_{15}(ju) = u^8 + ju^7 + ju^5 - u^4 - ju^3 - ju + 1.$$

The factorization of $u^{60} - 1$ in terms of its irreducible factors over CZ is now complete.

5.4 Extension fields of Cyclotomic Fields and Cyclotomic Polynomial Factorization

With the algebra of fields, it is clear that we can treat a cyclotomic field as a ground field and define an extension field in much the same way as a cyclotomic field was obtained as an extension field of the field of rational numbers. We hope that such as algebra, though tedious to represent, is conceptually clear to the readers at this stage. In this section, we focus on the cyclotomic field of direct interest in digital signal processing, the field of complex rationals, $CF(4)$ or CZ, and consider cyclotomic polynomial factorization of $u^n - 1$ in it.

Consider CZ as the ground field and its extension by an irreducible polynomial over CZ. We remain focused on cyclotomic polynomials. Therefore, the irreducible polynomial is a cyclotomic polynomial. If $(4, N) = d < 4$, then $C_N(v)$ is irreducible in CZ. The extension field, in this case, is the same as $CF(N')$, where $N' = (4/d) \cdot N$. The cyclotomic polynomial factorization over such fields has already been described earlier in Section 5.3.1 and need not be pursued again. The only difference is in the representation of the N'-th root of unity. In $CF(N')$, the N'-th root of unity is represented by jv if $d = 1$ and by jv^2 if $d = 2$.

If $(4, N) = 4$, then $C_N(v)$ factorizes further over CZ and the factors $C_{1,N}(v)$ and $C_{3,N}(v)$ are irreducible over CZ. Both of these factors contain the N-th root of unity and either one of them can be used as an extension polynomial to obtain an extension field starting with CZ as the ground field. Let us term the extension field thus obtained as the **complex cyclotomic field** $\mathsf{CCF}(N)$. This field contains the N-th root of unity. If $(n, N) = d$, then $\mathsf{CCF}(N) \cap \mathsf{CF}(n) = \mathsf{CF}(d)$ and $\mu = \phi(d)$. Thus each of $C_n(u)$ and $C_N(u)$ will factor into $\phi(d)$ factors in the fields $\mathsf{CCF}(N)$ and $\mathsf{CF}(n)$, respectively.

Example 5.7 Consider the factorization of $C_8(v) = v^4 + 1$ into two factors $v^2 + j$ and $v^2 - j$. Either one of these can be used to generate a complex cyclotomic field containing the 8-th primitive root of unity. The elements in this field are polynomials in v of degree up to 1 with their coefficients defined over the field of complex rational numbers, CZ. The computations in this extension field are defined modulo $v^2 + j$ or $v^2 - j$ and are simplified using the recursion

$$\begin{aligned} v^2 &= -j, \\ v^3 &= -j\,v, \\ v^4 &= -1, \\ v^5 &= -v, \\ v^6 &= j, \\ v^7 &= j\,v, \\ v^8 &= 1, \end{aligned}$$

if $v^2 + j$ is used, or

$$\begin{aligned} v^2 &= j, \\ v^3 &= j\,v, \\ v^4 &= -1, \\ v^5 &= -v, \\ v^6 &= -j, \\ v^7 &= -j\,v, \\ v^8 &= 1, \end{aligned}$$

if $v^2 - j$ is used as the modulo polynomial. It is not surprising that the coefficients of the various polynomials are as simple as $-1, 1, 0, j, -j$. A close examination of (5.3.8) to (5.3.12) reveals that this will indeed be the case.

The procedure to obtain the factors of $C_n(u)$ over $\mathsf{CCF}(N)$ is the same as the procedure to obtain the factors of $C_n(u)$ over $\mathsf{CF}(N)$. If 4 is not a factor of n, then the factorization of $C_n(u)$ over $\mathsf{CCF}(N)$ is the same as its factorization over

$CF(N)$. If 4 is a factor of n, then we proceed in the following manner: first, factorize $C_n(u)$ over $CF(N)$ and then use the recursion as outlined in Example 5.7 to represent the coefficients of the factor polynomials as elements in $CCF(N)$. Alternately, one can first factorize $C_n(u)$ over CZ into two factors, $C_{1,n}(u)$ and $C_{2,n}(u)$, and then factorize each of $C_{1,n}(u)$ and $C_{2,n}(u)$ further over $CCF(N)$. It is most interesting to observe here that the coefficients of the factor polynomials of $C_n(u)$ in $CCF(N)$ are *simple polynomials in the indeterminate* v having the form $\alpha \cdot v^t$, where α is one of $\{-1, 1, 0, j, -j\}$ and $t < deg(C_{1,n}(u)) = deg(C_{2,n}(u)) = \phi(N)/2$.

For example, let $n = d \cdot e$, $(d, e) = 1$, $(e, N) = 1$ and let d be an integer of the type $d = 4 \cdot f$, $(4, f) = 1$. In this case, we first factorize $C_n(u)$ into 2 factors over CZ, each of degree $\phi(f \cdot e)$. These polynomials are given in (5.3.13) and (5.3.14) and have coefficients in CZ. Each of these factors is then further split into $\phi(f)$ factors having coefficients in $CCF(N)$ using the expressions in (5.3.11).

Example 5.8 Let $n = 60$, $N = 12$. Then $d = 12$, $f = 3$, $e = 5$. We first factorize $C_{60}(u)$ into 2 factors of degree 8 each. Using (5.3.13) and (5.3.14), these factors are obtained as

$$C_{1,60}(u) = C_{15}(-j\,u),$$
$$C_{3,60}(u) = C_{15}(j\,u).$$

Using (5.3.11), we can factorize $C_{15}(u)$ over $CF(12)$ to get the two factors as

$$C_{1,15}(u) = v^{16}\,C_5(v^{-4}\,u) = u^4 + v^4\,u^3 - v^2\,u^2 + u + v^4,$$
$$C_{3,15}(u) = v^{32}\,C_5(v^{-8}\,u) = u^4 - v^2\,u^3 + v^4\,u^2 + u - v^2.$$

Replacing u by $j\,u$ and $-j\,u$, we get the four degree four factors of $C_{60}(u)$ over the complex cyclotomic field generated by either $C_{1,12}(v) = v^2 - j\,v - 1$ or $C_{3,12}(u) = v^2 + j\,v - 1$.

Example 5.9 This example uses the recursions in Example 5.7. Let $n = 24$ and $N = 8$. Then $C_N(v) = v^4 + 1 = (v^2 + j)\,(v^2 - j)$ and $C_n(u) = u^8 - u^4 + 1$. In this case, $d = (n, N) = 8$ and $C_n(u)$ will factorize into $\phi(d) = 4$ factors, each of degree 2. These factors are obtained using (5.3.11) as

$$C_{1,n}(u) = u^2 + v\,u + v^2,$$
$$C_{3,n}(u) = u^2 + v^3\,u - v^2,$$
$$C_{5,n}(u) = u^2 - v\,u + v^2,$$
$$C_{7,n}(u) = u^2 - v^3\,u - v^2.$$

The above factorization is over $\mathcal{CF}(8)$. It can be converted to a factorization over $\mathcal{CCF}(8)$ by using either one of the two factors of $v^4 + 1$ over $\mathcal{CZ}$, that is, $(v^2 + j)$ or $(v^2 - j)$. The recursions to be used in either case are given in example 5.7. Based on those recursions, the above factors take the following form over $\mathcal{CCF}(8)$:

$$C_{1,n}(u) = u^2 + v\,u - j,$$
$$C_{3,n}(u) = u^2 - j\,v\,u + j,$$
$$C_{5,n}(u) = u^2 - v\,u - j,$$
$$C_{7,n}(u) = u^2 + j\,v\,u + j$$

if $v^2 + j$ is used to generate $\mathcal{CCF}(8)$, and

$$C_{1,n}(u) = u^2 + v\,u + j,$$
$$C_{3,n}(u) = u^2 + j\,v\,u - j,$$
$$C_{5,n}(u) = u^2 - v\,u + j,$$
$$C_{7,n}(u) = u^2 - j\,v\,u - j$$

if $v^2 - j$ is used to generate $\mathcal{CCF}(8)$.

5.5 A Preview of Applications to Digital Signal Processing

We have learned the algebra of fields in sufficient detail at this stage. It is our strong belief that the polynomial algebra and the CRT, in both integer and polynomial forms, constitute the heart of most digital signal processing algorithms under discussion. This section is on some direct applications of the various concepts presented thus far in this chapter to digital signal processing algorithms.

5.5.1 Application to Simplifying Complex-Valued Computations

Consider a computation of the type

$$C(u) = A(u)\,B(u) \bmod C_n(u), \tag{5.5.1}$$

where all the polynomials $A(u)$, $B(u)$, and $C(u)$ are defined over $\mathcal{CZ}$ (same as $\mathcal{CF}(4)$), that is, they have complex rational numbers as coefficients. Such a computation will arise in the computation of cyclic convolution using the cyclotomic polynomials. In essence, the algebra of complex rationals is a polynomial extension of algebra of rational numbers where all the computations are performed modulo $(j^2 + 1)$. Let us express this interpretation of complex algebra *explicitly* and rewrite (5.5.1) as

$$C(u, j) = A(u, j)\, B(u, j) \bmod [j^2 + 1, C_n(u)]. \tag{5.5.2}$$

A direct approach to this computation is to represent it in terms of real and imaginary parts and write

$$\begin{aligned} \mathrm{Re}\{C(u, j)\} = {} & \mathrm{Re}\{A(u, j)\} \cdot \mathrm{Re}\{B(u, j)\} \\ & - \mathrm{Im}\{A(u, j)\} \cdot \mathrm{Im}\{B(u, j)\} \bmod C_n(u), \end{aligned}$$

$$\begin{aligned} \mathrm{Im}\{C(u, j)\} = {} & \mathrm{Im}\{A(u, j)\} \cdot \mathrm{Re}\{B(u, j)\} \\ & + \mathrm{Re}\{A(u, j)\} \cdot \mathrm{Im}\{B(u, j)\} \bmod C_n(u). \end{aligned}$$

Since a complex MULT can be performed in three real MULTs, the above computation is equivalent to three polynomial products of the type $C(u) = A(u)\, B(u) \bmod C_n(u)$ with all the polynomials having rational coefficients. Therefore, the direct approach breaks a complex-valued computation to three rational-valued computations. In the following, we describe two approaches that can be utilized to simplify a complex-valued computation to either two complex-valued computations involving smaller degree polynomials, or to two rational-valued computations.

In Section 5.3.2, complex cyclotomic polynomials were presented and it was shown that $C_n(u)$ factorizes over the field of complex rationals $\mathsf{CF}(4)$ *iff* $4 \mid n$. In earlier notation, $N = 4$. Let us assume that this condition is satisfied by n. Then

$$C_n(u, j) = C_{1,n}(u, j)\, C_{3,n}(u, j) \bmod (j^2 + 1). \tag{5.5.3}$$

Consequently, the computation in (5.5.2) can be reduced to the following three-step procedure.

Step 1. CRT-P reduction. Compute

$$\begin{aligned} A_i(u, j) &\equiv A(u, j) \bmod [C_{i,n}(u), j^2 + 1], \\ B_i(u, j) &\equiv B(u, j) \bmod [C_{i,n}(u), j^2 + 1], \quad i = 1, 3. \end{aligned}$$

Step 2. Small degree polynomial multiplication. Compute

$$C_i(u, j) \equiv A_i(u, j) B_i(u, j) \bmod [C_{i,n}(u), j^2 + 1], \quad i = 1, 3.$$

Step 3. CRT-P reconstruction. Employ (3.4.2) to obtain $C(u)$ from $C_i(u, j)$, $i = 1, 3$.

In the special case when n is a power of 2 and $n > 2$, the various polynomials are very simple in nature. In this case,

$$C_n(u) = u^{n/2} + 1 = C_{1,n}(u)C_{3,n}(u) = \left(u^{n/4} - j\right)\left(u^{n/4} + j\right). \tag{5.5.4}$$

The CRT-P reduction for $C_{1,n}(u, j)$, $deg(C_{1,n}(u)) = n / 4$, is performed by replacing $u^{n/4+k}$ by $j\ u^k$ in the polynomials $A(u)$ and $B(u)$ to get degree $n/4 - 1$ polynomials $A_1(u, j)$ and $B_1(u, j)$. Similarly, the CRT-P reduction for $C_{3,n}(u, j)$, $deg(C_{3,n}(u, j)) = n / 4$, is performed by replacing $u^{n/4+k}$ by $-j\ u^k$ in the polynomials $A(u)$ and $B(u)$ to get degree $n / 4 - 1$ polynomials $A_3(u, j)$ and $B_3(u, j)$. Note that such a reduction *does not require any MULT*, a result that holds in general even when n is not a power of 2. This aspect will be analyzed in later chapters. Once $C_1(u, j)$ and $C_3(u, j)$ are determined, the CRT-P reconstruction in Step 3 can be performed as follows:

$$C(u, j) = u^{n/4}(j/2)\,[C_3(u, j) - C_1(u, j)] + (1/2)\,[\,C_3(u, j) + C_1(u, j)]. \tag{5.5.5}$$

Once again, the reconstruction requires ADDs and *rotation* (MULT by j).

Using the Dedekind's reciprocity theorem (Theorem 5.5), we also know that the same condition implies that $C_N(v) = v^2 + 1$ factorizes in $\mathsf{CF}(n)$. Once again, it is more popular to use the indeterminate j in the context of complex arithmetic. Replacing v by j, we have the following factorization in $\mathsf{CF}(n)$:

$$j^2 + 1 = (j - u^{n/4}) \cdot (j + u^{n/4}) \bmod C_n(u). \tag{5.5.6}$$

Consequently, the computation in (5.5.1) can also be reduced to the following three-step procedure.

Step 1. CRT-P reduction. Compute

$$A_1(u, u^{n/4}) \equiv A(u, j) \bmod [j - u^{n/4}, C_n(u)],$$
$$B_1(u, u^{n/4}) \equiv B(u, j) \bmod [j - u^{n/4}, C_n(u)],$$

and

$$A_3(u, -u^{n/4}) \equiv A(u, j) \bmod [j + u^{n/4}, C_n(u)],$$
$$B_3(u, -u^{n/4}) \equiv B(u, j) \bmod [j + u^{n/4}, C_n(u)].$$

Step 2. Polynomial multiplication. Compute

$$C_i(u) \equiv A_i(u)\, B_i(u) \bmod \; C_n(u)],\;\; i = 1, 3.$$

Step 3. CRT-P reconstruction. $C(u)$ is obtained from $C_i(u)$, $i = 1, 3$, as follows:

$$C(u,j) = (1/2)\,[C_1(u) + C_2(u)] + (j/2)\, u^{n/4}\,[C_2(u) - C_1(u)]. \qquad (5.5.7)$$

Interestingly, CRT-P reduction and reconstruction require only ADDs and rotation. Between the two approaches outlined above, either one may be preferred depending on the organization of the overall computation.

5.5.2 Expressing an One-Dimensional Computation as a Two-Dimensional Computation

We demonstrate this application with the help of two examples.

Example 5.10 Let us say that we are interested in the computation

$$C(u) = A(u)\, B(u) \bmod u^{16} + 1, \qquad (5.5.8)$$

where $A(u)$, $B(u)$, and $C(u)$ are degree 15 polynomials. Let us express $A(u) = A_0 + A_1 u + ... + A_{15} u^{15}$, as a two-dimensional polynomial in the following manner:

$$A(u, v) = [A_0 + A_1 u + A_2 u^2 + A_3 u^3] + v\,[A_4 + A_5 u + A_6 u^2 + A_7 u^3]$$
$$+ v^2\,[A_8 + A_9 u + A_{10} u^2 + A_{11} u^3] + v^3\,[A_{12} + A_{13} u + A_{14} u^2 + A_{15} u^3], \qquad (5.5.9)$$

where $v = u^4$. Note that the two representations are equivalent in all respects. Therefore, a one-dimensional polynomial of degree 15 is represented as a two-dimensional polynomial with degree 3 in either dimension. Similar two-dimensional representations hold for the polynomials $B(u)$ and $C(u)$. The computation in (5.5.8) can now be expressed as

$$C(u, v) = A(u, v)\, B(u, v) \bmod [u^4 - v,\, v^4 + 1]. \qquad (5.5.10)$$

The modulo polynomials are obtained from the definitions $v = u^4$ and $u^{16} + 1$ or $v^4 + 1$. Suppressing the functional dependence $v = u^4$ from (5.5.10), we can express it as a *true* two-dimensional computation,

$$C(u, v) = A(u, v)\, B(u, v) \bmod v^4 + 1. \qquad (5.5.11)$$

Once the two-dimensional polynomial $C(u, v)$ is computed in (5.5.11), the one-dimensional polynomial is obtained by replacing v by u^4.

Now we make the crucial observation. The computation in (5.5.11) can be performed as a one-dimensional polynomial computation in the indeterminate u, where the coefficients of the polynomials are themselves polynomials in the indeterminate v that are defined modulo $v^4 + 1$. Thus, the coefficients of the one-dimensional polynomials $A(u, v)$ and $B(u, v)$, each of degree 3, in the indeterminate u are defined in $\mathcal{CF}(8)$ as $v^4 + 1 = C_8(v)$. Since $deg(A(u, v)) = deg(B(u, v)) = 3$ (along the u dimension), $deg(C(u, v)) = 6$ (in (5.5.11)). *Now the CRT-P can be used to compute this polynomial product as an acyclic convolution with the coefficients of the various polynomials defined in* $\mathcal{CF}(8)$. *The interesting part is that the modulo polynomial $P(u)$ used in the CRT-P is also defined in* $\mathcal{CF}(8)$. Consequently, there are far more choices available for such a polynomial. A direct computation of (5.5.8) would require that the modulo polynomial be defined in $\mathcal{Z}$ or $\mathcal{CF}(4)$ (= $\mathcal{CZ}$). Following the basic formulation of the CRT-P based acyclic convolution algorithm in Section 4.4.1, $P(u)$ has degree at least 7. One choice of a degree 7 polynomial with coefficients in $\mathcal{CF}(8)$ is

$$P(u) = u\,(u - 1)\,(u + 1)\,(u - v)\,(u + v)\,(u + v^2)\,(u - v^2). \tag{5.5.12}$$

The three-step, CRT-P based algorithm of the acyclic convolution in (5.5.11) that employs the above $P(u)$ as the modulo polynomial is clear at this stage. The CRT-P reduction with respect to all the 7 degree one factors can be performed with ADDs and shifts as v is just an indeterminate quantity and MULT by powers of v does not require MULT. Therefore, we have broken down the computation in (5.5.8) to 7 polynomial products, each defined modulo $v^4 + 1$. An order recursive approach to further break down each of these 7 polynomial products defined modulo $v^4 + 1$ to still smaller degree polynomial product is clear at this stage.

Another approach that may be preferred in some situations is to cast the computation in (5.5.11) as a cyclic convolution and then compute it using the DFT. Many fast algorithms are available for DFT computation, some of which will be described in Chapters 13 and 14. This is done by defining a DFT of length 8 in $\mathcal{CF}(8)$ with the 8-th root of unity being v. This corresponds to letting

$$P(u) = u^8 - 1 = \prod_{i=1}^{8}\left(u - v^i\right). \tag{5.5.13}$$

The DFT based algorithm of the cyclic convolution in (5.5.11) is clear at this stage. The computations for the DFT and the IDFT require only ADDs and shifts as v is just an indeterminate quantity. Note that we only need to define a length 7 DFT as $C(u)$ has degree 6. But there does not exist a 7-th root of unity in $\mathcal{CF}(8)$. Once again, we have broken down the computation in (5.5.8) to 8 polynomial products, each defined modulo $v^4 + 1$. An order recursive approach to further break down each of these 8 polynomial products defined modulo $v^4 + 1$ to still smaller degree polynomial product is clear at this stage. Also, one may work with only 7 of the 8 factors in (5.5.13).

Example 5.11 Let us say that we are interested in the computation

$$C(u) = A(u)\, B(u) \bmod u^{18} + u^9 + 1, \tag{5.5.14}$$

where $A(u)$, $B(u)$, and $C(u)$ are degree 17 polynomials. Let us express $A(u) = A_0 + A_1 u + ... + A_{15} u^{17}$ as a two-dimensional polynomial in the following manner:

$$A(u, v) = [A_0 + A_1 u + A_2 u^2] + v\,[A_3 + A_4 u + A_5 u^2] + ...$$

$$+ v^5 [A_{15} + A_{16} u + A_{17} u^2], \tag{5.5.15}$$

where $v = u^3$. Therefore, a one-dimensional polynomial of degree 17 is represented as a two-dimensional polynomial with degrees 3 and 5 in its dimensions. Similar two-dimensional representations hold for the polynomials $B(u)$ and $C(u)$. The computation in (5.5.14) can now be expressed in two-dimensions as

$$C(u, v) = A(u, v)\, B(u, v) \bmod [u^3 - v, v^6 + v^3 + 1]. \tag{5.5.16}$$

Suppressing the functional dependence $v = u^3$ from (5.5.16), we can express it as a *true* two-dimensional computation,

$$C(u, v) = A(u, v)\, B(u, v) \bmod v^6 + v^3 + 1. \tag{5.5.17}$$

The computation in (5.5.17) can be performed as a one-dimensional polynomial computation in the indeterminate u where the coefficients of the polynomials are themselves polynomials in the indeterminate v that are defined modulo $v^6 + v^3 + 1$. The coefficients of one-dimensional polynomials $A(u, v)$ and $B(u, v)$, each of degree 2, in the indeterminate u are defined in $\mathcal{CF}(9)$ as $v^6 + v^3 + 1 = C_9(v)$. Since $deg(A(u, v)) = deg(B(u, v)) = 2$ (along the u dimension), $deg(C(u, v)) = 4$ (in (5.5.17)). Now the CRT-P can be used to compute this polynomial product as an acyclic convolution of length 5. One choice for the degree 5 modulo polynomial with coefficients in $\mathcal{CF}(9)$ is

$$P(u) = u(u-1)(u+1)(u-v)(u+v). \tag{5.5.18}$$

Therefore, we have broken down the computation in (5.5.14) to 5 polynomial products, each defined modulo $v^6 + v^3 + 1$.

In the DFT approach, a DFT of length 6 in $CF(9)$ can be defined by noting that $-v^3$ is a 6-th root of unity. This corresponds to letting

$$P(u) = u^6 - 1 = \prod_{i=1}^{6}\left(u - \left(-v^3\right)^i\right). \tag{5.5.19}$$

The DFT based algorithm of the cyclic convolution in (5.5.17) is clear at this stage. The computations for the DFT and the IDFT require only ADDs and shifts as v is just an indeterminate quantity. Note that we only need to define a length 5 DFT as $C(u)$ has degree 4, but there does not exist a 5-th root of unity in $CF(9)$. However, a choice such that

$$P(u) = u(u+1)(u^3 - 1) = u(u+1)\prod_{i=0}^{2}\left(u - \left(v^3\right)^i\right)$$

can also be used. Finally, one can also convert this one-dimensional computation to a two-dimensional computation by letting $u^9 = v$ and then either perform a length 17 acyclic convolution or a length 18 cyclic convolution in $CF(3)$. The details of such a procedure are straightforward and left to the reader.

It is interesting to observe here that if the sequences in the first example were complex-valued in nature, then further savings could be realized by employing either one of the two approaches described in Section 5.5.1. We close this discussion by emphasizing that, though we end up increasing the length of convolution in order to cast it as a cyclic convolution and then compute it using DFTs, it may be preferred due to the availability of a wide range of FFT algorithms for the DFT computation.

Notes

The objective of this chapter was to analyze the factorization of the polynomial $u^n - 1$ and the associated fields. The procedure can be described as first seeking a factorization over an extension field in terms of degree one factors by defining an n-th root of unity and then combine just enough of those factors to obtain the factorization over the field of interest. We have provided proofs of all the results with the exception of irreducibility of cyclotomic polynomials over the field of rational numbers. In our analysis, this proof does not play a significant role, the important aspect being that even though the cyclotomic polynomials are irreducible over the field of rational numbers, they do factor in the cyclotomic extension fields. We have covered most fields of interest for one- and higher-dimensional digital signal processing algorithms. We must emphasize that all the different algorithms for digital signal processing employ two basic concepts; the first one is polynomial theory, and the second one is Chinese remainder theorem, both in integer and polynomial forms. It is the most sincere hope of the author that the readers will develop an appreciation for these basic concepts as an umbrella under which the majority of algorithms reside.

A number of texts were used to assimilate the material for this chapter. There are a number of mathematics books that deal with many of these ideas but with an entirely different focus. The ones that influenced our presentation in a significant manner are Hasse [1980], Washington [1982], Edwards [1984], Pohst and Zassenhaus [1990], Nagell [1951], and Sah [1967]. The last two references contain the proof of irreducibility of cyclotomic polynomials over the field of rational number systems. The statement of the Dedekind's reciprocity theorem is taken almost verbatim from Edwards [1984]. The earlier books in digital signal processing do not provide a comprehensive coverage of the topics presented in this chapter. We have used McClellan and Rader [1979] and Heideman [1988] to enhance our understanding of cyclotomic polynomials as applied to digital signal processing algorithms. The author came in contact with the cyclotomic polynomials during his research on the computational algorithms for cyclic convolution, some of which is being reported in this book.

Bibliography

[5.1] H. Hasse, *Number Theory*, Springer-Verlag, 1980.

[5.2] L.C. Washington, *Introduction to Cyclotomic Fields*, Springer-Verlag, 1984.

[5.3] H.M. Edwards, *Galois Theory*, Springer-Verlag, 1984.

[5.4] M. Pohst and H. Zassenhaus, *Algorithmic Algebraic Number Theory*, Cambridge University Press, 1990.

[5.5] T. Nagell, *Introduction to Number Theory*, John Wiley & Sons, Inc., 1951.

[5.6] C.-H. Sah, *Abstract Algebra*, Academic Press, 1967.

[5.7] J.H. McClellan and C.M. Rader, *Number Theory in Digital Signal Processing*, Prentice Hall, Inc., 1979.

[5.8] M.T. Heideman, *Multiplicative Complexity, Convolution, and the DFT*, Springer-Verlag, 1988.

Problems

5.1 Prove Lemma 5.1.

5.2 Cast Example 5.11 in Section 5.5.2 as a computation of polynomial product with coefficients in CF(3). Write the formulation in the form of an acyclic and cyclic convolution. Obtain cyclotomic factors of $u^{18} - 1$ in CF(3).

5.3 Obtain the cyclotomic polynomials for values of n equal to 9, 15, 16, 27, 55, 95, and 105.

5.4 Show that the coefficients of the cyclotomic polynomial $C_n(u)$ lie in the range $-(r - 1)$ to $(r - 1)$ if n contains at most three distinct odd primes p, q, and r in its standard factorization, $r < q < p$.

5.5 Show that $C_n(1)$ is equal to 0 if $n = 1$, p if n is a power of a prime, and 1 if n has two or more prime factors [McClellan and Rader 1979].

5.6 Consider the cyclotomic factors of $u^n - 1$ for $n = 30$. Is this factorization for $u^n - 1$ for $n = 30$ valid over $GF(7)$?

5.7 Show that the non-zero coefficients of $C_n(u)$ alternate between 1 and -1 if n has two odd prime numbers in its factorization.

5.8 Factorize $u^{40} - 1$ in CF(5) and CCF(8).

5.9 Consider cyclotomic extension of the field of real numbers Re by the polynomial $v^2 - (2 \cos \theta) v + 1$, where $\theta = 2\pi / n$, n being a power of 2. Show that the computation of $\cos \theta$ requires only the operation of square rooting. Express cyclotomic factorization of $u^n - 1$ in this field.

5.10 Factorize $u^{56} - 1$ over Z, CZ, CF(7), CF(8), CF(16), CF(56), CCF(7), CCF(8), CCF(28), CCF(56).

5.11 Consider the two-dimensional ($p \times p$) cyclic convolution expressed in polynomial form as the computation

$$Z(u,v) = X(u,v)Y(u,v) \bmod \left[u^p - 1, v^p - 1\right],$$

where p is a prime number. Use the cyclotomic factorization of $v^p - 1$ over Z and that of $u^p - 1$ over CF(p) to show that the entire computation can be performed using ($p + 1$) multiplications of the type $A(w)\ B(w) \bmod C_p(w)$ and some other simple products. Identify these simple products.

5.12 Consider the two-dimensional ($2 \cdot p \times p$) cyclic convolution expressed in polynomial form as the computation

$$Z(u, v) = X(u, v)\ Y(u, v) \bmod [u^{2p} - 1, v^p - 1],$$

where p is a prime number, $p > 2$. Use the cyclotomic factorization of $v^p - 1$ over Z and that of $u^{2 \cdot p} - 1$ over CF(p) to show that the entire computation can be performed using MULTs of the type $A(w)\ B(w) \bmod C_p(w)$ and some other simple products. How many such MULTs do we have? Identify these simple products.

5.13 Consider the two-dimensional ($p^2 \times p$) cyclic convolution expressed in polynomial form as the computation

$$Z(u, v) = X(u, v)\ Y(u, v) \bmod [\, u^{p^2} - 1, v^p - 1],$$

where p is a prime number, $p > 2$. Use the cyclotomic factorization of $v^p - 1$ over Z and that of $u^{p^2} - 1$ over CF(p) to simplify the entire computation to MULTs of the type $A(w)\ B(w) \bmod C_p(w)$.

5.14 This is a continuation of Problem 5.13. Now use the cyclotomic factorization of $u^{p^2} - 1$ over Z and that of $v^p - 1$ over CF(p^2) to simplify the two-dimensional ($p^2 \times p$) cyclic convolution to MULTs of the type $A(w)\ B(w) \bmod C_{p^2}(w)$.

5.15 Consider the complex cyclotomic polynomials $C_{1,n}(u)$ and $C_{3,n}(u)$, $n = 24$. If $X(u)$ has coefficients defined over Z, then show that $X_1(u) \equiv X(u) \bmod C_{1,n}(u)$ can be used directly to determine $X_3(u) \equiv X(u) \bmod C_{3,n}(u)$. Hence, devise a method to process two polynomials having coefficients in $\mathbf{Z}$ simultaneously. In other words, determine a way to obtain $X(u)$ mod $C_{1,n}(u)$, $X(u)$ mod $C_{3,n}(u)$, $Y(u)$ mod $C_{1,n}(u)$, and $Y(u)$ mod $C_{3,n}(u)$, $X(u)$, $Y(u)$ having coefficients in $\mathbf{Z}$ with one computation of the type $Z(u)$ mod $C_{1,n}(u)$ and $Z(u)$ mod $C_{3,n}(u)$, $Z(u)$ having coefficients in $\mathsf{CF}(4)$. This is known as the property of **conjugate symmetry**.

5.16 Write a program to determine all the cyclotomic polynomials of length up to 500.

Chapter 6

Cyclotomic Polynomial Factorization In Finite Fields

In this chapter, we describe factorization of $u^n - 1$ in terms of its cyclotomic factors over the finite fields $GF(p)$ and $GF(q)$, $q = p^m$. $GF(p)$ is the field with elements $\{0, 1, ..., p-1\}$, where the arithmetic operations defined mod p. $GF(q)$ is obtained from $GF(p)$ by defining as elements all polynomials of degree up to $m - 1$ with coefficients in $GF(p)$. The arithmetic operations of '+' and '·' are defined mod $P(v)$, $P(v)$ being an *irreducible polynomial over* $GF(p)$, $deg(P(v)) = m$. Thus, $GF(q)$ is a polynomial extension of $GF(p)$. Note that we have changed the indeterminate from u to v in order to distinguish the polynomial representation of the elements of $GF(q)$ from the remainder of our discussion on polynomial algebra. $GF(q)$ exists for all values of prime p and m. In other words, irreducible polynomials of degree m exist over $GF(p)$ for all prime p and $m > 0$. In fact, for a finite field having a elements, a must have the form $a = p^m$.

6.1 Cyclotomic Polynomial Factorization

For the case that n is a multiple of p, $n = c \cdot p$, we have

$$u^n - 1 = (u^c - 1)^p. \qquad (6.1.1)$$

If c is still a multiple of p, then (6.1.1) can be repeated until they are relatively prime. Therefore, we need to focus only on the case $(n, p) = 1$, that is, n and p are coprime.

Theorem 6.1 If $(a, b) = 1$, then there always exists an integer ϕ such that

$$b \mid (a^{\phi} - 1).$$

Proof: This theorem follows directly from Theorem 2.2, the Euler's theorem, when ϕ is chosen to be the Euler's totient function of b.

Since $(n, p) = 1$, Theorem 6.1 can be used to find an integer such that $n \mid (p^m - 1)$. We shall use the smallest value of m for which $n \mid (p^m - 1)$ in order to factorize $u^n - 1$ over $GF(p)$ and $GF(p^m)$. It is clear from Theorem 6.1 that $m \mid \phi$.

Theorem 6.2 Let A be any non-zero element in $GF(q)$, $q = p$ for $GF(p)$ and $q = p^m$ for $GF(p^m)$. Then

$$A^{q-1} = 1. \tag{6.1.2}$$

Proof: Let $A_1, A_2, \ldots, A_{q-1}$ be the $(q - 1)$ non-zero elements of $GF(q)$. Consider the elements in $GF(q)$ defined as

$$B_i = A \cdot A_i, \ \ A_i \neq 0, \ \ i = 1, 2, \ldots, q - 1.$$

Clearly $B_i \in GF(q)$, $B_i \neq 0$, and $B_i \neq B_j$, $i \neq j$. Therefore, $B_1, B_2, \ldots, B_{q-1}$ are also the $(q - 1)$ non-zero elements of $GF(q)$. Thus,

$$\prod_{i=1}^{q-1} B_i = \prod_{i=1}^{q-1} \left(A \cdot A_i\right) = \prod_{i=1}^{q-1} A_i . \tag{6.1.3}$$

Multiplying both sides by the multiplicative inverse of $\prod_{i=1}^{q-1} A_i$, we get the result.

Theorem 6.3 A unique factorization of $u^{q-1} - 1$ over $GF(q)$ is given by

$$u^{q-1} - 1 = \prod_{i=1}^{q-1} \left(u - A_i\right), \ \ A_i \in GF(q), \ \ A_i \neq 0. \tag{6.1.4}$$

Proof: From Theorem 6.2, every non-zero element $A_i \in GF(q)$ is a root of $u^{q-1} - 1$. Therefore, $(u - A_i)$, $i = 1, 2, \ldots, q - 1$, is a factor of $u^{q-1} - 1$. Since there are $q - 1$ non-zero elements in $GF(q)$, we have

$$\prod_{i=1}^{q-1}(u - A_i) \Big| \left(u^{q-1} - 1\right).$$

Both polynomials $\prod_{i=1}^{q-1}(u - A_i)$ and $\left(u^{q-1} - 1\right)$ are monic and of degree $q - 1$. The theorem follows.

Theorem 6.3 gives a complete factorization of the polynomial $\left(u^{q-1} - 1\right)$ over $GF(q)$, $q = p^m$. Given the above factorization over $GF(p^m)$, we will explore the following two issues associated with this factorization:

1. Factorization of $u^n - 1$ over $GF(p)$, $n \mid (p - 1)$, and
2. Factorization of $u^n - 1$ over $GF(p)$ using $GF(p^m)$, $n \mid (p^m - 1)$.

Definition 6.1 Given a non-zero element A in $GF(q)$, the smallest non-zero value of n such that $A^n = 1$ is called the *order* of A. It is clear from Theorem 6.2 that $n \le q - 1$.

Theorem 6.4 The order of every non-zero element in $GF(q)$ divides $q - 1$.
Proof: If n does not divide $q - 1$, then we write

$$q - 1 = a\,n + r, \quad 0 \le r < n.$$

Using Theorem 6.2, we get

$$A^{q-1} = 1 = A^{an+r} = (A^n)^a \cdot A^r = A^r, \quad 0 \le r < n.$$

However, it contradicts the definitions of the order of A. Therefore, the statement of the theorem is true.

Theorem 6.5

$$(u^a - 1, u^b - 1) = u^{(a,\, b)} - 1.$$

Proof: This is same as Theorem 3.4.

A direct consequence of this theorem is that if $b \mid a$, then $(u^b - 1) \mid (u^a - 1)$ and vice versa. Another consequence of this theorem is that for every factor d of $q - 1$, that is $d \mid (q - 1)$, $(u^d - 1) \mid (u^{q-1} - 1)$. Combining this with the factorization in Theorem 6.3, we get

$$u^{q-1}-1=\left(u^d-1\right)C(u)=\prod_{i=1}^{q-1}\left(u-A_i\right),\ deg(C(u))=q-1-d.$$

Since $deg(C(u)) = q - 1 - d$, it can have only $q - 1 - d$ roots in $GF(q)$. Therefore, there must be d elements among the $q - 1$ non-zero elements A_1, A_2, ..., A_{q-1} in $GF(q)$ that are roots of $u^d - 1$. Let these elements be B_1, B_2, ..., B_d. This gives factorization of $u^d - 1$ for every $d \mid (q - 1)$ over $GF(q)$ as

$$u^d-1=\prod_{i=1}^{d}\left(u-B_i\right),\ B_i\in GF(q),\ B_i\neq 0. \tag{6.1.5}$$

The above factorization also implies that the element B_i has order equal to d or a factor of d. This is consistent with Theorem 6.4. There are exactly d elements that have d or a factor of d as their order. Now, consider the factorization of $q - 1$ as

$$q-1=r_1^{a_1}\cdot r_2^{a_2}\cdots r_l^{a_l}=q_1 q_2\cdots q_l$$

and let $d = 1$. There is one element in $GF(q)$ of order = 1. This element is 1. Let $d = r_1$. From (6.1.5), we see that there are r_1 distinct elements in $GF(q)$ having order r_1 or a factor of r_1. Since r_1 is prime, there must be exactly $r_1 - 1$ elements in $GF(p)$ of order r_1. Similarly, there are $r_1^i - r_1^{i-1}$ elements in $GF(p)$ of order r_1^i, i = 1, 2, ..., a_1. All of these $r_1^i - r_1^{i-1}$ elements are roots of the polynomial $u^{r_1^i}-1$ but not of $u^{r_1^{i-1}}-1$. Let A_j denote an element in $GF(q)$ of order $q_j = r_j^{a_j}$, j = 1, 2, ..., l. A fundamental property of these elements is stated in the following.

Theorem 6.6 In $GF(q)$, there is an element of order $q - 1$.

Proof: Consider the element $A = A_1 \cdot A_2$. Since $A_1^{q_1}=1$, $A_2^{q_2}=1$, we have $A_1^{q_1 q_2}=1$ and $A_2^{q_1 q_2}=1$. Therefore, $\left(A_1 A_2\right)^{q_1 q_2}=1$ or $A^{q_1 q_2}=1$. Let n be the order of the element A, that is $A^n = 1$. Since $A^{q_1 q_2}=1$, we have $n \mid (q_1 \cdot q_2)$. $A^n = 1$ also leads to $A^{nq_1}=1$ and $A^{nq_2}=1$. Replacing A by $A_1 \cdot A_2$ in these equations and realizing that $A_1^{q_1}=1$ and $A_2^{q_2}=1$, we get $A_2^{nq_1}=1$ and $A_1^{nq_2}=1$. These equations imply that $q_1 \mid n$ and $q_2 \mid n$. Since $(q_1, q_2) = 1$, we have $(q_1 \cdot q_2) \mid n$. The two relations $n \mid (q_1 \cdot q_2)$ and $(q_1 \cdot q_2) \mid n$ lead to n

$= q_1 \cdot q_2$. Since the element $A_1 \cdot A_2$ has order $q_1 \cdot q_2$, by induction, the element $A = A_1 \cdot A_2 \cdots A_l$ has order $\prod_{i=1}^{l} r_i^{a_i} = q-1$.

This theorem plays a critical role in the process of factorizing $u^n - 1$.

Definition 6.2 An element in $GF(q)$ having order equal to $q - 1$ is called a **primitive element** of the field.

Theorem 6.6 guarantees that every $GF(q)$ has at least one primitive element. Let A denote one such primitive element. Since A has order $q - 1$, every non-zero element A_i in $GF(q)$ can be expressed as

$$A_i = A^i, \; i = 1, 2, ..., q-1.$$

Based on (6.1.4), this leads to the factorization of $u^{q-1} - 1$ over $GF(q)$ as

$$u^{q-1} - 1 = \prod_{i=1}^{q-1} \left(u - A^i\right) \tag{6.1.6}$$

or

$$u^{q-1} - 1 = \prod_{i=0}^{q-2} \left(u - A^i\right), \tag{6.1.7}$$

since $A^{q-1} = A^0 = 1$. In terms of the element A, the factorization of $u^d - 1$, $d \mid (q-1)$ over $GF(q)$ can be expressed as

$$u^d - 1 = \prod_{i=1}^{d} \left(u - B^i\right) \tag{6.1.8}$$

or

$$u^d - 1 = \prod_{i=0}^{d-1} \left(u - B^i\right), \tag{6.1.9}$$

where

$$B = A^{(q-1)/d}. \tag{6.1.10}$$

Once again, (6.1.9) follows from (6.1.8) as $B^0 = B^d = 1$. In digital signal processing, the emphasis is on processing polynomials defined over either $GF(p)$ or $Z(M)$. Therefore, we need to further analyze the factorizations in (6.1.6) to (6.1.9). This is done in the following.

6.2 Factorization of $(u^n - 1)$ Over $GF(p)$

A direct consequence of (6.1.6) to (6.1.10) is that the factorization of $u^{p-1} - 1$ in $GF(p)$ or $GF(p^m)$ is given by

$$u^{p-1} - 1 = \prod_{i=1}^{p-1} \left(u - A^i\right) \tag{6.2.1}$$

or

$$u^{p-1} - 1 = \prod_{i=0}^{p-2} \left(u - A^i\right), \tag{6.2.2}$$

since $A^0 = A^{p-1} = 1$. Here A is a primitive element in $GF(p)$ or equivalently an element of order $(p - 1)$ in $GF(p^m)$. Also, for $d \mid (p - 1)$, the factorization of $u^d - 1$ over $GF(p)$ or $GF(p^m)$ is given by

$$u^d - 1 = \prod_{i=1}^{d} \left(u - B^i\right) \tag{6.2.3}$$

or

$$u^d - 1 = \prod_{i=0}^{d-1} \left(u - B^i\right), \tag{6.2.4}$$

where

$$B = A^{(p-1)/d}. \tag{6.2.5}$$

Example 6.1 Let $p = 11$. In $GF(11)$, the orders of 1, 2, 3, 4, 5, 6, 7, 8, 9, 10 are 1, 10, 5, 5, 5, 10, 10, 10, 5, 2, respectively. The orders are determined by noting that order of any element in $GF(11)$ divides $q - 1 = 11 - 1 = 10$. Therefore, we only need to test if the order of any element is either one of 1, 2, 5, 10. The integers 2, 6, 7, and 8 are primitive elements. All the elements A_i in $GF(11)$ can be expressed as powers of A, where A is one of 2, 6, 7, and 8. Let us select $A = 2$. Then we have

$$u^{10} - 1 = \prod_{i=0}^{9} \left(u - 2^i\right).$$

The integers 3, 4, 5, and 9 have order equal to 5. Therefore, any one of them can be used to obtain the factorization of $u^5 - 1$. In general, we pick $A^{(q-1)/d}$ as an element of order d. Here $d = 5$ and we can show that 3, 4, 5, or 9 can be expressed as A^2. Let us select $B = 4$ as the integer of order 5. Then we have

$$u^5 - 1 = \prod_{i=0}^{4} \left(u - 4^i\right).$$

The selection of the integer to be used in the factorization may depend ultimately on the ease of hardware realization of the various algorithms. We summarize the above procedure for factorizing $u^n - 1$ over $GF(p)$, $n|(q-1)$ in Table 6.1.

Table 6.1 Factorization of $u^n - 1$ Over $GF(p)$

	Let A be a primitive element in $GF(p)$.
$n = p - 1$	$u^n - 1 = \prod_{i=0}^{n-1} \left(u - A^i\right)$
$n \mid (p-1)$	$u^n - 1 = \prod_{i=0}^{n-1} \left(u - B^i\right)$ $B = A^{(p-1)/d}$
	Note: A primitive integer in $GF(p)$ always exists.

6.2.1 Returning Back to $GF(p)$

Given the factorization of $u^{q-1} - 1$ over $GF(q)$, $q = p^m$, in (6.1.6), we partition the set $\Gamma = \{1, 2, ..., q-1\}$ containing the powers of the non-zero elements in $GF(q)$ into subsets $\Gamma_{j_1}, \Gamma_{j_2}, \ldots$. A **cyclotomic set** Γ_j begins with j, where j is the smallest power of A not included in the preceding subsets. Other elements in the subset Γ_j are obtained as

$$\Gamma_j = \{j, j\,p, j\,p^2, j\,p^3, ...\}. \tag{6.2.6}$$

Since $A^{q-1} = 1$, the powers of A are defined mod $q - 1 = p^m - 1$. Also, $A^{p^m - 1} = 1$ implies that $A^{jp^m} = A^j$. Therefore, there are at most m elements in each Γ_j. It is easy to verify that no elements in two different cyclotomic sets are equal. Let Ψ be the set of indices j_1, j_2, ... Based on this partitioning and (6.2.6), we write the factorization of $u^{q-1} - 1$ as

$$u^{q-1} - 1 = \prod_{j \in \Psi} \left\{ \prod_{\theta \in \Gamma_j} \left(u - A^{\theta}\right) \right\} = \prod_{j \in \Psi} Q_j(u). \tag{6.2.7}$$

The polynomial $Q_j(u)$ is defined as

$$Q_j(u) = \left(u - A^{j}\right)\left(u - A^{jp}\right)\left(u - A^{jp^2}\right)\cdots\left(u - A^{jp^{l-1}}\right) \tag{6.2.8}$$

such that

$$jp^l \equiv j \bmod (p^m - 1). \tag{6.2.9}$$

The last step in this factorization process is to show that $Q_j(u)$ has coefficients over $GF(p)$ and that it is irreducible. This is established in the following.

Theorem 6.7 The polynomial $Q_j(u)$ as defined in (6.2.8) has coefficients over $GF(p)$.

Proof: Consider the polynomial

$$Q_j^p(u) = \left\{ \prod_{i=0}^{l-1} \left(u - A^{jp^i}\right) \right\}^p = \prod_{i=0}^{l-1} \left(u - A^{jp^i}\right)^p$$

$$= \prod_{i=0}^{l-1} \left(u^p - A^{jp^{i+1}}\right) = \left\{ \prod_{i=0}^{l-2} \left(u^p - A^{jp^{i+1}}\right) \right\} \left(u^p - A^{jp^l}\right).$$

Recalling (6.2.9), the above equation simplifies to

$$\left\{ \prod_{i=1}^{l-1} \left(u^p - A^{jp^i}\right) \right\} \left(u^p - A^{j}\right) = \prod_{i=0}^{l-1} \left(u^p - A^{jp^l}\right).$$

Thus, $Q_j^p(u) = Q_j\left(u^p\right)$. If $Q_j(u) = F_0 + F_1 u + \cdots + F_l u^l$, then

$$Q_j^p(u) = \left\{ \sum_{i=0}^{l} F_i u^i \right\}^p = \sum_{i=0}^{l} F_i^p u^{ip}$$

and

$$Q_j\left(u^p\right) = \sum_{i=0}^{l} F_i u^{ip} .$$

The condition $Q_j^p(u) = Q_j\left(u^p\right)$ and the above expressions lead to the property that if F_i is the i-th coefficient of the polynomial $Q_j(u)$ defined over $GF(p^m)$, then $F_i^p = F_i$. In other words, $F_i = 0$ or F_i is a root of the polynomial $u^{p-1} - 1$. The polynomial $u^{p-1} - 1$ has $p - 1$ non-zero elements of $GF(p)$ as its roots. Consequently, $F_i \in GF(p)$.

A result useful in establishing the irreducibility of the $Q_j(u)$ polynomial is as follows.

Theorem 6.8 If a polynomial $Q(u)$ defined over $GF(p)$ has $\beta \in GF(p^m)$ as a root, then β^p is also a root of $Q(u)$.
Proof: If β is a root of the polynomial

$$Q(u) = \sum_{i=0}^{l} F_i u^i , \quad F_i \in GF(p),$$

then $Q(\beta) = 0$ or

$$\sum_{i=0}^{l} F_i \beta^i = 0 .$$

Raising both sides to the power p and recalling that, since $F_i \in GF(p)$, $F_i^p = F_i$, we get

$$0 = \left\{ \sum_{i=0}^{l} F_i \beta^i \right\}^p = \sum_{i=0}^{l} F_i^p \beta^{pi} = \sum_{i=0}^{l} F_i \left(\beta^p\right)^i ,$$

or β^p is a root of $Q(u)$.

Applying this theorem recursively we see that if a polynomial defined over $GF(p)$ has $\beta \in GF(p^m)$ as a root, then it also has β^p, β^{p^2}, ... as roots. Since the polynomial $Q_j(u)$ as defined in (6.2.8) has only A^j, A^{jp}, A^{jp^2}, ... as its roots and has coefficients in $GF(p)$ (Theorem 6.7), it is the polynomial of the lowest degree in $GF(p)$ having A^j as its roots. Therefore, it must be irreducible. The polynomial $Q_j(u)$ having coefficients in $GF(p)$ and A^j, A^{jp}, A^{jp^2}, ... as its roots is called the **minimum polynomial** of its roots. Above analysis completes factorization of $u^n - 1$ over $GF(p)$, $n = p^m - 1$.

Example 6.2 Consider the factorization of $u^{15} - 1$ over $GF(2)$. Here $15 = 2^4 - 1$. We begin the factorization process by first considering the factorization of $u^{15} - 1$ over $GF(16)$. $GF(16)$ is created as an extension of $GF(2)$ by an irreducible polynomial of degree 4 over $GF(2)$. Thus, $p = 2$, $q = 16$, $m = 4$. One such polynomial is given by $v^4 + v^3 + 1$. If A is a primitive element in $GF(2^4)$, then the non-zero elements in $GF(2^4)$ can be expressed as A^i, $i = 1, 2, ..., q - 1$. A partitioning of the set $\Gamma = \{1, 2, ..., 15\}$ into subsets $\Gamma_{j_1}, \Gamma_{j_2}, \ldots$, is given by

$$\Gamma_1 = \{1, 2, 4, 8\},$$
$$\Gamma_3 = \{3, 6, 12, 9\},$$
$$\Gamma_5 = \{5, 10\},$$
$$\Gamma_7 = \{7, 14, 13, 11\},$$
$$\Gamma_{15} = \{15\}.$$

Note that each of the cyclotomic sets contains a listing of the power of A and since $A^{15} = 1$, the elements in Γ_j are defined mod 15. The set Ψ is given by $\{1, 3, 5, 7, 15\}$ and the factorization of $u^{15} - 1$ is given by

$$\begin{aligned} u^{15} - 1 &= \prod_{j \in \Psi} \left\{ \prod_{\theta \in \Gamma_j} \left(u - A^{\theta}\right) \right\} \\ &= \left\{(u - A)\left(u - A^2\right)\left(u - A^4\right)\left(u - A^8\right)\right\} \cdot \left\{\left(u - A^3\right)\left(u - A^6\right)\left(u - A^{12}\right)\left(u - A^9\right)\right\} \\ &\quad \cdot \left\{\left(u - A^5\right)\left(u - A^{10}\right)\right\} \cdot \left\{\left(u - A^7\right)\left(u - A^{14}\right)\left(u - A^{13}\right)\left(u - A^{11}\right)\right\} \cdot \left\{u - A^{15}\right\} \\ &= Q_1(u) Q_3(u) Q_5(u) Q_7(u) Q_{15}(u). \end{aligned} \tag{6.2.10}$$

Each of the polynomials $Q_i(u)$, $i = 1, 3, 5, 7, 15$, has coefficients over $GF(2)$ and is irreducible. We shall return to this example a little later and determine the exact form of the Q polynomials.

For the case $d \mid (p^m - 1)$, a factorization of $u^d - 1$ over $GF(p)$ begins with the factorization in (6.1.8) and (6.1.10) reproduced below:

$$u^d - 1 = \prod_{i=1}^{d} \left(u - B^i\right), \tag{6.2.11}$$

$$B = A^{(q-1)/d}. \quad (6.2.12)$$

Clearly, $B^d = 1$. The process of partitioning the set $\Gamma = \{1, 2, ..., d\}$ into cyclotomic sets remains the same as before. The cyclotomic set Γ_j is obtained as

$$\Gamma_j = \left\{ j, jp, jp^2, \ldots \right\}. \quad (6.2.13)$$

In this case, $B^d = 1$ and, therefore, the powers of B are defined mod d. Once again, there are at most m elements in each Γ_j. The rest of the procedure for computing the irreducible factors of $u^d - 1$ over $GF(p)$ remains the same. We summarize the procedure for factorizing $u^n - 1$ over $GF(p)$ using the above analysis in Table 6.2 along with the salient features of factor polynomials.

Example 6.3 Consider the factorization of $u^{13} - 1$ over $GF(3)$. Here $3^3 - 1 = 26 = 2 \cdot 13$. Therefore, we begin the factorization process by first considering factorization of $u^{13} - 1$ over $GF(27)$. $GF(27)$ is created by an extension of $GF(3)$ by an irreducible polynomial of degree 3 over $GF(3)$. Thus, $p = 3$, $q = 27$, $m = 3$. One such polynomial is given by $v^3 + 2v^2 + 1$. If A is a primitive element in $GF(27)$, then all the non-zero elements in $GF(3^3)$ can be expressed as A^i, $i = 1, 2, ..., q - 1$. Let B be an element in $GF(3^3)$ of order 13. We may express B as $B = A^2$. A partitioning of the set $\Gamma = \{1, 2, ..., 13\}$ into subsets $\Gamma_{j_1}, \Gamma_{j_2}, \ldots$, is given by

$$\Gamma_1 = \{1, 3, 9\},$$
$$\Gamma_2 = \{2, 6, 5\},$$
$$\Gamma_4 = \{4, 12, 10\},$$
$$\Gamma_7 = \{7, 8, 11\},$$
$$\Gamma_{13} = \{13\}.$$

Note that each of the cyclotomic sets contains a listing of the power of B and since $B^{13} = 1$, the elements in Γ_j are defined mod 13. The set Ψ is given by $\{1, 2, 4, 7, 13\}$ and the factorization of $u^{13} - 1$ is given by

$$\begin{aligned} u^{13} - 1 &= \prod_{j \in \Psi} \left\{ \prod_{\theta \in \Gamma_j} \left(u - B^{\theta} \right) \right\} \\ &= \left\{ (u - B)\left(u - B^3\right)\left(u - B^9\right) \right\} \cdot \left\{ \left(u - B^2\right)\left(u - B^6\right)\left(u - B^5\right) \right\} \\ &\cdot \left\{ \left(u - B^4\right)\left(u - B^{12}\right)\left(u - B^{10}\right) \right\} \cdot \left\{ \left(u - B^7\right)\left(u - B^8\right)\left(u - B^{11}\right) \right\} \cdot \left\{ u - B^{13} \right\} \end{aligned}$$

$$= Q_1(u)Q_2(u)Q_4(u)Q_7(u)Q_{13}(u). \quad (6.2.14)$$

Each of the polynomials $Q_i(u)$, $i = 1, 2, 4, 7, 13$, has coefficients in $GF(3)$ and is irreducible. We shall return to this example a little later and determine the exact form of the Q polynomials.

Table 6.2 Factorization of $u^n - 1$ Over $GF(p)$ Using $GF(p^m)$

	Let A be a primitive element in $GF(p^m)$.
$n = (q-1)$ $q = p^m$	$u^n - 1 = \prod_{j\in\Psi} Q_j(u)$ $Q_j(u) = \prod_{i=0}^{l-1}\left(u - A^{jp^i}\right), l \le m.$ Ψ is the set of the indices of the cyclotomic sets Γ_j, $\Gamma_j = \{j, jp, jp^2, \ldots\}$. Here j is the smallest integer not appearing in $\Gamma_1, \ldots, \Gamma_{j-1}$. The elements of Γ_j are defined modulo $(p^m - 1)$.
$n \mid (q-1)$	Let $B = A^{(q-1)/n}$. $u^n - 1 = \prod_{j\in\Psi} Q_j(u)$ $Q_j(u) = \prod_{i=0}^{l-1}\left(u - B^{jp^i}\right), l \le m$ Ψ is the set of the indices of the cyclotomic sets Γ_j, $\Gamma_j = \{j, jp, jp^2, \ldots\}$. Here j is the smallest integer not appearing in $\Gamma_1, \ldots, \Gamma_{j-1}$. The elements of Γ_j are defined modulo n.
$(n, p) = 1$	Find the smallest integer m such that $n \mid (p^m - 1)$. Apply the previous case to factorize $u^n - 1$.
$n = cp$ $(c, p) = 1$	$u^n - 1 = (u^c - 1)^p$ Factorize $u^c - 1$ using the techniques described above.
	Note: A primitive element in $GF(p^m)$ always exists. The polynomials $Q_j(u)$ are irreducible.

6.3 Primitive Polynomials Over $GF(p)$

The structure of $GF(p^m)$ was described at the beginning of this chapter. It consists of all the polynomials of degrees up to $m - 1$. The arithmetic operations of '+' and '·' in $GF(p^m)$ are defined mod $P(v)$, $P(v)$ being a monic degree m irreducible polynomial over $GF(p)$. All the computations over $GF(p^m)$ can be performed once $P(v)$ is specified. However, it is clear from the analysis in Section 6.2 that we encounter primitive elements of $GF(p^m)$ repeatedly in our factorization process. In this section, we further study the algebraic properties of the primitive elements of $GF(p^m)$.

A primitive element A of $GF(p^m)$ has order $q - 1$, $q = p^m$. Thus, $A^i \neq 1$, $i < q - 1$, and $A^{q-1} = 1$. Also, from (6.2.7) and (6.2.8), we see that A is a root of $Q_1(v)$, $deg(Q_1(v)) = m$. Note that $Q_1(v)$ is an irreducible polynomial over $GF(p)$ and is given by

$$Q_1(v) = \prod_{\theta=0}^{m-1} \left(v - A^{p^\theta}\right). \qquad (6.3.1)$$

Definition 6.3 Primitive Polynomial. A degree m irreducible polynomial over $GF(p)$ having a primitive element A of $GF(p^m)$ as its root is termed as a **primitive polynomial.**

By definition, every primitive polynomial is irreducible while the converse is not true. Since A is a root of $v^{q-1} - 1$ and $Q_1(v)$, $Q_1(v)$ being the smallest degree polynomial in $GF(p)$ having A as a root,

$$Q_1(v) \mid (v^{q-1} - 1).$$

In fact, the smallest value of i for which $Q_1(v) \mid (v^i - 1)$ is given by $i = q - 1$ or else A is not a primitive element. This analysis can be used to determine if a given degree m irreducible polynomial $Q(v)$ over $GF(p)$ is primitive or not. If the smallest value of i for which $Q_1(v) \mid (v^i - 1)$ is $i = q - 1$, $q = p^m$, then $Q(v)$ is primitive, otherwise it is not.

Thus far we have developed two distinct representations for the elements in $GF(p^m)$, one where all the elements are represented as elements in the set

$$\{0, A^i,\ i = 1, 2, ..., q - 1, A \text{ a primitive element}\},$$

and the second where all the elements are represented as polynomials in the indeterminate v having degrees up to $m - 1$. We need to bring the two representations together in order to complete this analysis. Let

$$Q_1(v) = \sum_{l=0}^{m-1} F_l v^l + v^m. \qquad (6.3.2)$$

Setting $v = A$ in (6.3.2) and using the fact that A is a root of $Q_1(v)$, we get

$$A^m = -\sum_{l=0}^{m-1} F_l A^l . \qquad (6.3.3)$$

This expression provides a basis for representing the elements A^i, $i = 1, 2, ..., q - 1$ as polynomials in A of degree up to $m - 1$. It can be used recursively to develop an one-to-one equivalence between the polynomial representation and the representation as powers of A^i. The equation (6.3.3), in essence, implies that the polynomial representation of A^i is obtained as $A^i \bmod Q_1(A)$. Thus

$$A^i \leftrightarrow A^i \bmod Q_1(A), \quad i = 1, 2, ..., q - 1. \qquad (6.3.4)$$

In the *rhs* of (6.3.4), the elements of $GF(p^m)$ are expressed as polynomials in A of degree up to $m - 1$ with the arithmetic operations defined mod $Q_1(A)$. The *lhs* is the corresponding representation as a power of A. In general, it is easier to perform '+' when two elements in $GF(p^m)$ are expressed in polynomial form. Similarly, it is easier to perform '·' when two elements in $GF(p^m)$ are expressed as a power of A. Since the elements of $GF(p^m)$ are expressed as polynomials in A of degree up to $m - 1$, they can equivalently be represented as vectors of length m having elements in $GF(p)$. Consequently, in many instances the extension field $GF(p^m)$ is also termed the **vector extension** of the **scalar field** $GF(p)$.

The algebra of finite fields and factorizations that we have described here are complete when a primitive polynomial $Q_1(A)$ is specified. We observe that primitive polynomials exist for all values of prime p and m. Furthermore, their choice is non-unique, that is, there exists more than one primitive polynomial for all values of p and m. Tables are available in books that list primitive polynomials for a large selection of p and m. In the following, we present primitive polynomials over $GF(p)$ for $p = 2, 3, 5$, and 7, and $m = 2, 3, 4$, and 5.

$p = 2$

$m = 2, \quad Q_1(A) = 1 + A + A^2$
$m = 3, \quad Q_1(A) = 1 + A + A^3$
$m = 4, \quad Q_1(A) = 1 + A^3 + A^4$
$m = 5, \quad Q_1(A) = 1 + A^3 + A^5,$

$p = 3$

$m = 2, \quad Q_1(A) = 2 + A + A^2$
$m = 3, \quad Q_1(A) = 1 + 2A^2 + A^3$

$m = 4, \quad Q_1(A) = 2 + A^3 + A^4$
$m = 5, \quad Q_1(A) = 1 + 2\,A^4 + A^5,$

$p = 5$

$m = 2, \quad Q_1(A) = 3 + 3\,A + A^2$
$m = 3, \quad Q_1(A) = 3 + 4\,A^2 + A^3$
$m = 4, \quad Q_1(A) = 3 + 3\,A^2 + A^3 + A^4$
$m = 5, \quad Q_1(A) = 3 + 2\,A^2 + A^5,$

$p = 7$

$m = 2, \quad Q_1(A) = 5 + 5\,A + A^2$
$m = 3, \quad Q_1(A) = 4 + 5\,A^2 + A^3$
$m = 4, \quad Q_1(A) = 3 + 3\,A^2 + 2\,A^3 + A^4$
$m = 5, \quad Q_1(A) = 1 + 2\,A + A^5.$

We end this section by stating that all the results derived here are also valid for studying the factorization of $u^n - 1$ over $GF(q)$ when n does not divide $q - 1$, $q = p^m$. In this case, we first seek factorization of $u^n - 1$ over the field $GF(q^a)$, a being the smallest integer for which $n \mid (q^a - 1)$. Here, $GF(q)$ is treated as the ground field and $GF(q^a)$ as an extension field. $GF(q^a)$ is obtained by defining an irreducible polynomial of degree a over $GF(p^m)$. Such polynomials exist for all values of p, m, and a. The analysis can be duplicated for this case by replacing p by q, and m by a. The details are straightforward and are omitted here. We reassert that rarely would one encounter data sequences in digital signal processing that are defined over $GF(p^m)$, $m > 1$.

Example 6.4 Let $p = 2$ and $m = 4$. $GF(2^4)$ is created by taking the primitive polynomial of degree 4 over $GF(2)$. Let us use $Q_1(A) = 1 + A^3 + A^4$. The recursion $A^i \leftrightarrow A^i \bmod Q_1(A)$, $i = 1, 2, ..., q - 1$, gives rise to Table 6.3.

Example 6.5 Let $p = 3$ and $m = 3$. The $GF(3^3)$ is created by taking the primitive polynomial of degree 3 over $GF(3)$. Let us use $Q_1(A) = 1 + 2A^2 + A^3$. Then the recursion $A^i \leftrightarrow A^i \bmod Q_1(A)$, $i = 1, 2, ..., q - 1$, gives rise to Table 6.4.

Example 6.6 Let us return to Example 6.2. Using the polynomial representation of the various elements in Table 6.3 and simplifying, we get the following polynomials as factors of $u^{15} - 1$ over $GF(2)$,

$Q_1(u) = 1 + u^3 + u^4,$
$Q_3(u) = 1 + u + u^2 + u^3 + u^4,$
$Q_5(u) = 1 + u^2,$

$Q_7(u) = 1 + u + u^4,$
$Q_{15}(u) = 1 + u.$

All the polynomials are irreducible over $GF(2)$. In addition, since A and A^7 are primitive elements, $Q_1(u)$ and $Q_7(u)$ are primitive polynomials over $GF(2)$.

Table 6.3 Equivalent Representations of the Elements of $GF(2^4)$

Polynomial representation	Vector representation	Power representation
0	(0 0 0 0)	$0 = A^{-\infty}$
A	(0 1 0 0)	A^1
A^2	(0 0 1 0)	A^2
A^3	(0 0 0 1)	A^3
$1 + A^3$	(1 0 0 1)	A^4
$1 + A + A^3$	(1 1 0 1)	A^5
$1 + A + A^2 + A^3$	(1 1 1 1)	A^6
$1 + A + A^2$	(1 1 1 0)	A^7
$A + A^2 + A^3$	(0 1 1 1)	A^8
$1 + A^2$	(1 0 1 0)	A^9
$A + A^3$	(0 1 0 1)	A^{10}
$1 + A^2 + A^3$	(1 0 1 1)	A^{11}
$1 + A$	(1 1 0 0)	A^{12}
$A + A^2$	(0 1 1 0)	A^{13}
$A^2 + A^3$	(0 0 1 1)	A^{14}
1	(1 0 0 0)	A^{15}

Example 6.7 Let us return to Example 6.3. Using the polynomial representation of the various elements in Table 6.4 and simplifying, we get the following polynomials as factors of $u^{13} - 1$ over $GF(3)$:

$Q_1(u) = 2 + 2u + u^2 + u^3,$
$Q_2(u) = 2 + 2u + u^3,$
$Q_4(u) = 2 + u + u^2 + u^3,$
$Q_7(u) = 1 + u^2 + u^3,$
$Q_{15}(u) = 2 + u.$

All the polynomials are irreducible over $GF(3)$.

Table 6.4 Equivalent Representations of the Elements of $GF(3^3)$

Polynomial representation	Vector representation	Power representation
0	(0 0 0)	$0 = A^{-\infty}$
A	(0 1 0)	A^1
A^2	(0 0 1)	A^2
$2 + A^2$	(2 0 1)	A^3
$2 + 2A + A^2$	(2 2 1)	A^4
$2 + 2A$	(2 2 0)	A^5
$2A + 2A^2$	(0 2 2)	A^6
$1 + A^2$	(1 0 1)	A^7
$2 + A + A^2$	(2 1 1)	A^8
$2 + 2A + 2A^2$	(2 2 2)	A^9
$1 + 2A + A^2$	(1 2 1)	A^{10}
$2 + A$	(2 1 0)	A^{11}
$2A + A^2$	(0 2 1)	A^{12}
2	(2 0 0)	A^{13}
$2A$	(0 2 0)	A^{14}
$2A^2$	(0 0 2)	A^{15}
$1 + 2A^2$	(1 0 2)	A^{16}
$1 + A + 2A^2$	(1 1 2)	A^{17}
$1 + A$	(1 1 0)	A^{18}
$A + A^2$	(0 1 1)	A^{19}
$2 + 2A^2$	(2 0 2)	A^{20}
$1 + 2A + 2A^2$	(1 2 2)	A^{21}
$1 + A + A^2$	(1 1 1)	A^{22}
$2 + A + 2A^2$	(2 1 2)	A^{23}
$1 + 2A$	(1 2 0)	A^{24}
$A + 2A^2$	(0 1 2)	A^{25}
1	(1 0 0)	A^{26}

6.4 Complex Finite Fields and Cyclotomic Polynomial Factorization

The complex finite field, if it exists, is a generalization of $GF(p)$. It consists of all complex integers of the type

$$W = U + j\,V, \tag{6.4.1}$$

where $U, V \in GF(p)$, and

$$j^2 \equiv -1 \bmod p. \tag{6.4.2}$$

The arithmetic operations of '+' and '·' in this number system are same as those for complex numbers with the exception that both real and imaginary parts are defined modulo p. A useful interpretation of this number system is that its elements can be expressed as a polynomial of degree 1 over $GF(p)$. The arithmetic operations of '+' and '·' between two elements in this number system can equivalently be represented as operations between two polynomials in j of degree 1 modulo $(j^2 + 1)$. Thus, all sequences defined over such a number system essentially involve arithmetic operations modulo $(j^2 + 1)$. It is clear from our discussion in the earlier sections of this chapter that if $(j^2 + 1)$ is irreducible in $GF(p)$, then this number system constitutes a field, $GF(p^2)$ to be precise, the elements of which are represented as in (6.4.1). Similarly, if $(j^2 + 1)$ factorizes in terms of two mutually coprime factors over $GF(p)$, then any computation modulo $(j^2 + 1)$ can be broken down into two computations by using the CRT-P. In the following, we pursue both of these aspects.

6.4.1 Factorization of $(j^2 + 1)$ Over $GF(p)$

For $p = 2$, $(j^2 + 1) = (j + 1)^2$. Therefore, $(j^2 + 1)$ does not factor into two distinct factors over $GF(2)$. However, it does factor which implies that it is not irreducible over $GF(2)$ and the complex integers over $GF(2)$ along with the arithmetic operations of '+' and '·' modulo $(j^2 + 1)$ do not constitute a finite field.

Now consider the factorization of $(j^2 + 1)$ over $GF(p)$, $p \neq 2$. The order of every non-zero element in $GF(p)$ divides $p - 1$. Since $p - 1$ is always even, factorization of $j^2 - 1$ always exists and is given by $j^2 - 1 = (j + 1)\,(j - 1)$. Therefore, the factorization of $(j^2 + 1)$ over $GF(p)$ can be determined from the factorization of $(j^2 + 1)\,(j^2 - 1) = j^4 - 1$. The polynomial $j^2 + 1$ factors over $GF(p)$ *iff* $j^4 - 1$ has four distinct factors, that is, there is an element of order 4 in $GF(p)$. Recalling that the order of every non-zero element in $GF(p)$ divides $p - 1$, we get the following theorem.

Theorem 6.9 For $p \neq 2$, $j^2 + 1$ is *irreducible* over $GF(p)$ *iff* 4 is not a factor of $(p - 1)$ or $p = 4L + 3$ or $p \equiv 3 \bmod 4$. It factors into the product of two relatively prime polynomials over $GF(p)$ *iff* $4 \mid (p - 1)$ or $p = 4L + 1$ or $p \equiv 1 \bmod 4$.

For example, $j^2 + 1$ is irreducible over $GF(3)$, and, therefore, the complex integers defined in (6.4.1) constitute the finite field $GF(3^2)$. Over $GF(5)$, $j^2 + 1 = (j + 2)(j + 3)$, which implies that the complex integers defined mod 5 do not constitute the finite field $GF(5^2)$.

Definition 6.4 Mersenne prime. For a prime number p, if the number $2^p - 1$ is also *prime*, then it is called a **Mersenne prime**.

Definition 6.5 Fermat Prime. If the number $2^a + 1$, $a = 2^t$, is *prime*, then it is called a **Fermat prime**.

The Mersenne and Fermat prime numbers play a special role in digital signal processing as their form makes it easier to realize the associated arithmetic. It is known that the five values of t, that is, $t = 0, 1, 2, 3, 4$, give rise to Fermat prime numbers. The next Fermat prime remains unknown to date. Let us apply Theorem 6.9 to the form of Mersenne and Fermat primes. Taking modulo 4 of their representations, we get $2^p - 1 \equiv -1 \bmod 4 \equiv 3 \bmod 4$. Similarly, $2^a + 1 \equiv 1 \bmod 4$, $a > 1$; for $a = 1$, $2^a + 1 \equiv 3 \bmod 4$. In summary, we get the following two theorems.

Theorem 6.10 $j^2 + 1$ is *irreducible* over $GF(r)$ if r is a Mersenne prime. Therefore, complex integers defined modulo a Mersenne prime constitute the finite field $GF(r^2)$.

Theorem 6.11 $j^2 + 1$ is *reducible* over $GF(r)$ if r is a Fermat prime. Therefore, complex integers defined modulo a Fermat prime do not constitute the finite field $GF(r^2)$. The only exception to the statement is the first Fermat prime that is equal to 3.

We draw an important distinction between the irreducibility of $j^2 + 1$ over $GF(p)$, $p = 4L + 3$, and the repeated factors of $j^2 + 1$ over $GF(2)$. In the former case, the complex integers over $GF(p)$ constitute a finite field while in the latter case, no such property holds. This plays a crucial role in the factorization of polynomials over complex integer fields and in the design of fast algorithms for computing convolutions. We now turn to the form of computations involving such a number system.

6.4.2 Special Cases

Case 1. $p = 2$. For $p = 2$, $j^2 + 1 = (j + 1)^2$. In this case, the algorithms for computing convolutions involving the complex arithmetic are same as the algorithms defined over $GF(2)$. The input and output quantities are complex-valued and the arithmetic operations are now defined modulo $(j^2 + 1)$. No further computational savings can be realized.

Case 2. $p = 4L + 1$. In this case, $j^2 + 1 = (j - B)(j - C) \bmod p$, $B \neq C$. Therefore, a polynomial, $Z(u, j)$ having complex-valued coefficients (j is shown explicitly here), can equivalently be represented as two polynomials $Z_1(u)$ and $Z_2(u)$ having coefficients defined over $GF(p)$. These polynomials are obtained as

$$Z_1(u) \equiv Z(u, j) \bmod (j - B) = Z(u, B), \tag{6.4.3}$$
$$Z_2(u) \equiv Z(u, j) \bmod (j - C) = Z(u, C). \tag{6.4.4}$$

This representation can be used to express all complex-valued polynomial manipulations as equivalent to 2 polynomial manipulations with the coefficients defined over $GF(p)$. Readers will immediately recognize this formulation as CRT-P in $GF(p)$ with the modulo polynomial $j^2 + 1$. Consequently, given $Z_1(u)$ and $Z_2(u)$, the polynomial $Z(u)$ can be obtained using the CRT-P reconstruction expressed as

$$Z(u) = [B\,Z_2(u) - C\,Z_1(u)]/(B - C) + j\,[Z_1(u) - Z_2(u)]/(B - C). \tag{6.4.5}$$

Factorization properties of polynomials over $GF(p^2)$, $p = 4L + 3$, forms the topic of the following section.

6.4.3 Cyclotomic Factorization Over Complex Finite Fields

The following discusion is valid only when $p \equiv 3 \bmod 4$, that is, the complex integers constitute the finite field $GF(p^2)$. The procedure to obtain the factorization of $u^n - 1$ over the complex finite field, $GF(p^2)$, is similar to the procedure to obtain the factorization of $u^n - 1$ over the finite field, $GF(p)$. In this case, we seek the smallest integer m such that $n \mid (p^{2m} - 1)$, factorize $u^n - 1$ over $GF(p^{2m})$ first and then combine those degree one factors that belong to the same cyclotomic set to obtain factors over $GF(p^2)$. The entire procedure requires a primitive polynomial over $GF(p^2)$. The details of the procedure are rather straightforward and the derivations need not be presented as they are similar to the earlier derivations. Tables 6.5 and 6.6 contain the summary of the entire procedure for the factorization of $u^n - 1$ over the complex finite field, $GF(p^2)$.

Example 6.8 Consider the case $p = 3$. The polynomial $j^2 + 1$ is irreducible in $GF(3)$ and, therefore, complex integers of the type $U + jV$, $U, V \in GF(3)$ constitute a finite field. This complex finite field is same as $GF(3^2)$. The order of every element in this field divides $9 - 1 = 8$. It can be verified that the element $1 + j\,0$ has order 1, the element $2 + j\,0$ has order 2, the elements $0 + j$ and $0 + 2j$ have order 4, and the elements $1 + j$, $1 + 2j$, $2 + j$, and $2 + 2j$ have order 8. The last four elements are also primitive in this field. Let us pick $1 + j$ as the primitive element. The factorization of $u^8 - 1$ can be written as

$$u^8 - 1 = \prod_{i=1}^{8}\left[u - (1+j)^i\right]. \tag{6.4.6}$$

The factorization of $u^4 - 1$ is trivial as it is the same as the one over the field of complex numbers.

Table 6.5 Factorization of $u^n - 1$ in $GF(p^2)$

	Let A be a primitive element in $GF(p^2)$.
$n = p^2 - 1$	$u^n - 1 = \prod_{i=0}^{n-1}\left(u - A^i\right)$
$n \mid (p^2 - 1)$	$u^n - 1 = \prod_{i=0}^{n-1}\left(u - B^i\right)$ $B = A^{(p^2-1)/d}$
	Note: A primitive integer in $GF(p^2)$ always exists. All elements of $GF(p^2)$ are represented as complex integers, $U + jV;\ U, V \in GF(p)$.

Table 6.6 Factorization of $u^n - 1$ in $GF(p^2)$ Using $GF(p^{2m})$

	Let A be a primitive element in $GF(p^{2m})$.
$n = (q-1)$ $q = p^{2m}$	$u^n - 1 = \prod_{j \in \Psi} Q_j(u)$ $Q_j(u) = \prod_{i=0}^{l-1}\left(u - A^{jp^{2i}}\right), l \leq m.$ Ψ is the set of the indices of the cyclotomic sets Γ_j, $\Gamma_j = \{j, j\,p^2, j\,p^4, \ldots\}$. Here j is the smallest integer not appearing in $\Gamma_1, \ldots, \Gamma_{j-1}$. The elements of Γ_j are defined modulo $(p^{2m} - 1)$.
$n \mid (q-1)$	Let $B = A^{(q-1)/n}$. $u^n - 1 = \prod_{j \in \Psi} Q_j(u)$ $Q_j(u) = \prod_{i=0}^{l-1}\left(u - B^{jp^{2i}}\right), l \leq m$ Ψ is the set of the indices of the cyclotomic sets Γ_j, $\Gamma_j = \{j, j\,p^2, j\,p^4, \ldots\}$. Here j is the smallest integer not appearing in $\Gamma_1, \ldots, \Gamma_{j-1}$. The elements of Γ_j are defined modulo n.
$(n, p) = 1$	Find the smallest integer m such that $n \mid (p^{2m} - 1)$. Apply the previous case to factorize $u^n - 1$.
$n = c \cdot p$ $(c, p) = 1$	$u^n - 1 = (u^c - 1)^p$ Factorize $u^c - 1$ using the techniques described above.
	Note: A primitive element in $GF(p^{2m})$ always exists. The polynomials $Q_j(u)$ are irreducible. Elements of $GF(p^2)$ are represented as complex integers, $U + j\,V;\ U, V \in GF(p)$.

Example 6.9 Consider the extension of the complex field $GF(3^2)$ by the polynomial $A^2 + 2\,A + (2 + 2\,j)$. Using techniques outlined earlier, it can be verified that it is indeed primitive. This polynomial gives rise to an extension field, $GF(3^4)$, where all the elements are polynomials of degree up to 1 having coefficients in $GF(3^2)$. The order of every non-zero element in the extension

field divides $3^4 - 1 = 80$. Therefore, the polynomials of the type $u^d - 1$ will factor as a product of degree-one factors for d equal to 1, 2, 4, 5, 8, 10, 16, 20, 40, and 80. Now, consider the factorization of $u^5 - 1$ over $GF(3^2)$. Let B be an element in $GF(3^4)$ of order 5. We may express B as $B = A^{16}$. A partitioning of the set $\Gamma = \{1, 2, 3, 4, 5\}$ into subsets $\Gamma_{j_1}, \Gamma_{j_2}, \ldots,$ is given by

$$\Gamma_1 = \{1, 4\},$$
$$\Gamma_2 = \{2, 3\},$$
$$\Gamma_5 = \{5\}.$$

Note that each of the cyclotomic sets contains a listing of the power of B and since $B^{16} = 1$, the elements in Γ_j are defined mod 16. The set Ψ is given by $\{1, 2, 5\}$ and the factorization of $u^5 - 1$ is given by

$$u^{13} - 1 = \prod_{j \in \Psi} \left\{ \prod_{\theta \in \Gamma_j} \left(u - B^{\theta} \right) \right\}$$
$$= \left\{ \left(u - B^1 \right)\left(u - B^4 \right) \right\} \cdot \left\{ \left(u - B^2 \right)\left(u - B^3 \right) \right\} \cdot \left\{ u - B^5 \right\}$$
$$= Q_1(u) Q_2(u) Q_5(u). \tag{6.4.7}$$

Each of the polynomials $Q_i(u)$, $i = 1, 2, 5$ has coefficients over $GF(3^2)$ and is irreducible. The exact form of these factors is determined by noting that (i) $B = A^{16}$, and (ii) A satisfies the recursion $A^2 \equiv A + (1 + j)$. Thus, the elements A^{16}, A^{32}, A^{48}, and A^{64} are obtained as

$$A^{16} = 2A + (1 + j),$$
$$A^{32} = (2 + j)A + 1,$$
$$A^{48} = (1 + 2j)A + j,$$
$$A^{64} = A + j.$$

Substituting in (6.4.7), we get the factor polynomials as

$$Q_1(u) = u^2 + (2 + j)u + 1,$$
$$Q_2(u) = u^2 + (2 + 2j)u + 1,$$
$$Q_5(u) = u + 2.$$

The factorization of $u^d - 1$ for d equal to 1, 2, 4, 5, 8, 10, 16, 20, 40, and 80 can be obtained in a similar manner.

Notes

The author came in contact with finite field algebra 15 years ago during his graduate studies at Concordia University. It is his understanding that this topic, even though of fundamental importance to the design and analysis of discrete time systems, is not well studied from a digital signal processing point of view. One of the objectives of this book is to rectify this situation. Factorization of $u^n - 1$ over $GF(p)$ is crucial to digital signal processing of sequences defined over $GF(p)$, its extension fields, the finite integer rings, and their polynomial extensions. This chapter is a collection of all those results that pertain to this factorization over finite fields. The book by Peterson and Weldon [1972] has influenced our understanding of algebras of finite fields and polynomials significantly. The primitive polynomials listed in Section 6.3 are taken from Lidl and Niederreiter [1983] that contains a more extensive list of the same. The graduate students who have taken courses on digital signal processing, coding theory, and digital communications with the author over the past fifteen years have made an indirect yet significant contribution to the overall presentation.

We hope that this book will put the algebra of finite fields in a proper perspective within the framework of digital signal processing.

Bibliography

[6.1] W.W. Peterson and E.J. Weldon, Jr., *Error Correcting Codes*, II edition, MIT Press, 1972.

[6.2] R. Lidl and H. Niederreiter, *Finite Fields*, Addison-Wesley Publishing Co., 1983.

Problems

6.1 Show that $j^2 + 1$ can never be a primitive polynomial over $GF(p)$ even though it is irreducible over those $GF(p)$ for which $p \equiv 3 \bmod 4$.

6.2 Show that all non-zero elements in $GF(2^3)$, $GF(2^5)$, and $GF(2^7)$ are primitive elements (except unity, of course). Can you generalize this result to $GF(2^P)$, where $2^P - 1$ is a Mersenne prime?

6.3 Compute the cyclotomic factors of $u^{16} - 1$ over $GF(3^2)$, the field of complex integers.

6.4 Find minimal polynomials of A^6, A^7, A^{13}, and A^{21} in $GF(2^6)$. Are these polynomials irreducible? primitive?

6.5 Factorize $u^{23} - 1$ over $GF(2)$.

6.6 What is the number of primitive elements in $GF(q)$? Justify.

6.7 Factorize $u^{16} - 1$ over $GF(5)$.

6.8 Find the order of all the non-zero elements in $GF(7^2)$, the field of complex integers. How many elements of each order does it contain?

6.9 Factorize $u^9 - 1$ over $GF(2^4)$ and hence over $GF(2)$.

6.10 Factorize $u^{13} - 1$ over $GF(2)$.

6.11 For this problem, let $n = 15$. We are given that a polynomial $X(u)$ defined over $GF(2^4)$ having degree up to $n - 1$, has only one non-zero coefficient. Also, we are given the values of $X(u)$ for $u = A$, and $u = A^2$, A being a primitive element in $GF(2^4)$. Derive an algorithm for determining $X(u)$. Will this procedure work if $n > 15$, everything else remaining the same?
Note: This problem corresponds to single-error-correcting (SEC) Reed-Solomon (RS) codes.

6.12 A generalization of Problem 6.11 can be made in the following manner. Let $n = 2^m - 1$. We are given that a polynomial $X(u)$ defined over $GF(2^m)$ having degree up to $n - 1$ has, at most, t non-zero coefficients. Also, we are given the values of $X(u)$ for $u = A$, $u = A^2$, ..., $u = A^{2t}$, A being a primitive element in $GF(2^4)$. Derive an algorithm for determining $X(u)$.
Note: This problem corresponds to t error-correcting RS codes.

6.13 Solve Problem 6.12 for $t = 2$.

6.14 Let us define complex arithmetic where the modulo polynomial is $j^2 + j + 1$ (instead of $j^2 + 1$). Derive conditions under which this can be used to obtain the extension field $GF(p^2)$ from the ground field $GF(p)$. Does $p = 5$ satisfy the conditions?

6.15 Repeat the above problem if the complex arithmetic is defined modulo $j^2 - j + 1$.

6.16 Consider the two-dimensional (31×31) cyclic convolution over $GF(2)$ expressed in polynomial form as the computation

$$Z(u,v) = X(u,v)Y(u,v) \bmod \left[u^{31} - 1, v^{31} - 1\right].$$

Use the cyclotomic factorization of $v^{31} - 1$ over $GF(2)$ and that of $u^n - 1$ over $GF(2^7)$ to show that the entire computation is equivalent to several one-dimensional polynomial multiplication of the type $A(w)$ $B(w)$ mod $Q(w)$, where $Q(w)$ is a primitive polynomial, $deg(Q(w)) = 5$, and some other simple products. Identify these simple products.

Chapter 7

Finite Integer Rings: Polynomial Algebra and Cyclotomic Factorization

It was shown in Section 2.3 that the set of integers $\{0, 1, ..., M-1\}$ forms the finite integer ring $Z(M)$ when the arithmetic operation defined modulo M. If M is prime, then we have a finite field $GF(p)$. Finite fields and the cyclotomic polynomial factorization in them were studied in Chapter 6. In this chapter, we consider the last type of number system that is the finite integer ring $Z(M)$ and its extension rings, where M is composite. In essence, we employ the CRT-I and the finite field properties to obtain various results in finite integer rings.

7.1 Polynomial Algebra Over a Ring

The set of all polynomials with coefficients defined over the ring $Z(M)$ constitutes a polynomial ring. Given a degree n polynomial $M(u)$ over $Z(M)$, the set of polynomials of degree up to $n-1$ over $Z(M)$ also constitutes a polynomial ring. Once again, the operations of '+' and '·' between two polynomials in such a polynomial ring are defined modulo $M(u)$. A close examination reveals that in order to perform the modulo $M(u)$ operation, the multiplicative inverse of the leading coefficient of $M(u)$ must exist. In other words, if

$$M(u)=\sum_{i=0}^{n} M_i u^i ,$$

then M_n^{-1} must exist and be unique in $Z(M)$. This requires that $(M_n, M) = 1$. Without any loss in generality, we may assume that $M_n = 1$. Thus, $M(u)$ is a *monic* polynomial. All the modulo polynomials $M(u)$ that characterize the polynomial rings under study will be assumed to be monic throughout this work.

Consider the ring of all polynomials of degrees up to $n - 1$ over $Z(M)$ with operations defined modulo $M(u)$. Given the standard factorization of M in (2.2.15), this ring can also be expressed as a direct sum of t polynomial sub-rings each defined over $Z\left(p_i^{a_i}\right)$, $i = 1, 2, ..., t$. Each of the t polynomial sub-rings is a polynomial ring of all polynomials of degrees up to $n - 1$ over $Z\left(p_i^{a_i}\right)$ with operations defined modulo $M^{(i)}(u)$, where

$$M^{(i)}(u) \equiv M(u) \bmod p_i^{a_i}, \quad i = 1, 2, \ldots, t.$$

In other words, given a polynomial $A(u)$ with coefficients over $Z(M)$, $M = p_1^{a_1} p_2^{a_2} \cdots p_t^{a_t}$, it can be represented as a direct sum of polynomials $A^{(i)}(u)$, $i = 1, 2, ..., t$, where

$$A^{(i)}(u) \equiv A(u) \bmod p_i^{a_i}. \tag{7.1.1}$$

We can also write

$$A(u) = A^{(1)}(u) \oplus A^{(2)}(u) \oplus \ldots \oplus A^{(t)}(u). \tag{7.1.2}$$

The above statements follow directly from the CRT-I and the definition of a polynomial. Therefore, analysis for polynomials over the ring $Z(M)$ can be performed once the mathematical properties of polynomials over integer rings of the type $Z(p^a)$ are known. In summary, we will focus our attention on the algebraic structure of the rings of polynomials of degrees up to $n - 1$ over $Z(p^a)$ where the operations are performed modulo a degree n monic polynomial $M(u)$ over $Z(p^a)$. Next, we present the p-adic form of a polynomial defined over $Z(p^a)$.

Definition 7.1 Let $A(u)=\sum_{i=0}^{n-1} A_i u^i$, $A_i \in Z(p^a)$, be a polynomial of degree $n - 1$ over $Z(p^a)$. The p-adic representation of $A(u)$ is

$$A(u) = \sum_{j=0}^{a-1} A_j(u) p^j, \tag{7.1.3}$$

where

$$A_j(u) = \sum_{i=0}^{n-1} A_{i,j} u^i, \; A_{i,j} \in GF(p). \tag{7.1.4}$$

The above p-adic representation of $A(u)$ is obtained by noting that the constant $A_i \in Z(p^a)$ can be represented in its p-adic form as

$$A_i = \sum_{j=0}^{a-1} A_{i,j} p^j, \; i = 0,1,\ldots,n-1, \; A_{i,j} \in GF(p). \tag{7.1.5}$$

Substituting in $A(u)$ and interchanging the summations, we get

$$A(u) = \sum_{i=0}^{n-1} \left(\sum_{j=0}^{a-1} A_{i,j} p^j \right) u^i = \sum_{j=0}^{a-1} \left(\sum_{i=0}^{n-1} A_{i,j} u^i \right) p^j. \tag{7.1.6}$$

In this work, we express a ring of polynomials of degrees up to $n - 1$ over $Z(p^a)$ as a direct sum in a manner analogous to (3.4.9). The various issues are

(i) Factorization of the modulo polynomial $M(u)$ over $Z(p^a)$,
(ii) CRT-P based direct sum representation of the polynomial ring based on factorization of $M(u)$, and
(iii) Computational methods to convert from direct sum representation to the original polynomial representation and vice versa.

This leads to generalization of the CRT-P over finite integer rings and their extensions. This generalization will be termed the **CRT-PR (Chinese remainder theorem for polynomials in rings)**. NTTs are shown to be a special case of the general approach being taken here.

Definition 7.2 Let $M(u)$ be a polynomial over $Z(M)$. A polynomial $A(u)$ over $Z(M)$ has $M(u)$ as its *factor* if $A(u)$ and $M(u)$ can be written as $A(u) = M(u)\, Q(u)$ mod M, where $Q(u)$ is also a polynomial over $Z(M)$. We also write $M(u) \mid A(u)$.

7.1.1 Roots of a Polynomial Over $Z(M)$

Given $A(u)$ over the ring $Z(M)$ and a scalar $E \in Z(M)$, consider dividing $A(u)$ by $(u - E)$ to get

$$A(u) = Q(u)\,(u - E) + R(u),$$

where $0 \le deg(R(u)) < deg(u - E) = 1$. Therefore, $R(u) = R$, $R \in Z(M)$. Setting $u = E$ in both sides of the above equation, we get

$$A(E) = R.$$

The scalar E is said to be a root of polynomial $A(u)$ if $A(E) = 0$ or equivalently $R = 0$. We see that if E is a root of $A(u)$, then $(u - E) \mid A(u)$. Therefore, if E_1 and E_2 are two roots of a polynomial $A(u)$, $E_1 \neq E_2$, then $(u - E_1) \mid A(u)$ and $(u - E_2) \mid A(u)$ over $Z(M)$. However, unlike a field, it does *not* imply that $[(u - E_1) \cdot (u - E_2)] \mid A(u)$ over $Z(M)$. Consequently, Theorem 3.2 does not hold for a polynomial defined over $Z(M)$. The number of roots of a polynomial over $Z(M)$ may exceed its degree. For example, the polynomial $u^2 + u$ has 0, 2, 3, and 5 as roots over $Z(6)$.

Let us examine the behavior of these roots further. Expressions (3.2.3) to (3.2.6) remain valid except that they are defined mod M in this case. The congruence $R_2 \cdot (E_2 - E_1) \equiv 0 \bmod M$ does not imply that $R_2 = 0$. However, if $(E_2 - E_1, M) = 1$, then $R_2 = 0$ and

$$[(u - E_1) \cdot (u - E_2)] \mid A(u)$$

from (3.2.5). This analysis leads to the following theorem.

Theorem 7.1 If $E_1, E_2, \ldots, E_k$ are roots of a polynomial $A(u)$ over $Z(M)$ and

$$(E_i - E_j, M) = 1, \quad 1 \le i, j \le k, \quad i \neq j, \tag{7.1.7}$$

then we have

$$\left[\prod_{i=1}^{k} (u - E_i)\right] \Big| A(u). \tag{7.1.8}$$

Note that $(E_2 - E_1, M) = 1$ is only a sufficient condition for $R_2 \cdot (E_2 - E_1) = 0$ to imply that $R_2 = 0$. It is *not* a necessary condition. For example, the polynomial $u^3 + u^2 + 2\,u + 2$ has $E_1 = 2$ and $E_2 = 4$ as roots and $(u + 2) \cdot (u + 4)$ as a factor over $Z(6)$ even though $(E_2 - E_1, M) = (2, 6) = 2 \neq 1$.

Definition 7.3 Differentially coprime integers. Integers $A_1, A_2, ..., A_k$ that satisfy $(A_i - A_j, M) = 1$, $1 \le i, j \le k$, $i \ne j$, are said to be **pairwise differentially coprime** to M.

We observe that the condition $(A_i - A_j, M) = 1$, $1 \le i, j \le k$, $i \ne j$, does not imply that $(A_i, M) = 1$, $i = 1, 2, ..., k$, in any way. For example, if $M = 35$, one can have $A_1 = 3$, $A_2 = 4$, $A_3 = 5$, $A_4 = 6$, and $A_5 = 7$. Finally, given the standard factorization of M in (2.2.15), it is easy to establish that the largest number of integers that are pairwise differentially coprime to M is p,

$$p = min\{p_1, p_2, ..., p_t\}.$$

7.2 Lagrange Interpolation

In this section, we describe the Lagrange interpolation (LI) algorithm for evaluating polynomials when they are defined over $Z(M)$. Subsequently, the LI will be shown to be a special case of the CRT-PR defined over $Z(M)$. Secondly, NTTs are demonstrated to be a special case of LI if all the interpolating points are in a special geometric order. It is worthwhile to mention here that NTTs are nothing but a DFT defined in a finite integer ring.

Let $X(u)$ be a degree $n - 1$ polynomial defined over $Z(M)$ and represented as

$$X(u) = \sum_{i=0}^{n-1} X_i u^i .$$

LI deals with the problem of interpolating $X(u)$ through n points. This suggests a method for determining the coefficients of $X(u)$ based on the values of $X(u)$ evaluated at n points A_i, $i = 1, 2, ..., n$; namely, $X(A_1), X(A_2), ..., X(A_n)$. Since

$$X(A_i) = X(u) \bmod (u - A_i),$$

LI is same as CRT-PR when the modulo polynomial $M(u)$ is set equal to the product of n degree-one polynomials $(u - A_i)$. In other words,

$$m_i(u) = (u - A_i), \quad i = 1, 2, ..., n.$$

One has to be careful in making these statements when the quantities are defined over a ring as the multiplicative inverse of a non-zero scalar does not exist all the time. $X(u)$ can be reconstructed using LI as follows:

$$X(u) = \sum_{i=1}^{n} X(A_i) \prod_{\substack{j=1 \\ j \neq i}}^{n} \frac{(u - A_j)}{(A_i - A_j)}. \tag{7.2.1}$$

Recall that the CRT-P requires that all the factor polynomials $m_i(u) = (u - A_i)$ be pairwise relatively prime. If A_i, $i = 1, 2, \ldots, n$ are elements of a field, the only constraint for a unique solution is $A_i \neq A_j$, $i \neq j$.

Alternatively, the LI algorithm can be represented as a Vandermonde system of linear equations,

$$\begin{bmatrix} 1 & A_1 & A_1^2 & \cdots & A_1^{n-1} \\ 1 & A_2 & A_2^2 & \cdots & A_2^{n-1} \\ \vdots & \vdots & \vdots & \ddots & \vdots \\ 1 & A_n & A_n^2 & \cdots & A_n^{n-1} \end{bmatrix} \begin{bmatrix} X_0 \\ X_1 \\ \vdots \\ X_{n-1} \end{bmatrix} = \begin{bmatrix} X(A_1) \\ X(A_2) \\ \vdots \\ X(A_n) \end{bmatrix}. \tag{7.2.2}$$

Let the Vandermonde matrix be denoted by $\mathbf{V}$. The determinant of $\mathbf{V}$ is given by

$$det(\mathbf{V}) = \prod_{i=1}^{n-1} \prod_{j=i+1}^{n} (A_j - A_i). \tag{7.2.3}$$

Over a field, the condition for the existence of a unique solution to the Vandermonde system shown in (7.2.2) is $det(\mathbf{V}) \neq 0$ or $A_i \neq A_j$, $i \neq j$.

Consider the system of linear congruences in (7.2.2) defined mod M. The form of the solution in (7.2.1) remains valid in this case as well. For all the quantities to be uniquely defined over $Z(M)$, the multiplicative inverse of all the terms in the denominator of (7.2.1) must exist and be unique. Therefore, a unique solution to the Vandermonde system exists *iff*

$$(A_i - A_j, M) = 1, \quad 1 \leq i, j \leq n, \; i \neq j. \tag{7.2.4}$$

Recall from Section 7.1 that $A_1, A_2, \ldots, A_n$ that satisfy (7.2.4) are said to be pairwise differentially coprime to M. Thus, we conclude the following.

Theorem 7.2 The n coefficients of a degree $n - 1$ polynomial $X(u)$ can be uniquely determined over $Z(M)$ from the values of $X(u)$ evaluated at $u = A_i$, $i = 1, 2, \ldots, n$, *iff* these n integers are pairwise differentially coprime to M.

The Vandermonde system of linear equations can be solved by either the LI in (7.2.1) or order-recursive algorithms designed for them. The order-recursive algorithm is applicable in the present situation as well, since all the principal matrices of **V** are also Vandermonde matrices having determinants that are coprime to M. The largest number of integers that are pairwise differentially coprime to M is p, $p = min\{p_1, p_2, ..., p_t\}$. Thus, the largest value of n for a uniquely solvable Vandermonde system of linear congruences is $n = p$. Also, the condition in (7.2.4) for the existence of a unique solution simplifies to

$$(A_i - A_j, p_k) = 1, \ \ k = 1, 2, ..., t. \tag{7.2.5}$$

This approach can also be used to compute the acyclic convolution of two sequences over $Z(M)$ in a computationally efficient manner. This algorithm will be described in Chapter 8; however, the readers will indeed recognize similarity of this approach over a ring and earlier approaches over fields. In polynomial notation, we can equivalently express this LI based approach as a direct sum over a ring of polynomials defined over $Z(M)$:

$$\langle X(u)\rangle_{M(u)} = \langle X(u)\rangle_{m_1(u)} \oplus \langle X(u)\rangle_{m_2(u)} \oplus \ldots \oplus \langle X(u)\rangle_{m_n(u)} \tag{7.2.6}$$

$$m_i(u) = u - A_i, \ \ i = 1, 2, \ldots, n \tag{7.2.7}$$

and

$$M(u) = \prod_{i=1}^{n} m_i(u). \tag{7.2.8}$$

This is analogous to the direct sum in (3.4.9) valid for the ring of polynomials defined over a field.

7.2.1 A Special Case of Lagrange Interpolation

In many instances, there are computational benefits to expressing the interpolating points as powers of a single value. Let us say that the scalars $A_i \in Z(M)$ are to be chosen such that

$$A_i = G^i \bmod M, \ \ 1 \le i \le n. \tag{7.2.9}$$

Once again, $n \le min\{p_1, p_2, ..., p_t\}$. Applying (7.2.4) to the above case, we see that a unique solution exists, *iff*

$$(G, M) = 1 \tag{7.2.10}$$

and

$$(G^i - 1, M) = 1, \;\; i = 1, 2, ..., n - 1. \tag{7.2.11}$$

Finding an integer G over $Z(M)$, $M = \prod_{j=1}^{t} p_j^{a_j}$, is equivalent to finding an integer G_j over $Z\left(p_j^{a_j}\right)$ that satisfies

$$\left(G_j, p_j^{a_j}\right) = 1 \tag{7.2.12}$$

and

$$\left(G_j^i - 1, p_j^{a_j}\right) = 1, \;\; i = 1, 2, ..., n - 1; \; j = 1, 2, ..., t, \tag{7.2.13}$$

and then using the CRT-I to get G over $Z(M)$. The conditions in (7.2.12) and (7.2.13) further simplify to

$$(G_j, p_j) = 1 \tag{7.2.14}$$

and

$$\left(G_j^i - 1, p_j\right) = 1, \;\; i = 1, 2, ..., n - 1; \; j = 1, 2, ..., t. \tag{7.2.15}$$

Therefore, one has to compute an integer G over $Z(p^a)$ such that

$$(G, p) = 1 \tag{7.2.16}$$

and

$$(G^i - 1, p) = 1, \;\; i = 1, 2, ..., n - 1, \;\; n \leq p. \tag{7.2.17}$$

Consider an integer $A \in Z(p^a)$ such that $(A, p) = 1$. We can write $\alpha A + \beta p = 1$. Also, dividing A by p, we have $A = Q p + A'$, $A' \in GF(p)$. Combining the two expressions leads to $\alpha' A' + \beta' p = 1$, that is, $(A', p) = 1$. An important consequence is that finding an integer G over $Z(p^a)$ that satisfies (7.2.16) and (7.2.17) is same as finding such an integer over $GF(p)$. Expressed over $GF(p)$, (7.2.16) and (7.2.17) become

$$(G, p) = 1, \;\; G \in GF(p) \tag{7.2.18}$$

and

$$(G^i - 1, p) = 1, \;\; i = 1, 2, ..., n - 1; \;\; n \le p. \tag{7.2.19}$$

Since $G < p$, (7.2.18) is trivially satisfied for $G \neq 0$. Also, $G^i - 1 \bmod p < p$ and, therefore, (7.2.19) is satisfied as well except when $G^i - 1 \equiv 0 \bmod p$ or $G^i \equiv 1 \bmod p$, for any value of $i \le n - 1$. Therefore, G can only take those values that have order $\ge n$ over $GF(p)$. In $GF(p)$, there exist integers of order up to $p - 1$. Consequently, a value can always be assigned to G provided that $n \le p - 1$. This discussion leads to an important theorem.

Theorem 7.3 All the n interpolating points can be chosen as powers of an integer G in $Z(M)$, $M = \prod_{j=1}^{t} p_j^{a_j}$ provided that G has order at least n in each of $GF(p_j)$, $n \le p_j - 1$.

The interpolating polynomials can now be written as

$$m_i(u) = u - G^i, \;\; i = 1, 2, \ldots, n \tag{7.2.20}$$

$$M(u) = \prod_{i=1}^{n} m_i(u). \tag{7.2.21}$$

7.3 Number Theoretic Transforms

A special case of the above approach to computing acyclic convolution gives rise to NTTs. In addition to the constraints in (7.2.12) and (7.2.13) for a unique solution to the Vandermonde system of linear congruences, NTTs further require that

$$M(u) = u^n - 1. \tag{7.3.1}$$

It is clear that such a modulo polynomial will lead to *cyclic convolution* in $Z(M)$. Since G is a root of $M(u)$, $G^n - 1 \equiv 0 \bmod M$, or G has order n over $Z(M)$. Therefore, necessary and sufficient conditions for NTTs to exist are

1. $(G, M) = 1$,

2. $(G^i - 1, M) = 1, \;\; i = 1, 2, ..., n - 1,$

and

3. $G^n = 1 \bmod M.$ (7.3.2)

Equivalently, G has order n over $Z\left(p_j^{a_j}\right)$, $j = 1, 2, \ldots, t$. Also, according to (7.2.13), order of G cannot be less than n or else, if order of $G = b$, $b < n$, then $G^b - 1 \equiv 0 \bmod p_j^{a_j}$ and (7.2.13) is violated. Thus, the integer G is an element of order exactly equal to n over $Z\left(p_j^{a_j}\right)$, $j = 1, 2, \ldots, t$. An element of order n over $Z(p^a)$ also implies that it has order n over $GF(p)$. Since the order of every element in $GF(p)$ divides $p - 1$ (Theorem 3.6), the value of n gets further constrained to $n \mid (p_j - 1), j = 1, 2, \ldots, t$.

Given a primitive element G_1 over $GF(p)$, an element of order n over $Z(p^a)$ can be obtained as

$$G \equiv G_1^{p^{a-1}(p-1)/n} \bmod p^a. \tag{7.3.3}$$

This statement can be proven as follows. Since $(G_1, p) = 1$, we have $(G_1, p^a) = 1$ and

$$G^n \equiv G_1^{p^{a-1}(p-1)} \bmod p^a \equiv 1 \bmod p^a$$

(Euler's theorem, Theorem 2.2). Also, if the order of G is b, $G^b \equiv 1 \bmod p^a$, $b < n$, then

$$G_1^{bp^{a-1}(p-1)/n} \equiv 1 \bmod p^a. \tag{7.3.4}$$

Taking mod p of both sides leads to

$$G_1^{bp^{a-1}(p-1)/n} \equiv 1 \bmod p. \tag{7.3.5}$$

Over $GF(p)$, $G_1^{p-1} \equiv 1 \bmod p$ or $G_1^p \equiv G_1 \bmod p$. Applying this argument to the above equation, we get

$$G_1^{b(p-1)/n} \equiv 1 \bmod p.$$

Since G_1 is a primitive element, we must have $b = n$ as $G_1^i \neq 1 \bmod p$ for $i < p - 1$. This discussion leads to the definition and existence of NTTs which is summarized in the following.

Theorem 7.4 An NTT of length n exists over $Z(M)$, $M = \prod_{j=1}^{t} p_j^{a_j}$ provided that

$$n \mid (p_i - 1), \quad i = 1, 2, ..., t,$$

or, equivalently,

$$n \mid gcd(p_1 - 1, p_2 - 1, ..., p_t - 1). \tag{7.3.6}$$

The value of the generator G over $Z(M)$ can be computed using the CRT-I as

$$G = G_1 \oplus G_2 \oplus ... \oplus G_t,$$

$$G_i = A_i^{\theta_i},$$

$$\theta_i = p_i^{a_i - 1}(p_i - 1)/n, \tag{7.3.7}$$

A_i being a primitive element in $GF(p_i)$. In terms of M, the conditions for NTTs to exist are

$$(G, M) = 1,$$

$$(G^i - 1, M) = 1, \quad i = 1, 2, ..., n - 1,$$

and

$$G^n = 1 \bmod M.$$

The constraint on the length of the NTTs in (7.3.6) is a very severe one. For example, for even values of M, the maximum value of $n = 1$. This constraint has restricted the usefulness of NTTs for computing the cyclic and acyclic convolution. This is also known as the **word sequence length constraint** (WSLC) problem. This problem will be studied in Chapters 8, 9, and 10.

The matrix-vector representation of the NTTs is identical to that of the DFT (refer to (4.1.2)) with G replacing the n-th root of unity. Consequently, the expression for the inverse NTT is the same as in (4.1.5). In the present situation, n^{-1} is defined as an integer A that satisfies the congruence $A \cdot n \equiv 1 \bmod M$.

In the following, we describe two special cases of NTTs that are of significant interest in digital signal processing due to the simple nature of their generator and the ease of hardware realization of the associated arithmetic. These are the Mersenne number transform (MNT) and the Fermat number transform (FNT). Some basic properties of numbers used to establish the structure of MNT and FNT are described in the following.

Property A. Any two consecutive odd integers are relatively coprime.

Property B. $(2^a - 1, 2^b - 1) = 2^{(a,\, b)} - 1$.

Property C. If $(A, B) = D_1$ and $(A, C) = D_2$ for two integers B and C such that $(B, C) = 1$, then $(A, B \cdot C) = D_1 \cdot D_2$.

7.3.1 Mersenne Number Transforms

Mersenne number transforms (MNTs) are a special case of NTTs when M is a Mersenne number of the type

$$M = 2^p - 1, \tag{7.3.8}$$

where p is a prime number. When M is prime (called the Mersenne prime), then $Z(M)$ is same as $GF(M)$, and NTTs exist for all values of n that divide $M - 1$. This is the case when $p = 2, 3, 5, 7, 11$, and so on.

When M is composite, let us test the conditions (7.3.2) for the NTT to exist for $G = 2$ and $n = p$. It is seen that the first two conditions are satisfied since $(2, M) = 1$ and $(2^a - 1, 2^p - 1) = 1$ for $a < p$ (Property B). Finally, $G^n = 2^p \bmod (2^p - 1) = 1$. Consequently, all three conditions for the NTT to exist are satisfied and $G = 2$ gives rise to a NTT of length p.

Now, test the conditions (7.3.2) for $G = -2$ and $n = 2 \cdot p$. The first condition is satisfied trivially. The third condition leads to $(-2)^{2p} = (2^p)^2 \bmod (2^p - 1) = 1$. For the second condition, we partition the values of i in three parts, (i) i: even, (ii) $i = p$, and (iii) i: odd except $i = p$. For the first part, $(-2)^i - 1 = 2^i - 1$, which leads to

$$((-2)^i - 1, 2^p - 1) = (2^i - 1, 2^p - 1) = 1 \text{ (Property B)}.$$

For the second part, $(-2)^p - 1 = -2^p - 1$ which leads to

$$((-2)^p - 1, 2^p - 1) = (-2^p - 1, 2^p - 1) = (2^p + 1, 2^p - 1) = 1 \text{ (Property A)}.$$

For the third part, let $A = 2^p - 1$, $B = 2^i - 1$, $C = 2^i + 1$ in Property C. This leads to

$$(A, B \cdot C) = (2^p - 1, 2^{2i} - 1) = 2^{(p,\, 2i)} - 1 = 2^{(p,\, i)} - 1 = 1.$$

Consequently, all three conditions for the NTT to exist are satisfied, and $G = -2$ gives rise to a NTT of length $2 \cdot p$. Combining this analysis with the condition in (7.3.6) for an NTT to exist, we see that $2 \cdot p$ divides $p_i - 1$, p_i being any one of the prime factors in the standard factorization of $2^p - 1$. Thus, every prime factor of $2^p - 1$ is of the form $2 \cdot L \cdot p + 1$.

NTTs when M is a Mersenne number and $n = p$ or $n = 2 \cdot p$ are popularly known as the Mersenne number transforms (MNTs). MNTs are considered attractive due to the ease of hardware realization of arithmetic modulo $2^p - 1$.

7.3.2 Fermat Number Transforms

Fermat number transforms (FNTs) are a special case of NTTs when M (also denoted by F_t) is a Fermat number of the type

$$M = 2^{2^t} + 1, \tag{7.3.9}$$

where t takes on integer values. The first five Fermat numbers 3, 5, 17, 257, and 65537 are prime. The next Fermat prime remains unknown to date. When M is prime (called the Fermat prime), then $Z(M)$ is same as $GF(M)$, and NTTs exist for all values of n such that $n \mid (M-1)$. This is the case when M is one of 3, 5, 17, 257, and 65537 and NTTs of length n will exist for $n \mid 2$, $n \mid 4$, $n \mid 16$, $n \mid 256$, and $n \mid 65536$, respectively. This is highly desirable in digital signal processing as the length of the NTT can be chosen to be a power of 2 and the overall arithmetic realized mod $2^{2^t} + 1$, a simple task.

When M is composite, let us test the conditions (7.3.2) for the NTT to exist for $G = 2$ and $n = 2^{t+1}$. It is seen that the first condition is trivially satisfied since $(2, M) = 1$. The third condition is satisfied as

$$2^{2^{t+1}} \bmod \left(2^{2^t} + 1\right) = (-1)^2 = 1.$$

Note that

$$2^{2^t} \bmod \left(2^{2^t} + 1\right) = -1.$$

For $i < 2^{t+1}$,

$$\left(2^i - 1, 2^{2^t} - 1\right) = 2^{\left(i, 2^t\right)} - 1.$$

Also,

$$\left(2^i - 1, 2^{2^{t+1}} - 1\right) = 2^{\left(i, 2^{t+1}\right)} - 1 = 2^{\left(i, 2^t\right)} - 1, \quad i < 2^{t+1}.$$

Use Property C with $A = 2^i - 1$, $B = 2^{2^t} - 1$, $C = 2^{2^t} + 1$, and $D_1 = 2^{\left(i, 2^t\right)} - 1$. We recognize that $(A, B) = D_1$, $(B, C) = 1$, $(A, B \cdot C) = D_1$. Thus,

$$\left(2^i - 1, 2^{2^t} + 1\right) = (A, C) = 1.$$

This establishes the second condition. Consequently, all three conditions for the NTT to exist are satisfied and $G = 2$ gives rise to a NTT of length 2^{t+1}. In a similar manner, it can be shown that -2 gives rise to a NTT of length 2^{t+1}.

If G were selected to be

$$2^{2^{t-2}}\left(2^{2^{t-1}} - 1\right),$$

then it can be shown that $G^2 \equiv 2 \bmod M$. Therefore, G is a square root of 2, and gives rise to a NTT of length 2^{t+2}. It is straightforward to verify that the three conditions in (7.3.2) for NTT to exist are satisfied by this root for $n = 2^{t+2}$. Combining this analysis with the condition in (7.3.6) for an NTT to exist, we see that 2^{t+2} divides $p_i - 1$, p_i being any one of the prime factors in the standard factorization of $2^{2^t} + 1$. Thus, every prime factor of $2^{2^t} + 1$ is of the form $L\ 2^{t+2} + 1$.

The NTTs when M is a Fermat number and n is a power of 2 are popularly known as the Fermat number transforms (FNTs). FNTs are considered attractive due to the length of transform being a power of 2 and the ease of hardware realization of arithmetic modulo $2^{2^t} + 1$.

Up to this point, we have been seeking a DFT defined over $Z(M)$. The DFT, whether defined over $\mathcal{C}$, $GF(q)$, or $Z(M)$, may be interpreted as a special case of the CRT-P with the modulo polynomial $u^n - 1$ factorizing into n degree-one polynomials of the type $u - \omega^i$, where ω is the n-th root of unity in the given number system. This interpretation is the major impetus for the remainder of this chapter which focuses on finding the CRT-P for polynomials with coefficients defined over $Z(M)$, thereby generalizing the NTTs.

It is clear from CRT-I that a computation defined mod M can be written as a direct sum of t computations defined mod $p_i^{a_i}$, $i = 1, 2, ..., t$, where the standard factorization of M is given by $M = \prod_{i=1}^{t} p_i^{a_i}$. Consequently, we only consider the structure of computations over integer rings of the type $Z(p^a)$. In the next few sections, we develop a systematic approach to (a) polynomial factorization over the ring $Z(p^a)$ and its polynomial extensions, and (b) CRT-P over such rings. The basic approach in establishing the various results over a given finite integer ring (which is either $Z(p^a)$ or its polynomial extension) is to express it in its p-adic expansion form, establish similar results over the underlying finite field (which is either $GF(p)$ or its extension), and then generalize them to the ring. This approach is adopted in order to exploit fully the algebraic properties of finite fields. One of the major elements of CRT-P is factorization of the modulo polynomial $M(u)$. We begin the next section by developing techniques for monic polynomial factorization over integer rings.

7.4 Monic Polynomial Factorization

In this section, we focus on factorization of monic polynomials over $Z(p^a)$. Later CRT-I is used to obtain a non-unique factorization over $Z(M)$ based on the factors over the sub-rings. Results of this section will be utilized in developing subsequent sections.

The following three theorems establish relationships between polynomials defined over $Z(p^a)$ and the corresponding polynomials defined over $GF(p)$. Theorem 7.5 is one of the versions of the Hensel's theorem. This version plays a key role in our analysis. Theorems 7.6 and 7.7 can be used to determine the irreducibility and relatively prime properties for polynomials over $Z(p^a)$. The definitions of factor polynomial, irreducible polynomial, and relatively prime polynomials are given in definitions 7.4, 7.5, and 7.6, respectively. Definition 7.4 is the same as Definition 7.2 stated earlier in this chapter. Interestingly, these determinations are made in $GF(p)$. In this regard, many fruitful results that have been developed in the context of a field thus far are applied to integer rings.

Definition 7.4 Let $f(u)$ be a polynomial over the ring $Z(M)$. A polynomial $h(u)$ defined over $Z(M)$ has $f(u)$ as a *factor* whenever $h(u)$ and $f(u)$ can be written as $h(u) = f(u)\, q(u) \bmod M$, where $q(u)$ is a polynomial defined over $Z(M)$.

For example,

$$(3\,u^2 + 5\,u + 1) \cdot (6\,u^2 + 5\,u + 5) = u^2 + 3\,u + 5 = (u + 4) \cdot (u + 8) \bmod 9.$$

This demonstrates that many results in polynomial algebra over a field are no longer valid over a ring. The sum of the degrees of the factor polynomials may not be equal to the degree of the product, and, in general, the factorization of the polynomial over $Z(M)$ may be non-unique. Furthermore, we cannot even test if $3u^2 + 5u + 1$ divides $u^2 + 3u + 5$ over $Z(9)$ or not. To circumvent these problems, we will *limit our discussion to the factorization of monic polynomials over $Z(p^a)$ into factor polynomials that are monic as well.* However, we choose to retain the term monic whenever it is needed for the sake of readability.

Definition 7.5 A monic polynomial is termed *irreducible* if it has no other monic polynomial except itself as its factor.

Definition 7.6 If two monic polynomials have no common monic factors over $Z(p^a)$, then they are said to be *relatively prime* (or coprime).

It is verified that for monic polynomials $f(u)$ and $h(u)$, if $h(u) = f(u) \cdot q(u) \bmod M$, then

(i) $q(u)$ is monic, and
(ii) $deg(h(u)) = deg(f(u)) + deg(q(u))$.

Also, given $f(u)$ and $h(u)$, it is straightforward to verify if $f(u)$ is a factor of $h(u)$ or not.

Theorem 7.5 Hensel's Theorem. Let $M(u)$ be a polynomial defined over $Z(p^a)$ and the factorization

$$M(u) = \prod_{i=1}^{\lambda} m_{0,i}(u) \bmod p$$

be known, where $m_{0,i}(u)$ and $m_{0,j}(u)$, $i \neq j$, are monic coprime polynomials defined over $GF(p)$. Then, there exists λ monic polynomials over $Z(p^a)$, denoted by $m_i(u)$, such that

$$m_i(u) \equiv m_{0,i}(u) \bmod p$$

and

$$M(u) = \prod_{i=1}^{\lambda} m_i(u) \bmod p .$$

Proof: We prove this theorem by describing a procedure to obtain these factor polynomials. According to Definition 7.1, $M(u)$ can be expressed in its p-adic form as shown in (7.1.3) and (7.1.4). Define a series of polynomials as

$$A_k(u) = M(u) \bmod p^{k+1}, \quad 0 \le k \le a-1. \tag{7.4.1}$$

Clearly, $A_0(u) = M_0(u)$ and $A_{a-1}(u) = M(u)$. Now, we introduce a recursive procedure that utilizes all the $A_k(u)$ polynomials to factorize $M(u)$. We start with $A_0(u)$ and $A_1(u)$. Assume that the set of relatively prime factors of $A_0(u)$ over $GF(p)$ is given by $\{B_{0,1}(u), B_{0,2}(u), ..., B_{0,\lambda}(u)\}$. Since $M_0(u)$ has $m_{0,1}(u), ..., m_{0,\lambda}(u)$ as its factors, we assign

$$m_{0,i}(u) = B_{0,i}(u), \quad i = 1, 2, ..., \lambda.$$

In the most general case, these polynomials are either irreducible or power of an irreducible polynomial over $GF(p)$. Also, they are unique and pairwise relatively prime in that situation. However, in the present situation, we only require that the factor polynomials be relatively prime over $GF(p)$. Hence

$$A_0(u) = \prod_{i=1}^{\lambda} B_{0,i}(u) \bmod p. \tag{7.4.2}$$

From (7.4.1), $A_1(u)$ equals

$$A_1(u) = M_0(u) + pM_1(u). \tag{7.4.3}$$

On the other hand, if the statement of the theorem is true, $A_1(u)$ may be expressed as

$$A_1(u) = \prod_{i=1}^{\lambda} \left[B_{0,i}(u) + pQ_{1,i}(u)\right] \bmod p^2. \tag{7.4.4}$$

Equating (7.4.3) and (7.4.4), and rearranging terms, we have

$$M_0(u) + pM_1(u) = [A_0(u) + pA_0'(u)] + p\left[\sum_{i=1}^{\lambda} Q_{1,i}(u) \prod_{\substack{j=1 \\ j \ne i}}^{\lambda} B_{0,j}(u)\right] \bmod p^2, \tag{7.4.5}$$

where $A_0'(u)$ is determined by

$$pA_0'(u) = \left[\left\{\prod_{i=1}^{\lambda} B_{0,i}(u)\right\} - A_0(u)\right] \bmod p^2. \tag{7.4.6}$$

Canceling p on both sides and taking mod $B_{0,l}(u)$, $l = 1, 2, ..., \lambda$ in (7.4.5), we get the following λ congruences:

$$[M_1(u) - A_0'(u)] \equiv Q_{1,l}(u) \prod_{\substack{j=1 \\ j \neq l}}^{\lambda} B_{0,j}(u) \bmod (B_{0,l}(u), p), \tag{7.4.7}$$

where $A_0'(u)$ is given in (7.4.6), and all $B_{0,j}(u)$ and $M_1(u)$ are known. Thus, l-th congruence shown in (7.4.7) will give a closed-form solution for $Q_{1,l}(u)$. Note that the solution to congruences in (7.4.7) is unique. This is due to the reason that, first, the term $\prod_{\substack{j=1 \\ j \neq l}}^{\lambda} B_{0,j}(u)$ and $B_{0,l}(u)$ are coprime mod p; second, $deg(Q_{1,l}(u)) < deg(B_{0,l}(u))$. Once we have solved for all $Q_{1,l}(u)$, (7.4.4) reveals that $A_1(u)$, defined over $Z(p^2)$ has λ factors, namely,

$[B_{0,i}(u) + pQ_{1,i}(u)]$, $i = 1, 2, ..., \lambda$.

Let $B_{1,i}(u) = B_{0,i}(u) + pQ_{1,i}(u)$ be the i-th factor of $A_1(u)$. Also, $B_{1,i}(u) \equiv B_{0,i}(u) \bmod p$. Hence,

$$A_1(u) = \prod_{i=1}^{\lambda} B_{1,i}(u) \bmod p^2. \tag{7.4.8}$$

Repeating the same analysis for $A_1(u)$ and $A_2(u)$, we have the following λ congruences:

$$[M_2(u) - A_1'(u)] \equiv Q_{2,l}(u) \prod_{\substack{j=1 \\ j \neq l}}^{\lambda} B_{1,j}(u) \bmod (B_{0,l}(u), p), \tag{7.4.9}$$

where $A_1'(u)$ is computed from

$$p^2 A_1'(u) = \left[\left\{ \prod_{i=1}^{\lambda} B_{1,i}(u) \right\} - A_1(u) \right] \bmod p^3 . \tag{7.4.10}$$

Once again, (7.4.9) and (7.4.10) give polynomial factorization of $A_2(u)$ over $Z(p^3)$. The factors

$$B_{2,i}(u) = B_{1,i}(u) + p^2 Q_{2,i}(u), \quad i = 1, 2, ..., \lambda$$

are used to construct the factors of $A_3(u)$. The same analysis repeats until $A_{a-1}(u) = M(u)$ is factored. The factors of $A_{a-1}(u)$ are given by $B_{a-1,1}(u)$, ..., $B_{a-1,\lambda}(u)$. Setting $m_i(u) = B_{a-1,i}(u)$, $i = 1, 2, ..., \lambda$ completes the procedure to compute the factors and the proof of the theorem.

Theorem 7.6 Two monic polynomials defined over $Z(p^a)$, denoted by $C_1(u)$ and $C_2(u)$, have no monic factors in common mod p^a if they have no monic factors in common mod p.

Proof: Proving this theorem is equivalent to proving the statement: "If $C_1(u)$ and $C_2(u)$ have at least one monic factor in common mod p^a, say $q(u)$, then they have at least one monic factor in common mod p." Let $C_1(u)$ and $C_2(u)$ be expressed as

$$\begin{aligned} C_1(u) &= q(u)\, A(u) \bmod p^a, \\ C_2(u) &= q(u)\, B(u) \bmod p^a. \end{aligned} \tag{7.4.11}$$

After taking mod p in (7.4.11), the result is

$$\begin{aligned} C_1(u) &= q_0(u)\, A_0(u) \bmod p, \\ C_2(u) &\equiv q_0(u)\, B_0(u) \bmod p, \end{aligned} \tag{7.4.12}$$

where $A(u)$, $B(u)$ and $q(u)$ satisfy the congruences

$$\begin{aligned} A(u) &\equiv A_0(u) \bmod p, \\ B(u) &\equiv B_0(u) \bmod p, \\ q(u) &\equiv q_0(u) \bmod p. \end{aligned} \tag{7.4.13}$$

Obviously (7.4.12) tells that $q_0(u)$ is the common monic factor between $C_1(u)$ and $C_2(u)$ mod p. This completes the proof.

We observe that Theorem 7.6 is only a sufficient condition. If $C_1(u)$ and $C_2(u)$ defined mod p^a have a monic factor $q_0(u)$ in common mod p, then we may conclude that $C_1(u)$ and $C_2(u)$ have monic factors $q_1(u)$ and $q_2(u)$ mod p^a, respectively, such that $q_1(u) \equiv q_0(u) \bmod p$ and $q_2(u) \equiv q_0(u) \bmod p$. It is not necessary that $q_1(u) = q_2(u) \bmod p^a$.

Theorem 7.7 A monic polynomial $M(u)$ defined over $Z(p^a)$ is irreducible *iff* $M_0(u) \equiv M(u) \bmod p$ is irreducible over $GF(p)$.
Proof: We prove sufficiency first. Assume that $M(u)$ is reducible polynomial over $Z(p^a)$ and has two monic factors; namely, $A(u)$ and $B(u)$. Therefore, $M(u) = A(u)\ B(u)$. Applying mod p operation, we have $M(u) \equiv A_0(u)\ B_0(u) \bmod p$. Obviously, $M(u)$ has $A_0(u)$ and $B_0(u)$ as its factors mod p, and, therefore, is not an irreducible polynomial over $GF(p)$. Secondly, we prove the necessity. Let $M_0(u) \equiv M(u) \bmod p$. Assume that $M_0(u)$ is a reducible monic polynomial defined over $GF(p)$ and has at least two relatively prime monic factors; namely, $A_0(u)$ and $B_0(u)$. By applying Theorem 7.5, $M(u) = A(u)\ B(u) \bmod p^a$, where $A(u)$ and $B(u)$, both defined over $Z(p^a)$, satisfy $A_0(u) \equiv A(u) \bmod p$ and $B_0(u) \equiv B(u) \bmod p$, respectively. Hence, $M(u)$ is reducible monic polynomial over $Z(p^a)$. This proves the theorem.

Theorems 7.5, 7.6 and 7.7 serve as a bridge connecting $GF(p)$ and $Z(p^a)$. We plan to present various results on $Z(p^a)$ as a natural extension to the corresponding results on $GF(p)$. This is desirable as the algebraic structure of finite fields has already been studied extensively by researchers. Methods are well known for selecting relatively prime factors of a polynomial in a way that the multiplicative complexity of the CRT-P based algorithms for computing convolutions is as low as possible. The following theorem states the unique factorization property of the above procedure.

Theorem 7.8 Unique factorization property. A monic polynomial defined over $Z(p^a)$ has a unique factorization into a product of relatively prime monic polynomials that are either irreducible or power of an irreducible polynomial over $GF(p)$.
Proof: Let $M(u)$ be a monic polynomial defined over $Z(p^a)$ and $M_0(u) \equiv M(u) \bmod p$. It is well known that the prime factorization of a polynomial with coefficients from a field satisfies the uniqueness property. Based on this statement, $M_0(u)$ has a unique factorization into pairwise relatively prime factors,

$$M_0(u) = \prod_{i=1}^{\lambda} B_{0,i}(u) = B_{0,1}(u) B_0'(u), \tag{7.4.14}$$

where $B_0'(u) = \prod_{i=2}^{\lambda} B_{0,i}(u)$. Suppose that the theorem fails, which means that the polynomial $M(u)$ has two distinct factorizations over $Z(p^a)$. Thus

$$M(u) = B_1(u)\,B'(u) = C_1(u)\,C'(u) \bmod p^a. \tag{7.4.15}$$

Using the p-adic form and taking mod p^2 on (7.4.15), we have

$$\begin{aligned}&[B_{0,1}(u) + pQ_{1,1}(u)][B_0'(u) + pQ_1'(u)]\\ &\quad = [C_{0,1}(u) + pR_{1,1}(u)][C_0'(u) + pR_1'(u)] \bmod p^2.\end{aligned} \tag{7.4.16}$$

Expanding (7.4.16), substituting $B_{0,1}(u) = C_{0,1}(u)$ and $B_0'(u) = C_0'(u)$ (recall that $M_0(u)$ has unique factorization over $GF(p)$), and canceling common terms, we get

$$B_{0,1}(u)[Q_1'(u) - R_1'(u)] = B_0'(u)[R_{1,1}(u) - Q_{1,1}(u)] \bmod p. \tag{7.4.17}$$

First, $B_{0,1}(u)$ and $B_0'(u)$ are coprime; second, $deg(B_{0,1}(u)) > deg(R_{1,1}(u))$ and $deg(Q_{1,1}(u))$. Therefore, $B_{0,1}(u)$ does not divide $B_0'(u)[R_{1,1}(u) - Q_{1,1}(u)]$. Similarly, $B_0'(u)$ does not divide $B_{0,1}(u)[Q_1'(u) - R_1'(u)]$. Hence, the only solution to (7.4.17) is that $Q_1'(u) = R_1'(u)$ and $Q_{1,1}(u) = R_{1,1}(u)$.

Repeating the same analysis up to mod p^a in (7.4.15), we can conclude that the pairs $[B_1(u), C_1(u)]$ and $[B'(u), C'(u)]$ have the same p-adic form; therefore, they are equal. This proves the theorem.

It is worthwhile to note that each of the relatively prime monic factors in the unique factorization over $Z(p^a)$ is either irreducible over $GF(p)$ or is a power of an irreducible polynomial over $GF(p)$. Those monic factor polynomials over $Z(p^a)$ that are powers of an irreducible polynomial over $GF(p)$ may further factor over $Z(p^a)$. This further factorization may be non-unique and is not of interest as it does not contribute to extensions of CRT-P over finite integer rings.

Example 7.1 Let $M(u) = u^7 + 2u^4 + 2u^3 \bmod 4$. The prime field is $GF(2)$ and $M_0(u) \equiv M(u) \bmod 2 = u^7$. Since $M_0(u)$ is a power of an irreducible polynomial $GF(2)$, factorization of $M(u)$ cannot continue further on the basis of the factorization algorithm outlined above. However,

$$\begin{aligned}M(u) &= u^7 + 2u^4 + 2u^3 \bmod 4\\ &= u^3(u^4 + 2u + 2) \bmod 4\\ &= (u^3 + 2)(u^4 + 2) \bmod 4.\end{aligned}$$

In the next subsection, we summarize the procedure by outlining steps involving closed-form equations of all intermediate polynomials. Later, we give examples to illustrate the concepts.

7.4.1 A Systematic Procedure

Based on the concepts developed here, we summarize the procedure to be followed in order to factorize a given monic polynomial, $M(u)$, defined over $Z(p^a)$.

1. According to the p-adic form of $M(u)$, we obtain $M_i(u)$, $i = 0, 1, ..., a - 1$. Note that all the coefficients of $M_i(u)$ are defined over $GF(p)$, even though $M(u)$ is defined over $Z(p^a)$.

2. Form the following polynomials based on $M_i(u)$ obtained in the previous step:

$$A_k(u) = \sum_{i=0}^{k} p^i M_i(u), \; k = 0,1,...,a-1. \tag{7.4.18}$$

   Clearly, $A_0(u) = M_0(u)$ and $A_{a-1}(u) = M(u)$. In order to factorize $M(u)$, we restate that $M_0(u) \equiv M(u) \bmod p$ has λ relatively prime factors over $GF(p)$, denoted by $B_{0,i}(u)$, $i = 1, 2, ..., \lambda$. Hence, the factorization of $A_0(u)$, which equals $M_0(u)$, is shown in (7.4.2). In the following, we present a recursive method such that factorization of $A_k(u)$ can be computed from the factorization of $A_{k-1}(u)$.

3. Let $k = 1$.

4. Compute $A'_{k-1}(u)$ based on $A_{k-1}(u)$ and λ factors $B_{k-1,i}(u)$, $i = 1, 2, ..., \lambda$,

$$p^k A'_{k-1}(u) = \left[\left(\prod_{i=1}^{\lambda} B_{k-1,i}(u)\right) - A_{k-1}(u)\right] \bmod p^{k+1}. \tag{7.4.19}$$

5. Solve for $Q_{k,l}(u)$, $l = 1, 2, ..., \lambda$. The l-th one is based on the congruence shown below.

$$\left[M_k(u) - A'_{k-1}(u)\right] \equiv Q_{k,l}(u) \prod_{\substack{i=1 \\ i \neq l}}^{\lambda} B_{k-1,i}(u) \bmod\left(B_{k-1,l}(u), p\right). \tag{7.4.20}$$

6. The factorization of $A_k(u)$ is available and can be written as

$$A_k(u) = \prod_{i=1}^{\lambda} B_{k,i}(u) \bmod p^{k+1},$$

where its i-th factor equals

$$B_{k,i}(u) = [B_{k-1,i}(u) + p^k Q_{k,i}(u)] \bmod p^{k+1}. \tag{7.4.21}$$

7. Increase k by one. If $k = a$, then stop. Otherwise, go to Step 4.

The following three examples are given to illustrate working of the procedure. Especially, we focus on polynomials having the form $u^n \pm 1$ for their obvious importance in digital signal processing. It is seen that, even though two polynomials may have the same factorization over $GF(p)$, their individual factorization over $Z(p^a)$ may be distinct. This reinforces the uniqueness property established in Theorem 7.8.

Example 7.2 Factorize $M(u) = u^7 - 1$ over $Z(2^3)$. Here, $p = 2$, $a = 3$. In Step 1, we get the p-adic form of $M(u)$: $M_0(u) = u^7 + 1$, $M_1(u) = 1$ and $M_2(u) = 1$. In Step 2, based on (7.4.18), we construct $A_1(u) = u^7 + 3$ and $A_2(u) = M(u) = u^7 + 7$. Also, factorization of $M(u)$ over $GF(2)$ is obtained using the techniques described in Chapter 6. It is given by

$$M_0(u) = (u + 1)(u^3 + u + 1)(u^3 + u^2 + 1).$$

Namely, $B_{0,1}(u) = u + 1$, $B_{0,2}(u) = u^3 + u + 1$, $B_{0,3}(u) = u^3 + u^2 + 1$. We start the recursive procedure, outlined from Step 4 to Step 6, at $k = 1$. According to (7.4.19), $A_0'(u) = u^6 + u^5 + u^2 + u$. Next, we solve the congruences in (7.4.20) for $Q_{1,l}(u)$, $l = 1, 2, 3$. They are

$$\begin{cases} Q_{1,1}(u) = 1 \bmod B_{0,1}(u) \\ Q_{1,2}(u) = u^2 + 1 \bmod B_{0,2}(u) \\ Q_{1,3}(u) = u^2 + u + 1 \bmod B_{0,3}(u). \end{cases} \tag{7.4.22}$$

In Step 6, based on (7.4.21), the factors of $A_1(u)$ (defined mod 4) are $B_{1,1}(u) = u + 3$, $B_{1,2}(u) = u^3 + 2u^2 + u + 3$, $B_{1,3}(u) = u^3 + 3u^2 + 2u + 3$.

Repeating Step 4 through Step 6 for $k = 2$, we obtain

$$\begin{cases} Q_{2,1}(u) = 1 \bmod B_{0,1}(u) \\ Q_{2,2}(u) = u^2 + u + 1 \bmod B_{0,2}(u) \\ Q_{2,3}(u) = 1 \bmod B_{0,3}(u), \end{cases} \tag{7.4.23}$$

and $B_{2,1}(u) = u + 7$, $B_{2,2}(u) = u^3 + 6u^2 + 5u + 7$, $B_{2,3}(u) = u^3 + 3u^2 + 2u + 7$. Since $k + 1 = a$, the recursive procedure stops. We write the polynomial factorization of $M(u)$ over $Z(8)$ as

$$M(u) = (u + 7)(u^3 + 6u^2 + 5u + 7)(u^3 + 3u^2 + 2u + 7) \bmod 8.$$

Example 7.3 Factorize $M(u) = u^7 + 1$ over $Z(2^3)$. The p-adic form of $M(u)$ gives $M_0(u) = u^7 + 1$, $M_1(u) = 0$ and $M_2(u) = 0$. Note that $M_0(u) = u^7 + 1 = u^7 - 1$ over $GF(2)$, which is the same as the polynomial in Example 7.2; therefore, $M_0(u)$ has the same factors: $B_{0,1}(u) = u + 1$, $B_{0,2}(u) = u^3 + u + 1$, $B_{0,3}(u) = u^3 + u^2 + 1$. Similarly, based on (7.4.18), we construct $A_1(u) = u^7 + 1$ and $A_2(u) = M(u) = u^7 + 1$. The intermediate polynomials are:

$$\begin{cases} Q_{1,1}(u) = 0 \bmod B_{0,1}(u) \\ Q_{1,2}(u) = u^2 \bmod B_{0,2}(u) \\ Q_{1,3}(u) = u \bmod B_{0,3}(u), \end{cases} \tag{7.4.24}$$

$B_{1,1}(u) = u + 1$, $B_{1,2}(u) = u^3 + 2u^2 + u + 1$, $B_{1,3}(u) = u^3 + u^2 + 2u + 1$,

$$\begin{cases} Q_{2,1}(u) = 0 \bmod B_{0,1}(u) \\ Q_{2,2}(u) = u \bmod B_{0,2}(u) \\ Q_{2,3}(u) = u^2 \bmod B_{0,3}(u), \end{cases} \tag{7.4.25}$$

and $B_{2,1}(u) = u + 1$, $B_{2,2}(u) = u^3 + 2u^2 + 5u + 1$, $B_{2,3}(u) = u^3 + 5u^2 + 2u + 1$. We write the polynomial factorization of $M(u)$ over $Z(8)$ as

$$M(u) = (u + 1)(u^3 + 2u^2 + 5u + 1)(u^3 + 5u^2 + 2u + 1) \bmod 8.$$

Example 7.4 Factorize $M(u) = u^8 - 1$ over $Z(3^2)$. Here, $p = 3$, $a = 2$. The p-adic form of $M(u)$ gives $M_0(u) = u^8 + 2$, $M_1(u) = 2$. Since $a = 2$, in Step 2, based on (7.4.18), we construct $A_0(u) = M_0(u)$, $A_1(u) = M(u)$. Also, factorization of $M(u)$ over $GF(3)$ is obtained using the techniques described in Chapter 6. It is given by

$$M_0(u) = (u + 1)(u + 2)(u^2 + 1)(u^2 + 2u + 2)(u^2 + u + 2).$$

Namely, $B_{0,1}(u) = u + 1$, $B_{0,2}(u) = u + 2$, $B_{0,3}(u) = u^2 + 1$, $B_{0,4}(u) = u^2 + 2u + 2$, $B_{0,5}(u) = u^2 + u + 2$, and $\lambda = 5$. We start the recursive procedure, outlined from Step 4 to Step 6, at $k = 1$. According to (7.4.19), $A_0'(u) = 2u^7 + 2u^4 + 2u^2 + 2u + 2$. Next, we solve the congruences in (7.4.20) for $Q_{1,l}(u)$, $l = 1, \ldots, 5$. They are

$$\begin{cases} Q_{1,1}(u) = 0 \bmod B_{0,1}(u) \\ Q_{1,2}(u) = 2 \bmod B_{0,2}(u) \\ Q_{1,3}(u) = 0 \bmod B_{0,3}(u) \\ Q_{1,4}(u) = u + 2 \bmod B_{0,4}(u) \\ Q_{1,5}(u) = u + 2 \bmod B_{0,5}(u). \end{cases} \tag{7.4.26}$$

Finally, based on (7.4.21), the monic factors of $M(u)$ over $Z(9)$ are $B_{1,1}(u) = u + 1$, $B_{1,2}(u) = u + 8$, $B_{1,3}(u) = u^2 + 1$, $B_{1,4}(u) = u^2 + 5u + 8$, $B_{1,5}(u) = u^2 + 4u + 8$.

Once the monic polynomial factorizations over the rings of integers $Z(p^a)$ are completed, we extend the results further by employing CRT-I to obtain factorization over $Z(M)$. This general factorization over $Z(M)$ is non-unique in many respects and is studied in the following subsection.

7.4.2 Monic Polynomial Factorization Over $Z(M)$

The factorization of a monic polynomial $M(u)$ of degree n over $Z(M)$ is considered. Recall that since M is composite, the standard factorization can be applied to M such that $M = \prod_{i=1}^{t} p_i^{a_i}$. Further, we assume that factorization over the individual sub-rings $Z(p^a)$ has been completed. Namely, they are

$$\begin{cases} M(u) \equiv m_1^{(1)}(u) \cdot m_2^{(1)}(u) \cdots m_{\gamma_1}^{(1)}(u) \bmod p_1^{a_1} \\ M(u) \equiv m_1^{(2)}(u) \cdot m_2^{(2)}(u) \cdots m_{\gamma_2}^{(2)}(u) \bmod p_2^{a_2} \\ \vdots \\ M(u) \equiv m_1^{(t)}(u) \cdot m_2^{(t)}(u) \cdots m_{\gamma_t}^{(t)}(u) \bmod p_t^{a_t}. \end{cases} \tag{7.4.27}$$

The main idea is to combine the coefficients of the factor polynomials shown in (7.4.27) using the CRT-I to form factorization over $Z(M)$. Assume that $M(u)$ has γ factors in $Z(M)$; the goal is to find these γ factors. We will soon see that γ can take more than one value; therefore, by setting $\gamma = 2$ tentatively, the main idea can be conceptually illustrated. Later, γ can be generalized to any arbitrary integer value. The first step is to divide γ_i factors in $Z\left(p_i^{a_i}\right)$ into two groups. The product of polynomial factors of each group give rise to two factors in $Z\left(p_i^{a_i}\right)$, denoted by $m_1'^{(i)}(u)$ and $m_2'^{(i)}(u)$. Then (7.4.27) becomes

$$\begin{cases} M(u) \equiv m_1'^{(1)}(u) \cdot m_2'^{(1)}(u) \bmod p_1^{a_1} \\ M(u) \equiv m_1'^{(2)}(u) \cdot m_2'^{(2)}(u) \bmod p_2^{a_2} \\ \quad \vdots \\ M(u) \equiv m_1'^{(t)}(u) \cdot m_2'^{(t)}(u) \bmod p_t^{a_t}. \end{cases} \tag{7.4.28}$$

By claiming that

$$m_k(u) \equiv m_k'^{(i)}(u) \bmod p_i^{a_i}, \quad k = 1, 2; \; i = 1, 2, \ldots, t,$$

the coefficients of the factors $m_1(u)$ and $m_2(u)$ can be computed using CRT-I. The same analysis can be extended easily to γ by dividing γ_i factors in $Z\left(p_i^{a_i}\right)$ into γ groups. In most cases, $\gamma_1, \gamma_2, \ldots, \gamma_t$ are unequal. If $\gamma > \gamma_i$, then additional 1 (degree 0 polynomial) can be appended. It is clear from this description that the factorization over $Z(M)$ is non-unique. One reason is that γ can be chosen arbitrarily; the other reason is that there are many ways to divide groups in the first step. If we remove the freedom in selecting γ by restricting $\gamma = max\{\gamma_1, \gamma_2, \ldots, \gamma_t\}$, then the total number of different ways of combination (or factorization) is

$$(\gamma!)^{t-1} \Big/ \prod_{i=1}^{t} (\gamma - \gamma_i)!\,.$$

A very special case arises when γ_i, $i = 1, 2, \ldots, t$ are all equal to γ and all $m_k'^{(i)}(u)$, $i = 1, 2, \ldots, t$; $k = 1, 2, \ldots, \gamma$, are degree-one factors. In this special case, there are a total of $(\gamma!)^{t-1}$ different factorizations possible over $Z(M)$. Example 7.5 is given to illustrate the procedure.

Example 7.5 Factorize $M(u) = u^7 - 1$ over $Z(2^3\,3^2) = Z(72)$. Clearly, $p_1 = 2$, $a_1 = 3$, and $p_2 = 2$, $a_2 = 2$. In Example 7.2, we have obtained the polynomial factorization over $Z(8)$:

$$M(u) = (u + 7)\,(u^3 + 6\,u^2 + 5\,u + 7)\,(u^3 + 3\,u^2 + 2\,u + 7) \bmod 8.$$

Following the systematic procedure described in Section 7.4, we obtain factorization over $Z(9)$ as

$$M(u) = (u + 8)\,(u^6 + u^5 + u^4 + u^3 + u^2 + u + 1) \bmod 9.$$

Next we propose two factorizations over $Z(72)$ by different combinations of the factors in $Z(2^3)$ and $Z(3^2)$. Because CRT-I is used to compute the coefficients of the factors in $Z(72)$, the following information about the scalars T_i in (2.4.2) is necessary: $T_1 = 1$, $T_2 = 8$. If we combine $(u + 7) \in Z(8)$ and $(u + 8) \in Z(9)$ together, a factor in $Z(72)$ is generated as $(u + 71)$. Further, we combine $(u^3 + 6\,u^2 + 5\,u + 7)\,(u^3 + 3\,u^2 + 2\,u + 7) \in Z(8)$ and $(u^6 + u^5 + u^4 + u^3 + u^2 + u + 1) \in Z(9)$ to form the second factor $(u^6 + u^5 + u^4 + u^3 + u^2 + u + 1) \in Z(72)$. Hence, one factorization of $M(u)$ that has two factors is

$$M(u) = (u + 71)\,(u^6 + u^5 + u^4 + u^3 + u^2 + u + 1) \bmod 72.$$

Another factorization which has three factors can be found by employing the following combination: $((u + 7) \in Z(8)$ and $(u + 8) \in Z(9))$, $((u^3 + 6\,u^2 + 5\,u + 7) \in Z(8)$ and $(u^6 + u^5 + u^4 + u^3 + u^2 + u + 1) \in Z(9))$, $((u^3 + 3\,u^2 + 2\,u + 7) \in Z(8)$ and $1 \in Z(9))$. It leads to

$$M(u) = (u + 71)\,(64\,u^6 + 64\,u^5 + 64\,u^4 + u^3 + 46\,u^2 + 37\,u + 55) \cdot (9\,u^3 + 27\,u^2 + 18\,u + 55) \bmod 72.$$

7.5 Extension of CRT-P Over Finite Integer Rings

In the previous section, we introduced a systematic procedure to factorize a monic polynomial over $Z(p^a)$. This methodology is primarily based on the known factorization of the monic polynomial over $GF(p)$. In this section, we present an extension of CRT-P when the coefficients are defined over $Z(p^a)$. This extension will be termed as the **CRT-PR (CRT for polynomial rings)**. It has also been termed as the **AICE-CRT (American-Indian-Chinese extension of CRT-P)** in recent literature. We demonstrate that the results obtained in the previous section play a fundamental role in deriving the CRT-PR. As its predecessors, we expect that the CRT-PR will find a large number of applications in designing efficient algorithms for many computationally intensive tasks involving arithmetic over integer rings. Later sections of this chapter will be oriented towards deriving the corresponding versions of CRT-PR over complex finite integer rings and finite integer polynomial rings.

Let $X(u)$ and $M(u)$ be two polynomials defined over $Z(p^a)$. The polynomial $X(u)$ is an arbitrary polynomial over $Z(p^a)$ having degree at most $n - 1$. It is necessary and sufficient that $M(u)$ be monic and of degree greater than $n - 1$. Assume that factorization of $M(u)$ over $Z(p^a)$ has been completed by following the steps listed in the previous section; that is,

$$M(u) = \prod_{i=1}^{\lambda} m_i(u) \bmod p^a . \tag{7.5.1}$$

Recall that the monic factors $m_i(u)$ and $m_j(u)$, $i \neq j$, $1 \leq i, j \leq \lambda$ are relatively prime. This important property leads to CRT-PR. Define a vector of polynomials, $[x_1(u)\ x_2(u)\ ...\ x_\lambda(u)]$ with the component $x_i(u)$ computed using

$$x_i(u) \equiv X(u) \bmod (m_i(u), p^a), \quad i = 1, 2, ..., \lambda. \tag{7.5.2}$$

The CRT-PR delivers a unique relationship between the polynomial $X(u)$ and its vector representation. Theorem 7.9 plays a preliminary role in proving the one-to-one correspondence. The CRT-PR is described in Theorem 7.10.

Theorem 7.9 Consider the polynomials $X(u)$, $m_1(u)$, and $m_2(u)$, all defined over $Z(p^a)$. Let $m_1(u)$ and $m_2(u)$ be monic and coprime mod p. If $m_1(u)$ is a factor of $X(u)$ and $m_2(u)$ is a factor of $X(u)$, then the statement "$m_1(u)\ m_2(u)$ divides $X(u)$" is true.

Proof: Since $m_1(u)$ is a factor of $X(u)$, it follows,

$$X(u) = m_1(u)\ Q(u) \bmod p^a. \tag{7.5.3}$$

Dividing $Q(u)$ by $m_2(u)$ (note that such a division is possible as $m_2(u)$ is monic), we may write

$$Q(u) = m_2(u)\ Q'(u) + R(u) \bmod p^a, \tag{7.5.4}$$

where $R(u)$ is the remainder, $deg(R(u)) < deg(m_2(u))$. Substituting (7.5.4) into (7.5.3), we obtain,

$$X(u) = m_1(u)\ m_2(u)\ Q'(u) + m_1(u)\ R(u) \bmod p^a. \tag{7.5.5}$$

Since $m_2(u)$ divides $X(u)$, (7.5.5) reveals that

$$R(u)\ m_1(u) \equiv 0 \bmod (m_2(u), p^a). \tag{7.5.6}$$

Obviously, (7.5.6) holds true mod p; therefore,

$$R(u)\ m_1(u) \equiv 0 \bmod (m_2(u), p). \tag{7.5.7}$$

Because $m_1(u)$ and $m_2(u)$ are coprime mod p, in order for (7.5.7) to be true, we must have $R(u) \equiv 0 \bmod p$ or

$$R(u) = p\, R_1(u). \tag{7.5.8}$$

Substituting (7.5.8) into (7.5.6), and canceling out common terms, we have

$$R_1(u)\, m_1(u) \equiv 0 \bmod (m_2(u), p^{a-1}). \tag{7.5.9}$$

This analysis is repeated until the arithmetic operations reduce to modulo p. Finally, we have

$$R_{a-1}(u)\, m_1(u) \equiv 0 \bmod (m_2(u), p), \tag{7.5.10}$$

where $R_{a-1}(u)$ satisfies $R(u) = p^{a-1} R_{a-1}(u) \bmod p^a$. Also, $deg(R_{a-1}(u)) = deg(R(u)) < deg(m_2(u))$. Based on these statements, along with the fact that $m_1(u)$ and $m_2(u)$ are coprime, the only solution to (7.5.10) is $R_{a-1}(u) = 0$ or $R(u) = 0$. From (7.5.5), if $R(u) = 0$, then $m_1(u)\, m_2(u)$ is a factor of $X(u)$.

Theorem 7.10 CRT-P for polynomial rings (CRT-PR). There exists an one-to-one correspondence between $X(u)$ and the vector representation $[x_1(u)\ x_2(u)\ \ldots\ x_\lambda(u)]$, as long as $deg(X(u)) < deg(M(u))$ and the monic factors of $M(u)$, denoted by $m_1(u), m_2(u), \ldots, m_\lambda(u)$, are relatively prime mod p. Here the entire theorem is defined over the ring $Z(p^a)$.

Proof: Assume that the theorem is false. This implies that there exists another polynomial of degree less than n defined over $Z(p^a)$ such that it has the same representation as the one for $X(u)$. Let this polynomial be $Y(u)$ and $X''(u) = X(u) - Y(u)$. Then, $X''(u)$ has all zeros in its vector representation. In other words, the following equation holds for all the monic factors of $M(u)$:

$$m_i(u) \,|\, X''(u) \bmod p^a, \quad i = 1, 2, \ldots, \lambda. \tag{7.5.11}$$

According to Theorem 7.9, (7.5.11) indicates that $\left[\prod_{i=1}^{\lambda} m_i(u)\right] \Big| X''(u)$. This is not possible as $M(u)$ is a monic polynomial of degree n while $deg(X''(u)) < n$.

On the basis of the CRT-PR, one can form an alternative system for representing polynomials with coefficients from an integer ring. In fact, we can now express the direct sum property of the polynomials over $Z(p^a)$ as

$$\langle X(u)\rangle_{M(u)} = \langle X(u)\rangle_{m_1(u)} \oplus \langle X(u)\rangle_{m_2(u)} \oplus \cdots \oplus \langle X(u)\rangle_{m_\lambda(u)} \tag{7.5.12}$$

in a manner analogous to (3.4.9). As its two counterparts (the CRT-I and CRT-P), we expect the CRT-PR to find applications as a tool to partition a large size computational task into a number of smaller but independent subtasks which may be performed in parallel. We will exploit this belief in succeeding work.

The CRT-PR has two phases. In Theorem 7.10, we have proven that the solution to the set of congruences shown in (7.5.2) is unique. Next we describe a procedure to find it.

7.5.1 The CRT-PR Reconstruction

Assume that a monic polynomial $M(u)$ defined over $Z(p^a)$ has λ relatively prime factors, $m_1(u)$, $m_2(u)$, ..., $m_\lambda(u)$. These factors are obtained based on the systematic procedure introduced in Section 7.4. Every polynomial $X(u)$ with degree less than the degree of $M(u)$ can be represented by an unique vector $\underline{X}(u)$ having λ components $x_1(u), x_2(u), ..., x_\lambda(u)$,

$$X(u) \leftrightarrow \underline{X}(u) = [x_1(u)\ \ x_2(u)\ \ \ldots\ \ x_k(u)],$$

where $x_i(u)$, the i-th residue polynomial corresponding to $m_i(u)$, is given in (7.5.2). According to CRT-PR, if $\underline{X}(u)$ is given, then the corresponding polynomial $X(u)$ can be uniquely determined by solving λ congruences in (7.5.2). A closed form expression for such a solution for $X(u)$ is given by

$$X(u) \equiv \sum_{i=1}^{\lambda} x_i(u) T_i(u) \left[\frac{M(u)}{m_i(u)}\right] \operatorname{mod}\left(M(u), p^a\right). \tag{7.5.13}$$

The polynomials $T_i(u)$, $i = 1, 2, ..., \lambda$, satisfy the congruences,

$$T_i(u)\left[\frac{M(u)}{m_i(u)}\right] \equiv 1 \operatorname{mod}\left(m_i(u), p^a\right). \tag{7.5.14}$$

We now propose a procedure for reconstructing the polynomial $X(u)$ over $Z(p^a)$ from its corresponding residue vector. The goal is to find the polynomials $T_i(u)$, $i = 1, 2, ..., \lambda$, such that (7.5.14) is satisfied. Let $M_i(u) = M(u) / m_i(u)$. Therefore, we need to solve recurrences of the type $T_i(u)\ M_i(u) \equiv 1 \bmod (m_i(u), p^a)$. First, we represent $T_i(u)$ and $M_i(u)$ in the p-adic form and then take mod p on both sides. Then (7.5.14) becomes

$$T_{0,i}(u)\, M_{0,i}(u) \equiv 1 \bmod (m_i(u), p), \tag{7.5.15}$$

or $T_{0,i}(u)$ is the inverse of $M_{0,i}(u)$. A solution to (7.5.15) exists because $M_{0,i}(u)$ and $m_i(u)$ are coprime. This solution can be obtained using the *Euclid's algorithm*. If we perform mod p^2 operation on the p-adic form, it becomes

$$[T_{0,i}(u) + p\, T_{1,i}(u)]\, [M_{0,i}(u) + p\, M_{1,i}(u)] \equiv 1 \bmod (m_i(u), p^2). \tag{7.5.16}$$

Canceling out the product and substituting

$$T_{0,i}(u)\, M_{0,i}(u) \equiv 1 + p\, Q_{1,i}(u) \bmod (m_i(u), p^2), \tag{7.5.17}$$

we get

$$T_{0,i}(u)\, M_{1,i}(u) \equiv -[T_{1,i}(u)\, M_{0,i}(u) + Q_{1,i}(u)] \bmod (m_i(u), p). \tag{7.5.18}$$

Hence, $T_{1,i}(u)$ can be solved as

$$T_{1,i}(u) = -T_{0,i}(u)\, [T_{0,i}(u)\, M_{1,i}(u) + Q_{1,i}(u)] \bmod (m_i(u), p). \tag{7.5.19}$$

The same analysis can be repeated until $T_{2,i}(u)$, ..., $T_{a-1,i}(u)$ are computed. In summary, we may express the closed-form solutions for finding $T_{k,i}(u)$, $k = 1, 2, ..., a-1$, as

$$T_{k,i}(u) = -T_{0,i}(u)\, [T_{0,i}(u)\, M_{k,i}(u) + Q_{k,i}(u)] \bmod (m_i(u), p). \tag{7.5.20}$$

Here $Q_{k,i}(u)$ satisfies the congruence

$$\left(\sum_{l=0}^{k-1} p^l T_{l,i}(u)\right)\left(\sum_{l=0}^{k-1} p^l M_{l,i}(u)\right) = 1 + p^k Q_{k,i}(u) \bmod \left(m_i(u), p^{k+1}\right). \tag{7.5.21}$$

Once $T_{k,i}(u)$ are computed, they can be used to construct $T_i(u)$ as

$$T_i(u) = \sum_{k=0}^{a-1} p^k T_{k,i}(u) \bmod p^a,\ i = 1, 2, ..., \lambda. \tag{7.5.22}$$

The polynomials $T_i(u)$, $i = 1, 2, ..., \lambda$, *unique* over $Z(p^a)$. Following example is given to illustrate the reconstruction procedure.

Example 7.6 Let $M(u) = u^7 + 1 \bmod 2^3$. Based on factorization derived in Example 7.3, the monic factors of $M(u)$ over $Z(8)$ are $m_1(u) = u + 1$, $m_2(u) =$

$u^3 + 2u^2 + 5u + 1$ and $m_3(u) = u^3 + 5u^2 + 2u + 1$. Consequently, $M_1(u) = u^6 + 7u^5 + u^4 + 7u^3 + u^2 + 7u + 1$, $M_2(u) = u^4 + 6u^3 + 7u^2 + 3u + 1$ and $M_3(u) = u^4 + 3u^3 + 7u^2 + 6u + 1$. Based on their p-adic forms, we have

$$\begin{cases} M_{1,0}(u) = u^6 + u^5 + u^4 + u^3 + u^2 + u + 1 \\ M_{1,1}(u) = u^5 + u^3 + u \\ M_{1,2}(u) = u^5 + u^3 + u, \end{cases}$$

$$\begin{cases} M_{2,0}(u) = u^4 + u^2 + u + 1 \\ M_{2,1}(u) = u^3 + u^2 + u \\ M_{2,2}(u) = u^3 + u^2, \end{cases}$$

$$\begin{cases} M_{3,0}(u) = u^4 + u^3 + u^2 + 1 \\ M_{3,1}(u) = u^3 + u^2 + u \\ M_{3,2}(u) = u^2 + u. \end{cases}$$

First, we solve for $T_{0,1}(u)$, $T_{0,2}(u)$, and $T_{0,3}(u)$ according to (7.5.15). They are $T_{0,1}(u) = T_{0,2}(u) = 1$, $T_{0,3}(u) = u^2 + 1$. Then, we compute $Q_{1,i}(u)$, for $i = 1, 2, 3$, from (7.5.21) to get $Q_{1,1}(u) = 0$, $Q_{1,2}(u) = u + 1$ and $Q_{1,3}(u) = u^2$. Substituting into (7.5.20), we solve for $T_{1,i}(u)$, $i = 1, 2, 3$, $T_{1,1}(u) = u$, $T_{1,2}(u) = u^2 + u$, and $T_{1,3}(u) = u^2$. Finally, repeating the same procedure for $k = 2$, we get, $Q_{2,1}(u) = 1$, $Q_{2,2}(u) = 0$, $Q_{2,3}(u) = 0$; and $T_{2,1}(u) = 0$, $T_{2,2}(u) = u^2 + u + 1$, $T_{2,3}(u) = u + 1$. By (7.5.22), $T_1(u) = 7$, $T_2(u) = 6u^2 + 6u + 5$, $T_3(u) = 3u^2 + 4u + 5$.

Consider the polynomial $X(u) = u^6 + 3u^5 + 7u^3 + u^2 + 6u + 3$ defined over $Z(8)$. Its residues corresponding to $m_1(u)$, $m_2(u)$, and $m_3(u)$ are $x_1(u) = 5$, $x_2(u) = 5u^2 + 2u + 4$ and $x_3(u) = u^2 + 2u + 1$, respectively. The CRT-PR can be verified by computing $X(u)$ from (7.5.13).

7.5.2 CRT-PR in Matrix Form

It is clear from the form of CRT-PR that it is identical to the CRT-P, the only difference being that all the quantities are defined over the ring $Z(p^a)$ for CRT-PR instead of a field for CRT-P. However, it is the field that acts as a key to obtaining various results over the ring. The CRT-P was cast in a matrix-vector form in Section 3.5. This form is valid in the case of CRT-PR. In this case, expressions in (3.5.1) to (3.5.5) are defined modulo p^a. We present an example to illustrate the matrix-vector form.

Example 7.7 Let $M(u) = u^7 + 1 \bmod 2^3$. Factors of $M(u)$ over $Z(8)$ are $m_1(u) = u + 1$, $m_2(u) = u^3 + 2u^2 + 5u + 1$, and $m_3(u) = u^3 + 5u^2 + 2u + 1$. The polynomials $M_i(u)$ and $T_i(u)$, $i = 1, 2, 3$ have been found in Example 7.6. Based on the factors $m_1(u)$, $m_2(u)$ and $m_3(u)$, matrices $\mathbf{A}_i$, $i = 1, 2, 3$ can be computed. The joint matrix $\mathbf{A}$ is seen to be

$$\mathbf{A} = \begin{bmatrix} \mathbf{A}_1 \\ \mathbf{A}_2 \\ \mathbf{A}_3 \end{bmatrix} = \begin{bmatrix} 1 & 7 & 1 & 7 & 1 & 7 & 1 \\ \hline 1 & 0 & 0 & 7 & 2 & 1 & 5 \\ 0 & 1 & 0 & 3 & 1 & 7 & 2 \\ 0 & 0 & 1 & 6 & 7 & 3 & 1 \\ \hline 1 & 0 & 0 & 7 & 5 & 1 & 2 \\ 0 & 1 & 0 & 6 & 1 & 7 & 5 \\ 0 & 0 & 1 & 3 & 7 & 6 & 1 \end{bmatrix}.$$

Similarly, based on the polynomials $T_i(u)$ and $M_i(u)$, matrices $\mathbf{B}_i(u)$, $i = 1, 2, 3$ can be computed. The joint matrix $\mathbf{B}$ is seen to be

$$\mathbf{B} = \begin{bmatrix} \mathbf{B}_1 & \mathbf{B}_2 & \mathbf{B}_3 \end{bmatrix} = \left[\begin{array}{c|ccc|ccc} 7 & 5 & 2 & 6 & 5 & 5 & 3 \\ 1 & 5 & 5 & 2 & 2 & 5 & 5 \\ 7 & 3 & 5 & 5 & 6 & 2 & 5 \\ 1 & 2 & 3 & 5 & 5 & 6 & 2 \\ 7 & 3 & 2 & 3 & 6 & 5 & 6 \\ 1 & 2 & 3 & 2 & 5 & 6 & 5 \\ 7 & 6 & 2 & 3 & 3 & 5 & 6 \end{array}\right].$$

It can be verified that the product of matrices $\mathbf{A}$ and $\mathbf{B}$ over $Z(8)$ gives the identity matrix having dimension 7.

Above matrix representation over $Z(p^a)$ can be extended to matrix representation over $Z(M)$ using the CRT-I and the direct sum approach. This aspect will be analyzed in later sections. We end this section by stating that the NTTs are a very special case of the CRT-PR as described in this section and can be so derived by setting $M(u) = u^n - 1$ and placing the constraint that $u^n - 1$ factorize into n relatively prime degree-one factors over $Z(p^a)$.

7.6 Polynomial Algebra and CRT-PR: The Complex Case

This section focuses on the extension of CRT-PR for processing complex-valued integer sequences. Complex NTTs (CNTTs) turn out to be a special case of this general case and will be so described in Section 7.7. Once again, the analysis is based on factorization properties of polynomials defined over complex finite fields and their p-adic extension integer rings. The readers are referred to Section 6.4 for complex finite fields as they are fundamental to analyzing the structure of the complex integer rings.

The complex integer ring $CZ(M)$ is a generalization of the $Z(M)$. It consists of all complex integers of the type

$$W = U + j\,V, \tag{7.6.1}$$

where $U, V \in Z(M)$, and

$$j^2 \equiv -1 \bmod M. \tag{7.6.2}$$

The arithmetic operations of '+' and '·' in $CZ(M)$ are the same as those for complex numbers with the exception that both real and the imaginary parts are defined modulo M. The $CZ(M)$ can be expressed as the direct sum

$$CZ(M) = \sum_{\substack{\oplus \\ i=1}}^{t} CZ(q_i), \tag{7.6.3}$$

where $M = \prod_{i=1}^{t} q_i, q_i = p_i^{a_i}$, the standard factorization of M. There is a one-to-one correspondence between the complex integers $W \in CZ(M)$ and a vector of residues $W = [W_1\ W_2\ \ldots\ W_t]$, where

$$W_i \equiv W \bmod q_i,\ \ i = 1, 2, \ldots, t. \tag{7.6.4}$$

A useful interpretation of this description of $CZ(M)$ is that every element in $CZ(M)$ can be expressed as a polynomial of degree 1 over $Z(M)$. Here, j is an indeterminate. The arithmetic operations between two elements of $CZ(M)$ can equivalently be represented as operations between two polynomials in j of degree 1 modulo $(j^2 + 1, M)$. Similarly, any computation in $CZ(M)$ can be represented as a computation in $Z(M)$, where all the scalars are replaced by polynomials of degree 1 in j. Using the direct sum property of $CZ(M)$ in (7.6.3), we see that the overall computation in $CZ(M)$ can be written as a direct sum of computations in $CZ(q_i)$, $i = 1, 2, ..., t$. Consequently, in the remainder of this section, we focus on developing polynomial algebra when the various quantities are defined modulo $(j^2 + 1, q)$, that is, they are defined in $CZ(q)$, $q = p^a$.

We observe here that any algorithm designed for a computation over $Z(M)$ may also be used for the computation over $CZ(M)$. The main thrust of this section is in deriving polynomial algebra results that will exploit the number theoretic properties of $CZ(M)$ and thereby require fewer computations than the algorithms originally designed over $Z(M)$.

7.6.1 Computations Over Complex Integer Rings

In this section, we analyze factorization of $j^2 + 1$ over $Z(q)$, $q = p^a$, and the resulting CRT-PR based simplifications. Consider any computation defined in $CZ(q)$, where all the quantities are expressed as polynomials in j of degree 1 over $Z(q)$ modulo $(j^2 + 1, q)$. If $j^2 + 1$ factors into the product of two relatively prime polynomials over $Z(q)$, say, $j^2 + 1 = (j - A)(j - B) \bmod q$, then, based on CRT-PR, the computation can be performed in three steps:

1. Reduce the computation over $CZ(q)$ by evaluating the various quantities for $j = A$ and $j = B$. This results in two computations over $Z(q)$.

2. Perform the two computations over $Z(q)$.

3. Reconstruct the original computation over $CZ(q)$ from the two computations in $Z(q)$.

These three steps are based on a straightforward application of CRT-PR to evaluating computations defined mod $(j^2 + 1, q)$. The factorization of $j^2 + 1$ over $Z(q)$ begins with its factorization over $GF(p)$ and employs the Hensel's theorem (Theorem 7.5) in a recursive manner. Two degree-one polynomials $j - A$ and $j - B$ are relatively prime over $Z(q)$ (or equivalently over $GF(p)$) *iff* $(A - B, q) = 1$ (or equivalently $A \bmod p \neq B \bmod p$, that is, the factors are distinct mod p). Using the CRT-I, we get the following lemma.

Lemma 7.1 If there exist two integers A and B in $Z(M)$, such that $(A - B, M) = 1$ and $j^2 + 1 = (j - A)(j - B) \bmod M$, then a computation in $CZ(M)$ can be performed as a direct sum of two computations in $Z(M)$ obtained by setting $j = A$ and $j = B$ in the original computation.

As an example, let $M = 2^a + 1$, a being even. It is seen that $A = 2^{a/2}$, $B = -2^{a/2}$ satisfy the condition in Lemma 7.1. Therefore, any computation in $CZ(2^a + 1)$

can be written as a direct sum of two computations in $Z(2^a + 1)$. Using the CRT-PR, a polynomial $X(u)$ in $CZ(2^a + 1)$ can be written as

$$X(u, j) = j\,[-X_1(u) + X_2(u)]\,2^{a/2-1} - [X_1(u) + X_2(u)]\,2^{a-1},$$

where $X_1(u)$ and $X_2(u)$ are polynomials in $Z(M)$ obtained after substituting $j = A$ and $j = B$ in $X(u, j)$, respectively.

7.6.2 Factorization of $j^2 + 1$ Over $GF(p)$

For $p = 2$, $j^2 + 1 = (j + 1)^2$. Therefore, $j^2 + 1$ does not factor into two distinct factors over $GF(2)$. However, it does factor which implies that it is not irreducible over $GF(2)$ and the complex integers over $GF(2)$ along with the arithmetic operations defined mod $(j^2 + 1)$ do not constitute a finite field.

Now consider the factorization of $j^2 + 1$ over $GF(p)$, $p \neq 2$. The order of every non-zero element in $GF(p)$ divides $p - 1$. Since $p - 1$ is always even, factorization of $j^2 - 1$ always exists and is given by $j^2 - 1 = (j - 1)(j + 1)$. Therefore, factorization of $j^2 + 1$ over $GF(p)$ can be determined from the factorization of $(j^2 + 1)(j^2 - 1) = j^4 - 1$. The polynomial $j^2 + 1$ factors over $GF(p)$ *iff* $j^4 - 1$ has four distinct factors, that is, there is an element of order 4 in $GF(p)$. Recalling that the order of every non-zero element in $GF(p)$ divides $p - 1$, we get the following lemma.

Lemma 7.2 For $p \neq 2$, $j^2 + 1$ is irreducible over $GF(p)$ *iff* 4 is not a factor of $p - 1$ or $p = 4L + 3$ or $p \equiv 3 \bmod 4$. It factors into two relatively prime polynomials over $GF(p)$ *iff* $4 \mid (p - 1)$ or $p = 4L + 1$ or $p \equiv 1 \bmod 4$.

For $p = 4L + 1$, $j^2 + 1$ factorizes uniquely over $GF(p)$ and $Z(p^a)$ as

$$j^2 + 1 = (j - A_0)(j + A_0) \bmod p$$
$$j^2 + 1 = (j - A)(j + A) \bmod p^a.$$

Here A_0 and A are elements of order 4 in $GF(p)$ and $Z(p^a)$, respectively. If π is a primitive element in $GF(p)$, then A_0 and A are given by

$$A_0 = \pi^{(p-1)/4} \bmod p$$
$$A = A_0^{p^{a-1}} \bmod p^a.$$

For example, $j^2 + 1$ is irreducible over $GF(3)$ and, therefore, over $Z(3^a)$ as well. Over $GF(5)$, 2 is a primitive element and

$$j^2 + 1 = (j + 2)(j + 3) \bmod 5$$

which leads to

$$j^2 + 1 = (j + 7)(j + 18) \bmod 25$$

using (7.6.7) and (7.6.8). The factorization of $j^2 + 1$ over $GF(p)$, or equivalently over $Z(q)$, is summarized as follows.

$p = 2$	$j^2 + 1 = (j + 1)^2$
$p = 4L + 3$	$j^2 + 1$ is irreducible over $GF(p)$ and $Z(p^a)$.
$p = 4L + 1$	$j^2 + 1 = (j - A)(j - B)$, $A, B \in Z(p^a)$

We draw an important distinction between the irreducibility of $j^2 + 1$ over $GF(p)$, $p = 4L + 3$, and repeated factors of $j^2 + 1$ over $GF(2)$. In the former case, complex integers over $GF(p)$ constitute a finite field while in the latter case, no such property holds. This plays a crucial role in the factorization of polynomials over complex integer rings and in the design of fast algorithms for computing convolutions. Factorization properties of polynomials over $CZ(q)$, $q = p^a$, $p = 4L + 3$, CRT-PR and corresponding algorithms for computing convolutions form the topic of the following sections.

7.6.3 Factorization Over a Complex Integer Ring

Throughout this section, we assume that the polynomials under discussion are defined over $CZ(q)$, $q = p^a$, $p = 4L + 3$. In this case, $j^2 + 1$ is irreducible (not primitive) over $GF(p)$ and, therefore, $CZ(p)$ is same as $GF(p^2)$.

Definition 7.6 Let $f(u) = \sum_{i=0}^{n} c_i u^i$, $c_i \in CZ(q)$, $q = p^a$, a polynomial of degree n. The p-adic representation of $f(u)$ is given by

$$f(u) = \sum_{i=0}^{a-1} f_i(u) p^i, \tag{7.6.5}$$

where

$$f_i(u) = \sum_{k=0}^{n} c_{i,k} u^k, \; c_{i,k} \in GF(p^2). \tag{7.6.6}$$

This definition is based on the p-adic representation of a scalar $c \in CZ(q)$ as

$$c = c_R + jc_I = \sum_{i=0}^{a-1} \left(c_{i,R} + jc_{i,I}\right) p^i . \tag{7.6.7}$$

Definition 7.7 Given two polynomials $f(u)$ and $h(u)$, $h(u)$ has $f(u)$ as its *factor* whenever $h(u)$ and $f(u)$ can be written as $h(u) = f(u)\, q(u)$.

Definition 7.8 If two monic polynomials have no common monic factors, they are said to be *relatively prime* (or *coprime*).

Theorem 7.11 Let $M(u)$ be a polynomial defined over $CZ(p^a)$ and

$$M(u) = \prod_{i=1}^{\lambda} m_{0,i}(u) \bmod p$$

be known, where $m_{0,i}(u)$ and $m_{0,j}(u)$, $i \neq j$, are monic coprime polynomials defined over $GF(p^2)$. Then, there exists λ monic polynomials over $CZ(p^a)$, denoted by $m_i(u)$, such that

$$m_i(u) \equiv m_{0,i}(u) \bmod p$$

and

$$M(u) = \prod_{i=1}^{\lambda} m_i(u) \bmod p .$$

This theorem may be treated as an extension of the Hensel's theorem as the original theorem deals with polynomials defined over $Z(q)$. For the proof, the readers are referred to the proof of Hensel's theorem (Theorem 7.5). We observe that if a polynomial defined over $CZ(q)$ is reduced mod p, it results in a polynomial defined over $GF(p^2)$.

Theorem 7.12 Two monic polynomials defined over $CZ(p^a)$, denoted by $C_1(u)$ and $C_2(u)$, have no monic factors in common mod p^a if they have no monic factors in common mod p.

Theorem 7.13 A monic polynomial $M(u)$ defined over $CZ(p^a)$ is an irreducible polynomial *iff* $M_0(u) \equiv M(u) \bmod p$ is an irreducible polynomial over the finite field $GF(p^2)$.

We omit proofs of the theorem and discussion of the definitions as they are similar to the case when all the quantities are defined over $Z(q)$. The following theorem states the unique factorization property of the procedure described above.

Theorem 7.14 Unique factorization property. A monic polynomial defined over $CZ(p^a)$ has a unique factorization into a product of relatively prime monic polynomials that are either irreducible or power of an irreducible polynomial over $GF(p^2)$.

The procedure to obtain the factors of a given polynomial over $CZ(q)$ begins with its representation in p-adic form, computing its factorization $GF(p^2)$ (or equivalently, it may be given), and then apply Theorem 7.11 in a recursive manner. The procedure described in Section 7.4.1 remains valid here with the minor exception that all the computations need to be performed in $GF(p^2)$ instead of $GF(p)$. The coefficients of various polynomials in the p-adic representation are elements of $GF(p^2)$.

Example 7.7 Factorize $M(u) = u^5 - 1$ over $CZ(3^2)$. Clearly, $p = 3$ and $a = 2$. In Step 1, we get the p-adic form of $M(u)$: $M_0(u) = u^5 + 2$ and $M_1(u) = 2$. In Step 2, we get $A_0(u) = M_0(u) = u^5 + 2$ and $A_1(u) = M(u) = u^5 + 8$. The factorization of $M(u)$ over $GF(3^2)$ is given by (obtained using the techniques in Section 6.4)

$$M_0(u) = (u + 2)\,(u^2 + (2 + j)\,u + 1)\,(u^2 + (2 + 2j)\,u + 1).$$

Thus, $\lambda = 3$ and $B_{0,1}(u) = u + 2$, $B_{0,2}(u) = u^2 + (2 + j)\,u + 1$, $B_{0,3}(u) = u^2 + (2 + 2j)\,u + 1$. For $k = 1$, we get $A_0'(u) = (2 + j)\,u^4 + (1 + j)\,u^3 + (1 + 2j)\,u^2 + 2j\,u$ in Step 4. In Step 5, the three congruences to be solved are

$$\begin{cases} Q_{1,1}(u)(1+j)(1+2j) \equiv 1 \bmod (u+2,3) \\ Q_{1,2}(u)(u+2j) \equiv (1+2j)u + 2 \bmod \left(u^2 + (2+j)u + 1, 3\right) \\ Q_{1,3}(u)(u+j) \equiv ju + (2+j) \bmod \left(u^2 + (2+2j)u + 1, 3\right). \end{cases}$$

The solutions are $Q_{1,1}(u) = 2$, $Q_{1,2}(u) = u$, and $Q_{1,3}(u) = (1 + 2j)\,u$. Finally, the factorization of $A_1(u) = M(u)$ over $CZ(9)$ is given by (Step 6),

$$M(u) = (u + 8)\,(u^2 + (5 + j)\,u + 1)(u^2 + (5 + 8j)\,u + 1)$$

7.6.4 CRT-PR Over Complex Integer Rings

In this section, we present the complex version of CRT-PR. Once again, we assume that all the polynomials under discussion are defined over $CZ(q)$, $q = p^a$, $p = 4L + 3$.

Let $X(u)$ and $M(u)$ be two polynomials defined over $CZ(p^a)$. The polynomial $X(u)$ is an arbitrary polynomial over $CZ(p^a)$ having degree at most $n - 1$. It is necessary and sufficient that $M(u)$ be monic and of degree greater than $n - 1$. Assume that factorization of $M(u)$ over $CZ(p^a)$ has been done by following the steps listed in the previous section; that is,

$$M(u) = \prod_{i=1}^{\lambda} m_i(u) \bmod p^a. \qquad (7.6.8)$$

Recall that the monic factors $m_i(u)$ and $m_j(u)$, $i \neq j$, $1 \leq i, j \leq \lambda$, are relatively prime. This important property will lead to CRT-PR. Define a vector of polynomials, $[x_1(u)\ x_2(u)\ \ldots\ x_\lambda(u)]$ with the component $x_i(u)$ computed using

$$x_i(u) \equiv x(u) \bmod (m_i(u), p^a), \quad i = 1, 2, \ldots, \lambda. \qquad (7.6.9)$$

The CRT-PR delivers a unique relationship between the polynomial $X(u)$ and its vector representation. Theorem 7.15 plays a preliminary role in proving the one-to-one correspondence. The CRT-PR is stated in Theorem 7.16. The proofs of these theorems are the same as the ones for Theorems 7.9 and 7.10, respectively, and are omitted here.

Theorem 7.15 Consider the polynomials $X(u)$, $m_1(u)$ and $m_2(u)$, all defined over $CZ(p^a)$. Let $m_1(u)$ and $m_2(u)$ be monic and coprime mod p. If $m_1(u)$ is a factor of $X(u)$ and $m_2(u)$ is a factor of $X(u)$, then the statement "$m_1(u)\ m_2(u)$ divides $X(u)$" is true.

Theorem 7.16 CRT-PR. There exists an one-to-one correspondence between $X(u)$ and the vector representation $[x_1(u)\ x_2(u)\ \ldots\ x_\lambda(u)]$, as long as $deg(X(u)) < deg(M(u))$ and the monic factors of $M(u)$, denoted by $m_1(u), m_2(u), \ldots, m_\lambda(u)$, are relatively prime mod p. Here the entire theorem is defined over $CZ(p^a)$.

We can now express the direct sum property of the polynomials over $CZ(p^a)$ as

$$\langle X(u)\rangle_{M(u)} = \langle X(u)\rangle_{m_1(u)} \oplus \langle X(u)\rangle_{m_2(u)} \oplus \cdots \oplus \langle X(u)\rangle_{m_\lambda(u)}. \qquad (7.6.10)$$

The CRT-PR reconstruction

The procedure for CRT-PR reconstruction for polynomials defined over $CZ(q)$ remains the same as the one for polynomials defined over $Z(q)$ as described in Section 7.5.1. The following example is given to illustrate the reconstruction procedure.

Example 7.9 The monic factors of $u^5 - 1$ over $CZ(9)$ are $m_1(u) = u - 1$, $m_2(u) = u^2 + (5 + j)\, u + 1$, and $m_3(u) = u^2 + (5 + 8\, j)\, u + 1$. Therefore,

$$\begin{cases} M_1(u) = u^4 + u^3 + u^2 + u + 1 \\ M_2(u) = u^3 + (4+8j)u^2 + (5+j)u + 8 \\ M_3(u) = u^3 + (4+j)u^2 + (5+8j)u + 8. \end{cases}$$

Based on their p-adic forms, we have

$$\begin{cases} M_{0,1}(u) = u^4 + u^3 + u^2 + u + 1,\; M_{1,1}(u) = 0 \\ M_{0,2}(u) = u^3 + (1+2j)u^2 + (2+j)u + 2,\; M_{1,2}(u) = (1+2j)u^2 + u + 2 \\ M_{0,3}(u) = u^3 + (1+j)u^2 + (2+2j)u + 2,\; M_{1,3}(u) = u^2 + (1+2j)u + 2. \end{cases}$$

Equations (7.5.15), (7.5.17), and (7.5.19) are used to obtain $T_{0,i}(u)$, $Q_{1,i}(u)$ and $T_{1,i}(u)$, respectively, for $i = 1, 2, 3$. In this case, we get

$$i = 1 \begin{cases} T_{0,1}(u) = 2 \\ Q_{1,1}(u) = 0 \\ T_{1,1}(u) = 0, \end{cases}$$

$$i = 2 \begin{cases} T_{0,2}(u) = (2+j)u + 2 \\ Q_{1,2}(u) = (1+j)u + (1+j) \\ T_{1,2}(u) = (2+2j)u + 1, \end{cases}$$

$$i = 3 \begin{cases} T_{0,3}(u) = (2+j)u + 2 \\ Q_{1,3}(u) = ju + 1 \\ T_{1,3}(u) = 2u + 1. \end{cases}$$

The reconstruction polynomials are obtained as

$$T_1(u) = T_{0,1}(u) + 3\,T_{1,1}(u) = 2$$
$$T_2(u) = T_{0,2}(u) + 3\,T_{1,2}(u) = (8 + 7\,j)\,u + 5$$
$$T_3(u) = T_{0,3}(u) + 3\,T_{1,3}(u) = (8 + 2\,j)\,u + 5.$$

Finally,

$$T_1(u)\,M_1(u) = 2\,u^4 + 2\,u^3 + 2\,u^2 + 2\,u + 2$$
$$T_2(u)\,M_2(u) = (8 + 7\,j)\,u^4 + (8 + 2\,j)\,u^3 + (8 + 2\,j)\,u^2 + (8 + 7\,j)\,u + 4$$
$$T_3(u)\,M_3(u) = (8 + 2\,j)\,u^4 + (8 + 7\,j)\,u^3 + (8 + 7\,j)\,u^2 + (8 + 2\,j)\,u + 4.$$

It is easily verified that $T_i(u)\,M_i(u) \equiv 1 \bmod m_i(u)$, $i = 1, 2, 3$.

7.7 Number Theoretic Transforms: The Complex Case

We have seen that depending on the form of prime numbers in the standard factorization of M, there can be three ways to process polynomials defined over $CZ(M)$. If $p = 2$ is a factor, then algorithms for processing polynomials defined over $CZ(2^a)$ are the same as those for processing polynomials defined over $Z(2^a)$. Only the input and output quantities are defined over $CZ(2^a)$ instead of $Z(2^a)$. When $p = 4L + 1$ or $p \equiv 1 \bmod 4$, then a complex polynomial defined over $CZ(p^a)$ can be expressed as a direct sum of two polynomials defined over $Z(p^a)$. Last, when $p = 4L + 3$ or $p \equiv 3 \bmod 4$, then one can employ $GF(p^2)$ to carry out the factorization. In the following, we establish the complex NTTs (CNTTs) as a special case of CRT-PR when M is such that all the prime numbers in its standard factorization are of the form $p = 4L + 3$.

CNTTs arise when $M(u) = u^n - 1$ and we want $M(u)$ to factor into n relatively prime factors (each of degree 1) over $CZ(q)$, $q = p^a$, $p = 4L + 3$. Following the analysis in Section 7.6, such a factorization begins with the factorization of $M(u)$ over $GF(p^2)$. $M(u)$ factors into n degree one relatively prime polynomials *iff* there is an element of order n in $GF(p^2)$. This leads to the following theorem in a straightforward manner.

Theorem 7.17 Existence of CNTTs. A CNTT of length n exists in the complex integer ring $CZ(M)$ *iff* all the prime integers in the standard factorization of M are the type $p = 4L + 3$ and n satisfies the condition

$$n \mid (p^2 - 1),\; q = p^a. \qquad (7.7.1)$$

The remainder of this section is devoted to further connecting the factorization over $GF(p^2)$ and $CZ(q)$ and extending the Fermat's and Euler's theorems to $CZ(q)$ when $p = 4L + 3$.

Theorem 7.18 Generalization of Fermat's theorem. Let $A_0 = A_{0R} + jA_{0I}$ be an element of complex integer field $GF(p^2)$, $A_0 \neq 0$. Then,

$$A_0^{p^2-1} = 1 \bmod\left(j^2 + 1, p\right). \tag{7.7.2}$$

Proof: There are $p^2 - 1$ non-zero elements in $GF(p^2)$. Since the order of every non-zero element in $GF(p^2)$ divides $p^2 - 1$, we have the statement of the theorem.

This is generalization of the Fermat's theorem for integers in $GF(p)$.

Let us now analyze the elements in $CZ(q)$, $q = p^a$. There are q^2 such complex integers. Consider all those elements A in $CZ(q)$, $A = A_R + j\,A_I$, that have square of their absolute values coprime to p, that is, $\left(A_R^2 + A_I^2, p\right) = 1$. It is clear that if the p-adic expansion of A is

$$A = \sum_{k=0}^{a-1}\left[A_{kR} + jA_{kI}\right]p^k,\ A_{kR}, A_{kI} \in GF(p), \tag{7.7.3}$$

then $\left(A_R^2 + A_I^2, p\right) = \left(A_{0R}^2 + A_{0I}^2, p\right)$ which is either 1 or a multiple of p in which case $A_{0R}^2 + A_{0I}^2 \equiv 0 \bmod p$. Since $A_{0R}^2 + A_{0I}^2 = \left(A_{0R} + jA_{0I}\right)\left(A_{0R} - jA_{0I}\right)$, it is a product of two elements in $GF(p^2)$. Consequently, $A_{0R}^2 + A_{0I}^2 \equiv 0 \bmod p$ holds *iff* $A_{0R} = A_{0I} = 0$. Therefore, coefficient of p^0 is non-zero in the p-adic expansion of all such integers that have square of their absolute values coprime to p. There are $p^{2a} - p^{2(a-1)}$ such integers. It is straightforward to verify that the set of all such complex integers is closed under the operation of multiplication. This leads to the following theorem.

Theorem 7.19 Generalization of Euler's theorem. If $p = 4L + 3$, then for every complex integer A in $CZ(q)$, $q = p^a$, such that $\left(|A|^2, q\right) = 1$, we have

$$A^{p^{2(a-1)}\left(p^2-1\right)} \equiv 1 \bmod\left(j^2 + 1, p^a\right). \tag{7.7.4}$$

Expressions in (7.7.3) and (7.7.4) can be used to identify elements in $CZ(q)$ that have a given order starting from elements of the same order in $GF(p^2)$.

Let π be a primitive element in $GF(p^2)$. Then the following factorization holds in $GF(p^2)$,

$$u^n - 1 = \prod_{i=0}^{n-1}\left(u - B^i\right), \tag{7.7.5}$$

where

$$B = \pi^{\left(p^2-1\right)/n}. \tag{7.7.6}$$

This gives rise to a DFT (or an NTT) of length n in $GF(p^2)$. This statement combined with the extension of the Euler's theorem, gives rise to the following theorem.

Theorem 7.20 Given a DFT of length n, defined over $GF(p^2)$, p prime, there is always a CNTT of length n defined over $CZ(q)$, $q = p^a$, $p = 4L + 3$, $n \mid (p^2 - 1)$. This CNTT is based on the factorization

$$u^n - 1 = \prod_{i=0}^{n-1}\left(u - G^i\right), \tag{7.7.7}$$

where

$$G = \pi^{p^{2(a-1)}\left(p^2-1\right)/n} \bmod\left(j^2 + 1, p^a\right), \tag{7.7.8}$$

π being a primitive element in $GF(p^2)$. The overall root for the CNTT defined over $CZ(M)$ is given by

$$G = \underset{\oplus}{\sum_{i=1}^{t}} G_i, \tag{7.7.9}$$

where G_i is the root for $CZ(q_i)$, $p_i = 4L_i + 3$, $i = 1, 2, \ldots, t$. CNTTs exist for all values of n such that $n \left| \left(p_i^2 - 1\right)\right.$, $i = 1, 2, \ldots, t$.

Given a root G that generates a CNTT of length n over $CZ(q)$, G^k, $(k, n) = 1$, can also be used as a root for the CNTT of length n. Therefore, there are $\phi(n)$ possible values for G. For a CNTT of length n defined over $CZ(M)$, there are $\phi(n)^t$ values possible for the generator G. Starting from all the possible values for π, the primitive element in $GF(p^2)$, expressions in (7.7.8) and (7.7.9) may also be used to identify all the possible values for the root that can be used to generate CNTT of a specified length n over $CZ(M)$. This is very helpful in identifying those roots whose structure may lead to a simple hardware realization. We leave the determination of all the roots as an exercise for the readers.

Example 7.10 This example may be treated as a continuation of Example 6.8. Let $p = 3$, $q = 27$, and $n = 8$. The element $1 + j$ is primitive in $GF(3^2)$. Thus, $\pi = 1 + j$ and $a = 3$. Substituting these values in (7.7.8), we get $G = 16\ (1 + j)$ as a root that can be used to generate length 8 CNTT over $CZ(27)$. Similarly, for $q = 9$, $p = 3$, and $n = 8$, we have $G = 7\ (1 + j)$ as a root that can be used to generate length 8 CNTT over $CZ(9)$. Instead, if $2 + 2\,j$ is used as the primitive element in $GF(3^2)$, then we obtain $G = 2 + 2\,j$ and $G = 11 + 11\,j$ as roots that can be used to generate a length 8 CNTT over $CZ(27)$ and $CZ(9)$, respectively.

We end this section by stating that all Mersenne primes q (primes of the type $q = 2^p - 1$) satisfy the relation $q = 4L + 3$. Therefore, $CZ(q)$ is the same as $GF(p^2)$ for all Mersenne primes q. In this finite field, for $p > 2$, $2\,j$ and $1 + j$ are elements of order $4\,p$ and $8\,p$, respectively. Consequently, these roots can be used to define a CNTT of length $4\,p$ and $8\,p$, respectively.

7.8 Pseudo Number Theoretic Transforms

In many digital signal processing applications, it is easier to design hardware realization when the algebra is defined modulo an integer having the form $2^a + 1$ or $2^a - 1$. However, the standard factorization of M having this form may severely limit the length of the NTTs that exist. The pseudo NTTs are a way to circumvent this problem. Given an NTT or CNTT defined mod M, all the intermediate quantities may be processed mod $A \cdot M$ (a multiple of M) and then the final result may be reduced mod M. Consequently, except for the final stage, the entire hardware can be realized mod $A \cdot M$. The integer $A \cdot M$ can be chosen to have the form $2^a + 1$ or $2^a - 1$ or some other desired form. For example, for $M = 31 \cdot 151$, A can be chosen as 7 to get $A \cdot M = 2^{15} - 1$. Thus, a length 30 NTT defined mod $31 \cdot 151$ may be obtained in which all the intermediate quantities are processed mod $2^{15} - 1$. Note that if M itself were chosen to be $2^{15} - 1$, then the maximum length possible for a NTT is only 6. Similarly, if $M = 15{,}790{,}321$, then choosing $A = 17$ leads to $A \cdot M = 2^{28} + 1$.

Since M is prime, NTTs defined mod 15,790,321 exist for all values of length n such that $n \mid 15{,}790{,}320$. If M were chosen to be $2^{28} + 1$, then the length of the NTT gets limited to $n \mid 16$. Similar results hold for $2^{31} + 1 = 3 \cdot 715{,}827{,}883$.

In the case of CNTT, pseudo CNTTs may be defined to take advantage of the situation in many ways. Previous discussion in the context of NTTs is valid in case of CNTTs as well. In addition, there is another way to exploit the idea. Recall that if the prime numbers in the standard factorization of M are of the type $4L + 1$, then the entire computation over $CZ(M)$ can be carried out as a direct sum of two computations over $Z(M)$. Similarly, if the prime numbers in the standard factorization of M are of the type $4L + 3$, then CNTTs exist for all lengths n that divide $n \mid (p^2 - 1)$, p being any one of the primes that constitute M. Both of these situations may be desirable in certain computational environments. In most cases, the standard factorization of integers of the type $2^a + 1$ or $2^a - 1$ have primes of both types in their factorization. Exception to this statement are the integers of the type $2^a + 1$, a even, in which case all the prime factors are of the type $4L + 1$. In case the standard factorization consists of both types of primes, one may define M to be an integer consisting of purely one type of primes and define A as the product of the remaining primes. For example, using $2^{16} - 1 = 3 \cdot 5 \cdot 17 \cdot 257$, one may set $M = 5 \cdot 17 \cdot 257$ and $A = 3$ to express the entire computation over $CZ(5 \cdot 17 \cdot 257)$ as a direct sum of two computations over $Z(5 \cdot 17 \cdot 257)$ and carry it out mod $2^{16} - 1$ with the final reduction carried out mod $5 \cdot 17 \cdot 257$. Another example of the same type of formulation is $2^{32} - 1 = 3 \cdot 5 \cdot 17 \cdot 257 \cdot 65{,}537$, where $M = 5 \cdot 17 \cdot 257 \cdot 65{,}537$ and $A = 3$ can be used. The factorization

$$2^{42} - 1 = 3 \cdot 3 \cdot 7 \cdot 7 \cdot 43 \cdot 127 \cdot 337 \cdot 5{,}419$$

may be used with $M = 3 \cdot 3 \cdot 7 \cdot 7 \cdot 43 \cdot 127 \cdot 5{,}419$ and $A = 337$ to obtain the prime factors of the type $4L + 3$ and then the corresponding CNTT.

We end this section by stating that all the results stated in the context of pseudo NTTs and CNTTs are also valid in the context of CRT-PR and pseudo CRT-PR may be defined in a similar manner.

7.9 Polynomial Algebra and Direct Sum Properties in Integer Polynomial Rings

In previous sections, we have described the generalization of NTTs defined in finite integer rings. This is based on the CRT-PR, an extension of the CRT-P. The main idea is to express the given finite integer ring $Z(M)$ ($CZ(M)$) as a direct sum of finite integer rings $Z(q_i)$ $(CZ(q_i))$, $M = \prod_{i=1}^{t} p_i^{a_i} = \prod_{i=1}^{t} q_i$, the standard factorization of M. The CRT-PR is employed to process a sequence in $Z(M)$ $(CZ(M))$ by first expressing it as a direct sum of sequences in each of $Z(q_i)$ $(CZ(q_i))$, $i = 1, 2, \ldots, t$. All sequences are represented as polynomials for the ease of notation. This section builds further on the results by extending them to the domain of integer polynomial rings. The theory of polynomial factorization and the resulting direct sum property are studied in depth. Emphasis is on the theory and computational algorithms for processing sequences defined in finite integer and complex integer rings. Factorization properties of polynomials in extension rings are further developed. This leads to extensions of CRT-PR in integer polynomial rings. Several properties of algebraic interest are also established for the rings under study.

7.9.1 Mathematical Preliminaries

In this section, we are interested in developing the algebra for processing integer valued data sequences. The data sequences we encounter in this regard are defined in finite integer rings $Z(M)$ and $CZ(M)$. Given the standard factorization of M as

$$M = \prod_{i=1}^{t} p_i^{a_i} = \prod_{i=1}^{t} q_i, \tag{7.9.1}$$

S ($\mathsf{S} = Z(M)$ or $CZ(M)$) can be expressed as the direct sum of sub-rings

$$\mathsf{S} = \sum_{\substack{\oplus \\ i=1}}^{t} \mathsf{S}_i, \tag{7.9.2}$$

where $\mathsf{S}_i = Z(q_i)$ for $\mathsf{S} = Z(M)$ and $\mathsf{S}_i = CZ(q_i)$ for $\mathsf{S} = CZ(M)$. The CRT-I provides a basis for (7.9.2). Based on (7.9.1) and (7.9.2), a polynomial in S may be expressed as a direct sum of polynomials $X^{(i)}(u)$ in the following manner:

$$X(u) = \sum_{\substack{\oplus \\ i=1}}^{t} X^{(i)}(u), \tag{7.9.3}$$

where

$$X^{(i)}(u) \equiv X(u) \bmod q_i \tag{7.9.4}$$

for $\mathsf{S} = Z(M)$ and

$$X^{(i)}(u) \equiv X(u) \bmod (j^2 + 1, q_i) \tag{7.9.5}$$

for $\mathsf{S} = CZ(M)$. The complex polynomial is shown explicitly as a polynomial mod $j^2 + 1$ in (7.9.5).

The finite integer rings that we study in this section are the rings obtained as a polynomial extension of S_i. In this case, the elements of E_i are given by

$$\mathsf{E}_i = \left\{ A_i(\theta), A_i(\theta) = \sum_{k=0}^{m_i-1} A_{i,k}\theta^k, A_{i,k} \in \mathsf{S}_i \right\}. \tag{7.9.6}$$

The arithmetic operations of '·' and '+' between two elements in E_i are defined modulo a *monic* degree m_i polynomial $Q_i(\theta)$ having the form

$$Q_i(\theta) = \theta^{m_i} + \sum_{k=0}^{m_i-1} Q_{i,k}\theta^k, Q_{i,k} \in S_i\,. \tag{7.9.7}$$

E_i will be denoted by $Z_i(q, Q(\theta))$ for $\mathsf{S}_i = Z(q_i)$ and by $CZ_i(q, Q(\theta))$ for $\mathsf{S}_i = CZ(q_i)$. Note that, in general,

$$Z(M, Q(\theta)) \neq \sum_{\substack{\oplus \\ i=1}}^{t} Z_i(q, Q(\theta)) \tag{7.9.8}$$

$$Q(\theta) = \sum_{\substack{\oplus \\ i=1}}^{t} Q_i(\theta)\,. \tag{7.9.9}$$

The equality in (7.9.8) holds *iff*

$$m = m_1 = m_2 = \ldots = m_t. \tag{7.9.10}$$

In this case, $Q(\theta)$ in (7.9.9) is a monic degree m polynomial. In addition, if (7.9.10) is satisfied, then we can write $\mathsf{E} = Z(M, Q(\theta))$. It is clear from this description that $CZ(M)$ is a polynomial extension of $Z(M)$ with $Q(\theta) = \theta^2 + 1$.

In the remainder of this section, we derive CRT-PR in $Z_i(q, Q(\theta))$ and $CZ_i(q, Q(\theta))$ which may be employed as a tool to design algorithms for one- and higher-dimensional data sequences. This is based on extending various results described in earlier sections that are valid in $Z(q_i)$ and $CZ(q_i)$ to obtain new results in $Z_i(q, Q(\theta))$ and $CZ_i(q, Q(\theta))$. In the following, we first treat the case $\mathsf{E}_i = Z_i(q, Q(\theta))$ and then apply the results to analyze the case $\mathsf{E}_i = CZ_i(q, Q(\theta))$. Also the subscript i will be deleted. This will simplify notation without creating any confusion.

There are three types of polynomials that we encounter in (7.9.1) to (7.9.10):

(i) Monic polynomial $Q(\theta)$ with coefficients in $Z(q)$ or $CZ(q)$,
(ii) Monic polynomial $M(u)$ with coefficients in $Z(q, Q(\theta))$ or $CZ(q, Q(\theta))$, and
(iii) Input polynomial $X(u)$ with coefficients in $Z(q, Q(0))$ or $CZ(q, Q(\theta))$.

These polynomials play distinct roles in the direct sum and their structure is analyzed separately in the following subsections.

7.9.2 Polynomial Rings and Fields

In this section, we analyze the structure of the rings $Z(q, Q(\theta))$ and $CZ(q, Q(\theta))$, $deg(Q(\theta)) = m$. The mathematical properties of the monic polynomial $Q(\theta)$ are based on its p-adic expansion given by

$$Q(\theta) = \sum_{l=0}^{a-1} Q_l(\theta) p^l \tag{7.9.11}$$

where

$$Q_l(0) - \sum_{i=0}^{m} q_{l,i} 0^i, q_{l,i} \in GF(p), \tag{7.9.12}$$

$q_{l,i}$ being the l-th coefficient in the p-adic expansion of Q_i. Also, let

$$Q^{(k)}(\theta) = Q(\theta) \bmod p^k = \sum_{l=0}^{k-1} Q_l(\theta) p^l, \; k = 1,2,\ldots,a. \tag{7.9.13}$$

It is clear that $Q^{(a)}(\theta) = Q(\theta)$. Since $Q(\theta)$ is monic, so are the polynomials $Q_0(\theta)$, …, $Q^{(k)}(\theta)$, $k = 1, 2, \ldots, a$. In addition, $m = deg(Q(\theta)) = deg(Q_0(\theta)) = deg(Q^{(a)}(\theta))$. Also, $deg(Q_l(\theta)) < m$, $l = 1, 2, \ldots, a - 1$.

Any field is a unique factorization domain. Let $Q_0(\theta)$ have the unique factorization

$$Q_0(\theta) = \prod_{i=1}^{\gamma} \left[r_{0,i}(\theta) \right]^{b_i} \bmod p = \prod_{i=1}^{\gamma} Q_{0,i}(\theta) \bmod p, \tag{7.9.14}$$

where $r_{0,i}(\theta)$ are irreducible in $GF(p)$. The factors $Q_{0,i}(\theta)$, $i = 1, 2, ..., \gamma$, are assumed to be monic without any loss in generality. Unique factorization of $Q(\theta)$ (recall Hensel's Theorem) is given by

$$Q(\theta) = \prod_{i=1}^{\gamma} Q_i(\theta) \bmod q, \tag{7.9.15}$$

where $Q_i(\theta)$ are monic and

$$Q_i(\theta) \equiv Q_{0,i}(\theta) \bmod p. \tag{7.9.16}$$

Unique factorization of $Q(\theta)$ in $Z(q)$ must be interpreted carefully. All the polynomials are monic in (7.9.15) and (7.9.16). In addition, $Q_i(\theta)$ is unique as it satisfies (7.9.16). Those $Q_i(\theta)$ that are obtained from $Q_{0,i}(\theta)$ with $b_i > 1$ may factor further in $Z(q)$ and such factorization is non-unique in general. Therefore, we restrict our attention to factorization properties only within the framework of (7.9.14) to (7.9.16). An algorithm to compute $Q_i(\theta)$ starting from $Q_{0,i}(\theta)$ is a recursive method based on Hensel's Theorem.

Given the factorization of $Q(\theta)$ in (7.9.15), the CRT-PR studied in sections 7.5 and 7.6 establishes a one-to-one correspondence between $Z(q, Q(\theta))$ and $Z(q, Q_i(\theta))$, $i = 1, 2, ..., \gamma$, that is,

$$Z(q, Q(\theta)) = \sum_{\substack{\oplus \\ i=1}}^{\gamma} Z(q, Q_i(\theta)). \tag{7.9.17}$$

This will be employed in later chapters to derive fast algorithms for computing convolutions in $Z(q)$ and subsequently in $Z(M)$. If $p \equiv 3 \bmod 4$, then $CZ(p)$ is the same as $GF(p^2)$ and we have the following direct sum:

$$CZ(q, Q(\theta)) = \sum_{\substack{\oplus \\ i=1}}^{\gamma} CZ(q, Q_i(\theta)). \tag{7.9.18}$$

In (7.9.18), $Q(\theta)$ is a monic polynomial with coefficients in $CZ(q)$. The algorithm to compute the factorization in $CZ(q)$ starting from the factorization in $CZ(p)$ has already been summarized in the earlier sections of this chapter.

These direct sums imply that we only need to study $Q(\theta)$ having the form $Q(\theta) \equiv Q_0(\theta) \bmod p = [r_0(\theta)]^b \bmod p$. In the present context, we are interested in the lowest degree polynomial $Q(\theta)$ for given values of q and n which result in the direct sum property. This leads to $b = 1$ or the condition that $Q_0(\theta)$ be irreducible over $GF(p)$. Consequently, $Q(\theta)$ is irreducible over $Z(q)$. Throughout the remainder of this section, $Q(\theta)$ will be assumed to be a *monic irreducible* degree m polynomial. Some properties associated with $Z(p, Q_0(\theta))$ that will be used in the sequel are as follows:

Property 1. $Z(p,Q_0(\theta))$ is same as $GF(r)$, $r = p^m$, $deg(Q_0(\theta)) = m$.

Property 2. The order of every non-zero element in $Z(p, Q_0(\theta))$ divides $r - 1$. Also if $n \mid (r - 1)$, then there is an element of order n in $Z(p, Q_0(\theta))$.

Property 3. $Q_0(\theta)$ can be expressed as

$$Q_0(\theta) = \prod_{i=0}^{m-1} \left(\theta - \beta^{p^i}\right). \tag{7.9.19}$$

$\theta = \beta$ is a root of $Q_0(\theta)$ in $Z(p, Q_0(\theta))$ and an element of order n, $n \mid (r - 1)$. θ and β will be used interchangeably in our analysis. $Q_0(\theta)$ is also called the *minimum polynomial* of β.

Property 4.

$$Q_0(\theta) \mid (\theta^n - 1),\ Q_0(\theta) \nmid (\theta^i - 1),\ i < n. \tag{7.9.20}$$

Property 5. A primitive element α in $Z(p, Q_0(\theta))$ has order $r - 1$ and β, an element of order n, can be written as $\beta = \alpha^a$, $a = (r - 1)/n$. Every element in $Z(p, Q_0(\theta))$ can be expressed as a polynomial in θ of degree less than m with coefficients in $Z(p)$. If $\beta = \alpha$ or $\beta = \alpha^i$, $(i, r - 1) = 1$, then β is a primitive element and $Q_0(\theta)$ is a primitive polynomial.

Property 6. Given β, β^{p^i}, $i = 1, 2, ..., m-1$, are known as its conjugates.

Property 7.

$$u^{r-1}-1=\prod_{i=1}^{r-1}\left(u-\alpha^{i}\right)\bmod\left(Q_0(\theta),p\right) \tag{7.9.21}$$

$$u^{n}-1=\prod_{i=1}^{n}\left(u-\theta^{i}\right)\bmod\left(Q_0(\theta),p\right). \tag{7.9.22}$$

$Z(p, Q_0(\theta))$, being a finite field, is a unique factorization domain. If $M_0(u)$ is a polynomial with coefficients in $Z(p, Q_0(\theta))$, then $M_0(u)$ has unique factorization

$$M_0(u)=\prod_{i=1}^{\lambda}\left[M_{0,i}(u)\right]^{b_i}=\prod_{i=1}^{\lambda}m_{0,i}(u)\bmod\left(Q_0(\theta),p\right). \tag{7.9.23}$$

Given $Q_0(\theta)$ and the associated $Z(p, Q_0(\theta))$, we now turn to $Q(\theta)$ in (7.9.16) and consider finding the remaining polynomials in (7.9.16). If choice of $M(u)$ is arbitrary, it does not matter how $Q_k(\theta)$, $k = 1, 2, ..., a-1$ are chosen. In some applications, we are interested in factorization of $u^n - 1$ over $Z(q, Q(\theta))$. Based on (7.9.21), (7.9.22), and Property 4, we conclude that in such situations, $Q(\theta) \mid (\theta^n - 1) \bmod q$. This is accomplished by using Hensel's Theorem. In this case, we first write $(\theta^n - 1) = Q_0(\theta)\, T_0(\theta) \bmod p$ and then use Hensel's Theorem.

Now consider $CZ(q)$. In the p-adic expansion in $CZ(q)$, the coefficients take values in $GF(p^2)$, $p \equiv 3 \bmod 4$. All the equations are valid in $CZ(q)$ except that instead of mod p, mod q, and mod $(p, Q_0(\theta))$, they are defined mod $(j^2 + 1, p)$, mod $(j^2 + 1, q)$, and mod $(j^2 + 1, Q_0(\theta), q)$, respectively. In this case, $Q_0(\theta)$ is a degree m irreducible polynomial in $CZ(p^2)$ and gives rise to $CZ(q, Q(\theta))$ such that $\theta^n - 1 = Q_0(\theta)\, T_0(\theta) \bmod (p, j^2 + 1)$, $n \mid (r-1)$, $r = p^{2m}$. The following three examples are presented to illustrate the ideas presented in this section.

Example 7.11 Consider the factorization of

$$A(\theta) = \theta^5 + 2\,\theta^4 + 6\,\theta^3 + 3\,\theta^2 + 3\,\theta + 7 \bmod 8.$$

Here $p = 2$, $a = 3$, and $q = 2^3$. The factorization of $A(\theta)$ in $GF(2)$ (which is assumed to be known) is given by

$$A_0(\theta) = \theta^5 + \theta^2 + \theta + 1 = A_{0,1}(\theta)\, A_{0,2}(\theta) = (\theta^2 + 1)\,(\theta^3 + \theta + 1) \bmod 2.$$

In this case,

$A_1(\theta) = \theta^4 + \theta^3 + \theta^2 + \theta + 1,$
$A^{(2)}(\theta) = \theta^5 + 2\,\theta^4 + 2\,\theta^3 + 3\,\theta^2 + 3\,\theta + 3,$
$A_2(\theta) = \theta^3 + 1.$

In order to obtain the factors in $Z(4)$, we substitute for the various quantities in the procedure described in Section 7.4.1 and solve the resulting congruences. This gives $A_{1,1}(\theta) = \theta + 1$ and $A_{1,2}(\theta) = \theta$. Thus,

$$A_1^{(2)}(\theta) = A_{0,1}(\theta) + 2A_{1,1}(\theta) = \theta^2 + 2\theta + 3$$
$$A_2^{(2)}(\theta) = A_{0,2}(\theta) + 2A_{1,2}(\theta) = \theta^3 + 3\theta + 1\,.$$

It is verified that

$$A^{(2)}(\theta) = A(\theta)\bmod 4 = A_1^{(2)}(\theta)A_2^{(2)}(\theta)\bmod 4 = \theta^5 + 2\theta^4 + 2\theta^3 + 3\theta^2 + 3\theta + 3\,.$$

Finally,

$$A^{(3)}(\theta) = A(\theta) = A_1^{(3)}(\theta)A_2^{(3)}(\theta) = \left(\theta^2 + 2\theta + 3\right)\left(\theta^3 + 3\theta + 5\right)\bmod 8\,.$$

Example 7.12 Consider the polynomial rings $Z(p, \theta^2 + 1)$, $Z(\theta^2 + \theta + 1)$, and $CZ(p, \theta^2 + \theta + 1)$. As stated earlier, $Z(q, \theta^2 + 1)$ is the same as $CZ(q)$ and $\theta^2 + 1$ is irreducible over $Z(q)$ *iff* $p \equiv 3 \bmod 4$. Consider $Z(p, \theta^2 + \theta + 1)$. Since $\theta^3 - 1 = (\theta - 1)(\theta^2 + \theta + 1)$ for every field, $\theta^2 + \theta + 1$ is irreducible over $Z(q)$ *iff* there is no element in $GF(p)$ of order 3. The order of every element in $GF(p)$ divides $p - 1$. Therefore, $\theta^2 + \theta + 1$ is irreducible over $Z(q)$ *iff* $3 \nmid (p-1)$ or $p \equiv 2 \bmod 3$. Note that for $p = 3$, $\theta^2 + \theta + 1 = (\theta + 2)^2 \bmod 3$. When they exist, $Z(p, \theta^2 + 1)$ and $Z(\theta^2 + \theta + 1)$ are finite fields with the order of every element a factor of $p^2 - 1$. Interestingly, for $p > 3$, $24 \mid (p^2 - 1)$. In both cases, $Q(\theta) = Q_0(\theta)$ and $Q_l(\theta) = 0$, $l = 1, 2, ..., a - 1$. Now, consider $CZ(p, \theta^2 + \theta + 1)$. For $p = 2$, $j^2 + 1 = (j + 1)^2 \bmod 2$ and for $p = 3$, $\theta^2 + \theta + 1 = (\theta + 2)^2 \bmod 3$. Therefore, we are restricted to $p \neq 2, 3$. $CZ(p)$ is the same as $GF(p^2)$ for $p \equiv 3 \bmod 4$ with order of every element a factor of $p^2 - 1$. $(\theta^2 + \theta + 1)$ is irreducible over $GF(p^2)$ *iff* there is no element in $GF(p^2)$ having order 3. For $p \neq 2, 3$, $3 \mid (p^2 - 1)$. Therefore, $(\theta^2 + \theta + 1)$ is always reducible over $GF(p^2)$, $p \neq 2, 3$. Consequently, we derive *no advantage* by extending $CZ(q)$ with $Q(\theta) = \theta^2 + \theta + 1$ or by extending $Z(q, j^2 + j + 1)$ with $Q(\theta) = \theta^2 + 1$. None of the extension rings correspond to an extension of a finite field.

Example 7.13 Consider $GF(2^4) = Z(2, Q_0(\theta))$, $Q_0(\theta) = \theta^4 + \theta + 1$. Since $Q_0(\theta)$ is a primitive polynomial in $GF(2)$, $Q_0(\theta) \mid (\theta^{15} - 1) \bmod 2$ and $Q_0(\theta) \nmid (\theta^i - 1) \bmod 2$, $i < 15$. This results in the factorization

$$u^{15} - 1 = \prod_{i=1}^{15} \left(u - \theta^i\right) \bmod \left(\theta^4 + \theta + 1, 2\right). \tag{7.9.24}$$

In order to write a similar factorization of $u^{15} - 1$ in $Z(4, Q(\theta))$, we must have $Q(\theta) \mid (\theta^{15} - 1) \bmod 4$. This is accomplished by first writing

$$\theta^{15} - 1 = Q_0(\theta)\, T_0(\theta) \bmod 2.$$

Given $Q_0(\theta) = \theta^4 + \theta + 1$, $T_0(\theta)$ is found by polynomial division in $GF(2)$ as

$$T_0(\theta) = \theta^{11} + \theta^8 + \theta^7 + \theta^5 + \theta^3 + \theta^2 + \theta + 1.$$

Based on Hensel's Theorem, we write

$$\theta^{15} - 1 = Q(\theta)\, T(\theta) \bmod 4 = [Q_0(\theta) + 2\, Q_1(\theta)]\, [T_0(\theta) + 2\, T_1(\theta)] \bmod 4.$$

This leads to

$$Q_0(\theta)T_1(\theta) + Q_1(\theta)T_0(\theta) = \left\{2^{-1}\left[\left(\theta^{15} - 1\right) - Q_0(\theta)T_0(\theta)\right]\right\} \bmod 2 = \sum_{i=0}^{12} \theta^i - \theta^{10}. \tag{7.9.25}$$

Given $Q_0(\theta)$ and $T_0(\theta)$, Euclid's algorithm is used to obtain

$$Q_1(\theta) = \theta^2 + \theta,$$
$$T_1(\theta) = \theta^9 + \theta^7 + \theta^4 + \theta^3 + \theta + 1,$$

and

$$Q(\theta) = \theta^4 + 2\theta^2 + 3\theta + 1,$$
$$T(\theta) = \theta^{11} + 2\theta^9 + \theta^8 + 3\theta^7 + \theta^5 + 2\theta^4 + 3\theta^3 + \theta^2 + 3\theta + 3.$$

Thus we can write

$$u^{15} - 1 = \prod_{i=1}^{15} \left(u - \theta^i\right) \bmod \left(\theta^4 + 2\theta^2 + 3\theta + 1, 4\right). \tag{7.9.26}$$

Arithmetic in $Z(q, Q(\theta))$. There are three arithmetic operations in $Z(q, Q(\theta))$, namely,

(1) '+' (or subtraction),

(2) '·' and

(3) division.

The operations of '+' and '·' are straightforward as they are defined mod $(q, Q(\theta))$. Division in $Z(q, Q(\theta))$ may be defined as: given $A(\theta) \in Z(q, Q(\theta))$, find $B(\theta)$ such that

$$A(\theta)\, B(\theta) \equiv 1 \bmod (Q(\theta), q). \tag{7.9.27}$$

Consider the general congruence in $Z(q, Q(\theta))$ given by

$$A(\theta)\, B(\theta) \equiv C(\theta) \bmod (Q(\theta), q) \tag{7.9.28}$$

for given $A(\theta)$ and $C(\theta)$ in $Z(q, Q(\theta))$. Taking mod p of both sides in (7.9.28) we get

$$A_0(\theta)\, B_0(\theta) \equiv C_0(\theta) \bmod (Q_0(\theta), p). \tag{7.9.29}$$

Therefore, a necessary condition for $B(\theta)$ to exist is that $A_0(\theta) \neq 0$ if $C_0(\theta) \neq 0$. In general, if $A_i(\theta) = 0$ for $i < b$, $A_b(\theta) \neq 0$ and $C_j(\theta) = 0$ for $j < c$, $C_c(\theta) \neq 0$, then a necessary condition for the solution to exist is that $b \leq c$. This is also a sufficient condition for the solution. Such a solution is unique in $Z(q, Q(\theta))$ if $b = 1$. For the inverse of $A(\theta) \in Z(q, Q(\theta))$, a necessary and sufficient condition is $A_0(\theta) \neq 0$. This discussion can be summarized in the form of a lemma as follows.

Lemma 7.3 A unique solution to the congruence $A(\theta)\, B(\theta) \equiv C(\theta)$ mod $(Q(\theta), q)$ exists *iff* $A_0(\theta) \neq 0$. If $C(\theta) = 1$, then $B(\theta)$ is called the inverse of $A(\theta)$. The inverse is unique whenever $(A_0(\theta) \neq 0)$ exists. These statements are also valid in $CZ(q, Q(\theta))$.

In the following, we describe a method to solve the congruence in (7.9.28) when $A_0(\theta) \neq 0$. Taking mod p, we get (7.9.29) which is solved using the Euclid's algorithm in $GF(p)$ by finding $B_0(\theta)$ and $D_0(\theta)$ such that

$$A_0(\theta)\, B_0(\theta) + D_0(\theta)\, Q_0(\theta) = C_0(\theta) \bmod p. \tag{7.9.30}$$

Note that $B_0(\theta)$ and $D_0(\theta)$ always exist. Based on (7.9.30), we write the solution mod p^{k+1} as

$$A^{(k+1)}(\theta)\, B^{(k+1)}(\theta) + D^{(k+1)}(\theta)\, Q^{(k+1)}(\theta) = C^{(k+1)}(\theta) \bmod p^{k+1},\quad k = 1, 2, ..., a-1$$

or

$$[A^{(k)}(\theta) + p^k A_k(\theta)]\, [B^{(k)}(\theta) + p^k B_k(\theta)]$$
$$+ [D^{(k)}(\theta) + p^k D_k(\theta)]\, [Q^{(k)}(\theta) + p^k Q_k(\theta)] = C^{(k)}(\theta) + p^k C_k(\theta) \bmod p^{k+1}, \tag{7.9.31}$$

where $B_k(\theta)$ and $D_k(\theta)$ are the unknown quantities. Expanding and rearranging the terms, we get

$$A_0(\theta)\, B_k(\theta) + D_k(\theta)\, Q_0(\theta) = E_k(\theta) \bmod p, \tag{7.9.32}$$

where

$$E_k(\theta) = \{-A_k(\theta)\, B_0(\theta) - D_0(\theta)\, Q_k(\theta) + C_k(\theta)$$
$$+ \{p^{-k}\, [C^{(k)}(\theta) - A^{(k)}(\theta)\, B^{(k)}(\theta) - D^{(k)}(\theta)\, Q^{(k)}(\theta)]\}\} \bmod p.$$

Once again, (7.9.32) can be solved uniquely for $B_k(\theta)$ and $D_k(\theta)$ using Euclid's algorithm. Note that all the polynomials are defined modulo $Q(\theta)$ and therefore have degree less than m.

7.9.3 Monic Polynomial Factorization in $Z(q, Q(\theta))$

In this section, we describe factorization properties of a monic degree n polynomial $M(u)$ defined in $Z(q, Q(\theta))$ starting from its monic factorization in $GF(r) = Z(p, Q_0(\theta))$, $r = p^m$, $q = p^a$, $deg(Q_0(\theta)) = m$. We begin by defining the various terms of interest.

Definition 7.9 Let $F(u) = \sum_{i=0}^{n} F_i u^i, F_i \in Z(q, Q(\theta))$, a polynomial of degree n. The p-adic expansion of $F(u)$ is given by

$$F(u) = \sum_{i=0}^{n} F_i u^i = \sum_{l=0}^{a-1} F_l(u) p^l, \tag{7.9.33}$$

where

$$F_l(u) = \sum_{i=0}^{n} f_{l,i} u^i, f_{l,i} \in Z(q, Q_0(\theta)). \tag{7.9.34}$$

This is based on the p-adic expansion of F_i as $F_i = \sum_{l=0}^{a-1} f_{l,i} p^l$. Also define $F^{(k)}(u)$ as

$$F^{(k)}(u) = F(u) \bmod p^k = \sum_{l=0}^{k-1} F_l(u) p^l, k = 1,2,\ldots,a. \tag{7.9.35}$$

Definition 7.10 Given two polynomials $F(u)$ and $H(u)$, $H(u)$ has $F(u)$ as its *factor* whenever $H(u)$ and $F(u)$ can be written as $H(u)=F(u)\,G(u)$.

Definition 7.11 A monic polynomial is termed *irreducible* if it has no other monic polynomial as its factor except itself.

Definition 7.12 If two monic polynomials have no common monic factors, they are said to be *relatively prime* or coprime.

Theorem 7.21 Two monic polynomials $F(u)$ and $H(u)$ defined in $Z(q, Q(\theta))$ have no monic factors in common in $Z(q, Q(\theta))$ if $F_0(u)$ and $H_0(u)$ have no monic factors in common in $Z(q, Q(\theta))$.

Proof: We prove the theorem by contradiction. If $A(u)$ is a common factor between $F(u)$ and $H(u)$, then

$$F(u) = A(u)\, R(u) \bmod (Q(\theta), q),$$
$$H(u) = A(u)\, S(u) \bmod (Q(\theta), q).$$

Taking mod $(Q_0(\theta), p)$ of the above equation implies that $A_0(u) \neq 0$ is a common factor between $F_0(u)$ and $H_0(u)$. This establishes the theorem.

This theorem establishes only a sufficient condition. $A_0(u) \neq 0$ being a common factor between $F_0(u)$ and $H_0(u)$ does not imply that $A(u)$ is a common factor between $F(u)$ and $H(u)$. In fact, $A(u)$ may not be a factor of either $F(u)$ or $H(u)$.

Theorem 7.22 A monic polynomial $M(u)$ defined in $Z(q,Q(\theta))$ is irreducible *iff* $M_0(u)$ is irreducible in $Z(p, Q_0(\theta))$.

Proof: If $M(u)$ factorizes in $Z(q, Q(\theta))$ in terms of monic factors as

$$M(u) = A(u)\, B(u) \bmod (Q(\theta), q),$$

then

$$M_0(u) = A_0(u)\, B_0(u) \bmod (Q_0(\theta), p),$$

thereby implying that $M_0(u)$ factorizes in $Z(p, Q_0(\theta))$. If $M_0(u)$ factorizes in $Z(p, Q_0(\theta))$ in terms of $A_0(u)$ and $B_0(u)$ as shown, then by extended Hensel's Theorem (proved in Theorem 7.23 in the following), $M(u)$ factorizes in $Z(q, Q(\theta))$ as well. This proves the theorem.

Theorem 7.23 Extended Hensel's Theorem. Let $M(u)$ be a monic polynomial defined in $Z(q, Q(\theta))$ and let $M_0(u)$ factorize as

$$M_0(u) = A_0(u)\, B_0(u) \bmod (Q_0(\theta), p) \tag{7.9.36}$$

where $A_0(u)$ and $B_0(u)$ are monic relatively prime polynomial in $Z(p, Q_0(\theta))$. Then there exist unique monic polynomials $A^{(k+1)}(u)$ and $B^{(k+1)}(u)$ in $Z(p^{k+1}, Q^{(k+1)}(\theta))$ such that

$$M^{(k+1)}(u) = A^{(k+1)}(u)\, B^{(k+1)}(u) \bmod (Q^{(k+1)}(\theta),\ p^{k+1}). \tag{7.9.37}$$

These polynomials satisfy the congruence

$$A^{(k+1)}(u) = A_0(u) \bmod (Q_0(\theta), p), \tag{7.9.38}$$
$$B^{(k+1)}(u) = B_0(u) \bmod (Q_0(\theta), p). \tag{7.9.39}$$

Proof: We first describe a procedure for obtaining $A^{(k+1)}(u)$ and $B^{(k+1)}(u)$ in (7.9.38) and (7.9.39) starting from $A_0(u)$ and $B_0(u)$ in (7.9.36). Let us express intermediate polynomials required in the algorithm as

$$\begin{aligned}
M^{(k+1)}(u) &\equiv M(u) \bmod p^{k+1} = M^{(k)}(u) + p^k M_k(u),\\
A^{(k+1)}(u) &\equiv A(u) \bmod p^{k+1} = A^{(k)}(u) + p^k A_k(u),\\
B^{(k+1)}(u) &\equiv B(u) \bmod p^{k+1} = B^{(k)}(u) + p^k B_k(u),\\
Q^{(k+1)}(\theta) &\equiv Q(\theta) \bmod p^{k+1} = Q^{(k)}(\theta) + p^k Q_k(\theta),\ \ k = 1, 2, \ldots, a-1.
\end{aligned} \tag{7.9.40}$$

Given (7.9.36) and (7.9.40) we are interested in the solution of (7.9.37) expressed as

$$\begin{aligned}
M^{(k)}(u) + p^k M_k(u) &= [A^{(k)}(u) + p^k A_k(u)]\, [B^{(k)}(u) + p^k B_k(u)]\\
&\qquad \bmod (Q^{(k)}(\theta) + p^k Q_k(\theta), p^{k+1}),
\end{aligned} \tag{7.9.41}$$

based on the solution at the previous stage

$$M^{(k)}(u) = A^{(k)}(u)\, B^{(k)}(u) \bmod (Q^{(k)}(\theta), p^k). \tag{7.9.42}$$

Expanding (7.9.41) and leaving $A_k(u)$ and $B_k(u)$ in the *rhs*, we get

$$M^{(k)}(u) + p^k M_k(u) - A^{(k)}(u)\, B^{(k)}(u)$$
$$= p^k [A^{(k)}(u)\, B_k(u) + A_k(u)\, B^{(k)}(u)] \bmod (Q^{(k)}(\theta) + p^k Q_k(\theta), p^{k+1}) \quad (7.9.43)$$

We now determine

$$A^{(k)}(u)\, B^{(k)}(u) \bmod (Q^{(k+1)}(\theta), p^{k+1})$$

from

$$A^{(k)}(u)\, B^{(k)}(u) \bmod (Q^{(k)}(\theta), p^k)$$

in (7.9.42). Equation (7.9.42) implies that there exists a polynomial $D^{(k)}(u)$ such that

$$A^{(k)}(u)\, B^{(k)}(u) + D^{(k)}(u)\, Q^{(k)}(\theta) = M^{(k)}(u) \bmod p^k. \quad (7.9.44)$$

Above equation can be interpreted as a set of n simultaneous congruences in polynomials in θ. Based on (7.9.44), we express

$$A^{(k)}(u)\, B^{(k)}(u) + [D^{(k)}(u) + p^k S_k(u)][Q^{(k)}(\theta) + p^k Q_k(\theta)]$$
$$= M^{(k)}(u) + p^k W_k(u) \bmod p^{k+1}, \quad (7.9.45)$$

where $S_k(u)$ and $W_k(u)$ are the two unknown polynomials in (7.9.45). Simplifying (7.9.45), we get

$$-Q_0(\theta)\, S_k(u) + W_k(u) = \{[A^{(k)}(u)\, B^{(k)}(u)$$
$$+ D^{(k)}(u)\, Q^{(k)}(\theta) - M^{(k)}(u)]\, p^{-k} + D_0(u)\, Q_k(\theta)\} \bmod p. \quad (7.9.46)$$

All the polynomials in the *rhs* are known and (7.9.46) can be solved for $S_k(u)$ and $W_k(u)$. Let $A_k'(u)$ denote the *rhs* of (7.9.46). Then

$$W_k(u) \equiv A_k'(u) \bmod (p, Q_0(\theta)) \quad (7.9.47)$$

and

$$S_k(u) = \{-[Q_0(\theta)]^{-1}[A_k'(u) - W_k(u)]\} \bmod p\,. \quad (7.9.48)$$

Thus we have

$$A^{(k)}(u)\,B^{(k)}(u) = M^{(k)}(u) + p^k\,W_k(u) \bmod (Q^{(k+1)}(\theta), p^{k+1}). \tag{7.9.49}$$

Substituting (7.9.49) in (7.9.43), it simplifies to

$$M_k(u) - W_k(u) = A_0(u)\,B_k(u) + A_k(u)\,B_0(u) \bmod (Q_0(\theta), p). \tag{7.9.50}$$

This is solved using the Euclid's algorithm to get $A_k(u)$ and $B_k(u)$ and thus $A^{(k+1)}(u)$ and $B^{(k+1)}(u)$.

We now prove the uniqueness of the factorization for $k = 1, 2, ..., a-1$. Starting from (7.9.36), it is shown that the polynomials $A^{(k+1)}(u)$ and $B^{(k+1)}(u)$ in (7.9.37) are unique. This can be established using mathematical induction as follows. If there are two sets of polynomials, say $\{A1^{(k+1)}(u)$ and $B1^{(k+1)}(u)\}$ and $\{A2^{(k+1)}(u)$ and $B2^{(k+1)}(u)\}$, that satisfy (7.9.37) while there is only one set $\{A^{(k)}(u)$ and $B^{(k)}(u)\}$ that satisfies (7.9.42), then

$$\begin{aligned} A1^{(k+1)}(u) &= A^{(k)}(u) + p^k A1_k(u),\\ A2^{(k+1)}(u) &= A^{(k)}(u) + p^k A2_k(u),\\ B1^{(k+1)}(u) &= B^{(k)}(u) + p^k B1_k(u),\\ B2^{(k+1)}(u) &= B^{(k)}(u) + p^k B2_k(u), \end{aligned} \tag{7.9.51}$$

with $A1_k(u) \neq A2_k(u)$ and $B1_k(u) \neq B2_k(u)$. Also (7.9.37) leads to

$$A1^{(k+1)}(u)\,B1^{(k+1)}(u) = A2^{(k+1)}(u)\,B2^{(k+1)}(u) \bmod (Q^{(k+1)}(\theta), p^{k+1}).$$

Substituting for various polynomials from (7.9.51), we get

$$A_0(u)\,[B1_k(u) - B2_k(u)] = B_0(u)\,[A2_k(u) - A1_k(u)] \bmod (Q_0(\theta), p).$$

This is not possible as $(A_0(u), B_0(u)) = 1$ and $deg(A_0(u)) > deg(A2_k(u))$ or $deg(A1_k(u))$. A similar statement also holds for $B_0(u)$, $B1_k(u)$, and $B2_k(u)$. Note that the validity of above equation must be tested in the finite field $Z(p, Q_0(\theta))$. The proof of the theorem is complete.

The various steps in the computation of $A^{(k+1)}(u)$ and $B^{(k+1)}(u)$ are given below.

Step 1. Initialization $A^{(1)}(u) = A_0(u)$, $B^{(1)}(u) = B_0(u)$, $M^{(1)}(u) = M_0(u)$, and $Q^{(1)}(\theta) = Q_0(\theta)$.

For $k = 1, 2, ..., a-1$, do

Step 2. Compute $D^{(k)}(u)$ in (7.9.44) as

$$D^{(k)}(u) = \left[Q^{(k)}(\theta)\right]^{-1}\left\{M^{(k)}(u) - A^{(k)}(u)B^{(k)}(u)\right\} \bmod p^k .$$

Step 3. Compute $W_k(u)$ using (7.9.47) with $A'_k(u)$ the *rhs* of (7.9.46). Note that by definition $D_0(u) = D^{(1)}(u)$ in (7.9.43).

Step 4. Solve (7.9.50) using the Euclid's algorithm for $A_k(u)$ and $B_k(u)$.

Step 5. Construct $A^{(k+1)}(u)$ and $B^{(k+1)}(u)$ using (7.9.40).

Theorem 7.23 establishes the procedure to obtain the factorization of $M(u)$ in $Z(q, Q(\theta))$ starting from its factorization in the finite field $Z(p, Q_0(\theta))$. The extension of these expressions to more than two factors is straightforward. This discussion and Theorem 7.23 also lead to the following result.

Theorem 7.24 The factorization of a monic degree n polynomial $M(u)$ in $Z(q, Q(\theta))$ in terms of its monic irreducible factors is unique and can be obtained from the factorization of $M_0(u)$ in $Z(p, Q_0(\theta))$.

Example 7.14 Consider the factorization of $u^5 - 1$ in $Z(8, \theta^2 + 7\theta + 5)$. It begins with the factorization of $u^5 - 1$ in $Z(2, \theta^2 + \theta + 1)$. Here $p = 2$, $a = 3$, $q = 2^3$, $Q(\theta) = \theta^2 + 7\theta + 5$. The factorization of $u^5 - 1$ in $Z(2, \theta^2 + \theta + 1)$ is well known and is given by

$$u^5 - 1 = (u + 1)\,(u^2 + \theta\, u + 1)\,(u^2 + (1 + \theta)\, u + 1) \bmod (\theta^2 + \theta + 1, 2).$$

In this case, $M_0(u) = u^5 + 1$, $M_1(u) = M_2(u) = 1$, $Q_0(\theta) = \theta^2 + \theta + 1$, $Q_1(\theta) = \theta$, and $Q_2(\theta) = \theta + 1$. The initialization is given by

$$\begin{aligned}
&A^{(1)}(u) = A_0(u) = u + 1,\\
&B^{(1)}(u) = B_0(u) = u^2 + 0\, u + 1,\\
&C^{(1)}(u) = C_0(u) = u^2 + (1 + \theta)\, u + 1,\\
&M^{(1)}(u) = M_0(u) = u^5 + 1 \text{ and } Q^{(1)}(\theta) = Q_0(\theta) = \theta^2 + \theta + 1.
\end{aligned}$$

For $k = 1$, various polynomials are

$$\begin{aligned}
&D^{(1)}(u) = u^3 + u^2,\\
&A'_1(u) = (1+\theta)u^4 + (\theta+\theta^2)u^3 + (\theta+\theta^2)u^2 + (1+\theta)u,\\
&W_1(u) = (1 + \theta)\, u^4 + u^3 + u^2 + (1 + \theta)\, u,\\
&A_1(u) = 1,\ B_1(u) = u,\ C_1(u) = (1 + \theta)\, u,\\
&A^{(2)}(u) = u + 3,\ B^{(2)}(u) = u^2 + (2 + \theta)\, u + 1,
\end{aligned}$$

and

$$C^{(2)}(u) = u^2 + (3 + 3\,\theta)\,u + 1.$$

For $k = 2$, the various polynomials are

$$D^{(2)}(u) = u^3 + 3\,u^2,$$
$$A_2'(u) = \theta u^4 + (1 + \theta + \theta^2)u^3 + (1 + \theta + \theta^2)u^2 + \theta u,$$
$$W_2(u) = \theta\,u^4 + \theta\,u,$$
$$A_2(u) = 1,\; B_2(u) = (1 + \theta)\,u,\; C_2(u) = 0,$$
$$A^{(3)}(u) = u + 7,\; B^{(3)}(u) = u^2 + (6 + 5\,\theta)\,u + 1,$$

and

$$C^{(3)}(u) = u^2 + (3 + 3\,\theta)\,u + 1.$$

It may be verified that

$$M^{(2)}(u) = u^5 + 3 = A^{(2)}(u)\,B^{(2)}(u)\,C^{(2)}(u) \bmod (\theta^2 + 3\,\theta + 1, 4),$$

and

$$M^{(3)}(u) = u^5 + 7 = A^{(3)}(u)\,B^{(3)}(u)\,C^{(3)}(u) \bmod (\theta^2 + 7\,\theta + 5, 8).$$

The last result that we need in order to establish the CRT-PR is stated in the following theorem.

Theorem 7.25 If two monic polynomials $A(u)$ and $B(u)$ divide a monic polynomial $M(u)$ in $Z(q, Q(\theta))$ and $(A_0(u), B_0(u)) = 1$ in $Z(p, Q_0(\theta))$, then $[A(u)\,B(u)] \mid M(u)$.

Proof: $B(u) \mid M(u)$ implies that $M(u) = R(u)\,B(u) \bmod (q, Q(\theta))$. Dividing $R(u)$ by $A(u)$, we get

$$R(u) = S(u)\,A(u) + V(u) \bmod (q, Q(\theta)),\;\; deg(V(u)) < deg(A(u)). \qquad (7.9.52)$$

Thus $M(u)$ can be expressed as

$$M(u) = S(u)\,A(u)\,B(u) + V(u)\,B(u) \bmod (Q(\theta), q).$$

Since $A(u) \mid M(u)$, we must have $A(u) \mid [V(u)\,B(u)]$ or $V(u)\,B(u) = U(u)\,A(u) \bmod (Q(\theta), q)$. Taking mod p on both sides,

$$V_0(u)\,B_0(u) = U_0(u)\,A_0(u) \bmod (Q_0(\theta), p). \qquad (7.9.53)$$

Recalling that $(A_0(u), B_0(u)) = 1$ and $deg(V(u)) < deg(A(u))$, (7.9.53) leads to $V_0(u) = U_0(u) = 0$. Using the same argument in a recursive manner, we get $V(u) = 0$. The theorem follows.

We end this section by stating that the results of this section are valid in $CZ(q, Q(\theta))$, $p \equiv 3 \bmod 4$. In this case all modulo operations will include mod $j^2 + 1$. The remaining algebraic manipulations remain the same. In the following, we state and prove the CRT-PR for polynomials defined in $Z(q, Q(\theta))$ and $CZ(q, Q(\theta))$.

7.9.4 CRT-PR in $Z(q, Q(\theta))$ and $CZ(q, Q(\theta))$

Throughout this section, $M(u)$ denotes a degree n polynomial in $Z(q, Q(\theta))$, $q = p^a$, $Q(\theta)$ being a monic irreducible polynomial in θ with coefficients in $Z(q)$. Also let $M(u)$ factorize as

$$M(u) = \prod_{i=1}^{\lambda} m_i(u) \text{mod}(Q(\theta), q) \tag{7.9.54}$$

where $m_i(u)$ are monic and $(m_{0,i}(u), m_{0,j}(u)) = 1$, $i \neq j$. The factorization techniques of the previous section are used to obtain (7.9.54) once the factorization of $M_0(u)$ is known in the finite field $Z(p, Q_0(\theta))$.

Theorem 7.26 CRT-PR for integer polynomial rings. Given an arbitrary polynomial $X(u)$ in $Z(q, Q(\theta))$, $deg(X(u)) < n$, there is a one-to-one correspondence between $X(u)$ and its residues $x_i(u)$, $i = 1, 2, ..., \lambda$ defined as

$$x_i(u) \equiv X(u) \bmod (m_i(u), Q(\theta), q), \quad i = 1, 2, ..., \lambda. \tag{7.9.55}$$

Proof: It is clear from (7.9.55) that $x_i(u)$, $i = 1, 2, ..., \lambda$ can be determined uniquely from $X(u)$ for given monic polynomials $m_i(u)$, $i = 1, 2, ..., \lambda$. To prove the uniqueness of $X(u)$ for a given set of $x_i(u)$, $i = 1, 2, ..., \lambda$, assume that there are two polynomials $X(u)$ and $Y(u)$ having the same residues. Then

$$X(u) - Y(u) \equiv 0 \bmod (m_i(u), Q(\theta), q), \quad i = 1, 2, ..., \lambda. \tag{7.9.56}$$

Theorem 7.25 and (7.9.56) imply that

$$X(u) - Y(u) = S(u)\prod_{i=1}^{\lambda} m_i(u) \text{mod}(Q(\theta), q) = S(u)M(u) \text{mod}(Q(\theta), q) \tag{7.9.57}$$

for some arbitrary polynomial $S(u)$. Since $deg(X(u) - Y(u)) < deg(M(u))$, (7.9.57) is possible *iff* $S(u) = 0$ or $X(u) = Y(u)$.

A. The CRT-PR Reconstruction. Given the residues of $X(u)$ in (7.9.55), CRT-PR reconstruction consists in obtaining $X(u)$ from its residues as follows:

$$X(u) \equiv \sum_{i=1}^{\lambda} x_i(u) T_i(u) \left\{ \prod_{\substack{j=1 \\ j \neq i}}^{\lambda} m_j(u) \right\} \bmod (M(u), Q(\theta), q), \tag{7.9.58}$$

where $deg(T_i(u)) < deg(m_i(u))$, and $T_i(u)$ satisfies the congruence

$$T_i(u) \left\{ \prod_{\substack{j=1 \\ j \neq i}}^{\lambda} m_j(u) \right\} \equiv 1 \bmod (m_i(u), Q(\theta), q). \tag{7.9.59}$$

The validity of (7.9.58) is verified by substituting for $X(u)$ from (7.9.58) in (7.9.55) and using (7.9.59) to simplify the *rhs*. In essence, the procedure to compute $T_i(u)$ is a special case of solving the polynomial congruence for $B(u)$,

$$A(u)\, B(u) \equiv C(u) \bmod (F(u), Q(\theta), q) \tag{7.9.60}$$

for given $A(u)$, $C(u)$, and $F(u)$ such that $A_0(u)$ and $F_0(u)$ are non-zero and $(A_0(u), F_0(u)) = 1$. Taking mod p on both sides of (7.9.60), we get

$$A_0(u)\, B_0(u) \equiv C_0(u) \bmod (F_0(u), Q_0(\theta), p). \tag{7.9.61}$$

Since $(A_0(u), F_0(u)) = 1$, $B_0(u)$ always exists and can be determined using Euclid's algorithm by expressing

$$A_0(u)\, B_0(u) + E_0(u)\, F_0(u) = C_0(u) \bmod (Q_0(\theta), p). \tag{7.9.62}$$

Now, assume that the solution to the congruence

$$A^{(k)}(u)\, B^{(k)}(u) \equiv C^{(k)}(u) \bmod (F^{(k)}(u), Q^{(k)}(\theta), p^k) \tag{7.9.63}$$

is known and we have to find the solution to

$$A^{(k+1)}(u)\, B^{(k+1)}(u) \equiv C^{(k+1)}(u) \bmod (F^{(k+1)}(u), Q^{(k+1)}(\theta), p^{k+1}). \tag{7.9.64}$$

Or, given

$$A^{(k)}(u)\, B^{(k)}(u) + E^{(k)}(u)\, F^{(k)}(u) = C^{(k)}(u) \bmod (Q^{(k)}(\theta), p^k) \tag{7.9.65}$$

we have to find $A_k(u)$ and $B_k(u)$ such that

$$A^{(k+1)}(u)\, B^{(k+1)}(u) + E^{(k+1)}(u)\, F^{(k+1)}(u) = C^{(k+1)}(u) \bmod (Q^{(k+1)}(\theta), p^{k+1}). \tag{7.9.66}$$

Based on (7.9.65), we write (7.9.66) as

$$\begin{aligned}&[A^{(k)}(u) + p^k A_k(u)]\,[B^{(k)}(u) + p^k B_k(u)]\\ &\quad + [E^{(k)}(u) + p^k E_k(u)]\,[F^{(k)}(u) + p^k F_k(u)]\\ &\quad = [C^{(k)}(u) + p^k C_k(u)] \bmod (Q^{(k+1)}(\theta), p^{k+1}).\end{aligned} \tag{7.9.67}$$

Expanding and simplifying, we get

$$\begin{aligned}p^k [A_0(u)\, B_k(u) + E_k(u)\, F_0(u)] &= \{[C^{(k)}(u) - A^{(k)}(u)\, B^{(k)}(u) - E^{(k)}(u)\, F^{(k)}(u)]\\ &\quad + p^k [C_k(u) - E_0(u)\, F_k(u) - A_k(u)\, B_0(u)]\} \bmod (Q^{(k+1)}(\theta), p^{k+1}).\end{aligned} \tag{7.9.68}$$

Following the same analysis as in (7.9.41) to (7.9.50), it can be shown that

$$A^{(k)}(u)\, B^{(k)}(u) + E^{(k)}(u)\, F^{(k)}(u) = C^{(k)}(u) + p^k\, W_k(u) \bmod (Q^{(k+1)}(\theta), p^{k+1}), \tag{7.9.69}$$

where

$$W_k(u) \equiv A_k'(u) \bmod (Q_0(\theta), p), \tag{7.9.70}$$

$$\begin{aligned}A_k'(u) = \{&[A^{(k)}(u) B^{(k)}(u) + E^{(k)}(u) F^{(k)}(u) + D^{(k)}(u) Q^{(k)}(\theta) - C^{(k)}(u)]\, p^{-k}\\ &+ D_0(u) Q_k(\theta)\} \bmod p\end{aligned} \tag{7.9.71}$$

with $D^{(k)}(u)$ satisfying (refer to (7.9.65))

$$A^{(k)}(u)\, B^{(k)}(u) + E^{(k)}(u)\, F^{(k)}(u) + D^{(k)}(u)\, Q^{(k)}(\theta) = C^{(k)}(u) \bmod p^k. \tag{7.9.72}$$

Substituting from (7.9.69) in (7.9.68), it simplifies to

$$\begin{aligned}A_0(u)\, B_k(u) + E_k(u)\, F_0(u) &= [C_k(u) - W_k(u) - E_0(u)\, F_k(u) - A_k(u)\, B_0(u)]\\ &\quad \bmod (Q_0(\theta), p).\end{aligned} \tag{7.9.73}$$

The above equation is solved using Euclid's algorithm to get $B_k(u)$ and $E_k(u)$ and thus $B^{(k+1)}(u)$. Furthermore, it can be shown that the solution is *unique* in all instances. The various steps in the computation of $B(u)$, the solution to the polynomial congruence in (7.9.60), are as follows.

Step 1. Initialization. Use Euclid's algorithm to find $B_0(u)$ and $E_0(u)$ in (7.9.61). The initializations are

$A^{(1)}(u) = A_0(u)$, $B^{(1)}(u) = B_0(u)$,
$C^{(1)}(u) = C_0(u)$, $F^{(1)}(u) = F_0(u)$,
$E^{(1)}(u) = E_0(u)$, and $Q^{(1)}(\theta) = Q_0(\theta)$.

For $k = 1, 2, ..., a - 1$, do

Step 2. Compute $D^{(k)}(u)$ in (7.9.72) as

$$D^{(k)}(u) = \left\{\left[Q^{(k)}(\theta)\right]^{-1}\left[C^{(k)}(u) - A^{(k)}(u)B^{(k)}(u) - E^{(k)}(u)F^{(k)}(u)\right]\right\} \bmod p^k .$$

Step 3. Compute $W_k(u)$ using (7.9.70) with $A_k'(u)$ obtained using (7.9.71). Note that by definition $D_0(u) = D^{(1)}(u)$ in (7.9.71).

Step 4. Solve (7.9.73) using Euclid's algorithm for $B_k(u)$ and $E_k(u)$.

Step 5. Compute $B^{(k+1)}(u) = B^{(k)}(u) + p^k B_k(u)$ and $E^{(k+1)}(u) = E^{(k)}(u) + p^k E_k(u)$.

In summary, under appropriate conditions as described here, the polynomial ring $Z(q, Q(\theta), M(u))$ satisfies the fundamental direct sum property,

$$Z(q, Q(\theta), M(u)) = \sum_{\substack{\oplus \\ i=1}}^{\lambda} Z(q, Q(\theta), m_i(u)). \tag{7.9.74}$$

The CRT-PR in $Z(q, Q(\theta))$ establishes the existence and uniqueness of such a direct sum. All the results of this section are also valid in $CZ(q, Q(\theta))$, $p \equiv 3 \bmod 4$. All the modulo operations in various equations will include $j^2 + 1$ and the congruences are to be solved in $Z(j^2 + 1, Q_0(\theta), p) = GF(p^{2m})$. The remaining algebraic manipulations remain the same. We end this section with an example of CRT-PR reconstruction.

Example 7.15 Consider computation of CRT-PR reconstruction polynomials for the factorization of $u^5 - 1$ in $Z(8, \theta^2 + 7\theta + 5)$. The following recurrences must be solved for CRT-PR reconstruction:

$$R_1(u)\,[u^4 + u^3 + u^2 + u + 1] \equiv 1 \bmod (u + 7, \theta^2 + 7\theta + 5, 8)$$

or

$$R_1(u)\, 5 \equiv 1 \bmod (u + 7, \theta^2 + 7\theta + 5, 8), \tag{7.9.75}$$

$$R_2(u)\,[u^3 + (2 + 3\theta)\,u^2 + (6 + 5\theta)\,u + 7] \equiv 1 \bmod (u^2 + (6 + 5\theta)\,u + 1, \theta^2 + 7\theta + 5, 8)$$

or

$$R_2(u)\,[(3 + 7\theta)\,u + (3 + 2\theta)] \equiv 1 \bmod (u^2 + (6 + 5\theta)\,u + 1, \theta^2 + 7\theta + 5, 8), \tag{7.9.76}$$

and

$$R_3(u)\,[u^3 + (5 + 5\theta)u^2 + (3 + 3\theta)\,u + 7] \equiv 1 \bmod (u^2 + (3 + 3\theta)\,u + 1, \theta^2 + 7\theta + 5, 8)$$

or

$$R_3(u)\,[(2 + \theta)\,u + (5 + 6\theta)] \equiv 1 \bmod (u^2 + (3 + 3\theta)\,u + 1, \theta^2 + 7\theta + 5, 8). \tag{7.9.77}$$

$R_1(u)$ is trivially solved as $R_1(u) = 5$. The steps in the solution to (7.9.76) are shown below. Note that the notation is slightly different in order to make it consistent with (7.9.60) to (7.9.73). For (7.9.76)

$$C_0(u) = 1,\; C_1(u) = 0,\; C_2(u) = 0,$$
$$A_0(u) = (1 + \theta)\,u + 1,\; A_1(u) = (1 + \theta)\,u + (1 + \theta),\; A_2(u) = \theta\,u,$$
$$F_0(u) = u^2 + \theta\,u + 1,\; F_1(u) = u,\; \text{and } F_2(u) = (1 + \theta)\,u.$$

The initialization is given by

$$A^{(1)}(u) = A_0(u) = (1 + \theta)\,u + 1,\; B^{(1)}(u) = B_0(u) = \theta\,u,$$
$$C^{(1)}(u) = C_0(u) = 1,\; F^{(1)}(u) = F_0(u) = u^2 + \theta\,u + 1,$$
$$E^{(1)}(u) = E_0(u) = 1,\; \text{and } Q^{(1)}(\theta) = Q_0(\theta) = \theta^2 + \theta + 1.$$

Note that $B_0(u)$ and $E_0(u)$ are obtained by solving (7.9.62) using Euclid's algorithm. For $k = 1$, the polynomials are

$$D^{(1)}(u) = u^2,$$
$$A_1'(u) = \left(1 + \theta^2\right)u^2 + \theta u,$$
$$W_1(u) = \theta\,u^2 + \theta\,u,$$

$B_1(u) = (1 + \theta)\, u + 1,$
$E_1(u) = 1,$
$B^{(2)}(u) = (2 + 3\,\theta)\, u + 2$, and $E^{(2)}(u) = 3.$

For $k = 2$, the polynomials are

$$D^{(2)}(u) = 3\,u^2 + 2\,u,$$
$$A_2'(u) = \left(\theta + \theta^2\right)u^2 + (1 + \theta)u + \theta,$$
$$W_2(u) = u^2 + (1 + \theta)\, u + \theta,$$
$$B_2(u) = \theta\, u + 1,$$
$$E_2(u) = 1 + \theta,$$
$$B^{(3)}(u) = (2 + 7\,\theta)\, u + 6, \text{ and } E^{(3)}(u) = 7 + 4\,\theta.$$

Therefore, $R_2(u) = (2 + 7\,\theta)\, u + 6$. In a similar manner it can be shown that $R_3(u) = (1 + \theta)\, u + 6$.

Notes

This chapter focused on bringing the richness of results in polynomial algebra in finite fields to polynomial algebra in finite integer rings and their extensions. All the computations in a ring are converted to equivalent computations in a finite field in order to demonstrate the existence and uniqueness of results. It is our belief that one must understand the algebra of finite fields in order to fully appreciate the algebra of finite integer rings and their extensions. An extension to the Chinese remainder theorem is described which is of fundamental importance to designing new methods for processing data sequences defined over finite integer rings and their extensions. The application of the results reported here to the design of computationally efficient algorithms for computing the one- and higher-dimensional convolutions will be described in later chapters. We expect that the mathematical framework established in this chapter will find diverse applications in computer arithmetic, digital signal processing, and other related areas. Several results presented here were derived by the author and his associates in their research work. The results on number theoretic transforms are well known. Many results on complex number theoretic transforms are cast under a different light. The generalization of the Euler's theorem may prove to be quite valuable in many situations.

Bibliography

[7.1] C.M. Rader, "Discrete Convolutions via Mersenne Transforms," *IEEE Transactions on Computers*, Vol. C-21, pp. 1269-1273, 1972.

[7.2] R.C. Agarwal and C.S. Burrus, "Fast Convolution using Fermat Number Transforms with Applications to Digital Filtering," *IEEE Transactions on Acoustics, Speech, and Signal Processing*, Vol. ASSP-22, pp. 87-97, 1974.

[7.3] R.C. Agarwal and C.S. Burrus, "Number Theoretic Transforms to Implement Fast Digital Convolution," *Proceeding IEEE*, Vol. 63, pp. 550-560, 1975.

[7.4] K.Y. Lin, H. Krishna, and B. Krishna, "Rings, Fields, the Chinese Remainder Theorem and an Extension, Part I: Theory," *IEEE Transactions on Circuits and Systems-II*, Vol. 41, pp. 641-655, 1994.

[7.5] K.Y. Lin, H. Krishna, and B. Krishna, "Rings, Fields, the Chinese Remainder Theorem and an Extension, Part II: Applications to Digital Signal Processing," *IEEE Transactions on Circuits and Systems-II*, Vol. 41, pp. 656-668, 1994.

[7.6] H. Krishna, B. Krishna, and K.Y. Lin, "The AICE-CRT and Digital Signal Processing Algorithms: The Complex Case," *Circuits, Systems, and Signal Processing*, Vol. 14, pp. 69-95, 1995.

[7.7] E. Dubois and A.N. Venetsanopoulos, "The Discrete Fourier Transform Over Finite Rings with Applications to Fast Convolution," *IEEE Transactions on Computers*, Vol. C-27, pp. 586-593, 1978.

[7.8] E. Dubois and A.N. Venetsanopoulos, "The Generalized Discrete Fourier Transform in Rings of Algebraic Integers," *IEEE Transactions on Acoustics, Speech, and Signal Processing*, Vol. ASSP-28, pp. 169-175, 1980.

[7.9] J.B. Martens and M.C. Vanwormhoudt, "Convolutions of Long Integer Sequences by Means of Number Theoretic Transforms over Residue Class Polynomial Rings," *IEEE Transactions on Acoustics, Speech, and Signal Processing*, Vol. ASSP-31, pp. 1125-1134, 1983.

[7.10] R.J. McElice, *Finite Fields for Computer Scientists and Engineers*, Kluwer Academic Publishers, 1987.

[7.11] W.W. Peterson and E.J. Weldon, Jr., *Error Correcting Codes*, MIT Press, 1978.

Problems

7.1 Show that all Mersenne numbers are relatively prime.

7.2 Show that all Fermat numbers are relatively prime.

7.3 Show that whenever Mersenne numbers are prime, they are of the type $4L+3$.

7.4 For prime Mersenne numbers, show that the numbers $2j$ and $(1+j)$ can be used to generate a CNTT of length $4p$ and $8p$, respectively.

7.5 Show that -2 can be used to define a length 2^{t+1} NTT when the arithmetic is defined modulo the Fermat number F_t.

7.6 Show that $2^{2^{t-2}}\left(2^{2^{t-1}}-1\right)$ can be used to define a length 2^{t+2} NTT when the arithmetic is defined modulo the Fermat number F_t.

7.7 Show that the factorization of u^n+1 over any field can be obtained from the factorization of u^n-1 and $u^{2n}-1$.

7.8 Using the statement in Problem 7.7 or otherwise, show that u^n+1 factors into the product of n degree-one relatively prime polynomials of the type $(u-G_i)$ over a field *iff* there is an element of order $2 \cdot n$ in that field. Can we express G_i as a power of a single element of the field? Justify.

7.9 Show that n is a factor of $(p-1)/2$ if the field under consideration in Problem 7.8 is $GF(p)$. Can the same statement be made when the factorization is considered over $Z(p^a)$?

7.10 Show that n is a factor of $(p^2-1)/2$ if the field under consideration in Problem 7.9 is $GF(p^2)$. Under what conditions can the same statement be made when the factorization is considered over $CZ(p^a)$?

7.11 Derive all the pertinent results with regard to the factorization of u^n+1 in terms of n degree-one factors over the ring $Z(M)$ and $CZ(M)$.

7.12 This problem deals with identifying all the possible values of the generator G for an NTT. Consider the length 30 NTT defined over $Z(31)$. Identify all the possible values for G by first obtaining a primitive element in $GF(31)$. You may write a computer program to carry out the necessary computation. How many values are possible for G? Now repeat the same procedure for a length 15 NTT defined over $Z(31)$. Think of criteria for suitability of a given generator for hardware implementation and identify those generators that best fit it.

7.13 Identify all the possible values for the generator G for NTTs of lengths 30 and 15 defined over $Z(2^{25} - 1)$. How many values for G are possible? You may write a computer program to carry out the necessary computation. Once again, discuss their suitability for hardware implementation. Note that $2^{25} - 1 = 31 \cdot 601 \cdot 1801$.

7.14 This problem deals with identifying all the possible values of the generator G for a CNTT. Consider the length 16 CNTT defined over $CZ(7)$. Identify all the possible values for G by first obtaining a primitive element in $GF(7^2)$. You may write a computer program to carry out the necessary computation. How many values are possible for G? Now repeat the same procedure for a length 24 NTT defined over $CZ(7)$. Identify those generators that best fit the criteria for suitability of a given generator for hardware implementation.

7.15 Identify all the possible values for the generator G for CNTT of length 16 defined over $CZ(2^{15} - 1)$. How many values for G are possible? You may write a computer program to carry out the necessary computation. Once again, discuss their suitability for hardware implementation. Note that $2^{15} - 1 = 7 \cdot 31 \cdot 151$.

7.16 Explain the reasons why rings of the type $Z(2^a + 1)$ or $Z(2^a - 1)$ are attractive from the point of view of hardware realization. Can you think of other rings that may also be attractive, for example, the ring $Z(2^a)$?

7.17 It has been stated that the polynomials used in the CRT-PR reconstruction for the CRT-PR defined in $Z(p^a)$, $CZ(p^a)$, $Z(p^a, Q(\theta))$ and $CZ(p^a, Q(\theta))$ are unique. Establish this result.

7.18 Consider the ring $CZ(2^a)$ for a even. It has been stated that $A = 2^{a/2}$ and $B = -2^{a/2}$ satisfy the conditions in Lemma 7.1. Are these values unique? Show that there exist other pairs of values for A and B that also satisfy those conditions. Specifically, let $a = 4$ and find all the different pairs of values. Note that the factorization of any monic polynomial is unique in the ring $Z(p^a)$, but not in the ring $Z(M)$.

7.19 Consider the congruence $A \cdot X \equiv 1 \bmod (M, j^2 + 1)$ defined over $CZ(M)$. In essence, we are interested in finding the inverse of a complex integer

A in $CZ(M)$. Derive the conditions under which a unique solution to this congruence exists. Note that this congruence may also be treated as a polynomial extension of the scalar congruence $A \cdot X \equiv 1 \bmod M$ or a (2×2) system of linear equations in $Z(M)$. One way to solve this problem could be to treat A and X as complex numbers, obtain X in $\mathbf{C}$, and then try to find the corresponding X in $CZ(M)$.

7.20 Develop a LI approach for polynomials defined over $CZ(M)$. What is the largest degree polynomial over $CZ(p^a)$ that can be determined using LI if $p = 4L + 3$ or if $p = 4L + 1$?

7.21 Specialize the LI approach developed in Problem 7.20 to the case when all the interpolating points are to be expressed as a power of a single number.

7.22 Now establish the conditions for the existence of CNTT as a special case of Problem 7.20.

7.23 Factorize $u^3 - 1$ over $Z(2^3)$. Are these factors unique?

7.24 Factorize $u^6 - 1$ over $Z(2^3)$? Are these factors unique?

7.25 Compute all the relevant quantities (factorization, reconstruction polynomials, etc.) for the CRT-RP for the polynomial $u^7 - 1$ over $Z(2^4)$. Cast the CRT-PR in the matrix-vector form.

7.26 Compute all the relevant quantities (factorization, reconstruction polynomials, etc.) for the CRT-RP for the polynomial $u^5 - 1$ over $CZ(3^3)$. Cast the CRT-PR in the matrix-vector form.

7.27 In this chapter, a generalization of the Euler's theorem was described for $CZ(p^a)$, $p = 4L + 3$. Can you obtain a similar generalization for $Z(p^a, Q(\theta))$ and $CZ(p^a, Q(\theta))$?

7.28 Obtain the factorization of $u^7 - 1$ over $GF(2, Q_0(\theta))$, $Q_0(\theta) = \theta^3 + \theta + 1$. What is $Q(\theta)$ if we wish to write a similar factorization of $u^7 - 1$ over $Z(8, Q(\theta))$?

7.29 Compute all the relevant quantities (factorization, reconstruction polynomials, etc.) for the CRT-RP for the polynomial $u^{10} - 1$ over $Z(3^2, Q(\theta))$, $Q_0(\theta) = \theta^2 + \theta + 2$. Cast the CRT-PR in the matrix-vector form.

7.30 Compute all the relevant quantities (factorization, reconstruction polynomials, etc.) for the CRT-RP for the polynomial $u^9 - 1$ over $Z(2^3, Q(\theta))$, $Q_0(\theta) = \theta^3 + \theta^2 + 1$. Cast the CRT-PR in matrix-vector form.

7.31 This problem deals with the derivation of NTTs over the integer polynomial ring $Z(p^a, Q(\theta))$, $deg(Q(\theta)) = m$, $Q_0(\theta)$ being a primitive polynomial in $GF(p)$. Show that NTTs exist in this ring for all values of n such that $n \mid (p^m - 1)$.

7.32 This problem deals with the derivation of NTTs over the integer polynomial ring $Z(M, Q(\theta))$, $M = \prod_{i=1}^{t} p_i^{a_i}$, $deg(Q(\theta)) = m$, $Q(\theta)$ being a monic polynomial such that $Q_0^{(i)}(\theta) \equiv Q(\theta) \bmod p_i$, $i = 1, 2, ..., t$ is a primitive polynomial in $GF(p_i)$. Show that NTTs exist in this ring for all values of n such that $n \mid \left(p_i^m - 1\right)$, $i = 1, 2, ..., t$.

7.33 This problem deals with the derivation of CNTTs over the integer polynomial ring $CZ(M,Q(\theta))$, $M = \prod_{i=1}^{t} p_i^{a_i}$, $deg(Q(\theta)) = m$, $Q(\theta)$ being a monic polynomial such that $Q_0^{(i)}(\theta) \equiv Q(\theta) \bmod p_i$, $i = 1, 2, ..., t$ is a primitive polynomial in $GF\left(p_i^2\right)$. Show that CNTTs exist in this ring for all values of n such that $n \mid \left(p_i^{2m} - 1\right)$, $i = 1, 2, ..., t$.

7.34 Develop a computer program that will compute all the relevant quantities (factorization, reconstruction polynomials, etc.) for the CRT-RP for the polynomial $u^n - 1$ over $Z(p^a, Q(\theta))$ and $CZ(p^a, Q(\theta))$ when n, p, a and m ($= deg(Q(\theta))$) are specified.

7.35 In many cases, one needs to solve the congruence

$$A(u)\, X(u) \equiv C(u) \bmod F(u)$$

over a ring $Z(q, Q(\theta))$ or $CZ(q, Q(\theta))$. Here $A(u)$ and $F(u)$ are relatively coprime factors of $u^n - 1$, $(n, q) = 1$. Show that it is possible to get a closed form expression for $C(u)$ in terms of the formal derivative of $F(u)$ and other related quantities.
Note: Write $u^n - 1$ in terms of $A(u)$ and $F(u)$, and take formal derivative.

PART II

CONVOLUTION ALGORITHMS

And Some More

Thoughts on Part II

In this part, we describe fast computationally efficient algorithms for the following tasks:

1. One-Dimensional Acyclic Convolution of Length $n = d + e - 1$:

$$Z(u) = X(u)\, Y(u), \tag{ii.1}$$

$deg(X(u)) = d - 1$, $deg(Y(u)) = e - 1$, and $deg(Z(u)) = n - 1$.

2. One-Dimensional Cyclic Convolution of Length n:

$$Z(u) = X(u)\, Y(u) \bmod (u^n - 1), \tag{ii.2}$$

$deg(X(u)) = deg(Y(u)) = deg(Z(u)) = n - 1$.

3. Two-Dimensional Acyclic Convolution of Length $n_1 \times n_2$, $n_1 = d_1 + e_1 - 1$, $n_2 = d_2 + e_2 - 1$:

$$Z(u, v) = X(u, v)\, Y(u, v), \tag{ii.3}$$

$deg(X(u, v)) = d_1 - 1$ in u,
$deg(X(u, v)) = d_2 - 1$ in v,
$deg(Y(u, v)) = e_1 - 1$ in u,
$deg(Y(u, v)) = e_2 - 1$ in v,
$deg(Z(u, v)) = n_1 - 1$ in u,
$deg(Z(u, v)) = n_2 - 1$ in v.

4. Two-Dimensional Cyclic Convolution of Length $n_1 \times n_2$:

$$Z(u, v) = X(u, v) Y(u, v) \bmod \left(u^{n_1} - 1, v^{n_2} - 1\right), \tag{ii.4}$$

$deg(X(u, v)) = deg(Y(u, v)) = deg(Z(u, v)) = n_1 - 1$ in u,
$deg(X(u, v)) = deg(Y(u, v)) = deg(Z(u, v)) = n_2 - 1$ in v.

5. Multidimensional Acyclic and Cyclic Algorithms. The expressions for multidimensional acyclic and cyclic convolutions can be defined in a manner similar to (ii.3) and (ii.4), respectively.

There are two distinct approaches to the one-dimensional computations in (ii.1) and (ii.2). The first approach consists in breaking down a large size one-dimensional convolution into a number of smaller size one-dimensional convolutions by using the CRT-P which are then evaluated independently. The second approach consists in converting a large size one-dimensional convolution into a multidimensional convolution. Algorithms for multidimensional convolution are then used for the overall computation. These multidimensional algorithms are collectively referred to as **nesting algorithms** and include **split and recursive nesting**.

It is important to observe here that certain tools, used to obtain the algorithms, are independent of the number system over which the resulting algorithms are valid. Thus, these tools remain valid for processing sequences defined over any number system. The CRT-I is one such tool. Another such tool is the conversion of one-dimensional convolution to multidimensional convolution which is based on the fact that the indices of the sequences are always integer-valued regardless of the number system over which the sequences themselves are defined. On the other hand, there are tools that are sensitive to the number system and, therefore, the resulting algorithms may be valid only over certain specified number systems.

Given a number of different algorithms to carry out the same computation, the question arises as to how one compares them to judge their usefulness. The single most important criterion in this regard is the time taken by an algorithm to compute all the required quantities in a specified computational environment. All other criteria that are used lead to this single criterion. Though it is easy to state, the measurement of time may become quite subjective and may depend on far too many parameters. In this work, we will measure the quality of an algorithm in terms of its computational complexity, that is, the number of arithmetic operations, ADDs and MULTs, it requires. The number of ADDs will be referred to as the **additive complexity** and the number of MULTs will be referred to as the **multiplicative complexity**. This is based on the notion that the arithmetic operations dominate the time that a computational algorithm requires. An algorithm will be called **computationally efficient** or **fast** if it has the least computational complexity of various algorithms. One also tends to focus more deeply on the multiplicative complexity for two reasons. First, MULTs require far more time than ADDs in most traditional machines. Second, in situations where an algorithm for solving a large size problem is built from an algorithm for solving a small size problem, it is the multiplicative complexity of the small size algorithm that plays the key role in determining the overall computational complexity. Finally, it is worthwhile stating that in many situations, various algorithms establish a trade off between the additive and the multiplicative complexities. Table II.1 lists the computational complexity of the various convolutions listed in (ii.1) to (ii.4) if they were computed directly.

Table II.1 Computational Complexity of Direct Computation of Various Convolutions

Computation	Additive Complexity	Multiplicative Complexity
(ii.1)	$(d-1)\cdot(e-1)$	$d\cdot e$
(ii.2)	$n\cdot(n-1)$	n^2
(ii.3)	$(e_1-1)\cdot(d_1-1)\cdot d_2\cdot e_2$ $+ (d_1+e_1-1)(d_2-1)(e_2-1)$ or $(e_2-1)\cdot(d_2-1)\cdot d_1\cdot e_1$ $+ (d_2+e_2-1)(d_1-1)(e_1-1)$	$d_1\cdot d_2\cdot e_1\cdot e_2$
(ii.4)	$n_1n_2\cdot(n_1n_2-1)$	$n_1^2 n_2^2$

Note: The two entries in the additive complexity column for (ii.3) correspond to the two orders in which the computations can be carried out.

The computational complexity of the direct approach for convolution is quite high and provides the impetus for search of efficient algorithms. In all cases, the algorithms can be cast as the bilinear form

$$\underline{Z} = \mathbf{C}\,[\mathbf{A}\,\underline{X} \otimes \mathbf{B}\,\underline{Y}],$$

where $\underline{X}$ and $\underline{Y}$ are the input vectors, $\underline{Z}$ is the output vector and **A**, **B**, and **C** are matrices of constants (elements of the field over which the algorithm is defined).

We now illustrate the above points with the help of the one-dimensional cyclic convolution algorithm of length 3. The cyclic convolution of length 3 can be computed using the algorithm represented in the bilinear form as

$$\begin{bmatrix} z_0 \\ z_1 \\ z_2 \end{bmatrix} = \frac{1}{3}\begin{bmatrix} 1 & 3 & 0 & -1 \\ 1 & -3 & -3 & 2 \\ 1 & 0 & 3 & -1 \end{bmatrix}\left\{\begin{bmatrix} 1 & 1 & 1 \\ 1 & 0 & -1 \\ 0 & 1 & -1 \\ 1 & 1 & -2 \end{bmatrix}\begin{bmatrix} x_0 \\ x_1 \\ x_2 \end{bmatrix} \otimes \begin{bmatrix} 1 & 1 & 1 \\ 1 & 0 & -1 \\ 0 & 1 & -1 \\ 1 & 1 & -2 \end{bmatrix}\begin{bmatrix} y_0 \\ y_1 \\ y_2 \end{bmatrix}\right\}.$$

This algorithm is based on the factorization $u^3 - 1 = (u-1)\,(u^2 + u + 1)$, CRT-P and the small degree polynomial MULT algorithm

$$(a_0 + a_1u)\cdot(b_0 + b_1u)$$

$$= a_0 \cdot b_0 + \{(a_0 + a_1) \cdot (b_0 + b_1) - a_1 \cdot b_1 - a_0 \cdot b_0\}\, u + a_1 \cdot b_1 u^2.$$

It is clear that there could be more than one bilinear form for the same convolution. One way to organize the evaluation of the above bilinear form is:

Step 1. Input ADDs

$$r_1 = (x_0 + x_1 + x_2),\ \ s_1 = (y_0 + y_1 + y_2)/3,$$
$$r_2 = x_0 - x_2,\ \ s_2 = y_0 - y_2,$$
$$r_3 = x_1 - x_2,\ \ s_3 = y_1 - y_2,$$
$$r_4 = r_2 + r_3,\ \ s_4 = (s_2 + s_3)/3.$$

Step 2. (General) MULTs

$$m_i = r_i \cdot s_i,\ \ i = 1, 2, 3, 4.$$

Step 3. Output ADDs

$$t_1 = m_2 - m_4,\ \ t_2 = m_3 - m_4,$$
$$z_0 = m_1 + t_1,$$
$$z_1 = m_1 - t_1 - t_2,$$
$$z_2 = m_1 + t_2.$$

The above organization requires 4 MULTs, 16 ADDs, and 2 scaling (by 3) operations. This is to be compared with the direct evaluation that requires 9 MULTs and 6 ADDs. It is of fundamental importance to note that the direct evaluation is valid over all fields and rings while the above algorithm is valid for those number systems over which 3^{-1} exists. Thus, it is valid over all fields except $GF(3)$ and all rings except those that are p-adic expansion rings of $GF(3^m)$. With some minor modification to the above three steps, one may use the bilinear form to process integer-valued sequences in a way that all the intermediate quantities are also integer valued.

In many instances arising in digital signal processing, one of the sequences corresponds to the *finite impulse response* (FIR) of the digital filter and, therefore, is fixed. In such instances, the arithmetic operations performed on the FIR sequence can be ignored as they need to be performed once only. In the above example, if the y sequence corresponds to the FIR filter, then the length 3 cyclic convolution algorithm requires 4 MULTs and 11 ADDs.

The last aspect that we discuss here is the acyclic convolution of an infinitely long sequence with a finite length sequence. This is relevant when an FIR filter is to be used to process an infinitely long input sequence. In the present framework, let x be the infinitely long sequence and y be the FIR sequence of length e. There are two approaches adopted in this case. They are termed as *overlap-add* and *overlap-save* and are described in the following.

Overlap-add technique. This technique consists in partitioning input x into non-overlapping blocks of length d. Let the j-th block be denoted by X_j. It is a vector of length d. Thus we have

$$X(u) = \sum_{i=0} x_i u^i = \sum_{j=0} X_j(u) u^{jd}, \tag{ii.5}$$

where $X_j(u)$ is the generating function of X_j and a polynomial of degree $d - 1$. The acyclic convolution of $X(u)$ and $Y(u)$ is given by

$$Z(u) = X(u)Y(u) = \sum_{j=0} \left[X_j(u) Y(u) \right] u^{jd}. \tag{ii.6}$$

Fast algorithms are employed to compute length $n = d + e - 1$ acyclic convolution $X_j(u)\ Y(u)$, $j = 0, 1, \ldots$. The length of $X_j(u)\ Y(u)$ is $n = d + e - 1$ while successive blocks $X_j(u)\ Y(u)$ are separated by d places. Therefore, there is an overlap of $n - d$ points among the adjacent blocks. The overlapping portions of successive blocks must be added in order to construct $Z(u)$ from $X_j(u)\ Y(u)$, $j = 0, 1, \ldots$. Consequently, the overlap-add technique requires one acyclic convolution of length n and $n - d$ ADDs per d output points.

Overlap-save technique. This technique consists in partitioning input x into overlapping blocks of length n. The j-th block overlaps with the $(j - 1)$th block in $(n - d)$ places, $n = d + e - 1$. Thus the j-th block is given by

$$X_j(u) = \sum_{i=0}^{n-1} x_{i+jd} u^i .$$

Fast algorithms are employed to compute cyclic convolution of length n

$$Z_j(u) = X_j(u)\ Y(u) \bmod (u^n - 1),\ j = 0, 1, \ldots .$$

Let

$$Z_j(u) = \sum_{i=0}^{n-1} z_{j,i} u^i .$$

Then

$$z_{j,i} = \sum_{k=0}^{n-1} x_{i-k+jd} y_k = \sum_{k=0}^{n-d} x_{i-k+jd} y_k ,$$

where the last equality is obtained by noting that $y_k = 0$ for $k = n - d + 1, \ldots, n - 1$. A comparison of this cyclic convolution with $X(u)\ Y(u)$ reveals that

$$z_{i+jd} = z_{j,i}, \quad i = n - d, \ldots, n - 1, \quad j = 0, 1, \ldots,$$

which implies that its last d points are the same as the output points, while the first $n - d$ points are to be discarded. The above technique does not compute the first d points of the output, that is, $z_0, \ldots, z_{d-1}$, which are to be computed separately, if required. Finally, we observe that one does not need to compute the entire cyclic convolution in this technique, only the last d points. This fact may be used to further reduce the computational complexity of the overlap-save technique.

Chapter 8

Fast Algorithms for Acyclic Convolution

In this chapter, we develop fast algorithms for computing one- and higher-dimensional acyclic convolution of discrete sequences. Techniques will also be described for converting one-dimensional acyclic convolution into multi-dimensional acyclic convolution. A worthwhile objective is to design algorithms with as low a multiplicative complexity as possible. We analyze the computational complexity of various algorithms and use it as a benchmark to compare them. Let us first consider one-dimensional convolution of two sequences expressed as the polynomial product,

$$Z(u) = X(u)\, Y(u), \tag{8.0.1}$$

where $deg(X(u)) = d - 1$, $deg(Y(u)) = e - 1$, and $deg(Z(u)) = n - 1$, $n = d + e - 1$. In a similar manner, two-dimensional acyclic convolution can be expressed as the polynomial product

$$Z(u, v) = X(u, v)\, Y(u, v). \tag{8.0.2}$$

There are two techniques that can be used to compute such products. The first technique employs the CRT-P and represents a large degree polynomial product as a number of small degree polynomial products. The second technique first converts a one-dimensional polynomial product into a multidimensional polynomial product and then evaluates the multidimensional polynomial product in an efficient manner. In essence, it computes a larger size convolution by employing algorithms for smaller size convolution. In the following, we describe both of these techniques in complete detail.

8.1 CRT-P Based Fast Algorithms for One-Dimensional Acyclic Convolution

Since $deg(Z(u))$ is $n - 1$, $Z(u)$ is unchanged if it is defined modulo any polynomial $P(u)$ of degree at least n, that is,

$$Z(u) \equiv X(u)\, Y(u) \bmod P(u), \quad deg(P(u)) \geq n. \tag{8.1.1}$$

If $P(u)$ is the product of λ relatively prime polynomials $p_i(u)$, $i = 1, 2, ..., \lambda$, that is,

$$P(u) = \prod_{i=1}^{\lambda} p_i(u), \tag{8.1.2}$$

then $Z(u)$ can be computed by the following CRT-P based three-step procedure:

Step 1. CRT-P reduction. Compute

$$\begin{aligned} X_i(u) &\equiv X(u) \bmod p_i(u) \\ Y_i(u) &\equiv Y(u) \bmod p_i(u), \quad i = 1, 2, ..., \lambda. \end{aligned} \tag{8.1.3}$$

Step 2. Small degree polynomial MULT. Compute

$$Z_i(u) \equiv Z(u) \bmod p_i(u) \equiv X_i(u)\, Y_i(u) \bmod p_i(u), \quad i = 1, 2, ..., \lambda. \tag{8.1.4}$$

Step 3. CRT-P reconstruction. Compute

$$Z(u) \equiv \sum_{i=1}^{\lambda} Z_i(u) T_i(u) [P(u)/p_i(u)] \bmod P(u). \tag{8.1.5}$$

If $p_i(u)$ is an irreducible polynomial, then one way to perform the computation of small degree polynomial product in Step 2 is:

1. Compute the acyclic convolution $V(u) = X_i(u)\, Y_i(u)$, and
2. Reduce $V(u) \bmod p_i(u)$.

Note that this is one of many possible ways to handle the computation of Step 2. It is neither necessary nor the best way (from a computational complexity point of view). There may be other ways. All of these different ways will have the same multiplicative complexity, however, they will differ in their additive complexity and the overall organization of the computation. Figure 8.1 shows the configuration of the CRT-P based fast algorithm for acyclic convolution.

Example 8.1 Consider the computation

$$\begin{aligned} Z(u) &= (x_0 + x_1 u)(y_0 + y_1 u) \bmod (u^2 + u + 1) \\ &= (x_0 y_0 - x_1 y_1) + (x_0 y_1 + x_1 y_0 - x_1 y_1)\, u. \end{aligned}$$

If ordinary polynomial MULT $(x_0 + x_1 u)(y_0 + y_1 u)$ is performed as

$$\begin{aligned} V(u) &= v_0 + v_1 u + v_2 u^2 \\ &= (x_0 y_0) + [(x_0 + x_1)(y_0 + y_1) - x_0 y_0 - x_1 y_1]\, u + x_1 y_1 u^2, \end{aligned}$$

and $Z(u)$ is obtained as $Z(u) = (v_0 - v_2) + (v_1 - v_2)\, u$, then the entire computation would cost 3 MULTs and 6 ADDs. The entire computation when the two parts are written together has the form

$$\begin{aligned} &(x_0 + x_1 u)(y_0 + y_1 u) \bmod (u^2 + u + 1) \\ &\quad = (x_0 y_0 - x_1 y_1) + [(x_0 + x_1)(y_0 + y_1) - x_0 y_0 - 2\, x_1 y_1]\, u. \end{aligned}$$

A method requiring 3 MULTs and 4 ADDs is obtained by expressing

$$\begin{aligned} &(x_0 + x_1 u)(y_0 + y_1 u) \bmod (u^2 + u + 1) \\ &\quad = (x_0 y_0 - x_1 y_1) + [x_0 y_0 - (x_0 - x_1)(y_0 - y_1)]\, u. \end{aligned}$$

This is based on the ordinary polynomial MULT

$$\begin{aligned} V(u) &= v_0 + v_1 u + v_2 u^2 \\ &= (x_0 y_0) + [x_0 y_0 + x_1 y_1 - (x_0 - x_1)(y_0 - y_1)]\, u + x_1 y_1 u^2. \end{aligned}$$

Interestingly, if we were to multiply polynomials modulo $u^2 - u + 1$, then the first approach would be preferable. There are other ways that cannot be obtained just by using these two stages. For example,

$$(x_0 + x_1 u)(y_0 + y_1 u) \bmod (u^2 + u + 1)$$

can also be computed as

$$[(x_0 - x_1)\, y_0 + x_1 (y_0 - y_1)] + [x_0 y_1 + x_1 (y_0 - y_1)]\, u.$$

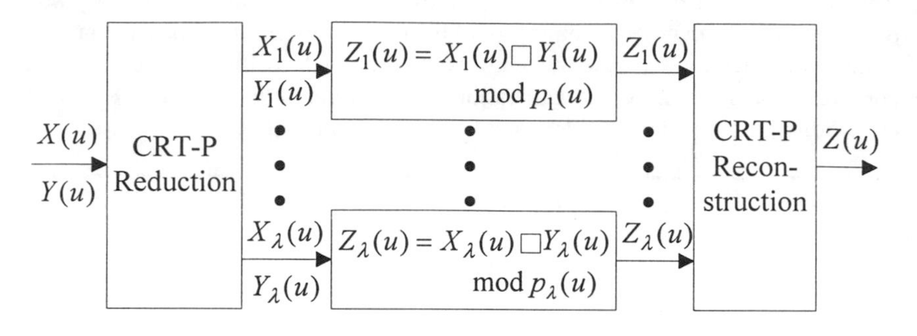

□ is either + or •

$$P(u) = \prod_{i=1}^{\lambda} p_i(u),\ \big(p_i(u),\ p_j(u)\big) = 1$$

$X(u) \longrightarrow (\square) \longrightarrow Z(u) = X(u)\ \square\ Y(u) \bmod P(u)$, with $Y(u)$ entering the $\square$ node from below.

Note: For acyclic convolution □ = •.

Figure 8.1 Configuration of CRT-P Based Algorithm for Acyclic Convolution

---

In the above procedure, Steps 1 and 3 require only ADDs. If $M(d_i)$ denotes the multiplicative complexity to compute the i-th product in Step 2, that is,

$$Z_i(u) \equiv X_i(u)\ Y_i(u) \bmod p_i(u),\ \ deg(p_i(u)) = d_i,$$

then the multiplicative complexity of the procedure is

$$M(n) = \sum_{i=1}^{\lambda} M(d_i). \tag{8.1.6}$$

Note that the multiplicative complexity is denoted by $M(d_i)$ for convenience; it depends on the precise form of $p_i(u)$. Though Step 3 of CRT-P based procedure requires no MULT, its additive complexity can be quite high. One may attempt to reduce it by experimenting with different ways to compute the CRT-P reconstruction. For example, if $P(u)$ has three factors, say $p_1(u)$, $p_2(u)$, and $p_3(u)$, then one can first combine the residues polynomials for $p_1(u)$ and $p_2(u)$ using the CRT-P reconstruction first and then combine the resulting polynomial (which is defined mod $q_1(u) = p_1(u)\, p_2(u)$) with the residue polynomial for $p_3(u)$. Other ways could also be tested in order to select the one corresponding to the least additive complexity. For example, we have found that if

$$P(u) = u\,(u^2 + 1)\,(u - 1)\,(u^2 + u + 1),$$

then it is better to combine the residues for $(u - 1)$ and $(u^2 + u + 1)$ first giving rise to a polynomial mod $u^3 - 1$ and then combine it with the other two residues.

The objective is to develop algorithms with as low a computational complexity as possible for a given value of n. Therefore, for a given degree of the polynomial $P(u)$, factors $p_i(u)$ are selected in such a way that the complexity of the computation $X(u)\, Y(u)$ mod $P(u)$ is as low as possible. In general, a large number of relatively prime factors leads to a computationally efficient procedure. With this in mind, the following criteria should be satisfied:

(A) $p_i(u)$ and $p_j(u)$ are relatively prime, $1 \le i < j \le \lambda;\ i \ne j$,

(B) $\sum_{i=1}^{\lambda} deg(p_i(u)) = deg(P(u))$, and

(C) Each of the polynomials $p_i(u)$ has as low a degree as possible.

Note that the form of $p_i(u)$ depends on the field of constants, F. For example, if all the factors are to be of degree one over Z, then there are infinitely many choices such as u, $u - 1$, $u + 1$, $u + 2$, $u - 2$, and so on, while over $GF(p)$, there are only p distinct choices. Thus, knowledge about the field of constants is incorporated directly into the design of algorithm. It is worthwhile to mention here that a CRT-P based acyclic convolution algorithm is known as the Cook-Toom algorithm when all the factor polynomials are of degree 1 and is the same as LI.

8.1.1 Improvements in the Basic Procedure

The complexity of the algorithm can be further reduced by modifying the basic procedure described above to permit *intentional* **wraparound** of the polynomial coefficients. For example, if

$$M(n - 1) < M(n) - 1,$$

then it is more efficient to calculate $Z(u)$ from the product

$$X(u)\ Y(u) \text{ modulo } P_1(u), \quad deg(P_1(u)) = n - 1,$$

with one extra MULT; that is, $x_{d-1} \cdot y_{e-1}$. Similarly, if

$$M(n-2) < M(n-1) - 2 < M(n) - 3,$$

it may be preferable to compute $Z(u)$ from the product

$$X(u)\ Y(u) \text{ modulo } P_2(u), \quad deg(P_2(u)) = n - 2,$$

with three more MULTs. Also, if $p_i(u)$ is of the form $\left(u - a_i\right)^{d_i}$, where a_i is an element of the number system, the product

$$X_i(u)\ Y_i(u) \bmod \left(u - a_i\right)^{d_i}$$

is relatively less complex to compute, in general, as compared to $X_i(u)\ Y_i(u)$ for several values of d_i. The modified algorithm to compute the product $X(u)\ Y(u)$ can be described in three schemes, as given below.

Scheme 1. Computation of $X(u)\ Y(u)$, $deg(X(u)) = d - 1$, $deg(Y(u)) = e - 1$. If $d = e = 1$, the result is obtained in one MULT; otherwise, select an integer s to minimize the total number of MULTs required to compute

(i) $X(u)\ Y(u) \bmod P(u)$, $deg(P(u)) = n - s$, and
(ii) $\overline{X}(u)\overline{Y}(u) \bmod u^s$ using Scheme 3, where

$$\overline{X}(u) = X\left(u^{-1}\right)u^{d-1},$$
$$\overline{Y}(u) = Y\left(u^{-1}\right)u^{e-1}.$$

Scheme 2. Computation of $X_i(u)\ Y_i(u) \bmod p_i(u)$. If $p_i(u) = \left(u - a_i\right)^{d_i}$, use Scheme 3 to compute $X_i(u)\ Y_i(u) \bmod p_i(u)$; otherwise, compute the ordinary product $X_i(u)\ Y_i(u)$ and reduce mod $p_i(u)$.

Scheme 3. Computation of $X(u)\ Y(u) \bmod p(u)$, $p(u) = (u - a)^d$. If $a = 0$, use one of the following two methods which is least computationally complex:

(i) Use Scheme 2.
(ii) Define

$$m_{l_1 l_2} = \left\{ \sum_{i=l_1}^{l_2} A_i \right\} \left\{ \sum_{j=l_1}^{l_2} B_j \right\}.$$

Then we have

$$A_{l_1} B_{l_2} = m_{l_1 l_2} + m_{l_1+1, l_2-1} - m_{l_1, l_2-1} - m_{l_1+1, l_2}.$$

It can be shown that the number of MULTs required to compute the product

$$C(u) = A(u)\, B(u) \bmod p(u)$$

is given by $f(f+1) - 1$ for $d = 2f - 1$ and $f(f+2)$ for $d = 2f$. If $a \neq 0$, define $\tilde{A}(u) = A(u+a)$ and $\tilde{B}(u) = B(u+a)$ and compute

$$\tilde{C}(u) = \tilde{A}(u)\tilde{B}(u) \bmod u^d$$

as described above. The product C(u) is then given by $C(u) = \tilde{C}(u-a)$.

By using bilinear small degree polynomial MULT algorithms, some of which are listed in **Appendix A**, and the procedure described above, we can design bilinear algorithms for larger values of n. Such a procedure is quite straightforward and requires no further elaboration.

8.2 Casting the Algorithm in a Bilinear Form

Let the coefficient vectors of $X(u)$, $Y(u)$, $Z(u)$, $X_i(u)$, $Y_i(u)$, and $Z_i(u)$ be $\underline{X}$, $\underline{Y}$, $\underline{Z}$, $\underline{X}_i$, $\underline{Y}_i$, and $\underline{Z}_i$, respectively. The computation of $\underline{Z}_i$ in (8.1.4) has the form

$$\underline{Z}_i = \mathbf{D}_i \left[\mathbf{E}_i \underline{X}_i \otimes \mathbf{F}_i \underline{Y}_i \right], \quad i = 1, 2, ..., \lambda + 1. \qquad (8.2.1)$$

Here the last block for $i = \lambda + 1$ corresponds to the computation of wraparound. The various polynomial degrees are $deg(X(u)) = d - 1$, $deg(Y(u)) = e - 1$, $deg(Z(u)) = n - 1$, $deg(p_i(u)) = n_i$ and $n_1 + n_2 + ... + n_\lambda = n - s$. According to (8.1.3), there exist matrices $\mathbf{G}_i$ and $\mathbf{K}_i$ of dimensions $n_i \times d$ and $n_i \times e$, respectively, such that $\underline{X}$, $\underline{Y}$, $\underline{X}_i$, $\underline{Y}_i$, $\mathbf{G}_i$, and $\mathbf{K}_i$ are related as

$$\underline{X}_i = \mathbf{G}_i \underline{X}$$
$$\underline{Y}_i = \mathbf{K}_i \underline{Y}, \quad i = 1, 2, ..., \lambda + 1. \qquad (8.2.2)$$

The matrices $\mathbf{G}_i$, $i = 1, 2, ..., \lambda$ depend only on the monic factor polynomial $p_i(u)$ while $\mathbf{G}_{i+1}$ depends on the wraparound. All of these are known *a priori*. The CRT-P reconstruction (along with wraparound) in (8.1.5) can be represented in the matrix form as

$$\underline{Z} = \sum_{i=1}^{\lambda+1} \mathbf{H}_i \underline{Z}_i \,. \tag{8.2.3}$$

Substituting (8.2.2) in (8.2.1) and replacing $\underline{Z}_i$ in (8.2.3) by the resulting expressions, we get

$$\underline{Z} = \mathbf{C}\,[\mathbf{A}\,\underline{X} \otimes \mathbf{B}\,\underline{Y}], \tag{8.2.4}$$

where

$$\mathbf{A} = \begin{bmatrix} \mathbf{E}_1 & & \cdots & \\ & \mathbf{E}_2 & \cdots & \\ & & \ddots & \\ & & \cdots & \mathbf{E}_{\lambda+1} \end{bmatrix} \begin{bmatrix} \mathbf{G}_1 \\ \mathbf{G}_2 \\ \vdots \\ \mathbf{G}_{\lambda+1} \end{bmatrix}$$

$$= \mathbf{E}\,\mathbf{G}, \tag{8.2.5}$$

$$\mathbf{B} = \begin{bmatrix} \mathbf{F}_1 & & \cdots & \\ & \mathbf{F}_2 & \cdots & \\ & & \ddots & \\ & & \cdots & \mathbf{F}_{\lambda+1} \end{bmatrix} \begin{bmatrix} \mathbf{K}_1 \\ \mathbf{K}_2 \\ \vdots \\ \mathbf{K}_{\lambda+1} \end{bmatrix}$$

$$= \mathbf{F}\,\mathbf{K}, \tag{8.2.6}$$

and

$$\mathbf{C} = \begin{bmatrix} \mathbf{H}_1 & \mathbf{H}_2 & \cdots & \mathbf{H}_{\lambda+1} \end{bmatrix} \begin{bmatrix} \mathbf{D}_1 & & \cdots & \\ & \mathbf{D}_2 & \cdots & \\ & & \ddots & \\ & & \cdots & \mathbf{D}_{\lambda+1} \end{bmatrix}$$

$$= \mathbf{H}\,\mathbf{D}. \tag{8.2.7}$$

The matrices **G** and **K** will be referred to as the CRT-P *reduction* matrices, while **H** will be referred to as the CRT-P *reconstruction* matrix. The matrix-vector products $\mathbf{A}\,\underline{X}$ and $\mathbf{B}\,\underline{Y}$ correspond to pre-ADDs while the matrix-vector product $\mathbf{C}\,\underline{M}$, $\underline{M} = [\mathbf{A}\,\underline{X} \otimes \mathbf{B}\,\underline{Y}]$ corresponds to post-ADDs. Combined together, they lead to the additive complexity of the overall algorithm. We will set up a bilinear form for fast algorithms and then attempt to identify a suitable organization of the computation that minimizes the additive complexity. It was mentioned earlier that one may attempt to reduce the additive complexity by experimenting with different ways to compute the CRT-P reconstruction. In essence this corresponds to different factorizations for the first λ blocks (the ones that correspond to the CRT-P) of the overall reconstruction matrix $\mathbf{H}$. Finally, this bilinear formulation remains valid for fast algorithms for the computation of cyclic convolution.

Example 8.2 Let us derive the bilinear form for acyclic convolution of length $n = 10$. It does not matter what the values of d and e are as long as $n = d + e - 1$. We choose

$$P(u) = u\,(u^2 + 1)\,(u - 1)\,(u + 1)\,(u^2 + u + 1)\,(u^2 - u + 1)$$

and a wraparound of $s = 1$. Let

$$\begin{aligned} p_1(u) &= u, \\ p_2(u) &= u^2 + 1, \\ p_3(u) &= u - 1, \\ p_4(u) &= u + 1, \\ p_5(u) &= u^2 + u + 1, \\ p_6(u) &= u^2 - u + 1. \end{aligned}$$

The various matrices of the above formulation are found from the following matrices:

$$\mathbf{G}' = \left[\begin{array}{cccccccccc} 1 & 0 & 0 & 0 & 0 & 0 & 0 & 0 & 0 & 0 \\ \hdashline 1 & 0 & -1 & 0 & 1 & 0 & -1 & 0 & 1 & 0 \\ 0 & 1 & 0 & -1 & 0 & 1 & 0 & -1 & 0 & 1 \\ \hdashline 1 & 1 & 1 & 1 & 1 & 1 & 1 & 1 & 1 & 1 \\ \hdashline 1 & -1 & 1 & -1 & 1 & -1 & 1 & -1 & 1 & -1 \\ \hdashline 1 & 0 & -1 & 1 & 0 & -1 & 1 & 0 & -1 & 1 \\ 0 & 1 & -1 & 0 & 1 & -1 & 0 & 1 & -1 & 0 \\ \hdashline 1 & 0 & -1 & -1 & 0 & 1 & 1 & 0 & -1 & -1 \\ 0 & 1 & 1 & 0 & -1 & -1 & 0 & 1 & 1 & 0 \\ \hdashline 0 & 0 & 0 & 0 & 0 & 0 & 0 & 0 & 0 & 1 \end{array}\right], \qquad (8.2.8)$$

$$
\mathbf{E} = \mathbf{F} = \begin{bmatrix}
1 & & & & & & & & & \\
& 1 & 0 & & & & & & & \\
& 0 & 1 & & & & & & & \\
& 1 & 1 & & & & & & & \\
& & & 1 & & & & & & \\
& & & & 1 & & & & & \\
& & & & & 1 & 0 & & & \\
& & & & & 0 & 1 & & & \\
& & & & & 1 & -1 & & & \\
& & & & & & & 1 & 0 & \\
& & & & & & & 0 & 1 & \\
& & & & & & & 1 & 1 & \\
& & & & & & & & & 1
\end{bmatrix}, \tag{8.2.9}
$$

$$
\mathbf{D} = \begin{bmatrix}
1 & & & & & & & & & & & & \\
& 1 & -1 & 0 & & & & & & & & & \\
& -1 & -1 & 1 & & & & & & & & & \\
& & & & 1 & & & & & & & & \\
& & & & & 1 & & & & & & & \\
& & & & & & 1 & -1 & 0 & & & & \\
& & & & & & -1 & -1 & 1 & & & & \\
& & & & & & & & & 1 & -1 & 1 & \\
& & & & & & & & & -1 & 0 & 1 & \\
& & & & & & & & & & & & 1
\end{bmatrix}, \tag{8.2.10}
$$

$$
\mathbf{H} = \left[\begin{array}{r|rr|r|r|rr|rr|r}
1 & 0 & 0 & 0 & 0 & 0 & 0 & 0 & 0 & 0 \\
0 & 0 & 1/2 & 1/12 & -1/12 & 1/6 & 1/6 & -1/6 & 1/6 & -1 \\
1 & -1/2 & 0 & 1/12 & 1/12 & -1/3 & 1/6 & -1/3 & -1/6 & 0 \\
0 & 0 & 0 & 1/6 & -1/6 & 1/3 & -1/6 & -1/3 & -1/6 & -1 \\
0 & 0 & 0 & 1/6 & 1/6 & -1/6 & 1/3 & -1/6 & -1/3 & 0 \\
0 & 0 & 0 & 1/6 & -1/6 & -1/6 & -1/6 & 1/6 & -1/6 & 0 \\
-1 & 0 & 0 & 1/6 & 1/6 & 1/3 & -1/6 & 1/3 & 1/6 & 0 \\
0 & 0 & -1/2 & 1/12 & -1/12 & -1/3 & 1/6 & 1/3 & 1/6 & 1 \\
-1 & 1/2 & 0 & 1/12 & 1/12 & 1/6 & -1/3 & 1/6 & 1/3 & 0 \\
\hline
0 & 0 & 0 & 0 & 0 & 0 & 0 & 0 & 0 & 1
\end{array}\right]. \tag{8.2.11}
$$

The part of **H** in the northwest of the partition marked by solid lines (let us call it **H**′) corresponds to CRT-P reconstruction while the remaining parts correspond to wraparound. There are two types of blocks in **G**′; the first λ blocks correspond to modulo polynomials while the last one corresponds to wraparound. We have shown **G**′ as a $(n \times n)$ matrix. However, the CRT-P reduction matrices **G** and **K** have dimensions $(n \times d)$ and $(n \times e)$, respectively. The matrix **G** can be obtained from **G**′ by deleting the last $(n - d)$ columns from the first t partitions and the first $(n - d)$ columns from the last partition. The matrix **K** is obtained from **G**′ by deleting $(n - e)$ columns in a similar manner. For example, if $d = 6$ and $e = 5$, then

$$\mathbf{G} = \left[\begin{array}{cccccc} 1 & 0 & 0 & 0 & 0 & 0 \\ \hline 1 & 0 & -1 & 0 & 1 & 0 \\ 0 & 1 & 0 & -1 & 0 & 1 \\ \hline 1 & 1 & 1 & 1 & 1 & 1 \\ \hline 1 & -1 & 1 & -1 & 1 & -1 \\ \hline 1 & 0 & -1 & 1 & 0 & -1 \\ 0 & 1 & -1 & 0 & 1 & -1 \\ \hline 1 & 0 & -1 & -1 & 0 & -1 \\ 0 & 1 & 1 & 0 & -1 & 1 \\ \hline 0 & 0 & 0 & 0 & 0 & 1 \end{array}\right] \tag{8.2.12}$$

$$\mathbf{K} = \left[\begin{array}{ccccc} 1 & 0 & 0 & 0 & 0 \\ \hline 1 & 0 & -1 & 0 & 1 \\ 0 & 1 & 0 & -1 & 0 \\ \hline 1 & 1 & 1 & 1 & 1 \\ \hline 1 & -1 & 1 & -1 & 1 \\ \hline 1 & 0 & -1 & 1 & 0 \\ 0 & 1 & -1 & 0 & 1 \\ \hline 1 & 0 & -1 & -1 & 0 \\ 0 & 1 & 1 & 0 & -1 \\ \hline 0 & 0 & 0 & 0 & 1 \end{array}\right]. \tag{8.2.13}$$

In the rest of this example, we assume that $d = 6$ and $e = 5$.

Given $z_1(u)$, $z_2(u)$, and $z_3^{(1)}(u)$, the residues of a degree 8 polynomial modulo u, (u^2+1), and (u^6-1), respectively, the CRT-P reconstruction matrix is given by

$$\mathbf{R} = \left[\begin{array}{c|cc|cccccc} 1 & 0 & 0 & 0 & 0 & 0 & 0 & 0 & 0 \\ 0 & 0 & 1/2 & 0 & 1/2 & 0 & 1/2 & 0 & -1/2 \\ 1 & -1/2 & 0 & -1/2 & 0 & 1/2 & 0 & 1/2 & 0 \\ 0 & 0 & 0 & 0 & 0 & 0 & 1 & 0 & 0 \\ 0 & 0 & 0 & 0 & 0 & 0 & 0 & 1 & 0 \\ 0 & 0 & 0 & 0 & 0 & 0 & 0 & 0 & 1 \\ -1 & 0 & 0 & 1 & 0 & 0 & 0 & 0 & 0 \\ 0 & 0 & -1/2 & 0 & 1/2 & 0 & -1/2 & 0 & 1/2 \\ 1 & 1/2 & 0 & 1/2 & 0 & 1/2 & 0 & -1/2 & 0 \end{array}\right]. \tag{8.2.14}$$

Similarly, given the residues of a degree 5 polynomial modulo $u-1$, $u+1$, u^2+u+1, u^2-u+1, CRT-P reduction and reconstruction matrices are given by

$$\mathbf{P} = \left[\begin{array}{cccccc} 1 & 1 & 1 & 1 & 1 & 1 \\ \hline 1 & -1 & 1 & -1 & 1 & -1 \\ \hline 1 & 0 & -1 & 1 & 0 & -1 \\ 0 & 1 & -1 & 0 & 1 & -1 \\ \hline 1 & 0 & -1 & -1 & 0 & 1 \\ 0 & 1 & 1 & 0 & -1 & -1 \end{array}\right],$$

$$\mathbf{S} = \frac{1}{6}\left[\begin{array}{c|c|cc|cc} 1 & 1 & 2 & -1 & 2 & 1 \\ 1 & -1 & -1 & 2 & 1 & 2 \\ 1 & 1 & -1 & -1 & -1 & 1 \\ 1 & -1 & 2 & -1 & -2 & -1 \\ 1 & 1 & -1 & 2 & -1 & -2 \\ 1 & -1 & -1 & -1 & 1 & -1 \end{array}\right]. \tag{8.2.15}$$

Based on $\mathbf{S}$, form the matrix $\mathbf{T}$ as

$$\mathbf{T} = \left[\begin{array}{c|c|c} \mathbf{I}_1 & & \\ \hline & \mathbf{I}_2 & \\ \hline & & \mathbf{S} \end{array}\right] \tag{8.2.16}$$

It is easy to verify that $\mathbf{H}' = \mathbf{R}\,\mathbf{T}$. These factorizations will prove to be of value in reducing the additive complexity of the CRT-P reconstruction. We also note that **P** and **S** correspond to the CRT-P reduction and reconstruction for a cyclic convolution of length 6 as

$$(u-1)(u+1)(u^2+u+1)(u^2-u+1) = u^6 - 1.$$

Therefore, other simpler forms of **S**, and thus $\mathbf{H}'$, are possible in this case that can be obtained using Theorem 4.8. One of the organizations for the computation of length 10 convolution based on the choice of $P(u)$ as stated, that reduces additive complexity can be described as follows. The derivation includes every detail in order to assist the reader to exploit similar ideas in other situations.

Based on (8.2.14), we observe that if, instead, we are given $z_1(u)$, $(1/2)\,z_2(u)$, and $(1/2)z_3^{(1)}(u)$, the CRT-P reconstruction matrix simplifies to

$$\mathbf{R} = \left[\begin{array}{c|cc|cccccc} 1 & 0 & 0 & 0 & 0 & 0 & 0 & 0 & 0 \\ 0 & 0 & 1 & 0 & 1 & 0 & 1 & 0 & -1 \\ 1 & -1 & 0 & -1 & 0 & 1 & 0 & 1 & 0 \\ 0 & 0 & 0 & 0 & 0 & 0 & 2 & 0 & 0 \\ 0 & 0 & 0 & 0 & 0 & 0 & 0 & 2 & 0 \\ 0 & 0 & 0 & 0 & 0 & 0 & 0 & 0 & 2 \\ -1 & 0 & 0 & 2 & 0 & 0 & 0 & 0 & 0 \\ 0 & 0 & -1 & 0 & 1 & 0 & -1 & 0 & 1 \\ 1 & 1 & 0 & 1 & 0 & 1 & 0 & -1 & 0 \end{array}\right]. \qquad (8.2.17)$$

Let

$$z_1(u) = z_{10}$$
$$(1/2)\,z_2(u) = z_{20} + z_{21}\,u$$

and

$$(1/2)z_3^{(1)}(u) = z_{30}^{(1)} + z_{31}^{(1)}u + z_{32}^{(1)}u^2 + z_{33}^{(1)}u^3 + z_{34}^{(1)}u^4 + z_{35}^{(1)}u^5.$$

Note that $(1/2)z_3^{(1)}(u)$ is an auxiliary polynomial and appears only as an intermediate quantity in the computation. Based on (8.2.17), the CRT-P reconstruction can be written as

$$
\begin{aligned}
z_0 &= z_{10} \\
z_1 &= z_{21} + z_{31}^{(1)} + \left(z_{33}^{(1)} - z_{35}^{(1)} \right) \\
z_2 &= z_{32}^{(1)} + \left(z_{10} - z_{20} - z_{30}^{(1)} + z_{34}^{(1)} \right) \\
z_3 &= z_{33}^{(1)} + z_{33}^{(1)} \\
z_4 &= z_{34}^{(1)} + z_{34}^{(1)} \\
z_5 &= z_{35}^{(1)} + z_{35}^{(1)} \\
z_6 &= -z_{10} + z_{30}^{(1)} + z_{30}^{(1)} \\
z_7 &= -z_{21} + z_{31}^{(1)} - \left(z_{33}^{(1)} - z_{35}^{(1)} \right) \\
z_8 &= z_{32}^{(1)} - \left(z_{10} - z_{20} - z_{30}^{(1)} + z_{34}^{(1)} \right).
\end{aligned}
\tag{8.2.18}
$$

The ADDs in parentheses are common and need to be performed only once. The expressions for z_1 and z_7, and for z_2 and z_8 share common ADDs. Clearly, the computational complexity of this reconstruction, if performed as above, is 15 ADDs.

Let us now turn to each of these residues. The computation

$$z_1(u) \bmod u = X(u)\ Y(u) \bmod u = X_0\ Y_0$$

requires only 1 MULT. The computation

$$
\begin{aligned}
(1/2)\ z_2(u) &= (1/2)\ Z(u) \bmod (u^2 + 1) \\
&= \{X(u) \bmod u^2 + 1\}\ \{(1/2)\ Y(u) \bmod u^2 + 1\}
\end{aligned}
$$

can be performed as

$$
\begin{aligned}
&a_0 = x_0 - x_2 + x_4, \quad b_0 = (y_0 - y_2 + y_4)/2, \\
&a_1 = x_1 - x_3 + x_5, \quad b_1 = (y_1 - y_3)/2, \\
&a_2 = a_0 + a_1, \quad b_2 = b_0 + b_1;
\end{aligned}
$$

$$K_i = a_i \cdot b_i, \ \ i = 0, 1, 2;$$

$$
\begin{aligned}
z_{20} &= K_0 - K_2, \\
z_{21} &= -K_0 - K_1 + K_2.
\end{aligned}
$$

This computation requires 5 ADDs for the reduction of $X(u)$, 4 ADDs for reduction of $Y(u)$, 3 MULTs, 2 scaling operations (by $1/2$), and 3 ADDs for reconstruction of $(1/2)\ z_2(u)$. Note that we count the number of ADDs for reduction of $X(u)$ and $Y(u)$ separately. This is to make the analysis valid for the situation when both $X(u)$ and $Y(u)$ are arbitrary as well as for the situation when one of the sequences ($Y(u)$ in the present set-up) is fixed. In the latter case, ADDs for the reduction of $Y(u)$ are to be ignored. The readers will notice that we put all the scaling operations on the y sequence for the same reason.

The computation for $(1/2)z_3^{(1)}(u)$ can be expressed as

$$(1/2)z_3^{(1)}(u) = X(u)\{(1/2)Y(u)\}\bmod\left(u^6 - 1\right).$$

Let the bilinear forms for the computations

$$\{a_0 + a_1 u\}\ \{b_0 + b_1 u\} \bmod u^2 + u + 1$$

and

$$\{a_0 + a_1 u\}\ \{b_0 + b_1 u\} \bmod u^2 - u + 1$$

be given by

$$\begin{bmatrix} 1 & -1 & 0 \\ 1 & 0 & -1 \end{bmatrix} \left\{ \begin{bmatrix} 1 & 0 \\ 0 & 1 \\ 1 & -1 \end{bmatrix} \begin{bmatrix} a_0 \\ a_1 \end{bmatrix} \otimes \begin{bmatrix} 1 & 0 \\ 0 & 1 \\ 1 & -1 \end{bmatrix} \begin{bmatrix} b_0 \\ b_1 \end{bmatrix} \right\}$$

and

$$\begin{bmatrix} 1 & -1 & 0 \\ -1 & 0 & 1 \end{bmatrix} \left\{ \begin{bmatrix} 1 & 0 \\ 0 & 1 \\ 1 & 1 \end{bmatrix} \begin{bmatrix} a_0 \\ a_1 \end{bmatrix} \otimes \begin{bmatrix} 1 & 0 \\ 0 & 1 \\ 1 & 1 \end{bmatrix} \begin{bmatrix} b_0 \\ b_1 \end{bmatrix} \right\}.$$

Based on (8.2.15) and the above bilinear forms, the length-6 cyclic convolution of two polynomials $X(u)$ and $(1/2)\ Y(u)$ can be computed as

$$\underline{Z}^{(1)} = \mathbf{C}'\ [\mathbf{A}'\ \underline{X} \otimes \mathbf{B}'\ (1/2)\ \underline{Y}]$$

where

$$\mathbf{A}' = \mathbf{B}' = \begin{bmatrix} 1 & & & & & \\ & 1 & & & & \\ & & 1 & 0 & & \\ & & 0 & 1 & & \\ & & 1 & -1 & & \\ & & & & 1 & 0 \\ & & & & 0 & 1 \\ & & & & 1 & 1 \end{bmatrix} \mathbf{P}$$

$$\mathbf{C}' = \mathbf{S} \begin{bmatrix} 1 & & & & & & & \\ & 1 & & & & & & \\ & & 1 & -1 & 0 & & & \\ & & 1 & 0 & -1 & & & \\ & & & & & 1 & -1 & 0 \\ & & & & & -1 & 0 & 1 \end{bmatrix}.$$

Using Theorem 4.8, we can also compute $(1/2)z_3^{(1)}(u)$ as the bilinear form

$$Z^{(1)} = (\mathbf{J}\,\mathbf{B}')\,[\mathbf{A}\,\underline{X} \otimes (\mathbf{C}'\,\mathbf{J})\,(1/2)\,\underline{Y}].$$

We adopt the latter form in the following. This leads to the following organization for $(1/2)z_3^{(1)}(u)$:

$$\begin{aligned}
&a_0 = (x_0 + x_2 + x_4) + (x_1 + x_3 + x_5),\quad b_0 = \{(y_3 + y_1) + (y_4 + y_2 + y_0)\}/12,\\
&a_1 = (x_0 + x_2 + x_4) - (x_1 + x_3 + x_5),\quad b_1 = \{(y_3 + y_1) - (y_4 + y_2 + y_0)\}/12,\\
&a_2 = x_0 + x_3 - (x_2 + x_5),\quad b_2 = \{(y_3 + y_0) - y_2 - (y_2 - (y_4 + y_1))\}/12,\\
&a_3 = x_1 + x_4 - (x_2 + x_5),\quad b_3 = \{(y_3 + y_0) - y_4 - y_1 - (y_2 - (y_4 + y_1))\}/12,\\
&a_4 = a_2 - a_3,\quad b_4 = -b_2 - b_3,\\
&a_5 = x_0 - x_3 - (x_2 - x_5),\quad b_5 = -\{-y_2 + (-y_2 + (y_4 - y_1)) - (y_3 - y_0)\}/12,\\
&a_6 = x_1 - x_4 + (x_2 - x_5),\quad b_6 = -\{y_4 - y_1 + (-y_2 + (y_4 - y_1)) + (y_3 - y_0)\}/12,\\
&a_7 = a_5 + a_6,\quad b_7 = -b_5 - b_6;
\end{aligned}$$

$$K_i = a_i \cdot b_i,\quad i = 0, \ldots, 7.$$

$$\begin{aligned}
z_{30}^{(1)} &= (K_0 - (K_2 + K_4) - (K_3 - K_4)) - (K_1 - (K_5 + K_7) + (K_6 + K_7)),\\
z_{31}^{(1)} &= (K_0 + (K_3 - K_4)) + (K_1 - (K_6 + K_7)),\\
z_{32}^{(1)} &= (K_0 + (K_2 + K_4)) - (K_1 + (K_5 + K_7)),\\
z_{33}^{(1)} &= (K_0 - (K_2 + K_4) - (K_3 - K_4)) + (K_1 - (K_5 + K_7) + (K_6 + K_7)),
\end{aligned}$$

$$z_{34}^{(1)} = (K_0 + (K_3 - K_4)) - (K_1 - (K_6 + K_7)),$$
$$z_{35}^{(1)} = (K_0 + (K_2 + K_4)) + (K_1 + (K_5 + K_7)).$$

Once again, ADDs in parentheses are common and need to be performed only once. This computation requires 18 ADDs for reduction of $X(u)$, 21 ADDs for reduction of $Y(u)$, 8 MULTs, 6 scaling operations (by $1/12$), and 18 ADDs for reconstruction of $(1/2)z_3^{(1)}(u)$. Finally, the recovery of the degree 9 polynomial from its residue mod $u\,(u^2+1)\,(u^6-1)$ and the wraparound computation (which requires 1 MULT) takes 3 ADDs.

Summing up the various quantities, we see that acyclic convolution of length 10 requires 23 ADDs for reduction of $X(u)$, 25 ADDs for reduction of $Y(u)$, 13 MULTs, 2 scaling operations by $1/2$, 6 scaling operations by $(1/12)$, and 39 ADDs for reconstruction. If computations performed on the y sequence are to be ignored, then the computational complexity of length 10 algorithm is 62 ADDs and 13 MULTs. However, if all operations are to be counted, then the computational complexity is 87 ADDs, 13 MULTs, and 8 scaling operations. This is to be compared with the direct approach that requires 30 MULTs and 20 ADDs.

If one were to compute the same convolution using the Cook-Toom algorithm with the modulo polynomial as

$$u\,(u-1)\,(u+1)\,(u-2)\,(u+2)\,(u-3)\,(u+3)\,(u-4)\,(u+4)$$

and a wraparound of 1, then the mulitplicative complexity will be 8 MULTs; however, it would require division by constants which are quite large in magnitude. We end this example by stating that the above algorithm is valid for all those number systems over which 12^{-1} is defined. Therefore, it is valid over all fields and rings except $GF(2^m)$ and $GF(3^m)$ and the integer rings that are p-adic expansions of them.

Example 8.3 Consider the design of length 6 acyclic convolution over $Z(2^2)$. Let $k = 4$ and $e = 3$. We know that the set of coprime polynomials over $GF(2)$ can be used in the present situation as well. Therefore, we choose

$$P(u) = u\,(u^2+1)\,(u^2+u+1)$$

and $s = 1$. The resulting bilinear form is given by

$$\begin{bmatrix} z_0 \\ z_1 \\ z_2 \\ z_3 \\ z_4 \\ z_5 \end{bmatrix} = \begin{bmatrix} 1 & 0 & 0 & 0 & 0 & 0 & 0 & 0 \\ 1 & -1 & 0 & 1 & 1 & 1 & 0 & 1 \\ 2 & 0 & -1 & 2 & 0 & 1 & 1 & 1 \\ 1 & 0 & -1 & 2 & 1 & 1 & 0 & 2 \\ 1 & -1 & -1 & 1 & 0 & 1 & 1 & 1 \\ 0 & 0 & 0 & 0 & 0 & 0 & 0 & 1 \end{bmatrix}$$

$$\cdot \left[\begin{bmatrix} 1 & 0 & 0 & 0 \\ 1 & 0 & -1 & 0 \\ 1 & 1 & -1 & -1 \\ 0 & 1 & 0 & -1 \\ 1 & 0 & -1 & 1 \\ -1 & 1 & 0 & -1 \\ 0 & 1 & -1 & 0 \\ 0 & 0 & 0 & 1 \end{bmatrix} \begin{bmatrix} x_0 \\ x_1 \\ x_2 \\ x_3 \end{bmatrix} \otimes \begin{bmatrix} 1 & 0 & 0 \\ 1 & 0 & -1 \\ 1 & 1 & -1 \\ 0 & 1 & 0 \\ 1 & 0 & -1 \\ 1 & -1 & 0 \\ 0 & 1 & -1 \\ 0 & 0 & 1 \end{bmatrix} \begin{bmatrix} y_0 \\ y_1 \\ y_2 \end{bmatrix} \right].$$

8.3 Multidimensional Approaches to One-Dimensional Acyclic Convolution

In this approach, one-dimensional polynomial product is computed by converting it to a multidimensional polynomial product. This technique can alternatively be interpreted as an algorithm that uses small degree polynomial products recursively to compute product of larger degree polynomials. The acyclic convolution algorithm based on CRT-P (described in previous section) and the multidimensional convolution algorithm described here may be compared in terms of the associated multiplicative complexity. For a given value of d and e, the algorithm having the smaller multiplicative complexity may be preferred. A very crucial feature of this approach is that it is valid for all types of number systems for which data sequences can be defined.

Let us assume that we wish to compute the polynomial product,

$$Z(u) = X(u)\,Y(u)$$

where

$$Z(u) = \sum_{i=0}^{n-1} z_i u^i ,$$

$$X(u) = \sum_{j=0}^{d-1} x_j u^j ,$$

and

$$Y(u) = \sum_{l=0}^{e-1} y_l u^l .$$

Let d and e be composite integers having a common factor, that is, d and e have the form $d = d_1\, c$ and $e = e_1\, c$. Then $X(u)\; Y(u)$ may be transformed into a two-dimensional polynomial product in the following manner. Define the quantities,

$$j = c\, j_2 + j_1,\; j_2 = 0, 1, ..., d_1 - 1;\; l_1, j_1 = 0, 1, ..., c - 1,$$

$$l = c\, l_2 + l_1,\; l_2 = 0, 1, ..., e_1 - 1,$$

$$u_1 = u^c, \tag{8.3.1}$$

and let $X(u, u_1)$, $Y(u, u_1)$, and $Z(u, u_1)$ denote the two-dimensional polynomials corresponding to the polynomials $X(u)$, $Y(u)$, and $Z(u)$, respectively. Using the mapping in (8.3.1), the polynomials $X(u, u_1)$ and $Y(u, u_1)$ can be expressed as

$$X(u) = X(u, u_1) = \sum_{j_1=0}^{c-1} X_{j_1}(u_1) u^{j_1}$$

and

$$Y(u) = Y(u, u_1) = \sum_{l_1=0}^{c-1} Y_{l_1}(u_1) u^{l_1} , \tag{8.3.2}$$

where

$$X_{j_1}(u) = \sum_{j_2=0}^{d_1-1} x_{cj_2+j_1} u_1^{j_2}$$

and

$$Y_{l_1}(u) = \sum_{l_2=0}^{e_1-1} y_{cl_2+l_1} u_1^{l_2} . \tag{8.3.3}$$

Thus, $X(u, u_1)$ is a $(c - 1)$ degree polynomial in u, where each coefficient, in turn, is a polynomial of degree $(d_1 - 1)$ in u_1. Similarly, $Y(u, u_1)$ is a $(c - 1)$ degree polynomial in u, where each coefficient, in turn, is a polynomial of degree $(e_1 - 1)$ in u_1. The product of $Z(u, u_1)$ and $Y(u, u_1)$ is a two-dimensional polynomial $Z(u, u_1)$, and is given by

$$Z(u,u_1) = X(u,u_1)Y(u,u_1) = \sum_{l_1=0}^{c-1}\sum_{j_1=0}^{c-1} X_{j_1}(u_1)Y_{l_1}(u_1)u^{j_1+l_1}. \tag{8.3.4}$$

The polynomial $Z(u)$ may be obtained from $Z(u, u_1)$ by replacing u_1 with u^c.

Alternatively, using the same quantities as in (8.3.1), $X(u, u_1)$ and $Y(u, u_1)$ can be expressed as

$$X(u) = X(u,u_1) = \sum_{j_2=0}^{d_1-1} X_{j_2}(u)u_1^{j_2}$$

and

$$Y(u) = Y(u,u_1) = \sum_{l_2=0}^{e_1-1} Y_{l_2}(u)u_1^{l_2}, \tag{8.3.5}$$

where

$$X_{j_2}(u) = \sum_{j_1=0}^{c-1} x_{cj_2+j_1}u^{j_1}$$

and

$$Y_{l_2}(u) = \sum_{l_1=0}^{c-1} y_{cl_2+l_1}u^{l_1}. \tag{8.3.6}$$

Here, $X(u, u_1)$ is a $(d_1 - 1)$ degree polynomial in u_1, where each coefficient, in turn, is a polynomial of degree $(c - 1)$ in u. Similarly, $Y(u, u_1)$ is a $(e_1 - 1)$ degree polynomial in u_1, where each coefficient, in turn, is a polynomial of degree $(c - 1)$ in u. The product of $Z(u, u_1)$ and $Y(u, u_1)$ is a two-dimensional polynomial $Z(u, u_1)$, and is given by

$$Z(u,u_1)=X(u,u_1)Y(u,u_1)=\sum_{l_2=0}^{e_1-1}\sum_{j_2=0}^{d_1-1}X_{j_2}(u)Y_{l_2}(u)u_1^{j_2+l_2}. \tag{8.3.7}$$

Once again, $Z(u)$ may be obtained from $Z(u, u_1)$ simply by replacing u_1 with u^c. It is worthwhile to mention that this approach is also based on algorithms for small length acyclic convolution. However, in order to convert a one-dimensional acyclic convolution into a multidimensional acyclic convolution, the lengths of the sequences must be composite with a common factor. This condition is not a restrictive one. In many cases such factors exist. In other cases, the sequences may be appended by 0s at the end (*zero-padding*) in order to increase the lengths in a way so as to have a common factor. The approach to convert a one-dimensional acyclic convolution to a higher-dimensional acyclic convolution is straightforward at this stage.

Let us now analyze the computational complexity of this approach. Converting the one-dimensional polynomials $X(u)$ and $Y(u)$ to two-dimensional polynomials $X(u, u_1)$ and $Y(u, u_1)$ as in (8.3.2), (8.3.3), (8.3.5), and (8.3.6), requires no ADD and MULT. The computational complexity of two-dimensional acyclic convolution algorithm to obtain $Z(u, u_1)$ in (8.3.4) or (8.3.7) will be analyzed in the next section.

Let us examine the process of recovery of $Z(u)$ from $Z(u, u_1)$ defined as

$$Z(u) = Z(u, u^c). \tag{8.3.8}$$

In (8.3.4), $Z(u, u_1)$ can be written as

$$Z(u, u_1) = Z_0(u_1) + Z_1(u_1)\, u + Z_2(u_1)\, u^2 + \ldots + Z_{2c-2}(u_1)\, u^{2c-2}, \tag{8.3.9}$$

where each of $Z_k(u_1)$ is a polynomial in u_1 of degree $d_1 + e_1 - 2$. Setting $u_1 = u^c$, we observe that the first $d_1 + e_1 - 2$ coefficients of $Z_c(u_1)$ overlap with the last $d_1 + e_1 - 2$ coefficients of $Z_0(u_1)$, and so on. Each overlap leads to one ADD. Thus, reconversion of the two-dimensional polynomial back to one-dimensional polynomial requires $(d_1 + e_1 - 2)$ $(c - 1)$ ADDs. In (8.3.7), $Z(u, u_1)$ can be written as

$$Z(u,u_1)=Z_0(u)+Z_1(u)u_1+Z_2(u)u_1^2+\ldots+Z_{d_1+e_1-2}(u)u_1^{d_1+e_1-2}, \tag{8.3.10}$$

where each of $Z_k(u)$ is a polynomial in u of degree $2c - 2$. Setting $u_1 = u^c$, we observe that the first $c - 1$ coefficients of $Z_1(u)$ overlap with the last $c - 1$ coefficients of $Z_0(u)$, and so on. Each overlap leads to one ADD. Thus, re-conversion of the two-dimensional polynomial back to a one-dimensional polynomial requires $(d_1 + e_1 - 2)(c - 1)$ ADDs. We observe that reconversion requires the same number of ADDs regardless of the order in which the computation of two-dimensional acyclic convolution is carried out.

Several other instances of converting an one-dimensional computation of the type $X(u)\,Y(u) \bmod P(u)$ were described in Section 5.5.2. The reader is strongly urged to visit that section one more time and understand the approach taken to carry out such conversions. Of most interest would be the cases when $P(u)$ is a cyclotomic polynomial (Example 5.11) is of the type $u^n + 1$ (Example 5.10) or $u^n - 1$. The case of $P(u) = u^n - 1$ will be studied extensively in the next chapter.

8.4 Multidimensional Acyclic Convolution Algorithms

Consider the two-dimensional acyclic convolution of length $n_1 \times n_2$, $n_1 = d_1 + e_1 - 1$, $n_2 = d_2 + e_2 - 1$:

$$Z(u,v) = \sum_{i=0}^{n_1-1}\sum_{j=0}^{n_2-1} z_{i,j}u^i v^j$$

$$= X(u,v)Y(u,v) = \left[\sum_{i=0}^{d_1-1}\sum_{j=0}^{d_2-1} x_{i,j}u^i v^j\right]\left[\sum_{i=0}^{e_1-1}\sum_{j=0}^{e_2-1} y_{i,j}u^i v^j\right], \qquad (8.4.1)$$

$deg(X(u, v)) = d_1 - 1$ in u,
$deg(X(u, v)) = d_2 - 1$ in v,
$deg(Y(u, v)) = e_1 - 1$ in u,
$deg(Y(u, v)) = e_2 - 1$ in v,
$deg(Z(u, v)) = n_1 - 1$ in u,
$deg(Z(u, v)) = n_2 - 1$ in v.

This computation may arise independently in multidimensional digital signal processing or, due to conversion of one-dimensional acyclic convolution, to two-dimensional acyclic convolution. Starting from (8.4.1), we observe that $Z(u, v)$ can be expressed as

$$Z(u,v) = X(u,v)Y(u,v) = \left[\sum_{i=0}^{d_1-1}\left\{\sum_{j=0}^{d_2-1} x_{i,j}v^j\right\}u^i\right]\left[\sum_{i=0}^{e_1-1}\left\{\sum_{j=0}^{e_2-1} y_{i,j}v^j\right\}u^i\right]$$

$$= \left[\sum_{i=0}^{d_1-1} X_i(v) u^i \right] \left[\sum_{i=0}^{e_1-1} Y_i(v) u^i \right]. \tag{8.4.2}$$

Based on (8.4.2), one may compute $Z(u, v)$ by employing the one-dimensional acyclic convolution algorithm for $X(u)\ Y(u)$, $deg(X(u)) = d_1 - 1$, $deg(Y(u)) = e_1 - 1$ in the following manner. In this algorithm,

(i) every pre-ADD for $X(u)$ and $Y(u)$ is replaced by polynomial ADDs of degree $d_2 - 1$ and $e_2 - 1$, respectively;

(ii) every MULT is replaced by the one-dimensional acyclic convolution algorithm for $A(v)\ B(v)$, $deg(A(v)) = d_2 - 1$, $deg(B(v)) = e_2 - 1$; and

(iii) every post-ADD is replaced by polynomial ADDs of degree $d_2 + e_2 - 2$.

In the one-dimensional acyclic convolution algorithm for $X(u)\ Y(u)$, let us define various quantities as

$A_{\text{I},1,x}$: number of pre-ADDs for $X(u)$,
$A_{\text{I},1,y}$: number of pre-ADDs for $Y(u)$,
$A_{\text{O},1}$: number of post-ADDs,
M_1: number of MULTs.

Also, let A_2 and M_2 be the number of ADDs and MULTs required by the one-dimensional acyclic convolution algorithm for $A(v)\ B(v)$. The computational complexity of the two-dimensional convolution algorithm is given by

$$M = M_1 M_2,$$
$$A = A_{\text{I},1,x}\, d_2 + A_{\text{I},1,y}\, e_2 + M_1 A_2 + A_{\text{O},1}\, (d_2 + e_2 - 1).$$

If one-dimensional acyclic convolution algorithms for computing $X(u)\ Y(u)$, $deg(X(u)) = d_1 - 1$, $deg(Y(u)) = e_1 - 1$, and $A(v)\ B(v)$, $deg(A(v)) = d_2 - 1$, $deg(B(v)) = e_2 - 1$, have the bilinear forms

$$\underline{Z}_1 = \mathbf{C}_1\, [\mathbf{A}_1\, \underline{X} \otimes \mathbf{B}_1\, \underline{Y}]$$

and

$$\underline{Z}_2 = \mathbf{C}_2\, [\mathbf{A}_2\, \underline{X} \otimes \mathbf{B}_2\, \underline{Y}],$$

respectively, then the bilinear form for the two-dimensional acyclic convolution algorithm is given by

$$\underline{Z} = \mathbf{C}\,[\mathbf{A}\,\underline{X} \otimes \mathbf{B}\,\underline{Y}]. \tag{8.4.3}$$

Here, the matrices **A**, **B**, and **C** are given by the Kronecker products

$$\mathbf{A} = \mathbf{A}_1 \,\kappa\, \mathbf{A}_2,$$
$$\mathbf{B} = \mathbf{B}_1 \,\kappa\, \mathbf{B}_2,$$

and

$$\mathbf{C} = \mathbf{C}_1 \,\kappa\, \mathbf{C}_2.$$

The I-th element of vector $\underline{X}$ is same as $x_{i,j}$, $I = i\, d_2 + j$. Similar statements also hold for the vectors $\underline{Y}$ and $\underline{Z}$.

Alternatively, $Z(u, v)$ can be expressed as

$$Z(u,v) = X(u,v)Y(u,v) = \left[\sum_{j=0}^{d_2-1}\left\{\sum_{i=0}^{d_1-1} x_{i,j}u^i\right\}v^j\right]\left[\sum_{j=0}^{e_2-1}\left\{\sum_{i=0}^{e_1-1} y_{i,j}u^i\right\}v^j\right]$$

$$= \left[\sum_{j=0}^{d_2-1} X_j(u)v^i\right]\left[\sum_{j=0}^{e_2-1} Y_j(u)v^i\right]. \tag{8.4.4}$$

Based on (8.4.4), $Z(u, v)$ can be computed by employing one-dimensional acyclic convolution algorithm for $X(v)\ Y(v)$, $deg(X(v)) = d_2 - 1$, $deg(Y(v)) = e_2 - 1$ in the following manner. In this algorithm,

(i) every pre-ADD for $X(v)$ and $Y(v)$ is replaced by polynomial ADDs of degree $d_1 - 1$ and $e_1 - 1$, respectively;

(ii) every MULT is replaced by the one-dimensional acyclic convolution algorithm for $A(u)\ B(u)$, $deg(A(u)) = d_1 - 1$, $deg(B(u)) = e_1 - 1$; and

(iii) every post-ADD is replaced by polynomial ADDs of degree $d_1 + e_1 - 2$.

In this case, the computational complexity is given by

$$M = M_1\, M_2,$$
$$A = A_{I,2,x}\, d_1 + A_{I,2,y}\, e_1 + M_2\, A_1 + A_{O,2}\,(d_1 + e_1 - 1),$$

where various quantities are defined analogously. We observe that number of MULTs do not depend on the order of computation, but the number of ADDs do.

The bilinear form for the overall computation can be obtained in a manner similar to (8.4.3). A fundamental feature of this approach is that, as long as the constituent algorithms are valid over the given number system, the overall algorithm is valid, too. We illustrate these ideas in the following example.

Example 8.4 Let $d_1 = e_1 = 2$, $d_2 = e_2 = 3$. The one-dimensional acyclic convolution algorithms of lengths 3 and 5 that we use here are

Length 3:

$$\begin{bmatrix} z_0 \\ z_1 \\ z_2 \end{bmatrix} = \begin{bmatrix} 1 & 0 & 0 \\ -1 & -1 & 1 \\ 0 & 1 & 0 \end{bmatrix} \left\{ \begin{bmatrix} 1 & 0 \\ 0 & 1 \\ 1 & 1 \end{bmatrix} \begin{bmatrix} x_0 \\ x_1 \end{bmatrix} \otimes \begin{bmatrix} 1 & 0 \\ 0 & 1 \\ 1 & 1 \end{bmatrix} \begin{bmatrix} y_0 \\ y_1 \end{bmatrix} \right\}. \tag{8.4.5}$$

Length 5:

$$\begin{bmatrix} z_0 \\ z_1 \\ z_2 \\ z_4 \\ z_5 \end{bmatrix} = \begin{bmatrix} 1 & 0 & 0 & 0 & 0 & 0 \\ -1 & -1 & 0 & 1 & 0 & 0 \\ -1 & 1 & -1 & 0 & 0 & 1 \\ 0 & -1 & -1 & 0 & 1 & 0 \\ 0 & 0 & 1 & 0 & 0 & 0 \end{bmatrix} \left\{ \begin{bmatrix} 1 & 0 & 0 \\ 0 & 1 & 0 \\ 0 & 0 & 1 \\ 1 & 1 & 0 \\ 0 & 1 & 1 \\ 1 & 0 & 1 \end{bmatrix} \begin{bmatrix} x_0 \\ x_1 \\ x_2 \end{bmatrix} \otimes \begin{bmatrix} 1 & 0 & 0 \\ 0 & 1 & 0 \\ 0 & 0 & 1 \\ 1 & 1 & 0 \\ 0 & 1 & 1 \\ 1 & 0 & 1 \end{bmatrix} \begin{bmatrix} y_0 \\ y_1 \\ y_2 \end{bmatrix} \right\}. \tag{8.4.6}$$

In this case, the various quantities are

$$A_{I,1,x} = A_{I,1,y} = 1,\ A_{O,1} = 2,\ M_1 = 3,$$
$$A_{I,2,x} = A_{I,2,y} = 3,\ A_{O,2} = 7,\ M_2 = 6.$$

If the first approach is used, then the computation is performed according to (8.4.5) by replacing x_0 and x_1 with $X_0(v)$ and $X_1(v)$, y_0 and y_1 with $Y_0(v)$ and $Y_1(v)$, and using (8.4.6) every time a MULT is encountered. The computational complexity of the algorithm is given by $M = 18$, $A = 55$. If the second approach is used, then the computation is performed according to (8.4.6) by replacing x_j with $X_j(u)$, y_j with $Y_j(u)$, $j = 0, 1, 2$, and using (8.4.5) every time a MULT is encountered. The computational complexity of the algorithm is given by $M = 18$, $A = 57$. If this approach is used to compute the one-dimendional acyclic convolution of two length 6 sequences (length of convolution = 11), then an additional $(d_1 + e_1 - 2)(c - 1) = 4$ ADDs are required. Finally, note that the algorithm presented here is *valid for all number systems* as it requires no division.

8.5 Nesting and Split Nesting Algorithms for Multidimensional Convolution

In the previous section, we described fast algorithms for multidimensional acyclic convolution that employ algorithms for one-dimensional acyclic convolution. The approach adopted is termed as the *nesting approach* whereby one-dimensional convolutions are nested together in order to perform higher order convolution. In many instances, each of the one-dimensional algorithms is based on CRT-P. In this case, one can realize additional computational savings by breaking the computations further. This is known as **split-nesting**. Consider the following formulation. Say, we are interested in the computation,

$$Z(u, v) = X(u, v)\, Y(u, v) \bmod [P(u), Q(v)], \tag{8.5.1}$$

where $deg(P(u)) = N_1$, $deg(Q(v)) = N_2$. Let these polynomials factor as

$$P(u) = A(u)\, B(u)$$

and

$$Q(v) = C(v)\, D(v),$$

with $deg(A(u)) = a$, $deg(B(u)) = b$, $deg(C(v)) = c$, and $deg(D(v)) = d$.

A straightforward approach would be to use the algorithm for computing the polynomial product $X(u)\, Y(u) \bmod P(u)$ in the following manner. Replace each ADD with ADD of vectors of length N_2 and MULT with a polynomial product of the type $X(v)\, Y(v) \bmod Q(v)$. The computational complexity of this approach is given by

$$M = M_1 M_2$$
$$A = A_1 N_2 + M_1 A_2. \tag{8.5.2}$$

Alternatively, one can employ the following CRT-P based algorithm for computing $Z(u, v)$ in (8.5.1):

Step 1 CRT-P Reduction.

Step 1 (a): Compute:

A1 $X(u, v) \bmod [A(u), Q(v)]$,
A2 $X(u, v) \bmod [B(u), Q(v)]$,
A3 $Y(u, v) \bmod [A(u), Q(v)]$,
A4 $Y(u, v) \bmod [B(u), Q(v)]$,

Step 1 (b): Compute:

A5 $X(u, v) \bmod [A(u), C(v)]$,
A6 $X(u, v) \bmod [A(u), D(v)]$,

from $X(u, v) \bmod [A(u), Q(v)]$;

A7 $X(u, v) \bmod [B(u), C(v)]$,
A8 $X(u, v) \bmod [B(u), D(v)]$,

from $X(u, v) \bmod [B(u), Q(v)]$;

A9 $Y(u, v) \bmod [A(u), C(v)]$,
A10 $Y(u, v) \bmod [A(u), D(v)]$,

from $Y(u, v) \bmod [A(u), Q(v)]$;

A11 $Y(u, v) \bmod [B(u), C(v)]$,
A12 $Y(u, v) \bmod [B(u), D(v)]$,

from $Y(u, v) \bmod [B(u), Q(v)]$.

Step 2. Perform four polynomial products.

A13 $Z(u, v) \bmod [A(u), C(v)] = X(u, v)\ Y(u, v) \bmod [A(u), C(v)]$,
A14 $Z(u, v) \bmod [A(u), D(v)] = X(u, v)\ Y(u, v) \bmod [A(u), D(v)]$,
A15 $Z(u, v) \bmod [B(u), C(v)] = X(u, v)\ Y(u, v) \bmod [B(u), C(v)]$,
A16 $Z(u, v) \bmod [B(u), D(v)] = X(u, v)\ Y(u, v) \bmod [B(u), D(v)]$,.

Step 3. CRT-P Reconstruction.

Step 3(a): Obtain:

A17 $Z(u, v) \bmod [A(u), Q(v)]$

from

$Z(u, v) \bmod [A(u), C(v)]$, $Z(u, v) \bmod [A(u), D(v)]$;

A18 $Z(u, v) \bmod [B(u), Q(v)]$

from

$Z(u, v) \bmod [B(u), C(v)]$, $Z(u, v) \bmod [B(u), D(v)]$;

Step 3(b): Obtain:

A19 $Z(u, v) \bmod [P(u), Q(v)]$

from

$Z(u, v) \bmod [A(u), Q(v)]$, $Z(u, v) \bmod [B(u), Q(v)]$.

Equivalently, the algorithm to compute the product $X(u)\ Y(u) \bmod P(u)$ has the following form:

Step 1 CRT-P Reductions. Compute

B1 $X(u) \bmod A(u)$,
B2 $X(u) \bmod B(u)$,
B1 $Y(u) \bmod A(u)$,
B2 $Y(u) \bmod B(u)$.

Step 2. Perform the two polynomial products:

B3 $Z(u) \bmod A(u) = X(u)\ Y(u) \bmod A(u)$,
B4 $Z(u) \bmod B(u) = X(u)\ Y(u) \bmod B(u)$.

Step 3. CRT-P Reconstruction. Obtain

B5 $Z(u) \bmod P(u)$

from

$Z(u) \bmod A(u)$ and $Z(u) \bmod B(u)$.

Let the additive complexity of Steps B1 to B5 be $A_{1,1,1}$, $A_{1,1,2}$, $A_{1,2,1}$, $A_{1,2,2}$, and $A_{1,3}$. Similarly, the multiplicative complexity (only Steps B3 and B4 require MULTs) is represented by $M_{1,2,1}$ for Step B3 and $M_{1,2,2}$ for Step B4. The form to compute the product $X(v)\ Y(v) \bmod Q(v)$ can be described in a similar manner. The corresponding numbers for this algorithm are represented by replacing the first subscript with 2. Here, the first subscript is 1 for the algorithm for $X(u)\ Y(u) \bmod P(u)$ and 2 for $X(v)\ Y(v) \bmod Q(v)$); the second subscript denotes the step in the algorithm and the third the computation with regard to the factor polynomials, 1 for mod $A(u)$ or $C(v)$ and 2 for mod $B(u)$ or $D(v)$.

We now analyze the computational complexity of the algorithm described in Steps A1 to A19. This analysis is based on the assumption that $X(u, v)$ mod

$[A(u), Q(v)]$ is computed by replacing scalar quantities in the computation $X(u)$ mod $A(u)$ with vectors of length N_2. Similarly, the computation

$$Z(u, v) \bmod [A(u), C(v)] = X(u, v)\, Y(u, v) \bmod [A(u), C(v)]$$

is performed by replacing (i) each scalar ADD in the computation $X(u)\, Y(u)$ mod $A(u)$ with ADD of two vectors of length c, and (ii) each scalar MULT in the computation $X(u)\, Y(u)$ mod $A(u)$ with the algorithm to compute the product $X(v)\, Y(v)$ mod $C(v)$.

The computational complexity of Steps A1 to A19 is shown in Table 8.1.

Table 8.1 Computational Analysis of the Split Nesting Algorithm

Steps	Additive Complexity	Multiplicative Complexity
A1 and A2	$A_{1,1,1}\, N_2$	0
A3 and A4	$A_{1,1,2}\, N_2$	0
A5 and A9	$A_{2,1,1}\, a$	0
A6 and A10	$A_{2,1,2}\, a$	0
A7 and A11	$A_{2,1,1}\, b$	0
A8 and A12	$A_{2,1,2}\, b$	0
A13	$A_{1,2,1}\, c + M_{1,2,1} \cdot A_{2,2,1}$	$M_{1,2,1} \cdot M_{2,2,1}$
A14	$A_{1,2,1}\, d + M_{1,2,1} \cdot A_{2,2,2}$	$M_{1,2,1} \cdot M_{2,2,2}$
A15	$A_{1,2,2}\, c + M_{1,2,2} \cdot A_{2,2,1}$	$M_{1,2,2} \cdot M_{2,2,1}$
A16	$A_{1,2,2}\, d + M_{1,2,2} \cdot A_{2,2,2}$	$M_{1,2,2} \cdot M_{2,2,2}$
A17	$A_{2,3}\, a$	0
A18	$A_{2,3}\, b$	0
A19	$A_{1,3}\, N_2$	0
A1 - A4	$A_{1,1}\, N_2$	0
A5 - A12	$A_{2,1}\, N_1$	0
A13 - A16	$A_{1,2}\, N_2 + A_{2,2} \cdot M_{1,2}$	$M_{1,2} \cdot M_{2,2}$
A17 - A19	$A_{2,3}\, N_1 + A_{1,3}\, N_2$	0
Total:	$N_2\, A_1 + N_1\, A_2 + (M_{1,2} - N_1)\, A_{2,2}$	$M_{1,2} \cdot M_{2,2}$

Comparing the computational cost in the last row of Table 8.1 with (8.5.2), we see that split nesting results in savings of

$$(M_1 - N_1)\,(A_2 - A_{2,2}) = (M_1 - N_1)\,(A_{2,1} - A_{2,3}) \text{ ADD}$$

over the nesting approach. Note that $M_1 = M_{1,2}$ as per notation. There are no savings in the number of MULTs.

The message of the above analysis and all the other techniques is clear. One obtains computational savings by

(i) Breaking down a large computation into as many small pieces as possible;
(ii) Computing them independently; and
(iii) Combining them to reconstruct the overall result.

8.6 Acyclic Convolution Algorithms Over Finite Fields and Rings

The previously described approaches to deriving fast algorithms for computing acyclic convolution of data sequences are valid over all number systems. For example, the various factors of modulo polynomial need to be relatively co-prime over the given number system for CRT-P based approach to work. There can be two ways one can go about designing algorithms over finite fields and rings. One is a direct method in which an algorithm is designed directly in the field and takes into account its algebraic structure. The second is an indirect method and basically consists in designing the algorithm over the field of rational numbers and then judging its validity over the finite field or ring of interest. The crucial thing to observe in judging validity is the existence of multiplicative inverses of the integers in the denominator of the rational-valued elements in **A**, **B**, and **C** matrices.

8.6.1 Algorithms Over Finite Fields

In the following, we present two examples to illustrate the salient features of the algorithms over finite fields.

Example 8.5 Consider the design of length 6 acyclic convolution over $GF(2)$. Let $k = 4$ and $e = 3$. We choose $P(u) = u\,(u^2 + 1)\,(u^2 + u + 1)$ and $s = 1$. The resulting bilinear form is given by

$$\begin{bmatrix} z_0 \\ z_1 \\ z_2 \\ z_3 \\ z_4 \\ z_5 \end{bmatrix} = \begin{bmatrix} 1 & 0 & 0 & 0 & 0 & 0 & 0 & 0 \\ 1 & 1 & 0 & 1 & 1 & 1 & 0 & 1 \\ 0 & 0 & 1 & 0 & 0 & 1 & 1 & 1 \\ 1 & 0 & 1 & 0 & 1 & 1 & 0 & 0 \\ 1 & 1 & 1 & 1 & 0 & 1 & 1 & 1 \\ 0 & 0 & 0 & 0 & 0 & 0 & 0 & 1 \end{bmatrix} \cdot \left[\begin{bmatrix} 1 & 0 & 0 & 0 \\ 1 & 0 & 1 & 0 \\ 1 & 1 & 1 & 1 \\ 0 & 1 & 0 & 1 \\ 1 & 0 & 1 & 1 \\ 1 & 1 & 0 & 1 \\ 0 & 1 & 1 & 0 \\ 0 & 0 & 0 & 1 \end{bmatrix} \begin{bmatrix} x_0 \\ x_1 \\ x_2 \\ x_3 \end{bmatrix} \otimes \begin{bmatrix} 1 & 0 & 0 \\ 1 & 0 & 1 \\ 1 & 1 & 1 \\ 0 & 1 & 0 \\ 1 & 0 & 1 \\ 1 & 1 & 0 \\ 0 & 1 & 1 \\ 0 & 0 & 1 \end{bmatrix} \begin{bmatrix} y_0 \\ y_1 \\ y_2 \end{bmatrix} \right]$$

Example 8.6 Consider the design of length 16 acyclic convolution over $GF(2)$. Let $k = 8$ and $e = 9$. We choose

$$P(u) = u^3\,(u^2 + 1)\,(u^3 + u + 1)\,(u^3 + u^2 + 1)\,(u^2 + u + 1)$$

and $s = 3$. The resulting bilinear form is characterized by

$$
\mathbf{C}^t = \begin{bmatrix}
1 & 1 & 1 & 1 & 1 & 1 & 0 & 1 & 1 & 1 & 1 & 1 & 1 & 0 & 0 & 0 \\
0 & 1 & 1 & 0 & 1 & 1 & 0 & 0 & 1 & 1 & 0 & 1 & 1 & 0 & 0 & 0 \\
0 & 0 & 1 & 0 & 0 & 1 & 0 & 0 & 0 & 1 & 0 & 0 & 0 & 0 & 0 & 0 \\
0 & 1 & 0 & 0 & 1 & 0 & 0 & 0 & 1 & 0 & 0 & 1 & 1 & 0 & 0 & 0 \\
0 & 0 & 1 & 0 & 0 & 1 & 0 & 0 & 0 & 1 & 0 & 0 & 0 & 0 & 0 & 0 \\
0 & 0 & 0 & 1 & 1 & 1 & 0 & 0 & 0 & 0 & 1 & 1 & 1 & 0 & 0 & 0 \\
0 & 0 & 0 & 1 & 1 & 1 & 0 & 0 & 0 & 0 & 1 & 1 & 1 & 0 & 0 & 0 \\
0 & 0 & 0 & 1 & 0 & 1 & 1 & 1 & 1 & 1 & 0 & 1 & 1 & 0 & 0 & 0 \\
0 & 0 & 0 & 0 & 1 & 1 & 0 & 0 & 0 & 0 & 0 & 1 & 1 & 0 & 0 & 0 \\
0 & 0 & 0 & 1 & 0 & 1 & 0 & 0 & 0 & 0 & 1 & 1 & 0 & 0 & 0 & 0 \\
0 & 0 & 0 & 1 & 1 & 0 & 0 & 0 & 0 & 1 & 0 & 1 & 0 & 0 & 0 & 0 \\
0 & 0 & 0 & 0 & 1 & 1 & 1 & 1 & 0 & 1 & 0 & 1 & 1 & 0 & 0 & 0 \\
0 & 0 & 0 & 0 & 0 & 1 & 1 & 1 & 1 & 0 & 1 & 0 & 0 & 0 & 0 & 0 \\
0 & 0 & 0 & 1 & 0 & 0 & 0 & 1 & 1 & 1 & 1 & 1 & 1 & 0 & 0 & 0 \\
0 & 0 & 0 & 0 & 1 & 0 & 0 & 0 & 1 & 1 & 1 & 1 & 1 & 0 & 0 & 0 \\
0 & 0 & 0 & 1 & 1 & 0 & 0 & 1 & 0 & 0 & 0 & 0 & 0 & 0 & 0 & 0 \\
0 & 0 & 0 & 1 & 0 & 1 & 1 & 0 & 0 & 1 & 0 & 1 & 1 & 0 & 0 & 0 \\
0 & 0 & 0 & 0 & 0 & 0 & 0 & 1 & 0 & 1 & 1 & 0 & 1 & 0 & 0 & 0 \\
0 & 0 & 0 & 1 & 0 & 0 & 0 & 0 & 1 & 0 & 0 & 0 & 1 & 0 & 0 & 0 \\
0 & 0 & 0 & 1 & 1 & 0 & 1 & 0 & 0 & 1 & 1 & 0 & 0 & 0 & 0 & 0 \\
0 & 0 & 0 & 0 & 1 & 1 & 1 & 1 & 1 & 0 & 0 & 0 & 0 & 0 & 0 & 0 \\
0 & 0 & 0 & 0 & 1 & 0 & 1 & 0 & 1 & 1 & 1 & 0 & 1 & 0 & 0 & 0 \\
0 & 0 & 0 & 1 & 0 & 1 & 0 & 1 & 1 & 1 & 1 & 0 & 0 & 0 & 0 & 0 \\
0 & 0 & 0 & 1 & 1 & 1 & 1 & 1 & 1 & 0 & 1 & 1 & 1 & 1 & 1 & 1 \\
0 & 0 & 0 & 0 & 1 & 0 & 0 & 1 & 0 & 0 & 0 & 1 & 0 & 0 & 1 & 0 \\
0 & 0 & 0 & 0 & 1 & 0 & 0 & 1 & 0 & 0 & 0 & 1 & 0 & 0 & 1 & 0 \\
0 & 0 & 0 & 1 & 0 & 0 & 1 & 0 & 0 & 0 & 1 & 0 & 0 & 1 & 0 & 0 \\
0 & 0 & 0 & 1 & 0 & 0 & 1 & 0 & 0 & 0 & 1 & 0 & 0 & 1 & 0 & 0
\end{bmatrix}
$$

$$
\mathbf{A} = \begin{bmatrix}
1 & 0 & 0 & 0 & 0 & 0 & 0 & 0 \\
0 & 1 & 0 & 0 & 0 & 0 & 0 & 0 \\
0 & 0 & 1 & 0 & 0 & 0 & 0 & 0 \\
1 & 1 & 0 & 0 & 0 & 0 & 0 & 0 \\
1 & 0 & 1 & 0 & 0 & 0 & 0 & 0 \\
1 & 0 & 1 & 0 & 1 & 0 & 1 & 0 \\
0 & 1 & 0 & 1 & 0 & 1 & 0 & 1 \\
1 & 1 & 1 & 1 & 1 & 1 & 1 & 1 \\
1 & 0 & 1 & 1 & 0 & 1 & 1 & 0 \\
0 & 1 & 1 & 0 & 1 & 1 & 0 & 1 \\
1 & 1 & 0 & 1 & 1 & 0 & 1 & 1 \\
1 & 0 & 0 & 1 & 0 & 1 & 1 & 1 \\
0 & 1 & 0 & 1 & 1 & 1 & 0 & 0 \\
0 & 0 & 1 & 0 & 1 & 1 & 1 & 0 \\
1 & 1 & 0 & 0 & 1 & 0 & 1 & 1 \\
0 & 1 & 1 & 1 & 0 & 0 & 1 & 0 \\
1 & 0 & 1 & 1 & 1 & 0 & 0 & 1 \\
1 & 0 & 0 & 1 & 1 & 1 & 0 & 1 \\
0 & 1 & 0 & 0 & 1 & 1 & 1 & 0 \\
0 & 0 & 1 & 1 & 1 & 0 & 1 & 0 \\
1 & 1 & 0 & 1 & 0 & 0 & 1 & 1 \\
0 & 1 & 1 & 1 & 0 & 1 & 0 & 0 \\
1 & 0 & 1 & 0 & 0 & 1 & 1 & 1 \\
0 & 0 & 0 & 0 & 0 & 0 & 0 & 1 \\
0 & 0 & 0 & 0 & 0 & 0 & 1 & 1 \\
0 & 0 & 0 & 0 & 0 & 0 & 1 & 0 \\
0 & 0 & 0 & 0 & 0 & 1 & 0 & 1 \\
0 & 0 & 0 & 0 & 0 & 1 & 0 & 0
\end{bmatrix}
$$

$$\mathbf{B} = \begin{bmatrix} 1 & 0 & 0 & 0 & 0 & 0 & 0 & 0 & 0 \\ 0 & 1 & 0 & 0 & 0 & 0 & 0 & 0 & 0 \\ 0 & 0 & 1 & 0 & 0 & 0 & 0 & 0 & 0 \\ 1 & 1 & 0 & 0 & 0 & 0 & 0 & 0 & 0 \\ 1 & 0 & 1 & 0 & 0 & 0 & 0 & 0 & 0 \\ 1 & 0 & 1 & 0 & 1 & 0 & 1 & 0 & 1 \\ 0 & 1 & 0 & 1 & 0 & 1 & 0 & 1 & 0 \\ 1 & 1 & 1 & 1 & 1 & 1 & 1 & 1 & 1 \\ 1 & 0 & 1 & 1 & 0 & 1 & 1 & 0 & 1 \\ 0 & 1 & 1 & 0 & 1 & 1 & 0 & 1 & 1 \\ 1 & 1 & 0 & 1 & 1 & 0 & 1 & 1 & 0 \\ 1 & 0 & 0 & 1 & 0 & 1 & 1 & 1 & 0 \\ 0 & 1 & 0 & 1 & 1 & 1 & 0 & 0 & 1 \\ 0 & 0 & 1 & 0 & 1 & 1 & 1 & 0 & 0 \\ 1 & 1 & 0 & 0 & 1 & 0 & 1 & 1 & 1 \\ 0 & 1 & 1 & 1 & 0 & 0 & 1 & 0 & 1 \\ 1 & 0 & 1 & 1 & 1 & 0 & 0 & 1 & 0 \\ 1 & 0 & 0 & 1 & 1 & 1 & 0 & 1 & 0 \\ 0 & 1 & 0 & 0 & 1 & 1 & 1 & 0 & 1 \\ 0 & 0 & 1 & 1 & 1 & 0 & 1 & 0 & 0 \\ 1 & 1 & 0 & 1 & 0 & 0 & 1 & 1 & 1 \\ 0 & 1 & 1 & 1 & 0 & 1 & 0 & 0 & 1 \\ 1 & 0 & 1 & 0 & 0 & 1 & 1 & 1 & 0 \\ 0 & 0 & 0 & 0 & 0 & 0 & 0 & 0 & 1 \\ 0 & 0 & 0 & 0 & 0 & 0 & 0 & 1 & 1 \\ 0 & 0 & 0 & 0 & 0 & 0 & 0 & 1 & 0 \\ 0 & 0 & 0 & 0 & 0 & 0 & 1 & 0 & 1 \\ 0 & 0 & 0 & 0 & 0 & 0 & 1 & 0 & 0 \end{bmatrix}.$$

8.6.2 Algorithms Over Finite Integer Rings

The structure of the algorithms over $Z(M)$ and $CZ(M)$ has the form

$$\underline{Z} = \mathbf{C}\,[\mathbf{A}\,\underline{X} \otimes \mathbf{B}\,\underline{Y}] \bmod M.$$

A direct application of the CRT-I implies that

$$\underline{Z} = \sum_{\substack{\oplus \\ i=1}}^{t} \underline{Z}^{(i)},$$

$$\underline{X} = \sum_{\substack{\oplus \\ i=1}}^{t} \underline{X}^{(i)},$$

$$\underline{Y} = \sum_{\substack{\oplus \\ i=1}}^{t} \underline{Y}^{(i)},$$

$$\underline{Z}^{(i)} = \mathbf{C}^{(i)}\,[\,\mathbf{A}^{(i)}\,\underline{X}^{(i)} \otimes \mathbf{B}^{(i)}\,\underline{Y}^{(i)}],$$

$$\mathbf{A}^{(i)} \equiv \mathbf{A} \bmod p_i^{a_i},$$
$$\mathbf{B}^{(i)} \equiv \mathbf{B} \bmod p_i^{a_i},$$
$$\mathbf{C}^{(i)} \equiv \mathbf{C} \bmod p_i^{a_i},$$

and

$$\mathbf{A} = \sum_{\substack{\oplus \\ i=1}}^{t} \mathbf{A}^{(i)},$$

$$\mathbf{B} = \sum_{\substack{\oplus \\ i=1}}^{t} \mathbf{B}^{(i)},$$

$$\mathbf{C} = \sum_{\substack{\oplus \\ i=1}}^{t} \mathbf{C}^{(i)}.$$

As a result, the task of deriving an algorithm for computing any bilinear form over $Z(M)$ and $CZ(M)$ (which consists in determining the matrices **A**, **B**, and **C**) is reduced to finding an algorithm for computing the same bilinear form over

$Z\left(p_i^{a_i}\right)$ and $CZ\left(p_i^{a_i}\right)$, $i = 1, 2, ..., t$. The matrices **A**, **B**, and **C** over $Z(M)$ and $CZ(M)$ are a direct sum of the corresponding matrices over $Z\left(p_i^{a_i}\right)$ and $CZ\left(p_i^{a_i}\right)$, respectively. We, therefore, focus on constructing algorithms over $Z(p^a)$ and $CZ(p^a)$.

For the CRT-PR based algorithm for computing acyclic convolution over $\mathbf{Z}(p^a)$, the modulo polynomial $P(u)$ is selected in terms of its factor polynomials that are relatively coprime over $GF(p)$. We illustrate the approach with the help of an example.

Example 8.7 Let $n = d + e - 1 = 10$ and $M = 3^2 \cdot 5^3 = 1{,}125$. The irreducible factor polynomials of $P(u)$ over $GF(3)$ are chosen to be

$$P_1(u) = u,$$
$$P_2(u) = u + 1,$$
$$P_3(u) = u + 2,$$
$$P_4(u) = u^2 + 1,$$
$$P_5(u) = u^2 + u + 2,$$

and

$$P_6(u) = u^2 + 2\,u + 2.$$

Clearly, $deg(P(u)) = 9$. Note that $\prod_{i=1}^{6} P_i(u) = u^9 - u$. In this case, we have a wraparound of 1 coefficient as $d = 9$. In the same way, the irreducible factor polynomials of $P(u)$ over $GF(5)$ are chosen to be

$$P_1(u) = u,$$
$$P_2(u) = u + 1,$$
$$P_3(u) = u + 2,$$
$$P_4(u) = u + 3,$$
$$P_5(u) = u + 4,$$
$$P_6(u) = u^2 + 2,$$

and

$$P_7(u) = u^2 + 3.$$

Let $d = 6$ and $e = 5$. The various $\mathbf{A}^{(i)}$, $\mathbf{B}^{(i)}$, and $\mathbf{C}^{(i)}$ matrices for computing the acyclic convolution over $Z\left(p_i^{a_i}\right)$, $i = 1, 2$ are given by

$$
\mathbf{A}^{(1)} = \left[\begin{array}{cccccc}
1 & 0 & 0 & 0 & 0 & 0 \\ \hdashline
1 & 8 & 1 & 8 & 1 & 8 \\ \hdashline
1 & 1 & 1 & 1 & 1 & 1 \\ \hline
1 & 1 & 8 & 8 & 1 & 1 \\
1 & 0 & 8 & 0 & 1 & 0 \\
0 & 1 & 0 & 8 & 0 & 1 \\ \hdashline
1 & 1 & 5 & 3 & 8 & 8 \\
1 & 0 & 1 & 4 & 8 & 0 \\
0 & 1 & 0 & 8 & 0 & 8 \\ \hdashline
1 & 1 & 6 & 4 & 8 & 8 \\
1 & 0 & 1 & 5 & 8 & 0 \\
0 & 1 & 5 & 8 & 0 & 8
\end{array}\right],
$$

$$
\mathbf{A}^{(2)} = \left[\begin{array}{cccccc}
1 & 0 & 0 & 0 & 0 & 0 \\ \hdashline
1 & 124 & 2 & 124 & 1 & 124 \\ \hdashline
1 & 1 & 1 & 1 & 1 & 1 \\ \hdashline
1 & 68 & 124 & 57 & 1 & 68 \\ \hdashline
1 & 57 & 124 & 68 & 1 & 57 \\ \hdashline
1 & 1 & 68 & 68 & 124 & 124 \\
1 & 0 & 68 & 0 & 124 & 0 \\
0 & 1 & 0 & 68 & 0 & 124 \\ \hdashline
1 & 1 & 57 & 57 & 124 & 124 \\
1 & 0 & 57 & 0 & 124 & 0 \\
0 & 1 & 0 & 57 & 0 & 124
\end{array}\right],
$$

$$\mathbf{B}^{(1)} = \begin{bmatrix} 1 & 0 & 0 & 0 & 0 \\ \hline 1 & 8 & 1 & 8 & 1 \\ \hline 1 & 1 & 1 & 1 & 1 \\ \hline 1 & 1 & 8 & 8 & 1 \\ 1 & 0 & 8 & 0 & 1 \\ 0 & 1 & 0 & 8 & 0 \\ \hline 1 & 1 & 5 & 3 & 8 \\ 1 & 0 & 1 & 4 & 8 \\ 0 & 1 & 4 & 8 & 0 \\ \hline 1 & 1 & 6 & 4 & 8 \\ 1 & 0 & 1 & 5 & 8 \\ 0 & 1 & 5 & 8 & 0 \end{bmatrix},$$

$$\mathbf{B}^{(2)} = \begin{bmatrix} 1 & 0 & 0 & 0 & 0 \\ \hline 1 & 124 & 1 & 124 & 1 \\ \hline 1 & 1 & 1 & 1 & 1 \\ \hline 1 & 68 & 124 & 57 & 1 \\ \hline 1 & 57 & 124 & 68 & 1 \\ \hline 1 & 1 & 68 & 68 & 124 \\ 1 & 0 & 68 & 0 & 124 \\ 0 & 1 & 0 & 68 & 0 \\ \hline 1 & 1 & 57 & 57 & 124 \\ 1 & 0 & 57 & 0 & 124 \\ 0 & 1 & 0 & 57 & 0 \end{bmatrix},$$

$$
\mathbf{C}^{(1)} = \left[\begin{array}{c|c|c|ccc|ccc|ccc}
1 & 0 & 0 & 0 & 0 & 0 & 0 & 0 & 0 & 0 & 0 & 0 \\
0 & 1 & 8 & 7 & 2 & 2 & 7 & 7 & 7 & 7 & 7 & 6 \\
0 & 8 & 8 & 0 & 2 & 7 & 4 & 5 & 3 & 5 & 4 & 2 \\
0 & 1 & 8 & 2 & 7 & 7 & 0 & 5 & 4 & 0 & 5 & 5 \\
0 & 8 & 8 & 0 & 7 & 2 & 4 & 3 & 5 & 5 & 6 & 4 \\
0 & 1 & 8 & 7 & 2 & 2 & 2 & 2 & 2 & 2 & 2 & 3 \\
0 & 8 & 8 & 0 & 2 & 7 & 5 & 4 & 6 & 4 & 5 & 7 \\
0 & 1 & 8 & 2 & 7 & 7 & 0 & 4 & 5 & 0 & 4 & 4 \\
0 & 8 & 8 & 0 & 7 & 2 & 5 & 6 & 4 & 4 & 3 & 5
\end{array}\right],
$$

and

$$
\mathbf{C}^{(2)} = \left[\begin{array}{c|c|c|c|c|ccc|ccc}
1 & 0 & 0 & 0 & 0 & 0 & 0 & 0 & 0 & 0 & 0 \\
0 & 78 & 47 & 54 & 71 & 94 & 31 & 31 & 94 & 31 & 31 \\
0 & 47 & 47 & 78 & 78 & 0 & 108 & 94 & 0 & 17 & 94 \\
0 & 78 & 47 & 71 & 54 & 108 & 17 & 17 & 17 & 108 & 108 \\
0 & 47 & 47 & 47 & 47 & 0 & 31 & 108 & 0 & 31 & 17 \\
0 & 78 & 47 & 54 & 71 & 31 & 94 & 94 & 31 & 94 & 94 \\
0 & 47 & 47 & 78 & 78 & 0 & 17 & 31 & 0 & 108 & 31 \\
0 & 78 & 47 & 71 & 54 & 17 & 108 & 108 & 108 & 17 & 17 \\
124 & 47 & 47 & 47 & 47 & 0 & 94 & 17 & 0 & 94 & 108
\end{array}\right].
$$

Notes

The objective of this chapter was to present fast algorithms for computing one- and higher-dimensional acyclic convolution of data sequences defined over a given number system. There are two methods for accomplishing this. One can employ the CRT-P directly or first convert the one-dimensional task to a multi-dimensional one and then apply CRT-P along each dimension. All these algorithms may vastly differ in their computational efficiencies and structural properties that may be important from the point of view of hardware realization. One may also wish to compare the algorithms in terms of their additive complexity and the nature of constants that show up as elements of the various matrices in the bilinear form. In that sense, one should be able to establish tradeoffs between multiplicative and additive complexities and the overall realization.

We have made a conscious effort to keep the description of algorithms over finite fields and rings separate from the description of algorithms over infinite fields. It is our hope that it will assist the readers in understanding the workings and derivations in all instances.

Bibliography

[8.1] J.H. McClellan and C.M. Rader, *Number Theory in Digital Signal Processing*, Prentice Hall, Inc., 1979.

[8.2] H.J. Nussbaumer, *Fast Fourier Transform and Convolution Algorithms*, Springer-Verlag, 1981.

[8.3] R.E. Blahut, *Fast Algorithms for Digital Signal Processing*, Addison-Wesley Publishing Co., 1984.

[8.4] M.S. Soderstrand, W.K. Jenkins, G.A. Jullien, and F.J. Taylor, *Residue Number System Arithmetic: Modern Applications in Digital Signal Processing*, IEEE Press, 1986.

[8.5] H. Krishna, B. Krishna, K.-Y. Lin, and J.-D Sun, *Computational Number Theory and Digital Signal Processing*, CRC Press, 1994.

[8.6] M.T. Heideman, *Multiplicative Complexity, Convolution, and the DFT*, Springer-Verlag, 1988.

[8.7] D.G. Myers, *Digital Signal Processing, Efficient Convolution and Fourier Transform Techniques*, Prentice Hall, Inc., 1990.

[8.8] R. Tolimieri, M. An, and C. Lu, *Algorithms for Discrete Fourier Transfrom and Convolution*, Springer-Verlag, 1989.

Problems

8.1 Derive a CRT-P based algorithm for acyclic convolution of length $n = 5$ by using $P(u) = u(u+1)(u+2)(u+3)$ and $s = 1$. Derive the matrices in the bilinear form and describe an organization of the computation. Evaluate the multiplicative and additive complexities. Comment on the validity of the resulting algorithm over finite fields and rings.

8.2 Repeat Problem 8.1 with $P(u) = u(u+1)(u-1)(u+2)$.

8.3 Repeat Problem 8.1 with $P(u) = u(u+1)(u^2+1)$.

8.4 Comment on the nature of the constants (elements of the **A**, **B**, and **C** matrices in the bilinear form) that are encountered in Problems 8.1, 8.2, and 8.3.

8.5 We wish to convolve two integer sequences of lengths 6 and 5, respectively. It is known that the elements of the two sequences lie in the range 0 to 2. Design a suitable scheme using CRT-P over the field of rational numbers, $\mathbf{Z}$.

8.6 Cast the computation in Problem 8.5 in a finite field $GF(p)$ by first identifying a suitable p. Now design a suitable scheme using the CRT-P over $GF(p)$.

8.7 This problem compares the two approaches for deriving fast algorithms. Let $d = e = 6$. Design a length $n = 11$ acyclic convolution algorithm (i) using the CRT-P and (ii) using the two-dimensional approach. Compare the two algorithms.

8.8 Design a multidimensional approach to computing acyclic convolution of two sequences of lengths $d = e = 2^a$. Show that this task can be accomplished in 3^a MULTs.

8.9 The two-dimensional acyclic convolution algorithms for $X(u, v)\ Y(u, v)$ described in this chapter employ two one-dimensional acyclic convolution algorithms for polynomial products of the type $A(u) \cdot B(u)$ and $C(v) \cdot D(v)$. Define $A(u) \cdot B(u) \bmod P(u)$ and $C(v) \cdot D(v) \bmod Q(v)$. Let $P(u) = P_1(u) \cdot P_2(u)$, $Q(v) = Q_1(v) \cdot Q_2(v)$, $(P_1(u), P_2(u)) = 1$ and $(Q_1(v), Q_2(v)) = 1$. Show that an alternate way to compute the acyclic convolution is to first compute the polynomial products

$$X(u, v)\ Y(u, v) \bmod (P_1(u), Q_1(v)),$$
$$X(u, v)\ Y(u, v) \bmod (P_1(u), Q_2(v)),$$
$$X(u, v)\ Y(u, v) \bmod (P_2(u), Q_1(v)),$$
$$X(u, v)\ Y(u, v) \bmod (P_2(u), Q_2(v)),$$

and then to reconstruct $X(u, v)\ Y(u, v)$ from it.

8.10 Let $d_1 = d_2 = 2$, $e_1 = e_2 = 3$. Design two-dimensional acyclic convolution algorithms using the method described in the Section 8.4 and the method in Problem 8.9.

8.11 We explore the idea of Problem 8.9 further. One way to compute $X(u, v)\ Y(u, v) \bmod (P(u), Q(v))$ is to employ the one-dimensional algorithms for computing $A(u)\ B(u) \bmod P(u)$ and $A(v)\ B(v) \bmod Q(v)$.

Another way could be to seek further factorization of $P(u)$ or $Q(v)$ in an extension field if either one of them is irreducible. Let $P(u) = u^2 + 1$, $Q(v) = v^2 + 1$. Describe the algorithm to compute $X(u, v)\ Y(u, v) \bmod (u^2 + 1, v^2 + 1)$ using two one-dimensional algorithms and evaluate its computational complexity. Now express $Q(v) = (v + u)(v - u) \bmod P(u)$ and use it to express $X(u, v)\ Y(u, v) \bmod (u^2 + 1, v^2 + 1)$ as a direct sum of $X(u, -u)\ Y(u, -u) \bmod u^2 + 1$ and $X(u, u)\ Y(u, u) \bmod u^2 + 1$. Describe the resulting algorithm and evaluate its computational complexity. Which of the two approaches is computationally superior?

8.12 Can you employ the approach described in Problems 8.9 and 8.11 to describe the complete algorithm for computing the two-dimensional acyclic convolution?

8.13 Use the idea of Problem 8.11 with $P(u) = u^a + 1$, $Q(v) = u^a + 1$ with $a = 2^b$.

8.14 Use the idea of Problem 8.11 with $P(u) = u^a + 1$, $Q(v) = u^b + 1$, $a = 2^c$, $b = 2^d$, $c > d$.

8.15 Use the idea of Problem 8.11 with $P(u) = u^a + 1$, $Q(v) = u^b + 1$, $a = 2^c$, $b = 2^d$, $c < d$.

8.16 Compare the results of Problems 8.14 and 8.15 in terms of computational complexity and decide which of the two approaches is better.

8.17 Design an acyclic convolution algorithm for $d = e = 10$ using the two-dimensional approach.

8.18 Examine the validity of the algorithm derived in Problem 8.17 over finite fields and integer rings.

8.19 Examine the validity of the algorithms derived in Problem 8.7 over finite fields and integer rings.

8.20 Design a length $n = 11$ acyclic convolution algorithm over $GF(3)$, $GF(5)$, $Z(8)$, $CZ(8)$.

8.21 Compare the algorithms in Problems 8.7 and 8.20.

Chapter 9

Fast One-Dimensional Cyclic Convolution Algorithms

In the next two chapters, we describe fast algorithms for computing one- and two-dimensional cyclic convolution of discrete sequences. The present chapter deals with one-dimensional algorithms and techniques for converting one-dimensional cyclic convolution to two- or higher-dimensional convolution. Chapter 10 is on fast algorithms for two- and higher-dimensional algorithms for cyclic convolution. In addition, we attempt to obtain closed form expressions for the polynomials in CRT-P reconstruction and various matrices that arise in the bilinear formulation of fast algorithms. The number systems included in this chapter are the fields of complex, real and rational numbers, finite fields, and finite integer rings. The analysis presented here is expected to be useful in understanding the structure of cyclic convolution algorithms in different number systems, organization of various steps in the computation, and overall realization of the algorithms.

This chapter deals with one-dimensional cyclic convolution of length n defined as the computation

$$Z(u) = X(u)\, Y(u) \bmod u^n - 1, \tag{9.0.1}$$

where $X(u)$ and $Y(u)$ are polynomials of degree $n - 1$. A direct computation of (9.0.1) requires n^2 MULTs and $n^2 - n$ ADDs. It is this computational cost that we seek to reduce.

CRT-P forms the backbone of all fast algorithms for cyclic convolution in a manner similar to fast algorithms for acyclic convolution described in Chapter 8. The results of Chapter 8 are also valid here with the fundamental difference that the modulo polynomial is restricted to be

$$P(u) = u^n - 1. \tag{9.0.2}$$

It could be any arbitrary polynomial of degree n for acyclic convolution. Consequently, we have to use the factorization of $u^n - 1$ in CRT-P formulation for computing cyclic convolution. This factorization has been studied in depth in Chapter 5 over the fields of complex, real, and rational numbers; in Chapter 6 over finite fields; and in Chapter 7 over finite integer rings.

The cyclotomic factorization studied in Chapters 5, 6, and 7, along with the CRT-P can be used to obtain a fast algorithm for computing one-dimensional cyclic convolution of data sequences. The objective of this chapter is to go one step further and analyze the structure of various matrices arising in the bilinear forms associated with fast algorithms based on CRT-P.

9.1 Bilinear Forms and Cyclic Convolution

Based on CRT-P, the fast algorithm can be outlined as follows. Given

$$u^n - 1 = \prod_{i=1}^{\lambda} p_i(u), \tag{9.1.1}$$

$(p_i(u), p_j(u)) = 1$, $i \neq j$, the algorithm involves the following five steps:

Step 1. CRT-P reduction. Compute

$$\begin{aligned} X_i(u) &\equiv X(u) \bmod p_i(u) \\ Y_i(u) &\equiv Y(u) \bmod p_i(u), \quad i = 1, 2, \ldots, \lambda. \end{aligned} \tag{9.1.2}$$

Step 2. Small degree polynomial MULT. Compute

$$Z_i(u) \equiv X_i(u)\, Y_i(u) \bmod p_i(u), \quad i = 1, 2, \ldots, \lambda. \tag{9.1.3}$$

Step 3. CRT-P reconstruction. Compute

$$A_i(u) \equiv Z_i(u)\, t_i(u) \bmod p_i(u), \quad i = 1, 2, \ldots, \lambda. \tag{9.1.4}$$

Step 4. CRT-P reconstruction-continued. Compute

$$B_i(u) \equiv A_i(u)\, P_i(u) \bmod (u^n - 1), \quad i = 1, 2, \ldots, \lambda. \tag{9.1.5}$$

Step 5. CRT-P recovery. Compute

$$Z(u) = \sum_{i=1}^{\lambda} B_i(u) \bmod (u^n - 1). \tag{9.1.6}$$

Here $P_i(u)$ and $t_i(u)$ are defined as

$$P_i(u) = (u^n - 1)/p_i(u) \tag{9.1.7a}$$
$$t_i(u)\, P_i(u) \equiv 1 \bmod p_i(u). \tag{9.1.7b}$$

Several observations are in order here.

Observation 1. Since the modulo polynomials $p_i(u)$, $i = 1, 2, \ldots, \lambda$ are monic, Steps 1, 4, and 5 do not involve any divisions.

Observation 2. The computation in Steps 2, 3, and 4 need not be performed mod $p_i(u)$ and $u^n - 1$. They are shown here for the sake of completeness. The final result is correct as long as (9.1.6) is defined mod $(u^n - 1)$.

Observation 3. As long as the coefficients of $p_i(u)$, $P_i(u)$, and $t_i(u)$ are simple, Steps 1, 3, 4, and 5 are assumed to involve ADDs only. The only MULTs required are in Step 2 to compute the acyclic convolution

$$Z_i(u) \equiv X_i(u)\, Y_i(u) \bmod p_i(u), \quad i = 1, 2, \ldots, \lambda.$$

The multiplicative complexity of the CRT-P cyclic convolution algorithm is given by

$$M(n) = \sum_{i=1}^{\lambda} M_a(d_i),$$

where $d_i = deg(p_i(u))$. Here $M_a(d_i)$ is multiplicative complexity of the acyclic convolution algorithm for multiplying two degree d_i polynomials mod $p_i(u)$. Note that $M_a(d_i)$ depends on the precise form of $p_i(u)$. In many cases, $M_a(d_i)$ turns out to be the multiplicative complexity of acyclic convolution algorithm for two polynomials of degree d_i each, which makes it independent of $p_i(u)$. In general, a large number of factors $p_i(u)$ lead to reduction in the number of MULTs. We will, therefore, focus exclusively on factorization of $u^n - 1$ in terms of its monic *irreducible* factors. The resulting factors are the cyclotomic polynomials that were studied in Chapters 5, 6, and 7. They can be determined in closed form and their structure depends on the number system of interest.

It will be assumed that the small degree acyclic convolution in Step 2 can be computed using pre-designed algorithms or, alternatively, CRT-P can be used one more time. Thus, the structure of cyclic convolution algorithm can be analyzed completely except for the **reconstruction polynomials (RPs)** $t_i(u)$, $i = 1, 2, ..., \lambda$. It is clear that $(p_i(u), P_i(u)) = 1$. The *Euclid's algorithm* consists in expressing the *gcd* $d(u)$ of two polynomials $a(u)$ and $b(u)$ as

$$d(u) = c_1(u)\, a(u) + c_2(u)\, b(u). \tag{9.1.8}$$

Setting $a(u) = p_i(u)$ and $b(u) = P_i(u)$, we get

$$d(u) = 1 = c_1(u)\, p_i(u) + c_2(u)\, P_i(u). \tag{9.1.9}$$

Therefore,

$$t_i(u) \equiv c_2(u) \bmod p_i(u). \tag{9.1.10}$$

Recall once again that the reduction mod $p_i(u)$ need not be preformed in Steps 2 and 3 of the algorithm. Only the final result needs to be computed mod $(u^n - 1)$. For the sake of analysis being pursued here, it is desirable to have a closed form expression for the RPs $t_i(u)$. We now proceed to establish such a closed form.

Theorem 9.1 The RPs $t_i(u)$ in CRT-P based cyclic convolution algorithm are given by

$$t_i(u) = \left[\frac{u}{n}\frac{d}{du}p_i(u)\right] \bmod p_i(u),\ i = 1,2,...,\lambda\,. \tag{9.1.11}$$

Proof: From the definitions of $p_i(u)$ and $P_i(u)$, we have

$$u^n - 1 = p_i(u)\, P_i(u). \tag{9.1.12}$$

Differentiating both sides (formal derivative for finite fields and rings), we get

$$nu^{n-1} = P_i(u)\frac{d}{du}p_i(u) + p_i(u)\frac{d}{du}P_i(u)\,. \tag{9.1.13}$$

Multiplying (9.1.13) by u/n and subtracting (9.1.12) from it, we get

$$1 = \left[\frac{u}{n}\frac{d}{du}p_i(u)\right]P_i(u) + \left[\frac{u}{n}\frac{d}{du}P_i(u) - P_i(u)\right] \bmod p_i(u). \tag{9.1.14}$$

The statement of the theorem follows.

Theorem 9.1 is useful in many ways. One, it bypasses Euclid's algorithm completely. Two, we can analytically compute all the quantities involved in the computation. Three, it indicates that the only division required in the algorithm is by n. In other words, n^{-1} must exist over the number system of interest. For **I**, **Z**, **Re**, and **C**, this condition is trivially satisfied. For finite fields and finite integer rings such as $GF(q)$, $Z(M)$, and $CZ(M)$, this translates into $(n, q) = 1$ and $(n, M) = 1$, respectively. We may also interpret that if n is ignored from the denominator in (9.1.11), then it is equivalent to computing the output with a gain of n. In such a case, no division is required.

Let $deg(p_i(u)) = d_i$, $\sum_{i=1}^{\lambda} d_i = n$. In matrix-vector notation, any bilinear form (of which cyclic convolution is a special case) can be expressed as

$$\underline{Z} = \mathbf{C}\,[\mathbf{A}\,\underline{X} \otimes \mathbf{B}\,\underline{Y}], \tag{9.1.15}$$

where $\otimes$ denotes component-by-component MULT of two vectors. The dimensions of **A**, **B**, and **C** matrices are $M(n) \times n$, $M(n) \times n$, and $n \times M(n)$, respectively. Recall that $M(n)$ denotes the multiplicative complexity of cyclic convolution algorithm. The elements of **A**, **B**, and **C** are scalars of the number system over which the algorithm is defined.

It is important to establish the forms of **A**, **B**, and **C** matrices for the five-step algorithm described above. In matrix-vector form, Step 1 becomes

$$\underline{X}_i = \mathbf{R}_i\,\underline{X},$$
$$\underline{Y}_i = \mathbf{R}_i\,\underline{Y}, \quad i = 1, 2, \ldots, \lambda. \tag{9.1.16}$$

Following (9.1.15), Step 2 can be expressed as (ignoring mod $p_i(u)$)

$$\underline{Z}_i = \mathbf{C}_i\,[\mathbf{A}_i\,\underline{X}_i \otimes \mathbf{B}_i\,\underline{Y}_i], \quad i = 1, 2, \ldots, \lambda. \tag{9.1.17}$$

The dimensions of $\mathbf{A}_i$, $\mathbf{B}_i$, and $\mathbf{C}_i$ matrices are $M_a(d_i) \times d_i$, $M_a(d_i) \times d_i$, and $(2d_i - 1) \times M_a(d_i)$, respectively. Ignoring mod $p_i(u)$ in (9.1.4), Steps 3 and 4 can be expressed as

$$\underline{A}_i = \mathbf{T}_i\,\underline{Z}_i \tag{9.1.18}$$

$$\underline{B}_i = \mathbf{P}_i\, \underline{A}_i, \quad i = 1, 2, \ldots, \lambda. \tag{9.1.19}$$

The dimensions of $\mathbf{T}_i$ and $\mathbf{P}_i$ matrices are $(3d_i - 2) \times (2d_i - 1)$ and $n \times (3d_i - 2)$, respectively. Combining (9.1.16) to (9.1.19) with (9.1.6) leads to the matrices **A**, **B**, and **C** in the algorithm as expressed in (9.1.15). It is straightforward to verify that

$$\mathbf{A} = \begin{bmatrix} \mathbf{A}_1\mathbf{R}_1 \\ \hline \mathbf{A}_2\mathbf{R}_2 \\ \hline \vdots \\ \hline \mathbf{A}_\lambda\mathbf{R}_\lambda \end{bmatrix} = \begin{bmatrix} \mathbf{A}_1 & & & \\ & \mathbf{A}_2 & & \\ & & \ddots & \\ & & & \mathbf{A}_\lambda \end{bmatrix} \begin{bmatrix} \mathbf{R}_1 \\ \hline \mathbf{R}_2 \\ \hline \vdots \\ \hline \mathbf{R}_\lambda \end{bmatrix} = \mathbf{A}'\mathbf{R}, \tag{9.1.20}$$

$$\mathbf{B} = \begin{bmatrix} \mathbf{B}_1\mathbf{R}_1 \\ \hline \mathbf{B}_2\mathbf{R}_2 \\ \hline \vdots \\ \hline \mathbf{B}_\lambda\mathbf{R}_\lambda \end{bmatrix} = \begin{bmatrix} \mathbf{B}_1 & & & \\ & \mathbf{B}_2 & & \\ & & \ddots & \\ & & & \mathbf{B}_\lambda \end{bmatrix} \begin{bmatrix} \mathbf{R}_1 \\ \hline \mathbf{R}_2 \\ \hline \vdots \\ \hline \mathbf{R}_\lambda \end{bmatrix} = \mathbf{B}'\mathbf{R}, \tag{9.1.21}$$

and

$$\begin{aligned} \mathbf{C} &= \left[\mathbf{P}_1\mathbf{T}_1\mathbf{C}_1 \mid \mathbf{P}_2\mathbf{T}_2\mathbf{C}_2 \mid \cdots \mid \mathbf{P}_\lambda\mathbf{T}_\lambda\mathbf{C}_\lambda\right] \\ &= \left[\mathbf{P}_1 \mid \mathbf{P}_2 \mid \cdots \mid \mathbf{P}_\lambda\right] \begin{bmatrix} \mathbf{T}_1 & & & \\ & \mathbf{T}_2 & & \\ & & \ddots & \\ & & & \mathbf{T}_\lambda \end{bmatrix} \begin{bmatrix} \mathbf{C}_1 & & & \\ & \mathbf{C}_2 & & \\ & & \ddots & \\ & & & \mathbf{C}_\lambda \end{bmatrix} = \mathbf{PTC}'. \end{aligned} \tag{9.1.22}$$

Above partitioned forms will be useful in subsequent analyses. These forms show explicitly the origin of various computations. Also, the structure of **R**, **T**, and **P** matrices can be analyzed independently. We assume that the small degree acyclic convolution algorithms are pre-designed. Consequently, we have little control over the forms of **A**′, **B**′, and **C**′ matrices. However, we further assume that their elements are simple in nature.

Example 9.1 Let $n = 6$ and the field of interest be Z. Over Z,

$$u^6 - 1 = (u - 1)\,(u + 1)\,(u^2 + u + 1)\,(u^2 - u + 1).$$

Let $p_1(u) = u - 1$, $p_2(u) = u + 1$, $p_3(u) = u^2 + u + 1$, and $p_4(u) = u^2 - u + 1$. Also, $P_i(u) = (u^6 - 1)/p_i(u)$ and we have

$$P_1(u) = u^5 + u^4 + u^3 + u^2 + u + 1,$$
$$P_2(u) = u^5 - u^4 + u^3 - u^2 + u - 1,$$
$$P_3(u) = u^4 - u^3 + u - 1,$$
$$P_4(u) = u^4 + u^3 - u - 1.$$

Using (9.1.11), the RPs, $t_i(u)$, $i = 1, 2, ..., 4$, are computed as,

$$t_1(u) = (1/6),$$
$$t_2(u) = -(1/6),$$
$$t_3(u) = (1/6)\,(-u - 2),$$
$$t_4(u) = (1/6)\,(u - 2).$$

The matrices **T** and **P** are obtained from polynomials $t_i(u)$ and $P_i(u)$. The various blocks are given by

$$\mathbf{T}_1 = \mathbf{T}_2 = (1/6)[1]$$

$$\mathbf{T}_3 = (1/6)\begin{bmatrix} -2 & 0 & 0 \\ -1 & -2 & 0 \\ 0 & -1 & -2 \\ 0 & 0 & -1 \end{bmatrix}$$

$$\mathbf{T}_4 = (1/6)\begin{bmatrix} -2 & 0 & 0 \\ 1 & -2 & 0 \\ 0 & 1 & -2 \\ 0 & 0 & 1 \end{bmatrix}$$

$$\mathbf{P}_1 = [1 \;\; 1 \;\; 1 \;\; 1 \;\; 1 \;\; 1]^t$$
$$\mathbf{P}_2 = [-1 \;\; 1 \;\; -1 \;\; 1 \;\; -1 \;\; 1]^t$$

$$\mathbf{P}_3 = \begin{bmatrix} -1 & 0 & 1 & -1 \\ 1 & -1 & 0 & 1 \\ 0 & 1 & -1 & 0 \\ -1 & 0 & 1 & -1 \\ 1 & -1 & 0 & 1 \\ 0 & 1 & -1 & 0 \end{bmatrix}$$

$$\mathbf{P}_4 = \begin{bmatrix} -1 & 0 & 1 & 1 \\ -1 & -1 & 0 & 1 \\ 0 & -1 & -1 & 0 \\ 1 & 0 & -1 & -1 \\ 1 & 1 & 0 & -1 \\ 0 & 1 & 1 & 0 \end{bmatrix}.$$

The product of two degree 0 polynomials $X(u) = x_0$ and $Y(u) = y_0$ can be computed in 1 MULT as $Z(u) = x_0\, y_0$. The product of two degree 1 polynomials $X(u) = x_0 + x_1 u$ and $Y(u) = y_0 + y_1 u$ is computed in 3 MULTs and 4 ADDs as

$$Z(u) = m_0 + (-m_0 + m_1 - m_2)\, u + m_1 u^2,$$

$$m_0 = x_0\, y_0,$$
$$m_1 = (x_0 + x_1)\,(y_0 + y_1),$$
$$m_2 = x_1\, y_1.$$

The corresponding blocks of the $\mathbf{A}'$, $\mathbf{B}'$, and $\mathbf{C}'$ matrices are given by

$$\mathbf{A}_1 = \mathbf{B}_1 = \mathbf{A}_2 = \mathbf{B}_2 = [1]$$

$$\mathbf{A}_3 = \mathbf{B}_3 = \mathbf{A}_4 = \mathbf{B}_4 = \begin{bmatrix} 1 & 1 & 0 \\ 0 & 1 & 1 \end{bmatrix}^t$$

$$\mathbf{C}_1 = \mathbf{C}_2 = [1]$$

$$\mathbf{C}_3 = \mathbf{C}_4 = \begin{bmatrix} 1 & 0 & 0 \\ -1 & 1 & -1 \\ 0 & 0 & 1 \end{bmatrix}.$$

Finally, the blocks of the reduction matrix $\mathbf{R}$ computed using $p_i(u)$, $i = 1, 2, \ldots, 4$ are given by

$$\mathbf{R}_1 = [\,1 \;\; 1 \;\; 1 \;\; 1 \;\; 1 \;\; 1\,]$$
$$\mathbf{R}_2 = [\,1 \;\; -1 \;\; 1 \;\; -1 \;\; 1 \;\; -1\,]$$

$$\mathbf{R}_3 = \begin{bmatrix} 1 & 0 & -1 & 1 & 0 & -1 \\ 0 & 1 & -1 & 0 & 1 & -1 \end{bmatrix}$$

$$\mathbf{R}_4 = \begin{bmatrix} 1 & 0 & -1 & -1 & 0 & 1 \\ 0 & 1 & 1 & 0 & -1 & -1 \end{bmatrix}.$$

It is clear that these matrices are also valid over all those $GF(q)$, $Z(M)$, and $CZ(M)$ for which $(6, q) = (6, M) = 1$.

There are three motivations in presenting the above example in such depth. First, once the number system N is chosen and the factorization of $u^n - 1$ over N is known, all the matrices can be determined in closed form. Second, is the appearance of numbers {0, 1, −1} in **R** and **P** matrices which is a desirable feature from the standpoint of the organization of computations and computational complexity of the algorithm. Let U = {0, 1, −1}. Third, is the structure of the $\mathbf{T}_i$ matrices that have entries other than U. All three motivations are further pursued. Cyclotomic polynomials are analyzed and conditions are derived for which the elements of **R** and **P** matrices belong to U. Also, the elements of the **T** matrix are analyzed for their nature and a simple bound is derived for their values.

We end this section by stating that the elements of **A′**, **B′**, and **C′** matrices depend on the small degree acyclic convolution algorithm and it is desirable that they take values in the set U. Several such algorithms have been described in Appendix A. Furthermore, it may be desirable to compute the product $Z_i(u) \equiv X_i(u)\, Y_i(u) \bmod p_i(u)$ in Step 2 of the algorithm directly instead of computing it as $X_i(u)\, Y_i(u)$ and then reducing it mod $p_i(u)$. Readers may refer to the statements in Section 8.1 in this regard.

9.2 Cyclotomic Polynomials and Related Algorithms Over Re and C

We begin with factorization of $u^n - 1$ over C. Over C, $u^n - 1$ is factored as

$$u^n - 1 = \prod_{i=1}^{n} \left(u - \omega^i\right) \qquad (9.2.1)$$

where

$$\omega = \exp(-2\pi j/n). \qquad (9.2.2)$$

Here, ω is the n-th root of unity in C. Thus, we have

$$p_i(u) = u - \omega^i,$$

$$P_i(u) = (u^n - 1)/(u - \omega^i) = \omega^{-i} \sum_{l=0}^{n-1} (\omega^{-i} u)^l ,$$

$$t_i(u) = \frac{1}{n}\left[u \frac{d}{du}(u - \omega^i) \right] \bmod (u - \omega^i) = \frac{\omega^i}{n},$$

and

$\mathbf{A}' = \mathbf{B}' = \mathbf{C}' = \mathbf{I}$ (Identity matrix).

The CRT-P reduction matrix $\mathbf{R}$ is the well-known DFT matrix and the reconstruction matrix $\mathbf{P}\ \mathbf{T}$ is the IDFT matrix. Computation of DFT and IDFT using the FFT is a topic that will be studied in Chapters 13 and 14. Above factorization and the resulting expressions are valid for any field that possesses an n-th root of unity. Such roots exist in $GF(q)$ for values of n that divide $q - 1$.

Over Re, $u^n - 1$ factors as

$$u^n - 1 = \begin{cases} (u-1)(u+1)\prod_{i=1}^{n/2-1}(u^2 - 2\cos\theta_i u + 1) & n: even \\ (u+1)\prod_{i=1}^{(n-1)/2}(u^2 - 2\cos\theta_i u + 1) & n: odd \end{cases}$$

where

$$\cos\theta_i = \frac{(\omega^i + \omega^{-i})}{2} = \cos\left(\frac{2\pi i}{n}\right).$$

Thus,

$p_1(u) = u - 1,$

$$P_1(u) = (u^n - 1)/(u - 1) = \sum_{l=0}^{n-1} u^l ,$$

$$t_1(u) = \frac{1}{n}\left[u \frac{d}{du}(u - 1) \right] \bmod (u - 1) = \frac{1}{n},$$

$p_i(u) = u^2 - 2\cos\theta_i + 1,$

$$t_i(u) = \frac{1}{n}\left[u\frac{d}{du}\left(u^2 - 2\cos\theta_i u + 1\right)\right]\mathrm{mod}\left(u^2 - 2\cos\theta_i u + 1\right)$$
$$= \frac{1}{n}\left[2\cos\theta_i u - 1\right],$$
$$P_i(u) = \left(u^n - 1\right)/\left(u^2 - 2\cos\theta_i u + 1\right) = -\sum_{k=0}^{n-1}\sum_{l=0}^{\infty} u^{l+k}\omega^{i(l-k)},$$

$i = 1, 2, \ldots, (n/2 - 1)$ (n even), $i = 1, 2, \ldots, (n-1)/2$ (n odd). Finally, for n even, let $l = n/2 + 1$. Then

$$p_l(u) = u + 1,$$
$$P_l(u) = \left(u^n - 1\right)/(u+1) = -\sum_{k=0}^{n-1}(-u)^k,$$
$$t_l(u) = \frac{1}{n}\left[u\frac{d}{du}(u+1)\right]\mathrm{mod}(u+1) = -\frac{1}{n}.$$

9.3 Cyclotomic Polynomials and Related Algorithms Over Z

The factorization of $u^n - 1$ in terms of its irreducible factor polynomials over I and Z is the same and is given by

$$u^n - 1 = \prod_{d|n} C_d(u), \tag{9.3.1}$$

where

$$C_d(u) = \prod_{\substack{i \le d \\ (i,d)=1}}\left(u - \omega^i\right), \tag{9.3.2}$$

ω being the d-th root of unity in (9.3.2). The number of cyclotomic factors of $u^n - 1$ is equal to the number of divisors of n including 1 and n, and the degree of $C_d(u)$ is equal to $\phi(d)$, the Euler's totient function of d.

There are a number of important properties associated with cyclotomic polynomials.

(1) Their coefficients are integers.
(2) They are irreducible over I and Z.
(3) Their coefficients take values in U when $\alpha \leq 2$.

Recall that

$$U = \{0, 1, -1\}.$$

In property (3) above and the remainder of this section, α denotes the number of odd prime factors of d. Also, β denotes the total number of prime factors of d. Clearly, $\beta = \alpha + 1$ or $\beta = \alpha$ depending on whether 2 is a factor of d or not. Cyclotomic polynomials are computed using various expressions given in Chapter 5. There are some additional properties that follow from a closer examination of (5.2.8) to (5.2.32).

(4) $deg(C_d(u)) = \phi(d)$, which is even except for $d = 1, 2$.
(5) They are symmetric except $C_1(u)$ that is asymmetric. A polynomial $X(u) = \sum_{i=0}^{r} x_i u^i$ is called *symmetric* if $x_i = x_{r-i}$, and *asymmetric* if $x_i = -x_{r-i}$.
(6) For $\alpha = \beta = 1$, the coefficients of $C_d(u)$ belong to the set $\{0, 1\}$.
(7) For $\beta \leq 2$, the non-zero coefficients of $C_d(u)$ alternate in sign between 1 and -1, starting with the coefficient of u^0 which is 1.
(8) The number of non-zero coefficients in $C_d(u)$, $d = p^a q^b r^c \ldots$ is the same as the number of non-zero coefficients in $C_t(u)$, $t = pqr\ldots$. This is also true for $C_{2m}(u)$ and $C_m(u)$, m odd.

One might wonder why these properties are important. Properties (1) to (7) will be used to establish the nature of the elements of CRT-P reduction and reconstruction matrices. Property (8), along with others, can be used in determining the additive complexity of the cyclic convolution algorithm. The number of non-zero coefficients in $C_{pq}(u)$ can be evaluated numerically for all values of p and q that one is likely to encounter in digital signal processing.

Given the cyclotomic factorization of $u^n - 1$, we now turn to CRT-P reduction and reconstruction matrices **R**, **T**, and **P**. Each of these matrices has a block structure having D blocks, D being the number of factors of n or the number of cyclotomic factors of $u^n - 1$. Elements in the d-th block depend only on $C_d(u)$. Given $C_d(u)$, we will derive closed form expressions for these elements for the most frequently occurring values of d. These values are $d = 2^m$, p^k, $2^m p^k$, $p^k q^l$, and $2^m p^k q^l$. Let

$$C_d(u) = \sum_{i=0}^{\phi(d)} c_{i,d} u^i . \tag{9.3.3}$$

Clearly, $c_{i,d} \in \mathrm{U}$ for $\alpha \leq 2$. The corresponding RP $t_d(u)$ as obtained from (9.1.11) is given by

$$t_d(u) = \left[\frac{u}{n} \frac{d}{du} C_d(u) \right] \operatorname{mod} C_d(u) = \frac{1}{n} \sum_{i=0}^{\phi(d)} i c_{i,d} u^i \operatorname{mod} C_d(u). \tag{9.3.4}$$

Since $deg(C_d(u)) = \phi(d)$, $u^{\phi(d)} \bmod C_d(u) = u^{\phi(d)} - C_d(u)$ and (9.3.4) simplifies to

$$t_d(u) = \frac{1}{n} \sum_{i=0}^{\phi(d)-1} \{i - \phi(d)\} c_{i,d} u^i . \tag{9.3.5}$$

For $d = 1, 2$, $C_d(u) = u + (-1)^d$ and $t_d(u) = (1/n)(-1)^{d+1}$. For $d = 2^m$,

$$C_d(u) = u^{2^{m-1}} + 1 \tag{9.3.6}$$

and

$$t_d(u) = -(1/n) 2^{m-1}. \tag{9.3.7}$$

For $d = p$,

$$t_d(u) = \frac{1}{n} \sum_{i=0}^{p-2} (i - p + 1) u^i . \tag{9.3.8}$$

For $d = p \cdot q$,

$$t_d(u) = \frac{1}{n} \left[\sum_{i=0}^{\infty} \sum_{l=0}^{p-1} \{lq + ip - \phi(pq)\} u^{lq+ip} - u \sum_{i=0}^{\infty} \sum_{l=0}^{p-1} \{lq + ip + 1 - \phi(pq)\} u^{lq+ip} \right]. \tag{9.3.9}$$

The expressions in (9.3.7), (9.3.8), and (9.3.9) are computed using (9.3.5) for the forms of $C_d(u)$ for various values of d. For $d = p^k q^l r^j \ldots$, let $\theta = p^{k-1} q^{l-1} r^{j-1} \ldots$, $d' = pqr\ldots$. Then we have the following results:

$$t_d(u) = \theta t_{d'}(v) \Big|_{v=u^\theta} \tag{9.3.10}$$

and for $d = 2m$, $(2, m) = 1$, $m \geq 3$,

$$t_{2m}(u) = t_m(-u). \tag{9.3.11}$$

These expressions are obtained by a straightforward application of the forms of $C_d(u)$ to (9.1.11). Let

$$t_d(u) = \frac{1}{n}\sum_{i=0}^{\gamma} t_{i,d}u^i,\ \gamma = \phi(d) - 1, \tag{9.3.12}$$

then $t_{i,d} = [i - \phi(d)] \cdot c_{i,d}$. Thus, the signs of the non-zero coefficients of $t_d(u)$ and $C_d(u)$ are opposite in nature. If $\alpha \leq 2$, $c_{i,d} \in \mathsf{U}$ and we have the following lemma.

Lemma 9.1 If d has at most two odd prime factors, then the largest absolute value of any non-zero coefficient of the RP $t_d(u)$ is $\phi(d)$ (ignoring n in the denominator).

The d-th block of the **T** matrix has dimensions $(3\phi(d) - 2) \times (2\phi(d) - 1)$ and is given by

$$\mathbf{T}_d = \frac{1}{n}\begin{bmatrix} t_{0,d} & & & \\ t_{1,d} & t_{0,d} & & \\ \vdots & t_{1,d} & \ddots & \\ t_{\gamma,d} & \vdots & \ddots & t_{0,d} \\ & t_{\gamma,d} & & t_{1,d} \\ & & \ddots & \vdots \\ & & & t_{\gamma,d} \end{bmatrix}. \tag{9.3.13}$$

Now, let us analyze the other RPs associated with $C_d(u)$. We have

$$P_d(u) = (u^n - 1)/C_d(u). \tag{9.3.14}$$

For $d = 1$, $C_d(u) = u - 1$, and

$$P_1(u) = \frac{u^n - 1}{u - 1} = \sum_{i=0}^{n-1} u^i. \tag{9.3.15}$$

For $d = p^k$, n must have the form $n = N \cdot p^a$, $(N, p) = 1$, $a \geq k$. In this case,

$$C_d(u) = \frac{\left(u^{p^k} - 1\right)}{\left(u^{p^{k-1}} - 1\right)} \tag{9.3.16}$$

and

$$P_d(u) = \left(u^{Np^a} - 1\right)\frac{\left(u^{p^{k-1}} - 1\right)}{\left(u^{p^k} - 1\right)} = \sum_{i=0}^{Np^{a-k}-1}\left[u^{(ip+1)p^{k-1}} - u^{ip^k}\right]. \tag{9.3.17}$$

We observe that $P_d(u)$, $d = p^k$ has $2N \cdot p^{a-k}$ non-zero coefficients that alternate in value between -1 and 1 starting with -1. For $d = p^k q^l$, n must have the form $n = N \cdot p^a q^b$, $(N, p) = (N, q) = 1$, $a \geq k$ and $b \geq l$. In this case,

$$C_d(u) = C_{pq}\left(u^{p^{k-1}q^{l-1}}\right) = \frac{\left(u^{p^k q^l} - 1\right)}{\left(u^{p^{k-1}q^l} - 1\right)}\frac{\left(u^{p^{k-1}q^{l-1}} - 1\right)}{\left(u^{p^k q^{l-1}} - 1\right)} \tag{9.3.18}$$

and

$$P_d(u) = \frac{\left(u^{Np^a q^b} - 1\right)}{\left(u^{p^k q^l} - 1\right)}\frac{\left(u^{p^k q^{l-1}} - 1\right)}{\left(u^{p^{k-1}q^{l-1}} - 1\right)}\left(u^{p^{k-1}q^l} - 1\right). \tag{9.3.19}$$

Without any loss in generality, assume that $q > p$. Then (9.3.19) simplifies to

$$P_d(u) = \sum_{i=0}^{Np^{a-k}q^{b-l}-1}\sum_{j=0}^{p-1}\left[u^{ip^k q^l + jp^{k-1}q^{l-1} + p^{k-1}q^l} - u^{ip^k q^l + jp^{k-1}q^{l-1}}\right]. \tag{9.3.20}$$

It can be shown that no two terms cancel out or overlap each other in (9.3.20). Thus, $P_d(u)$, $d = p^k q^l$ has $2N \cdot p^{a-k+1}q^{b-l}$, $q > p$, non-zero coefficients which are either 1 or -1. The following theorem establishes a more general result.

Theorem 9.2 RPs satisfy the following relationships:

$$P_d(u)\Big|_{\substack{d=2^m\cdot M, M>3\\ n=N\cdot 2^c\\ M|N\\ (M,2)=(N,2)=1\\ c\geq m}} = \left\{\prod_{i=1}^{c-m}\left(v^{2^{c-i}}+1\right)\right\}\left(1+w^N\right)P_d(w)\Big|_{\substack{d=M\\ n=N\\ w=-u^a\\ a=2^{m-1}\\ v=u^N}} \tag{9.3.21}$$

and

$$P_d(u)\Big|_{\substack{d=2^m\\ n=N\cdot 2^c\\ (N,2)=1\\ c\geq m}} = \left\{\prod_{i=1}^{c-m}\left(v^{2^{c-i}}+1\right)\right\}\left(w^N-1\right)\sum_{j=0}^{N-1}\left(-w^j\right)\Big|_{\substack{w=u^a\\ a=2^{m-1}\\ v=u^N}}. \tag{9.3.22}$$

Proof: For $d = 2^m \cdot M$, n must have the form $n = 2^c \cdot N$, $c \geq m$, $(2, M) = (2, N) = 1$ and $M \mid N$. Also,

$$C_d(u) = C_M\left(-u^{2^{m-1}}\right), \quad M \geq 3 \tag{9.3.23}$$

and

$$P_d(u) = \left(u^n - 1\right)/C_d(u) = \left(u^{N2^c} - 1\right)/C_M\left(-u^{2^{m-1}}\right). \tag{9.3.24}$$

Define a change of variables as

$$v = u^N \tag{9.3.25}$$

and write

$$u^{N2^c} - 1 = \left(u^N\right)^{2^c} - 1 = \left\{\prod_{i=1}^{c-m}\left(v^{2^{c-i}}+1\right)\right\}\left(v^{2^m}-1\right). \tag{9.3.26}$$

As a result, (9.3.24) becomes

$$P_d(u) = \left\{\prod_{i=1}^{c-m}\left(v^{2^{c-i}}+1\right)\right\}\left(v^{2^m}-1\right)\Big/C_M\left(-u^{2^{m-1}}\right). \tag{9.3.27}$$

In order to simplify (9.3.27), we write

$$v^{2^m} - 1 = u^{N2^m} - 1 = \left(1 - u^{N2^{m-1}}\right)\left(-1 - u^{N2^{m-1}}\right). \tag{9.3.28}$$

Substituting (9.3.28) into (9.3.27), we get

$$P_d(u) = \left\{\prod_{i=1}^{c-m}\left(v^{2^{c-i}} + 1\right)\right\}\left(1 + w^N\right)\left[\left(-1 + w^N\right)/C_M(w)\right]\Bigg|_{\substack{w=-u^a \\ a=2^{m-1}}}. \tag{9.3.29}$$

It is clear from the definition of $P_d(u)$ that the term in square parentheses in (9.3.29) is equal to $P_d(w)$ for $d = M$ and $n = N$. Therefore, in general,

$$P_d(u)\Big|_{\substack{d=2^m\cdot M, M>3 \\ n=N\cdot 2^c \\ M|N \\ (M,2)=(N,2)=1 \\ c\geq m}} = \left\{\prod_{i=1}^{c-m}\left(v^{2^{c-i}} + 1\right)\right\}\left(1 + w^N\right)P_d(w)\Big|_{\substack{d=M \\ n=N \\ w=-u^a \\ a=2^{m-1} \\ v=u^N}}. \tag{9.3.30}$$

Since $deg(C_d(u)) = \phi(d)$, we have $deg(P_d(u)) = n - \phi(d)$. Therefore,

$$\deg(P_d(w))\Big|_{\substack{d=M \\ n=N \\ w=-u^a \\ a=2^{m-1}}} = 2^{m-1}\{N - \phi(M)\} < N2^{m-1}. \tag{9.3.31}$$

Let the coefficients of $P_d(u)$ take values in a set S and t be the number of non-zero coefficients in $P_d(u)$, $d = M$, $n = N$. Then, the polynomial in (9.3.31) has degree less than $N \cdot 2^{m-1}$ with t non-zero coefficients taking values in S. Therefore, the polynomial

$$\left(1 + w^N\right)P_d(w)\Big|_{\substack{d=M \\ n=N \\ w=-u^a \\ a=2^{m-1} \\ v=u^N}}$$

in (9.3.30) has degree less than $N \cdot 2^m$ with $2t$ non-zero coefficients taking values in S. Finally, since $v = u^N$, $P_d(u)$, $d = 2^m \cdot M$, $n = 2^c \cdot N$, has degree less than

$2^c \cdot N$ (in fact, the exact value is $n - \phi(d) = 2^c \cdot N - 2^{m-1}\phi(M)$) and has $2^{c-m+1}t$ non-zero coefficients taking values in S.

For $M = p^k$, $N = J \cdot p^a$, $(J, p) = 1$, $a \geq k$, $t = 2J \cdot p^{a-k}$ and S = U. Similarly, for $M = p^k q^l$, $N = J \cdot p^a q^b$, $(J, p) = (J, q) = 1$, $a \geq k$, $b \geq l$ and $q > p$, $t = 2J \cdot p^{a-k+1} q^{b-l}$ and S = U.

A similar analysis for the case $d = 2^m$, $n = N \cdot 2^c$, $c \geq m$, $(N, 2) = 1$ leads to

$$P_d(u) = \left\{ \prod_{i=1}^{c-m} \left(v^{2^{c-i}} + 1 \right) \right\} \left(w^N - 1 \right) \sum_{j=0}^{N-1} \left(-w^j \right) \Bigg|_{\substack{w=u^a \\ a=2^{m-1} \\ v=u^N}} . \tag{9.3.32}$$

There are $N \cdot 2^{c-m+1}$ non-zero coefficients in $P_d(u)$. Once again, S = U. This completes our analysis for the case $d = 2^m \cdot M$.

We infer from (9.3.17), (9.3.20), (9.3.21), and (9.3.22) that the coefficients of the RP $P_d(u)$ take values in U if $\alpha \leq 2$. The number of non-zero coefficients can be determined in closed form.

For the case, $d = p^k q^l r^j \ldots$, let $\theta = p^{k-1} q^{l-1} r^{j-1} \ldots$ and $d' = pqr\ldots$. Then n must have the form $n = N \cdot p^a q^b r^c \ldots$, $(N, p) = (N, q) = (N, r) = \ldots = 1$, $a \geq k$, $b \geq l$, $c \geq j$, … . In this case,

$$P_d(u) = P_{d'}(u) \Big|_{\substack{n = Np^{a-k+1} q^{b-l+1} r^{c-j+1} \\ v = u^\theta}} . \tag{9.3.33}$$

Given $P_d(u)$ as

$$P_d(u) = \sum_{i=0}^{\gamma} p_{i,d} u^i, \quad \gamma = n - \phi(d),$$

the corresponding d-th block of the **P** matrix has dimension $n \times (3\phi(d) - 2)$ and is given by

$$\mathbf{P}_d = \begin{bmatrix} p_{0,d} & 0 & 0 & 0 & p_{\gamma,d} & \\ p_{1,d} & p_{0,d} & 0 & \vdots & 0 & \\ \vdots & p_{1,d} & \ddots & 0 & \vdots & \\ p_{\gamma,d} & \vdots & & p_{0,d} & 0 & \cdots \\ 0 & p_{\gamma,d} & & \vdots & p_{0,d} & \\ \vdots & \vdots & \ddots & p_{\gamma-1,d} & \vdots & \\ 0 & 0 & 0 & p_{\gamma,d} & p_{\gamma-1,d} & \end{bmatrix}. \tag{9.3.34}$$

In other words, first column of $\mathbf{P}_d$ is obtained by listing the coefficients of $P_d(u)$. Subsequently, i-th column of $\mathbf{P}_d$ is obtained by circularly shifting the $(i-1)$-th column, $i = 2, 3, ..., 3\phi(d) - 2$. This analysis on the RP $P_d(u)$ can be summarized in the form of a theorem.

Theorem 9.3 If d has at most two odd prime factors, then coefficients of the RP $P_d(u)$ and elements of the corresponding block $\mathbf{P}_d$ in the reconstruction matrix $\mathbf{P}$ associated with $C_d(u)$ take values in U.

In the following, we analyze the structure of CRT-P reduction matrix $\mathbf{R}$. The computation of $X_d(u)$ in (9.1.2) (or equivalently, $Y_d(u)$) is performed as

$$X_d(u) = X(u) \bmod C_d(u), \tag{9.3.35}$$

where $deg(X(u)) = n - 1$ and $deg(C_d(u)) = \phi(d)$. Let

$$X(u) = \sum_{i=0}^{n-1} x_i u^i . \tag{9.3.36}$$

Substituting (9.3.36) in (9.3.35), we get

$$X_d(u) = \sum_{i=0}^{n-1} x_i u^i \bmod C_d(u) = \sum_{i=0}^{n-1} \left(u^i \bmod C_d(u)\right) x_i . \tag{9.3.37}$$

Therefore, in matrix-vector form, the i-th column of the block $\mathbf{R}_d$ of $\mathbf{R}$ is obtained by simply listing the coefficients of the polynomial

$$r_{i,d}(u) = u^i \bmod C_d(u), \quad i = 0, 1, ..., n-1. \tag{9.3.38}$$

Since $deg(C_d(u)) = \phi(d)$, we have

$$r_{i,d}(u) = u^i,\ \ i = 0, 1, ..., \phi(d) - 1. \tag{9.3.39}$$

The cyclotomic polynomial $C_d(u)$ contains all roots of unity of order d as its roots. They are also contained in $u^d - 1$. Therefore,

$$u^d \equiv 1 \bmod C_d(u). \tag{9.3.40}$$

In fact, d is the smallest value of i for which $u^i \equiv 1 \bmod C_d(u)$ or else $C_d(u)$ contains roots of unity other than the roots of unity of order d. The above equation also leads to

$$u^{d+i} \equiv u^i \bmod C_d(u). \tag{9.3.41}$$

Thus, the $(i + d)$-th column of $\mathbf{R}_d$ is same as the i-th column of $\mathbf{R}_d$ for all i. Combining (9.3.41) with (9.3.39) we see that $r_{i,d}(u)$ needs to be evaluated only for $i = \phi(d), \phi(d) + 1, ..., d - 1$. This forms a total of $d - \phi(d)$ values. In the following, we evaluate these polynomials for cases when d has at most two odd prime factors. We begin with $d = p$.

For $d = p$, $\phi(d) = p - 1$ and we need to evaluate only u^{p-1}. In this case,

$$C_p(u) = \sum_{j=0}^{p-1} u^j$$

and

$$u^{p-1} = -\sum_{j=0}^{p-2} u^j \bmod C_p(u). \tag{9.3.42}$$

The corresponding column vector contains all -1s and the $(p - 1) \times n$ dimensional block $\mathbf{R}_p$ of the reduction matrix is given by

$$\mathbf{R}_p = \begin{bmatrix} & -1 & & -1 & \\ & -1 & & -1 & \\ \mathbf{I}_{p-1} & \vdots & \mathbf{I}_{p-1} & \vdots & \cdots \\ & \vdots & & \vdots & \\ & -1 & & -1 & \end{bmatrix}. \tag{9.3.43}$$

For $d = pq$, $\phi(d) = (p-1)(q-1)$ and we need to evaluate $u^i \bmod C_{pq}(u)$, $i = (p-1)(q-1), \ldots, pq-1$. Recall that $C_{pq}(u)$ is a symmetric polynomial. Once again, we assume that $q > p$. Since

$$C_{pq}(u) = \frac{\left(u^{pq}-1\right)}{\left(u^{p}-1\right)} \frac{(u-1)}{\left(u^{q}-1\right)}, \tag{9.3.44}$$

it is clear that

$$\frac{\left(u^{pq}-1\right)}{\left(u^{q}-1\right)} \equiv 0 \bmod C_{pq}(u)$$

or

$$u^{q(p-1)} + u^{q(p-2)} + \ldots + u^q + 1 \equiv 0 \bmod C_{pq}(u). \tag{9.3.45}$$

For $q > p$, $q(p-1) > (q-1)(p-1)$ while $q(p-2) < (q-1)(p-1)$. Therefore, we may alternatively write (9.3.45) as

$$u^{q(p-1)} \bmod C_{pq}(u) = -\sum_{j=0}^{p-2} u^{qj}.$$

In fact, one also obtains

$$u^{q(p-1)+l} \bmod C_{pq}(u) = -\sum_{j=0}^{p-2} u^{qj+l}, \ l = 0,1,\ldots,q-p+1. \tag{9.3.46}$$

Given (9.3.46), we only need to evaluate $u^i \bmod C_{pq}(u)$ for $i = (p-1)(q-1), \ldots, pq-q-1$ and $i = pq-(p-1), \ldots, pq-1$. We first evaluate $u^i \bmod C_{pq}(u)$ for $i = (p-1)(q-1), \ldots, pq-q-1$. If

$$C_{pq}(u) = \sum_{j=0}^{\phi(pq)} c_j u^j, \tag{9.3.47}$$

then (5.2.23) reveals that $c_0 = c_1 = -1$, $c_j = 0$, $j = 2, \ldots, p-1$ and $c_p = 1$ (remember $q > p$). Since $C_{pq}(u)$ is symmetric, we have

$$c_{\phi(pq)} = c_0 = 1, \tag{9.3.48}$$

$$c_{\phi(pq)-1} = c_1 = -1, \tag{9.3.49}$$

and

$$c_{\phi(pq)-p} = c_p = 1. \tag{9.3.50}$$

Based on (9.3.47) - (9.3.50), we get

$$u^{\phi(pq)} = C_{pq}(u) + (u^{\phi(pq)} - C_{pq}(u)) \tag{9.3.51}$$

and, therefore,

$$u^{\phi(pq)} \bmod C_{pq}(u) = (u^{\phi(pq)} - C_{pq}(u)). \tag{9.3.52}$$

An important observation is that $u^{\phi(pq)} - C_{pq}(u)$ is a polynomial of degree equal to $\phi(pq) - 1$ and has the highest degree coefficient equal to 1 (from (9.3.49)) while the next non-zero coefficient corresponds to $u^{\phi(pq)-p}$ (from (9.3.50)). This implies that for $j \leq p - 1$, $(u^{\phi(pq)} - C_{pq}(u))u^j$ is a polynomial of degree equal to $\phi(pq) + j - 1$ with the coefficient of $u^{\phi(pq)+j-1}$ equal to 1 while the next non-zero coefficient corresponds to $u^{\phi(pq)-p+j}$. Note that $\phi(pq) - p + j < \phi(pq)$ for $j \leq p - 1$. This analysis combined with (9.3.51) leads to

$$u^{\phi(pq)+1} = uC_{pq}(u) + (u^{\phi(pq)+1} - uC_{pq}(u))$$

$$= (u + 1)C_{pq}(u) + (u^{\phi(pq)+1} - (u + 1)C_{pq}(u)).$$

This is not a trivial identity as the second term in the *rhs* has degree less than $\phi(pq)$ which implies

$$u^{\phi(pq)+1} \bmod C_{pq}(u) = u^{\phi(pq)+1} - (u + 1)C_{pq}(u).$$

Using the same argument in a recursive manner, it is seen that

$$u^{\phi(pq)+j} \bmod C_{pq}(u) = u^{\phi(pq)+j} - (u^j + u^{j-1} + \ldots + u + 1)C_{pq}(u), \quad j = 0, 1, \ldots, p - 1. \tag{9.3.53}$$

Clearly, (9.3.53) evaluates $u^i \bmod C_d(u)$ for $i = \phi(pq), \ldots, \phi(pq) + p - 1 = pq - q$. It remains to be shown that the coefficients of the *rhs* polynomial in (9.3.53) take values in U. Substituting for $C_{pq}(u)$ in (9.3.53) from (5.2.23), we get

$$u^{\phi(pq)+j} \bmod C_{pq}(u) = u^{\phi(pq)+j} - \left\{\sum_{k=0}^{j} u^k\right\} \sum_{l=0}^{p-1}\sum_{m=0} u^{lq+mp}(1-u)$$

$$= u^{\phi(pq)+j} - \left[\sum_{l=0}^{p-1}\sum_{m=0} u^{lq+mp} - \sum_{l=0}^{p-1}\sum_{m=0} u^{lq+mp+j+1}\right],$$

$$j = 0, 1, ..., p-1. \tag{9.3.54}$$

Once again, since $lq + mp$ takes on distinct values for $l < p$, the first term in the square parenthesis of (9.3.54) is a polynomial with coefficient values in the set $\{0, 1\}$. It is clear that the coefficients of $u^i \bmod C_{pq}(u)$ take values in U, $i = (p-1)(q-1), ..., pq - q - 1$.

The analysis for $u^i \bmod C_{pq}(u)$, $i = pq - (p-1), ..., pq - 1$, is similar to the analysis for $u^i \bmod C_{pq}(u)$, $i = \phi(pq), ..., pq - q - 1$, and proceeds as follows. Since $u^d \equiv 1 \bmod C_d(u)$, we have

$$u^{pq-j} \equiv u^{-j} \bmod C_{pq}(u). \tag{9.3.55}$$

Therefore, analysis of $u^i \bmod C_{pq}(u)$, $i = pq - (p-1), ..., pq - 1$, is identical to the analysis for $u^{-j} \bmod C_{pq}(u)$, $j = 1, 2, ..., p-1$. Based on (9.3.47) to (9.3.50), it can be shown that

$$u^{-1} = u^{-1}C_{pq}(u) + (1 - C_{pq}(u))u^{-1}. \tag{9.3.56}$$

Using the same arguments as those to obtain (9.3.53), we get

$$u^{-j} \bmod C_{pq}(u) = u^{-j} - \left(\sum_{k=0}^{j} u^{-k}\right) C_{pq}(u) \tag{9.3.57}$$

$$= u^{-j} - u^{-j}\left[\sum_{l=0}^{p-1}\sum_{m=0} u^{lq+mp} - \sum_{l=0}^{p-1}\sum_{m=0} u^{lq+mp+j}\right],$$

$$j = 1, 2, ..., p-1. \tag{9.3.58}$$

It is clear from (9.3.58) that the coefficients of $u^i \bmod C_{pq}(u)$ take values in U, $i = pq - (p-1), ..., pq - 1$. A comparison of (9.3.54) and (9.3.58) reveals that $u^i \bmod C_{pq}(u)$, $i = \phi(pq), ..., pq - q - 1$, can be obtained directly from $u^i \bmod C_{pq}(u)$, $i = pq - (p-1), ..., pq - 1$.

The case of $u^i \bmod C_{2d}(u)$, $i = 0, 1, ..., 2d - 1$, $(d, 2) = 1$, $d \geq 3$, needs to be analyzed separately. In this case, let us assume that $r_{j,d}(u) = u^j \bmod C_d(u)$, $j = 0, 1, ..., d - 1$ are already known. For example, if $d = p$ or $d = pq$, then $u^j \bmod C_d(u)$, $j = 0, 1, ..., d - 1$ are known from the above analysis. In this case, $C_{2d}(u) = C_d(-u)$, which leads to

$$u^i \bmod C_{2d}(u) = u^i \bmod C_d(-u) = (-v)^i \bmod C_d(v)\Big|_{v=-u},$$

$$i = 0, 1, ..., 2d - 1. \tag{9.3.59}$$

Since $v^j \bmod C_d(v)$ is known only for $j = 0, 1, \ldots, d-1$, we split the interval $i = 0, 1, \ldots, 2d-1$, into two parts as $i_1 = 0, 1, \ldots, d-1$, and $i_2 = d, \ldots, 2d-1$. For the first interval,

$$(-v)^{i_1} \bmod C_d(v)\Big|_{v=-u} = \begin{cases} r_{i_1,d}(-u), & i_1 : even \\ -r_{i_1,d}(-u), & i_1 : odd \end{cases}. \tag{9.3.60}$$

For the second interval, let $i_2 = d + k$, $k = 0, 1, \ldots, d-1$. Recalling that $(-v)^d \equiv -1 \bmod C_d(v)$, we get

$$(-v)^{i_2} \bmod C_d(v)\Big|_{v=-u} = (-v)^{d+k} \bmod C_d(v) = -(-v)^k \bmod C_d(v).$$

Therefore, for the second interval,

$$(-v)^{i_2} \bmod C_d(v)\Big|_{v=-u} = \begin{cases} -r_{k,d}(-u), & k{:}even \\ r_{k,d}(-u), & k{:}odd \end{cases}, \tag{9.3.61}$$

where $k = i_2 - d$. Since d is odd, k being even implies that i_2 is odd and vice versa. As a result, (9.3.60) and (9.3.61) can be combined together as

$$u^i \bmod C_{2d}(u) = -u^{i+d} \bmod C_{2d}(u) = (-1)^i r_{i,d}(-u) \tag{9.3.62}$$

for $i = 0, 1, \ldots, d-1$. This discussion is summarized in the form of a lemma in the following.

Lemma 9.2 The coefficients of the polynomials $u^i \bmod C_{2d}(u)$, $i = 0, 1, \ldots, 2d-1$ and $u^j \bmod C_d(u)$, $j = 0, 1, \ldots, d-1$ take values in the same set, $(2, d) = 1$, $d \geq 3$.

Finally, we establish the following theorem for the form of the reduction polynomials.

Theorem 9.3 The coefficients of the polynomials $u^i \bmod C_{d_e}(u)$, $i = 0, 1, \ldots, d_e - 1$ take values in the same set as the coefficients of the polynomials $u^j \bmod C_d(u)$, $j = 0, 1, \ldots, d-1$, where,

$$d_e = 2^m \prod_{k=1}^{l} p_k^{\theta_k},$$

(standard factorization of d_e) and

$$d = \prod_{k=1}^{l} p_k \,.$$

Proof: From the properties of cyclotomic polynomials,

$$C_{d_e}(u) = C_{2d}\left(u^a\right),$$

where

$$a = 2^{m-1} \prod_{k=1}^{l} p_k^{\theta_k - 1} \,.$$

We have to examine $u^i \bmod C_{d_e}(u), i = 0,1,\ldots,d_e - 1$. Dividing i by a, we may write $i = i_1 a + i_2$, $i_2 < a$, and $i_1 < 2d$. Therefore,

$$u^i \bmod C_{d_e}(u) = u^{i_1 a + i_2} \bmod C_{2d}\left(u^a\right)$$
$$= \left\{ v^{i_1} \bmod C_{2d}(v) \Big|_{v=u^a} \cdot u^{i_2} \right\} \bmod C_{2d}\left(u^a\right).$$

The polynomial $v^{i_1} \bmod C_{2d}(v)$, $v = u^a$ has degree at most $a(\phi(d) - 1)$. As a result, polynomial in the curly parentheses has degree at most $a \cdot \phi(d) - 1$. Since $deg(C_{2d}(u^a)) = a \cdot \phi(d)$, the above expression simplifies to

$$u^i \bmod C_{d_e}(u) = v^{i_1} \bmod C_{2d}(v) \Big|_{v=u^a} \cdot u^{i_2} \,. \tag{9.3.63}$$

Combining this with Lemma 9.2, the statement of the theorem follows in a straightforward manner. In addition, for $d = 2^m$, we have

$$u^i \bmod C_d(u) = u^i, \; i = 0, 1, \ldots, 2^{m-1} - 1 \tag{9.3.64}$$

and

$$u^i \bmod C_d(u) = -u^{i-2^{m-1}}, \quad i = 2^{m-1},\ldots,2^m - 1. \tag{9.3.65}$$

The result of this theorem is valid for odd d_e also, in which case $m = 0$.

Combining the results of Theorem 9.3, Lemma 9.2, and the analysis preceding it, we get the following theorem.

Theorem 9.4 If d has at most two odd prime factors, the coefficients of the polynomials $u^i \bmod C_d(u)$, $i = 0, 1, \ldots, d-1$, take values in U. The elements of the corresponding block $\mathbf{R}_d$ of CRT-P reduction matrix $\mathbf{R}$ also take values in U.

We present the following two examples to illustrate various results developed in this section.

Example 9.2 Let $n = 60 = 2^2 \cdot 3 \cdot 5$. n has two odd prime factors, namely, 3 and 5. The factors of n are 1, 2, 3, 4, 5, 6, 10, 12, 15, 20, 30, and 60, and the factorization of $u^{60} - 1$ over Z is given by

$$u^{60} - 1 = \prod_{d|n} C_d(u).$$

The various cyclotomic polynomials are

$$\begin{aligned}
C_1(u) &= u - 1,\\
C_2(u) &= u + 1,\\
C_3(u) &= u^2 + u + 1,\\
C_4(u) &= u^2 + 1,\\
C_5(u) &= u^4 + u^3 + u^2 + u + 1,\\
C_6(u) &= C_2(u^3)/C_2(u) = u^2 - u + 1,\\
C_{10}(u) &= C_5(-u) = u^4 - u^3 + u^2 - u + 1,\\
C_{12}(u) &= C_3(-u^2) = u^4 - u^2 + 1,\\
C_{15}(u) &= C_3(u^5)/C_3(u) = u^8 - u^7 + u^5 - u^4 + u^3 - u + 1,\\
C_{20}(u) &= C_5(-u^2) = u^8 - u^6 + u^4 - u^2 + 1,\\
C_{30}(u) &= C_{15}(-u) = u^8 + u^7 - u^5 - u^4 - u^3 + u + 1,\\
C_{60}(u) &= C_{15}(-u^2) = u^{16} + u^{14} - u^{10} - u^8 - u^6 + u^2 + 1.
\end{aligned}$$

The CRT–P RPs $t_d(u)$ are obtained as

$$\begin{aligned}
t_1(u) &= (1/60),\\
t_2(u) &= -(1/60),\\
t_3(u) &= (1/60)(-u - 2),\\
t_4(u) &= (1/60)2,\\
t_5(u) &= (1/60)(-u^3 - 2\,u^2 - 3\,u - 4),\\
t_6(u) &= (1/60)(u - 2),\\
t_{10}(u) &= (1/60)(u^3 - 2\,u^2 + 3\,u - 4),
\end{aligned}$$

$t_{12}(u) = (1/60)(2\,u^2 - 4),$
$t_{15}(u) = (1/60)(u^7 - 3\,u^5 + 4\,u^4 - 5\,u^3 + 7\,u - 8),$
$t_{20}(u) = (1/60)(2\,u^6 - 4\,u^4 + 6\,u^2 - 8),$
$t_{30}(u) = (1/60)(-u^7 + 3\,u^5 + 4\,u^4 + 5\,u^3 - 7\,u - 8),$
$t_{60}(u) = (1/60)(-2\,u^{14} + 6\,u^{10} + 8\,u^8 + 10\,u^6 - 14\,u^2 - 16).$

These polynomials can either be computed using (9.3.5) directly or by using the appropriate expressions in (9.3.6) to (9.3.11). For example, $t_{12}(u) = 2t_6(u^2) = 2t_3(-u^2) = (1/60)(2u^2 - 4)$ from (9.3.10) and (9.3.11). The corresponding blocks $\mathbf{T}_d$ of the reconstruction matrix $\mathbf{T}$ are obtained by using (9.3.13). Note that $(1/60)$ is a common factor present throughout and the maximum absolute value of any element in $\mathbf{T}_d$ is $\phi(d)$.

The RP $P_1(u)$ is given in (9.3.15). The RP's $P_3(u)$ and $P_5(u)$ are computed using (9.3.17); $P_6(u)$, $P_{10}(u)$, $P_{12}(u)$, $P_{20}(u)$, $P_{30}(u)$, and $P_{60}(u)$ using (9.3.21); $P_{15}(u)$ using (9.3.20); and $P_2(u)$ and $P_4(u)$ using (9.3.22). A representative polynomial in each case is given below.

$$P_5(u) = \sum_{i=0}^{11}\left[u^{5i+1} - u^{5i}\right],$$

$$P_{15}(u) = \sum_{i=0}^{3}\sum_{j=0}^{2}\left[u^{15i+j+5} - u^{15i+j}\right],$$

$$P_{60}(u) = \left(1 + w^5\right)P_{15}(w)\Big|_{\substack{d=15\\ n=15\\ w=-u^2}}.$$

Evaluating $P_{15}(u)$ for $d = 15$, $n = 15$ using (9.3.20), and substituting above, we get

$$P_{60}(u) = \left(1 + w^5\right)\sum_{j=0}^{j+5}\left(w^{j+5} - w^j\right)\Bigg|_{w=-u^2},$$

$$P_4(u) = \left(u^{30} - 1\right)\left[\sum_{j=0}^{14}\left(-u^2\right)^j\right].$$

The block $\mathbf{R}_\mathrm{d}$ of the reduction matrix $\mathbf{R}$ is $(\phi(d) \times n)$ dimensional and has the structure

$$\mathbf{R}_d = \left[\mathbf{Q}_d \;\vdots\; \mathbf{Q}_d \;\vdots\; \cdots \;\vdots\; \mathbf{Q}_d\right],$$

where $\mathbf{Q}_d$ is a $\phi(d) \times d$ matrix having identity matrix of dimension $\phi(d)$ in front. The remaining columns of $\mathbf{Q}_d$ are obtained by evaluating $u^i \bmod C_d(u)$, $i = \phi(d), ..., d-1$. For $d = 1$, $\mathbf{R}_d$ is a row vector having all elements equal to unity. For $d = 2, 3$, and 5, $\mathbf{R}_d$ is given in (9.3.43). For $d = 15$, $u^i \bmod C_d(u)$, $i = 8, ..., 15$ is evaluated using (9.3.54) for $i = 8, 9, 10$, (9.3.46) for $i = 10, ..., 13$, and (9.3.58) for $i = 14, 15$. The $\mathbf{Q}_{15}$ matrix is obtained as

$$\mathbf{Q}_{15} = \left[\mathbf{I}_8 \begin{matrix} -1 & -1 & -1 & 0 & 0 & 1 & 1 \\ 1 & 0 & 0 & -1 & 0 & -1 & 0 \\ 0 & 1 & 0 & 0 & -1 & 0 & -1 \\ -1 & -1 & 0 & 0 & 0 & 0 & 1 \\ 1 & 0 & 0 & 0 & 0 & -1 & -1 \\ -1 & 0 & -1 & 0 & 0 & 1 & 0 \\ 0 & -1 & 0 & -1 & 0 & 0 & 1 \\ 1 & -1 & 0 & 0 & -1 & -1 & -1 \end{matrix} \right].$$

The block $\mathbf{Q}_4$ for $d = 4$ is obtained using (9.3.64) and (9.3.65) as

$$\mathbf{Q}_4 = \left[\mathbf{I}_2 \begin{matrix} -1 & 0 \\ 0 & -1 \end{matrix} \right].$$

Blocks $\mathbf{Q}_d$ for $d = 6, 10$, and 30 are obtained by first obtaining the blocks for $d = 3, 5$, and 15 and then applying (9.3.62). Given $\mathbf{Q}_5$ in (9.3.43), $\mathbf{Q}_{10}$ can be shown to be

$$\mathbf{Q}_{10} = \left[\mathbf{I}_4 \begin{matrix} -1 \\ 1 \\ -1 \\ 1 \end{matrix} \; -\mathbf{I}_4 \begin{matrix} 1 \\ -1 \\ 1 \\ -1 \end{matrix} \right].$$

The block $\mathbf{Q}_d$ for $d = 12, 20$, and 60 are obtained by first using the blocks for $d = 6, 10$, and 30 and then applying (9.3.63). Given

$$\mathbf{Q}_6 = \left[\mathbf{I}_2 \begin{matrix} -1 \\ 1 \end{matrix} \; -\mathbf{I}_2 \begin{matrix} 1 \\ -1 \end{matrix} \right],$$

$\mathbf{Q}_{12}$ is obtained as

$$\mathbf{Q}_{12} = \begin{bmatrix} \mathbf{I}_4 & \begin{matrix} -1 & 0 \\ 0 & -1 \\ 1 & 0 \\ 0 & 1 \end{matrix} & -\mathbf{I}_4 & \begin{matrix} 1 & 0 \\ 0 & 1 \\ -1 & 0 \\ 0 & -1 \end{matrix} \end{bmatrix}.$$

Other blocks may be obtained in a similar manner.

Example 9.3 Let $n = 105 = 3 \cdot 5 \cdot 7$. The factors of n are 1, 3, 5, 15, 21, 35, and 105. The CRT-P reduction matrix **R** and the reconstruction matrices **T** and **P** have 7 blocks. Since integers 1, 3, 5, 7, 15, 21, and 35 have at most two odd prime factors, coefficients of cyclotomic polynomials $C_1(u)$, $C_3(u)$, $C_5(u)$, $C_7(u)$, $C_{15}(u)$, $C_{21}(u)$, and $C_{35}(u)$ take values in U. The RPs $t_d(u)$ may be determined using (9.3.9). Only $C_{105}(u)$ has two coefficients that take the value −2. In the case of $C_{105}(u)$, the absolute maximum value of the coefficients of $t_{105}(u)$ is 82.

The blocks of the reduction matrix **R** and the reconstruction matrix **P** corresponding to d = 1, 3, 5, 7, 15, 21, and 35 can be determined using the closed form expressions derived here. The elements in all these blocks take values in U. Only one of the blocks in **R** and **P** matrices corresponding to d = 105 may have elements taking values other than 0, 1, and −1.

It is observed that all the results derived in this section pertain to values of n that have at most two odd primes in their standard factorization. In **Appendix B**, results are presented on the classification of the cyclotomic polynomial $C_n(u)$, where n is the product of three distinct odd primes p, q, and r, that is, $n = p \cdot q \cdot r$. We present classification of the coefficients for $C_n(u)$, $P_n(u)$ and the remainders $u^i \bmod C_n(u)$, $i = 1, 2, \ldots, \phi(n) - 1$. The cases considered cover most values of length n that are of interest from the point of view of cyclic convolution algorithms.

9.4 Other Considerations

The multiplicative complexity of CRT-P based cyclic convolution algorithm is given by

$$M(N) = \sum_{d|n} M_a(\phi(d)).$$

Therefore, it is important to design efficient acyclic convolution algorithms of length $\phi(d)$. If $\phi(d)$ is small, say $\phi(d) \le 10$, such algorithms may be designed using ad-hoc approaches. For values of $\phi(d)$ in a moderate-to-large range, for example, for $n = d = 105$, $\phi(d) = 48$, one may design acyclic convolution algorithm using the CRT-P, again with the important difference that the modulo polynomial is not restricted to be $u^n - 1$. This provides considerable flexibility that may be exploited to reduce the computational burden. A second approach to designing acyclic convolution algorithms is to use multidimensional techniques as described in Section 8.3.

We feel that the closed form expression for the RPs $t_i(u)$ in (9.1.11) is a very valuable one in many respects. It leads to the following lemma.

Lemma 9.3 The cyclotomic polynomials based CRT-P algorithm for computing cyclic convolution of length n can be used for data sequences defined over $\mathbf{I}$, $\mathbf{Z}$, $\mathbf{Re}$, and $\mathbf{C}$ and those finite fields and rings over which n^{-1} exists. For the case of $GF(q)$, $Z(M)$ and $CZ(M)$, it is required that $(q, n) = (M, n) = 1$ for n^{-1} to exist.

Interestingly, as long as $(M, n) = 1$, cyclic convolution algorithm over $Z(M)$ is no more complex computationally than the cyclic convolution algorithm over $\mathbf{Z}$. In fact, the algebraic properties of $Z(M)$ may lead to cyclic convolution algorithms that require an even fewer number of arithmetic operations than those algorithms designed over $\mathbf{Z}$. The reason for this can be attributed to the fact that cyclotomic polynomials defined over $\mathbf{Z}$ are irreducible over $\mathbf{Z}$ but may factor further over $Z(M)$. These aspects will be explored in Section 9.7.

It is possible to write closed form expressions for the RPs $t_i(u)$ in a manner similar to (9.1.11) in some other cases as well. If the convolution is performed modulo $u^n - A$ instead of $u^n - 1$, it is seen that

$$t_i(u) = \left[\frac{u}{An} \frac{d}{du} p_i(u) \right] \bmod p_i(u). \tag{9.4.1}$$

For $A = -1$, $p_i(u)$ are also cyclotomic polynomials and can be obtained from the cyclotomic factors of $u^{2n} - 1$.

Consider the case when the computation in (9.0.1) is defined modulo a polynomial $P(u)$ which can be expressed as

$$P(u) = \prod_{j=1}^{l} C_{i_j}(u), \tag{9.4.2}$$

where $C_{i_j}(u)$ is the i_j-th cyclotomic polynomial. For example, this may be one way to compute acyclic convolution. Since $C_{i_j}(u)$ divides $\left(u^{i_j}-1\right)$ but not $u^k - 1$, $k < i_j$, we have

$$P(u) \mid (u^n - 1), \tag{9.4.3}$$

where

$$n = lcm(n_1, n_2, ..., n_l). \tag{9.4.4}$$

Therefore, there exists a polynomial with integer coefficients, say $\theta(u)$, such that

$$u^n - 1 = \theta(u)\prod_{j=1}^{l} C_{i_j}(u). \tag{9.4.5}$$

Comparing (9.4.5) with (9.3.1), we see that $\theta(u)$ is the product of those cyclotomic polynomials that are factors of $u^n - 1$ and are not included in $P(u)$. In this case, RPs are given by

$$t_i(u) = \left[\frac{u\theta(u)}{n}\frac{d}{du}C_{i_j}(u)\right] \bmod C_{i_j}(u). \tag{9.4.6}$$

Once again, we note that the only division that may be required is by n. For example, if $P(u) = C_1(u) \cdot C_2(u) \cdot C_5(u)$, then $n = 10$ and $\theta(u) = C_{10}(u)$. The RPs are computed using (9.4.6). The only division required in this case is by 10. With $P(u)$ as defined in (9.4.2), the elements of the reconstruction matrix will be integers. Their values will depend on cyclotomic polynomials included in (9.4.2). The analysis performed in the previous section can be used here to determine the elements of the l blocks in the reduction matrix **R** as long as i_j, $j = 1, 2, ..., l$ has at most two odd prime factors. These elements take values in U.

In some situations, it is easier to process one of the input sequences. For example, in digital filter implementation, $\underline{X}$ could be the FIR sequence and, therefore, known *a priori*. In such situations, **A** $\underline{X}$ could be computed in advance and stored. Therefore, it is desirable that **C** and **B** be matrices having simple elements preferably in U. For the sake of discussion, recall that the matrices **A′**, **B′**, and **C′** are obtained from small length acyclic convolution algorithms and are assumed to have simple elements. The way in which the cyclic convolution algorithm is analyzed here, the blocks of the reduction matrix **R** have elements in U as long as the cyclotomic polynomials have at most two odd prime factors.

Even though elements of the block $\mathbf{P}_d$ of the reconstruction matrix are in U, if d has at most two odd prime factors, the structure of the overall reconstruction matrix is not simple. This is due to the matrix $\mathbf{T}$ for the cyclic convolution algorithm. It was shown in Theorem 4.8 that if $\underline{Z} = \mathbf{C}\ [\mathbf{A}\ \underline{X} \otimes \mathbf{B}\ \underline{Y}]$ is a bilinear form for computing the cyclic convolution, then one also has

$$\underline{Z} = (\mathbf{J}\ \mathbf{B}^t)\ [\mathbf{A}\ \underline{X} \otimes (\mathbf{C}^t\ \mathbf{J})\ \underline{Y}]. \tag{9.4.7}$$

This immediately leads to the following lemma.

Lemma 9.4 If one of the inputs to the cyclic convolution algorithm is known *a priori* (equivalently, $(\mathbf{C}^t\ \mathbf{J})\ \underline{Y}$ is determined *a priori*), then the remaining matrices, namely, $(\mathbf{J}\ \mathbf{B}^t)$ and $\mathbf{A}$ consist of factor matrices having simple elements.

9.5 Complex Cyclotomic Polynomials and Related Algorithms Over CZ

The $\mathbf{R}$, $\mathbf{T}$, and $\mathbf{P}$ matrices involved in CRT-P reduction and reconstruction for cyclic convolution are highly structured. The elements of those blocks of $\mathbf{R}$ and $\mathbf{P}$ matrices that correspond to $C_d(u)$ with d having at most two odd prime factors take values in the set U, $\mathsf{U} = \{0, 1, -1\}$. This is a highly desirable property to be pursued further and generalized over CZ in this section.

The factorization over CZ is important and useful in many respects. It is expected to result in cyclic convolution algorithms having lower multiplicative complexity for processing data sequences defined over I, Z, CI, CZ, Re, and C. A necessary and sufficient condition for $C_d(u)$ to factorize further over CZ is that $4 \mid d$. Therefore, only one-quarter of all cyclotomic polynomials satisfy this condition. However, if $4 \mid n$, that is, n has the form $n = 2^c \cdot N$, $c \geq 2$, $(2, N) = 1$, then it is seen that the fraction of cyclotomic polynomials that constitute $u^n - 1$ and further factorize over CZ is given by $(c-1)/(c+1)$. This includes $C_n(u)$ that has the highest degree. Thus, of all the cyclotomic polynomials that constitute $u^n - 1$, either all are irreducible over CZ or at least one-third ($c \geq 2$) are reducible into product of two complex conjugate factors of equal degree. In digital signal processing applications, the condition that $4 \mid n$ may not be a restrictive condition at all. Those cyclotomic polynomials that factorize further over CZ are termed as the *complex cyclotomic polynomials*.

Throughout this section, it is assumed that the condition $4 \mid n$ is satisfied for complex cyclotomic polynomials to exist. Given the cyclotomic factorization of $u^n - 1$ over CZ, we now turn to the CRT-P reduction and reconstruction matrices **R**, **T**, and **P**. Each of these matrices has a block structure with D blocks, D being the number of complex cyclotomic factors of $u^n - 1$. Here, $D = 2c/(c + 1)$ times the number of factors of n, $n = 2^c \cdot N$, $(2, N) = 1$.

The elements in the d-th block of **R**, **T**, and **P** matrices depend only on $C_d(u)$, 4 not a factor of d. They have already been studied in Section 9.3. Therefore, in the following, we focus only on the complex cyclotomic polynomials $C1_d(u)$ and $C2_d(u)$ and the corresponding blocks of **R**, **T**, and **P** matrices, $4 \mid d$. These blocks will be labeled as d,1-th and d,2-th blocks, respectively.

Let $\phi'(d) = deg(C1_d(u)) = deg(C2_d(u)) = \phi(d)/2$. The RPs $t1_d(u)$ and $t2_d(u)$ are obtained as

$$t1_d(u) = \left[\frac{u}{n}\frac{d}{du}C1_d(u)\right] \mod C1_d(u)\,.$$

Substituting for $C1_d(u)$ from (5.3.15) and simplifying, we get

$$t1_d(u) = 2^{m-2}(-j)^{\phi(M)}\, t_M\!\left(ju^{2^{m-2}}\right), \tag{9.5.1}$$

$t_M(u)$ being the RP for $C_M(u)$. Replacing j by $-j$ in (9.5.1), we get an expression for $t2_d(u)$. The following lemma is obtained from (9.5.1) and the analysis for $t_M(u)$ in Section 9.3.

Lemma 9.5 The coefficients of the RPs $t1_d(u)$ and $t2_d(u)$ are alternately either integers or purely imaginary integers. If d has at most two odd prime factors, then the largest absolute value of any non-zero coefficient of the polynomials $t1_d(u)$ and $t2_d(u)$ is $\phi'(d)$, (ignoring n in the denominator).

The d,1-th and d,2-th blocks of the **T** matrix have dimensions $(3\phi'(d) - 2) \times (2\phi'(d) - 1)$. Their structure is analogous to the structure of $\mathbf{T}_d$ corresponding to $t_d(u)$ in (9.3.13).

Now, let us analyze the RPs associated with $C1_d(u)$ and $C2_d(u)$. By definition,

$$P1_d(u) = \left(u^n - 1\right)\!/C1_d(u)\,.$$

For $d = 2^m M$, $(2, M) = 1$, $m \geq 2$, n must have the form $n = 2^c \cdot N$, $M \mid N$, $(2, N) = 1$ and $c \geq m$. Thus $P1_d(u)$ can be expressed as:

$$P1_d(u) = \left(u^{2^c N} - 1\right) \Big/ \left[(-j)^{\phi(M)} C_M\left(ju^{2^{m-2}}\right)\right]. \tag{9.5.2}$$

Setting $ju^{2^{m-2}} = t$ in (9.5.2), we get

$$P1_d(u) = j^{\phi(M)} \left(t^{N.2^{c-m+2}} - 1\right) \Big/ C_M(t) = j^{\phi(M)} P_d(t) \Big|_{\substack{d=M \\ n=N.2^{c-m+2} \\ (2,N)=(2,M)=1 \\ M|n}}$$

Following the same analysis as in Theorem 9.2, we have

$$P1_d(u) \Big|_{\substack{d=2^m.M \\ n=N.2^c \\ M|n \\ (M,2)=(N,2)=1 \\ c\geq m\geq 2}} = j^{\phi(M)} \left\{ \prod_{i=1}^{c-m+2-i} \left(v^{2^{c-m+2-i}} + 1\right) \right\} P_d(t) \Big|_{\substack{d=M \\ n=N \\ v=t^N \\ t=ju^a \\ a=2^{m-2}}} . \tag{9.5.3}$$

The expression for $P2_d(u)$ is obtained by replacing j by $-j$ in the *rhs* of (9.5.3). It is clear from (9.3.17), (9.3.20), (9.3.21), (9.3.22), and (9.5.3) above that the coefficients of the RPs, $P1_d(u)$ and $P2_d(u)$, take values in the set U' if d has at most two odd prime factors. In these cases, the number of non-zero coefficients can be determined in closed form. The structure of the d,1-th and d,2-th blocks of the **P** matrix is analogous to the structure of the $\mathbf{P}_d$ corresponding to $P_d(u)$ in (9.3.34). The dimension of these blocks is $n \times (3\phi'(d) - 2)$. Here, U' is defined as the set

$$\mathsf{U}' = \{0, 1, -1, j, -j\}.$$

Summarizing this analysis, we get the following lemma.

Lemma 9.6 The coefficients of the RPs, $P1_d(u)$ and $P2_d(u)$, are alternately either integers or purely imaginary integers. If d has at most two odd prime factors, then the coefficients take values in U'.

In the following, we analyze the structure of the reduction matrix **R**. The computation of $X1_d(u)$ is performed as

$$X1_d(u) = X(u) \bmod C1_d(u),$$

where $deg(X(u)) = n - 1$ and $deg(C1_d(u)) = \phi'(d) = \phi(d)/2$. Following (9.3.37), (9.3.38), we conclude that the i-th column of the block $\mathbf{R}1_d$ of the reduction matrix is obtained by listing the elements of the polynomial

$$r1_{i,d}(u) \equiv u^i \bmod C1_d(u), \quad i = 0, 1, ..., n - 1. \tag{9.5.4}$$

Once again, $C1_d(u)$ consists of all the d-th roots of unity. Therefore, $u^d \equiv 1 \bmod C1_d(u)$, while $u^i \not\equiv 1 \bmod C1_d(u)$ for $i < d$. Thus, $\mathbf{R}1_d$ is periodic with $(d + i)$-th column the same as the i-th column. Also,

$$r1_{i,d}(u) = u^i, \quad i < \phi'(d). \tag{9.5.5}$$

For $\phi'(d) \leq i < d$, we get

$$r1_{i,d}(u) \equiv u^i \bmod\left[(-j)^{\phi(M)} C_M\left(ju^{2^{m-2}}\right)\right] \tag{9.5.6}$$

$$= u^i \bmod C_M\left(ju^{2^{m-2}}\right). \tag{9.5.7}$$

Writing $i = i_1 2^{m-2} + i_2$, $i_2 < 2^{m-2}$, $i_1 < 4M$, we get

$$r1_{i,d}(u) = (-j)^{i_1}\left[t^{i_1} \bmod C_M(t)\right]_{\substack{t = ju^a \\ a = 2^{m-2}}} \cdot u^{i_2} \tag{9.5.8}$$

$$= (-j)^{i_1} r_{i_1,M}(t).u^{i_2}\Big|_{\substack{t = ju^a \\ a = 2^{m-2}}}. \tag{9.5.9}$$

Replacing j by $-j$, we get an expression for $r2_{i,d}(u)$. The polynomial $r_{i_1,M}(t)$ is periodic as $r_{i_1+M,M}(t) = r_{i_1,M}(t)$. This analysis leads to the following lemma.

Lemma 9.7 The coefficients of the polynomials $u^i \bmod C1_d(u)$, $i = 0, 1, ..., d - 1$ are obtained from $u^i \bmod C_M(u)$, $i = 0, 1, ..., M - 1$, $d = 2^m M$, $(2, M) = 1$. They occur alternately as either integers or purely imaginary integers. They take values in U′, if d has at most two odd prime factors.

We illustrate the results of this section with the following example.

Example 9.4 Let $n = 60 = 2^2 \cdot 3 \cdot 5$. In this case, $c = 2$, $N = 15$. The cyclotomic polynomials that factorize over **CZ** are $C_4(u)$, $C_{12}(u)$, $C_{20}(u)$, and $C_{60}(u)$. The corresponding complex cyclotomic polynomials are

$$\begin{aligned}
C_4(u) &= C1_4(u) \cdot C2_4(u) = (u - j) \cdot (u + j),\\
C_{12}(u) &= C1_{12}(u) \cdot C2_{12}(u) = (u^2 - ju - 1) \cdot (u^2 + ju - 1),\\
C_{20}(u) &= C1_{20}(u) \cdot C2_{20}(u) = (u^4 - ju^3 - u^2 + ju + 1) \cdot (u^4 + ju^3 - u^2 - ju + 1),\\
C_{60}(u) &= C1_{60}(u) \cdot C2_{60}(u)\\
&= (u^8 + ju^7 + ju^5 - u^4 - ju^3 - ju + 1) \cdot (u^8 - ju^7 - ju^5 - u^4 + ju^3 + ju + 1).
\end{aligned}$$

The RPs for the complex cyclotomic polynomials are given by

$$\begin{aligned}
t1_4(u) &= (1/60)(-j),\\
t1_{12}(u) &= j^2 t_3(ju) = (1/60)\, ju + 2,\\
t1_{20}(u) &= t_5(ju) = (1/60)[ju^3 + 2u^2 - 3ju - 4],\\
t1_{60}(u) &= t_{15}(ju) = (1/60)[-ju^7 - 3ju^5 + 4u^4 + 5ju^3 + 7ju - 8].
\end{aligned}$$

These polynomials are obtained by using (9.5.1) and Example 9.2. The RPs, $P1_d(u)$, $d = 4, 12, 20$, and 60, are obtained by setting $M = 1, 2, 5$, and 15, respectively in (9.5.3). In each case $c = m = 2$. Thus,

$$P1_d(u)\Big|_{\substack{d=4.M\\ n=4\cdot 15}} = j^{\phi(M)}\left\{\prod_{i=1}^{2}\left[v^{2^{2-i}}+1\right]\right\} P_d(t)\Bigg|_{\substack{d=M\\ n=15\\ v=t^{15}\\ t=ju}}$$

which leads to:

$$\begin{aligned}
P1_4(u) &= j\left(-u^{30}+1\right)\left(-ju^{15}+1\right).\sum_{i=0}^{14}(ju)^i,\\
P1_{12}(u) &= -\left(-u^{30}+1\right)\left(-ju^{15}+1\right).\sum_{i=0}^{4}\left[(ju)^{3i+1}-(ju)^{3i}\right],\\
P1_{20}(u) &= \left(-u^{30}+1\right)\left(-ju^{15}+1\right).\sum_{i=0}^{2}\left[(ju)^{5i+1}-(ju)^{5i}\right],\\
P1_{60}(u) &= \left(-u^{30}+1\right)\left(-ju^{15}+1\right).\sum_{i=0}^{2}\left[(ju)^{5+i}-(ju)^{i}\right].
\end{aligned}$$

The block $\mathbf{R}_{d,1}$ of the reduction matrix $\mathbf{R}$ is $\phi'(d) \times n$ dimensional and has a structure:

$$\mathbf{R}_{d,1} = \left[\mathbf{Q}_{d,1} \mid \mathbf{Q}_{d,1} \mid \cdots \mid \mathbf{Q}_{d,1}\right],$$

where $\mathbf{Q}_{d,1}$ is a $\phi'(d) \times d$ matrix having identity matrix of dimension $\phi'(d)$ in front. For $d = 4$, $\phi'(d) = 1$ and the $\mathbf{R}_{4,1}$ is a row vector with $\mathbf{Q}_{4,1} = [1\ \ -j\ \ -1\ \ j]$. In this case, $m = 2$, therefore, $i_2 = 0$ in (9.5.7) which leads to $i = i_1$ and

$$r1_{i,d}(u) = (-j)^i r_{i,M}(t)\Big|_{t=ju} . \qquad (9.5.10)$$

For $d = 20$, $M = 5$, we have $u^i \bmod C_5(u) = u^i$, $i = 0, \ldots, 3$, and $u^4 \bmod C_5(u) = -u^3 - u^2 - u - 1$. Combining this with (9.5.10), we have

$$\mathbf{Q}_{20,1} = \left[\begin{array}{c|c|c|c|c|c|c|c} & -1 & & j & & 1 & & -j \\ \mathbf{I}_4 & -j & -j\mathbf{I}_4 & -1 & -\mathbf{I}_4 & j & j\mathbf{I}_4 & 1 \\ & 1 & & -j & & -1 & & j \\ & -j & & -1 & & j & & -1 \end{array}\right].$$

Similar forms are obtained for $\mathbf{Q}_{12,1}$ and $\mathbf{Q}_{60,1}$ starting from $\mathbf{Q}_3$ and $\mathbf{Q}_{15}$. The expressions corresponding to $C2_d(u)$ are obtained by replacing j with $-j$ in the above expressions.

We emphasize that the coefficients of complex cyclotomic polynomials and elements of the CRT-P matrices **R**, **T**, and **P** always alternate between integer and purely imaginary integer values (ignoring n in the denominator for **T**). In addition, those coefficients and the elements of those blocks of the **R** and **P** matrices for which d has at most two odd prime factors, take values in U'.

9.5.1 Conjugate Symmetry Property of Complex Cyclotomic Polynomials

In the following, we analyze the conjugate symmetry property associated with complex cyclotomic polynomials and its applications for processing real-valued sequences. It is well known that the DFT, X_k, $k = 0, 1, \ldots, n - 1$, of a real-valued sequence, x_l, $l = 0, 1, \ldots, n - 1$, satisfies the conjugate symmetry property

$$X_k = X^*_{n-k}, \quad k = 1, 2, \ldots, n-1 . \qquad (9.5.11)$$

This property (to be established in Chapter 13) can be exploited in many ways to reduce the computational complexity of the DFT algorithms.

For complex cyclotomic polynomials, the CRT-P reduction consists in the computations $X1_d(u) \equiv X(u) \bmod C1_d(u)$ and $X2_d(u) \equiv X(u) \bmod C2_d(u)$. Since $C1_d(u) = C2^*_d(u)$, for real $X(u)$, we have

$$X1_d(u) = X2^*_d(u) \tag{9.5.12}$$

This is a generalization of (9.5.11). Thus, once $X1_d(u)$ is known, $X2_d(u)$ is obtained as well. This conjugate symmetry property can also be exploited in the same way. As an example, consider the following.

Simultaneous processing of two real-valued sequences. Given two real-valued sequences $A(u)$ and $B(u)$, a complex-valued sequence $D(u)$ is formed as

$$D(u) = A(u) + jB(u).$$

$D(u)$ is reduced mod $C1_d(u)$ and $C2_d(u)$ to get $D1_d(u) \equiv D(u) \bmod C1_d(u)$ and $D2_d(u) \equiv D(u) \bmod C2_d(u)$. It is easy to show that

$$A1_d(u) \equiv A(u) \bmod C1_d(u) = (1/2)\left[D1_d(u) + D2^*_d(u)\right]$$

and

$$B1_d(u) \equiv B(u) \bmod C2_d(u) = (1/2j)\left[D1_d(u) - D2^*_d(u)\right].$$

Similar expressions may be written for CRT-P reconstruction. The above expressions can also be employed to simultaneously compute two real cyclic convolutions, $A(u)B(u) \bmod (u^n - 1)$ and $D(u)\,E(u) \bmod (u^n - 1)$, with one complex cyclic convolution using the complex cyclotomic polynomials.

9.5.2 Computational Complexity Comparisons

It is seen that complex cyclotomic polynomials lead to a reduction in the number of MULTs for computing the cyclic convolution of complex-valued sequences as compared to the cyclotomic polynomials. For example, a cyclotomic polynomial based cyclic convolution algorithm for $n = 60$ requires

$$M(60) = 2M_a(1) + 3M_a(2) + 3M_a(4) + 3M_a(8) + M_a(16) = 200 \text{ MULTs}.$$

For complex-valued sequences, these MULTs are complex as well. A complex cyclotomic polynomial based cyclic convolution algorithm for $n = 60$ requires

$$M(60) = 4M_a(1) + 4M_a(2) + 4M_a(4) + 4M_a(8) = 133 \text{ MULTs}.$$

Here, we assume that $M_a(2^c) = 3^c$ for simplicity. For real-valued sequences, both approaches appear to require the same number of arithmetic operations.

9.6 The Agarwal-Cooley Algorithm

The basic idea of the Agarwal-Cooley algorithm is to convert one-dimensional cyclic convolution into a multidimensional cyclic convolution. In essence, a one-dimensional cyclic convolution of length n, where $n = n_1 \cdot n_2$, $(n_1, n_2) = 1$, can be expressed as a two-dimensional cyclic convolution of lengths n_1 and n_2, respectively. The extension of the idea to convert one-dimensional cyclic convolution to a d-dimensional cyclic convolution when n has d relatively co-prime factors, that is, $n = n_1 \cdot n_2 \cdots n_d$, $(n_i, n_j) = 1$, $i \neq j$, is straightforward. The Agarwal-Cooley algorithm consists in the application of CRT-I to the indices of sequences being convoluted. Therefore, it is valid for data sequences defined over any arbitrary number system. A major advantage of the Agarwal-Cooley algorithm is that large length cyclic convolution algorithms can be constructed from smaller length cyclic convolution algorithms.

9.6.1 Conversion from One Dimension to Higher Dimensions

Consider the cyclic convolution of length n expressed as

$$z_k = \sum_{i=0}^{n-1} x_{(k-i) \bmod n} y_i \,, \quad k = 0, 1, \ldots, n-1. \tag{9.6.1}$$

In matrix-vector form, we have

$$\begin{bmatrix} z_0 \\ z_1 \\ z_2 \\ \vdots \\ z_{n-2} \\ z_{n-1} \end{bmatrix} = \begin{bmatrix} x_0 & x_{n-1} & x_{n-2} & \cdots & x_2 & x_1 \\ x_1 & x_0 & x_{n-1} & \cdots & x_3 & x_2 \\ x_2 & \ddots & \ddots & \ddots & x_4 & x_3 \\ \ddots & \ddots & \ddots & \ddots & \ddots & \ddots \\ x_{n-2} & x_{n-3} & \cdots & x_1 & x_0 & x_{n-1} \\ x_{n-1} & x_{n-2} & \cdots & x_2 & x_1 & x_0 \end{bmatrix} \begin{bmatrix} y_0 \\ y_1 \\ y_2 \\ \vdots \\ y_{n-2} \\ y_{n-1} \end{bmatrix}. \tag{9.6.2}$$

Let n be the product of two relatively prime numbers n_1 and n_2, that is, $n = n_1 \cdot n_2$, $(n_1, n_2) = 1$. Using the CRT-I, we can write

$$\begin{aligned} k &= a\, k_1 + b\, k_2 \bmod n, \\ i &= a\, i_1 + b\, i_2 \bmod n, \end{aligned} \tag{9.6.3}$$

where

$$k_1 \equiv k \bmod n_1, \;\; k_1 = 0, 1, ..., n_1 - 1,$$
$$k_2 \equiv k \bmod n_2, \;\; k_2 = 0, 1, ..., n_2 - 1,$$
$$i_1 \equiv i \bmod n_1, \;\; i_1 = 0, 1, ..., n_1 - 1,$$
$$i_2 \equiv i \bmod n_2, \;\; i_1 = 0, 1, ..., n_2 - 1,$$

and

$$a = T_1\, n_2, \;\; b = T_2\, n_1.$$

Substituting in (9.6.1), we get

$$z_{ak_1+bk_2} = \sum_{i_2=0}^{n_2-1}\sum_{i_1=0}^{n_1-1} x_{a(k_1-i_1) \bmod n_1 + b(k_2-i_2) \bmod n_2}\, y_{ai_1+bi_2}\,. \tag{9.6.4}$$

In the above expression, terms $(k_1 - i_1)$ and $(k_2 - i_2)$ are defined mod n_1 and n_2, respectively. Define a two-dimensional sequence w_{j_1,j_2} as $w_{j_1,j_2} = w_{aj_1+bj_2}$ for the sequences x, y, and z in (9.6.4). This results in

$$z_{k_1,k_2} = \sum_{i_2=0}^{n_2-1}\sum_{i_1=0}^{n_1-1} x_{(k_1-i_1) \bmod n_1, (k_2-i_2) \bmod n_2}\, y_{i_1,i_2}\,. \tag{9.6.5}$$

It is clear that the first dimension is defined mod n_1 and the second dimension is defined mod n_2. This expression demonstrates that the one-dimensional cyclic convolution of length n in (9.6.1) has been converted to a two-dimensional cyclic convolution of length $n_1 \times n_2$ in (9.6.5) for $n = n_1 \cdot n_2$, $(n_1, n_2) = 1$. In a similar manner, the one-dimensional cyclic convolution of length n in (9.6.1) can be converted to a two-dimensional cyclic convolution of length $n_2 \times n_1$ for $n = n_1 \cdot n_2$, $(n_1, n_2) = 1$.

Define one-dimensional vectors $\underline{w}_{j_1}$, $j_1 = 0, 1, ..., n_1 - 1$, as

$$\underline{w}_{j_1} = \begin{bmatrix} w_{aj_1} \\ w_{aj_1+b} \\ w_{aj_1+b2} \\ \vdots \\ w_{aj_1+b(n_2-1)} \end{bmatrix}, \; j_1 = 0, 1, ..., n_1 - 1. \tag{9.6.6}$$

Now, arrange two-dimensional sequences z_{j_1,j_2} and y_{j_1,j_2} as one-dimensional vectors in the following manner:

$$\underline{z} = \begin{bmatrix} \underline{z}_0 \\ \underline{z}_1 \\ \vdots \\ \underline{z}_{n_1-1} \end{bmatrix}, \ \underline{y} = \begin{bmatrix} \underline{y}_0 \\ \underline{y}_1 \\ \vdots \\ \underline{y}_{n_1-1} \end{bmatrix}. \tag{9.6.7}$$

In terms of vectors $\underline{z}$ and $\underline{y}$, two-dimensional cyclic convolution in (9.6.5) can be expressed as

$$\begin{bmatrix} \underline{z}_0 \\ \underline{z}_1 \\ \underline{z}_2 \\ \vdots \\ \underline{z}_{n_1-2} \\ \underline{z}_{n_1-1} \end{bmatrix} = \begin{bmatrix} \mathbf{X}_0 & \mathbf{X}_{n_1-1} & \mathbf{X}_{n_1-2} & \cdots & \mathbf{X}_2 & \mathbf{X}_1 \\ \mathbf{X}_1 & \mathbf{X}_0 & \mathbf{X}_{n_1-1} & \cdots & \mathbf{X}_3 & \mathbf{X}_2 \\ \mathbf{X}_2 & \ddots & \ddots & \ddots & \mathbf{X}_4 & \mathbf{X}_3 \\ \ddots & \ddots & \ddots & \ddots & \ddots & \ddots \\ \mathbf{X}_{n_1-2} & \mathbf{X}_{n_1-3} & \cdots & \mathbf{X}_1 & \mathbf{X}_0 & \mathbf{X}_{n_1-1} \\ \mathbf{X}_{n_1-1} & \mathbf{X}_{n_1-2} & \cdots & \mathbf{X}_2 & \mathbf{X}_1 & \mathbf{X}_0 \end{bmatrix} \begin{bmatrix} \underline{y}_0 \\ \underline{y}_1 \\ \underline{y}_2 \\ \vdots \\ \underline{y}_{n_1-2} \\ \underline{y}_{n_1-1} \end{bmatrix}. \tag{9.6.8}$$

Here, $\mathbf{X}_{j_1}$ is a $n_2 \times n_2$ matrix corresponding to a cyclic convolution of length n_2. The (k_2, i_2)th element of $\mathbf{X}_{j_1}$ is given by

$$\left[\mathbf{X}_{j_1}\right]_{k_2,i_2} = x_{j_1+b(k_2-i_2)\bmod n_2}. \tag{9.6.9}$$

Let us cast these operations in polynomial form. Given a one-dimensional sequence w_i, $i = 0, 1, \ldots, n-1$, define two-dimensional polynomials as

$$W(u,v) = \sum_{j_1=0}^{n_1-1}\sum_{i_2=0}^{n_2-1} w_{aj_1+bj_2} u^{j_1} v^{j_2} = \sum_{j_1=0}^{n_1-1}\left[\sum_{i_2=0}^{n_2-1} w_{aj_1+bj_2} v^{j_2}\right] u^{j_1} = \sum_{j_1=0}^{n_1-1} w_{j_1}(v) u^{j_1}. \tag{9.6.10}$$

The one-dimensional cyclic convolution can now be expressed as a $(n_1 \times n_2)$ cyclic convolution given by

$$Z(u,v) = X(u,v)Y(u,v) \bmod \left[v^{n_2}-1, u^{n_1}-1\right]. \tag{9.6.11}$$

Example 9.6 Let $n = 10 = 2 \cdot 5$. Let us choose $n_1 = 2$, $n_2 = 5$. Using the Euclid's algorithm, we have $a = 5$ and $b = 6$. The various expressions in this case are

$$\begin{bmatrix} z_0 \\ z_1 \\ z_2 \\ \vdots \\ z_8 \\ z_9 \end{bmatrix} = \begin{bmatrix} x_0 & x_9 & x_8 & \cdots & x_2 & x_1 \\ x_1 & x_0 & x_9 & \cdots & x_3 & x_2 \\ x_2 & \ddots & \ddots & \ddots & x_4 & x_3 \\ \ddots & \ddots & \ddots & \ddots & \ddots & \ddots \\ x_8 & x_7 & \cdots & x_1 & x_0 & x_9 \\ x_9 & x_8 & \cdots & x_2 & x_1 & x_0 \end{bmatrix} \begin{bmatrix} y_0 \\ y_1 \\ y_2 \\ \vdots \\ y_8 \\ y_9 \end{bmatrix},$$

$$\underline{w}_0 = \begin{bmatrix} w_0 \\ w_6 \\ w_2 \\ w_8 \\ w_4 \end{bmatrix}, \quad \underline{w}_1 = \begin{bmatrix} w_5 \\ w_1 \\ w_7 \\ w_3 \\ w_9 \end{bmatrix},$$

$$\begin{bmatrix} \underline{z}_0 \\ \underline{z}_1 \end{bmatrix} = \begin{bmatrix} \mathbf{X}_0 & \mathbf{X}_1 \\ \mathbf{X}_1 & \mathbf{X}_0 \end{bmatrix} \begin{bmatrix} \underline{y}_0 \\ \underline{y}_1 \end{bmatrix},$$

$$\mathbf{X}_0 = \begin{bmatrix} x_0 & x_4 & x_8 & x_2 & x_6 \\ x_6 & x_0 & x_4 & x_8 & x_2 \\ x_2 & x_6 & x_0 & x_4 & x_8 \\ x_8 & x_2 & x_6 & x_0 & x_4 \\ x_4 & x_8 & x_2 & x_6 & x_0 \end{bmatrix}, \quad \mathbf{X}_1 = \begin{bmatrix} x_5 & x_9 & x_3 & x_7 & x_1 \\ x_1 & x_5 & x_9 & x_3 & x_7 \\ x_7 & x_1 & x_5 & x_9 & x_3 \\ x_3 & x_7 & x_1 & x_5 & x_9 \\ x_9 & x_3 & x_7 & x_1 & x_5 \end{bmatrix}.$$

This gives rise to a (2×5) two-dimensional cyclic convolution. We note that one could also choose $n_1 = 5$, $n_2 = 2$, and obtain a (5×2) two-dimensional formulation of the one-dimensional cyclic convolution of length 10.

9.6.2 Computing Higher-Dimensional Cyclic Convolutions

If $n = n_1 \cdot n_2$, $(n_1, n_2) = 1$, then length n one-dimensional cyclic convolution can be converted to a $(n_1 \times n_2)$ two-dimensional cyclic convolution. This is equivalent to computing a length n_1 cyclic convolution, where each of the scalar input is a vector of length n_2. Let us outline a basic procedure to compute this two-dimensional cyclic convolution. Given an algorithm to compute lengths n_1 and n_2 one-dimensional cyclic convolutions, $(n_1 \times n_2)$ two-dimensional cyclic convolution can be computed by employing the length n_1 algorithm. Every ADD in this algorithm is replaced by ADD of two length n_2 vectors. Similarly, every MULT is replaced by a cyclic convolution of length n_2 that is computed by using the cyclic convolution algorithm of length n_2.

Let us develop the bilinear forms representation of the algorithm described above. If

$$\mathbf{C}_1 [\mathbf{A}_1 \underline{x}_1 \otimes \mathbf{B}_1 \underline{y}_1], \tag{9.6.12}$$
$$\mathbf{C}_2 [\mathbf{A}_2 \underline{x}_2 \otimes \mathbf{B}_2 \underline{y}_2], \tag{9.6.13}$$
$$\mathbf{C} [\mathbf{A} \underline{x} \otimes \mathbf{B} \underline{y}], \tag{9.6.14}$$

denote the bilinear forms for cyclic convolution algorithms of lengths n_1, n_2, and n, respectively, then for the approach described above,

$$\begin{aligned} \mathbf{C} &= \mathbf{C}_1 \,\kappa\, \mathbf{C}_2 \\ \mathbf{A} &= \mathbf{A}_1 \,\kappa\, \mathbf{A}_2 \\ \mathbf{B} &= \mathbf{B}_1 \,\kappa\, \mathbf{B}_2, \end{aligned} \tag{9.6.15}$$

κ being the Kronecker product of two matrices. Let $A(n_1)$, $A(n_2)$, and $A(n)$ denote the additive complexities, and $M(n_1)$, $M(n_2)$, and $M(n)$ denote the multiplicative complexities of cyclic convolution algorithms of lengths n_1, n_2, and n, respectively. If the above described method is employed, then

$$M(n) = M(n_1) \cdot M(n_2) \tag{9.6.16}$$
$$A(n) = M(n_1) \cdot A(n_2) + A(n_1) \cdot n_2. \tag{9.6.17}$$

If $n = n_1 \cdot n_2$, $(n_1, n_2) = 1$, then length n one-dimensional cyclic convolution can also be converted to a $(n_2 \times n_1)$ two-dimensional cyclic convolution. This is equivalent to computing a length n_2 cyclic convolution where each of the scalar input is a vector of length n_1. Now, $(n_2 \times n_1)$ two-dimensional cyclic convolution can be computed by employing the length n_2 algorithm. Now, every ADD in this algorithm is replaced by ADD of two length n_1 vectors. Similarly, every MULT is replaced by a cyclic convolution of length n_1 that is computed using by the cyclic convolution algorithm of length n_1. In this case,

$$\begin{aligned} \mathbf{C} &= \mathbf{C}_2 \,\kappa\, \mathbf{C}_1 \\ \mathbf{A} &= \mathbf{A}_2 \,\kappa\, \mathbf{A}_1 \\ \mathbf{B} &= \mathbf{B}_2 \,\kappa\, \mathbf{B}_1. \end{aligned} \tag{9.6.18}$$

For this method we have

$$M(n) = M(n_1) \cdot M(n_2) \tag{9.6.19}$$
$$A(n) = M(n_2) \cdot A(n_1) + A(n_2) \cdot n_1. \tag{9.6.20}$$

It is seen that in either of the two approaches, the overall multiplicative complexity remains the same while the additive complexity depends on the order in which convolutions are performed. We would prefer the order that has smaller additive complexity. Thus, if

$$M(n_1) \cdot A(n_2) + A(n_1) \cdot n_2 < M(n_2) \cdot A(n_1) + A(n_2) \cdot n_1$$

or

$$(M(n_1) \; - n_1)/A(n_1) < (M(n_2) \; - n_2)/A(n_2),$$

then the first order is selected and if

$$(M(n_1) \; - n_1)/A(n_1) > (M(n_2) \; - n_2)/A(n_2),$$

then the second order is selected. Finally, it is worthwhile to mention here that one may realize further reduction in additive complexity by employing the split nesting algorithm described in Section 8.5 to compute the two- or higher-dimensional cyclic convolution in the Agarwal-Cooley algorithm.

Example 9.7 For $n = 10 = 2 \cdot 5$, we have $n_1 = 2$, $n_2 = 5$, $M(2) = 2$, $M(5) = 10$, $A(2) = 4$, $A(5) = 31$. In this case,

$$(M(n_1) \; - n_1)/A(n_1) = 0,$$
$$(M(n_2) \; - n_2)/A(n_2) = 5/31.$$

It is clear that converting a length 10 one-dimensional cyclic convolution to a 2 × 5 two-dimensional convolution is to be preferred from the point of view of computational complexity. The computational complexity of the Agarwal-Cooley technique based length 10 cyclic convolution algorithm is given by

$$M(10) = 20, \quad A(10) = 82.$$

9.7 Cyclic Convolution Algorithms Over Finite Fields and Rings

All the fast algorithms described in this chapter, including the Agarwal-Cooley algorithm, for computing cyclic convolution are valid over all number systems. In the present case, one seeks the cyclotomic factorization of $u^n - 1$ over the given finite field or the finite integer ring for the CRT-PR based algorithms. All other steps in the computation remain the same.

Over $Z(M)$ and $CZ(M)$, a cyclic convolution algorithm is a direct sum of the cyclic convolution algorithms over $Z\left(p_i^{a_i}\right)$ and $CZ\left(p_i^{a_i}\right)$, respectively, $M = \prod_{i=1}^{t} p_i^{a_i}$. Therefore, the multiplicative complexity of a cyclic convolution algorithm over $Z(M)$ and $CZ(M)$ is the maximum of the multiplicative complexities over $Z\left(p_i^{a_i}\right)$ and $CZ\left(p_i^{a_i}\right)$, respectively, $i = 1, 2, \ldots, t$. Interestingly, the computational complexity of the cyclic convolution algorithm over $Z(p^a)$ and $GF(p)$ are same. This leads to

$$M(n) = \max\left\{M_{GF(p_i)}(n), i = 1,2,\ldots,t\right\}.$$

We now turn to specific examples to illustrate this approach. The superscript (i) refers to the particular quantity over $Z\left(p_i^{a_i}\right)$ and $CZ\left(p_i^{a_i}\right)$, $i = 1, 2, \ldots, t$.

Example 9.8 Consider Example 9.1. The length 6 cyclic convolution algorithm is valid for all finite fields $GF(q)$ and finite integer rings $Z(M)$, $CZ(M)$ such that $(6, q) = 1$ and $(6, M) = 1$.

Example 9.9 Let $M = 15$ and $n = 8$. Thus $t = 2$, $p_1 = 3$, $p_2 = 5$, $a_1 = a_2 = 1$. This example demonstrates a simple algorithm obtained by using CRT-PR and the direct sum property. The factorizations of $P(u) = u^8 - 1$ over $GF(3)$ and $GF(5)$ are given by

$$u^8 - 1 = (u+1)(u+2)(u^2+1)(u^2+2u+2)(u^2+u+2) \bmod 3,$$
$$u^8 - 1 = (u+1)(u+4)(u+2)(u+3)(u^2+2)(u^2+3) \bmod 5.$$

The number of irreducible factor polynomials over $GF(3)$ and $GF(5)$ is 5 and 6, respectively. Based on these factor polynomials, $t_j^{(i)}(u)$ and $P_j^{(i)}(u)$ are computed as

$$P_1^{(1)}(u) = u^7 + 2u^6 + u^5 + 2u^4 + u^3 + 2u^2 + u + 2$$

$$t_1^{(1)}(u) = 1$$

$$P_2^{(1)}(u) = u^7 + u^6 + u^5 + u^4 + u^3 + u^2 + u + 1$$

$$t_2^{(1)}(u) = 2$$

$$P_3^{(1)}(u) = u^6 + 2u^4 + u^2 + 2$$

$$t_3^{(1)}(u) = 2$$

$$P_4^{(1)}(u) = u^6 + u^5 + 2u^4 + 2u^2 + 2u + 1$$

$$t_4^{(1)}(u) = 2u + 1$$

$$P_5^{(1)}(u) = u^6 + 2u^5 + 2u^4 + 2u^2 + u + 1$$

$$t_5^{(1)}(u) = u + 1$$

$$P_1^{(2)}(u) = u^7 + 4u^6 + u^5 + 4u^4 + u^3 + 4u^2 + u + 4$$

$$t_1^{(2)}(u) = 3$$

$$P_2^{(2)}(u) = u^7 + u^6 + u^5 + u^4 + u^3 + u^2 + u + 1$$

$$t_2^{(2)}(u) = 2$$

$$P_3^{(2)}(u) = u^7 + 3u^6 + 4u^5 + 2u^4 + u^3 + 3u^2 + 4u + 2$$

$$t_3^{(2)}(u) = 1$$

$$P_4^{(2)}(u) = u^7 + 2u^6 + 4u^5 + 3u^4 + u^3 + 2u^2 + 4u + 3$$

$$t_4^{(2)}(u) = 4$$

$$P_5^{(2)}(u) = u^6 + 3u^4 + 4u^2 + 2$$

$$t_5^{(2)}(u) = 2$$

$$P_6^{(2)}(u) = u^6 + 2u^4 + 4u^2 + 3$$

$$t_6^{(2)}(u) = 3$$

Consequently, the reconstruction matrices $\mathbf{P}^{(1)}\mathbf{T}^{(1)}$ and $\mathbf{P}^{(2)}\mathbf{T}^{(2)}$ are generated as

$$\mathbf{P}^{(1)}\mathbf{T}^{(1)} = \left[\begin{array}{c:c:cc:cc:cc} 2 & 2 & 1 & 0 & 1 & 2 & 1 & 1 \\ 1 & 2 & 0 & 1 & 1 & 1 & 2 & 1 \\ 2 & 2 & 2 & 0 & 0 & 1 & 0 & 2 \\ 1 & 2 & 0 & 2 & 1 & 0 & 2 & 0 \\ 2 & 2 & 1 & 0 & 2 & 1 & 2 & 2 \\ 1 & 2 & 0 & 1 & 2 & 2 & 1 & 2 \\ 2 & 2 & 2 & 0 & 0 & 2 & 0 & 1 \\ 1 & 2 & 0 & 2 & 2 & 0 & 1 & 0 \end{array}\right],$$

$$\mathbf{P}^{(2)}\mathbf{T}^{(2)} = \left[\begin{array}{c|c|c|c|cc|cc} 2 & 2 & 2 & 2 & 4 & 0 & 4 & 0 \\ 3 & 2 & 4 & 1 & 0 & 4 & 0 & 4 \\ 2 & 2 & 3 & 3 & 3 & 0 & 2 & 0 \\ 3 & 2 & 1 & 4 & 0 & 3 & 0 & 2 \\ 2 & 2 & 2 & 2 & 1 & 0 & 1 & 0 \\ 3 & 2 & 4 & 1 & 0 & 1 & 0 & 1 \\ 2 & 2 & 3 & 3 & 2 & 0 & 3 & 0 \\ 3 & 2 & 1 & 4 & 0 & 2 & 0 & 3 \end{array}\right].$$

In Step 1, we proceed with CRT-PR reduction to generate the matrices $\mathbf{R}^{(1)}$ and $\mathbf{R}^{(2)}$. They are given by

$$\mathbf{R}^{(1)} = \left[\begin{array}{cccccccc} 1 & 2 & 1 & 2 & 1 & 2 & 1 & 2 \\ \hdashline 1 & 1 & 1 & 1 & 1 & 1 & 1 & 1 \\ \hdashline 1 & 0 & 2 & 0 & 1 & 0 & 2 & 0 \\ 0 & 1 & 0 & 2 & 0 & 1 & 0 & 2 \\ \hdashline 1 & 0 & 1 & 1 & 2 & 0 & 2 & 2 \\ 0 & 1 & 1 & 2 & 0 & 2 & 2 & 1 \\ \hdashline 1 & 0 & 1 & 2 & 2 & 0 & 2 & 1 \\ 0 & 1 & 2 & 2 & 0 & 2 & 1 & 1 \end{array}\right],$$

and

$$\mathbf{R}^{(2)} = \left[\begin{array}{cccccccc} 1 & 4 & 1 & 4 & 1 & 4 & 1 & 4 \\ \hdashline 1 & 1 & 1 & 1 & 1 & 1 & 1 & 1 \\ \hdashline 1 & 3 & 4 & 2 & 1 & 3 & 4 & 2 \\ \hdashline 1 & 2 & 4 & 3 & 1 & 2 & 4 & 3 \\ \hdashline 1 & 0 & 3 & 0 & 4 & 0 & 2 & 0 \\ 0 & 1 & 0 & 3 & 0 & 4 & 0 & 2 \\ \hdashline 1 & 0 & 2 & 0 & 4 & 0 & 3 & 0 \\ 0 & 1 & 0 & 2 & 0 & 4 & 0 & 3 \end{array}\right].$$

The matrices $\mathbf{C}'^{(i)}$, $\mathbf{A}'^{(i)}$, and $\mathbf{B}'^{(i)}$, $i = 1, 2$, are given by

$$\mathbf{C}'^{(1)} = \begin{bmatrix} 1 & & & & & & & & & & \\ & 1 & & & & & & & & & \\ & & 0 & 1 & 2 & & & & & & \\ & & 1 & 2 & 2 & & & & & & \\ & & & & & 0 & 1 & 1 & & & \\ & & & & & 1 & 2 & 0 & & & \\ & & & & & & & & 0 & 1 & 1 \\ & & & & & & & & 1 & 2 & 1 \end{bmatrix},$$

$$\mathbf{C}'^{(2)} = \begin{bmatrix} 1 & & & & & & & & & \\ & 1 & & & & & & & & \\ & & 1 & & & & & & & \\ & & & 1 & & & & & & \\ & & & & 0 & 1 & 3 & & & \\ & & & & 1 & 4 & 4 & & & \\ & & & & & & & 0 & 1 & 2 \\ & & & & & & & 1 & 4 & 4 \end{bmatrix},$$

$$\mathbf{A}'^{(1)} = \mathbf{B}'^{(1)} = \begin{bmatrix} 1 & & & & & & & \\ & 1 & & & & & & \\ & & 1 & 1 & & & & \\ & & 1 & 0 & & & & \\ & & 0 & 1 & & & & \\ & & & & 1 & 1 & & \\ & & & & 1 & 0 & & \\ & & & & 0 & 1 & & \\ & & & & & & 1 & 1 \\ & & & & & & 1 & 0 \\ & & & & & & 0 & 1 \end{bmatrix},$$

$$
\mathbf{A}'^{(2)} = \mathbf{B}'^{(2)} = \begin{bmatrix}
1 & & & & & & & \\
 & 1 & & & & & & \\
 & & 1 & & & & & \\
 & & & 1 & & & & \\
 & & & & 1 & 1 & & \\
 & & & & 1 & 0 & & \\
 & & & & 0 & 1 & & \\
 & & & & & & 1 & 1 \\
 & & & & & & 1 & 0 \\
 & & & & & & 0 & 1
\end{bmatrix}.
$$

The matrices $\mathbf{A}^{(i)}$, $\mathbf{B}^{(i)}$, and $\mathbf{C}^{(i)} = \mathbf{P}^{(i)}\,\mathbf{T}^{(i)}\,\mathbf{C}'^{(i)}$, $i = 1, 2$, can now be computed as

$$
\mathbf{A}^{(1)} = \mathbf{B}^{(1)} = \mathbf{A}'^{(1)}\mathbf{R}^{(1)} = \left[\begin{array}{cccccccc}
1 & 2 & 1 & 2 & 1 & 2 & 1 & 2 \\ \hline
1 & 1 & 1 & 1 & 1 & 1 & 1 & 1 \\ \hline
1 & 1 & 2 & 2 & 1 & 1 & 2 & 2 \\
1 & 0 & 2 & 0 & 1 & 0 & 2 & 0 \\
0 & 1 & 0 & 2 & 0 & 1 & 0 & 2 \\ \hline
1 & 1 & 2 & 0 & 2 & 2 & 1 & 0 \\
1 & 0 & 1 & 1 & 2 & 0 & 2 & 2 \\
0 & 1 & 1 & 2 & 0 & 2 & 2 & 1 \\ \hline
1 & 1 & 0 & 1 & 2 & 2 & 0 & 2 \\
1 & 0 & 1 & 2 & 2 & 0 & 2 & 1 \\
0 & 1 & 2 & 2 & 0 & 2 & 1 & 1
\end{array}\right],
$$

$$
\mathbf{A}^{(2)} = \mathbf{B}^{(2)} = \mathbf{A}'^{(2)}\mathbf{R}^{(2)} = \left[\begin{array}{cccccccc}
1 & 4 & 1 & 4 & 1 & 4 & 1 & 4 \\ \hline
1 & 1 & 1 & 1 & 1 & 1 & 1 & 1 \\ \hline
1 & 3 & 4 & 2 & 1 & 3 & 4 & 2 \\ \hline
1 & 2 & 4 & 3 & 1 & 2 & 4 & 3 \\ \hline
1 & 1 & 3 & 3 & 4 & 4 & 2 & 2 \\
1 & 0 & 3 & 0 & 4 & 0 & 2 & 0 \\
0 & 1 & 0 & 3 & 0 & 4 & 0 & 2 \\ \hline
1 & 1 & 2 & 2 & 4 & 4 & 3 & 3 \\
1 & 0 & 2 & 0 & 4 & 0 & 3 & 0 \\
0 & 1 & 0 & 2 & 0 & 4 & 0 & 3
\end{array}\right],
$$

$$\mathbf{C}^{(1)} = \mathbf{P}^{(1)}\mathbf{T}^{(1)}\mathbf{C}'^{(1)} = \left[\begin{array}{c|c|ccc|ccc|ccc} 2 & 2 & 0 & 1 & 2 & 2 & 2 & 1 & 1 & 0 & 2 \\ 1 & 2 & 1 & 2 & 2 & 1 & 0 & 1 & 1 & 1 & 0 \\ 2 & 2 & 0 & 2 & 1 & 1 & 2 & 0 & 2 & 1 & 2 \\ 1 & 2 & 2 & 1 & 1 & 0 & 1 & 1 & 0 & 2 & 2 \\ 2 & 2 & 0 & 1 & 2 & 1 & 1 & 2 & 2 & 0 & 1 \\ 1 & 2 & 1 & 2 & 2 & 2 & 0 & 2 & 2 & 2 & 0 \\ 2 & 2 & 0 & 2 & 1 & 2 & 1 & 0 & 1 & 2 & 1 \\ 1 & 2 & 2 & 1 & 1 & 0 & 2 & 2 & 0 & 1 & 1 \end{array}\right],$$

and

$$\mathbf{C}^{(2)} = \mathbf{P}^{(2)}\mathbf{T}^{(2)}\mathbf{C}'^{(2)} = \left[\begin{array}{c|c|c|c|ccc|ccc} 2 & 2 & 2 & 2 & 0 & 4 & 2 & 0 & 4 & 3 \\ 3 & 2 & 4 & 1 & 4 & 1 & 1 & 4 & 1 & 1 \\ 2 & 2 & 3 & 3 & 0 & 3 & 4 & 0 & 2 & 4 \\ 3 & 2 & 1 & 4 & 3 & 2 & 2 & 2 & 3 & 3 \\ 2 & 2 & 2 & 2 & 0 & 1 & 3 & 0 & 1 & 2 \\ 3 & 2 & 4 & 1 & 1 & 4 & 4 & 1 & 4 & 4 \\ 2 & 2 & 3 & 3 & 0 & 2 & 1 & 0 & 3 & 1 \\ 3 & 2 & 1 & 4 & 2 & 3 & 3 & 3 & 2 & 2 \end{array}\right].$$

Finally, we combine $\mathbf{C}^{(1)}$ and $\mathbf{C}^{(2)}$, $\mathbf{A}^{(1)}$ and $\mathbf{A}^{(2)}$, and $\mathbf{B}^{(1)}$ and $\mathbf{B}^{(2)}$ component-by-component using the direct sum to get the algorithm for cyclic convolution over $Z(15)$. Note that proper zero appending may have to be done to match the dimensions of all of $\mathbf{A}^{(i)}$, $\mathbf{B}^{(i)}$, and $\mathbf{C}^{(i)}$ before direct sum is performed. The multiplicative complexity of computing a cyclic convolution of length 8 over $Z(15)$ is 11 MULTs.

Example 9.10 Let $M = 3^2 \cdot 5^3 = 1{,}125$ and $n = 8$. Thus $t = 2$, $p_1 = 3$, $p_2 = 5$, $a_1 = 2$, $a_2 = 3$. The factorizations of $P(u) = u^8 - 1$ over $Z(9)$ and $Z(125)$ are obtained by applying Hensel's theorem to the factorizations over $GF(3)$ and $GF(5)$, respectively. They are given by

$$u^8 - 1 = (u+1)(u+8)(u^2+1)(u^2+5u+8)(u^2+4u+8) \bmod 9,$$
$$u^8 - 1 = (u+1)(u+124)(u+57)(u+68)(u^2+57)(u^2+68) \bmod 125.$$

The number of irreducible factor polynomials over $GF(3)$ and $GF(5)$ is 5 and 6, respectively. Based on these factor polynomials, $t_j^{(i)}(u)$ and $P_j^{(i)}(u)$ are computed as

$$P_1^{(1)}(u) = u^7 + 8u^6 + u^5 + 8u^4 + u^3 + 8u^2 + u + 8$$

$$t_1^{(1)}(u) = 1$$

$$P_2^{(1)}(u) = u^7 + u^6 + u^5 + u^4 + u^3 + u^2 + u + 1$$

$$t_2^{(1)}(u) = 8$$

$$P_3^{(1)}(u) = u^6 + 8u^4 + u^2 + 8$$

$$t_3^{(1)}(u) = 2$$

$$P_4^{(1)}(u) = u^6 + 4u^5 + 8u^4 + 8u^2 + 5u + 1$$

$$t_4^{(1)}(u) = 5u + 7$$

$$P_5^{(1)}(u) = u^6 + 5u^5 + 8u^4 + 8u^2 + 4u + 1$$

$$t_5^{(1)}(u) = 4u + 7$$

$$P_1^{(2)}(u) = u^7 + 124u^6 + u^5 + 124u^4 + u^3 + 124u^2 + u + 124$$

$$t_1^{(2)}(u) = 78$$

$$P_2^{(2)}(u) = u^7 + u^6 + u^5 + u^4 + u^3 + u^2 + u + 1$$

$$t_2^{(2)}(u) = 47$$

$$P_3^{(2)}(u) = u^7 + 68u^6 + 124u^5 + 57u^4 + u^3 + 68u^2 + 124u + 57$$

$$t_3^{(2)}(u) = 71$$

$$P_4^{(2)}(u) = u^7 + 57u^6 + 124u^5 + 68u^4 + u^3 + 57u^2 + 124u + 68$$

$$t_4^{(2)}(u) = 54$$

$$P_5^{(2)}(u) = u^6 + 68u^4 + 124u^2 + 57$$

$$t_5^{(2)}(u) = 17$$

$$P_6^{(2)}(u) = u^6 + 57u^4 + 124u^2 + 68$$

$$t_6^{(2)}(u) = 108$$

Consequently, the reconstruction matrices $\mathbf{P}^{(1)}\,\mathbf{T}^{(1)}$ and $\mathbf{P}^{(2)}\,\mathbf{T}^{(2)}$ are generated as

$$\mathbf{P}^{(1)}\mathbf{T}^{(1)} = \left[\begin{array}{c|c|cc|cc|cc} 8 & 8 & 7 & 0 & 7 & 4 & 7 & 5 \\ 1 & 8 & 0 & 7 & 4 & 7 & 5 & 7 \\ 8 & 8 & 2 & 0 & 0 & 4 & 0 & 5 \\ 1 & 8 & 0 & 2 & 4 & 0 & 5 & 0 \\ 8 & 8 & 7 & 0 & 2 & 4 & 2 & 5 \\ 1 & 8 & 0 & 7 & 5 & 2 & 4 & 2 \\ 8 & 8 & 2 & 0 & 0 & 5 & 0 & 4 \\ 1 & 8 & 0 & 2 & 5 & 0 & 4 & 0 \end{array}\right],$$

$$\mathbf{P}^{(2)}\mathbf{T}^{(2)} = \left[\begin{array}{c|c|c|c|cc|cc} 47 & 47 & 47 & 47 & 94 & 0 & 94 & 0 \\ 78 & 47 & 54 & 71 & 0 & 94 & 0 & 94 \\ 47 & 47 & 78 & 78 & 108 & 0 & 17 & 0 \\ 78 & 47 & 71 & 54 & 0 & 108 & 0 & 17 \\ 47 & 47 & 47 & 47 & 31 & 0 & 31 & 0 \\ 78 & 47 & 54 & 71 & 0 & 31 & 0 & 31 \\ 47 & 47 & 78 & 78 & 17 & 0 & 108 & 0 \\ 78 & 47 & 71 & 54 & 0 & 17 & 0 & 108 \end{array}\right].$$

In Step 1, we proceed with CRT-PR reduction to generate the matrices $\mathbf{R}^{(1)}$ and $\mathbf{R}^{(2)}$. They are given by

$$\mathbf{R}^{(1)} = \left[\begin{array}{cccccccc} 1 & 8 & 1 & 8 & 1 & 8 & 1 & 8 \\ \hdashline 1 & 1 & 1 & 1 & 1 & 1 & 1 & 1 \\ \hdashline 1 & 0 & 8 & 0 & 1 & 0 & 8 & 0 \\ 0 & 1 & 0 & 8 & 0 & 1 & 0 & 8 \\ \hdashline 1 & 0 & 1 & 4 & 8 & 0 & 8 & 5 \\ 0 & 1 & 4 & 8 & 0 & 8 & 5 & 1 \\ \hdashline 1 & 0 & 1 & 5 & 8 & 0 & 8 & 4 \\ 0 & 1 & 5 & 8 & 0 & 8 & 4 & 1 \end{array}\right],$$

and

$$\mathbf{R}^{(2)} = \left[\begin{array}{cc:cc:cc:cc} 1 & 124 & 1 & 124 & 1 & 124 & 1 & 124 \\ \hdashline 1 & 1 & 1 & 1 & 1 & 1 & 1 & 1 \\ \hdashline 1 & 68 & 124 & 57 & 1 & 68 & 124 & 57 \\ \hdashline 1 & 57 & 124 & 68 & 1 & 57 & 124 & 68 \\ \hdashline 1 & 0 & 68 & 0 & 124 & 0 & 57 & 0 \\ 0 & 1 & 0 & 68 & 0 & 124 & 0 & 57 \\ \hdashline 1 & 0 & 57 & 0 & 124 & 0 & 68 & 0 \\ 0 & 1 & 0 & 57 & 0 & 124 & 0 & 68 \end{array}\right].$$

The matrices $\mathbf{C}'^{(i)}$, $\mathbf{A}'^{(i)}$, and $\mathbf{B}'^{(i)}$, $i = 1, 2$, are given by

$$\mathbf{C}'^{(1)} = \begin{bmatrix} 1 & & & & & & & & & & \\ & 1 & & & & & & & & & \\ & & 0 & 1 & 8 & & & & & & \\ & & 1 & 8 & 8 & & & & & & \\ & & & & & 0 & 1 & 1 & & & \\ & & & & & 1 & 8 & 3 & & & \\ & & & & & & & & 0 & 1 & 1 \\ & & & & & & & & 1 & 8 & 4 \end{bmatrix},$$

$$\mathbf{C}'^{(2)} = \begin{bmatrix} 1 & & & & & & & & & & \\ & 1 & & & & & & & & & \\ & & 1 & & & & & & & & \\ & & & 1 & & & & & & & \\ & & & & 0 & 1 & 68 & & & & \\ & & & & 1 & 124 & 124 & & & & \\ & & & & & & & 0 & 1 & 57 \\ & & & & & & & 1 & 124 & 124 \end{bmatrix}.$$

The matrices $\mathbf{A}'^{(i)}$ and $\mathbf{B}'^{(i)}$, $i = 1, 2$, remain the same as in the previous example. The matrices $\mathbf{A}^{(i)}$, $\mathbf{B}^{(i)}$, and $\mathbf{C}^{(i)} = \mathbf{P}^{(i)}\,\mathbf{T}^{(i)}\,\mathbf{C}'^{(i)}$, $i = 1, 2$, can now be computed as

$$\mathbf{A}^{(1)} = \mathbf{B}^{(1)} = \mathbf{A}'^{(1)}\mathbf{R}^{(1)} = \left[\begin{array}{cccccccc}
1 & 8 & 1 & 8 & 1 & 8 & 1 & 8 \\ \hdashline
1 & 1 & 1 & 1 & 1 & 1 & 1 & 1 \\ \hdashline
1 & 1 & 8 & 8 & 1 & 1 & 8 & 8 \\
1 & 0 & 8 & 0 & 1 & 0 & 8 & 0 \\
0 & 1 & 0 & 8 & 0 & 1 & 0 & 8 \\ \hdashline
1 & 1 & 5 & 3 & 8 & 8 & 4 & 6 \\
1 & 0 & 1 & 4 & 8 & 0 & 8 & 5 \\
0 & 1 & 4 & 8 & 0 & 8 & 5 & 1 \\ \hdashline
1 & 1 & 6 & 4 & 8 & 8 & 3 & 5 \\
1 & 0 & 1 & 5 & 8 & 0 & 8 & 4 \\
0 & 1 & 5 & 8 & 0 & 8 & 4 & 1
\end{array}\right],$$

$$\mathbf{A}^{(2)} = \mathbf{B}^{(2)} = \mathbf{A}'^{(2)}\mathbf{R}^{(2)} = \left[\begin{array}{cccccccc}
1 & 124 & 1 & 124 & 1 & 124 & 1 & 124 \\ \hdashline
1 & 1 & 1 & 1 & 1 & 1 & 1 & 1 \\ \hdashline
1 & 68 & 124 & 57 & 1 & 68 & 124 & 57 \\ \hdashline
1 & 57 & 124 & 68 & 1 & 57 & 124 & 68 \\ \hdashline
1 & 1 & 68 & 68 & 124 & 124 & 57 & 57 \\
1 & 0 & 68 & 0 & 124 & 0 & 57 & 0 \\
0 & 1 & 0 & 68 & 0 & 124 & 0 & 57 \\ \hdashline
1 & 1 & 57 & 57 & 124 & 124 & 68 & 68 \\
1 & 0 & 57 & 0 & 124 & 0 & 68 & 0 \\
0 & 1 & 0 & 57 & 0 & 124 & 0 & 68
\end{array}\right],$$

$$\mathbf{C}^{(1)} = \mathbf{P}^{(1)}\mathbf{T}^{(1)}\mathbf{C}'^{(1)} = \left[\begin{array}{c;{2pt/2pt}c;{2pt/2pt}ccc;{2pt/2pt}ccc;{2pt/2pt}ccc}
8 & 8 & 0 & 7 & 2 & 5 & 2 & 4 & 4 & 3 & 5 \\
1 & 8 & 7 & 2 & 2 & 7 & 6 & 7 & 7 & 7 & 6 \\
8 & 8 & 0 & 2 & 7 & 4 & 5 & 3 & 5 & 4 & 2 \\
1 & 8 & 2 & 7 & 7 & 0 & 4 & 4 & 0 & 5 & 5 \\
8 & 8 & 0 & 7 & 2 & 4 & 7 & 5 & 5 & 6 & 4 \\
1 & 8 & 7 & 2 & 2 & 2 & 3 & 2 & 2 & 2 & 3 \\
8 & 8 & 0 & 2 & 7 & 5 & 4 & 6 & 4 & 5 & 7 \\
1 & 8 & 2 & 7 & 7 & 0 & 5 & 5 & 0 & 4 & 4
\end{array}\right],$$

and

$$\mathbf{C}^{(2)} = \mathbf{P}^{(2)}\,\mathbf{T}^{(2)}\,\mathbf{C}'^{(2)}$$

$$
= \left[\begin{array}{c|c|c|c|ccc|ccc}
47 & 47 & 47 & 47 & 0 & 94 & 17 & 0 & 94 & 108 \\
78 & 47 & 54 & 71 & 94 & 31 & 31 & 94 & 31 & 31 \\
47 & 47 & 78 & 78 & 0 & 108 & 94 & 0 & 108 & 94 \\
78 & 47 & 71 & 54 & 108 & 17 & 17 & 17 & 108 & 108 \\
47 & 47 & 47 & 47 & 0 & 31 & 108 & 0 & 31 & 17 \\
78 & 47 & 54 & 71 & 31 & 94 & 94 & 31 & 94 & 94 \\
47 & 47 & 78 & 78 & 0 & 17 & 31 & 0 & 17 & 31 \\
78 & 47 & 71 & 54 & 17 & 108 & 108 & 108 & 108 & 17
\end{array}\right].
$$

Finally, we combine $\mathbf{C}^{(1)}$ and $\mathbf{C}^{(2)}$, $\mathbf{A}^{(1)}$ and $\mathbf{A}^{(2)}$, and $\mathbf{B}^{(1)}$ and $\mathbf{B}^{(2)}$ component by component using the direct sum to get the algorithm for cyclic convolution over $Z(1,125)$. Again, note that proper zero appending may have to be done to match the dimensions of all of $\mathbf{A}^{(i)}$, $\mathbf{B}^{(i)}$, and $\mathbf{C}^{(i)}$ before a direct sum is performed. The multiplicative complexity of computing a cyclic convolution of length 8 over $Z(1,125)$ is 11 MULTs.

Example 9.11 Consider a length 5 cyclic convolution over $CZ(9)$. The corresponding factorization is given in Example 7.9. Assuming that the product of two degree one polynomials is computed in 3 MULTs as

$$(x_0 + x_1 u)(y_0 + y_1 u) = x_0 y_0 + [(x_0 + x_1)(y_0 + y_1) - x_0 y_0 - x_1 y_1]\, u + x_1 y_1 u^2,$$

various matrices for length 5 cyclic convolution of integer sequences over $CZ(3^2)$ are found to be

$$
\mathbf{A} = \mathbf{B} = \left[\begin{array}{ccccc}
1 & 1 & 1 & 1 & 1 \\ \hline
1 & 0 & 8 & 5+j & 4+8j \\
0 & 1 & 4+8j & 5+j & 8 \\
1 & 1 & 3+8j & 1+2j & 3+8j \\ \hline
1 & 0 & 8 & 5+8j & 4+j \\
0 & 1 & 4+j & 5+8j & 8 \\
1 & 1 & 3+j & 1+7j & 3+j
\end{array}\right],
$$

$$\mathbf{C}=\left[\begin{array}{c|ccc|ccc} 2 & 5+2j & 4j & 8+7j & 5+7j & 5j & 8+2j \\ 2 & 4+7j & 4+7j & 4 & 4+2j & 4+2j & 4 \\ 2 & 4j & 5+2j & 8+7j & 5j & 5+7j & 8+2j \\ 2 & 0 & 5j & 8+2j & 0 & 4j & 8+7j \\ 2 & 5j & 0 & 8+2j & 4j & 0 & 8+7j \end{array}\right].$$

The partitions as marked in the above matrices correspond to the CRT-PR reduction and reconstruction with regard to the modulo polynomials $p_i(u)$, $i = 1, 2, 3$, respectively.

Notes

In this chapter, we have studied Chinese remainder theorem based fast algorithms for cyclic convolution as derived from the cyclotomic factorization of $u^n - 1$. Over CZ, the elements of the resulting matrices turn out to be either integers or purely imaginary integers. In many cases of interest, the matrices involved in the computation have simple elements such as 0, 1, −1, j, and $-j$. All the algorithms derived for computation over Z and CZ are also valid for computation over $Z(M)$ and $CZ(M)$ provided that $(n, M) = 1$. Complex integers (or more generally complex rational numbers) are also known as Gaussian numbers. Factorization of $u^n - 1$ over CZ or equivalently over CI has not received much attention in the digital signal processing literature. We expect this work to lead to better understanding and implementation of Chinese remainder theorem based algorithms for computing cyclic convolution. A large part of this chapter is based on Garg [1997] and Garg and Mendis [1997].

Bibliography

[9.1] J.H. McClellan and C.M. Rader, *Number Theory in Digital Signal Processing*, Prentice Hall, 1979.

[9.2] R.E. Blahut, *Fast Algorithms for Digital Signal Processing*, Addison Wesley, 1982.

[9.3] H.J. Nussbaumer, *Fast Fourier Transform and Convolution Algorithms*, Springer-Verlag, 1981.

[9.4] M.T. Heideman, *Multiplicative Complexity, Convolution and the DFT*, Springer-Verlag, 1988.

[9.5] R. Tolimieri, M. An, and C. Lu, *Algorithms for Discrete Fourier Transform and Convolution*, Springer-Verlag, 1989.

[9.6] D.G. Myers, *Digital Signal Processing, Efficient Convolution and Fourier Transform Techniques*, Prentice Hall, 1990.

[9.7] S.D. Morgera and H. Krishna, *Digital Signal Processing, Applications to Communications and Algebraic Coding Theories*, Academic Press, 1989.

[9.8] H. Krishna, B. Krishna, K.Y. Lin, and J.D. Sun, *Computational Number Theory and Digital Signal Processing*, CRC Press, 1994.

[9.9] H. Hasse, *Number Theory*, Springer-Verlag, 1980.

[9.10] L.C. Washington, *Introduction to Cyclotomic Fields*, Springer-Verlag, 1982.

[9.11] H.M. Edwards, *Galois Theory*, Springer-Verlag, 1984.

[9.12] M. Pohst and H. Zassenhaus, *Algorithmic Algebraic Number Theory*, Cambridge University Press, 1990.

[9.13] K.Y. Lin, B. Krishna, and H. Krishna, "Rings, Fields, the Chinese Remainder Theorem and an Extension, Part I: Theory," *IEEE Transactions on Circuits and Systems*, Vol. 41, pp. 641-655, October 1994.

[9.14] K.Y. Lin, B. Krishna, and H. Krishna, "Rings, Fields, the Chinese Remainder Theorem and an Extension, Part II: Applications to Digital Signal Processing," *IEEE Transactions on Circuits and Systems*, Vol. 41, pp. 656-668, October 1994.

[9.15] H. Krishna, B. Krishna, and K.Y. Lin, "The AICE-CRT and Algorithms for Digital Signal Processing: The Complex Case," *Circuits, Systems, and Signal Processing*, Vol. 14. no. 1, pp. 69-95, 1995.

[9.16] V.U. Reddy and N.S. Reddy, "Complex Rectangular Transforms," *Proceedings International Conference on Acoustics, Speech and Signal Processing*, 1979.

[9.17] N.S. Reddy and V.U. Reddy, "Complex Rectangular Transforms for Digital Convolution," *IEEE Transactions Acoustics, Speech and Signal Processing*, Vol. ASSP-28, No. 5, pp. 592-596, Oct. 1980.

[9.18] H. Krishna Garg, "Analysis of the Chinese Remainder Theorem and Cyclotomic Polynomials Based Algorithms for Cyclic Convolution, Part I: Rational Number System," *Circuits, Systems, and Signal Processing*, Vol. 16, pp. 569-594, 1997.

[9.19] H. Krishna Garg and F.V.C. Mendis, 'Analysis of the Chinese Remainder Theorem and Cyclotomic Polynomials Based Algorithms for Cyclic Convolution, Part II: Complex Rational Number System," *Circuits, Systems, and Signal Processing*, Vol. 16, pp. 595-610, 1997.

[9.20] L-K. Hua, *Introduction to Number Theory*, Springer-Verlag, 1982.

[9.21] E. Dubois and A.N. Venetsanopoulos, "Convolution using a Conjugate Symmetry Property for the Generalized Discrete Fourier Transform," *IEEE Transactions Acoustics, Speech and Signal Processing*, Vol. ASSP-26, pp. 165-170, 1978.

Problems

9.0 It is common to have a table of cyclic convolution algorithms over $\mathbf{Z}$ for small values of n. Create such a table over $\mathbf{Z}$ and $C\mathbf{Z}$ for $n = 2, 3, 4, 5, 6, 7, 8, 9$ using any technique and evaluate the complexities.

9.1 Derive a length 4 cyclic convolution algorithm based on the CRT-P for real, complex, rational, and complex rational sequences. Evaluate the computational complexity in all instances.

9.2 Repeat Problem 9.1 for length $n = 12$ and $n = 15$.

9.3 Repeat Problem 9.1 for $n = 20, 24, 30, 36, 60$.

9.4 In Section 9.1, expressions were derived for the polynomials $t_i(u)$ and $P_i(u)$ for the cases when n had at most 2 odd prime factors. However, it is to be observed that the CRT-P reconstruction requires only the product $t_i(u) \cdot P_i(u)$. Derive expressions for this product when n has at most 2 odd prime factors. Hence, derive the matrix formulation of CRT-P reconstruction based on this product.

9.5 In this chapter, we derived expressions when there were at most two odd prime factors in the standard factorization of n. Can you extend them to the case of at most three odd prime factors? For a start, let $n = pqr$, $p < q < r$ with $p = 3$.

9.6 Consider the cyclotomic factorization and the corresponding matrices for cyclic convolution of length n over CF(3), CF(5), CF(8), CCF(8). In each case, one needs to consider only those values of n for which the cyclotomic factors of $u^n - 1$ over Z factor further these extension fields.

9.7 Show that the polynomial product $A(u) \cdot B(u) \bmod u^n + 1$ can be converted to a cyclic convolution by replacing u by $u \cdot \omega$, ω being the $2n$-th root of unity. Also, if n is odd, then u may be replaced by $-u$.

9.8 Using the result of Problem 9.7, show that the cyclic convolution of length n (n being an even number) can be computed as two cyclic convolutions of length $n / 2$ each at a computational cost of approximately n MULTs.

9.9 Show that the polynomial product $A(u) \cdot B(u) \bmod u^n + 1$ can be converted to a two-dimensional convolution of the kind $A(u, v) \cdot B(u, v) \bmod [u^a - v, v^b + 1]$ for $n = a \cdot b$.

9.10 Show that for $n = a \cdot b$, $(a, b) \neq 1$, a length n cyclic convolution can be converted to a two-dimensional convolution where the convolution is cyclic along one dimension and acyclic along the second. Identify the length of convolution along each dimension.

9.11 Based on problems 9.9 and 9.10, show that polynomial product $A(u) \cdot B(u) \bmod u^n + 1$ can be converted to a two-dimensional convolutions of the kind $A(v) \cdot B(v) \bmod v^a + 1$ for $n = a^2$, a being a power of 2.

9.12 Describe how one may use the idea of Problem 9.11 in a recursive manner for $n = a^t$.

9.13 Describe how one may use the idea of Problem 9.11 in a recursive manner for $n = n_1 \cdot a^t$.

9.14 Derive algorithms for computing polynomial product $A(u) \cdot B(u) \bmod u^n + 1$, $n = 2, 4, 8, 16$.

9.15 On some occasions, it is possible to derive algorithms on an ad-hoc basis (say by some deep insight or observation into the computational task) that are computationally more efficient than the ones derived using the mathematical approaches. Derive (if possible) one such algorithm for the case $A(u) \cdot B(u) \bmod u^4 + 1$.

9.16 Derive an algorithm to compute $A(u) \cdot B(u) \bmod (u^6 - u^3 + 1)$. Such a computation arises in a CRT-P based cyclic convolution algorithm of length 18.

9.17 One way to design an algorithm for computing

$$A(u) \cdot B(u) \bmod (u^{p-1} + u^{p-2} + \ldots + 1),$$

p prime, is to compute the product $A(u) \cdot B(u)$ and then reduce it modulo $(u^{p-1} + u^{p-2} + \ldots + 1)$. Design such an algorithm for $p = 5$. Let modulo polynomial be $u\,(u + 1)(u^2 + 1)(u^2 + u + 1)$ with a wraparound of one coefficient.

9.18 Repeat Problem 9.17 for $p = 7$ with

$$u\,(u + 1)(u - 1)(u^2 + 1)(u^2 + u + 1)(u^2 - u + 1)$$

as the modulo polynomial with a wraparound of two coefficients.

9.19 Examine the validity of the algorithms derived in Problems 9.17 and 9.18 over finite fields and integer rings.

9.20 Repeat Problems 9.17 and 9.18 using the technique described in Problem 9.15.

9.21 Derive algorithms for Problems 9.17 and 9.18 using multidimensional approach and compare their computational complexity with the algorithms derived in Problems 9.17 and 9.18.

9.22 Design a cyclic convolution algorithm for $n = 15$ based on Agarwal-Cooley algorithm.

9.23 Compare the computational complexity of the two algorithms, one derived in Problem 9.2 and the second derived in Problem 9.22.

9.24 We wish to compute a length $n = 210 = 2 \cdot 3 \cdot 5 \cdot 7$ cyclic convolution using Agarwal-Cooley algorithm. Describe the optimal way (in terms of computational complexity) to accomplish this. Assume that the computational complexities of the individual algorithms are $M(2) = 2$, $A(2) = 4$, $M(3) = 4$, $A(3) = 11$, $M(5) = 10$, $A(5) = 31$, $M(7) = 16$, $A(7) = 70$.

9.25 Examine the validity of the cyclic convolution algorithms derived in Problems 9.1, 9.2, 9.13 over various finite fields and integer rings.

9.26 Derive a cyclic convolution algorithm for $n = 15$ over $GF(2)$ and $Z(2^3)$. Compare it with the algorithm in Problem 9.2, if it is valid over $GF(2)$.

9.27 Derive a cyclic convolution algorithm for $n = 16$ over $GF(17)$ and $Z(17^4)$.

9.28 Based on Section 8.5 describe a split nesting version of the Agarwal-Cooley algorithm for $n = 15$.

9.29 Generalize the split nesting version of the Agarwal-Cooley algorithm for any arbitrary n.

9.30 Derive a length $n = 210$ cyclic convolution algorithm over Z and $GF(11)$. Compare their computational complexities.

Chapter 10

Two- and Higher-Dimensional Cyclic Convolution Algorithms

In this chapter, we describe fast computationally efficient algorithms for computing two- and higher-dimensional cyclic convolution of sequences. These computational tasks may arise naturally in multidimensional digital signal processing or may be used as a tool to compute one-dimensional cyclic convolution via the Agarwal-Cooley algorithm as described in Section 9.6. The basic idea remains the same as in previous chapters. We are focused on exposing and exploiting the mathematical properties of the computational task in order to reduce its computational burden. Number-theoretic results and polynomial algebra form the core of these properties. Once again, results pertaining to finite fields and finite integer rings are presented in separate sections.

This chapter deals primarily with two-dimensional cyclic convolution of length $n_1 \times n_2$ defined as

$$z_{k_1,k_2} = \sum_{i_2=0}^{n_2-1}\sum_{i_1=0}^{n_1-1} x_{(k_1-i_1)\bmod n_1,(k_2-i_2)\bmod n_2}\, y_{i_1,i_2}\,, \qquad (10.0.1)$$

where x_{i_1,i_2} and y_{i_1,i_2} are the two-dimensional input sequences and z_{i_1,i_2} is the output sequence, $i_1 = 0, 1, ..., n_1 - 1$; $i_2 = 0, 1, ..., n_2 - 1$. All sequences have length $n_1 \times n_2$.

10.1 Polynomial Formulation and an Algorithm

Define the generating polynomial for a sequence a_{i_1,i_2}, $i_1 = 0, 1, ..., n_1 - 1$; $i_2 = 0, 1, ..., n_2 - 1$ as

$$A(u,v) = \sum_{i_2=0}^{n_2-1}\sum_{i_1=0}^{n_1-1} a_{i_1,i_2} u^{i_1} v^{i_2} = \sum_{i_2=0}^{n_2-1} A_{i_2}(u) v^{i_2} = \sum_{i_1=0}^{n_1-1} A_{i_1}(v) u^{i_1} . \quad (10.1.1)$$

Here,

$$A_{i_2}(u) = \sum_{i_1=0}^{n_1-1} a_{i_1,i_2} u^{i_1}$$

and

$$A_{i_1}(v) = \sum_{i_2=0}^{n_2-1} a_{i_1,i_2} v^{i_2} .$$

The two-dimensional cyclic convolution can be expressed in polynomial form as

$$Z(u, v) = X(u, v)\ Y(u, v) \bmod \left(u^{n_1} - 1, v^{n_2} - 1\right), \quad (10.1.2)$$

$deg(X(u, v)) = deg(Y(u, v)) = deg(Z(u, v)) = n_1 - 1$ in u and $deg(X(u, v)) = deg(Y(u, v)) = deg(Z(u, v)) = n_2 - 1$ in v. Based on (10.1.1), we may also write

$$Z_{k_2}(u) = \sum_{i_2=0}^{n_2-1} X_{(k_2-i_2) \bmod n_2}(u) Y_{i_2}(u) \bmod \left(u^{n_1} - 1\right) \quad (10.1.3)$$

or

$$Z_{k_1}(v) = \sum_{i_1=0}^{n_1-1} X_{(k_1-i_1) \bmod n_1}(v) Y_{i_1}(v) \bmod \left(v^{n_2} - 1\right). \quad (10.1.4)$$

A straightforward method to compute the two-dimensional cyclic convolution in (10.1.3) or (10.1.4) would be to employ one-dimensional cyclic convolution algorithms of lengths n_1 and n_2 in the following manner. For (10.1.3), the one-dimensional algorithm for cyclic convolution of length n_2 is used. Each ADD is replaced by ADD of two length n_1 vectors, and every MULT is replaced by a cyclic convolution of length n_1 which is computed by using the cyclic convolution algorithm of length n_1. For (10.1.4), the one-dimensional algorithm for cyclic convolution of length n_1 is used. Each ADD is replaced by

ADD of two length n_2 vectors, and every MULT is replaced by a cyclic convolution of length n_2 which is computed using the cyclic convolution algorithm of length n_2.

The associated computational complexities of the two approaches are

$$M(n) = M(n_1) \cdot M(n_2), \tag{10.1.5}$$
$$A(n) = M(n_2) \cdot A(n_1) + A(n_2) \cdot n_1 \tag{10.1.6}$$

for (10.1.3) based algorithm, and

$$M(n) = M(n_1) \cdot M(n_2), \tag{10.1.7}$$
$$A(n) = M(n_1) \cdot A(n_2) + A(n_1) \cdot n_2 \tag{10.1.8}$$

for (10.1.4) based algorithm. It is seen that the multiplicative complexity remains the same while the additive complexity depends on the order of computation. A comparison of (10.1.6) and (10.1.8) reveals that (10.1.3) is to be chosen if

$$(M(n_2) - n_2)/A(n_2) < (M(n_1) - n_1)/A(n_1),$$

and (10.1.4) is to be chosen if

$$(M(n_2) - n_2)/A(n_2) > (M(n_1) - n_1)/A(n_1).$$

Let us develop the bilinear forms representation of the algorithm described above. If

$$\mathbf{C}_1\,[\mathbf{A}_1\,\underline{x}_1 \otimes \mathbf{B}_1\,\underline{y}_1], \tag{10.1.9}$$
$$\mathbf{C}_2\,[\mathbf{A}_2\,\underline{x}_2 \otimes \mathbf{B}_2\,\underline{y}_2], \tag{10.1.10}$$
$$\mathbf{C}\,[\mathbf{A}\,\underline{x} \otimes \mathbf{B}\,\underline{y}], \tag{10.1.11}$$

denote the bilinear forms for cyclic convolution algorithms of lengths n_1, n_2, and n, then for (10.1.3) based algorithm,

$$\begin{aligned} \mathbf{C} &= \mathbf{C}_2\ \kappa\ \mathbf{C}_1 \\ \mathbf{A} &= \mathbf{A}_2\ \kappa\ \mathbf{A}_1 \\ \mathbf{B} &= \mathbf{B}_2\ \kappa\ \mathbf{B}_1 \end{aligned} \tag{10.1.12}$$

and

$$\begin{aligned} \mathbf{C} &= \mathbf{C}_1\ \kappa\ \mathbf{C}_2 \\ \mathbf{A} &= \mathbf{A}_1\ \kappa\ \mathbf{A}_2 \\ \mathbf{B} &= \mathbf{B}_1\ \kappa\ \mathbf{B}_2. \end{aligned} \tag{10.1.13}$$

for (10.1.4) based algorithm.

10.2 Improvements and Related Algorithms

Several improvements can be made in the algorithm described in the previous section. One, if any one of n_1 or n_2 factor further into relatively coprime factors, then Agarwal-Cooley algorithm can be employed to convert the two-dimensional cyclic convolution into higher-dimensional cyclic convolutions. Say, $n_1 = A \cdot B$, $(A, B) = 1$, then the two-dimensional cyclic convolution of length $n_1 \times n_2$ can be expressed as three-dimensional cyclic convolution of length $A \times B \times n_2$. Now, the algorithm of the previous section can be used to get an algorithm for this three-dimensional cyclic convolution. The extension to higher dimensions is quite straightforward.

10.2.1 Split Nesting Algorithms

One can also decompose the computation further by using polynomial factorization of each of the modulo polynomials. For sake of simplicity and effectiveness, say we can write

$$u^{n_1} - 1 = A(u)B(u) \tag{10.2.1}$$

and

$$v^{n_2} - 1 = P(v)Q(v). \tag{10.2.2}$$

Based on this factorization, we can break the two-dimensional cyclic convolution into four polynomial products:

$$X(u, v) \cdot Y(u, v) \bmod [A(u), P(v)],$$
$$X(u, v) \cdot Y(u, v) \bmod [B(u), P(v)],$$
$$X(u, v) \cdot Y(u, v) \bmod [A(u), Q(v)],$$

and

$$X(u, v) \cdot Y(u, v) \bmod [B(u), Q(v)]. \tag{10.2.3}$$

The construction of the overall algorithm based on CRT-P is clear at this stage. This split nesting approach has been analyzed in depth in Section 8.5 where it is shown that it leads to a reduction in additive complexity as compared to the algorithm in Section 10.1, while the multiplicative complexities of the two algorithms remain the same. Since we are interested in reducing the multiplicative complexity to the extent possible, we seek the factorizations of $u^{n_1} - 1$ and $v^{n_2} - 1$ in terms of the irreducible cyclotomic polynomials.

10.3 Discrete Fourier Transform Based Algorithms

We now describe a special case of split nesting algorithm when all the modulo polynomials are degree one polynomials. Using the factorization over the number system of interest,

$$u^n - 1 = \prod_{i=0}^{n-1}\left(u - \omega^i\right), \tag{10.3.1}$$

where ω is the n-th root of unity ($\omega = \exp(-2\pi j/n)$ for complex numbers), we can break the two-dimensional cyclic convolution into the following computation:

Step 1. Compute the two-dimensional DFT (CRT-P reduction):

$$\begin{aligned} X_{k_1,k_2} &= X(u,v)\bmod\left[\left(u-\omega_1^{k_1}\right)\left(v-\omega_2^{k_2}\right)\right] \\ &= X\left(\omega_1^{k_1},\omega_2^{k_2}\right) = \sum_{i_2=0}^{n_2-1}\sum_{i_1=0}^{n_1-1} x_{i_1,i_2}\,\omega_1^{i_1k_1}\,\omega_2^{i_2k_2}, \end{aligned} \tag{10.3.2}$$

$$\begin{aligned} Y_{k_1,k_2} &= Y(u,v)\bmod\left[\left(u-\omega_1^{k_1}\right)\left(v-\omega_2^{k_2}\right)\right], \\ &= Y\left(\omega_1^{k_1},\omega_2^{k_2}\right) = \sum_{i_2=0}^{n_2-1}\sum_{i_1=0}^{n_1-1} y_{i_1,i_2}\,\omega_1^{i_1k_1}\,\omega_2^{i_2k_2}, \end{aligned} \tag{10.3.3}$$

$$k_1 = 0, 1, \ldots, n_1 - 1,\ k_2 = 0, 1, \ldots, n_2 - 1,$$

where ω_1 and ω_2 are the n_1-th and n_2-th roots of unity. For complex numbers, they are given by

$$\omega_1 = \exp(-2\pi j/n_1),\ \omega_2 = \exp(-2\pi j/n_2).$$

Step 2. Compute the point-by-point product (small degree polynomial MULT in CRT-P):

$$Z_{k_1,k_2} = X_{k_1,k_2}\cdot Y_{k_1,k_2},\ k_1 = 0, 1, \ldots, n_1 - 1,\ k_2 = 0, 1, \ldots, n_2 - 1. \tag{10.3.4}$$

Note that

$$Z_{k_1,k_2} = Z(u,v)\bmod\left[\left(u-\omega_1^{k_1}\right)\left(v-\omega_2^{k_2}\right)\right]$$

$$= Z\left(\omega_1^{k_1}, \omega_2^{k_2}\right) = \sum_{i_2=0}^{n_2-1}\sum_{i_1=0}^{n_1-1} z_{i_1,i_2}\,\omega_1^{i_1k_1}\,\omega_2^{i_2k_2}$$

$$= \left\{X(u,v)\cdot Y(u,v)\right\} \bmod\left[\left(u-\omega_1^{k_1}\right),\left(v-\omega_2^{k_2}\right)\right]$$

$$= X(u,v) \bmod\left[\left(u-\omega_1^{k_1}\right),\left(v-\omega_2^{k_2}\right)\right]\cdot Y(u,v) \bmod\left[\left(u-\omega_1^{k_1}\right),\left(v-\omega_2^{k_2}\right)\right]$$

$$= \boldsymbol{X}_{k_1,k_2}\cdot \boldsymbol{Y}_{k_1,k_2}\,.$$

Step 3. Compute the two-dimensional IDFT (CRT-P reconstruction):

$$x_{i_1,i_2} = \left(1/\left(n_2 n_1\right)\right)\sum_{k_2=0}^{n_2-1}\sum_{k_1=0}^{n_1-1} \boldsymbol{Z}_{k_1,k_2}\,\omega_1^{-i_1k_1}\,\omega_2^{-i_2k_2}\,. \tag{10.3.5}$$

Observations are in order with regard to the above approach.

A. Computational burden has been shifted from Step 2 to Step 1 and Step 3.

B. It is valid over all number systems over which the appropriate quantities are defined.

C. For rational, real, and complex number sequences, this approach will involve complex arithmetic.

D. Computational procedures for computing two- and higher-dimensional DFT will be studied in Chapter 14.

E. The computation of DFT and IDFT is very similar in form and an algorithm for one will lead to an algorithm for the other.

We now turn to algorithms whose structure is extremely sensitive to the number system under consideration. In that regard, we will study extensively the rational and complex rational number systems, finite fields, and finite integer rings. The remainder of this chapter will be devoted to such algorithms.

10.4 Algorithms Based on Extension Fields

Let us return to the split nesting formulation of the algorithm for two-dimensional cyclic convolution as stated in (10.2.1) to (10.2.3). It is based on the factorizations $u^{n_1} - 1 = A(u)B(u)$ and $v^{n_2} - 1 = P(v)Q(v)$. However, instead the expressing the computation

$$X(u, v)\,Y(u, v) \bmod \left(u^{n_1} - 1, v^{n_2} - 1\right),$$

as a direct sum of four polynomial products in (10.2.3), we express it as a direct sum of

$$X(u, v)\,Y(u, v) \bmod \left(A(u), v^{n_2} - 1\right)$$

and

$$X(u, v)\,Y(u, v) \bmod \left(B(u), v^{n_2} - 1\right). \qquad (10.4.1)$$

The **crucial observation** which leads to new forms of algorithms that are superior computationally to the split nesting algorithms is as follows: "If $C(u)$ is an irreducible polynomial over the number system N under consideration, then mod $C(u)$ defines an extension field and, therefore, we may seek factorization of $v^{n_2} - 1$ in the extension field rather than N itself." Let us illustrate this statement with examples.

Example 10.1 Let $n_1 = n_2 = 4$. Write

$$u^4 - 1 = (u - 1)\,(u + 1)\,(u^2 + 1)$$

which implies that $X(u, v)\,Y(u, v) \bmod (u^4 - 1, v^4 - 1)$ can be written as a direct sum of

$$X(u, v)\,Y(u, v) \bmod (u - 1, v^4 - 1) = X(1, v)\,Y(1, v) \bmod v^4 - 1,$$
$$X(u, v)\,Y(u, v) \bmod (u + 1, v^4 - 1) = X(-1, v)\,Y(-1, v) \bmod v^4 - 1,$$

and

$$X(u, v)\,Y(u, v) \bmod (u^2 + 1, v^4 - 1).$$

The first two terms correspond to one-dimensional cyclic convolution of length 4 and are computed as such. The split nesting algorithm will proceed with expressing $X(u, v)\,Y(u, v) \bmod (u^2 + 1, v^4 - 1)$ as a direct sum of

$$X(u, v)\,Y(u, v) \bmod (u^2 + 1, v - 1) = X(u, 1)\,Y(u, 1) \bmod u^2 + 1,$$
$$X(u, v)\,Y(u, v) \bmod (u^2 + 1, v + 1) = X(u, -1)\,Y(u, -1) \bmod u^2 + 1,$$

and

$$X(u, v)\,Y(u, v) \bmod (u^2 + 1, v^2 + 1),$$

and then compute the last term by using the one-dimensional algorithm for computing $A(v)\,B(v) \bmod v^2 + 1$, where each scalar ADD is replaced by ADD of vectors of length 2 and each scalar MULT is replaced by polynomial product of the kind $C(u)\, D(u) \bmod u^2 + 1$. Such a computation of the last term will require 9 MULTs.

For the approach being described here, we now seek factorization of $v^4 - 1$ in the cyclotomic field CF(4) (readers may refer to Chapter 5 for such factorizations). This is given by

$$v^4 - 1 = (v - 1)\,(v + 1)\,(v - u)\,(v + u).$$

Now, we proceed with expressing $X(u, v)\, Y(u, v) \bmod (u^2 + 1, v^4 - 1)$ as a direct sum of

$$X(u, v)\, Y(u, v) \bmod (u^2 + 1, v - 1) = X(u, 1)\, Y(u, 1) \bmod u^2 + 1,$$
$$X(u, v)\, Y(u, v) \bmod (u^2 + 1, v + 1) = X(u, -1)\, Y(u, -1) \bmod u^2 + 1,$$
$$X(u, v)\, Y(u, v) \bmod (u^2 + 1, v - u) = X(u, u)\, Y(u, u) \bmod u^2 + 1,$$

and

$$X(u, v)\, Y(u, v) \bmod (u^2 + 1, v + u) = X(u, -u)\, Y(u, -u) \bmod u^2 + 1.$$

Once again, the one-dimensional algorithm for computing $A(u)\, B(u) \bmod u^2 + 1$ is used. In essence, we have expressed $X(u, v)\, Y(u, v) \bmod (u^2 + 1, v^2 + 1)$ as a direct sum of $X(u, u)\, Y(u, u) \bmod u^2 + 1$ and $X(u, -u)\, Y(u, -u) \bmod u^2 + 1$. Such a computation of the last term in (10.4.3) now requires 6 MULTs. It is of **fundamental importance** to note here that u and v are indeterminate quantities and computations of the type $X(u, v) \bmod (v - u) = X(v, u)$ do not require any MULT.

Example 10.2 Let $n_1 = 7$ and $n_2 = 7$. In this case, we express the computation $X(u, v)\, Y(u, v) \bmod (u^7 - 1, v^7 - 1)$ as a direct sum of

$$X(u, v)\, Y(u, v) \bmod (u^7 - 1, v - 1) = X(u, 1)\, Y(u, 1) \bmod u^7 - 1$$

and

$$X(u, v)\, Y(u, v) \bmod (u^7 - 1, v^7 + v^6 + v^5 + v^4 + v^3 + v^2 + v + 1).$$

We now compute the second term by using the following factorization over the cyclotomic field CF(7),

$$\left(u^7 - 1\right) = \prod_{i=0}^{6} \left(u - v^i\right),$$

which implies that

$$X(u, v)\, Y(u, v) \bmod (u^7 - 1, v^7 + v^6 + v^5 + v^4 + v^3 + v^2 + v + 1)$$

can be written as a direct sum of

$$X(v^i, v)\, Y(v^i, v) \bmod v^4 + v^3 + v^2 + v + 1, \quad i = 0, 1, \ldots, 6.$$

Once again, no MULT is required in computing $X(v^i, v)$ from $X(u, v)$, $i = 0, 1, \ldots, 6$. The cyclic convolution algorithm of length 7 requires 16 MULTs. Therefore, the two-dimensional cyclic convolution of length 7 × 7 requires 16 + 7 × 15 = 121 MULTs.

Example 10.3 Let $n_1 = 6$ and $n_2 = 3$. In this case, we express the computation $X(u, v)\, Y(u, v) \bmod (u^6 - 1, v^3 - 1)$ as a direct sum of

$$X(u, v)\, Y(u, v) \bmod (u^6 - 1, v - 1) = X(u, 1)\, Y(u, 1) \bmod u^6 - 1$$

and

$$X(u, v)\, Y(u, v) \bmod (u^6 - 1, v^2 + v + 1).$$

The first term is a one-dimensional cyclic convolution of length 6. The second term employs the following factorization over the cyclotomic field $CF(3)$:

$$u^6 - 1 = \prod_{i=0}^{5} \left(u - (-v)^i\right).$$

It implies that $X(u, v)\, Y(u, v) \bmod (u^6 - 1, v^2 + v + 1)$ can be written as a direct sum of

$$X((-v)^i, v)\, Y((-v)^i, v) \bmod v^2 + v + 1, \quad i = 0, 1, \ldots, 5.$$

The cyclic convolution algorithm of length 6 requires 8 MULTs. Therefore, the two-dimensional cyclic convolution of length 6 × 3 requires 8 + 6 × 3 = 26 MULTs.

Example 10.4 Let $n_1 = 7$ and $n_2 = 5$. In this case, we express the computation $X(u, v)\, Y(u, v) \bmod (u^7 - 1, v^5 - 1)$ as a direct sum of

$$X(u, v)\, Y(u, v) \bmod (u^7 - 1, v - 1) = X(u, 1)\, Y(u, 1) \bmod u^7 - 1,$$

$$X(u, v)\, Y(u, v) \bmod (u - 1, v^4 + v^3 + v^2 + v + 1)$$
$$= X(1, v)\, Y(1, v) \bmod v^4 + v^3 + v^2 + v + 1),$$

and

$$X(u, v)\, Y(u, v) \bmod (u^6 + u^5 + u^4 + u^3 + u^2 + 1, v^4 + v^3 + v^2 + v + 1).$$

Of the three terms, the last one is the most intensive computationally. We compute it by first computing

$X(u, v)\, Y(u, v) \bmod (u(u^{10} - 1), v^4 + v^3 + v^2 + v + 1),$

and then reducing the product mod $u^6 + u^5 + u^4 + u^3 + u^2 + 1$. This is computed by using the following factorization over the cyclotomic field $CF(5)$:

$$u\left(u^{10} - 1\right) = u\prod_{i=0}^{9}\left(u - (-v)^i\right),$$

which implies that

$X(u, v)\, Y(u, v) \bmod (u(u^{10} - 1), v^4 + v^3 + v^2 + v + 1)$

can be written as a direct sum of

$X(0, v)\, Y(0, v) \bmod v^4 + v^3 + v^2 + v + 1$

and

$X((-v)^i, v)\, Y((-v)^i, v) \bmod v^4 + v^3 + v^2 + v + 1, \; i = 0, 1, ..., 9.$

The cyclic convolution algorithm of length 7 requires 16 MULTs. Therefore, the two-dimensional cyclic convolution of length 7×5 requires $16 + 12 \times 9 = 124$ MULTs.

Example 10.5 Let $n_1 = n_2 = 12$. Based on

$v^{12} - 1 = C_{12}(v) \cdot C_6(v) \cdot C_4(v) \cdot C_3(v) \cdot C_2(v) \cdot C_1(v),$

we express the computation $X(u, v)\, Y(u, v) \bmod (u^{12} - 1, v^{12} - 1)$ as a direct sum of

I. $X(u, v)\, Y(u, v) \bmod (u^{12} - 1, v - 1) = X(u, 1)\, Y(u, 1) \bmod u^{12} - 1,$
II. $X(u, v)\, Y(u, v) \bmod (u^{12} - 1, v + 1) = X(u, -1)\, Y(u, -1) \bmod u^{12} - 1,$
III. $X(u, v)\, Y(u, v) \bmod (u^{12} - 1, v^2 + v + 1),$
IV. $X(u, v)\, Y(u, v) \bmod (u^{12} - 1, v^2 + 1),$
V. $X(u, v)\, Y(u, v) \bmod (u^{12} - 1, v^2 - v + 1),$

and

VI. $X(u, v)\, Y(u, v) \bmod (u^{12} - 1, v^4 - v^2 + 1).$

The computations in I and II correspond to one-dimensional cyclic convolution of length 12 and are performed in 20 MULTs each. To perform III, we seek factorization of $u^{12} - 1$ in $CF(3)$. Such a factorization is given by

$u^{12} - 1 = [(u^2 + v)(u^2 + v^2)] \cdot [(u + v)(u + v^2)] \cdot [u^2 + 1]$

$$\cdot [(u - v)(u - v^2)] \cdot [u + 1] \cdot [u - 1].$$

Readers will recognize that $C_{12}(u)$, $C_6(u)$, and $C_3(u)$ factorize further in $CF(3)$. Similarly, to perform IV, V, and VI, we seek factorization of $u^{12} - 1$ in $CF(4)$, $CF(6)$, and $CF(12)$, respectively. They are given by

$$u^{12} - 1 = [(u^2 + vu - 1)(u^2 - vu - 1)] \cdot [(u^2 - u + 1] \cdot [(u - v)(u + v)] \cdot [u^2 + u + 1] \cdot [u + 1] \cdot [u - 1],$$

$$u^{12} - 1 = [(u^2 - v)(u^2 + v^2)] \cdot [(u - v)(u + v^2)] \cdot [u^2 + 1] \cdot [(u + v)(u - v^2)] \cdot [u + 1] \cdot [u - 1],$$

$$u^{12} - 1 = \prod_{i=0}^{11} \left(u - v^i\right).$$

As a result, the computation in I, II, III, IV, V, and VI are a direct sum of 6, 6, 9, 8, 9, and 12 terms, respectively. If a 2×2 convolution is computed in 9 MULTs, then the overall multiplicative complexity is given by

$$20 + 20 + 45 + 48 + 45 + 108 = 286 \text{ MULTs}.$$

Example 10.6. Let $n_1 = n_2 = 9$. Based on

$$v^9 - 1 = C_9(v) \cdot C_3(v) \cdot C_1(v),$$

we express the computation $X(u, v)\, Y(u, v) \bmod (u^9 - 1, v^9 - 1)$ as a direct sum of

I. $X(u, v)\, Y(u, v) \bmod (u^9 - 1, v - 1) = X(u, 1)\, Y(u, 1) \bmod u^9 - 1$,
II. $X(u, v)\, Y(u, v) \bmod (u^9 - 1, v^2 + v + 1)$,

and

III. $X(u, v)\, Y(u, v) \bmod (u^9 - 1, v^6 + v^3 + 1)$.

The computation in I corresponds to one-dimensional cyclic convolution of length 9 and is performed in 19 MULTs. To perform II, we seek factorization of $u^9 - 1$ in $CF(3)$. Such a factorization is given by

$$u^9 - 1 = \left[\left(u^3 - v\right)\left(u^3 - v^2\right)\right] \cdot \left[\left(u - v\right)\left(u - v^2\right)\right] \cdot \left[u - 1\right].$$

Readers will recognize that $C_9(u)$ and $C_3(u)$ factorize further in $CF(3)$. Similarly, to perform III, we seek factorization of $u^9 - 1$ in $CF(9)$. It is given by

$$u^9 - 1 = \prod_{i=0}^{8}\left(u - v^i\right).$$

Computations mod $(v^6 + v^3 + 1)$, $(u^3 - v, v^2 + v + 1))$ and $(u^3 - v^2, v^2 + v + 1)$ can be performed by first formulating them as acyclic convolution of length 5 with coefficients defined mod $(v^2 + v + 1)$. Thus, each of them requires $5 \times 3 = 15$ MULTs. The computation in I, II, and III are a direct sum of 3, 5, and 9 terms, respectively. The overall multiplicative complexity is given by

$$19 + 39 + 135 = 193 \text{ MULTs}.$$

Example 10.7 There is an alternative formulation and perhaps a different organization of the computations involved in the previous example. Once again, let $n_1 = n_2 = 9$. Based on

$$v^9 - 1 = C_9(v) \cdot (v^3 - 1),$$

we express the computation $X(u, v)\ Y(u, v) \bmod (u^9 - 1, v^9 - 1)$ as a direct sum of

I. $X(u, v)\ Y(u, v) \bmod (u^9 - 1, v^3 - 1)$

and

II. $X(u, v)\ Y(u, v) \bmod (u^9 - 1, v^6 + v^3 + 1)$.

The computation in I corresponds to a two-dimensional cyclic convolution of length 9×3. To perform II, we seek factorization of $u^9 - 1$ in $CF(9)$. Such a factorization is given by

$$u^9 - 1 = \prod_{i=0}^{8}\left(u - v^i\right).$$

To perform I, we now express it as

$$X(u, v)\ Y(u, v) \bmod (v^3 - 1, u^9 - 1).$$

This is now written as a direct sum of

III. $X(u, v)\ Y(u, v) \bmod (v^3 - 1, u^3 - 1)$

and

IV. $X(u, v)\, Y(u, v) \bmod (v^3 - 1, u^6 + u^3 + 1)$.

The computation in IV can be written as a direct sum of 3 terms defined mod $u^6 + u^3 + 1$. These 3 terms are based on the factorization of $v^3 - 1$ in $CF(9)$ given by

$$v^3 - 1 = \prod_{i=0}^{2} \left[v - \left(u^3\right)^i \right].$$

Overall, we have expressed the 9 × 9 cyclic convolution as 9 + 3 = 12 polynomial products mod $C_9(u) = u^6 + u^3 + 1$ and a 3 × 3 cyclic convolution which can be computed in a similar manner. Computation mod $u^6 + u^3 + 1$ is performed by first formulating it as acyclic convolution of length 5 with coefficients defined mod $(v^2 + v + 1)$. Thus, each of them requires 5 × 3 = 15 MULTs. The overall multiplicative complexity is given by

12 × 15 + 13 = 193 MULTs.

Example 10.8 Let $n_1 = n_2 = 15$. Based on

$$v^{15} - 1 = C_{15}(v) \cdot C_5(v) \cdot C_3(v) \cdot C_1(v),$$

we express the computation $X(u, v)\, Y(u, v) \bmod (u^{15} - 1, v^{15} - 1)$ as a direct sum of

I. $X(u, v)\, Y(u, v) \bmod (u^{15} - 1, v - 1) = X(u, 1)\, Y(u, 1) \bmod u^{15} - 1$,

II. $X(u, v)\, Y(u, v) \bmod (u^{15} - 1, v^2 + v + 1)$,

III. $X(u, v)\, Y(u, v) \bmod (u^{15} - 1, v^5 + v^4 + v^3 + v^2 + 1)$,

and

IV. $X(u, v)\, Y(u, v) \bmod (u^{15} - 1, v^8 - v^7 + v^5 - v^4 + v^3 - v + 1)$.

The computation in I corresponds to a one-dimensional cyclic convolution of length 15 and is performed in 40 MULTs. To perform II, III, and IV, we seek factorization of $u^{15} - 1$ in $CF(3)$, $CF(5)$, and $CF(15)$, respectively. Such factorizations are given by

$$u^{15} - 1 = \left[\left(u^4 + vu^3 + v^2u^2 + u + v\right)\left(u^4 + v^2u^3 + vu^2 + u + v^2\right)\right]$$

$$\cdot \left[u^4 + u^3 + u^2 + u + 1\right] \cdot \left[(u - v)\left(u - v^2\right)\right] \cdot [u - 1],$$

$$u^{15}-1=\left[\left(u^2+vu+v^2\right)\left(u^2+v^2u+v^4\right)\left(u^2+v^3u+v\right)\left(u^2+v^4u+v^3\right)\right]$$

$$\cdot\left[\prod_{i=1}^{4}\left(u-v^i\right)\right]\cdot\left[u^2+u+1\right]\cdot[u-1],$$

$$u^{15}-1=\prod_{i=0}^{14}\left(u-v^i\right).$$

As a result, the computations in I, II, III, and IV are a direct sum of 4, 6, 10, and 15 terms, respectively. If length 3, 7, 15, 4 × 2 convolutions are computed in 3, 9, 27, and 27 MULTs, respectively, then the overall multiplicative complexity is given by

$$40 + 90 + 180 + 405 = 715 \text{ MULTs}.$$

Example 10.9 Let $n_1 = n_2 = 12$ and the field of computation be complex rational numbers, CZ. Based on

$$v^{12}-1 = C_{1,12}(v)\cdot C_{3,12}(v)\cdot C_6(v)\cdot C_{1,4}(v)\cdot C_{3,4}(v)\cdot C_3(v)\cdot C_2(v)\cdot C_1(v),$$

the computation $X(u, v)\,Y(u, v) \bmod (u^{12}-1,\, v^{12}-1)$ is expressed as a direct sum of

I. $X(u, v)\,Y(u, v) \bmod (u^{12}-1,\, v-1) = X(u, 1)\,Y(u, 1) \bmod u^{12}-1,$
II. $X(u, v)\,Y(u, v) \bmod (u^{12}-1,\, v+1) = X(u, -1)\,Y(u, -1) \bmod u^{12}-1,$
III. $X(u, v)\,Y(u, v) \bmod (u^{12}-1,\, v^2+v+1),$
IV. $X(u, v)\,Y(u, v) \bmod (u^{12}-1,\, v+j) = X(u, -j)\,Y(u, -j) \bmod u^{12}-1,$
V. $X(u, v)\,Y(u, v) \bmod (u^{12}-1,\, v-j) = X(u, j)\,Y(u, j) \bmod u^{12}-1,$
VI. $X(u, v)\,Y(u, v) \bmod (u^{12}-1,\, v^2-v+1),$
VII. $X(u, v)\,Y(u, v) \bmod (u^{12}-1,\, v^2-jv-1),$

and

VIII. $X(u, v)\,Y(u, v) \bmod (u^{12}-1,\, v^2+jv-1).$

The computations in I, II, IV, and V correspond to a one-dimensional cyclic convolution of length 12 and each is performed in 16 complex MULTs (CMULTs). To perform III, we seek factorization of $u^{12}-1$ in $CCF(3)$. Such a factorization is given by

$$u^{12}-1=\prod_{i=0}^{11}\left[\left(u-(jv)^{i}\right)\right].$$

Similarly, to perform VI, VII, and VIII, we seek factorization of $u^{12}-1$ in $CCF(6)$, $CCF(12)$, and $CCF(12)$, respectively. These factorizations are the same and are given by

$$u^{12}-1=\prod_{i=0}^{11}\left(u-v^{i}\right).$$

As a result, the computations in I, II, IV, and V are a direct sum of 8 terms and those in III, VI, VII, and VIII are a direct sum of 12 terms. If a length 2 convolution is computed in 3 CMULTs, then the overall multiplicative complexity is given by

$$16+16+36+16+16+36+36+36=208 \text{ CMULTs.}$$

In the following, a general methodology is presented for obtaining fast algorithms for a two-dimensional cyclic convolution based on factorization in extension fields. It is shown that in many cases of interest, it can equivalently be reduced to one-dimensional polynomial products that can then be computed using suitable algorithms.

Let us return to the computation

$$X(u,v)\,Y(u,v) \bmod \left(u^{n_1}-1, v^{n_2}-1\right).$$

Consider the case when we are interested in algorithms over the rational number system Z. There are two approaches (basically the same formulation but different routes) for proceeding from this point onwards.

Approach 1. Based on the CRT-P and cyclotomic factorization of $v^{n_2}-1$ as

$$v^{n_2}-1=\prod_{d_2|n_2}C_{d_2}(v), \tag{10.4.2}$$

we can express the two-dimensional cyclic convolution as a direct sum of

$$X(u,v)\,Y(u,v) \bmod \left(u^{n_1}-1, C_{d_2}(v)\right), \text{ for all } d_2 \mid n_2. \tag{10.4.3}$$

Now, we seek factorization of $u^{n_1}-1$ in cyclotomic fields $CF(d_2)$, for all $d_2 \mid n_2$. Such factorizations are described in Chapter 5. Let such factorizations be given by

$$u^{n_1}-1=\prod_{d_1|n_1} C_{d_1}(u)=\prod_{d_1|n_1}\left[\prod_{j} C_{j,d_1}(u)\right]\text{mod}C_{d_2}(v). \tag{10.4.4}$$

The coefficients of $C_{j,d_1}(u)$ are elements in $CF(d_2)$ with v as the indeterminate quantity. Now each of the terms in (10.4.3) can be further written as a direct sum of

$$X(u, v)\ Y(u, v) \text{ mod } \left(C_{j,d_1}(u), C_{d_2}(v)\right), \text{ for all } j,\ d_1, \text{ and } d_2. \tag{10.4.5}$$

Now, polynomial MULT algorithms in $CF(d_2)$ are used to compute

$$Z(u, v) \equiv X(u, v)\ Y(u, v) \text{ mod } \left(C_{j,d_1}(u), C_{d_2}(v)\right), \text{ for all } j,\ d_1, \text{ and } d_2, \tag{10.4.6}$$

and then CRT-P reconstruction in $CF(d_2)$ is used to get

$$Z(u, v) \equiv X(u, v)\ Y(u, v) \text{ mod } \left(u^{n_1}-1, C_{d_2}(v)\right), \text{ for all } d_2 \mid n_2. \tag{10.4.7}$$

Finally, CRT-P reconstruction over $\mathbf{Z}$ is used to get

$$Z(u, v) \equiv X(u, v)\ Y(u, v) \text{ mod } \left(u^{n_1}-1, v^{n_2}-1\right). \tag{10.4.8}$$

All the steps can be performed with only ADDs except the polynomial MULT in (10.4.5). All the examples presented in this section, except Example 10.7, are based on this approach. In Example 10.9, at first factorization is obtained over $CF(4)$, the field of complex rational numbers, and then over the appropriate CCF.

Approach 2. Based on CRT-P and the cyclotomic factorization of $v^{n_2}-1$ as

$$v^{n_2}-1=\prod_{d_2|n_2} C_{d_2}(v)=C_{n_2}(v)\prod_{\substack{d_2|n_2\\ d_2<n_2}} C_{d_2}(v)=C_{n_2}(v)A(v), \tag{10.4.9}$$

we can express the two-dimensional cyclic convolution as a direct sum of

$$\text{I. } X(u,v)\,Y(u,v) \bmod \left(u^{n_1}-1, C_{n_2}(v)\right) \tag{10.4.10}$$

and

$$\text{II. } X(u,v)\,Y(u,v) \bmod \left(u^{n_1}-1, A(v)\right). \tag{10.4.11}$$

In many cases, $A(v)$ has a desirable form such as $v^a - 1$ or $v^b + 1$, where b is a power of 2. Now, we seek factorization of $u^{n_1}-1$ in cyclotomic field $CF(n_2)$. Let this factorization be given by

$$u^{n_1}-1=\prod_{d_1|n_1} C_{d_1}(u)=\prod_{d_1|n_1}\left[\prod_{j} C_{j,d_1}(u)\right] \bmod C_{n_2}(v). \tag{10.4.12}$$

The coefficients of $C_{j,d_1}(u)$ are elements in $CF(n_2)$ with v as the indeterminate quantity. Now each of the terms in (10.4.10) can be further written as a direct sum of

$$X(u,v)\,Y(u,v) \bmod \left(C_{j,d_1}(u), C_{n_2}(v)\right), \text{ for all } j, \text{ and } d_1. \tag{10.4.13}$$

Polynomial MULT algorithms in $CF(n_2)$ are used to compute

$$Z(u,v) \equiv X(u,v)\,Y(u,v) \bmod \left(C_{j,d_1}(u), C_{n_2}(v)\right), \text{ for all } j, \text{ and } d_1, \tag{10.4.14}$$

and then CRT-P reconstruction in $CF(n_2)$ is used to get

$$Z(u,v) \equiv X(u,v)\,Y(u,v) \bmod \left(u^{n_1}-1, C_{n_2}(v)\right). \tag{10.4.15}$$

The computation in (10.4.11) is expressed as

$$X(u,v)\,Y(u,v) \bmod \left(A(v), u^{n_1}-1\right). \tag{10.4.16}$$

Based on CRT-P and cyclotomic factorization of $u^{n_1}-1$ as

$$u^{n_1}-1=\prod_{d_1|n_1} C_{d_1}(u)=C_{n_1}(u)\prod_{\substack{d_1|n_1\\ d_1<n_1}} C_{d_1}(u)=C_{n_1}(u)B(u), \tag{10.4.17}$$

we can express the two-dimensional computation in (10.4.16) as a direct sum of

$$\text{I. } X(u, v)\, Y(u, v) \bmod \left(A(v), C_{n_1}(u)\right) \tag{10.4.18}$$

and

$$\text{II. } X(u, v)\, Y(u, v) \bmod (A(v), B(u)). \tag{10.4.19}$$

In order to simplify (10.4.18), we now seek factorization of $A(v)$ in $CF(n_1)$. Let

$$A(v) = \prod_i A_i(v) \bmod C_{n_1}(u). \tag{10.4.20}$$

The coefficients of $A_i(v)$ are elements in $CF(n_1)$ and hence polynomials in the indeterminate u. Based on CRT-P in $CF(n_1)$ and the factorization in (10.4.20), we express the computation in (10.4.18) as a direct sum of

$$X(u, v)\, Y(u, v) \bmod \left(A_i(v), C_{n_1}(u)\right), \text{ for all } i. \tag{10.4.21}$$

In summary, the two-dimensional cyclic convolution is expressed as a direct sum of

A. $Z(u, v) \equiv X(u, v)\, Y(u, v) \bmod \left(C_{j,d_1}(u), C_{n_2}(v)\right)$, for all j, and d_1; (10.4.22)

B. $X(u, v)\, Y(u, v) \bmod \left(A_i(v), C_{n_1}(u)\right)$, for all i; (10.4.23)

and

C. $X(u, v)\, Y(u, v) \bmod (A(v), B(u))$. (10.4.24)

In many cases of interest, the polynomial $A(v)$ has the form $v^a - 1$, a a power of a prime p; or $v^b + 1$, b a power of 2. Similar statement holds for $B(u)$ as well. In such cases, the computation in (10.4.24) has the same form as the original two-dimensional cyclic convolution with the exception that the length is reduced. Examples 10.1, 10.2, 10.3, and 10.7 are based on this approach.

Both approaches are similar in their organization of computation and have certain attractive features. One, all two-dimensional computations are converted to equivalent one-dimensional computations which can be performed by using one basic algorithm in an iterative manner. Two, the resulting smaller size computational task has a form similar to the original computation. Three, in many instances, the factorizations used in (10.4.22) and (10.4.23) correspond to DFT in the appropriate cyclotomic field. Such factorizations and the resulting algorithms are termed as **polynomial transform** in the DSP literature. In such instances, the computations can be performed using FFT algorithms over the field of interest. In this regard, one may seek original factorizations over the field of complex numbers, CF(4), and then obtain further factorizations over suitable *CCF*. Four, the methodology applied to derive these algorithms is quite general and remains valid over finite fields and integer rings as well.

The message and the principle behind these algorithms are very clear. Computational efficiency is realized by expressing the large size computation as a direct sum of as many small size computations as possible. We now turn to certain special cases and establish the specific forms for factorizations and the resulting computations.

Case 1. $n_1 = p^a$, $n_2 = p^b$, $a < b$; p an odd prime. In this case, we use the factorization

$$\begin{aligned} v^{p^b} - 1 &= (v-1)\prod_{j=1}^{b}\left[\sum_{k=0}^{p-1} v^{kp^{b-j}}\right] = \left(v^{p^a} - 1\right)\prod_{j=1}^{b-a}\left[\sum_{k=0}^{p-1} v^{kp^{b-j}}\right] \\ &= \left(v^{p^a} - 1\right)\prod_{j=1}^{b-a} C_{p^{b-j+1}}(v). \end{aligned} \tag{10.4.25}$$

Using (10.4.25), the two-dimensional cyclic convolution can be expressed as a direct sum of

I. $X(u, v)\ Y(u, v) \bmod \left(u^{p^a} - 1, v^{p^a} - 1\right)$, which is a $p^a \times p^a$ two-dimensional cyclic convolution, and

II. $X(u, v)\ Y(u, v) \bmod \left(u^{p^a} - 1, C_{p^{b-j+1}}(v)\right), j = 1, 2, ..., b - a.$ (10.4.26)

The j-th computation in (10.4.26) can be converted to p^a one-dimensional polynomial products mod $C_{p^{b-j+1}}(v)$ by using the factorization

$$u^{p^a} - 1 = \prod_{i=0}^{p^a-1}\left[u - \left(v^{p^{b-a+1-j}}\right)^i\right] \bmod C_{p^{b-j+1}}(v), j = 1, 2, ..., b-a. \tag{10.4.27}$$

Each of these is a DFT in CF(p^{b-j+1}).

Case 2. $n_1 = 2^a$, $n_2 = 2^b$, $a < b$. In this case, we use the factorization

$$v^{2^b}-1=(v-1)\prod_{j=1}^{b}\left[1+v^{2^{b-j}}\right]=\left(v^{2^a}-1\right)\prod_{j=1}^{b-a}\left[1+v^{2^{b-j}}\right]$$

$$=\left(v^{2^a}-1\right)\prod_{j=1}^{b-a}C_{2^{b-j+1}}(v). \qquad (10.4.28)$$

Using (10.4.28), the two-dimensional cyclic convolution can be expressed as a direct sum of

I. $X(u, v)\ Y(u, v) \bmod \left(u^{2^a}-1, v^{2^a}-1\right)$, which is a $2^a \times 2^a$ two-dimensional cyclic convolution, and

II. $X(u, v)\ Y(u, v) \bmod \left(u^{2^a}-1, C_{2^{b-j+1}}(v)\right)$, $j = 1, 2, \ldots, b-a.$ (10.4.29)

The j-th computation in (10.4.29) can be converted to 2^a one-dimensional polynomial products mod $C_{2^{b-j+1}}(v)$ by using the factorization

$$u^{2^a}-1=\prod_{i=0}^{2^a-1}\left[u-\left(v^{2^{b-a+1-j}}\right)^i\right] \bmod C_{2^{b-j+1}}(v),\ j=1,2,\ldots,b-a. \qquad (10.4.30)$$

Each of these is a DFT in $CF(2^{b-j+1})$. Interestingly, because of the special form of $C_{2^{b-j+1}}(v)$, the j-th computation in (10.4.30) is a circular correlation of length 2^{b-j}, $j = 1, 2, \ldots, b-a$. We will study such a computational task in one of the problems at the end of this chapter.

Case 3. $n_1 = n_2 = p^a$; p an odd prime. In this case, we can use the factorization

$$v^{p^a}-1=(v-1)\prod_{j=1}^{a}\left[\sum_{k=0}^{p-1}v^{kp^{a-j}}\right]=(v-1)\prod_{j=1}^{a}C_{p^{a-j+1}}(v)$$

$$=C_{p^a}(v)\left\{v^{p^{a-1}}-1\right\} \qquad (10.4.31)$$

in two ways as per the two approaches described above.

Approach 1. Using (10.4.31), the two-dimensional cyclic convolution can be expressed as a direct sum of

I. $X(u, v)\, Y(u, v) \bmod \left(u^{p^a} - 1, v - 1\right)$, which is a length p^a one-dimensional cyclic convolution, and

II. $X(u, v)\, Y(u, v) \bmod \left(u^{p^a} - 1, C_{p^{a-j+1}}(v)\right)$, $j = 1, 2, \ldots, a.$ (10.4.32)

The j-th computation in (10.4.32) can be written as a direct sum of p^{a-j+1} polynomial products by using the factorization

$$u^{p^a} - 1 = \prod_{i=0}^{p^{a-j+1}-1} \left[u^{p^{j-1}} - v^i\right] \bmod C_{p^{a-j+1}}(v),\ j = 1, 2, \ldots, a \qquad (10.4.33)$$

in $CF(p^{a-j+1})$.

Approach 2. In this case, based on (10.4.31) and the analysis previously presented, we write the two-dimensional cyclic convolution as a direct sum of the following three computations:

A. $X(u, v)\, Y(u, v) \bmod \left(u^{p^a} - 1, C_{p^a}(v)\right)$, (10.4.34)

B. $X(u, v)\, Y(u, v) \bmod \left(v^{p^{a-1}} - 1, C_{p^a}(u)\right)$, (10.4.35)

and

C. $X(u, v)\, Y(u, v) \bmod \left(u^{p^{a-1}} - 1, v^{p^{a-1}} - 1\right)$. (10.4.36)

The computations in (10.4.34) and (10.4.35) are converted to lengths p^a and p^{a-1} DFT in $CF(p^a)$, respectively. The roots of unity are given by v and u^a, respectively. Note that the computations in **A** and **B** are now performed by first converting them to one-dimensional polynomial products.

Case 4. $n_1 = n_2 = 2^a$. In this case, we can use the factorization

$$v^{2^a} - 1 = (v-1)\prod_{j=1}^{a}\left[1 + v^{2^{a-j}}\right] = (v-1)\prod_{j=1}^{a} C_{2^{a-j+1}}(v) = C_{2^a}(v)\left\{v^{2^{a-1}} - 1\right\} \qquad (10.4.37)$$

in two ways as per the two approaches described above.

Approach 1. Using (10.4.37), the two-dimensional cyclic convolution can be expressed as a direct sum of

I. $X(u, v)\ Y(u, v) \bmod \left(u^{2^a} - 1, v - 1\right)$, which is a 2^a length one-dimensional cyclic convolution, and

II. $X(u, v)\ Y(u, v) \bmod \left(u^{2^a} - 1, C_{2^{a-j+1}}(v)\right)$, $j = 1, 2, \ldots, a$. (10.4.38)

The j-th computation in (10.4.38) can be written as a direct sum of 2^{a-j+1} polynomial products by using the factorization

$$u^{2^a} - 1 = \prod_{i=0}^{2^{a-j+1}-1} \left[u^{2^{j-1}} - v^i\right] \bmod C_{2^{a-j+1}}(v),\ j = 1, 2, \ldots, a \qquad (10.4.39)$$

in $CF(2^{a-j+1})$.

Approach 2. In this case, based on (10.4.37) and the analysis previously presented, we write the two-dimensional cyclic convolution as a direct sum of the following three computations:

A. $X(u, v)\ Y(u, v) \bmod \left(u^{2^a} - 1, C_{2^a}(v)\right)$, (10.4.40)

B. $X(u, v)\ Y(u, v) \bmod \left(v^{2^{a-1}} - 1, C_{2^a}(u)\right)$, (10.4.41)

and

C. $X(u, v)\ Y(u, v) \bmod \left(u^{2^{a-1}} - 1, v^{2^{a-1}} - 1\right)$. (10.4.42)

The computations in (10.4.40) and (10.4.41) are converted to lengths 2^a and 2^{a-1} DFT in $CF(2^a)$, respectively. The roots of unity are given by v and u^a, respectively. Note that the computations in **A** and **B** are now performed by first converting them to one-dimensional polynomial products which correspond to circular correlations of length 2^{a-1} because of the special form of $C_{2^a}(v)$.

There are many other cases of interest and algorithms that can be studied in the present mathematical framework. In the following, we list those cases that can follow from the above cases with suitable modifications.

Case 5. $n_1 = 2p$, $n_2 = p$; p an odd prime,

Case 6. $n_1 = 2p$, $n_2 = 2p$; p an odd prime,

Case 7. $n_1 = 2p^2$, $n_2 = 2p^2$; p an odd prime,

Case 8. $n_1 = 2p^a$, $n_2 = 2p^a$; p an odd prime,

Case 9. $n_1 = 4p^a$, $n_2 = 4p^a$; p an odd prime, factorizations over CFs,

Case 10. $n_1 = 4p^a$, $n_2 = 4p^a$; p an odd prime, factorizations over CCFs,

Case 11. $n_1 = p_1$, $n_2 = p_1 \times p_2$; p_1 and p_2 distinct odd primes,

Case 12. $n_1 = p_1 \times p_2$, $n_2 = p_1 \times p_2$; p_1 and p_2 distinct odd primes,

Case 13. $n_1 = 2^a \times p$, $n_2 = 2^a \times p$; p an odd prime, factorizations over CFs,

Case 14. $n_1 = 2^a \times p$, $n_2 = 2^a \times p$; p an odd prime, factorizations over CCFs.

These cases are left to the readers as problems at the end of this chapter.

10.5 Algorithms for Multidimensional Cyclic Convolution

A vast majority of the techniques and algorithms for processing multi-dimensional sequences follow from the two-dimensional methods. For example, three-dimensional cyclic convolution of length $n_1 \times n_2 \times n_3$ is defined as

$$z_{k_1,k_2,k_3} = \sum_{i_3=0}^{n_3-1}\sum_{i_2=0}^{n_2-1}\sum_{i_1=0}^{n_1-1} x_{(k_1-i_1)\bmod n_1,(k_2-i_2)\bmod n_2,(k_3-i_3)\bmod n_3}\, y_{i_1,i_2,i_3}, \tag{10.5.1}$$

where x_{i_1,i_2,i_3} and y_{i_1,i_2,i_3} are three-dimensional input sequences, and z_{i_1,i_2,i_3} is the output sequence, $i_1 = 0, 1, \ldots, n_1 - 1$; $i_2 = 0, 1, \ldots, n_2 - 1$; $i_3 = 0, 1, \ldots, n_3 - 1$. All sequences have length $n_1 \times n_2 \times n_3$.

Defining the generating polynomial in the usual way for three-dimensional sequences, the three-dimensional cyclic convolution can be expressed in polynomial form as

$$Z(u, v, w) = X(u, v, w)\, Y(u, v, w) \bmod \left(u^{n_1} - 1, v^{n_2} - 1, w^{n_3} - 1\right), \tag{10.5.2}$$

$$deg(X(u, v, w)) = deg(Y(u, v, w)) = deg(Z(u, v, w)) = n_1 - 1 \text{ in } u,$$
$$deg(X(u, v, w)) = deg(Y(u, v, w)) = deg(Z(u, v, w)) = n_2 - 1 \text{ in } v,$$

and

$$deg(X(u, v, w)) = deg(Y(u, v, w)) = deg(Z(u, v, w)) = n_3 - 1 \text{ in } w.$$

Based on (10.5.2), we may also write

$$Z_{k_2,k_3}(u) = \sum_{i_3=0}^{n_3-1}\sum_{i_2=0}^{n_2-1} X_{(k_2-i_2)\bmod n_2,(k_3-i_3)\bmod n_3}(u) Y_{i_2,i_3}(u) \bmod \left(u^{n_1}-1\right). \quad (10.5.3)$$

Thus, a length $n_1 \times n_2 \times n_3$ three-dimensional cyclic convolution can be imagined as two-dimensional cyclic convolution of length $n_2 \times n_3$, where each scalar is replaced by a polynomial of length n_1. Here, the choice and ordering of lengths is arbitrary and must be made keeping computational efficiency in mind. Therefore, a three-dimensional cyclic convolution algorithm can be obtained by replacing each ADD by a polynomial ADD and each MULT by a one-dimensional cyclic convolution in the two-dimensional cyclic convolution algorithm. Interestingly, three-dimensional cyclic convolution can also be imagined as one-dimensional cyclic convolution, where each scalar is replaced by two-dimensional polynomials. Once again, the choice and ordering of lengths is arbitrary and must be made keeping computational efficiency in mind. In such a case, the three-dimensional cyclic convolution algorithm can be obtained by replacing each ADD by a two-dimensional polynomial ADD and each MULT by a two-dimensional cyclic convolution in the one-dimensional cyclic convolution algorithm.

One can formulate expressions and algorithms for four- and higher-dimensional cyclic convolution as well. They follow a similar path. Two observations can be made at this stage.

Observation 1. A d-dimensional cyclic convolution of length $n_1 \times n_2 \times \cdots \times n_i \times \cdots \times n_d$ can be converted to a $(d + 1)$-dimensional cyclic convolution if one of the lengths, say n_i, factors into two relatively prime numbers. In other words, if $n_i = n_{i,1} \times n_{i,2}$, $(n_{i,1}, n_{i,2}) = 1$, then Agarwal-Cooley algorithm can be used to express a d-dimensional cyclic convolution as a $(d + 1)$-dimensional cyclic convolution of length $n_1 \times n_2 \times \cdots \times n_{i,1} \times n_{i,2} \times \cdots \times n_d$. This statement can be used in a recursive manner. For example, a 12×3 cyclic convolution can be converted to a 3-dimensional cyclic convolution of length $4 \times 3 \times 3$; similarly, a 10×10 cyclic convolution can be converted to a four-dimensional cyclic convolution of length $2 \times 2 \times 5 \times 5$.

Observation 2. One can use the factorizations in extension fields in order to compute a d-dimensional cyclic convolution, $d > 1$. Such an approach is straightforward to derive and is left as an exercise to the readers for many cases of interest. For example, a $p \times p \times p$ cyclic can be converted to a $p \times p$ cyclic convolution and p^2 one-dimensional polynomial products mod $C_p(u)$.

Once again, one may seek factorizations over the complex numbers and complex cyclotomic fields for complex valued sequences.

10.6 Algorithms for Two-Dimensional Cyclic Convolution in Finite Integer Rings

In this section, only those algorithms are described that are sensitive to the form of the finite integer rings. We describe algorithms for computing the two-dimensional cyclic convolution of length $n_1 \times n_2$ defined as

$$Z(u, v) = X(u, v)\, Y(u, v) \bmod (M,\ u^{n_1} - 1, v^{n_2} - 1) \tag{10.6.1}$$

in $Z(M)$ and

$$Z(u, v) = X(u, v)\, Y(u, v) \bmod (M, j^2 + 1,\ u^{n_1} - 1, v^{n_2} - 1) \tag{10.6.2}$$

in $CZ(M)$. Here, $deg(X(u, v)) = deg(Y(u, v)) = deg(Z(u, v)) = n_1$ in u, and $deg(X(u, v)) = deg(Y(u, v)) = deg(Z(u, v)) = n_2$ in v. Given the standard factorization of M in (2.2.15), the following direct sums are obtained using the CRT-I:

$$Z(M) = \sum_{\substack{\oplus\\ i=1}}^{t} Z(q_i)\,, \tag{10.6.3}$$

$$CZ(M) = \sum_{\substack{\oplus\\ i=1}}^{t} CZ(q_i)\,. \tag{10.6.4}$$

Based on (10.6.3) and (10.6.4), we only need to derive algorithms for the computations

$$Z(u, v) = X(u, v)\, Y(u, v) \bmod (q,\ u^{n_1} - 1, v^{n_2} - 1) \tag{10.6.5}$$

in $Z(q)$ and

$$Z(u, v) = X(u, v)\, Y(u, v) \bmod (q, j^2 + 1,\ u^{n_1} - 1, v^{n_2} - 1) \tag{10.6.6}$$

in $CZ(q)$, $q = p^a$.

Following the same arguments as in the case of one-dimensional algorithms, it is seen that

(i) for $p = 2$, the algorithm for (10.6.6) is the same as in (10.6.5), only the input and output quantities are defined in $CZ(2^a)$;

(ii) for $p \equiv 1 \bmod 4$, the computation in (10.6.6) can be written as a direct sum of two computations of the type as in (10.6.5); and

(iii) for $p \equiv 3 \bmod 4$, $CZ(p) = GF(p^2)$ and we employ the corresponding arithmetic to compute the quantities of interest.

Also, the above analysis is complete for all cases except $p \equiv 3 \bmod 4$ and in the context of algorithms in $CZ(q)$, we will assume that $p \equiv 3 \bmod 4$.

All the essential elements of the algorithm for cyclic convolution have already been described in earlier chapters. In this section, we bring them together to develop a two-dimensional cyclic convolution algorithm having the least computational complexity of all known algorithms. We begin with (10.6.5) first. In order to simplify our description and make it consistent with the earlier works, we assume that

$$(n_1, p) = (n_2, p) = 1. \tag{10.6.7}$$

A CRT-PR based algorithm for cyclic convolution over $Z(q)$ begins with the factorization of $u^{n_1} - 1$ (or $v^{n_2} - 1$) in $Z(q)$ as

$$u^{n_1} - 1 = \prod_{k=1}^{\lambda} p_k(u) \bmod q, \tag{10.6.8}$$

starting with its factorization in $GF(p)$. This factorization along with CRT-PR in $Z(q)$ implies that $Z(u, v)$ in (10.6.5) is a direct sum of $Z_k(u, v)$, $k = 1, 2, \ldots, \lambda$, where

$$Z_k(u, v) = X_k(u, v)\, Y_k(u, v) \bmod (q, p_k(u),\ v^{n_2} - 1),\ \ k = 1, 2, \ldots, \lambda. \tag{10.6.9}$$

We note that each of $p_k(u)$ in (10.6.9) is *irreducible* and a factor of $u^{n_1}-1$ in $Z(q)$. These are highly desirable properties.

The key to deriving a two-dimensional cyclic convolution algorithm is in observing that the computation in (10.6.9) can be treated as a one-dimensional cyclic convolution in integer polynomial ring $Z(p, p_k(u))$. Therefore, we seek the factorization of $v^{n_2}-1$ and the associated CRT-PR over $Z(p^a, p_k(u))$ in order to compute $Z_k(u, v)$, $k = 1, 2, \ldots, \lambda$, in (10.6.9) and then employ the CRT-PR in $Z(q)$ to obtain $Z(u, v)$ in (10.6.5). Over $CZ(q)$, we seek the factorization of $u^{n_1}-1$ (or $v^{n_2}-1$) in $CZ(q)$, $p \equiv 3 \bmod 4$ and the factorization of $v^{n_2}-1$ in $CZ(q, p_k(u))$, $p_k(u)$ being the k-th irreducible factor of $u^{n_1}-1$ in $CZ(q)$. The remaining structure of the two-dimensional cyclic convolution algorithm in $CZ(q)$ is similar to the structure of the algorithm in $Z(q)$. All the results described in Chapter 7 are valid and can be used here. The structure of the two-dimensional cyclic convolution algorithm of sequences defined in $Z(q)$ and $CZ(q)$ is straightforward and requires no further elaboration. In the following, we summarize some of the salient features of polynomial factorization over integer polynomial rings, and the structure of CRT-PR based algorithms for two-dimensional cyclic convolution algorithms in them. We begin with some examples.

Example 10.10 Let $n_1 = n_2 = 7$ and the field be $GF(2)$. The factorization of $v^7 - 1$ over $GF(2)$ is given by

$$v^7 - 1 = (v - 1)\,(v^3 + v^2 + 1)\,(v^3 + v + 1).$$

In this case, we express the computation $X(u, v)\,Y(u, v) \bmod (u^7 - 1, v^7 - 1)$ as a direct sum of

$$X(u, v)\,Y(u, v) \bmod (u^7 - 1, v - 1) = X(u, 1)\,Y(u, 1) \bmod u^7 - 1,$$
$$X(u, v)\,Y(u, v) \bmod (u^7 - 1, v^3 + v^2 + 1),$$

and

$$X(u, v)\,Y(u, v) \bmod (u^7 - 1, v^3 + v + 1).$$

The first term is a one-dimensional cyclic convolution of length 7. The second and third terms employ the following factorization over $GF(2)$:

$$u^7 - 1 = \prod_{i=0}^{6}\left(u - v^i\right) \operatorname{mod}\left(v^3 + v^2 + 1\right),$$

and

$$u^7 - 1 = \prod_{i=0}^{6} \left(u - v^i\right) \bmod \left(v^3 + v + 1\right),$$

which implies that $X(u, v)\ Y(u, v) \bmod (u^7 - 1, v^3 + v^2 + 1)$ and $X(u, v)\ Y(u, v) \bmod (u^7 - 1, v^3 + v + 1)$ can be written as a direct sum of

$$X(v^i, v)\ Y(v^i, v) \bmod v^3 + v^2 + 1,\ \ i = 0, 1, ..., 6$$

and

$$X(v^i, v)\ Y(v^i, v) \bmod v^3 + v + 1,\ \ i = 0, 1, ..., 6,$$

respectively. Noting that the cyclic convolution algorithm of length 7 over $GF(2)$ requires 13 MULTs, the overall two-dimensional cyclic convolution of length 7×7 over $GF(2)$ requires $13 + 14 \times 6 = 97$ MULTs.

Example 10.11 Let $n_1 = n_2 = 7$ and the ring be $Z(4)$. The factorization of $v^7 - 1$ over $Z(4)$ is given by

$$v^7 - 1 = (v - 1)\ (v^3 + 3v^2 + 2v + 3)\ (v^3 + 2v^2 + v + 3).$$

In this case, we express the computation $X(u, v)\ Y(u, v) \bmod (u^7 - 1, v^7 - 1)$ as a direct sum of

$$X(u, v)\ Y(u, v) \bmod (u^7 - 1, v - 1) = X(u, 1)\ Y(u, 1) \bmod u^7 - 1,$$
$$X(u, v)\ Y(u, v) \bmod (u^7 - 1, v^3 + 3v^2 + 2v + 3),$$

and

$$X(u, v)\ Y(u, v) \bmod (u^7 - 1, v^3 + 2v^2 + v + 3).$$

The first term is a one-dimensional cyclic convolution of length 7. The second and third terms employ the following factorization over $Z(4)$:

$$u^7 - 1 = \prod_{i=0}^{6} \left(u - v^i\right) \bmod \left(v^3 + 3v^2 + 2v + 3\right)$$

and

$$u^7 - 1 = \prod_{i=0}^{6} \left(u - v^i\right) \bmod \left(v^3 + 2v^2 + v + 3\right),$$

which implies that

$$X(u, v)\ Y(u, v) \bmod (u^7 - 1, v^3 + 3v^2 + 2v + 3)$$

and

$$X(u, v)\ Y(u, v) \bmod (u^7 - 1, v^3 + 2v^2 + v + 3)$$

can be written as a direct sum of

$$X(v^i, v)\ Y(v^i, v) \bmod v^3 + 3v^2 + 2v + 3, \ \ i = 0, 1, ..., 6$$

and

$$X(v^i, v)\ Y(v^i, v) \bmod v^3 + 2v^2 + v + 3, \ \ i = 0, 1, ..., 6,$$

respectively. Once, again, the overall two-dimensional cyclic convolution of length 7×7 over $Z(4)$ requires 97 MULTs.

Example 10.12 Let $q = 2^a$ and $n_1 = n_2 = 2^m - 1 = n$, a Mersenne prime integer. In this case $u^{n_1} - 1$ factors as product of $(u - 1)$ and $(n - 1)/m$ irreducible polynomials in $Z(q)$, each of degree m. All of these polynomials originate from *primitive polynomials* in $GF(2)$ as n is prime. Let $p(u)$ denote any one of these. For $(u - 1)$, (10.6.9) reduces to a one-dimensional cyclic convolution of length n in $Z(q)$ which is computed using CRT-PR based cyclic convolution algorithm. In all other cases, (10.6.9) is a computation of the type

$$Z(u, v) = X(u, v)\ Y(u, v) \bmod (q, p(u), v^n - 1). \tag{10.6.10}$$

The CRT-PR based algorithm for cyclic convolution for (10.6.10) is based on the following factorization of $v^n - 1$ in $Z(q, p(u))$,

$$v^n - 1 = \prod_{i=0}^{n-1} \left(v - u^i\right) \bmod (q, p(u)). \tag{10.6.11}$$

In all instances $p(u) \mid (u^n - 1)$ and $p_0(u)$ is primitive over $GF(2)$. Consequently, the two-dimensional cyclic convolution is computed as

1. One-dimensional cyclic convolution of length n in $Z(q)$, and
2. $(n - 1)/m$ NTTs each of length n in $Z(q, p_k(u))$ which in turn are computed as n polynomial products mod $p_k(u)$ in $Z(q)$.

If M_1 denotes the multiplicative complexity of the algorithm for computing a polynomial product mod $p_k(u)$ (which is equivalent to an acyclic convolution algorithm of length $2m - 1$ over $Z(q)$ followed by a reduction mod $p_k(u)$), the multiplicative complexity of the two-dimensional cyclic convolution algorithm is given by

$$M(n \times n) = \left(n^2 - 1\right)\frac{M_1}{m} + 1.$$

This corresponds to about M_1/m MULTs per point which is the same as if the entire computation were an one-dimensional acyclic convolution of length $2m-1$ over $Z(q)$.

10.6.1 Algorithms for Integer Polynomial Rings

In this section, algorithms are derived for computing the cyclic convolution

$$Z(u) = X(u)\,Y(u) \bmod (q, Q(\theta), u^n - 1) \tag{10.6.12}$$

and

$$Z(u) = X(u)\,Y(u) \bmod (q, j^2 + 1, Q(\theta), u^n - 1). \tag{10.6.13}$$

Once again $X(u)$, $Y(u)$, and $Z(u)$ are defined over $Z(q, Q(\theta))$ in (10.6.12) and over $CZ(q, Q(\theta))$ in (10.6.13). In (10.6.12), $Q(\theta)$ is such that $Q_0(\theta)$ is *irreducible* in $GF(p)$. Consequently, $Q(\theta)$ is irreducible over $Z(q)$. In the remainder of this section, $Q(\theta)$ is assumed to be a monic irreducible polynomial over $Z(q)$, $deg(Q(\theta)) = m$. Thus, $Z(q, Q(\theta))$ is a p-adic expansion ring of $Z(p, Q_0(\theta)) = GF(r)$, $r = p^m$. This analysis places no constraints on the structure of $Q_l(\theta)$, $l = 1, 2, \ldots, a-1$. Similarly, in (10.6.13) $Q(\theta)$ is assumed to be a monic irreducible polynomial over $CZ(q)$, $deg(Q(\theta)) = m$. $CZ(q, Q(\theta))$ is a p-adic expansion ring of $CZ(p, Q_0(\theta)) = GF(r)$, $r = p^{2m}$.

Let us analyze the computation in (10.6.12) first. In order to use CRT-PR over $Z(q, Q(\theta))$, we seek factorization of $u^n - 1$ in $Z(q, Q(\theta))$ in terms of relatively coprime factors. Such a factorization begins with the factorization of $u^n - 1$ in finite field $Z(p, Q_0(\theta)) = GF(r)$, $r = p^m$, and employs the extended Hensel's Theorem to obtain the factorization in $Z(q, Q(\theta))$ in an order-recursive manner. The factorization of $u^n - 1$ in $Z(p, Q_0(\theta))$ can be determined using the algebra of finite fields as described in Chapter 7. If $n = b \cdot p^c$ where $(b, p) = 1$, then

$$u^n - 1 = \left(u^b - 1\right)^{p^c} \bmod (p, Q_0(\theta)). \tag{10.6.14}$$

Therefore, we need to factorize $u^n - 1$ in $Z(p, Q_0(\theta))$ when $(n, p) = 1$. For $(n, p) = 1$, we find the positive integer a having the least value such that $n \mid (p^{ma} - 1)$. The procedure to factorize $u^n - 1$ can be divided into three categories as described briefly in the following.

Category 1. $n \mid (p-1)$. In this case, $u^n - 1$ is the product of n degree one polynomials

$$u^n - 1 = \prod_{i=0}^{n-1} (u - B^i) \bmod p,$$
$$B = A^{(p-1)/n}. \tag{10.6.15}$$

The integer A is a primitive integer in $GF(p)$. This factorization is also valid in $Z(p, Q_0(\theta))$.

Category 2. $n \mid (p^m - 1)$. In this case, $u^n - 1$ is the product of n degree one polynomials

$$u^n - 1 = \prod_{i=0}^{n-1} (u - B^i) \bmod (p, Q_0(\theta)),$$
$$B = A^{(p^m-1)/n}. \tag{10.6.16}$$

Here, A is a primitive element in $Z(p, Q_0(\theta))$ and a polynomial in θ.

Category 3. $n \mid (p^{ma} - 1)$, $a > 1$. Let $R_0(\gamma)$ be a degree a polynomial in γ with coefficients in $Z(q, Q_0(\theta))$. Also, $R_0(\gamma)$ is irreducible in $Z(p, Q_0(\theta))$. Following (10.6.16), we can write

$$u^n - 1 = \prod_{i=0}^{n-1} (u - B^i) \bmod (p, Q_0(\theta), R_0(\gamma)),$$
$$B = A^{(p^{ma}-1)/n}. \tag{10.6.17}$$

Here, A is a primitive element in $Z(p, Q_0(\theta), R_0(\gamma)) = GF(p^{ma})$ and a polynomial in γ. Let $S = \{1, 2, ..., n\}$. Partition S into subsets S_{i_1}, S_{i_2}, ...; $S_{i_k} = \{i_k, ri_k, r^2 i_k, \cdots\}$, where i_k is the smallest element in S not covered in earlier subsets. The elements in subsets S_{i_1}, S_{i_2}, ... are defined mod n. Each subset has a maximum of a elements. The factorization of $u^n - 1$ in (10.6.17) simplifies to

$$u^n - 1 = \prod_k \left\{ \left[\prod_{l \in S_{i_k}} \left(u - B^l\right) \right] \bmod R_0(\gamma) \right\} \bmod (p, Q_0(\theta))$$
$$= \prod_k p_k(u) \bmod (p, Q_0(\theta)). \tag{10.6.18}$$

Each of $p_k(u)$ in (10.6.18) is the polynomial of lowest degree having B^{i_k} and its conjugates in $GF(r^a)$ as its roots. Therefore, it is irreducible and has its coefficients in $GF(r)$.

The factorization of $u^n - 1$ in (10.6.15) (Category 1) and its corresponding factorization in $Z(q)$ lead to NTTs in the usual sense. The factorization of $u^n - 1$ in (10.6.16) (Category 2) and its corresponding factorization in $Z(q, Q(\theta))$ lead to NTT in integer polynomial rings. The factorization of u^n-1 in (10.6.18) (Category 3) along with the CRT-PR lead to new cyclic convolution algorithms. Note that Category 2 is a special case of Category 3 and Category 1 is a special case of Category 2. For algorithm for cyclic convolution in $CZ(q, Q(\theta))$, we seek factorization of $u^n - 1$ in $CZ(p, Q_0(\theta))$ first. As (10.6.14) also holds here, we assume that $(n, p) = 1$. Now, we seek the smallest a such that $n \mid (p^{2ma} - 1)$. The remaining analysis for $CZ(p, Q_0(\theta))$ is similar to the analysis for $Z(p, Q_0(\theta))$. Given the factorization of $u^n - 1$ in $Z(q, Q_0(\theta))$ and $CZ(q, Q_0(\theta))$, its factorization in $Z(q, Q(\theta))$ and $CZ(q, Q(\theta))$ is obtained by using the extended Hensel's Theorem. The application of CRT-PR to derive algorithms for cyclic convolution is straightforward once the factors of $u^n - 1$ are known in the finite integer ring of interest. The factorization in examples 10.10, 10.11, and 10.12 are all Category 2 factorizations.

Example 10.13 Consider the case $p = 2$ and $n = 5$. Since $5 \mid (2^4 - 1)$, let $m = 4$ and $Q_0(\theta) = \theta^4 + \theta + 1$, a degree 4 primitive polynomial in $GF(2)$. The factorization of $u^5 - 1$ falls into Category 2 and is given by

$$u^5 - 1 = \prod_{i=0}^{4} \left(u - \theta^{3i}\right) \bmod (2, \theta^4 + \theta + 1).$$

Similarly,

$$u^3 - 1 = \prod_{i=0}^{2} \left(u - \theta^{5i}\right) \bmod (2, \theta^4 + \theta + 1).$$

Both of these factorizations follow from

$$u^{15} - 1 = \prod_{i=0}^{14} \left(u - \theta^{i}\right) \bmod (2, \theta^4 + \theta + 1).$$

Combining above factorizations with (10.6.14), we see that

$$u^{30}-1=\prod_{i=0}^{14}\left(u-\theta^i\right)^2 \bmod (2,\theta^4+\theta+1)$$

$$u^{20}-1=\prod_{i=0}^{4}\left(u-\theta^i\right)^4 \bmod (2,\theta^4+\theta+1)\,.$$

The corresponding factorizations and CRT-PR in $Z(q, Q(\theta))$ follow in a straightforward manner.

Example 10.14 Consider the case $p = 2$, $m = 3$, $n = 7$, and $Q_0(\theta) = \theta^2 + \theta + 1$. In this case $7 \mid (2^6 - 1)$ and, therefore, $a = 3$. An irreducible polynomial in $Z(2, \theta^2 + \theta + 1)$ having degree 3 is $R_0(\gamma) = \gamma^3 + \theta\gamma^2 + \theta^2\gamma + \theta$. In fact, $R_0(\gamma)$ is also primitive. The partition of $S = \{1, 2, ..., 7\}$ is given by $S_1 = \{1, 4, 2\}$, $S_3 = \{3, 5, 6\}$ and $S_7 = \{7\}$. Thus, $u^7 - 1$ factorizes over $Z(2, \theta^2 + \theta + 1)$ as

$$\begin{aligned}u^7-1=\Big\{&\left[(u-B)\left(u-B^4\right)\left(u-B^2\right)\right]\left[\left(u-B^3\right)\left(u-B^5\right)\left(u-B^6\right)\right]\\&\cdot\left[u-B^7\right]\bmod R_0(\gamma)\Big\} \bmod (2,\theta^2+\theta+1),\end{aligned}$$

where $B = \gamma^9$. Simplifying,

$$u^7-1=\left(u^3+u+1\right)\left(u^3+u^2+1\right)(u+1) \bmod (2,\theta^2+\theta+1).$$

Example 10.15 Consider Example 10.14 with $n = 9$. $9 \mid (2^6 - 1)$. In this case, the partition of $S = \{1, 2, ..., 9\}$ is given by $S_1 = \{1, 4, 7\}$, $S_2 = \{2, 8, 5\}$, $S_3 = \{3\}$, $S_6 = \{6\}$ and $S_9 = \{9\}$. Thus, $u^9 - 1$ factorizes over $Z(2, \theta^2 + \theta + 1)$ as

$$\begin{aligned}u^9-1=\{&[(u-B)(u-B^4)(u-B^7)][(u-B^2)(u-B^8)(u-B^5)]\\&\cdot[u-B^3][u-B^6][u-B^9] \bmod R_0(\gamma)\} \bmod (2,\theta^2+\theta+1),\end{aligned}$$

where $B = \gamma^7$. Simplifying,

$$u^9-1=\left(u^3+\theta\right)\left(u^3+\theta^2\right)(u+\theta)\left(u+\theta^2\right)(u+1) \bmod (2,\theta^2+\theta+1).$$

Examples 10.14 and 10.15 fall into Category 3. In the next section, we address the issue of selecting $Q(\theta)$ and analyze the resulting factorizations of $u^n - 1$.

10.6.2 Selecting $Q(\theta)$ for Cyclic Convolution Algorithms

Up to this point, all the analysis is based on the assumption that $Q(\theta)$ is given. The only condition is that $Q_0(\theta)$ be irreducible over $GF(p)$. This is indeed true when $Q(\theta)$ arises as in two-dimensional cyclic convolution algorithm. However, in other situations, its choice may be arbitrary. In this section, we address the question of selecting $Q(\theta)$ in terms of its degree m, irreducible $Q_0(\theta)$ and the remaining polynomials in its p-adic expansion when only p and n are given. Keeping (10.6.14) in mind, it is further assumed that $(p, n) = 1$.

For $(p, n) = 1$, there always exists e (the smallest value), the order of p mod n, such that $p^e \equiv 1 \bmod n$ or $n \mid (p^e - 1)$. From Euler's theorem, $p^{\phi(n)} \equiv 1 \bmod n$, where $\phi(n)$ is the Euler's totient function of n. Therefore, $e \mid \phi(n)$. Since in general we seek the smallest values of a and m such that $n \mid (p^{ma} - 1)$ in order to factorize $u^n - 1$, we have $e \mid (ma)$. Therefore, if m is given, then a is obtained as

$$a = e/(e, m). \tag{10.6.19}$$

If $m = e$, then $a = 1$ and the Category 2 factorization of $u^n - 1$ in $Z(p, Q_0(\theta))$ follows. In essence, it leads to a length n one-dimensional cyclic convolution in $Z(q)$ being computed as n polynomial MULTs in $Z(q)$, each of degree e.

If $(e, m) = 1$, then $a = e$ and a Category 3 factorization follows. This is the same as the factorization of $u^n - 1$ in $GF(p)$ starting from its factorization in $GF(p^a)$. The resulting CRT-PR based algorithm for cyclic convolution over $Z(p, Q_0(\theta))$ is the same as CRT-PR based algorithm for cyclic convolution over $Z(q)$. Example 10.14 corresponds to this situation. It is clear that in this case, working in the polynomial extension provides no advantage whatsoever. Of all the values of m for which $(m, e) = 1$, it is best to select $m = 1$.

When e is prime, then we have either $m = e$ (or a multiple of e) or $(m, e) = 1$. Since we look for the smallest values of m and a in all instances, we consider only the situation $m = e$ or $m = 1$. If $m = e$, then $a = 1$ and the Category 2 factorization follows. If $m = 1$, then $a = e$. This also provides us with two distinct approaches to computing one-dimensional cyclic convolution of length n. Consider the following example.

Example 10.16 Let $p = 5$ and $n = 31$. In this case, $e = 3$. Let $Q_0(\theta)$ be a degree 3 primitive polynomial in $GF(5)$. Then the Category 2 factorization of $u^{31} - 1$ is given by

$$u^{31}-1=\prod_{i=0}^{30}\left(u-\theta^{4i}\right) \bmod (5, Q_0(\theta)). \qquad (10.6.20)$$

On the other hand, $u^{31}-1$ factorizes into 11 polynomials in $GF(5)$ (10 of degree 3 each and 1 of degree 1). Even though the two factorizations are closely related to each other, they may lead to algorithms for one-dimensional cyclic convolution in $Z(5^a)$ having vastly different computational and structural properties. A direct use of (10.6.20) leads to an algorithm that requires 186 MULTs (that may be reduced by exploiting the conjugate symmetry property as explained in Paper 3) while the second alternative leads to an algorithm that requires 61 MULTs.

In general, for a non-prime e, m and a must be selected to satisfy

$$ma = e.$$

Also, m and e should be as close to each other as possible. For example, if we let $m = 3$, $Q_0(\theta) = \theta^3 + \theta + 1$, and $a = 2$ in Example 10.15, then we get

$$u^9-1=\left(u^2+\theta u+1\right)\left(u^2+\theta^2 u+1\right)\left(u^2+u+1\right)\left(u^2+\theta^4 u+1\right)(u+1) \bmod (2, \theta^3+\theta+1).$$

A direct implementation of one-dimensional length 9 cyclic convolution based on the above factorization requires 58 MULTs while a similar implementation for the factorization in Example 10.15 requires 43 MULTs.

For the complex case, $Q_0(\theta)$ is an irreducible polynomial over $GF(p^2)$, $p \equiv 3 \bmod 4$. Here we seek the smallest value of a and m such that $n \mid (p^{2ma}-1)$. It is clear that $e \mid (2ma)$ and $a = (e/2)/(e/2, m)$ for e even and $a = e/(e, m)$ for e odd. Therefore, we derive an advantage only if e is even. Otherwise, one may simply use the factorization in $GF(p)$. If $e = 2e_1$, then $ma = e_1$. If e_1 is prime, then two alternatives follow (i) $m = e_1$, $a = 1$, and (ii) $m = 1$, $a = e_1$. The first alternative leads to the Category 2 factorization of $u^n - 1$ in $CZ(p, Q_0(\theta))$ while the second leads to the factorization of $u^n - 1$ in $CZ(p)$. In general, for an even e with non-prime $e/2$, m and a must be selected to satisfy $ma = e/2$, the values being as close to each other as possible.

Example 10.17 Given in the following are four factorizations of $u^{10} - 1$ in $GF(3)$ and related fields, all of which may be used to obtain algorithms for cyclic convolution in $Z(3^a)$ and $CZ(3^a)$ having different properties:

$$u^{10}-1=(u+2)(u+1)\left(u^4+u^3+u^2+u+1\right)\left(u^4+2u^3+u^2+2u+1\right) \bmod 3$$

$$=\prod_{i=0}^{9}\left(u-\theta^{8i}\right) \bmod (3,\theta^4+2\theta+2)$$

$$= (u+2)(u+1)\cdot [u^2+(2+\theta)u+1]\cdot[u^2+(2+2\theta)u+1]$$
$$\cdot [u^2+(1+2\theta)u+1]\cdot[u^2+(1+\theta)u+1] \bmod (3, \theta^2+1)$$

$$=\prod_{i=0}^{9}\left(u-\theta^{2i}\right) \bmod (3, j^2+1, \theta^2+(1+j)\theta+1).$$

The factorizations are in $GF(3)$, $Z(3, \theta^4 + 2\theta + 2)$, $Z(3, \theta^2 + 1)$, and $CZ(3, \theta^2 + (1 + j)\theta + 1)$, respectively. Note that by setting $j = \theta$ in the factorization over $Z(3, \theta^2 + 1)$, we get the factorization over $CZ(3)$.

Once m is selected, we have to find $Q_0(\theta)$, an irreducible polynomial in θ with coefficients in $GF(p)$ or $GF(p^2)$ for the algorithm for cyclic convolution over $Z(q)$ and $CZ(q)$, respectively, $deg(Q_0(\theta)) = m$. Such polynomials exist for all values of m and their choice is non-unique. Let us consider the case of $Q_0(\theta)$ defined over $GF(p)$ first.

Of all the possible choices for $Q_0(\theta)$, we are interested in those $Q_0(\theta)$ for which B in (10.6.16) is a simple polynomial in θ. This is done to keep the computational complexity of the algorithm for cyclic convolution as low as possible. The simplest form of B is $B = \theta^k$. The order of θ is defined as the lowest value of N for which $Q_0(\theta) \mid (\theta^N - 1)$. It is clear from (10.6.15) and (10.6.16) that for $B = \theta^k$ we must have $n \mid N$. Therefore, any one of the irreducible polynomials of degree m can be selected as long as the order of θ associated with it is a multiple of n. This statement is valid for $Q_0(\theta)$ in $GF(p^2)$ as well. Note that in general, $N \mid (p^m - 1)$ for $Q_0(\theta)$ in $GF(p)$ and $N \mid (p^{2m} - 1)$ for $Q_0(\theta)$ in $GF(p^2)$. If N is equal to the largest value possible, then the corresponding polynomials are *primitive*.

Once $Q_0(\theta)$ is selected, $Q_0(\theta) \mid (\theta^N - 1)$, $N \mid (p^m - 1)$, we must use the p-adic expansion in order to compute the factorization of $u^n - 1$ in $Z(q, Q(\theta))$. Once again, we would like to preserve the transform structure of the algorithm for cyclic convolution based on (10.6.15) and (10.6.16). This can be accomplished by selecting $Q_1(\theta), \ldots, Q_{a-1}(\theta)$ such that $Q(\theta) = \sum_{i=0}^{a-1} Q_i(\theta)$ divides $(\theta^N - 1)$ in

$Z(q)$, $q = p^a$. This is also accomplished by using the p-adic expansion. It is clear that if $Q(\theta) \mid (\theta^N - 1)$ in $Z(q)$, and (10.6.15) and (10.6.16) hold, then

$$u^n - 1 = \prod_{i=0}^{n-1} \left(u - \theta^{(N/n)i}\right) \bmod (q, Q(\theta)). \tag{10.6.21}$$

Since the choice of $Q_0(\theta)$ is non-unique, we have non-unique choices for $Q(\theta)$. All the possible $Q(\theta)$ lead to algorithm for cyclic convolution having the same multiplicative complexity. However, the additive complexity of the algorithm for cyclic convolution will depend on the exact form of $Q(\theta)$. The search for that particular $Q(\theta)$ that minimizes the additive complexity may turn out to be quite elaborate.

In summary, $Q(\theta)$ in $Z(q)$ is selected in a way that $Q(0) \mid (0^N - 1)$, $n \mid N$, N being the order of θ in $Q_0(\theta)$. Such polynomials exist for all values of N such that $N \mid (p^m - 1)$. Similarly, $Q(\theta)$ in $CZ(q)$ is chosen in a way that $Q(\theta) \mid (\theta^N - 1)$, $n \mid N$, N being the order of θ in $Q_0(\theta)$. Such polynomials exist for all values of N such that $N \mid (p^{2m} - 1)$.

Notes

We have focused on the algebra of fields, rings, and polynomial theory with a view to cast the computational problems within a unified mathematical framework. This has led to many interesting results which, when applied to the topic of computing discrete convolution, give rise to generalization of previously known algorithms in some instances and new algorithms in others. These algorithms are also the most efficient algorithms in terms of the number of arithmetic operations, multiplications, and additions. We hope that this mathematical framework will find applications in simplifying many other computationally intensive tasks as well. The idea of using cyclotomic fields to simplify the computation of two-dimensional cyclic convolution was first described by Nussbaumer [1980] and the resulting discrete Fourier transforms were termed as polynomial transforms. It was further generalized by Pitas and Strintzis [1987]. CRT-PR based fast algorithms for two-dimensional cyclic convolutions were first described in Garg [1996] and Garg and Ko [1997].

The algebra of polynomials defined over finite integer rings and their polynomial extensions has not been well studied in digital signal processing till recently. In our recent work, we have shown that this algebra is as rich in structure as the algebra of polynomials defined over finite fields. These results, besides being of significance from a mathematical point of view, lead to new computationally efficient algorithms for computing convolutions.

Bibliography

[10.1] H.J. Nussbaumer, *Fast Fourier Transform and Convolution Algorithms*, Springer-Verlag, 1981.

[10.2] D.G. Myers, *Digital Signal Processing: Efficient Convolution and Fourier Transform Techniques*, Prentice Hall, 1990.

[10.3] H.J. Nussbaumer, "Fast Polynomial Transform Algorithms for Digital Convolution," *IEEE Transactions on Acoustics, Speech and Signal Processing*, Vol. ASSP-28, pp. 205-215, 1980.

[10.4] J.H. McClellan and C.M. Rader, *Number Theory in Digital Signal Processing*, Prentice Hall, 1979.

[10.5] R.E. Blahut, *Fast Algorithms for Digital Signal Processing*, Addison-Wesley Publishing Co., 1984.

[10.6] T.K. Troung, I.S. Reed, R.G. Lipes, and C. Wu, "On the Application of a Fast Polynomial Transform and the Chinese Remainder Theorem to Compute a Two-Dimensional Convolution," *IEEE Transactions on Acoustics, Speech and Signal Processing*, Vol. ASSP-29, pp. 91-97, 1981.

[10.7] I. Pitas and M.B. Strintzis, "Multidimensional Cyclic Convolution Algorithms with Minimal Multiplicative Complexity," *IEEE Transactions on Acoustics, Speech and Signal Processing*, Vol. ASSP-35, pp. 384-390, 1987.

[10.8] H. Krishna Garg, "On Factorization of Polynomials and Direct Sum Properties of Integer Polynomial Rings," *Circuits, Systems, and Signal Processing*, Vol. 15, pp. 415-435, 1996.

[10.9] H. Krishna Garg and C.C. Ko, "Fast Algorithms for Computing Two-Dimensional Convolution in Integer Polynomial Rings," *Circuits, Systems, and Signal Processing*, Vol. 16, pp. 121-139, 1997.

[10.10] K.Y. Lin, B. Krishna, and H. Krishna, "Rings, Fields, the Chinese Remainder Theorem and an American-Indian-Chinese Extension, Part I: Theory," *IEEE Transactions on Circuits and Signal Processing*, Vol. 41, pp. 641-655, 1994.

[10.11] K.Y. Lin, H. Krishna, and B. Krishna, "Rings, Fields, the Chinese Remainder Theorem and an American-Indian-Chinese Extension, Part II: Applications to Digital Signal Processing," *IEEE Transactions on Circuits and Signal Processing*, Vol. 41, pp. 656-668, 1994.

[10.12] H. Krishna, B. Krishna, and K.Y. Lin, "The AICE-CRT and Algorithms for Digital Signal Processing: The Complex Case," *Circuits, Systems, and Signal Processing*, Vol. 14, pp. 69-95, 1995.

[10.13] E. Dubois and A.N. Venetsanopoulos, "Convolution Using a Conjugate Symmetry Property of the Generalized Discrete Fourier Transform," *IEEE Transactions on Acoustics, Speech and Signal Processing*, Vol. ASSP-26, pp. 165-170, 1978.

[10.14] J.B. Martens and M.C. Vanwormhoudt, "Convolution Using a Conjugate Symmetry Property for Number Theoretic Transforms Over Rings of Regular Integers," *IEEE Transactions on Acoustics, Speech and Signal Processing*, Vol. ASSP-31, pp. 1121-1124, 1983.

[10.15] J.B. Martens and M.C. Vanwormhoudt, "Convolution of Long Integer Sequences by Means of Number Theoretic Transforms Over Residue Class Polynomial Rings," *IEEE Transactions on Acoustics, Speech and Signal Processing*, Vol. ASSP-31, pp. 1125-1134, 1983.

[10.16] C.M. Rader, "Discrete Convolutions via Mersenne Transforms," *IEEE Transactions on Computers*, Vol. C-21, pp.1269-1273, 1972.

[10.17] R.C. Agarwal and C.S. Burrus, "Fast Convolution Using Fermat Number Transforms with Applications to Digital Filtering," *IEEE Transactions on Acoustics, Speech and Signal Processing*, Vol. ASSP-22, pp. 87-97, 1974.

[10.18] R.C. Agarwal and C.S. Burrus, "Number Theoretic Transforms to Implement Fast Digital Convolution," *Proceeding IEEE*, Vol. 63, pp.550-560, 1975.

[10.19] H.J. Nussbaumer, "Digital Filtering Using Pseudo Fermat Number Transforms," *IEEE Transactions on Acoustics, Speech and Signal Processing*, Vol. ASSP-25, pp. 79-83, 1977.

[10.20] M.C. Vanwormhoudt, "Structural Properties of Complex Residue Rings Number Theoretic Transforms," *IEEE Transactions on Acoustics, Speech and Signal Processing*, Vol. ASSP-26, pp. 99-104, 1978.

[10.21] E. Dubois and A.N. Venetsanopoulos, "The Discrete Fourier Transform Over Finite Rings with Applications to Fast Convolution," *IEEE Transactions on Computers*, Vol. C-27, pp. 586-593, 1978.

[10.22] I.S. Reed and T.K. Troung, "A Fast Computation of Complex Convolution using a Hybrid Transform," *IEEE Transactions on Acoustics, Speech and Signal Processing*, Vol. ASSP-26, pp. 566-570, 1978.

[10.23] I.S. Reed, T.K. Troung, B. Benjanthrit, and C. Wu, "A Fast Algorithm for Computing a Complex Number Theoretic Transform for Long Sequences," *IEEE Transactions on Acoustics, Speech and Signal Processing*, Vol. ASSP-29, pp. 22-24, 1983.

[10.24] E. Dubois and A.N. Venetsanopoulos, "The Generalized Discrete Fourier Transform in Rings of Algebraic Integers," *IEEE Transactions on Acoustics, Speech and Signal Processing*, Vol. ASSP-28, pp. 169-175, 1980.

[10.25] J.B. Martens, "Number Theoretic Transforms for the Calculation of Convolutions," *IEEE Transactions on Acoustics, Speech and Signal Processing*, Vol. ASSP-31, pp. 969-978, 1983.

[10.26] M. Bhattacharya and R.C. Agarwal, "Number Theoretic Techniques for computation of Digital Convolution," *IEEE Transactions on Acoustics, Speech and Signal Procesing*, Vol. ASSP-32, pp. 507-511, 1984.

[10.27] B. Arambepola and P.J.W. Rayner, "Discrete Transforms Over Polynomial Rings with Applications in Computing Multidimensional Convolutions," *IEEE Transactions on Acoustics, Speech and Signal Processing*, Vol. ASSP-28, pp. 407-414, 1980.

[10.28] R.H.V. Kraats and A.N. Venetsanopoulos, "Hardware for Two-Dimensional Digital Filtering using Fermat Number Transforms," *IEEE Transactions on Acoustics, Speech and Signal Processing*, Vol. ASSP-30, pp. 155-161, 1982.

[10.29] J.B. Martens, "Comments on 'Hardware for Two-Dimensional Digital Filtering using Fermat Number Transforms'," *IEEE Transactions on Acoustics, Speech and Signal Processing*, Vol. ASSP-31, pp. 1034-1037, 1983.

[10.30] J.B. Martens, "Two-Dimensional Convolutions by Means of Number Theoretic Transforms Over Residue Class Polynomial Rings," *IEEE*

Transactions on Acoustics, Speech and Signal Processing, Vol. ASSP-32, pp. 862-871, 1984.

[10.31] G. Martinelli, "Long Convolutions Using Number Theoretic and Polynomial Transforms," *IEEE Transactions on Acoustics, Speech and Signal Processing,* Vol. ASSP-32, pp. 1090-1092, 1984.

Problems

10.1 Derive computationally efficient algorithms for two-dimensional cyclic convolution for $n_1 = 2p$, $n_2 = p$; p an odd prime.

10.2 Derive computationally efficient algorithms for two-dimensional cyclic convolution for $n_1 = 2p$, $n_2 = 2p$; p an odd prime.

10.3 Derive computationally efficient algorithms for two-dimensional cyclic convolution for $n_1 = 2p^2$, $n_2 = 2p^2$; p an odd prime.

10.4 Derive computationally efficient algorithms for two-dimensional cyclic convolution for $n_1 = 2p^a$, $n_2 = 2p^a$; p an odd prime.

10.5 Derive computationally efficient algorithms for two-dimensional cyclic convolution for $n_1 = 4p^a$, $n_2 = 4p^a$; p an odd prime, using factorizations over cyclotomic fields.

10.6 Derive computationally efficient algorithms for two-dimensional cyclic convolution for $n_1 = 4p^a$, $n_2 = 4p^a$; p an odd prime, using factorizations over complex cyclotomic fields.

10.7 Derive computationally efficient algorithms for two-dimensional cyclic convolution for $n_1 = p_1$, $n_2 = p_1 \times p_2$; p_1 and p_2 distinct odd primes.

10.8 Derive computationally efficient algorithms for two-dimensional cyclic convolution for $n_1 = p_1 \times p_2$, $n_2 = p_1 \times p_2$; p_1 and p_2 distinct odd primes.

10.9 Derive computationally efficient algorithms for two-dimensional cyclic convolution for $n_1 = 2^a \times p$, $n_2 = 2^a \times p$; p an odd prime, using factorizations over cyclotomic fields.

10.10 Derive computationally efficient algorithms for two-dimensional cyclic convolution for $n_1 = 2^a \times p$, $n_2 = 2^a \times p$; p an odd prime, using factorizations over complex cyclotomic fields.

10.11 It has been claimed all along that the CRT-P reduction and CRT-P reconstruction are very simple for the two-dimensional cyclic convolution algorithms based on factorization in extension fields and do not involve large valued scalars for the computation. Verify this statement by explicitly deriving the expressions for CRT-P reduction and reconstruction for the fourteen cases discussed in Section 10.5 in the context of two-dimensional cyclic convolution algorithm. Compute the number of ADDs and scaling operations in each case.
Hint: You may use the results of Chapter 9.

10.12 Show that a computation $X(u, v)\ Y(u, v) \bmod (u^{n/2} - 1, v^n - 1)$ can be converted to a computation $X(u, v)\ Y(u, v) \bmod (u^{n/2} + 1, v^n - 1)$ by defining a transformation $u \leftarrow u \cdot \omega$, ω being an n-th root of unity in the field of computation. Describe how one may use the algorithm for the computation $X(u, v)\ Y(u, v) \bmod (u^{n/2} + 1, v^n - 1)$ to compute $X(u, v)$ $Y(u, v) \bmod (u^{n/2} - 1, v^n - 1)$.
Note: If $n/2$ is odd, then the transformation $u \leftarrow -u$ can be used.

10.13 Show that a computation $X(u, v)\ Y(u, v) \bmod (u^n - 1, v^n + 1)$ can be converted to a computation $X(u, v)\ Y(u, v) \bmod (u^n + 1, v^n + 1)$ by defining a transformation $u \leftarrow u \cdot v$. Describe how one may use the algorithm for the computation $X(u, v)\ Y(u, v) \bmod (u^n + 1, v^n + 1)$ to compute $X(u, v)\ Y(u, v) \bmod (u^n - 1, v^n + 1)$.

10.14 Show that the one-dimensional computation $X(u)\ Y(u) \bmod C_{p^2}(u)$ can be converted to a two-dimensional computation $X(u, v)\ Y(u, v) \bmod (u^p - v, C_p(v))$ which can be further written as a direct sum of $(2p - 1)$ one-dimensional computations of the type $X_i(v)\ Y_i(v) \bmod C_p(v)$.

10.15 This is a generalization of Problem 10.14. Show that the one-dimensional computation $X(u)\ Y(u) \bmod C_{p^b}(u)$ can be converted to a two-dimensional computation $X(u, v)\ Y(u, v) \bmod \left(u^{p^{b-a}} - v, C_{p^a}(v)\right)$, which can be further written as a direct sum of one-dimensional computations of the type $X_i(v)\ Y_i(v) \bmod C_{p^a}(u)$ under certain conditions.

10.16 Repeat Problem 10.15 for $p = 2$. In this case, we show that a large size cyclic correlation can be written as a direct sum of a number of smaller size cyclic correlations.

10.17 Write computationally efficient algorithms for computing three-dimensional cyclic convolution of length $p \times p \times p$, p an odd prime number.

10.18 Write computationally efficient algorithms for computing three-dimensional cyclic convolution of length $2p \times 2p \times 2p$, p an odd prime number.

10.19 Write computationally efficient algorithms for computing three-dimensional cyclic convolution of length $2^a \times 2^a \times 2^a$.

10.20 Write computationally efficient algorithms for computing two-dimensional cyclic convolution of length $(p_1 \cdot p_2) \times (p_1 \cdot p_2)$ by first converting it to a four-dimensional cyclic convolution of length $p_1 \times p_1 \times p_2 \times p_2$, p_1 and p_2 being distinct odd prime numbers.

10.21 Now specialize the method in Problem 10.20 for $p_1 = 3$, $p_2 = 5$.

10.22 Compare the algorithm in Problem 10.21 to the algorithm described in Example 10.8.

10.23 Derive computationally efficient algorithm for computing two-dimensional convolution of length 3×3 over $GF(2)$.

10.24 Derive computationally efficient algorithm for computing two-dimensional cyclic convolution of length 5×5 over $GF(2)$.

10.25 Derive computationally efficient algorithm for computing two-dimensional cyclic convolution of length 9×9 over $GF(2)$.

10.26 Derive computationally efficient algorithm for computing two-dimensional cyclic convolution of length 3×3 over $Z(8)$ and $Z(16)$.

10.27 Derive computationally efficient algorithm for computing two-dimensional cyclic convolution of length 5×5 over $Z(4)$ and $Z(32)$.

10.28 Derive computationally efficient algorithm for computing two-dimensional cyclic convolution of length 3×3 over $Z(2^a)$.

10.29 Derive computationally efficient algorithm for computing two-dimensional cyclic convolution of length 5×5 over $Z(2^a)$.

10.30 Derive computationally efficient algorithm for computing two-dimensional cyclic convolution of length 9×9 over $Z(2^a)$.

10.31 Derive computationally efficient algorithm for computing two-dimensional cyclic convolution of length 17×17 over $GF(2)$.
Hint: 17 is a factor of $255 = 2^8 - 1$.

10.32 Derive computationally efficient algorithm for computing two-dimensional cyclic convolution of length 73×73 over $GF(2)$.
Hint: 73 is a factor of $511 = 2^9 - 1$.

10.33 Derive computationally efficient algorithm for computing two-dimensional cyclic convolution of length 17×17 over $Z(2^a)$.

10.34 Derive computationally efficient algorithm for computing two-dimensional cyclic convolution of length 73×73 over $Z(2^a)$.

10.35 Derive computationally efficient algorithm for computing two-dimensional cyclic convolution of length 13×13 over $GF(2)$.
Hint: 13 is a factor of $4095 = 2^{12} - 1$.

10.36 Derive computationally efficient algorithm for computing two-dimensional cyclic convolution of length 13×13 over $Z(2^a)$.

10.37 Derive computationally efficient algorithm for computing two-dimensional cyclic convolution of length 13×13 over $Z(3^a)$.
Hint: 13 is a factor of $26 = 3^3 - 1$.

10.38 Derive computationally efficient algorithm for computing two-dimensional cyclic convolution of length 13×13 over $CZ(3^a)$.
Hint: 13 is a factor of $26 = 3^3 - 1$.

10.39 Derive computationally efficient algorithm for computing two-dimensional cyclic convolution of length 16×16 over $CZ(3^a)$.
Hint: 16 is a factor of $80 = 3^4 - 1$.

10.40 Derive computationally efficient algorithms for computing two-dimensional cyclic convolution of length 31×31 over $Z(5^a)$ and $CZ(5^a)$. This may be treated as a continuation of Example 10.16.
Hint: 31 is a factor of $124 = 5^3 - 1$.

Chapter 11

Validity of Fast Algorithms Over Different Number Systems

In this chapter, we analyze the algebraic structure of fast algorithms for computing one- and two-dimensional convolution of sequences defined over Z and CZ. These algorithms are based on factorization properties of polynomials and the direct sum property of modulo computation over such fields. Algorithms are described for cyclic as well as acyclic convolution. It is shown that under certain non-restrictive conditions, all the previously defined algorithms over Z and CZ are also valid over the rings of finite integers.

11.1 Introduction

Traditionally, the design of algorithms for computing convolution of sequences has been studied within the framework of infinite fields, examples of which are Z, Re, and C. This is followed by similar algorithms over finite fields. These algorithms include FFT algorithms for the computation of DFT. The mathematical tools for deriving these fast algorithms are taken from polynomial algebra. The concepts of roots, unique polynomial factorization, greatest common divisor (*gcd*), Euclid's algorithm, relatively coprime polynomials, modulo arithmetic, and CRT are well established in the algebra of polynomials defined over infinite and finite fields.

In Chapters 7 to 10, we have shown that the algebra of polynomials defined over finite integer rings and their polynomial extensions is as rich in structure as the algebra of polynomials defined over finite fields. These results, besides being of significance from a mathematical point of view, lead to new algorithms for computing convolutions. This is also our main objective in this chapter.

In this chapter, we undertake to further extend the results for computing cyclic and acyclic convolution of data sequences defined over finite integer rings. The algebraic structure of fast algorithms for computing one- and two-dimensional convolution of sequences defined over Z and CZ is analyzed. Algorithms are described for cyclic as well as acyclic convolution. Certain non-restrictive conditions are established under which all the previously defined algorithms over Z and CZ are also valid over the rings of finite integers. It also extends the CRT-P over finite integer rings.

11.2 Mathematical Preliminaries

The number systems that we encounter in this chapter are as follows:

1. Rings of integers, I, and complex integers, CI,
2. Fields of rational, real, complex, and complex rational numbers, Z, Re, C, and CZ,
3. Cyclotomic field CF, and
4. Finite integer and complex integer rings, $Z(M)$ and $CZ(M)$.

It is straightforward to see that $\mathsf{I} \subset \mathsf{Z} \subset \mathsf{CF}$, $\mathsf{I} \subset \mathsf{CI} \subset \mathsf{CZ}$, $\mathsf{Z} \subset \mathsf{CZ}$, $Z(M) \subset CZ(M)$. Given the standard factorization of M as

$$M = \prod_{i=1}^{t} p_i^{a_i} = \prod_{i=1}^{t} q_i \,, \tag{11.2.1}$$

we have the following direct sums:

$$Z(M) = \sum_{\substack{\oplus \\ i=1}}^{t} Z(q_i) \,, \tag{11.2.2}$$

$$CZ(M) = \sum_{\substack{\oplus \\ i=1}}^{t} CZ(q_i) \,. \tag{11.2.3}$$

(11.2.2) and (11.2.3) are based on CRT-I. A bilinear form over a commutative ring R is a computation of the type,

$$z_a = \sum_{\forall b} \sum_{\forall c} A_{a,b,c} x_b y_c, \; a = 0, 1, \ldots, \tag{11.2.4}$$

where x_b and y_c are the b-th and c-th elements of the input vectors $\underline{X}$ and $\underline{Y}$. All quantities in (11.2.4) are defined over N. Here, N is any one of $\mathcal{I}$, $\mathcal{CI}$, Z, Re, $\mathcal{C}$, $\mathcal{CZ}$, $Z(M)$, or $CZ(M)$.

In digital signal processing, two of the most widely occurring bilinear forms correspond to cyclic and acyclic convolution. In this chapter, we are interested in studying the interrelationship between structures of those algorithms for cyclic and acyclic convolution in which the elements of **A**, **B**, and **C** matrices belong to $\mathcal{I}$, $\mathcal{CI}$, Z, $\mathcal{CZ}$, $\mathcal{CF}$, $Z(M)$, and $CZ(M)$. We begin with a brief description of the CRT-P over $Z(M)$ and $CZ(M)$.

11.3 Chinese Remainder Theorem Over Finite Integer Rings

The results presented in this section are unconventional in the sense that CRT-P is shown to exist over the non-unique polynomial factorization domains. Unless otherwise mentioned, all polynomials in this section are assumed to be defined over either $Z(M)$ or $CZ(M)$.

Definition 11.1 A polynomial $A(u)$ is called a *factor* of $C(u)$ if we can write $C(u) = A(u)\, B(u)$.

Definition 11.2 Two monic polynomials $A(u)$ and $B(u)$ are said to be *relatively coprime* if $A_i(u)$ and $B_i(u)$ have no monic factors in common mod q_i, $i = 1, 2, \ldots, t$. Here, $A_i(u) = A(u) \bmod q_i$ and $B_i(u) = B(u) \bmod q_i$. In this case, we write $(A(u), B(u)) = 1$.

Lemma 11.1 Two *monic* polynomials $A(u)$ and $B(u)$ are relatively coprime *iff* we can write

$$D(u)\, A(u) + E(u)\, B(u) - 1 \mod M, \tag{11.3.1}$$

where $deg(D(u)) < deg(B(u))$, $deg(E(u)) < deg(A(u))$. In addition, $D(u)$ and $E(u)$ are unique (not necessarily monic).
Proof: The expression in (11.3.1) holds *iff*

$$D_i(u)\, A_i(u) + E_i(u)\, B_i(u) = 1 \mod q_i, \tag{11.3.2}$$

where $D_i(u) \equiv D(u) \bmod q_i$ and $E_i(u) \equiv E(u) \bmod q_i$. If (11.3.2) holds, then $A_i(u)$ and $B_i(u)$ have no monic factor in common as their common factor will also be a factor of $D_i(u)\, A_i(u) + E_i(u)\, B_i(u)$. It has been shown in Section 7.5.1 that if $A_i(u)$ and $B_i(u)$ have no factors in common mod q_i, then there exist polynomials $D_i(u)$ and $E_i(u)$ that satisfy (11.3.2).

To prove the uniqueness of $D(u)$ and $E(u)$, we proceed as follows. From the direct sum properties it is clear that

$$F(u) = \sum_{\substack{\oplus \\ i=1}}^{t} F_i(u), \tag{11.3.3}$$

where $F(u)$ is any one of $A(u)$, $B(u)$, $C(u)$, $D(u)$, and $E(u)$. Thus, $D(u)$ and $E(u)$ is unique *iff* $D_i(u)$ and $E_i(u)$ are unique. This was shown to be valid in Section 7.5.1.

Lemma 11.2 If two monic relatively coprime polynomials $A(u)$ and $B(u)$ are factors of $C(u)$, then $[A(u)\, B(u)\,] \mid C(u)$. Here, $C(u)$ is not necessarily monic.
Proof: Divide $C(u)$ by $A(u)$ to get

$$C(u) = Q(u)\, A(u). \tag{11.3.4}$$

Dividing $Q(u)$ by $B(u)$, we write $Q(u) = R(u)\, B(u) + S(u)$. Substituting in (11.3.4), we get

$$C(u) = R(u)\, B(u)\, A(u) + S(u)\, A(u) \bmod M \tag{11.3.5}$$

$deg(S(u)) < deg(B(u))$. Since $B(u)$ is a factor of $C(u)$, (11.3.5) leads to

$$S(u)\, A(u) = V(u)\, B(u) \bmod M \tag{11.3.6}$$

$deg(V(u)) < deg(A(u))$, or

$$S_i(u)\, A_i(u) = V_i(u)\, B_i(u) \bmod q_i. \tag{11.3.7}$$

Once again, it has been shown in Section 7.5 that (11.3.7) leads to $V_i(u) = S_i(u) = 0$, or $V(u) = S(u) = 0$. The lemma follows.

Theorem 11.1 CRT-P over finite integer rings. There is a one-to-one correspondence between a polynomial $X(u)$ mod $P(u)$ and its residues $x_j(u)$ defined as

$$x_j(u) = X(u) \bmod p_j(u), \quad j = 1, 2, \ldots, \lambda, \tag{11.3.8}$$

where $p_j(u)$ are monic relatively coprime factors of the monic polynomial $P(u)$. Given $x_j(u)$, $X(u)$ can be reconstructed as:

$$X(u) = \sum_{j=1}^{\lambda} x_j(u) T_j(u) P_j(u) \bmod P(u), \tag{11.3.9}$$

where

$$P_j(u) = \prod_{\substack{k=1 \\ k \neq j}}^{\lambda} p_k(u), \tag{11.3.10}$$

and the unique polynomials $T_j(u)$, $deg(T_j(u)) < deg(p_j(u))$ are obtained from

$$T_j(u)\, P_j(u) + E_j(u)\, p_j(u) = 1 \mod M. \tag{11.3.11}$$

All the polynomials in the above theorem are defined over either $Z(M)$ or $CZ(M)$. Lemmas 11.1 and 11.2 provide the basis for the above theorem. The one-to-one correspondence is established using Lemma 11.2 while the validity of (11.3.9) to (11.3.11) is verified using Lemma 11.1.

It has been shown that $Z(q)$, $q = p^a$, is a unique factorization domain for monic polynomials. This result and the accompanying analysis led to the CRT-PR. The above analysis shows that we require only the factorization of the monic polynomial $P(u)$ in terms of its monic relatively coprime factors. The CRT-P over finite integer rings holds even though this factorization may be non-unique. For example, over $Z(35)$, we have the following four factorizations for $u^6 - 1$:

$$u^6 - 1 = (u + 34)\,(u + 1)\,(u^2 + u + 1)\,(u^2 + 34u + 1) \bmod 35,$$

$$u^6 - 1 = (u + 29)\,(u + 6)\,(u^2 + u + 1)\,(u^2 + 34u + 1) \bmod 35,$$

$$u^6 - 1 = (u + 34)\,(u + 1)\,(u^2 + 6u + 1)\,(u^2 + 29u + 1) \bmod 35,$$

$$u^6 - 1 = (u + 29)\,(u + 6)\,(u^2 + 6u + 1)\,(u^2 + 29u + 1) \bmod 35.$$

Any one of the above factorizations can be used to compute a length 6 cyclic convolution over $Z(35)$. Even though the intermediate quantities will be different, the final result will be the same due to the uniqueness of the CRT-P over $Z(M)$ and $CZ(M)$.

Recall that over Z or CZ, the polynomials $T_j(u)$ are obtained by using the Euclid's algorithm to write the relation

$$T_j(u)\,P_j(u) + E_j(u)\,p_j(u) = 1. \qquad (11.3.12)$$

If A_j is *lcm* of the denominators of the coefficients of $T_j(u)$ and $E_j(u)$, then (11.3.12) can equivalently be written as

$$T_j'(u)\,P_j(u) + E_j'(u)\,p_j(u) = A_j, \qquad (11.3.13)$$

where $T_j'(u) = A_j\,T_j(u)$ and $E_j'(u) = A_j\,E_j(u)$. Clearly, $T_j'(u)$ and $E_j'(u)$ have integer coefficients. It is straightforward to see that (11.3.13) can be converted to an equivalent expression over $Z(M)$ or $CZ(M)$ *iff* $(A_j, M) = 1$, thereby giving rise to unique $T_j(u)$ over $Z(M)$ or $CZ(M)$. Since this chapter focuses on the validity of algorithms originally designed in Z and CZ over $Z(M)$ and $CZ(M)$, a major effort will be on establishing the value of A_j , $j = 1, 2, \ldots , \lambda$ and verifying the condition $(A_j, M) = 1$. No effort is made to exploit the algebraic structure of $Z(M)$ or $CZ(M)$ to directly design algorithms for cyclic and acyclic convolution over finite integer rings.

11.4 Interrelationships Among Algorithms Over Different Number Systems

Algorithms for cyclic and acyclic convolution in Z and CZ are described in chapters 8, 9, and 10. It is our purpose in this section to examine their validity over $Z(M)$ and $CZ(M)$. Let us begin with algorithms for cyclic convolution. These algorithms can be computed as the bilinear form $\underline{Z} = \mathbf{C}\,[\mathbf{A}\,\underline{X} \otimes \mathbf{B}\,\underline{Y}]$. Over Z and CZ, all the elements of **A**, **B**, and **C** matrices are of the type α/β, $\alpha \in \mathtt{I}$ or $\alpha \in \mathtt{CI}$ and $\beta \in \mathtt{I}$, $\beta \neq 0$. Let a, b, and c be the *lcm* of the elements of **A**, **B**, and **C** matrices, respectively. Then the algorithm for cyclic convolution can be cast as

$$\underline{Z}' = K\,\underline{Z} = \mathbf{C}'\,[\mathbf{A}'\,\underline{X} \otimes \mathbf{B}'\,\underline{Y}], \qquad (11.4.1)$$

where

$$\mathbf{C}' = c\,\mathbf{C}$$

$$\mathbf{B}' = b\,\mathbf{B}$$
$$\mathbf{A}' = a\,\mathbf{A}$$

and

$$K = a\,b\,c. \tag{11.4.2}$$

Clearly, the bilinear form $\mathbf{C}'\,[\mathbf{A}'\,\underline{X} \otimes \mathbf{B}'\,\underline{Y}]$ computes the output $\underline{Z}' = K\,\underline{Z}$ with $\mathbf{A}'$, $\mathbf{B}'$, and $\mathbf{C}'$ having elements in $\mathcal{I}$ or $C\mathcal{I}$. Also, it is valid over all commutative rings including $Z(M)$ and $CZ(M)$ as it involves only MULT and ADD. In essence, it computes $\underline{Z}'$ which is same as $\underline{Z}$ except for the gain of K. Over Z and $C\mathsf{Z}$, these statements are trivial. However, over $Z(M)$ and $CZ(M)$, $\underline{Z}$ must be uniquely recoverable from $\underline{Z}'$. Recalling that the congruence $a\,x \equiv d \bmod e$ has a unique solution in $Z(e)$ *iff* $(a, e) = 1$, we get the following result.

Theorem 11.2 All the bilinear algorithms valid over Z and $C\mathsf{Z}$ are valid over $Z(M)$ and $CZ(M)$ *iff* K as defined in (11.4.2) satisfies the condition $(K, M) = 1$. This includes algorithm for cyclic and acyclic convolution as special cases.

The respective matrices over $Z(M)$ and $CZ(M)$ are obtained by replacing each element of the original **A**, **B**, and **C** matrices by its equivalent in the finite integer ring. For $\alpha / \beta \in \mathsf{Z}$ or $C\mathsf{Z}$, its equivalent in $Z(M)$ or $CZ(M)$ is defined as $(\beta^{-1} \bmod M) \cdot \alpha$.

Example 11.1 Consider the length $n = 4$ algorithm for cyclic convolution over $C\mathsf{Z}$. The **A**, **B**, and **C** matrices are

$$\mathbf{A} = \mathbf{B} = \begin{bmatrix} 1 & 1 & 1 & 1 \\ 1 & -j & -1 & j \\ 1 & -1 & 1 & -1 \\ 1 & j & -1 & -j \end{bmatrix},$$

$$\mathbf{C} = \frac{1}{4}\mathbf{A}^*.$$

These matrices may be used to compute cyclic convolution over Z and $C\mathsf{Z}$. In this case, $a = b = 1$, $c = 4$, and $K = 4$. Therefore, these matrices may also be used to compute cyclic convolution over $Z(M)$ and $CZ(M)$ that satisfy $(4, M) = 1$ or $(2, M) = 1$. Clearly, it indicates the possibility of a wide range of values for M including integers of the type $2^r - 1$ (Mersenne numbers) and $2^r + 1$ (Fermat numbers).

Theorem 11.2 is simple to a great extent. It can be used on an ad-hoc basis as follows. Derive the bilinear form for the algorithm over Z or CZ and compute K in (11.4.2). These bilinear elements of the **A**, **B**, and **C** matrices are modified to reflect their structure over $Z(M)$ or $CZ(M)$.

In the remainder of this section, we consider specific methods to construct algorithms for cyclic and acyclic convolution over Z and CZ and analyze their validity over $Z(M)$ and $CZ(M)$. The important aspect of the convolution algorithms that are studied is *a priori* computation of the scalar K. It is not our aim to exhaustively analyze the structure of all the different methods for all possible values of n. We hope to show that the mathematical formulation of these methods is rich to the extent that their suitability can be judged *a priori* for given values of n and M. We begin with the analysis of algorithm for cyclic convolution.

11.4.1 Analysis of Cyclic Convolution Algorithms

These are three distinct approaches to designing algorithm for cyclic convolution over Z and CZ. They are

1. The direct approach,
2. The Agarwal-Cooley algorithm, and
3. Polynomial transforms or recursive nesting.

In the direct approach, $u^n - 1$ is factored over Z or CZ. The factor polynomials over Z (CZ) are known as the (complex) cyclotomic polynomials. They have been studied in Chapter 5. If $4 \mid n$, then $C_n(u)$ further factorizes in terms of complex cyclotomic polynomials.

Given the factorization of $u^n - 1 = \prod_{i=1}^{\lambda} p_i(u)$, the direct approach based algorithm for cyclic convolution over Z or CZ involves the following three steps:

(i) **CRT-P reduction.** Compute:

$$x_i(u) = X(u) \bmod p_i(u),$$
$$y_i(u) = Y(u) \bmod p_i(u). \tag{11.4.3}$$

(ii) **Small degree polynomial MULT:**

$$z_i'(u) = x_i(u)\, y_i(u), \tag{11.4.4}$$

$$z_i(u) = z_i'(u) \bmod p_i(u).$$

(iii) **CRT-P reconstruction.**

$$Z(u) = \sum_{i=1}^{\lambda} z_i(u)T_i(u)P_i(u) \bmod (u^n - 1). \tag{11.4.5}$$

It is known that the unique polynomials $T_i(u)$ are given by

$$T_i(u) = \frac{1}{n}T_i'(u) \tag{11.4.6}$$

$$T_i'(u) = \left[u\frac{d}{du}p_i(u)\right] \bmod p_i(u). \tag{11.4.7}$$

Since (complex) cyclotomic polynomials are monic with (complex) integer coefficients, Step (i) does not require any division. In Step (iii), the only division is by n as $T_i'(u)$ and $P_i(u)$ are polynomials with (complex) integer coefficients as well. Only Step (ii) may require division in the small length acyclic convolution for obtaining $z_i'(u)$ in (11.4.4).

In the Agarwal-Cooley algorithm, a length n cyclic convolution is computed by first converting it to a two- or higher-dimensional cyclic convolution. More precisely, if $n = n_1 \cdot n_2 \cdots n_d$, $(n_i, n_j) = 1$, $i \neq j$, then a length n cyclic convolution is first converted to a d-dimensional cyclic convolution, length being n_i in the i-th dimension, $i = 1, 2, \ldots, d$. This mapping is independent of the number system over which the data sequences are defined. It depends only on the number theoretic properties of the indices. In each of the d-dimensions, algorithm for cyclic convolution is obtained by applying the direct approach. Therefore, in the Agarwal-Cooley algorithm, the i-th dimension requires division by n_i. It is clear that the overall algorithm for cyclic convolution requires division by $n = n_1 \cdot n_2 \cdots n_d$.

In the polynomial transform or recursive nesting approach, the computation for Step 2 in (11.4.4) is written as

$$z(u) \equiv x(u)\, y(u) \bmod p_i(u). \tag{11.4.8}$$

This computation is converted to a two-dimensional convolution via suitable change of variables. This method is widely used for $n = 2^a$. Then $p_d(u) = u^{2^{d-1}} + 1$, $d \mid n$. In this case, the mapping is $u_1 = u^{2^k}$ and (11.4.8) is transformed into

$$z(u, u_1) \equiv x(u, u_1)\, y(u, u_1) \bmod (u^{2^k} - u_1, u_1^{2^{d-1-k}} + 1). \tag{11.4.9}$$

This involves computing polynomial product mod $(u^{2^k} - u_1)$ in the cyclotomic field $CF(2^{d-k})$. The computation in (11.4.9), in turn, is converted to a cyclic convolution as

$$z(u, u_1) \equiv x(u, u_1)\, y(u, u_1) \bmod (u_1^{2^{d-1-k}} + 1, u^{2^{k+1}} - 1). \tag{11.4.10}$$

Provided that $2^k < 2^{d-k}$ or $k < d/2$, the cyclic convolution in (11.4.10) is computed using the factorization of $(u^{2^{k+1}} - 1)$ over $CF(2^{d-k})$ as

$$u^{2^{k+1}} - 1 = \prod_{l=0}^{2^{k+1}-1} (u - u_1^{al}) \bmod (u_1^{2^{d-1-k}} + 1), \tag{11.4.11}$$

where

$$a = 2^{d-2k-1}. \tag{11.4.12}$$

This leads to a DFT type of algorithm for cyclic convolution of length 2^{k+1} in $CF(2^{d-k})$. In a manner similar to (11.4.6) and (11.4.7), it is easily seen that the only divisions this algorithm requires is by 2^{k+1} in the IDFT. Also, each of the DFT coefficients is a polynomial in u_1. In essence, the computation $x(u)\, y(u)$ mod $(u^{2^{d-1}} + 1)$ is expressed as 2^{k+1} computations of the type $x(u_1^{al}, u_1)\, y(u_1^{al}, u_1)$ mod $(u_1^{2^{d-1-k}} + 1)$. These 2^{k+1} computations can be further decomposed in a recursive manner. This recursive nesting approach leads to substantial savings when it is applied with exponentially increasing radices. Also, the entire algorithm can be based on a single small degree polynomial MULT of the type $x(u)\, y(u)$ mod $(u^2 + 1)$ which requires no division.

It is clear from the description of the three approaches that they all consist in reducing a large length cyclic convolution into a number of smaller length acyclic convolutions and in process require division by n. These smaller length acyclic convolutions are then computed in a computationally efficient manner. This discussion leads to the following result.

Theorem 11.3 The (complex) cyclotomic polynomial based direct method, the Agarwal-Cooley algorithm, and the recursive nesting method which are used to compute cyclic convolution over Z or CZ are also valid for computing cyclic convolution over $Z(M)$ and $CZ(M)$ *iff*

(1) $(n, M) = 1$, and
(2) The small length acyclic convolution are also valid over $Z(M)$ or $CZ(M)$.

It is relatively easier to verify the second condition in this theorem than the condition in Theorem 11.2. This is discussed further when acyclic convolution algorithms over **Z** and **CZ** are analyzed in order to establish their validity over $Z(M)$ and $CZ(M)$.

We end our discussion on algorithms for cyclic convolution by making a comment about the recursive nesting approach, $n = 2^a$. The factorization in (11.4.11) is valid over $Z(M)$ and $CZ(M)$ *iff* $(2^a, M) = 1$ or $(2, M) = 1$ with one important difference. The arithmetic mod $u_1^{2^a} + 1$ over **Z** gives rise to a cyclotomic field as $u_1^{2^a} + 1$ is irreducible over **Z**. Over $Z(M)$ and $CZ(M)$, the arithmetic mod $u_1^{2^a} + 1$ gives rise to a polynomial ring. However, all the computations including the algorithm for $x(u)\,y(u)$ mod $u^2 + 1$ are valid under the condition $(2, M) = 1$. Once again, the Mersenne and Fermat numbers trivially satisfy this constraint. This makes the recursive nesting approach a very powerful one for computing cyclic convolution over $Z(M)$ and $CZ(M)$ when M is either a Mersenne or a Fermat number.

11.4.2 Analysis of Acyclic Convolution Algorithms

There are two approaches to designing algorithms for acyclic convolution over **Z** and **CZ**. They are:

1. The direct method, and
2. The multidimensional approach.

In the direct method, a length n' acyclic convolution $Z(u) = X(u)\,Y(u)$ is expressed as

$$Z(u) = X(u)\,Y(u) \bmod P(u), \tag{11.4.13}$$

$deg(P(u)) = D$, $D \le n'$. Here, $P(u)$ is an arbitrary monic polynomial having the factorization

$$P(u) = \prod_{i=1}^{\lambda} p_i(u), \tag{11.4.14}$$

where $p_i(u)$ are monic and irreducible (or power of an irreducible) polynomials satisfying $(p_i(u), p_j(u)) = 1$, $i \neq j$ over $\mathbf{Z}$ or $C\mathbf{Z}$. The last $n' - D$ coefficients are recovered separately via the computation

$$\bar{z}(u) = \bar{x}(u)\bar{y}(u) \bmod u^{n'-D}, \tag{11.4.15}$$

called the *wraparound*. $\bar{A}(u)$ is obtained by listing the coefficients of $A(u)$ in the reverse order.

The computation in (11.4.13) is carried out using the CRT-P as in (11.4.3) to (11.4.5). Once again, the only division that may be required is due to the small length acyclic convolution required for (11.4.4) or the $T_i(u)$ polynomials in the CRT-P reconstruction in (11.4.5).

In the multidimensional approach, a length n' acyclic convolution, $n' = n_1 \cdot n_2 \cdots n_d$ is converted to a d-dimensional acyclic convolution via a suitable mapping of indices. This gives rise to

$$Z(u_1, u_2, \ldots, u_d) = x(u_1, u_2, \ldots, u_d)\, y(u_1, u_2, \ldots, u_d). \tag{11.4.16}$$

Algorithm for acyclic convolution in the i-th dimension is obtained by either the direct method or by trial and error if n_i is small, $i = 1, 2, \ldots, d$. The value of K for the multidimensional approach can, therefore, be determined once the divisors in the algorithm for acyclic convolution for each of the d-dimensions are known. These are determined by analyzing the direct method.

The modulo polynomials for acyclic convolution are far less structured due to the freedom in choosing $P(u)$. Therefore, they are also more difficult to analyze. However, we may still get a good insight into their behavior by discovering the exact source of division. The only division that may be required arises in either $T_i(u)$ or the small length acyclic convolution $z_i(u) = x_i(u)\, y_i(u)$. In most instances that we have come across, $n' - D$ is quite small, say 3 or 4, and the computation of wraparound coefficients in (11.4.15) requires no division. If the small length acyclic convolutions are computed using the CRT-P as well, then they can also be analyzed in a manner similar to the overall algorithm. In any case, the small length acyclic convolution algorithm is assumed to be known.

If $P(u)$ in (11.4.14) is chosen in a way that $p_i(u)$ is the d_i-th (complex) cyclotomic polynomial, then it is seen that over $\mathbf{Z}$ or $C\mathbf{Z}$,

$$T_i(u) = \frac{1}{n}\left[u\theta(u)\frac{d}{du}p_i(u)\right] \bmod p_i(u), \tag{11.4.17}$$

where $\theta(u)$ is such that

$$u^n - 1 = \theta(u)\, P(u) \tag{11.4.18}$$

and

$$n = lcm(d_1, d_2, \ldots). \tag{11.4.19}$$

In this case, the only division that may be required is by n. It is possible that the integer coefficients of the polynomial in parenthesis in (11.4.17) have $gcd\ a_i$ such that $(a_i, n) > 1$. In that case, the denominator in (11.4.17) is given by $n/(a_i, n)$.

It can be shown that two polynomials, $p_1^{\alpha_1}(u)$ and $p_2^{\alpha_2}(u)$, are relatively coprime over $Z(M)$ or $CZ(M)$ *iff* $p_1(u)$ and $p_2(u)$ are relatively coprime over $Z(M)$ or $CZ(M)$. In other words,

$$D_1(u)\ p_1^{\alpha_1}(u) + E_1(u)\ p_2^{\alpha_2}(u) = 1 \bmod M \tag{11.4.20}$$

iff

$$D_1'(u)\, p_1(u) + E_1'(u)\, p_2(u) = 1 \bmod M. \tag{11.4.21}$$

One of the commonly used factors in acyclic convolution is $p_1(u) = u^b$. For $(p_1(u), p_i(u)) = 1$ over **Z** or **CZ**, we have $p_0 \neq 0$, where p_0 is the coefficient of u^0 in $p_i(u)$. From (11.4.20) and (11.4.21), u^b is relatively coprime to $p_i^{\alpha_i}(u)$ *iff* u and $p_i(u)$ are relatively coprime. Let $p_i(u) = p_0 + p_1 u + \ldots + p_c u^c$, then

$$-(p_1 + p_2 u + \ldots + p_c u^{c-1})\, u + p_i(u) = p_0 \bmod M. \tag{11.4.22}$$

Thus, u^b and $p_i^{\alpha_i}(u)$ are relatively coprime over $Z(M)$ or $CZ(M)$ *iff* $(p_0, M) = 1$.

We summarize the above discussion in the following three lemmas.

Lemma 11.3 The acyclic convolution based on

$$P(u) = u^b \prod_{i=1}^{\lambda} [p_i(u)]^{\alpha_i}$$

is valid over $Z(M)$ or $CZ(M)$, *iff* the acyclic convolution based on the polynomial

$$P'(u) = u^{\delta(b)} \prod_{i=1}^{\lambda} p_i(u)$$

is valid over $Z(M)$ and $CZ(M)$. Here, $\delta(b) = 1$, $b > 0$, and $\delta(b) = 0$ for $b = 0$.

Lemma 11.4 The acyclic convolution based on

$$P(u) = u\prod_{i=1}^{\lambda} p_i(u) = uP'(u)$$

is valid over $Z(M)$ or $CZ(M)$ *iff* the acyclic convolution based on $P'(u)$ is valid over $Z(M)$ or $CZ(M)$ and $(p_0, M) = 1$, p_0 being the coefficient of u^0 in $P'(u)$.

Lemma 11.5 The acyclic convolution based on

$$P(u) = u^b \prod_{i=1}^{\lambda} [p_i(u)]^{\alpha_i}$$

$p_i(u)$ being the d_i-th (complex) cyclotomic polynomial, is valid over $Z(M)$ or $CZ(M)$ *iff* $(A, M) = 1$ where

$$A = lcm\left(\frac{n}{(a_1, n)}, \frac{n}{(a_2, n)}, \bullet\bullet\bullet\right)$$

$$n = lcm(d_1, d_2, \dots).$$

Here, a_i is *gcd* of the coefficients of the polynomial in parenthesis in (11.4.17). A sufficient condition is $(n, M) = 1$.

In Lemmas 11.3 to 11.5, the small length acyclic convolution algorithms originally designed in **Z** or ***CZ*** are assumed to be valid over $Z(M)$ or $CZ(M)$.

Example 11.2 Consider length 9 ($n' = 9$) acyclic convolution. Let us say that $D = 2$ and $P(u)$ is chosen to be

$$P(u) = u^2 (u - 1)^2 (u + 1) (u^2 + u + 1).$$

Based on Lemma 11.3, we need to test the validity of the acyclic convolution for $P(u) = u (u - 1) (u + 1) (u^2 + u + 1)$. Applying Lemma 11.4, we see that the condition $(p_0, M) = 1$ is trivially satisfied as $p_0 = -1$. Thus we need to test the validity of the algorithm only for $P(u) = (u - 1) (u + 1) (u^2 + u + 1)$. Recognizing $(u - 1)$, $(u + 1)$, and $(u^2 + u + 1)$ as the first, second, and third cyclotomic polynomials, we have $n = lcm(1, 2, 3) = 6$ and $\theta(u) = u^2 - u + 1$. Using (11.4.17), we have

$$T_1(u) = 1/6$$
$$T_2(u) = -1/2$$
$$T_3(u) = (1/3)(u - 1).$$

In this case, $a_1 = 1$, $a_2 = 3$, $a_3 = 2$, and $A = lcm(6, 2, 3) = 6$. Therefore, length 9 acyclic convolution algorithm based on $D = 2$ and $P(u) = u^2 (u - 1)^2 (u + 1)(u^2 + u + 1)$ is valid for all $Z(M)$ and $CZ(M)$ that satisfy $(6, M) = 1$. This algorithm requires 13 MULTs. If $P(u)$ is chosen to be $u^2 (u - 1)^2 (u + 1)(u^2 + 1)$, then the condition for validity becomes $(2, M) = 1$. The multiplicative complexity remains unchanged at 13.

11.5 Analysis of Two-Dimensional Cyclic Convolution Algorithms

In this section, we analyze the structure of algorithms for computing two-dimensional cyclic convolution over Z and CZ and examine their validity over $Z(M)$ and $CZ(M)$. The two-dimensional cyclic convolution of length $n_1 \times n_2$ is defined as

$$Z(u, v) = X(u, v)\, Y(u, v) \bmod \left(u^{n_1} - 1, v^{n_2} - 1\right). \tag{11.5.1}$$

There are two approaches to computing $Z(u, v)$. They are

1. The direct method, and
2. Polynomial transforms.

In the direct method, all polynomials are expressed in powers of u with the coefficients as polynomials in v. The method is essentially the same as the algorithm for length n_1 one-dimensional cyclic convolution in u where each scalar MULT is replaced by a one-dimensional cyclic convolution in v of length n_2. The following lemma follows in a straightforward manner.

Lemma 11.6 The two-dimensional algorithm for cyclic convolution of length $n_1 \times n_2$ over Z or CZ obtained using the direct method is valid over $Z(M)$ and $CZ(M)$ *iff* each of the constituent one-dimensional algorithm for cyclic convolution of length n_1 and n_2 are valid over $Z(M)$ and $CZ(M)$. In other words,

(i) $(n_1, M) = (n_2, M) = 1$, and

(ii) The small length acyclic convolution algorithms used in the computation of cyclic convolutions of length n_1 and n_2 are valid over $Z(M)$ or $CZ(M)$.

Polynomial transforms have been most widely studied for the case $n = n_1 = n_2$, with (a) $n = q$, q prime, and (b) $n = q^c$. It consists in converting a two-dimensional cyclic convolution into a number of one-dimensional convolutions. Consider the case $n = q^c$. In this case, we express

$$\left(u^{q^c} - 1\right) = \left(u^{q^{c-1}} - 1\right) C_{q^c}(u) . \tag{11.5.2}$$

In (11.5.2), $C_{q^c}(u)$ is a degree $q^c - q^{c-1}$ cyclotomic polynomial. Thus $Z(u, v)$ in (11.5.1) can be obtained from $Z_1(u, v)$ and $Z_2(u, v)$ by using the CRT-P corresponding to the factorization in (11.5.2), where

$$Z_1(u, v) = X(u, v)\, Y(u, v) \bmod \left(u^{q^{c-1}} - 1, v^{q^c} - 1\right) \tag{11.5.3}$$

and

$$Z_2(u, v) = X(u, v)\, Y(u, v) \bmod \left(C_{q^c}(u), v^{q^c} - 1\right). \tag{11.5.4}$$

Similarly, $Z_1(u, v)$ can be obtained from $Z_3(u, v)$ and $Z_4(u, v)$ using the CRT-P corresponding to the factorization in (11.5.2), where

$$Z_3(u, v) = X(u, v)\, Y(u, v) \bmod \left(u^{q^{c-1}} - 1, C_{q^c}(v)\right) \tag{11.5.5}$$

and

$$Z_4(u, v) = X(u, v)\, Y(u, v) \bmod \left(u^{q^{c-1}} - 1, v^{q^{c-1}} - 1\right). \tag{11.5.6}$$

Using the same analysis as in one-dimensional cyclic convolution, it is seen that obtaining $Z_1(u, v)$ from $Z_3(u, v)$ and $Z_4(u, v)$ and obtaining $Z(u, v)$ from $Z_1(u, v)$ and $Z_2(u, v)$ requires division by q^c.

The computations for $Z_2(u, v)$ in (11.5.4) and $Z_3(u, v)$ in (11.5.5) are performed using a DFT type algorithm corresponding to the factorization

$$v^{q^c} - 1 = \prod_{i=0}^{q^c - 1} (v - u^i) \bmod C_{q^c}(u) \tag{11.5.7}$$

and

$$u^{q^{c-1}} - 1 = \prod_{i=0}^{q^{c-1} - 1} (u - (v^q)^i) \bmod C_{q^c}(v) . \tag{11.5.8}$$

Once again, the CRT-P reconstruction corresponding to (11.5.7) and (11.5.8) require division by q^c and q^{c-1}, respectively. Based on (11.5.2) to (11.5.8), we see that a $q^c \times q^c$ cyclic convolution is computed in the following manner:

(i) q^c polynomial products mod $C_{q^c}(u)$ in (11.5.7),
(ii) q^{c-1} polynomial products mod $C_{q^c}(v)$ in (11.5.8), and
(iii) a two-dimensional cyclic convolution of length $q^{c-1} \times q^{c-1}$.

The two-dimensional cyclic convolution of length $q^{c-1} \times q^{c-1}$ can be computed using the above expressions in a recursive manner. This also leads to the following theorem.

Theorem 11.4 The polynomial transform based algorithm for computing two-dimensional cyclic convolution of length $q^c \times q^c$ is valid over $Z(M)$ and $CZ(M)$ *iff*

(i) $(M, q) = 1$, and
(ii) algorithms for computing products of the type $X(u) \cdot Y(u) \bmod C_{q^i}(u)$, $i = c, c-1, \ldots, 1$, are valid over $Z(M)$ and $CZ(M)$.

In most instances, the second condition in the above theorem also leads to the condition $(M, q) = 1$. More specifically, if $q = 2$, then $C_{2^i}(u) = u^{2^{i-1}} + 1$ and all the polynomial products of the type $X(u) \cdot Y(u) \bmod C_{q^i}(u)$ can be computed using recursive nesting. The resulting two-dimensional algorithm for cyclic convolution is valid over all $Z(M)$ and $CZ(M)$ that satisfy $(2, M) = 1$. Once again, this condition is satisfied trivially by Mersenne and Fermat numbers.

We conclude this section by observing that three- or higher-dimensional algorithm for cyclic convolution based on either the direct method or the polynomial transform approach can also be analyzed in a similar manner as the two-dimensional algorithm for cyclic convolution analyzed here.

Notes

In this work, we have shown that under certain non-restrictive conditions the algorithms for computing convolution over finite integer rings are the same as the algorithms for computing convolutions over fields of rational and complex rational numbers. All the different approaches for computing one-dimensional (cyclic and acyclic) convolution and two-dimensional cyclic convolution are analyzed. Thus, these algorithms are completely equivalent in terms of computational complexity, organization, and overall implementation. Interestingly, this result is established without any mathematical analysis of the structure of finite integer rings.

The results reported in this chapter were based on Garg et al [1996]. An example of the algorithm for a 3×3 cyclic convolution is described in Section 6.4.2 of Nussbaumer [1981] that requires division only by 9, thereby requiring $(M, 3) = 1$ for it to be valid over $Z(M)$ and $CZ(M)$.

Bibliography

[11.1] H. Krishna, B. Krishna, K.-Y. Lin, and J.-D. Sun, *Computational Number Theory and Digital Signal Processing*, CRC Press, Boca Raton, 1994.

[11.2] H.J. Nussbaumer, *Fast Fourier Transform and Convolution Algorithms*, Springer-Verlag, 1981.

[11.3] D.G. Myers, *Digital Signal Processing, Efficient Convolution and Fourier Transform Techniques*, Prentice Hall, 1990.

[11.4] J.H. McClellan and C.M. Rader, *Number Theory in Digital Signal Processing*, Prentice Hall, 1979.

[11.5] R.E. Blahut, *Fast Algorithms for Digital Signal Processing*, Addison-Wesley Publishing Co., 1984.

[11.6] T.K. Troung, I.S. Reed, R.G. Lipes, and C. Wu, "On the Application of a Fast Polynomial Transform and the Chinese Remainder Theorem to Compute a Two-Dimensional Convolution," *IEEE Transactions on Acoustics, Speech and Signal Processing*, Vol. ASSP-29, pp. 91-97, 1981.

[11.7] I. Pitas and M.G. Strintzis, "Multidimensional Cyclic Convolution Algorithms with Minimal Multiplicative Complexity," *IEEE Transactions on Acoustics, Speech & Signal Processing*, Vol. ASSP-35, pp. 384-390, 1987.

[11.8] G. Martinelli, "Long Convolutions Using Number Theoretic and Polynomial Transforms," *IEEE Transactions on Acoustics, Speech & Signal Processing*, Vol. ASSP-32, pp. 1090-1092, 1984.

[11.9] H. Krishna Garg, C.C. Ko, K.-Y. Lin, and H. Liu, "On Algorithms for Digital Signal Processing of Sequences," *Circuits, Systems, and Signal Processing*, Vol. 15, pp. 437-452, 1996.

Problems

11.1 Study the validity of the two-dimensional cyclic convolution algorithm of length 5×5 derived using cyclotomic polynomials over finite fields and integer rings.

11.2 Study the validity of the two-dimensional cyclic convolution algorithm of length 9×9 derived using cyclotomic polynomials over finite fields and integer rings.

11.3 Study the validity of the two-dimensional cyclic convolution algorithm of length 12×12 derived using cyclotomic and complex cyclotomic polynomials over finite fields and integer rings.

Chapter 12

Fault Tolerance for Integer Sequences

In many signal processing and computing systems, one is faced with the task of processing data which is inherently defined in $Z(M)$ or $CZ(M)$, or can equivalently be converted to one. Such arithmetic is based on the number theoretic properties of data sequences. Along other lines, coding theory has flourished as an art and science to protect numeric data from errors, thereby improving the performance of data processing systems. However, by and large, with certain exceptions, coding theory is developed as it applies to digital communication systems. This is evident from a number of texts on coding theory as it relates to communication theory. Exception to this is the application of coding theory concepts to fault tolerant computing systems. With the advent of *very large scale integration* (VLSI) design methodology and concepts of parallel processing, there has been tremendous thrust towards high performance computing systems that possess fault tolerance capability. The original idea of **algorithm based fault tolerance** (ABFT) is based on some of the classical results available in coding theory.

12.1 A Framework for Fault Tolerance

This chapter is an effort to bring together number theory and coding theory with the objective of developing a mathematical framework for building algorithm based fault tolerance capability in vectors defined over $Z(M)$ or $CZ(M)$. The impetus for this work is provided by the applications of integer arithmetic in certain computational environments. Lack of an underlying Galois field presents a unique challenge to this framework. We note here that the most well known application of coding theory is in digital communications where the data is defined in $GF(q)$, $q = p^m$. We develop the theory and algorithms for error control in data sequences defined over a ring of integers.

The finite integer ring $Z(M)$ is defined by a composite integer M having the standard factorization,

$$M = \prod_{i=1}^{t} p_i^{a_i} = \prod_{i=1}^{t} q_i , \tag{12.1.1}$$

where p_i is the i-th prime, a_i is a positive integer exponent, and $q_i = p_i^{a_i}$. We say that the ring $Z(M)$ has *degeneracy* $\{t, a\}$, where $a = max(a_1, a_2, \ldots, a_t)$. An integer X, $0 \le X < M$, can equivalently be represented by a residue vector $\underline{X}$ having t components $X_1, X_2, \ldots, X_t$ where

$$X_i \equiv X \bmod q_i, \quad i = 1, 2, \ldots, t. \tag{12.1.2}$$

Given a residue vector $\underline{X}$, the corresponding integer X in $Z(M)$ can be uniquely determined by solving t congruences in (12.1.2). The uniqueness of the integer X and a procedure to compute it is established by the CRT-I.

The ABFT technique is a (n, k) coding scheme **C** that consists in mapping a length k data vector $\underline{A}$ to a length n codevector $\underline{V}$. All the length k data vectors form a vector space of dimension k. Since the coding scheme **C** being introduced is linear in nature, the corresponding length n codevectors also form a vector space of dimension k. Such a coding scheme is completely characterized by k **linearly independent** (*l.i.*) codevectors $\underline{V}_i$, $i = 1, 2, \ldots, k$ associated with the k unit data vectors $\underline{E}_i$, $i = 1, 2, \ldots, k$. The unit data vector $\underline{E}_i$ is a vector having 1 in its i-th position and 0s elsewhere. Without any loss of generality, we assume that the codevectors $\underline{V}_i$ have the form

$$\underline{V}_i = [\underline{E}_i | \overbrace{**\cdots**}^{n-k}] . \tag{12.1.3}$$

The '*' in (12.1.3) denote the $(n - k)$ parity check digits to be assigned. As a result, the codevector for an arbitrary data vector $\underline{A} = [A_1\ A_2\ \ldots\ A_k]$ can be written as $\underline{V} = \underline{A}\,\mathbf{G}$, where, $\mathbf{G}$, a $(k \times n)$ matrix, is called the **generator matrix** of the code (in systematic form) and is given by

$$\mathbf{G} = \begin{bmatrix} \underline{V}_1 \\ \underline{V}_2 \\ \vdots \\ \underline{V}_k \end{bmatrix} = [\mathbf{I}_k \mid \mathbf{P}]. \tag{12.1.4}$$

Here $\mathbf{I}_k$ is the identity matrix of dimension k and $\mathbf{P}$ is a $k \times (n - k)$ matrix having elements in $Z(M)$. This is based on the linear nature of the coding scheme, modulo operation, and the decomposition of $\underline{A}$ as $\underline{A} = \sum_{i=1}^{k} A_i \underline{E}_i$. We also note that the codevector $\underline{V}$ has the form $\underline{V} = [\underline{A} \mid \underline{A}\,\mathbf{P}]$. The first k digits of $\underline{V}$ are the same as those of $\underline{A}$ and the remaining $(n - k)$ parity check digits are appended to provide for fault tolerance.

It is clear from the form of $\mathbf{G}$ in (12.1.4) that $rank(\mathbf{G}) = k$. Now consider the null space $\mathbf{U}$ of the vector space $\mathbf{V}$ created by the rows of the generator matrix $\mathbf{G}$ in (12.1.4). Let $\underline{H}_1$, $\underline{H}_2$, …, $\underline{H}_{n-k}$ be the $(n - k)$ *l.i.* vectors that span $\mathbf{U}$. It is clear that

$$\underline{V}\,\mathbf{H}^t = \underline{0}, \tag{12.1.5}$$

where $\mathbf{H}$, to be termed as the **parity check matrix** of the code $\mathbf{C}$, is a $(n - k) \times n$ matrix given by

$$\mathbf{H} = \begin{bmatrix} \underline{H}_1 \\ \underline{H}_2 \\ \vdots \\ \underline{H}_{n-k} \end{bmatrix}. \tag{12.1.6}$$

For the systematic form of $\mathbf{G}$, $\mathbf{H}$ can be written by inspection as

$$\mathbf{H} = [-\mathbf{P}^t \mid \mathbf{I}_{n-k}]. \tag{12.1.7}$$

The **minimum distance**, d, is a fundamental parameter associated with any random error correcting code and is defined as

$$d = min\{d(\underline{V}_i, \underline{V}_j): \underline{V}_i, \underline{V}_j \in \mathbf{C}, \underline{V}_i \neq \underline{V}_j\}.$$

Since the coding scheme being studied is lincar in nature, it is easily established that $d = min\{wt(\underline{V}), \underline{V} \in \mathbf{C}, \underline{V} \neq \underline{0}\}$. Here, $d(\underline{V}_i, \underline{V}_j)$ and $wt(\underline{V})$ denote the **Hamming distance** between the vectors $\underline{V}_i$ and $\underline{V}_j$ and the **Hamming weight** of the vector $\underline{V}$, respectively. The minimum distance, d, is intimately related to the fault detection and correction capability of the code **C**.

12.1.1 Fault Detection and Correction

Given a codevector $\underline{V}$, let $\underline{V}$ be changed to a vector $\underline{R}$ due to presence of faults in computation. In this case, we write $\underline{R} = \underline{V} + \underline{E}$ or

$$R_i \equiv V_i + E_i \bmod M, \quad i = 1, \ldots, n, \tag{12.1.8}$$

where $\underline{E}$ is the fault vector, $\underline{E} = [E_1\ E_2\ \ldots\ E_n]$. If $E_i \neq 0$, then a fault is said to have taken place in the i-th position. The Hamming weight of the fault vector $\underline{E}$, $wt(\underline{E})$, also denoted by γ, is called the number of faults in $\underline{R}$. Also, we have $d(\underline{R}, \underline{V}) = wt(\underline{E}) = \gamma$.

Fault detection consists in checking if $\underline{R}$ is a codevector in **C** or not. This can be accomplished by computing $\underline{R}\,\mathbf{H}^t$. Fault correction or decoding consists in computing the position of the faults and the fault magnitudes. The relationships between d and the fault detection and correction capabilities are straightforward to derive. The **fault detection capability** of a code **C**, defined as the largest value of γ for which $\underline{R}$ cannot be a codevector in **C**, is $d - 1$. The **fault correcting capability**, λ, of a code **C**, defined as the largest value of γ for which correct decoding takes place ($\hat{\underline{V}} = \underline{V}$), is $\lambda = \lfloor (d-1)/2 \rfloor$, where $\lfloor a \rfloor$ denotes the largest integer less than or equal to a. A code **C** can correct λ faults and simultaneously detect β faults, $\beta > \lambda$, if $d \geq \lambda + \beta + 1$.

Having established the general framework for ABFT techniques over $Z(M)$, we now analyze the mathematical structures of these codes in more detail.

12.2 Mathematical Structure of C Over $Z(M)$

Consider a length k data vector $\underline{A} = [A_1\ A_2\ \ldots\ A_k]$, $A_j \in Z(M)$. Based on 12.1.1 and the CRT-I, it can equivalently be represented by t vectors, each of length k, corresponding to t distinct primes that constitute M, that is,

$$\underline{A} \leftrightarrow [\underline{A}^{(1)}\ \underline{A}^{(2)}\ \ldots\ \underline{A}^{(t)}], \tag{12.2.1}$$

where

$$\underline{A}^{(i)} = \left[A_1^{(i)} \cdots A_k^{(i)}\right],\ A_j^{(i)} \in Z(q_i),\ j = 1, 2, \ldots, k. \tag{12.2.2}$$

This representation is obtained by writing $A_j \in Z(M)$ as

$$A_j \leftrightarrow \underline{A}_j = \left[A_j^{(1)} \cdots A_j^{(t)}\right] \tag{12.2.3}$$

$$A_j^{(i)} \equiv A_j \bmod q_i, \quad i = 1, 2, \ldots, t, \; j = 1, 2, \ldots, k. \tag{12.2.4}$$

Also, $A_j = 0$ implies that $A_j^{(i)} = 0$, $i = 1, 2, \ldots, t$, and vice versa, while $A_j \neq 0$ implies that at least one of $A_j^{(i)}$ is not equal to 0 and vice versa.

The codevector $\underline{V}$ associated with $\underline{A}$ is obtained as

$$\underline{V} = [V_1 \;\; V_2 \ldots \; V_n] = \underline{A}\,\mathbf{G}. \tag{12.2.5}$$

Using the property of linearity of the coding scheme **C** and the modulo operation, we may equivalently decompose the above equation mod q_i, $i = 1, 2, \ldots, t$ to get

$$\underline{V} \leftrightarrow [\underline{V}^{(1)} \;\; \underline{V}^{(2)} \;\; \ldots \;\; \underline{V}^{(t)}], \tag{12.2.6}$$

where using (12.2.5), we have

$$\underline{V}^{(i)} = \left[V_1^{(i)} \, V_2^{(i)} \cdots V_n^{(i)}\right] \tag{12.2.7}$$

$$\underline{V}^{(i)} = \underline{A}^{(i)}\,\mathbf{G}^{(i)} \bmod q_i, \quad i = 1, 2, \ldots, t. \tag{12.2.8}$$

$\mathbf{G}^{(i)}$ is the $(k \times n)$ sub-generator matrix obtained by reducing each element of $\mathbf{G}$ modulo q_i. Thus, the process of encoding $\underline{A}$ to get the codevector $\underline{V} = \underline{A}\,\mathbf{G}$ can be decomposed into t parallel processes as listed below:

1. Given a length k data vector $\underline{A}$, reduce each element of $\underline{A}$ mod q_i to get t data vectors $\underline{A}^{(1)}, \underline{A}^{(2)}, \ldots, \underline{A}^{(t)}$, each of length k, where

   $$\underline{A}^{(i)} = \left[A_1^{(i)} \, A_2^{(i)} \cdots A_k^{(i)}\right] \text{ and } A_j^{(i)} \equiv A_j \bmod q_i.$$

2. Encode each of these vectors independently as $\underline{V}^{(i)} \equiv \underline{A}^{(i)}\,\mathbf{G}^{(i)} \bmod q_i$, $i = 1, 2, \ldots, t$.
3. Use the CRT-I to combine $V_j^{(i)}$, $i = 1, 2, \ldots, t$ to get the j-th component, V_j, of the codevector $\underline{V}$, $j = 1, 2, \ldots, n$.

The expression in (12.2.8) is an important expression as it enables us to interpret the overall coding scheme **C** defined on $Z(M)$ to be constituted by t subcoding schemes $\mathbf{C}_i$ for each of the t primes that constitute M. Here, $\underline{A}^{(i)}$ is the data vector, $\underline{V}^{(i)}$ is the corresponding codevector, and $\mathbf{G}^{(i)}$ is the generator matrix associated with $\mathbf{C}_i$. The encoding method outlined above has the attractive feature that if the data is available in its reduced form modulo q_i, then the encoding can be performed directly to produce the corresponding codevector in reduced form. In such a case, Steps (1) and (3) are not required. The encoding method as outlined above is shown in Figure 12.1. It is clear from (12.1.4) that $\mathbf{G}^{(i)}$ is also in systematic form and is given by

$$\mathbf{G}^{(i)} = [\mathbf{I}_k \mid \mathbf{P}^{(i)}], \quad i = 1, 2, \ldots, t. \tag{12.2.9}$$

The parity check matrix, $\mathbf{H}^{(i)}$ for the subcode $\mathbf{C}_i$ can be expressed as

$$\mathbf{H}^{(i)} = [-\mathbf{P}^{(i)t} \mid \mathbf{I}_{n-k}]. \tag{12.2.10}$$

Based on this decomposition and the CRT-I, we make several observations with respect to the coding scheme $\mathbf{C}$. These are stated in the following.

Observation 1. A length n arbitrarily chosen vector $\underline{V}^{(i)}$ having components over $Z(q_i)$ is a codevector in the sub-code $\mathbf{C}_i$ *iff* it can be written in the form $\underline{V}^{(i)} = A^{(i)}\,\mathbf{G}^{(i)}$, where $A^{(i)}$ is a length k vector defined over $Z(q_i)$, or equivalently $\underline{V}^{(i)}\,\mathbf{H}^{(i)t} \equiv \underline{0} \bmod q_i$, $i = 1, 2, \ldots, t$.

Observation 2. A length n arbitrarily chosen $\underline{V}$ having components over $Z(M)$ is a codevector in the code $\mathbf{C}$ *iff* $\underline{V}^{(i)}$ (defined as $\underline{V}^{(i)} \equiv \underline{V} \bmod q_i$) is a codevector in the sub-code $\mathbf{C}_i$, $i = 1, 2, \ldots, t$.

The code $\mathbf{C}$ defined over $M = q_1 \cdot q_2 \cdots q_t$ is equivalent to t sub-codes $\mathbf{C}_i$, defined over q_i $i = 1, 2, \ldots, t$. An important question arises at this stage: how are the minimum distance d of the code $\mathbf{C}$ and the minimum distance d_i of the sub-code $\mathbf{C}_i$, $i = 1, 2, \ldots, t$ related to each other? We explore the structure and the decomposition of the $\mathbf{C}$ further to find an answer to this question. It is clear from the definition of *l.i.* of vectors that if a codevector $\underline{V}$ has non-zero component in positions $i_1, i_2, \ldots, i_b$ ($wt(\underline{V}) = b$), then the i_1-th, i_2-th, …, i_b-th rows of $\mathbf{H}^t$ are linearly dependent (*l.d.*) over $Z(M)$. Thus, if the minimum distance is d, no arbitrarily selected $d - 1$ or less rows of $\mathbf{H}^t$ are *l.d.* Combining this statement with the definition of rank of a matrix, we can state that the rank of the $(n - k) \times j$ matrix ($j \le d - 1$) obtained by arbitrarily selecting j rows of $\mathbf{H}^t$ is j.

THEOREM 12.1 If the minimum distance of a (n, k) code $\mathbf{C}$ defined over $Z(M)$ is d, $M = q_1 \cdot q_2 \cdots q_t$, then the minimum distance of the (n, k) sub-code $\mathbf{C}_i$ defined over $Z(q_i)$ is d_i, $d_i \ge d$. Also, there is at least one sub-code, say $\mathbf{C}_b$, such that $d_b = d$.

Proof: Consider the codeword $\underline{V}$ of Hamming weight d. When reduced mod q_i, it has either Hamming weight d or less. If its Hamming weight $= d$, the theorem is proved. If its Hamming weight when reduced mod q is less than d, say d', then consider the weight of the codeword in **C** obtained by combining $\underline{V}$ mod q_i with the all-zero codevectors for all other values of q_j, $j \neq i$. Such a codevector has Hamming weight $d' < d$. This violates the assumption that the minimum distance of the code is d.

Consider the task of combining independently designed (n, k) sub-codes $\mathbf{C}_i$ designed over $Z(q_i)$, $i = 1, 2, \ldots, t$ to obtain a (n, k) code **C** over $Z(M)$, $M = q_1 \cdot q_2 \cdots q_t$. If

$$\underline{V}^{(i)} = \left[\underline{V}_1^{(i)}\ \underline{V}_2^{(i)} \ldots \underline{V}_n^{(i)}\right]$$

is an arbitrary codevector in $\mathbf{C}_i$, $i = 1, 2, \ldots, t$ and $\underline{V} = [V_1\ V_2 \ldots V_n]$ is the corresponding codevector in **C**, then from the CRT-I, we have $V_j \neq 0$ if one or more of $V_j^{(i)} \neq 0$. Also, if $V_j^{(i)} = 0$, $i = 1, 2, \ldots, t$, then $V_j = 0$. Noting that each of the sub-codes $\mathbf{C}_i$ has the all-zero vector as a codevector, we get the following theorem.

THEOREM 12.2 If the minimum distance of the sub-code $\mathbf{C}_i$ designed over $Z(q_i)$ is d_i, $i = 1, 2, \ldots, t$, then the minimum distance of the code **C** defined over $Z(M)$ and obtained by combining $\mathbf{C}_i$ using the CRT-I, $M = q_1 \cdot q_2 \cdots q_t$, is d, where $d = min(d_1, d_2, \ldots, d_t)$.

We now turn to the final aspect of our general framework for fault control which is the design of decoding algorithms for codes defined over $Z(M)$.

12.2.1 Decoding Algorithms Over *Z(M)*

The above analysis (particularly, Theorems 12.1 and 12.2) also provides for a straightforward procedure for decoding the faulty vector in (12.1.8). Since the faults in (12.1.8) corrupt the codevector in an additive manner, we outline the following decoding procedure for such a fault model:

1. Given a length n vector $\underline{R}$ defined over $Z(M)$, reduce each element of $\underline{R}$ modulo q_i to get t vectors $\underline{R}^{(i)}$, defined over $Z(q_i)$, $i = 1, 2, \ldots, t$, each of length n.

2. Decode each of $\underline{R}^{(i)}$ independently to get $\underline{E}^{(i)}$ using the decoding algorithm for $\mathbf{C}_i$.

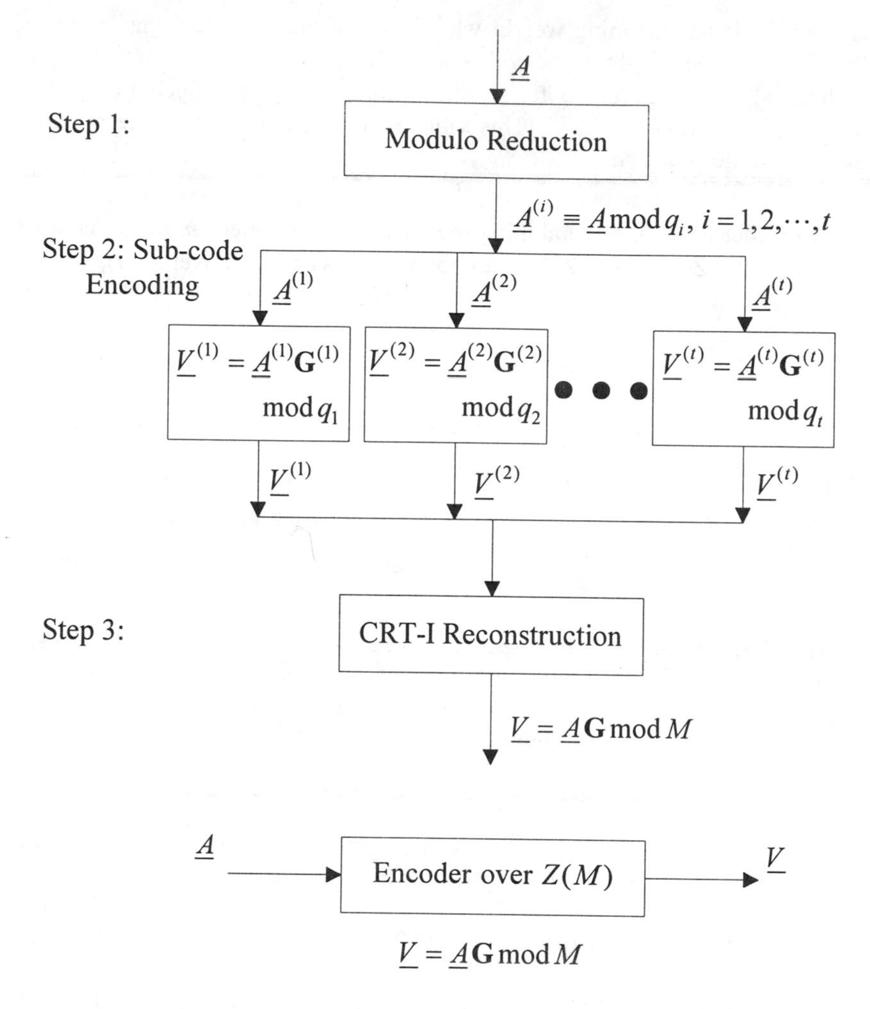

Figure 12.1 An Encoder Diagram for Encoding Over *Z(M)*

---

3. Combine $\underline{E}^{(i)}$, $i = 1, 2, \ldots, t$ using CRT-I to get the fault vector $\underline{E}$ which the decoder for $\mathbf{C}$ would equivalently compute.

4. Finally, $\underline{R}$ is decoded to $\hat{\underline{V}} = \underline{R} - \underline{E}$.

It is important to note here that the validity of the above decoding procedure is based on two facts, namely, (i) $wt(\underline{E}^{(i)}) \leq wt(\underline{E})$, and (ii) minimum distance d_i, of the sub-code $\mathbf{C}_i \geq d$, the minimum distance of the code $\mathbf{C}$ (Theorem 12.1).

Thus, the complete decoding procedure over $Z(M)$ can be implemented using t independent decoders defined over $Z(q_i)$ operating in parallel. This is a very interesting and useful feature of the coding scheme that we have introduced here. It exposes the inherent parallelism in the encoding and the decoding algorithms that may be exploited to obtain suitable VLSI architectures. A block diagram for the decoder is shown in Figure 12.2. Also, we note that each of the sub-codes $\mathbf{C}_i$ are defined over a much smaller size alphabet as compared to the original code.

Given a minimum distance d (n, k) code $\mathbf{C}$ defined over $Z(M)$, $M = q_1 \cdot q_2 \cdots q_t$, consider the problem of designing a minimum distance d code $\mathbf{C}'$ defined over $Z(N)$, $N = M \cdot q$, $q = p^a$, $(q_i, q) = 1$, $i = 1, 2, \ldots, t$. Based on Theorems 12.1 and 12.2, it is clear that we need to design a code $\mathbf{C}''$ over $Z(q)$ having minimum distance $d'' \geq d$. The generator matrix of the code $\mathbf{C}'$ is obtained by combining the generator matrices of the coding scheme $\mathbf{C}$ and $\mathbf{C}''$ using the CRT-I. The encoder for $\mathbf{C}'$ is obtained by including one more branch in parallel to the encoder to $\mathbf{C}$. Interestingly, the decoder for $\mathbf{C}'$ is obtained by combining the decoders for $\mathbf{C}$ and $\mathbf{C}''$ in parallel as well.

Since the design of a coding scheme over $Z(M)$ is equivalent to the design of t coding schemes $\mathbf{C}_i$ over $Z(q_i)$, $i = 1, 2, \ldots, t$, we now turn to the design of a coding scheme $\mathbf{C}$ over $Z(q)$, $q = p^a$. We reassert that $Z(q)$ is not the same as (or an automorphism of) $GF(q)$ due to modulo computation.

12.3 Coding Techniques Over $Z(q)$

For $a = 1$, $Z(q) = Z(p) = GF(p)$. Therefore, for $a = 1$, we can apply the standard coding theory techniques to the design of codes of interest here. As a result, we are motivated to establish a close correspondence between codes defined over $GF(p)$ and codes defined over a ring of integers, $Z(q)$, $q = p^a$. Such a correspondence is in two parts; part I is the characterization of codes in terms of the generator matrices and minimum distance and part II is the decoding algorithms.

Consider a (n, k) code $\mathbf{C}$ over $Z(q)$ having generator matrix $\mathbf{G}$, parity check matrix $\mathbf{H}$, and minimum distance d, and an equivalent code $\mathbf{C}_e$ over $GF(p)$ having generator matrix $\mathbf{G}_e$, parity check matrix $\mathbf{H}_e$, and minimum distance d_e. The equivalence between the codes $\mathbf{C}$ and $\mathbf{C}_e$ is defined in terms of their generator matrices $\mathbf{G}$ and $\mathbf{G}_e$ as

$$\mathbf{G}_e \equiv \mathbf{G} \bmod p. \tag{12.3.1}$$

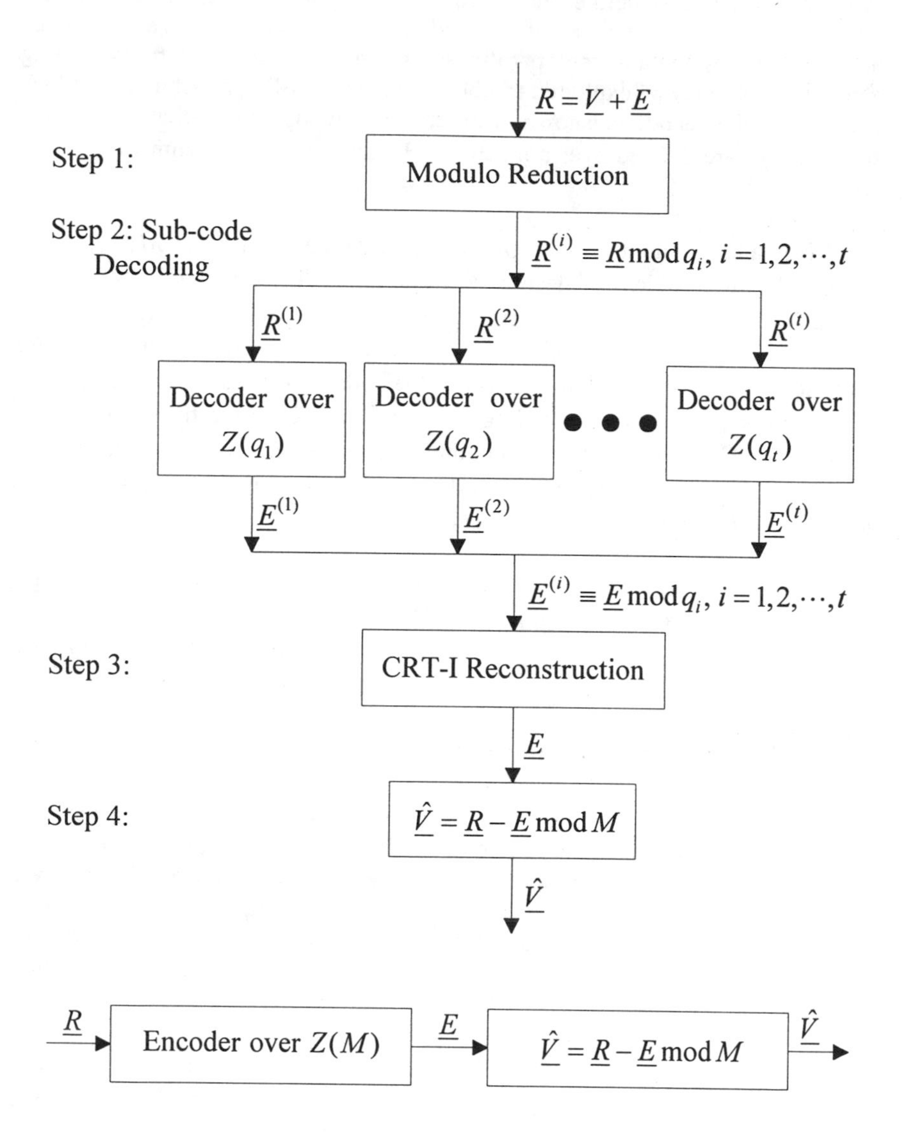

Figure 12.2 The Decoder Diagram for Decoding Over $Z(M)$

We now proceed to establish a relationship between d and d_e. It is established in the following theorem.

Theorem 12.3 The minimum distance of a code $\mathbf{C}$ in $Z(q)$ generated by $\mathbf{G}$ is the same as the minimum distance of a code $\mathbf{C}_e$ in $GF(p)$ generated by $\mathbf{G}_e$, where $\mathbf{G}_e \equiv \mathbf{G} \bmod p$.

Proof: The codevectors in $\mathbf{C}$ are obtained as $\underline{V} = \underline{A}\ \mathbf{G}$. Taking the p-adic expansion of the quantities, we get

$$\underline{V}_0 + p\,\underline{V}_1 + \ldots + p^{a-1}\,\underline{V}_{a-1} = \{\underline{A}_0 + p\,\underline{A}_1 + \ldots + p^{a-1}\,\underline{A}_{a-1}\} \cdot \{\mathbf{G}_0 + p\,\mathbf{G}_1 + \ldots + p^{a-1}\,\mathbf{G}_{a-1}\} \bmod p^a.$$

It is clear that $\mathbf{G}_0 = \mathbf{G}_e \bmod p$ is the generator matrix of the code $\mathbf{C}_e$ in $GF(p)$. Let $\underline{A}$ be a non-zero message vector and b be the smallest integer such that $\underline{A}_b \neq 0$, $b < a$. Such a b always exists or else $\underline{A} = \underline{0}$. Now, $\underline{V}_b = \underline{A}_b\,\mathbf{G}_0 \bmod p$, that is, $\underline{V}_b$ is a codevector in $\mathbf{C}_e$. Let d_e be the minimum distance of such a code. Thus,

$$wt(\underline{V}) \geq wt(\underline{V}_b) \geq d_e.$$

Let $\underline{B}$ be a message vector in $GF(p)$ such that the corresponding codevector $\underline{v}$ in $\mathbf{C}_e$ has Hamming weight equal to d_e. Then, for the message $\underline{A} = p^{a-1} \cdot \underline{B}$ in $Z(q)$, the corresponding codevector $\underline{V}$ $(= p^{a-1} \cdot \underline{v})$ has Hamming weight equal to d_e. This proves the statement of the theorem.

Using similar arguments as above, we can show that if we are given a coding scheme $\mathbf{C}_e$ over $GF(p)$, having minimum distance d, then the generator matrix for $\mathbf{C}_e$ can also be used to obtain $\mathbf{C}$ over $Z(q)$, $q = p^a$, having minimum distance d. A consequence of the analysis is that one needs to design only a code over $GF(p)$ that can be used over $Z(q)$. There is an abundance of good linear codes over $GF(p)$ available in coding theory texts and any of them can be used in the present context. Therefore, we do not need to pursue the topic of design of linear codes defined over $Z(q)$. Combining the above analysis and the analysis in Section 12.2, we get the following fundamental theorem.

Theorem 12.4 The generator matrix of a coding scheme defined over $Z(M)$, $M = q_1 \cdot q_2 \cdots q_t$, $q_i = p_i^{a_i}$, having minimum distance d is equivalent to the generator matrix of a coding scheme defined over $Z(M_e)$, $M_e = p_1 \cdot p_2 \cdots p_t$, having minimum distance d. The generator matrix of a coding scheme defined over $Z(M_e)$, M_e having minimum distance d is obtained by combining the generator matrices of coding schemes $\mathbf{C}_i$ over $GF(p_i)$ having minimum distances $d_i \geq d$ using the CRT-I. In addition, $d = min(d_1, d_2, \ldots, d_t)$.

In addition to finding the generator matrix of a coding scheme over $Z(q)$, we have the issue of the decoding algorithm for the code over $Z(q)$ that is originally designed over $GF(p)$. Once again, the decoding algorithms for linear codes over $GF(p)$ are well established. Consequently, there is a strong impetus to derive the decoding algorithm for the code over $Z(q)$ in terms of the decoding algorithm for the code over $GF(p)$. In the following, it is shown that the decoding algorithm over $GF(p)$ can be used recursively to decode a received vector over $Z(q)$. This is based on the observation that a linear code defined over $Z(q)$, $q = p^a$ is also a linear code over $Z(p^j)$, $j = 1, 2, \ldots, a$. It is worthwhile to mention here that the code over $Z(p^j)$ is obtained by reducing the code over $Z(p^i)$, $j < i$, modulo p^j.

12.3.1 Decoding Algorithms Over $Z(q)$

The decoding over $GF(p)$ is performed in three steps:

1. Compute the $(n - k)$ length vector called the syndrome as $\underline{s} = \underline{r}\,\mathbf{H}^t$;

2. Define a mapping called the standard array from the vector $\underline{s}$ over $GF(p)$ to a length n fault vector $\underline{e}$ over $GF(p)$; and

3. Obtain the estimate $\hat{\underline{v}}$ of the codevector as $\hat{\underline{v}} = \underline{r} - \underline{e}$.

Observe that the decoding procedure as outlined here does not require division, thereby alleviating the need for a field. Therefore, it may also be used to decode over $Z(q)$, where the operation of division by a non-zero scalar is not always available. However, except in certain simple cases, any practical realization of Step (2) must employ division in order to decode in a reasonable amount of time by computing $\underline{e}$ for a given $\underline{s}$ rather than storing $\underline{e}$ for every possible $\underline{s}$ in a memory. Note that the size of the memory for a memory-based realization of Step (2) is $p^{(n-k)}$. The previous statements are valid in the present context also, that is, given $\underline{S} = \underline{R}\,\mathbf{H}^t$ over $Z(q)$, we would like to compute $\underline{E}$ over $Z(q)$ in order to perform decoding over $Z(q)$. The size of the memory for a memory-based realization of Step (2) for a decoder over $Z(q)$ will be $q^{(n-k)}$. We will denote the decoder over $GF(p)$ by Đ.

Given $\underline{R} = \underline{V} + \underline{E}$ over $Z(q)$, we may write $R_i = V_i + E_i$, $i = 1, 2, \ldots, n$; R_i, V_i, $\underline{E}_i \in Z(q)$. Since R_i, V_i, and E_i are defined mod p^a, we can express them in the p-adic number representation as

$$R_i = \sum_{h=0}^{a-1} r_{i,h}\, p^h, \quad 0 \le r_{i,h} < p,$$

$$V_i = \sum_{h=0}^{a-1} v_{i,h} p^h, \quad 0 \le v_{i,h} < p,$$

$$E_i = \sum_{h=0}^{a-1} e_{i,h} p^h, \quad 0 \le e_{i,h} < p. \tag{12.3.2}$$

Note that

$$r_{i,0} \equiv v_{i,0} + e_{i,0} \bmod p, \tag{12.3.3}$$

while, in general, $r_{i,h} \neq v_{i,h} + e_{i,h} \bmod p$, $h > 0$. This is due to the non-linear nature of the arithmetic operation of '+'. Given $\underline{R}$, the recursive algorithm to compute $\underline{E}$ consists in computing $e_{i,h}$, $i = 1, \ldots, n$, $h = 0, 1, \ldots, a-1$ recursively by employing Đ. Based on (12.3.2) and the expression, $\underline{R} = \underline{V} + \underline{E}$, we can write

$$\underline{R} = \sum_{h=0}^{a-1} \underline{r}_h p^h = \sum_{h=0}^{a-1} \underline{v}_h p^h + \sum_{h=0}^{a-1} \underline{e}_h p^h \bmod p^a, \tag{12.3.4}$$

where the vectors $\underline{r}_h$, $\underline{v}_h$, and $\underline{e}_h$ are given by

$$\begin{aligned} \underline{r}_h &= [r_{1,h}\ r_{2,h}\ \ldots\ r_{n,h}], \\ \underline{v}_h &= [v_{1,h}\ v_{2,h}\ \ldots\ v_{n,h}], \\ \underline{e}_h &= [e_{1,h}\ e_{2,h}\ \ldots\ e_{n,h}],\ h = 0, 1, \ldots, a-1. \end{aligned} \tag{12.3.5}$$

As $\underline{V}$ is a code vector in a code over $Z(q)$,

$$\underline{V}_h \equiv \underline{V} \bmod p^{h+1},\ h = 0, 1, \ldots, a-1, \tag{12.3.6}$$

is a codevector in a code over $Z(p^{h+1})$. Corresponding to (12.3.6), we may write

$$\begin{aligned} \underline{R}_h &\equiv \underline{R} \bmod p^{h+1} \equiv \underline{V}_h + \underline{E}_h \bmod p^{h+1}, \\ \underline{E}_h &\equiv \underline{E} \bmod p^{h+1}. \end{aligned} \tag{12.3.7}$$

Recall that the first step in decoding $\underline{R}$ is computation of the syndrome as

$$\underline{S} \equiv \underline{R}\,\mathbf{H}^t = \underline{E}\,\mathbf{H}^t, \tag{12.3.8}$$

thereby implying that the syndrome $\underline{S}_h$ corresponding to the received vector $\underline{R}_h$ can be obtained as

$$\underline{S}_h \equiv \underline{R}\,\mathbf{H}^t \bmod p^{h+1} \equiv \underline{S} \bmod p^{h+1},\ h = 0, 1, 2, \ldots, a-1. \tag{12.3.9}$$

The decoder Ð over $GF(p)$ consists in a computational procedure to obtain $\underline{e}$, given the syndrome $\underline{s}$ and the relation $\underline{s} = \underline{e}\,\mathbf{H}'$ over $GF(p)$. We can convert the above equation to a relations of the type $\underline{s} = \underline{e}\,\mathbf{H}'$ over $GF(p)$ and then employ Ð as follows. Let

$$\underline{s}_0 = \underline{S}_0 \equiv \underline{R}_0\,\mathbf{H}' \bmod p = \underline{e}_0\,\mathbf{H}' \bmod p. \tag{12.3.10}$$

Given $\underline{s}_0$, Ð is used to find $\underline{e}_0$. We note that if $wt(\underline{E})$ is less than $\lfloor (d-1)/2 \rfloor$, the fault-correcting capability of the code over $Z(q)$, then $wt(\underline{E}_h)$ and $wt(\underline{e}_h)$ are less than the fault-correcting capability of the code over $Z(p^{h+1})$. This is due to the fact that $wt(\underline{e}_h) \le wt(\underline{E}_h) \le wt(\underline{E})$ while the fault correcting capability of the codes over $Z(q)$ and $Z(p^{h+1})$ are the same.

Once $\underline{e}_0$ is known, consider the expression for $\underline{S}_1$ as

$$\underline{S}_1 \equiv \underline{R}_1\,\mathbf{H}' \bmod p^2 \equiv \underline{E}_1\,\mathbf{H}' \bmod p^2 = (\underline{e}_0 + \underline{e}_1\,p)\,\mathbf{H}' \bmod p^2. \tag{12.3.11}$$

Rearranging various terms, we get

$$\underline{s}_1 = \{(\underline{S}_1 - \underline{e}_0\,\mathbf{H}')\,p^{-1}\} \bmod p \equiv \underline{e}_1\,\mathbf{H}' \bmod p.$$

Given $\underline{S}$ and $\underline{e}_0$, $\underline{s}_1$ can be computed, and Ð can be used to obtain $\underline{e}_1$. The recursive nature of the decoding algorithm is now clear. Let us say that $\underline{e}_0$, $\underline{e}_1$, ..., $\underline{e}_{h-1}$ are determined. The computations to determine $\underline{e}_h$ proceed as follows:

$$\underline{s}_h = \left\{\left(\underline{S}_h - \sum_{j=0}^{h-1} \underline{e}_j\,p^j\,\mathbf{H}'\right)p^{-h}\right\} \bmod p \equiv \underline{e}_h\mathbf{H}' \bmod p\,. \tag{12.3.12}$$

Given $\underline{s}_h$, Ð is employed to compute $\underline{e}_h$, $h = 1, \ldots, a-1$. Once the fault vectors $\underline{e}_0$, $\underline{e}_1$, ..., $\underline{e}_{a-1}$ (or equivalently, the vector $\underline{E}$) are determined, the decoder decodes $\underline{R}$ to a code vector $\hat{\underline{V}}$, where

$$\hat{\underline{V}} = \underline{R} - \underline{E} \bmod q\,. \tag{12.3.13}$$

A step-by-step description of the decoding algorithm over $Z(q)$ is given in the following.

1. Given $\underline{R}$, compute the syndrome $\underline{S} \equiv \underline{R}\,\mathbf{H}'$.
2. Let $\underline{s}_0 \equiv \underline{S} \bmod p = \underline{e}_0\,\mathbf{H}'$.
3. Employ Ð to determine $\underline{e}_0$.
4. Let $h \leftarrow 1$.

5. Compute $\underline{s}_h \equiv \underline{e}_h \mathbf{H}^t \bmod p$ by using (12.3.12).
6. Employ Đ to determine $\underline{e}_h$.
7. $h \leftarrow h + 1$. If $h < a$, go to Step (5).
8. Decode $\underline{R}$ to the codevector $\hat{\underline{V}}$ using (12.3.13).

A flowchart for this algorithm is shown in Figure 12.3. This algorithm leads to correct decoding provided that $wt(\underline{E}) \le \lfloor (d-1)/2 \rfloor$. A most interesting feature of this algorithm is that it leads to correct decoding for all those fault vectors $\underline{E}$ for which $wt(\underline{e}_h) \le \lfloor (d-1)/2 \rfloor$, $h = 0, 1, \ldots, a-1$, even though $wt(\underline{E})$ exceeds $\lfloor (d-1)/2 \rfloor$. Clearly, if $wt(\underline{E}) \le \lfloor (d-1)/2 \rfloor$, then $wt(\underline{e}_h) \le \lfloor (d-1)/2 \rfloor$, but if $wt(\underline{e}_h) \le \lfloor (d-1)/2 \rfloor$, then it *does not* imply that $wt(\underline{E}) \le \lfloor (d-1)/2 \rfloor$, $h = 0, 1, 2, \ldots, a-1$. We observe that one cannot directly decode the vector $\underline{r}_h$, the h-th coefficient in the p-adic representation of $\underline{R}$, $h = 0, 1, \ldots, a-1$, as the corresponding vector $\underline{v}_h$, though defined over $GF(p)$, is *not* a codevector in any code. Also, a divide-and-conquer technique based fast algorithm for decoding over $Z(q)$, $q = p^a$ is described in the following section.

12.3.2 A Fast Algorithm for Decoding Over $Z(q)$

In this subsection, we outline a divide-and-conquer technique based fast algorithm for decoding over $Z(q)$. It assumes that a is an integer of the type $a = 2^\beta$ and recursively decodes the received vector $\underline{R}$ over $Z(p^h)$, $h = 1, 2, 4, \ldots, 2^\beta$. Since this algorithm is similar in approach to the algorithm described in Section 12.3.1, we simply summarize it in the following.

Decoder for $h = 1$

$$\underline{s}_{h-1} = \underline{S} \bmod p^h.$$

Use decoder over $Z(p^h)$ to get $\underline{e}_0, \underline{e}_1, \ldots, \underline{e}_{h-1}$.

Decoder for $h = 2, 4, \ldots, 2^\beta$

$$\underline{s}_{h/2} = \left\{ \left(\underline{S}_h - \sum_{j=0}^{h/2-1} \underline{e}_j p^j \mathbf{H}^t \right) p^{-h/2} \right\} \bmod p^{h/2} = \left\{ \sum_{j=h/2}^{h-1} \underline{e}_j p^{j-h/2} \mathbf{H}^t \right\} \bmod p^{h/2}$$

Use decoder over $Z(p^{h/2})$) to get $\underline{e}_{h/2}, \ldots, \underline{e}_{h-1}$.

In most applications, a is a small integer. In such cases, the order-recursive algorithm described in Section 12.3.1 may perform better than this divide-and-conquer technique based algorithm due to its simple mathematical structure.

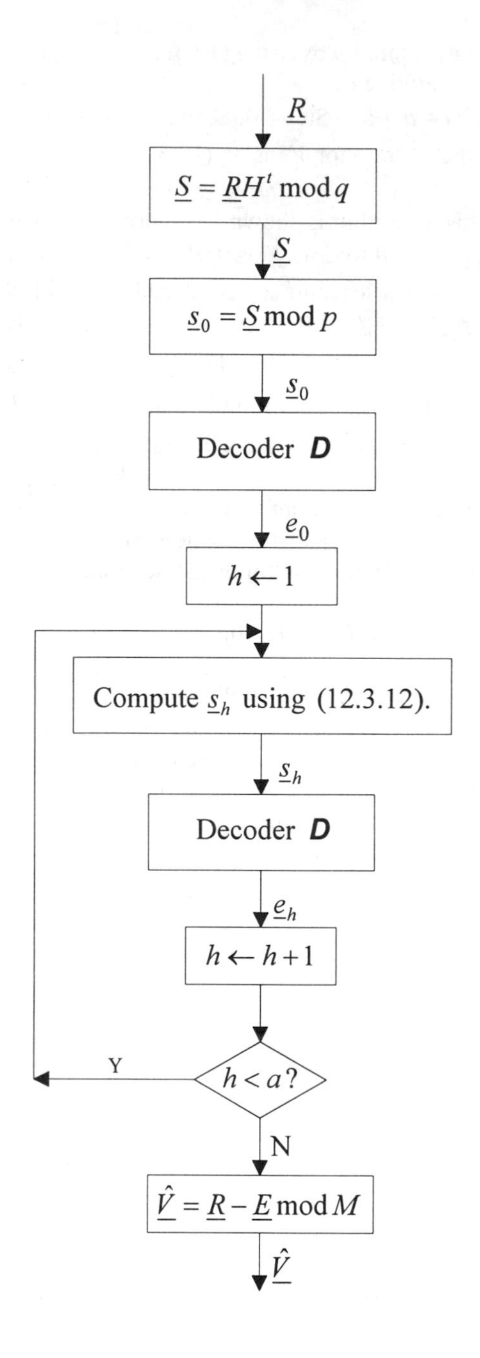

Figure 12.3 A Flowchart For the Decoder Over $Z(q)$, $q = p^a$

12.3.3 Some General Remarks

Sections 12.2 and 12.3 establish a complete mathematical framework for fault control techniques in data sequences defined over $Z(M)$. It is shown that the generator matrix **G** of a linear code defined over $Z(M)$, $M = q_1 \cdot q_2 \cdots q_t$, $q_i = p_i^{a_i}$ is equivalent to the generator matrix of a linear code defined over $Z(M_e)$, $M_e = p_1 \cdot p_2 \cdots p_t$. Also, the generator matrix of a linear code defined over $Z(M)$ is obtained by combining the generator matrices of linear codes $\mathbf{C}_i$ defined over $GF(p_i)$ using the CRT-I. The decoder over $Z(M)$ can be implemented using t decoders operating in parallel; here the i-th decoder performs decoding over $Z(q_i)$. The decoder over $Z(q_i)$ can be implemented by using recursively the decoder over $GF(p_i)$ a_i times. Consequently, the degeneracy of M, $\{t, a\}$ plays a vital role in establishing the overall realization and the decoding time, t being the number of parallel branches such a realization will have and a being the maximum number of recursions along each of the branches.

In some cases, it may not be possible to directly derive (n, k) codes over $Z(q_i)$ having minimum distance d. An easier task is to design (n_i, k) codes $\mathbf{C}_i$ over $Z(q_i)$ having minimum distance $d_i \geq d$. Then, a (n, k) code over $Z(M)$ having minimum distance d can be obtained such that $n = max(n_1, n_2, \ldots, n_t)$. This is accomplished by appending $n - n_i$ zeros to every codeword in $\mathbf{C}_i$ and then combining the resulting codewords using the CRT-I. The configuration of the decoder remains unaltered.

In this chapter, we have emphasized coding techniques over $Z(q)$. The same ideas can be used to design coding techniques over $CZ(q)$. In this case, the code over $CZ(q)$ (i) is the same as the code over $Z(q)$ if $p = 2$, and (ii) is a direct sum of two codes over $Z(q)$ if $p \equiv 1 \bmod 4$. If $p \equiv 3 \bmod 4$, then the code over $CZ(q)$ is obtained by defining its equivalent code in $GF(p^2)$. The decoder now employs the decoder over $GF(p^2)$ in order to perform decoding over $CZ(q)$. The overall analysis of the technique remains the same.

Finally, the readers are referred to Paper four in Part IV for a complete and mathematically rigorous treatment of **Bose-Chaudhary-Hoquenghem** (BCH) and **Reed-Solomon** (RS) codes in $Z(q)$ and $CZ(q)$.

We now present an example and specific details of minimum distance 4 single-fault-correcting, double-fault-detecting (SFC-DFD) codes.

12.4 Examples and SFC-DFD Codes

Let $M = 110{,}852{,}311 = 31^3 \cdot 61^2 = q_1 \cdot q_2$, $q_1 = 31^3$, $q_2 = 61^2$. Clearly, we have $p_1 = 31$, $a_1 = 3$, $p_2 = 61$, and $a_2 = 2$. Here, M has degeneracy $\{2, 3\}$. Suppose we are interested in designing a (n, k) minimum distance 3 SFC code over $Z(M)$. This is accomplished by designing (n, k) SFC codes over $Z(29{,}791)$ or $GF(31)$ and $Z(3{,}721)$ or $GF(61)$, respectively. A SFC code over $GF(p)$ is a Hamming code having the largest value of n as $(p^m - 1)/(p - 1)$, $m = n - k$. Thus, a SFC code over $Z(M)$ can have the largest value of n as

$$n = max(n_1, n_2, \ldots, n_t),$$
$$n_i \le (p_i^m - 1)/(p_i - 1), \quad i = 1, 2, \ldots, t,$$
$$m = n - k.$$

In our case, let $m = n - k = 2$. Then $n_1 = 32$, $n_2 = 62$, and $n = 62$. If we are interested in a (10, 8) SFC code over $Z(110{,}852{,}311)$, then we need to obtain the generator matrices $\mathbf{G}^{(1)}$ and $\mathbf{G}^{(2)}$ of the (10, 8) shortened Hamming codes over $GF(31)$ and $GF(61)$, respectively. One such set of generator matrices is

$$\mathbf{G}^{(1)} = \begin{bmatrix} & & & & \mathbf{I}_8 & & & \\ 1 & 1 & 1 & 1 & 1 & 1 & 1 & 1 \\ 1 & 2 & 3 & 4 & 5 & 6 & 7 & 8 \end{bmatrix}^t$$

and

$$\mathbf{G}^{(2)} = \begin{bmatrix} & & & & \mathbf{I}_8 & & & \\ 1 & 1 & 1 & 1 & 1 & 1 & 1 & 1 \\ 1 & 2 & 3 & 4 & 5 & 40 & 7 & 8 \end{bmatrix}^t . \qquad (12.4.1)$$

Combining $\mathbf{G}^{(1)}$ and $\mathbf{G}^{(2)}$ using the CRT-I, we get the overall generator matrix as

$$\mathbf{G} = \begin{bmatrix} & & & & \mathbf{I}_8 & & & \\ 1 & 1 & 1 & 1 & 1 & 1 & 1 & 1 \\ 1 & 2 & 3 & 4 & 5 & 77{,}158696 & 7 & 8 \end{bmatrix}^t .$$

Let the data vector be $\underline{A}$ = [50,000 29 60 2 5 6 110 25]. The corresponding code vector in $\mathbf{C}_1$ having $\mathbf{G}^{(1)}$ as the generator matrix is computed as $\underline{V}^{(1)} = \underline{A}^{(1)}\,\mathbf{G}^{(1)}$ mod 29,791, where $\underline{A}^{(1)} \equiv \underline{A}$ mod 29,791 = [20,209 29 60 2 5 6 110 25]. The codevector $\underline{V}^{(1)}$ = [20,209 29 60 2 5 6 110 25 20,446 21,486]. Similarly, $\underline{A}^{(2)} = \underline{A}$ mod 3,721 = [1,627 29 60 2 5 6 110 25] and $\underline{V}^{(2)}$ = [1,627 29 60 2 5 6 110 25 1,864 3,108]. The codevector $\underline{V}$ may be obtained by combining $\underline{V}^{(1)}$ and $\underline{V}^{(2)}$ using the CRT-I or directly by using the expression $\underline{V} = \underline{A}\,\mathbf{G}$. In this example, $\underline{V}$ = [50,000 29 60 2 5 6 110 25 50,237 19,594,173]. Let the vector $\underline{R}$ = [50,000 29 60 1,222 5 6 110 25 50,237 19,594,173]. Computing the syndrome, we get $\underline{S} = \underline{R}\,\mathbf{H}^t$ = [110,851,091

110,847,431]. Since $\underline{S} \neq \underline{0}$, we assume that one fault has taken place and proceed to find it. It can be shown that for a SFC code, $\underline{S} = \underline{E_i}\ \underline{H_i}$, where i is the fault location, E_i is the fault value, and $\underline{H_i}$ is the i-th row of the parity check matrix $\mathbf{H}'$. This relation can be employed directly to compute i and E_i as $i = 4$ and $E_i = 1{,}220$. However, we would like to demonstrate the parallel and recursive nature of the decoder.

Since the degeneracy of M is {2, 3}, the overall decoder can be implemented in parallel using two decoders over $Z(29{,}791)$ and $Z(3{,}721)$, respectively. The first step is to compute the syndrome over $Z(29{,}791)$ and $Z(3{,}271)$. They are given by

$$\underline{S}^{(1)} \equiv \underline{R}^{(1)}\ \mathbf{H}' \bmod 29{,}791 \equiv \underline{S} \bmod 29{,}791 = [28{,}571\ \ 24{,}911],$$

$$\underline{S}^{(2)} \equiv \underline{R}^{(2)}\ \mathbf{H}' \bmod 3{,}721 \equiv \underline{S} \bmod 3{,}721 = [2{,}501\ \ 2{,}562]. \tag{12.4.2}$$

Given $\underline{S}^{(1)}$ and $\underline{S}^{(2)}$, the decoder will now proceed to determine $\underline{E}^{(1)}$ and $\underline{E}^{(2)}$ using the decoders over $Z(29{,}751)$ and $Z(3{,}721)$, respectively, and the recursive approach as described in Section 12.3. At the heart of these decoders are two SFC decoders, one that operates in $GF(31)$ using $\mathbf{G}^{(1)}$ as the generator matrix and the second that operates in $GF(61)$ using $\mathbf{G}^{(2)}$ as its generator matrix.

In Table 12.1, we show the results of these recursive computations. Note that the first decoder performs three recursions while the second decoder performs two recursions. To simplify the notations, we delete the parenthesized superscript of (1) or (2) from the table. Based on Table 12.1, we conclude that the fault is in the fourth location. The fault magnitude is computed as follows: $E_4^{(1)} \equiv E_4 \bmod 29{,}791$. The coefficients in the 31-adic representation of $E_4^{(1)}$ are 11, 8, and 1, respectively, implying that

$$E_4^{(1)} = 11 + 8 \times 31 + 1 \times 31^2 = 1{,}220. \tag{12.4.3}$$

Similarly, $E_4^{(2)} \equiv E_4 \bmod 3{,}721$ and the coefficients in the 61-adic representation of $E_4^{(2)}$ are 0 and 20, respectively, implying that

$$E_4^{(2)} = 0 + 20 \times 61 = 1{,}220. \tag{12.4.4}$$

Combining (12.4.3) and (12.4.4) using the CRT-I, we get $E_4 = 1{,}220$. Finally, the decoder decodes $\underline{R}$ to $\underline{\hat{V}}$, where

$$\underline{\hat{V}} = R_i\,,\ \ i = 1, 2, \ldots, 10,\ \ i \neq 4$$

and

$$\hat{V}_4 = R_4 - E_4 = 2.$$

Table 12.1 Results of Computations for the Example in Section 12.4

Computational recursion	$\underline{S}_h$	$\underline{s}_h$	Decoder over $GF(31)$ Fault location, fault value
$h=0$	(20 18)	(20 18)	4, 11
$h=1$	(702 886)	(23 30)	4, 8
$h=2$	(28,751 24,911)	(30 24)	4, 1

A. Steps in decoding over $Z(29,751)$ $\underline{S}=[28,751 \;\; 224,911]$, $a=3$.

Computational recursion	$\underline{S}_h$	$\underline{s}_h$	Decoder over $GF(61)$ Fault location, fault value
$h=0$	(0 0)	(0 0)	no error, 0
$h=1$	(2,501 2,562)	(41 42)	4, 20

B. Steps in decoding over $Z(3,721)$ $\underline{S}=[2,501 \;\; 2,562]$, $a=2$.

A SFC-DFD code is obtained by adding an overall parity digit to every codevector in the SFC code. For example, a (11, 8) SFC-DFD code over Z(110,852,311) is obtained by combining the generator matrices of the SFC-DFD codes over $GF(31)$ and $GF(61)$, respectively. For SFC generator matrices in (12.4.1), the corresponding SFC-DFD generator matrices (in systematic form) are as follows:

$$\mathbf{G}^{(1)} = \begin{bmatrix} & & & \mathbf{I}_8 & & & & \\ 1 & 1 & 1 & 1 & 1 & 1 & 1 & 1 \\ 1 & 2 & 3 & 4 & 5 & 6 & 7 & 8 \\ 3 & 4 & 5 & 6 & 7 & 8 & 9 & 10 \end{bmatrix}^t$$

and

$$\mathbf{G}^{(2)} = \begin{bmatrix} & & & \mathbf{I}_8 & & & & \\ 1 & 1 & 1 & 1 & 1 & 1 & 1 & 1 \\ 1 & 2 & 3 & 4 & 5 & 40 & 7 & 8 \\ 3 & 4 & 5 & 6 & 7 & 42 & 9 & 10 \end{bmatrix}^t .$$

Decoding for SFC-DFD code is performed in a manner similar to the decoding for the SFC code. However, if any one of the t decoders operating in parallel detects the presence of two faults or the fault vector $\underline{E}$ has two non-zero components, then the decoder declares the presence of two faults.

Notes

In this work, our goal has been to bring together the richness of results available in the number theory and the theory of error control coding. The importance of this work lies in the suitability of modulo arithmetic in certain computational environments. A complete mathematical framework is described which may be used for designing schemes for algorithm based fault tolerant computing over a ring of integers. The readers are referred to Paper four of Part IV for a complete and mathematically rigorous treatment of BCH and RS codes in $Z(q)$ and $CZ(q)$.

Linearity of modulo operation and the coding scheme leads to very interesting and useful properties for both the encoding and the decoding algorithms. The encoding operation can be performed on the data vector directly or equivalently on modulo reduced vectors associated with it. Decoding is performed by combining several decoders operating in parallel. All the computations required for fault correction are converted to equivalent computations on a Galois field. This enables us to distinguish between the computations that may be performed in parallel and computations that are intrinsically serial in nature. Also, the emphasis in this chapter is on the mathematical structure of the various algorithms. The issues related to computational efficiency are not treated directly. We expect that this work will motivate further research on the application of coding theory techniques to algorithm based fault tolerant computing systems.

There are a number of texts on coding theory and its applications to communication systems. We have used the texts by Peterson and Weldon [1972], Blahut [1983], McElice [1977], MacWilliams and Sloane [1977], and Lin and Costello [1983] quite extensively. Application of coding theory concepts to fault tolerance in computing systems have been described in Wakerly [1978], Rao and Fujiwara [1989], Huang and Abraham [1984], Jou and Abraham [1986], Chen and Abraham [1986], and Anfinson and Luk [1988]. The earlier work on codes constructed over integer rings can be found in Blake [1972], Blake [1975], Spiegel [1977], and Spiegel [1978]. A more recent work on this topic can be found in Fang, Rao, and Kolluru [1994].

The original idea of algorithm based fault tolerance was introduced in Huang and Abraham [1984]. It is based on some of the classical results available in coding theory. Their proposed checksum scheme can detect and correct any failure within a single processor in a fault-tolerant multi-processor system under the assumption that at most one module is faulty within a given period of time. Subsequently, more applications of the algorithm based fault tolerance in signal processing and related tasks were proposed in Jou and Abraham [1986] and Chen and Abraham [1986]. A linear algebraic model of weighed checksum scheme has been introduced in Jou and Abraham [1986] and the techniques for correcting single faults were extended to correcting two faults in Anfinson and Luk [1988].

This chapter is based on the research work reported in Krishna [1994].

Bibliography

[12.1] W.W. Peterson and E.J. Weldon, Jr., *Error Correcting Codes*, II Edition, MIT Press, 1972.

[12.2] R.E. Blahut, *Theory and Practice of Error Control Codes*, Addison-Wesley Publishing Co., 1983.

[12.3] R.J. McElice, *The Theory of Information and Coding*, Addison-Wesley Publishing Co., 1977.

[12.4] F.J. MacWilliams and N.J.A. Sloane, *The Theory of Error Correcting Codes*, North Holland Publishing Co., 1997.

[12.5] S. Lin and D.J. Costello, Jr., *Error Control Coding: Fundamentals and Applications*, Prentice Hall, 1983.

[12.6] J. Wakerly, *Error Detecting Codes, Self-checking Circuits and Applications*, North Holland, 1978.

[12.7] T.R.N. Rao and E. Fujiwara, *Error Control Coding for Computer Systems*, Prentice Hall, 1989.

[12.8] M.A. Soderstrand, W.K. Jenkins, G.A. Jullien, and F.J. Taylor, *Residue Number System Arithmetic: Modern Applications in Digital Signal Processing*, IEEE Press, 1986.

[12.9] K.H. Huang and J.A. Abraham, "Algorithm-Based Fault Tolerance for Matrix Operations," *IEEE Transactions on Computers*, Vol. C-33, pp. 518-528, June 1984.

[12.10] J.Y. Jou and J.A. Abraham, "Fault-Tolerant Matrix Arithmetic and Signal Processing on Highly Concurrent Computing Structures," *Proceedings IEEE*, no. 5, pp. 721-741, May 1986.

[12.11] C.Y. Chen and J.A. Abraham, "Fault Tolerance Systems for the Computation of Eigen Values and Singular Values," *SPIE, Advanced Algorithms and Architectures for Signal Processing*, Vol. 696, pp. 228-237, 1986.

[12.12] C.J. Anfinson and F.T. Luk, "A Linear Algebraic Model of Algorithm Based Fault Tolerance," *IEEE Transactions on Computers*, Vol. 37, pp. 1599-1604, Dec. 1988.

[12.13] I.F. Blake, "Codes Over Certain Rings," *Information and Control*, Vol. 20, pp. 396-404, 1972.

[12.14] I.F. Blake, "Codes Over Integer Residue Rings," *Information and Control*, Vol. 29, pp. 295-300, 1975.

[12.15] E. Spiegel, "Codes Over Z_m," *Information and Control*, Vol. 35, pp. 48-51, 1977.

[12.16] E. Spiegel, "Codes Over Z_m, Revised," *Information and Control*, Vol. 37, pp. 100-104, 1978.

[12.17] H. Krishna, "A Mathematical Framework for Algorithm Based Fault Tolerant Computing Over a Ring of Integers," *Circuits, Systems and Signal Processing*, Vol. 13, pp. 625-653, 1994.

[12.18] G.L. Feng, T.R.N. Rao, and M.S. Kolluru, "Error Correcting Codes Over Z_{2^m} for Algorithm-Based Fault Tolerance," *IEEE Transactions on Computers*, Vol. 43, pp. 370-374, 1994.

Problems

12.1 Design a SFC code over $Z(4)$ starting from a SFC over $GF(2)$. Obtain the generator matrix and describe the decoding algorithm.

12.2 Design a SFC code over $Z(1,205)$ and write the decoding algorithm. What is the degeneracy of M?

12.3 One of the popular choices for M in digital signal processing is $M = 2^{16} + 1$. Describe a SFC-DFD code in terms of its generator matrix and the decoding algorithm.

12.4 If $V(u)$ and $A(u)$ are the generating polynomials of codevector $\underline{V}$ and message vector $\underline{A}$, respectively, then the (n, k) code is called **cyclic** if $V(u) = A(u) \cdot G(u)$, where the *monic* polynomial $G(u)$, termed the

generator polynomial, divides $u^n - 1$. For a cyclic code of length n generated by $G(u)$, if $G(u)$ does not divide $u^i - 1$, $i < n$, then show that this is a SFC code.

12.5 Consider $Z(q)$, $q = p^a$. If G is the generator of a length n NTT in $Z(q)$, then show that $G(u) = (u - G)(u - G^2)$ generates a SFC cyclic code of length n. Show that such SFC cyclic codes of length n exist only for $n \mid (p - 1)$.

12.6 Consider $Z(q)$, $q = p^a$. Let G be the generator of a length n NTT in $Z(q)$. Write the parity check matrix for the SFC cyclic code of length n generated by $G(u) = (u - G)(u - G^2)$. Now describe a decoding algorithm for single fault correction that directly operates in $Z(q)$.

12.7 Consider $Z(q)$, $q = p^a$. Let G be the generator of a length n NTT in $Z(q)$. Write the parity check matrix for the SFC cyclic code of length n having

$$G(u) = (u - G)(u - G^2)(u - G^3)\cdots(u - G^{d-1}).$$

12.8 Show that no matter how selected, up to $d - 1$, arbitrarily selected rows of $\mathbf{H}^t$ for the $\mathbf{H}$ matrix in Problem 12.7 are *l.i.*

12.9 Use the result in Problem 12.8 to show that the minimum distance of the cyclic code in Problem 12.7 is at least d.

12.10 Design a double-error correcting cyclic code for length $n = 16$ in $Z(17^2)$. Describe a decoding algorithm for such a code.

12.11 Design a SFC for fault control in $CZ(3^3 \cdot 7^2)$. Describe the generator matrix, the parity check matrix, and a decoding algorithm for the code.

12.12 Consider $CZ(q)$, $q = p^a, p \equiv 3 \bmod 4$. Let G be the generator of a length n NTT in $CZ(q)$. Write the parity check matrix for the SFC cyclic code of length n generated by $G(u) = (u - G)(u - G^2)$. Now describe a decoding algorithm for single fault correction that directly operates in $CZ(q)$. Show that such SFC cyclic codes of length n exist only for $n \mid (p^2 - 1)$.

12.13 Repeat Problems 12.7 and 12.8 for $CZ(q)$, $q = p^a, p \equiv 3 \bmod 4$.

12.14 Generalize Problems 12.7 and 12.8 to $Z(q, Q(\theta))$ and $CZ(q, Q(\theta))$.

PART III

FAST FOURIER TRANSFORM (FFT) ALGORITHMS

Thoughts on Part III

In this part, we describe fast computationally efficient algorithms for the following tasks:

1. One-Dimensional Discrete Fourier Transform (DFT) of Length n :

$$A_i = \sum_{l=0}^{n-1} A_l \alpha^{li}, \; i = 0, 1, \ldots, n-1, \qquad \text{(iii.1)}$$

where α is the n-th root of unity in the number system of interest. In the chapters that follow, α is the n-th root of unity in the complex number system C. Such a root is denoted by ω and is computed as

$$\omega = e^{-2\pi j/n}, \; j = \sqrt{-1}\,. \qquad \text{(iii.2)}$$

2. Two-Dimensional DFT of Length $n_1 \times n_2$:

$$A_{i_1,i_2} = \sum_{l_2=0}^{n_2-1}\sum_{l_1=0}^{n_1-1} A_{l_1,l_2}\,\omega_1^{l_1 i_1}\,\omega_2^{l_2 i_2},$$
$$i_1 = 0, 1, \ldots, n_1 - 1; \;\; i_2 = 0, 1, \ldots, n_2 - 1; \qquad \text{(iii.3)}$$

where ω_1 and ω_2 are the n_1-th and n_2-th roots of unity and are obtained by setting $n = n_1$ and $n = n_2$ in (iii.2), respectively.

3. Multidimensional DFTs. The expressions for multidimensional DFT can be defined in a manner similar to (iii.1) and (iii.3). The algorithms for computing multidimensional DFTs are similar in form to the algorithms for computing two-dimensional DFT.

The fast algorithms for the DFT computation are collectively known as fast Fourier transform (FFT) algorithms. We examine FFTs in great detail for all the possible cases of lengths. It will become obvious as Chapters 13 and 14 unfold that the number-theoretic properties of the indices play a fundamental role in the design of FFTs. In this regard, an effort is made to cover all possible cases of length n starting from $n = p$ to n as expressed in its standard factorization.

Given the one-dimensional DFT, the original sequence can be computed as

$$A_l = (1/n)\sum_{i=0}^{n-1} A_i \alpha^{-il}, \quad l = 0, 1, \ldots, n-1. \qquad \text{(iii.4)}$$

This is known as the inverse DFT (IDFT). Similarly, given the two-dimensional DFT, the original sequence can be computed as the IDFT:

$$A_{l_1,l_2} = (1/(n_1 n_2))\sum_{i_2=0}^{n_2-1}\sum_{i_1=0}^{n_1-1} A_{i_1,i_2}\,\omega_1^{-l_1 i_1}\,\omega_2^{-l_2 i_2},$$
$$l_1 = 0, 1, \ldots, n_1 - 1; \; l_2 = 0, 1, \ldots, n_2 - 1. \qquad \text{(iii.5)}$$

It is clear that if α is an n-th root of unity, then so is α^{-1}. Comparing (iii.1) and (iii.3) with (iii.4) and (iii.5), respectively, we see that DFT and IDFT computations are identical. Consequently, we need to study FFT algorithms only for DFTs. Algorithms for IDFT follow.

Table III.1 lists the computational complexity of the DFTs listed in (iii.1) and (iii.3), if they were computed directly.

Table III.1 Computational Complexity of Direct Computation of Various Discrete Fourier Transforms

Computation	Additive Complexity	Multiplicative Complexity
(iii.1)	$n \cdot (n-1)$	$(n-1)^2$
(iii.3)	$n_1 \cdot n_2(n_1 + n_2 - 2)$	$n_1 \cdot n_2(n_1 + n_2 - 4) + n_1 + n_2$

FFTs have been studied extensively for their computational features and hardware realizations. As time has progressed, scientists and researchers have discovered new and novel applications of FFTs. Today, one can find a very wide range of values of n, starting from small to large and going up to 50,000 or more, for which FFTs are being used in various applications. The readers will soon recognize that there exist a number of different FFTs for a given length n. They differ vastly in their computational complexity, organization, overheads, and hardware realization for serial and parallel computation. In that sense, the choice and the suitability of the FFT algorithms may depend on a range of factors. We are once again focused on the computational complexity aspects of FFT algorithms.

In general, DFTs are defined for complex-valued data. If the data is real then it is redundant as far as complex arithmetic is concerned. There are several ways to exploit this redundancy in FFT algorithms in order to reduce their computational complexity. Other symmetries present in the data may lead to simplifications as well. Also, DFTs exist in all fields and rings, infinite or finite, that contain the n-th root of unity. In most instances, FFTs are obtained by exploiting the number-theoretic properties of the indices that are always integers regardless of the number systems over which the sequences themselves are defined. Even though we focus on sequences defined in Re and C in our presentation, the readers are reminded to keep this in mind in order to appreciate the applicability of DFTs to a much wider range of applications. For example, by treating DFTs in finite fields $GF(p)$ and $GF(q)$, $q = p^m$, one can describe algebraic constructions of some of the most popular codes, namely Bose-Chaudhary-Hoquenghem (BCH) and Reed-Solomon (RS) codes, used extensively to improve the reliability of information processing systems.

Cyclic convolution algorithms and FFTs are intimately related. Such a relationship exists for one- and higher-dimensional data sequences. FFTs are a tool to compute cyclic convolution efficiently. Similarly, cyclic convolutions are a tool to obtain FFTs. This is a very interesting aspect of digital signal processing algorithms. We illustrate this with the help of an example of length 5 cyclic convolution and FFT.

Example iii.1 Let $n = 5$. Consider the length 5 cyclic convolution of two sequences defined in C:

$$Z(u) = X(u)\, Y(u) \bmod u^5 - 1.$$

A DFT based computation is as follows:

Step 1. Compute the DFT of x_l, y_l, $l = 0, 1, 2, 3, 4$, to get $\boldsymbol{X}_i$ and $\boldsymbol{Y}_i$, $i = 0, 1, 2, 3, 4$.

Step 2. Obtain the DFT of z_l, $l = 0, 1, 2, 3, 4$, as

$$\boldsymbol{Z}_i = \boldsymbol{X}_i \cdot \boldsymbol{Y}_i, \quad i = 0, 1, 2, 3, 4.$$

Step 3. Obtain the IDFT of $\boldsymbol{Z}_i$, $i = 0, 1, 2, 3, 4$, to get z_l, $l = 0, 1, 2, 3, 4$.

It is clear that a FFT algorithm for a length 5 DFT will result in a fast algorithm for cyclic convolution of length 5.

Now, express the DFT of a length 5 sequence as

$$\begin{bmatrix} A_0 \\ A_1 \\ A_2 \\ A_3 \\ A_4 \end{bmatrix} = \begin{bmatrix} 1 & 1 & 1 & 1 & 1 \\ 1 & \omega & \omega^2 & \omega^3 & \omega^4 \\ 1 & \omega^2 & \omega^4 & \omega & \omega^3 \\ 1 & \omega^3 & \omega & \omega^4 & \omega^2 \\ 1 & \omega^4 & \omega^3 & \omega^2 & \omega \end{bmatrix} \begin{bmatrix} A_0 \\ A_1 \\ A_2 \\ A_3 \\ A_4 \end{bmatrix}.$$

Note that the computation of A_0 involves only ADDs and can be done separately. Now consider the 4×4 core of the DFT pertaining to the computation

$$\begin{bmatrix} \boldsymbol{B}_1 \\ \boldsymbol{B}_2 \\ \boldsymbol{B}_3 \\ \boldsymbol{B}_4 \end{bmatrix} = \begin{bmatrix} \omega & \omega^2 & \omega^3 & \omega^4 \\ \omega^2 & \omega^4 & \omega & \omega^3 \\ \omega^3 & \omega & \omega^4 & \omega^2 \\ \omega^4 & \omega^3 & \omega^2 & \omega \end{bmatrix} \begin{bmatrix} A_1 \\ A_2 \\ A_3 \\ A_4 \end{bmatrix}. \qquad \text{(iii.6)}$$

Given $\boldsymbol{B}_i$, $\boldsymbol{A}_i$, $i = 1, 2, 3, 4$ can be recovered using one ADD per coefficient. Thus a total of 8 ADDs are required to compute $\boldsymbol{A}_i$, $i = 0, 1, 2, 3, 4$, once $\boldsymbol{B}_i$, $i = 1, 2, 3, 4$, are known.

Defining a permutation, we can alternately express (iii.6) as

$$\begin{bmatrix} \boldsymbol{B}_2 \\ \boldsymbol{B}_4 \\ \boldsymbol{B}_3 \\ \boldsymbol{B}_1 \end{bmatrix} = \begin{bmatrix} \omega & \omega^3 & \omega^4 & \omega^2 \\ \omega^2 & \omega & \omega^3 & \omega^4 \\ \omega^4 & \omega^2 & \omega & \omega^3 \\ \omega^3 & \omega^4 & \omega^2 & \omega \end{bmatrix} \begin{bmatrix} A_3 \\ A_4 \\ A_2 \\ A_1 \end{bmatrix}. \qquad \text{(iii.7)}$$

This matrix-vector product is seen to be a cyclic convolution of length 4 and can be expressed in polynomial form as

$$\begin{aligned}[\boldsymbol{B}_2 + \boldsymbol{B}_4\, u + \boldsymbol{B}_3\, u^2 + \boldsymbol{B}_1\, u^3] &= [\omega + \omega^2\, u + \omega^4\, u^2 + \omega^3\, u^3] \\ &\quad \cdot [A_3 + A_4\, u + A_2\, u^2 + A_1\, u^3] \bmod (u^4 - 1). \qquad \text{(iii.8)}\end{aligned}$$

This length 4 cyclic convolution can once again be computed by (i) taking the length 4 DFT of the input sequences, (ii) multiplying them together, and (iii) taking the IDFT. Alternately, any other cyclic convolution algorithm of length 4 can be used. The length 4 DFT corresponds to the DFT matrix having the form

$$\Lambda = \begin{bmatrix} 1 & 1 & 1 & 1 \\ 1 & -j & -1 & j \\ 1 & -1 & 1 & -1 \\ 1 & j & -1 & -j \end{bmatrix},$$

which implies that a length 4 DFT can be computed only in 8 ADDs. Alternately, the factorization $u^4 - 1 = (u - 1)(u + 1)(u^2 + 1)$ can be used to compute the quantities in (iii.8).

The link between the DFT and cyclic convolution will be further enhanced and strengthened in Chapters 13 and 14.

There are a number of approaches for computing one- and higher-dimensional DFT for a given length. We will present several of these in the chapters that follow. One interesting point about these algorithms is that many FFTs consist in expressing a d-dimensional large length DFT as a $(d + 1)$ or higher dimensional DFT of smaller length in one or more dimensions and compute them as such. Some of the most popular FFTs including radix-2 and radix-4 FFT fall into this category.

Example iii.2 Consider the length 6 DFT expressed as the matrix vector product

$$\begin{bmatrix} A_0 \\ A_1 \\ A_2 \\ A_3 \\ A_4 \\ A_5 \end{bmatrix} = \begin{bmatrix} 1 & 1 & 1 & 1 & 1 & 1 \\ 1 & \omega & \omega^2 & \omega^3 & \omega^4 & \omega^5 \\ 1 & \omega^2 & \omega^4 & 1 & \omega^2 & \omega^4 \\ 1 & \omega^3 & 1 & \omega^3 & 1 & \omega^3 \\ 1 & \omega^4 & \omega^2 & 1 & \omega^4 & \omega^2 \\ 1 & \omega^5 & \omega^4 & \omega^3 & \omega^2 & \omega \end{bmatrix} \begin{bmatrix} A_0 \\ A_1 \\ A_2 \\ A_3 \\ A_4 \\ A_5 \end{bmatrix},$$

where $\omega = exp(-j\, 2\pi / 6)$. Using the properties $6 = 2 \cdot 3$, $(2, 3) = 1$, this can be arranged as a two-dimensional DFT expressed as

$$\begin{bmatrix} A_0 & A_2 & A_4 \\ A_3 & A_5 & A_1 \end{bmatrix} = \begin{bmatrix} 1 & 1 \\ 1 & -1 \end{bmatrix} \begin{bmatrix} A_0 & A_4 & A_2 \\ A_3 & A_1 & A_5 \end{bmatrix} \begin{bmatrix} 1 & 1 & 1 \\ 1 & \omega_3 & \omega_3^2 \\ 1 & \omega_3^2 & \omega_3 \end{bmatrix},$$

where ω_3 is the third root of unity, $\omega_3 = exp(-j\, 2\, \pi / 3)$. The pre- and the post-MULT matrices are recognized as length 2 and length 3 DFT matrices. The entire computation has been recast as 3 DFTs of length 2 (DFT of each column) followed (or preceded) by 2 DFTs of length 3 (DFT of each row).

One has to be careful when counting MULTs and ADDs for FFT algorithms. Any MULT encountered in DFT is an expression of the type

$$A \cdot f(\omega), \tag{iii.9}$$

where A is data dependent and $f(\omega)$ is a pre-computed complex-valued function of ω. A straightforward computation of (iii.9) requires 4 real MULTs (RMULT) and 2 real ADDs (RADD). If A is real, then (iii.9) requires 2 RMULTs. If A is complex, let $A = a_R + ja_I$ and $f(\omega) = f_R + j f_I$. We can compute (iii.9) as

$$\begin{aligned} A \cdot f(\omega) &= (a_R + ja_I) \cdot (f_R + j f_I) \\ &= [(a_R + a_I) \cdot f_R - a_I \cdot (f_R + f_I)] \\ &\quad + j[(a_R + a_I) \cdot f_R + a_R(f_I - f_R)]. \end{aligned}$$

Thus a complex MULT (CMULT) for FFT can be realized as 3 RMULTs and 3 RADDs. A count of 3 RADDs for a CMULT is based on the assumption that only the data dependent RADD are included; 2 RADDs among the coefficients of $f(\omega)$ are ignored as they are pre-computed only once and remain fixed throughout. The count for RMULT and RADD can be further reduced for cases when $f(\omega) = \omega^i = \cos\theta + j\sin\theta$ and $\theta = \pm\,\pi/4$ or $\pm\,3\pi/4$. This corresponds to $n = 8$ or the 8th root of unity. In this case,

$$A \cdot (1 \pm j)/\sqrt{2} = (a_R - \pm a_I)/\sqrt{2} + j(\pm a_R + a_I)/\sqrt{2}.$$

This can now be computed in 2 RMULTs and 2 RADDs instead of either 4 RMULTs and 2 RADDs or 3 RMULTs and 3 RADDs as stated above. Finally, if A is real or purely imaginary valued, then (iii.9) can be computed in 2 RMULTs.

We end this discussion by stating that the treatment of DFT in the following chapters is a computational one. We remain focused on the FFT algorithms. There are a number of books that describe the application of DFT to science, engineering, and other areas of interest.

Chapter 13

Fast Fourier Transform: One-Dimensional Data Sequences

In this chapter, we study fast algorithms for computing the discrete Fourier transform (DFT). Collectively, they are known as fast Fourier transform (FFT) algorithms. This chapter deals with one-dimensional data sequences and the next chapter is focused on two- and higher-dimensional sequences. The nature of FFT algorithms depends on the number-theoretic properties of the data sequences and their indices. This fact will influence not only our presentation of the FFT but also the overall structure of the algorithms. We can partition the FFT into three categories: (a) FFT for complex-valued data sequences; (b) FFT for real-valued data sequences; and (c) FFT for data sequences defined in finite number systems (rings or fields). For the FFT belonging to category (a), one may exploit the number-theoretic properties of the indices only. For the FFT belonging to category (b), one may desire to exploit the real-valued nature of the data sequence and/or restrict the intermediate quantities to be real-valued to the extent possible. For the FFT belonging to category (c), there may be additional mathematical properties that could play a role in the derivation of an algorithm. All the quantities in this chapter are defined over a field F. We begin our description by defining and analyzing the mathematical properties of DFT.

13.1 The DFT: Definition and Properties

Definition 13.1 Discrete Fourier transform (DFT). Given a length n, data sequence A_l, $l = 0, 1, \ldots, n-1$ or, equivalently, a degree $n-1$ polynomial $A(u)$, the *discrete Fourier transform* (DFT) in a field F is defined as the sequence $\mathbf{A}_i$, $i = 0, 1, \ldots, n-1$, given by

$$\mathbf{A}_i = \sum_{l=0}^{n-1} A_l \alpha^{li}\ , \quad i = 0, 1, \ldots, n-1, \tag{13.1.1}$$

where α is an element of order n in F, that is, $\alpha^n = 1$ in F and $\alpha^i \neq 1$, $i < n$. Representing the DFT as a matrix-vector product, we have

$$\begin{bmatrix} \mathbf{A}_0 \\ \mathbf{A}_1 \\ \vdots \\ \mathbf{A}_{n-1} \end{bmatrix} = \begin{bmatrix} 1 & 1 & 1 & \cdots & 1 & 1 \\ 1 & \alpha & \alpha^2 & \cdots & \alpha^{n-2} & \alpha^{n-1} \\ 1 & \alpha^2 & \alpha^4 & \cdots & \alpha^{2(n-2)} & \alpha^{2(n-1)} \\ \vdots & \vdots & \vdots & \ddots & \vdots & \vdots \\ 1 & \alpha^{n-2} & \alpha^{(n-2)2} & \cdots & \alpha^{(n-2)(n-2)} & \alpha^{(n-2)(n-1)} \\ 1 & \alpha^{n-1} & \alpha^{(n-1)2} & \cdots & \alpha^{(n-1)(n-2)} & \alpha^{(n-1)(n-1)} \end{bmatrix} \begin{bmatrix} A_0 \\ A_1 \\ \vdots \\ A_{n-1} \end{bmatrix} \tag{13.1.2}$$

or

$$\underline{\mathbf{A}} = \Lambda\, \underline{A}. \tag{13.1.3}$$

Definition 13.2 Inverse discrete Fourier transform. Given the DFT $\mathbf{A}_i$, $i = 0, 1, \ldots, n-1$ of a sequence A_l, $l = 0, 1, \ldots, n-1$, the original data sequence A_l, $l = 0, 1, \ldots, n-1$ is said to constitute the *inverse discrete Fourier transform* (IDFT) of the DFT sequence $\mathbf{A}_i$, $i = 0, 1, \ldots, n-1$.

Theorem 13.1 Given the DFT $\mathbf{A}_i$, $i = 0, 1, \ldots, n-1$ of a sequence A_l, $l = 0, 1, \ldots, n-1$, the IDFT can be computed as

$$A_l = (1/n) \sum_{i=0}^{n-1} \mathbf{A}_i \alpha^{-il}\ , \quad l = 0, 1, \ldots, n-1, \tag{13.1.4}$$

which can alternately be represented as the matrix-vector product,

$$\underline{A} = \Lambda^{-1}\, \underline{\mathbf{A}}$$

or

$$\begin{bmatrix} A_0 \\ A_1 \\ \vdots \\ A_{n-1} \end{bmatrix} = \left(\frac{1}{n}\right) \begin{bmatrix} 1 & 1 & 1 & \cdots & 1 & 1 \\ 1 & \alpha^{-1} & \alpha^{-2} & \cdots & \alpha^{-(n-2)} & \alpha^{-(n-1)} \\ 1 & \alpha^{-2} & \alpha^{-4} & \cdots & \alpha^{-2(n-2)} & \alpha^{-2(n-1)} \\ \vdots & \vdots & \vdots & \ddots & \vdots & \vdots \\ 1 & \alpha^{-(n-2)} & \alpha^{-(n-2)2} & \cdots & \alpha^{-(n-2)(n-2)} & \alpha^{-(n-2)(n-1)} \\ 1 & \alpha^{-(n-1)} & \alpha^{-(n-1)2} & \cdots & \alpha^{-(n-1)(n-2)} & \alpha^{-(n-1)(n-1)} \end{bmatrix} \begin{bmatrix} A_0 \\ A_1 \\ \vdots \\ A_{n-1} \end{bmatrix}. \tag{13.1.5}$$

Proof: In other words, if the (l, k)-th element of the DFT matrix Λ is α^{lk}, then the (l, k)-th element of the IDFT matrix Λ^{-1} is $(1/n)\ \alpha^{-lk}$. We establish this result by showing that for the forms of Λ and Λ^{-1} matrices in (13.1.2) and (13.1.5), the product $[(\Lambda)\ (\Lambda^{-1})] = \mathbf{I}_{n\times n}$ is an identity matrix. In this case,

(i, k)-th element of $[(\Lambda)(\Lambda^{-1})]$

$$= \sum_{l=0}^{n-1} \big((i,l)\text{th element of } \Lambda\big) \cdot \big((l,k)\text{th element of } \Lambda^{-1}\big)$$

$$= \sum_{l=0}^{n-1} \alpha^{il} (1/n) \alpha^{-lk} = (1/n) \sum_{l=0}^{n-1} \alpha^{l(i-k)}. \tag{13.1.6}$$

For $i = k$, $i - k = 0$ and all the terms in the above sum become equal to 1. Therefore, the (i, k)-th element is 1 for $i = k$. For $i \neq k$, $i - k \neq 0$ and the sum corresponds to a geometric series with the first term = 1, number of terms equal to n, and the common ratio = $\alpha^{(i-k)}$. This leads to

$$\sum_{l=0}^{n-1} \alpha^{l(i-k)} = \left(1 - \alpha^{n(i-k)}\right) / \left(1 - \alpha^{(i-k)}\right) = 0. \tag{13.1.7}$$

The last equality holds due to the fact that α is an n-th root of unity in F and, therefore, $\alpha^n = 1$ and $\alpha^{(i-k)} \neq 1$, $0 < i, k < n$. We recognize that if α is a primitive element of the field F, then so is α^{-1}. Consequently, the (i, k)-th element of the matrix product $[(\Lambda)\ (\Lambda^{-1})]$ is 1 for $i = k$ and is 0 for $i \neq k$. This implies that $[(\Lambda)(\Lambda^{-1})] = \mathbf{I}_{n\times n}$, thereby proving the statement of the theorem.

The factor of $(1/n)$ in the IDFT matrix can also be distributed as $1/\sqrt{n}$ in the DFT matrix and $1/\sqrt{n}$ in the IDFT matrix. A comparison of (13.1.1) and (13.1.4) reveals an interesting aspect of the DFT and IDFT computation. The

IDFT computation is identical to the DFT computation with the α, the n-th root of unity in F replaced by α^{-1}. We recognize that if α is an n-th root of unity in the field F, then so is α^{-1}. The computation of DFT and IDFT can be performed in an identical manner. Consequently, we will discuss only algorithms for DFT.

Unless mentioned otherwise, this chapter deals with the DFT computation defined in C, the field of complex numbers. In C, the n-th root of unity (denoted by α thus far) is denoted by ω and is given by

$$\omega = e^{-2\pi j/n}, j = \sqrt{-1}. \qquad (13.1.8)$$

In other fields, the n-th root of unity may take different forms. For example, in the cyclotomic field $CF(7)$, that is the extension field of Z obtained by using the cyclotomic polynomial $1 + j + j^2 + j^3 + j^4 + j^5 + j^6$ (Problem 3.14), the 7-th root of unity is given by j which can be used to define a DFT of length 7 in this field. This aspect will be discussed thoroughly in Chapter 14. Unless otherwise stated, we assume that we are dealing with the DFT defined for data sequences in C.

Definition 13.3 Fast Fourier transform (FFT). The fast algorithms for the computation of DFT and IDFT are collectively known as the **fast Fourier transform** (FFT).

The computational complexity of FFTs, as measured in terms of the number of ADDs and MULTs required by the algorithm, forms the measure of their efficiency. In this regard, it is important to mention that the number of MULTs are measured by counting terms of the type $a \cdot b$, where a and b depend on the data sequence and ω, respectively. MULT by scalars such as 1, −1, 0, and so on are excluded from the count.

The DFT and FFT in two- and higher-dimensions will be studied in depth in Chapter 14.

13.1.1 Properties of DFT

In the following, we represent the original sequence A_l, $l = 0, 1, \ldots, n-1$, and its DFT $\boldsymbol{A}_i$, $i = 0, 1, \ldots, n-1$, in a compact form as

$$\underline{A}_l \xleftrightarrow{DFT} \underline{\boldsymbol{A}}_i. \qquad (13.1.9)$$

The indices l and i are defined mod n, as the n-th root of unity is defined mod n.

A. **Linearity.** Given (13.1.9) for two sequences A_l and B_l, the following holds:

$$a\underline{A}_l + b\underline{B}_l \xleftrightarrow{DFT} a\underline{\boldsymbol{A}}_i + b\underline{\boldsymbol{B}}_i \,. \tag{13.1.10}$$

B. **Time Shifting.** Given (13.1.9), the following holds:

$$\underline{A}_{l-m} \xleftrightarrow{DFT} \omega^{im}\,\underline{\boldsymbol{A}}_i \,. \tag{13.1.11}$$

C. **Duality.** If (13.1.9) holds, then

$$\underline{\boldsymbol{A}}_l \xleftrightarrow{DFT} n \cdot \underline{A}_{-i} \,. \tag{13.1.12}$$

Since all indices are defined mod n, A_{-i} is same as A_{n-i}.

D. **Frequency Shifting.** Given (13.1.9), the following holds:

$$\omega^{-lm}\,\underline{A}_l \xleftrightarrow{DFT} \underline{\boldsymbol{A}}_{i-m} \,. \tag{13.1.13}$$

E. **Permutation.** Recall that if $(a, n) = 1$, then one can always find b such that b is the inverse of a mod n, that is, $b = a^{-1}$ mod n. Also, $(b, n) = 1$. Thus, $c = a \cdot l$ mod n and $d = b \cdot i$ mod n define a permutation on l and i, respectively. In this case

$$\underline{A}_c \xleftrightarrow{DFT} \underline{\boldsymbol{A}}_d \,. \tag{13.1.14}$$

Proof: Let the DFT of the left-hand-side sequence be denoted by $\boldsymbol{B}_i$. Then, we have

$$\boldsymbol{B}_i = \sum_{l=0}^{n-1} A_{al}\,\omega^{li} = \sum_{c=0}^{n-1} A_c\,\omega^{bci} = \sum_{c=0}^{n-1} A_c\,\omega^{c(bi)} = \boldsymbol{A}_{bi} = \boldsymbol{A}_d \,.$$

The summation is changed from l to c by noting that as l goes from 0, 1, …, $n-1$, so does c and, therefore, the summation remains the same. Also, we have $l = a^{-1} \cdot c$ mod $n = b \cdot c$ mod n.

F. **Convolution.** If A_l is the cyclic convolution of two sequences B_l and C_l (in polynomial form, $A(u) = B(u) \cdot C(u)$ mod $u^n - 1$), then

$$\underline{A}_l \xleftrightarrow{DFT} \underline{\boldsymbol{B}}_i \cdot \underline{\boldsymbol{C}}_i \,. \tag{13.1.15}$$

G. **Product.** If $A_l = B_l \cdot C_l$, $l = 0, 1, \ldots, n-1$, then

$$\underline{A}_l \xleftrightarrow{DFT} \frac{1}{n}\sum_{k=0}^{n-1} \boldsymbol{B}_k \cdot \boldsymbol{C}_{i-k} \,. \tag{13.1.16}$$

H. **Parseval's Theorem.** Since DFT is an orthogonal transform,

$$\sum_{i=0}^{n-1} |\boldsymbol{A}_i|^2 = n \sum_{l=0}^{n-1} |A_l|^2 \,. \tag{13.1.17}$$

I. If (13.1.9) holds, then

$$\underline{A}_l^* \xleftrightarrow{DFT} \underline{\boldsymbol{A}}_{-i}^* \,. \tag{13.1.18}$$

J. If (13.1.9) holds, then

$$\underline{A}_{-l}^* \xleftrightarrow{DFT} \underline{\boldsymbol{A}}_i^* \,. \tag{13.1.19}$$

K. **Real Sequences.** If the sequence A_l is real-valued, then (13.1.18) implies that

$$\boldsymbol{A}_i = \boldsymbol{A}_{-i}^* = \boldsymbol{A}_{n-i}^* \,. \tag{13.1.20}$$

Thus, the DFT of a real sequence satisfies conjugate symmetry.

L. **Conjugate Symmetry.** If the sequence A_l satisfies conjugate symmetry, $A_l = A_l^*$, then its DFT is a real valued sequence. This follows from the duality property.

M. **Imaginary Sequences.** If the sequence A_l is a purely imaginary sequence, then

$$\boldsymbol{A}_i = -\boldsymbol{A}_{-i}^* = -\boldsymbol{A}_{n-i}^* \,. \tag{13.1.21}$$

In other words, it satisfies the property of conjugate anti-symmetry.

Properties K and M can be used to compute the DFT of two real-valued sequences, B_l and C_l, by first forming a complex-valued sequence $A_l = B_l + jC_l$ followed by the DFT of A_l. Finally, $\boldsymbol{B}_i$ and $\boldsymbol{C}_i$ are extracted from $\boldsymbol{A}_i$ by exploiting properties K and M. The details are straightforward and left as an exercise.

Combining properties K and L, we see that if the sequence A_l is real and satisfies conjugate symmetry, that is $A_l = A_{-l}$, then its DFT satisfies conjugate symmetry and is real. Therefore, $\boldsymbol{A}_i = \boldsymbol{A}_{-i}$.

The structure of FFTs in $\boldsymbol{C}$ depends on the number-theoretic properties of the indices. However, before we begin that path, we present a simple yet powerful idea for deriving FFTs that is valid for any n.

13.1.2 Bluestein's Algorithm

Based on the identity

$$l\,i = l^2/2 + i^2/2 - (i-l)^2/2,$$

we can express (13.1.1) as

$$\boldsymbol{A}_i = \omega^{i^2/2} \sum_{l=0}^{n-1} \left[A_l \omega^{l^2/2} \right] \omega^{-(i-l)^2/2}, \quad i = 0, 1, \ldots, n-1. \tag{13.1.22}$$

This is immediately recognized as convolution of two sequences, namely, $A_l \omega^{l^2/2}$ and $\omega^{-l^2/2}$, $l = 0, 1, \ldots, n-1$, followed by post-MULT by $\omega^{i^2/2}$. Forming $A_l \omega^{l^2/2}$ from A_l and post-MULT by $\omega^{i^2/2}$ requires at most n MULTs each. This aspect will be explored further in the problems at the end of the chapter. In general, the powers of ω are defined mod n. For n even, $-(l \pm n)^2/2 = -l^2/2$. Therefore, the powers of $\omega^{-l^2/2}$ are also defined mod n. This implies that for n even, the convolution in (13.1.22) is cyclic of length n. In a similar manner, it can be shown that for odd n, it is a cyclic convolution of length $2n$, where the last n coefficients of $A_l \omega^{l^2/2}$ are set equal to 0 and only the first n points in the convolution are computed. Any of the previously studied algorithms for cyclic convolution can be used to compute this convolution. The main reason for this algorithm to be so powerful is its applicability for any value of n. This technique for computing the DFT is also known as the *chirp-z* transform.

13.2 Rader's FFT Algorithm, $n = p$, p an Odd Prime

The FFT algorithm for $n = 2$ is trivial. Consider the case of $n = p$, p an odd prime. The FFT algorithm in this instance is known as the **Rader's algorithm**. In this case, we have

$$\boldsymbol{A}_0 = A_0 + A_1 + \ldots + A_{p-1} \tag{13.2.1}$$

and

$$\boldsymbol{A}_i - A_0 = \boldsymbol{B}_i = \sum_{l=1}^{p-1} A_l \omega^{li}, \quad i = 1, \ldots, p-1. \tag{13.2.2}$$

It is the term in (13.2.2) that the Rader's algorithm seeks to simplify using number theory. The computation of $\boldsymbol{B}_i$ can be interpreted as the computation of DFT of a sequence having first term equal to 0 and the first term of the output ignored. This is also known as the **reduced DFT**. Both the indices i and l in (13.2.2) are defined mod p and correspond to non-zero elements of $GF(p)$. In $GF(p)$, there is a primitive element of order $p-1$. Let α be one such element. Therefore, we can write

$$i = \alpha^g$$

and

$$l = \alpha^{-h}; \quad g, h = 0, 1, \ldots, p-2. \tag{13.2.3}$$

Substituting in (13.2.2), we get

$$\boldsymbol{B}_{\alpha^g} = \sum_{h=0}^{p-2} A_{\alpha^{-h}} \omega^{\alpha^{g-h}}; \quad g, h = 0, 1, \ldots, p-2. \tag{13.2.4}$$

This equation implies that the sequence $\boldsymbol{B}_{\alpha^g}$, $g = 0, 1, \ldots, p-2$ is cyclic convolution of $A_{\alpha^{-h}}$ and ω^{α^h}, $h = 0, 1, \ldots, p-2$. In polynomial form

$$\boldsymbol{B}(u) = \sum_{g=0}^{p-2} \boldsymbol{B}_{\alpha^g} u^g = \Delta(u) A(u) \bmod \left(u^{p-1} - 1\right)$$
$$= \left[\sum_{h=0}^{p-2} \omega^{\alpha^h} u^h\right] \cdot \left[\sum_{h=0}^{p-2} A_{\alpha^{-h}} u^h\right] \bmod \left(u^{p-1} - 1\right). \tag{13.2.5}$$

Once again, all the algorithms for computing the cyclic convolution can be used to compute $\boldsymbol{B}_i$, $i = 1, 2, \ldots, p-1$. Rader's algorithm is desirable as it expresses a length p DFT in terms of a length $p-1$ cyclic convolution. This cyclic convolution can either be computed using the FFT of length $p-1$ or any other cyclic convolution algorithm. It is important to note here that $p-1$ may turn out to be a highly composite integer and, therefore, there may exist a wide range of algorithms for the cyclic convolution of length $p - 1$. One must realize though that the input and the output sequences get permuted in the process of DFT computation using Rader's algorithm. Figure 13.1 depicts the process involved in the computation of $\boldsymbol{B}_i$ using the Rader's algorithm as described here.

There is a slight modification that one can make in the above formulation of Rader's algorithm. This is done as follows:

$$\boldsymbol{A}_0 = A_0 + A_1 + \ldots + A_{p-1} \tag{13.2.6}$$

and

$$\boldsymbol{A}_i - \boldsymbol{A}_0 = \boldsymbol{C}_i = \sum_{l=1}^{p-1} A_l\left(\omega^{li} - 1\right), \; i = 1, \ldots, p-1. \tag{13.2.7}$$

It is the term in (13.2.7) that one now simplifies. Following the same steps as in (13.2.1) to (12.2.5), we get

$$\boldsymbol{C}_{\alpha^g} = \sum_{h=0}^{p-2} A_{\alpha^{-h}}\left(\omega^{\alpha^{g-h}} - 1\right); \; g, h = 0, 1, \ldots, p-2. \tag{13.2.8}$$

This equation implies that the sequence $\boldsymbol{C}_{\alpha^g}$, $g = 0, 1, \ldots, p - 2$ is the cyclic convolution of $A_{\alpha^{-h}}$ and $\omega^{\alpha^h} - 1$, $h = 0, 1, \ldots, p - 2$, and can be computed accordingly.

Example 13.1 Let $p = 17$. The length 17 DFT is given by

$[\boldsymbol{A}_0\ \boldsymbol{A}_1\ \boldsymbol{A}_2\ \boldsymbol{A}_3\ \boldsymbol{A}_4\ \boldsymbol{A}_5\ \boldsymbol{A}_6\ \boldsymbol{A}_7\ \boldsymbol{A}_8\ \boldsymbol{A}_9\ \boldsymbol{A}_{10}\ \boldsymbol{A}_{11}\ \boldsymbol{A}_{12}\ \boldsymbol{A}_{13}\ \boldsymbol{A}_{14}\ \boldsymbol{A}_{15}\ \boldsymbol{A}_{16}]^t =$

$$\begin{bmatrix}
1 & 1 & 1 & 1 & 1 & 1 & 1 & 1 & 1 & 1 & 1 & 1 & 1 & 1 & 1 & 1 & 1 \\
1 & \omega^{1} & \omega^{2} & \omega^{3} & \omega^{4} & \omega^{5} & \omega^{6} & \omega^{7} & \omega^{8} & \omega^{9} & \omega^{10} & \omega^{11} & \omega^{12} & \omega^{13} & \omega^{14} & \omega^{15} & \omega^{16} \\
1 & \omega^{2} & \omega^{4} & \omega^{6} & \omega^{8} & \omega^{10} & \omega^{12} & \omega^{14} & \omega^{16} & \omega^{1} & \omega^{3} & \omega^{5} & \omega^{7} & \omega^{9} & \omega^{11} & \omega^{13} & \omega^{15} \\
1 & \omega^{3} & \omega^{6} & \omega^{9} & \omega^{12} & \omega^{15} & \omega^{1} & \omega^{4} & \omega^{7} & \omega^{10} & \omega^{13} & \omega^{16} & \omega^{2} & \omega^{5} & \omega^{8} & \omega^{11} & \omega^{14} \\
1 & \omega^{4} & \omega^{8} & \omega^{12} & \omega^{16} & \omega^{3} & \omega^{7} & \omega^{11} & \omega^{15} & \omega^{2} & \omega^{6} & \omega^{10} & \omega^{14} & \omega^{1} & \omega^{5} & \omega^{9} & \omega^{13} \\
1 & \omega^{5} & \omega^{10} & \omega^{15} & \omega^{3} & \omega^{8} & \omega^{13} & \omega^{1} & \omega^{6} & \omega^{11} & \omega^{16} & \omega^{4} & \omega^{9} & \omega^{14} & \omega^{2} & \omega^{7} & \omega^{12} \\
1 & \omega^{6} & \omega^{12} & \omega^{1} & \omega^{7} & \omega^{13} & \omega^{2} & \omega^{8} & \omega^{14} & \omega^{3} & \omega^{9} & \omega^{15} & \omega^{4} & \omega^{10} & \omega^{16} & \omega^{5} & \omega^{11} \\
1 & \omega^{7} & \omega^{14} & \omega^{4} & \omega^{11} & \omega^{1} & \omega^{8} & \omega^{15} & \omega^{5} & \omega^{12} & \omega^{2} & \omega^{9} & \omega^{16} & \omega^{6} & \omega^{13} & \omega^{3} & \omega^{10} \\
1 & \omega^{8} & \omega^{16} & \omega^{7} & \omega^{15} & \omega^{6} & \omega^{14} & \omega^{5} & \omega^{13} & \omega^{4} & \omega^{12} & \omega^{3} & \omega^{11} & \omega^{2} & \omega^{10} & \omega^{1} & \omega^{9} \\
1 & \omega^{9} & \omega^{1} & \omega^{10} & \omega^{2} & \omega^{11} & \omega^{3} & \omega^{12} & \omega^{4} & \omega^{13} & \omega^{5} & \omega^{14} & \omega^{6} & \omega^{15} & \omega^{7} & \omega^{16} & \omega^{8} \\
1 & \omega^{10} & \omega^{3} & \omega^{13} & \omega^{6} & \omega^{16} & \omega^{9} & \omega^{2} & \omega^{12} & \omega^{5} & \omega^{15} & \omega^{8} & \omega^{1} & \omega^{11} & \omega^{4} & \omega^{14} & \omega^{7} \\
1 & \omega^{11} & \omega^{5} & \omega^{16} & \omega^{10} & \omega^{4} & \omega^{15} & \omega^{9} & \omega^{3} & \omega^{14} & \omega^{8} & \omega^{2} & \omega^{13} & \omega^{7} & \omega^{1} & \omega^{12} & \omega^{6} \\
1 & \omega^{12} & \omega^{7} & \omega^{2} & \omega^{14} & \omega^{9} & \omega^{4} & \omega^{16} & \omega^{11} & \omega^{6} & \omega^{1} & \omega^{13} & \omega^{8} & \omega^{3} & \omega^{15} & \omega^{10} & \omega^{5} \\
1 & \omega^{13} & \omega^{9} & \omega^{5} & \omega^{1} & \omega^{14} & \omega^{10} & \omega^{6} & \omega^{2} & \omega^{15} & \omega^{11} & \omega^{7} & \omega^{3} & \omega^{16} & \omega^{12} & \omega^{8} & \omega^{4} \\
1 & \omega^{14} & \omega^{11} & \omega^{8} & \omega^{5} & \omega^{2} & \omega^{16} & \omega^{13} & \omega^{10} & \omega^{7} & \omega^{4} & \omega^{1} & \omega^{15} & \omega^{12} & \omega^{9} & \omega^{6} & \omega^{3} \\
1 & \omega^{15} & \omega^{13} & \omega^{11} & \omega^{9} & \omega^{5} & \omega^{7} & \omega^{3} & \omega^{1} & \omega^{16} & \omega^{14} & \omega^{12} & \omega^{10} & \omega^{8} & \omega^{6} & \omega^{4} & \omega^{2} \\
1 & \omega^{16} & \omega^{15} & \omega^{14} & \omega^{13} & \omega^{12} & \omega^{11} & \omega^{10} & \omega^{9} & \omega^{8} & \omega^{7} & \omega^{6} & \omega^{5} & \omega^{4} & \omega^{3} & \omega^{2} & \omega^{1}
\end{bmatrix}$$

$$\cdot [A_0\ A_1\ A_2\ A_3\ A_4\ A_5\ A_6\ A_7\ A_8\ A_9\ A_{10}\ A_{11}\ A_{12}\ A_{13}\ A_{14}\ A_{15}\ A_{16}]'.$$

Let $\alpha = 3$ be the primitive element in $GF(17)$ to be used in the computation. Then the sequences $\boldsymbol{B}_{\alpha^g}$, $g = 0, 1, \ldots, p-2$, $A_{\alpha^{-h}}$, and ω^{α^h}, $h = 0, 1, \ldots, p-2$, are given by $(\boldsymbol{B}_1\ \boldsymbol{B}_3\ \boldsymbol{B}_9\ \boldsymbol{B}_{10}\ \boldsymbol{B}_{13}\ \boldsymbol{B}_5\ \boldsymbol{B}_{15}\ \boldsymbol{B}_{11}\ \boldsymbol{B}_{16}\ \boldsymbol{B}_{14}\ \boldsymbol{B}_8\ \boldsymbol{B}_7\ \boldsymbol{B}_4\ \boldsymbol{B}_{12}\ \boldsymbol{B}_2\ \boldsymbol{B}_6)$, $(A_1\ A_6\ A_2\ A_{12}\ A_4\ A_7\ A_8\ A_{14}\ A_{16}\ A_{11}\ A_{15}\ A_5\ A_{13}\ A_{10}\ A_9\ A_3)$, and $(\omega^1\ \omega^3\ \omega^9\ \omega^{10}\ \omega^{13}\ \omega^5\ \omega^{15}\ \omega^{11}\ \omega^{16}\ \omega^{14}\ \omega^8\ \omega^7\ \omega^4\ \omega^{12}\ \omega^2\ \omega^6)$. In polynomial form,

$$\boldsymbol{B}_1 + \boldsymbol{B}_3 u + \boldsymbol{B}_9 u^2 + \boldsymbol{B}_{10} u^3 + \boldsymbol{B}_{13} u^4 + \boldsymbol{B}_5 u^5 + \boldsymbol{B}_{15} u^6 + \boldsymbol{B}_{11} u^7 + \boldsymbol{B}_{16} u^8 + \boldsymbol{B}_{14} u^9 + \boldsymbol{B}_8 u^{10} + \boldsymbol{B}_7 u^{11} + \boldsymbol{B}_4 u^{12} + \boldsymbol{B}_{12} u^{13} + \boldsymbol{B}_2 u^{14} + \boldsymbol{B}_6 u^{15}$$

$$= (A_1 + A_6 u + A_2 u^2 + A_{12} u^3 + A_4 u^4 + A_7 u^5 + A_8 u^6 + A_{14} u^7 + A_{16} u^8 + A_{11} u^9 + A_{15} u^{10} + A_5 u^{11} + A_{13} u^{12} + A_{10} u^{13} + A_9 u^{14} + A_3 u^{15})$$

$$\cdot (\omega^1 + \omega^3 u + \omega^9 u^2 + \omega^{10} u^3 + \omega^{13} u^4 + \omega^5 u^5 + \omega^{15} u^6 + \omega^{11} u^7 + \omega^{16} u^8 + \omega^{14} u^9 + \omega^8 u^{10} + \omega^7 u^{11} + \omega^4 u^{12} + \omega^{12} u^{13} + \omega^2 u^{14} + \omega^6 u^{15}) \bmod (u^{16} - 1).$$

In matrix-vector form

$$[\boldsymbol{B}_1\ \boldsymbol{B}_3\ \boldsymbol{B}_9\ \boldsymbol{B}_{10}\ \boldsymbol{B}_{13}\ \boldsymbol{B}_5\ \boldsymbol{B}_{15}\ \boldsymbol{B}_{11}\ \boldsymbol{B}_{16}\ \boldsymbol{B}_{14}\ \boldsymbol{B}_8\ \boldsymbol{B}_7\ \boldsymbol{B}_4\ \boldsymbol{B}_{12}\ \boldsymbol{B}_2\ \boldsymbol{B}_6]' =$$

$$\begin{bmatrix}
\omega^{1} & \omega^{6} & \omega^{2} & \omega^{12} & \omega^{4} & \omega^{7} & \omega^{8} & \omega^{14} & \omega^{16} & \omega^{11} & \omega^{15} & \omega^{5} & \omega^{13} & \omega^{10} & \omega^{9} & \omega^{3} \\
\omega^{3} & \omega^{1} & \omega^{6} & \omega^{2} & \omega^{12} & \omega^{4} & \omega^{7} & \omega^{8} & \omega^{14} & \omega^{16} & \omega^{11} & \omega^{15} & \omega^{5} & \omega^{13} & \omega^{10} & \omega^{9} \\
\omega^{9} & \omega^{3} & \omega^{1} & \omega^{6} & \omega^{2} & \omega^{12} & \omega^{4} & \omega^{7} & \omega^{8} & \omega^{14} & \omega^{16} & \omega^{11} & \omega^{15} & \omega^{5} & \omega^{13} & \omega^{10} \\
\omega^{10} & \omega^{9} & \omega^{3} & \omega^{1} & \omega^{6} & \omega^{2} & \omega^{12} & \omega^{4} & \omega^{7} & \omega^{8} & \omega^{14} & \omega^{16} & \omega^{11} & \omega^{15} & \omega^{5} & \omega^{13} \\
\omega^{13} & \omega^{10} & \omega^{9} & \omega^{3} & \omega^{1} & \omega^{6} & \omega^{2} & \omega^{12} & \omega^{4} & \omega^{7} & \omega^{8} & \omega^{14} & \omega^{16} & \omega^{11} & \omega^{15} & \omega^{5} \\
\omega^{5} & \omega^{13} & \omega^{10} & \omega^{9} & \omega^{3} & \omega^{1} & \omega^{6} & \omega^{2} & \omega^{12} & \omega^{4} & \omega^{7} & \omega^{8} & \omega^{14} & \omega^{16} & \omega^{11} & \omega^{15} \\
\omega^{15} & \omega^{5} & \omega^{13} & \omega^{10} & \omega^{9} & \omega^{3} & \omega^{1} & \omega^{6} & \omega^{2} & \omega^{12} & \omega^{4} & \omega^{7} & \omega^{8} & \omega^{14} & \omega^{16} & \omega^{11} \\
\omega^{11} & \omega^{15} & \omega^{5} & \omega^{13} & \omega^{10} & \omega^{9} & \omega^{3} & \omega^{1} & \omega^{6} & \omega^{2} & \omega^{12} & \omega^{4} & \omega^{7} & \omega^{8} & \omega^{14} & \omega^{16} \\
\omega^{16} & \omega^{11} & \omega^{15} & \omega^{5} & \omega^{13} & \omega^{10} & \omega^{9} & \omega^{3} & \omega^{1} & \omega^{6} & \omega^{2} & \omega^{12} & \omega^{4} & \omega^{7} & \omega^{8} & \omega^{14} \\
\omega^{14} & \omega^{16} & \omega^{11} & \omega^{15} & \omega^{5} & \omega^{13} & \omega^{10} & \omega^{9} & \omega^{3} & \omega^{1} & \omega^{6} & \omega^{2} & \omega^{12} & \omega^{4} & \omega^{7} & \omega^{8} \\
\omega^{8} & \omega^{14} & \omega^{16} & \omega^{11} & \omega^{15} & \omega^{5} & \omega^{13} & \omega^{10} & \omega^{9} & \omega^{3} & \omega^{1} & \omega^{6} & \omega^{2} & \omega^{12} & \omega^{4} & \omega^{7} \\
\omega^{7} & \omega^{8} & \omega^{14} & \omega^{16} & \omega^{11} & \omega^{15} & \omega^{5} & \omega^{13} & \omega^{10} & \omega^{9} & \omega^{3} & \omega^{1} & \omega^{6} & \omega^{2} & \omega^{12} & \omega^{4} \\
\omega^{4} & \omega^{7} & \omega^{8} & \omega^{14} & \omega^{16} & \omega^{11} & \omega^{15} & \omega^{5} & \omega^{13} & \omega^{10} & \omega^{9} & \omega^{3} & \omega^{1} & \omega^{6} & \omega^{2} & \omega^{12} \\
\omega^{12} & \omega^{4} & \omega^{7} & \omega^{8} & \omega^{14} & \omega^{16} & \omega^{11} & \omega^{15} & \omega^{5} & \omega^{13} & \omega^{10} & \omega^{9} & \omega^{3} & \omega^{1} & \omega^{6} & \omega^{2} \\
\omega^{2} & \omega^{12} & \omega^{4} & \omega^{7} & \omega^{8} & \omega^{14} & \omega^{16} & \omega^{11} & \omega^{15} & \omega^{5} & \omega^{13} & \omega^{10} & \omega^{9} & \omega^{3} & \omega^{1} & \omega^{6} \\
\omega^{6} & \omega^{2} & \omega^{12} & \omega^{4} & \omega^{7} & \omega^{8} & \omega^{14} & \omega^{16} & \omega^{11} & \omega^{15} & \omega^{5} & \omega^{13} & \omega^{10} & \omega^{9} & \omega^{3} & \omega^{1}
\end{bmatrix}$$

$$\cdot [A_1\ A_6\ A_2\ A_{12}\ A_4\ A_7\ A_8\ A_{14}\ A_{16}\ A_{11}\ A_{15}\ A_5\ A_{13}\ A_{10}\ A_9\ A_3]^t.$$

This cyclic convolution of length 16 can now be computed by using a length 16 FFT or any other technique for computing cyclic convolution.

13.3 Rader's FFT Algorithm, $n = p^c$, p an Odd Prime

Now, consider the case of $n = p^c$, p being an odd prime, and $c > 1$. The FFT algorithm in this instance follows the same ideas and formulation as the Rader's algorithm. All the lengths and exponents are defined mod p^c. In this case, we have

$$A_i = \sum_{l=0}^{p^c-1} A_l \omega^{li}, \quad i = 0, 1, \ldots, p^c - 1. \tag{13.3.1}$$

Let us partition the index i into two parts: part one corresponding to values of i that are divisible by p, and part two corresponding to values of i that are not divisible by p. In other words, we write i in terms of i_1 and i_2 such that

$$i = i_1 p + i_2, \ \ i_1 = 0, 1, \ldots, p^{c-1} - 1; \ \ i_2 = 0, 1, \ldots, p - 1. \tag{13.3.2}$$

The first part corresponds to $i_2 = 0$. For this part, we get

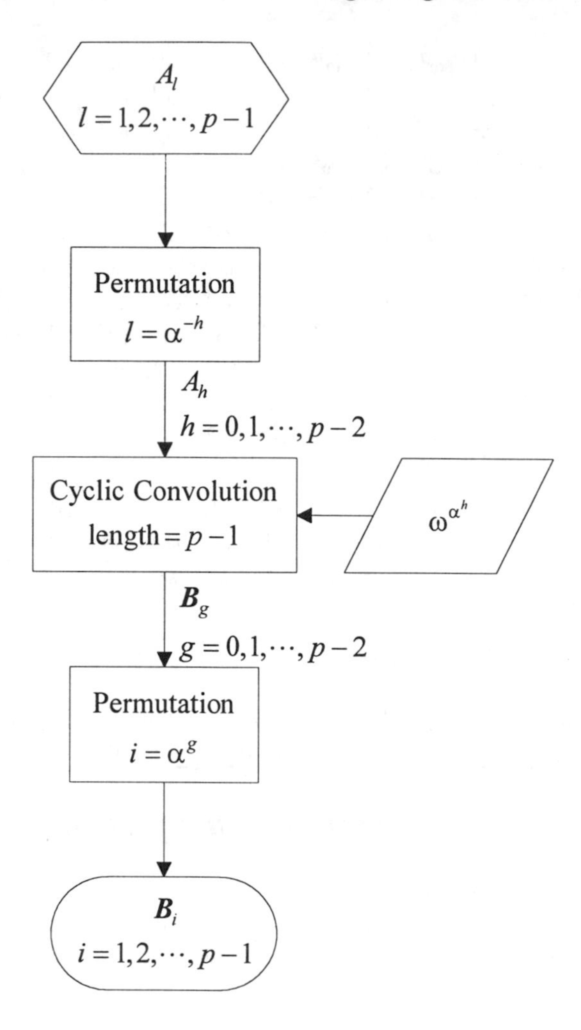

Figure 13.1 Formulation of Rader's Algorithm, $n = p$, p an Odd Prime

---

$$A_{i_1 p} = \sum_{l=0}^{p^c-1} A_l \omega^{pli_1}, \quad i_1 = 0, 1, ..., p^{c-1} - 1. \tag{13.3.3}$$

It is clear that ω^p is the p^{c-1}-th root of unity and, therefore, the second index term l in (13.3.3) can also be defined mod p^{c-1}. Thus, we write

$$l = l_2 p^{c-1} + l_1, \; l_1 = 0, 1, \ldots, p^{c-1} - 1; \; l_2 = 0, 1, \ldots, p - 1. \tag{13.3.4}$$

Substituting in (13.3.3), we get

$$A_{i_1 p} = \sum_{l=0}^{p^c-1} A_l \omega^{pli_1} = \sum_{l_1=0}^{p^{c-1}-1} \sum_{l_2=0}^{p-1} A_{l_2 p^{c-1}+l_1} \omega^{p\left(l_2 p^{c-1}+l_1\right) i_1}$$

$$= \sum_{l_1=0}^{p^{c-1}-1} \left[\sum_{l_2=0}^{p-1} A_{l_2 p^{c-1}+l_1} \right] \omega^{pl_1 i_1}, \quad i_1 = 0, 1, ..., p^{c-1} - 1. \tag{13.3.5}$$

Thus, for $i_2 = 0$, the FFT corresponds to a length p^{c-1} FFT of the data sequence

$$B_{l_1} = \sum_{l_2=0}^{p-1} A_{l_2 p^{c-1}+l_1}, \quad l_1 = 0, 1, ..., p^{c-1} - 1. \tag{13.3.6}$$

For the second part (i not divisible by p), we express the sum for $\boldsymbol{A}_i$ in (13.3.1) as a sum of two terms; first term corresponding to values of l that are divisible by p, and second term corresponding to values of i that are not divisible by p. In other words, we write l in terms of l_1 and l_2 such that

$$l = l_1 p + l_2, \; l_1 = 0, 1, \ldots, p^{c-1} - 1; \; l_2 = 0, 1, \ldots, p - 1. \tag{13.3.7}$$

The first term corresponds to $l_2 = 0$. Thus, (13.3.1) becomes

$$\boldsymbol{A}_i = \sum_{l_2=0} A_l \omega^{li} + \sum_{l_2 \neq 0} A_l \omega^{li} = \boldsymbol{C}_i + \boldsymbol{D}_i. \tag{13.3.8}$$

For the first term, we set $l = l_1 p$ and $i = i_2 p^{c-1} + i_1$, $i_2 = 0, 1, \ldots, p - 1$, $i_1 = 0, 1, \ldots, p^{c-1} - 1$, to get

$$\boldsymbol{C}_{i_2 p^{c-1}+i_1} = \sum_{l_1=0}^{p^{c-1}-1} A_{l_1 p} \omega^{pl_1 i_1}, \quad i_1 = 0, 1, ..., p^{c-1} - 1. \tag{13.3.9}$$

Once again, (13.3.9) defines a DFT of length p^{c-1}. Recalling that for this part, i is not divisible by p, we need to compute this DFT **only** for values of i_1 not divisible by p (*reduced DFT*).

Finally, let us turn to the computation of the second term in (13.3.8). Expressing it separately, we have

$$\boldsymbol{D}_i = \sum_{\substack{l \neq 0 \bmod p \\ i \neq 0 \bmod p}} A_l \omega^{li} . \tag{13.3.10}$$

It is this term that the Rader's algorithm seeks to simplify using number theory. Both indices i and l in (13.3.10) are defined mod p^c and are relatively prime to p^c. In $Z(p^c)$, there is a primitive element of order $p^c - p^{c-1}$ that generates all such values. Let α be one such element. Therefore, we can write

$$i = \alpha^g$$

and

$$l = \alpha^{-h}; \quad g, h = 0, 1, \ldots, p^c - p^{c-1} - 1.$$

Substituting in (13.3.10), we get

$$\boldsymbol{D}_{\alpha^g} = \sum_{h=0}^{p-2} A_{\alpha^{-h}} \omega^{\alpha^{g-h}} \; ; \; g, h = 0, 1, \ldots, p^c - p^{c-1} - 1. \tag{13.3.11}$$

This implies that the sequence $\boldsymbol{D}_{\alpha^g}$, $g = 0, 1, \ldots, p^c - p^{c-1} - 1$ is the cyclic convolution of $A_{\alpha^{-h}}$ and ω^{α^h}, $h = 0, 1, \ldots, p^c - p^{c-1} - 1$. In polynomial form

$$\boldsymbol{D}(u) = \sum_{g=0}^{p^c - p^{c-1} - 1} \boldsymbol{D}_{\alpha^g} u^g = \Delta(u) A(u) \bmod \left(u^{p^c - p^{c-1}} - 1\right)$$

$$= \left[\sum_{h=0}^{p^c - p^{c-1} - 1} \omega^{\alpha^h} u^h \right] \cdot \left[\sum_{h=0}^{p^c - p^{c-1} - 1} A_{\alpha^{-h}} u^h \right] \bmod \left(u^{p^c - p^{c-1}} - 1\right). \tag{13.3.12}$$

Once again, all the algorithms for computing the cyclic convolution can be used to compute $\boldsymbol{D}_i$. This FFT algorithm expresses a length p^c DFT in terms of two DFTs of length p^{c-1} and a length $p^c - p^{c-1}$ cyclic convolution. This cyclic convolution can either be computed using the FFT of length $p^c - p^{c-1}$ or any other cyclic convolution algorithm. The input and output sequences get permuted in this process of DFT computation. The polynomial

$$R(u) = \left[\sum_{h=0}^{p^c - p^{c-1} - 1} \omega^{\alpha^h} u^h \right] \tag{13.3.13}$$

in (12.3.12) is also known as the **generalized Rader's polynomial**.

Example 13.2 Let $n = 27 = 3^3$, $p = 3$, $c = 3$. Using the Rader's algorithm, the length 27 DFT can be computed as

(a) a length $9 = 3^2$ DFT as in (13.3.5), at an expense of $p^{c-1}\ (p-1)$ ADDs required to obtain the sequence in (13.3.6);
(b) a length $9 = 3^2$ DFT as in (13.3.9), the terms corresponding to $i_1 = 0, 3, 6$ are not computed;
(c) a length $18 = 3^3 - 3^2$ cyclic convolution as in (13.3.11) or (13.3.12).

A primitive root of order 18 in $Z(3^3)$ that can be used to obtain the cyclic convolution in (13.3.11) is given by $\alpha = 2$. The computation of the two length 9 DFTs can also be performed by expressing each of them as two DFTs of length 3 and a circular convolution of length 6.

13.4 Cooley-Tukey FFT Algorithm, $n = a \cdot b$

This case covers all possible values of n except for n equal to a prime. Collectively, the FFT algorithms of this section are called **splitting algorithms**. They are also known as the **Cooley-Tukey algorithms**. We will draw a distinction between the case $(a, b) = 1$ and $(a, b) > 1$ in our later analysis, but the present formulation is valid as long as n is a composite integer.

The DFT is defined as

$$\boldsymbol{A}_i = \sum_{l=0}^{n-1} A_l \omega^{li}, \quad i = 0, 1, \ldots, n-1. \qquad (13.4.1)$$

For $n = a \cdot b$, we express the indices as

$$l = a \cdot l_1 + l_2$$

$$i = b \cdot i_2 + i_1;\ l_1, i_1 = 0, 1, \ldots, b-1;\ l_2, i_2 = 0, 1, \ldots, a-1. \qquad (13.4.2)$$

Substituting for l and i in (13.4.1), we get

$$\boldsymbol{A}_{bi_2+i_1} = \boldsymbol{A}_{i_1,i_2} = \sum_{l_1=0}^{b-1}\sum_{l_2=0}^{a-1} A_{al_1+l_2}\,\omega^{(al_1+l_2)(bi_2+i_1)}. \qquad (13.4.3)$$

This expression, when simplified, results in an FFT formulation. These simplifications are

$$A_{bi_2+i_1} = \sum_{l_2=0}^{a-1} \omega^{bl_2 i_2} \omega^{l_2 i_1} \sum_{l_1=0}^{b-1} A_{l_1,l_2} \omega^{al_1 i_1},$$

$$i_1 = 0, 1, ..., b-1; i_2 = 0, 1, ..., a-1. \qquad (13.4.4)$$

This expression leads to some of the most widely studied FFT algorithms. Several observations are in order here.

Observation 1. It converts the one-dimensional input vector $\underline{A}$ into a two-dimensional array via the indices as defined in (13.4.2). The sequence A_l, $l = 0, 1, \ldots, a \cdot b - 1$ is arranged as $(b \times a)$ two-dimensional matrix **A** whose (l_1, l_2)-th element is given by $A_{al_1+l_2}$, $l_1 = 0, 1, \ldots, b-1$, $l_2 = 0, 1, \ldots, a-1$.

Observation 2. It converts the one-dimensional output vector $\underline{\boldsymbol{A}}$ into a two-dimensional array via the indices as defined in (13.4.2). The sequence $\boldsymbol{A}_i$, $i = 0, 1, \ldots, a \cdot b - 1$ is arranged as $(a \times b)$ two-dimensional matrix **D** whose (i_2, i_1)-th element is given by $\boldsymbol{A}_{bi_2+i_1}$, $i_1 = 0, 1, \ldots, b-1$, $i_2 = 0, 1, \ldots, a-1$.

Observation 3. ω^a and ω^b are the b-th and a-th roots of unity, respectively. The term $\omega^{l_2 i_1}$, $i_1 = 0, 1, \ldots, b-1$, $l_2 = 0, 1, \ldots, a-1$ is called the **twiddle factor** and we need to multiply by twiddle factors in appropriate places.

A computational procedure for the FFT algorithm based on (13.4.4) can be described as follows:

A Computational Method for Cooley-Tukey FFT

Step 1. Arrange the input sequence A_l, $l = 0, 1, \ldots, a \cdot b - 1$ into a $(b \times a)$ two-dimensional matrix **A** whose (l_1, l_2)-th element is given by $A_{al_1+l_2}$, $l_1 = 0, 1, \ldots, b-1$, $l_2 = 0, 1, \ldots, a-1$. This is accomplished by arranging the input on a row-by-row basis.

Step 2. Column DFTs. For each value of l_2, $l_2 = 0, 1, \ldots, a-1$, evaluate the sum

$$\boldsymbol{B}_{i_1,l_2} = \sum_{l_1=0}^{b-1} A_{l_1,l_2} \omega^{al_1 i_1}, \quad i_1 = 0, 1, \ldots, b-1.$$

This corresponds to taking a length b DFT of each of the a columns of $\boldsymbol{A}$. Let the resulting $(b \times a)$ matrix be denoted by **B**. The (i_1, l_2)-th element of **B** is given by $\boldsymbol{B}_{i_1,l_2}$.

Step 3. Multiplication by twiddle factors. Multiply the (i_1, l_2)-th element of **B**, that is $\boldsymbol{B}_{i_1,l_2}$, by the twiddle factor $\omega^{l_2 i_1}$, $i_1 = 0, 1, \ldots, b-1$, $l_2 = 0, 1, \ldots, a-1$. If either l_2 or i_1 is zero, then the twiddle factor is equal to unity. Thus, there are a total of $(a-1) \cdot (b-1)$ non-trivial MULTs by twiddle factors. Let the resulting $(i_1 \times l_2)$ matrix be denoted by **C**. The (i_1, l_2)-th element of **C** is given by $\boldsymbol{C}_{i_1,l_2}$, where

$$\boldsymbol{C}_{i_1,l_2} = \omega^{l_2 i_1} \cdot \boldsymbol{B}_{i_1,l_2}, \; i_1 = 0, 1, \ldots, b-1, l_2 = 0, 1, \ldots, a-1.$$

Step 4. Row DFTs. For each value of i_1, $i_1 = 0, 1, \ldots, b-1$, evaluate the sum

$$\boldsymbol{A}_{bi_2+i_1} = \sum_{l_2=0}^{a-1} \omega^{bl_2 i_2} \boldsymbol{C}_{i_1,l_2}, \; i_2 = 0, 1, \ldots, a-1.$$

This corresponds to taking a length a DFT of each of the b rows of **C**.

Step 5. As the DFT of the rows of **C** is undertaken, the resulting length a vectors are then arranged as columns of the matrix **D** (this involves permutation or **scrambling** of data). Recall that the $(a \times b)$ matrix **D** contains the one-dimensional DFT. Finally, $i = b\, i_2 + i_1$ is used to recover $\boldsymbol{A}_i$ from **D**. Thus, the DFT coefficients are read on a row-by-row basis from **D**. We can also interpret this as reading the DFT coefficients on a column-by-column basis after the row DFTs in Step 4 are performed.

End of Method

It is clear that this FFT algorithm requires (i) the computation of a DFTs of length b, (ii) b DFTs of length a, (iii) n MULTs by twiddle factors, and (iv) scrambling to obtain the DFT coefficients. This FFT algorithm does not require any additional storage beyond what is needed to store the input sequence. Once the input sequence is arranged as elements of **A**, each of the columns of **A** can be overwritten by their DFT to get the matrix **B**. Similarly, each of the elements of **B** can be multiplied by the twiddle factors and stored back in the same location to get the matrix **C**. Finally, each of the rows of **C** can be overwritten by their DFT and then scrambled to get the output DFT coefficients. Such a FFT algorithm is quite desirable and is termed as an **in-place** algorithm. Figure 13.2 depicts the process involved in the Cooley-Tukey algorithm as described here.

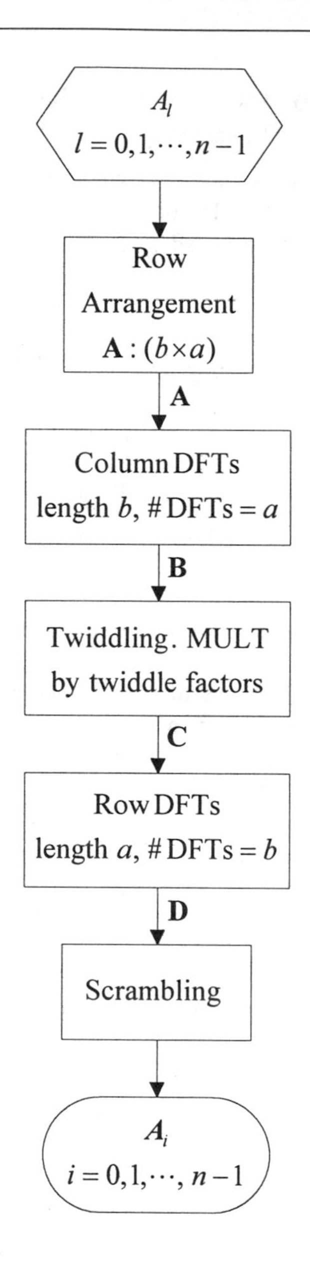

Figure 13.2 Formulation of Cooley-Tukey Splitting FFT Algorithms

The FFT algorithm as described above requires scrambling in the end. Alternatively, we can design a FFT algorithm that requires scrambling in the beginning. This is accomplished by defining the indices in the following manner:

$$l = l_1 + b \cdot l_2,$$

$$i = i_2 + a \cdot i_1, \; l_1, \; i_1 = 0, 1, \ldots, b-1; \; l_2, i_2 = 0, 1, \ldots, a-1. \tag{13.4.5}$$

Now, the input is arranged as $(b \times a)$ matrix $\mathbf{A}$ having its (l_1, l_2)-th element equal to A_l. This corresponds to arranging the input on a column-by-column basis. The derivation of the FFT algorithm is identical to the steps described above. Finally, the output is read on a row-by-row basis.

If $M(n)$ and $A(n)$ denote the multiplicative and additive complexities of a length n FFT, then for Cooley-Tukey FFT algorithm, we have

$$M(n) = b \cdot M(a) + a \cdot M(b) + (a-1) \cdot (b-1) \tag{13.4.6}$$

and

$$A(n) = b \cdot A(a) + a \cdot A(b). \tag{13.4.7}$$

The length a and length b DFTs can also be computed using the same approach if a and b are composite or other FFTs may be used when they are prime or a power of prime. Some simplifications may be obtained by noting that many of the twiddle factors take simple values such as ± 1, $\pm j$, $(\pm 1 \pm j)/\sqrt{2}$. If $n = p^c$, then we have a number of different choices for a and b. The factor a could go from p, $p^2, \ldots, p^{c-1}$.

13.5 FFT Algorithms for *n* a Power of 2

In this section, FFT algorithms are described for $n = 2^c$. These algorithms can be considered a special case of the splitting algorithms of the previous section. The primary motivation for describing them in a separate section is their popularity among the scientific community and simplifications that occur due to the special nature of n.

13.5.1 Radix-2 FFT

Consider the expression for Cooley-Tukey FFT in (13.4.4) for $a = 2$ and $b = n/2 = 2^{c-1}$. The resulting FFT algorithm is known as the **radix-2 FFT** as one of the factors is equal to 2. Noting that $\omega^{2^{c-1}} = -1$, (13.4.4) simplifies to

$$A_{2^{c-1}i_2+i_1} = \sum_{l_2=0}^{1} (-1)^{l_2 i_2} \omega^{l_2 i_1} \sum_{l_1=0}^{2^{c-1}-1} A_{l_1,l_2} \omega^{2 l_1 i_1},$$

$$i_1 = 0, 1, ..., 2^{c-1} - 1;\ i_2 = 0, 1. \tag{13.5.1}$$

For $i_2 = 0$,

$$A_{i_1} = \sum_{l_2=0}^{1} \omega^{l_2 i_1} \sum_{l_1=0}^{2^{c-1}-1} A_{2l_1+l_2} \omega^{2 l_1 i_1}$$

$$= \sum_{l_1=0}^{2^{c-1}-1} A_{2l_1} \omega^{2 l_1 i_1} + \omega^{i_1} \sum_{l_1=0}^{2^{c-1}-1} A_{2l_1+1} \omega^{2 l_1 i_1},\quad i_1 = 0, 1, ..., n/2 - 1. \tag{13.5.2}$$

Similarly, for $i_2 = 1$,

$$A_{2^{c-1}+i_1} = \sum_{l_1=0}^{2^{c-1}-1} A_{2l_1} \omega^{2 l_1 i_1} - \omega^{i_1} \sum_{l_1=0}^{2^{c-1}-1} A_{2l_1+1} \omega^{2 l_1 i_1},\quad i_1 = 0, 1, ..., n/2 - 1. \tag{13.5.3}$$

In summary, the FFT algorithm based on the factors $a = 2$ and $b = 2^{c-1}$ can be described in the following manner:

Step 1. Partition the input sequence into two sequences B_l and C_l, $l = 0, 1, \ldots, n/2 - 1$, each of length $n/2$. Here,

$$B_l = A_{2l}, \tag{13.5.4}$$

$$C_l = A_{2l+1},\quad l = 0, 1, \ldots, n/2 - 1. \tag{13.5.5}$$

Clearly, B_l sequence contains the even index terms and C_l sequence contains the odd index terms of the input.

Step 2. Take the length $n/2$ DFT of each of these sequences to get $\boldsymbol{B}_i$ and $\boldsymbol{C}_i$, $i = 0, 1, \ldots, n/2 - 1$.

Step 3. The first half of the original DFT sequence is given by

$$\boldsymbol{A}_i = \boldsymbol{B}_i + \omega^i \cdot \boldsymbol{C}_i,\quad i = 0, 1, \ldots, n/2 - 1, \tag{13.5.6}$$

as in (13.5.2), and the second half is given by

$$\boldsymbol{A}_{n/2+i} = \boldsymbol{B}_i - \omega^i \cdot \boldsymbol{C}_i,\quad i = 0, 1, \ldots, n/2 - 1, \tag{13.5.7}$$

as in (13.5.3).

The computation of each of the two length $n/2$ DFTs required in Step 2 can be expressed in terms of two length $n/4$ DFTs, and so on. This leads to an FFT algorithm known as the **decimation-in-time** FFT (DIT-FFT) algorithm. DIT-FFT algorithm requires that the input sequence be split into its even and odd index terms to get B_l and C_l, $l = 0, 1, \ldots, n/2 - 1$. These need to be split further into even and odd index terms if DIT-FFT is to be used again to compute the DFT of B_l and C_l, $l = 0, 1, \ldots, n/2 - 1$, and so on. This is called **bit-reversal**. There are numerous fast algorithms designed to perform bit-reversal for DIT-FFT. Finally, we observe that DIT-FFT is an *in-place algorithm* as $\boldsymbol{A}_i$ and $\boldsymbol{A}_{n/2+i}$ can be stored in the same locations as $\boldsymbol{B}_i$ and $\boldsymbol{C}_i$. Figure 13.3 shows the process of computing $\boldsymbol{A}_i$ and $\boldsymbol{A}_{n/2+i}$ in (13.5.6) and (13.5.7), once $\boldsymbol{B}_i$ and $\boldsymbol{C}_i$ are known. This process, called the **Butterfly**, constitutes the heart of radix-2 FFT algorithm. The figure is self-explanatory. Based on the butterfly operation, Figure 13.4 shows the overall configuration of the radix-2 DIT-FFT. Figures 13.5, 13.6, and 13.7 show the DIT-FFT algorithm structure for $n = 2$, 4, and 8, respectively. It is very interesting and important to note that the two DFTs of length 4 in Figure 13.7 need not be computed by DIT-FFT. Any other FFT is equally valid. In general, the computation of length $n/2$ DFTs $\boldsymbol{B}_i$ and $\boldsymbol{C}_i$, $i = 0, 1, \ldots, n/2 - 1$ as required in the DIT-FFT expressions (13.5.6) and (13.5.7) need not be performed using DIT-FFT. This introduces an additional degree of freedom in the FFT design.

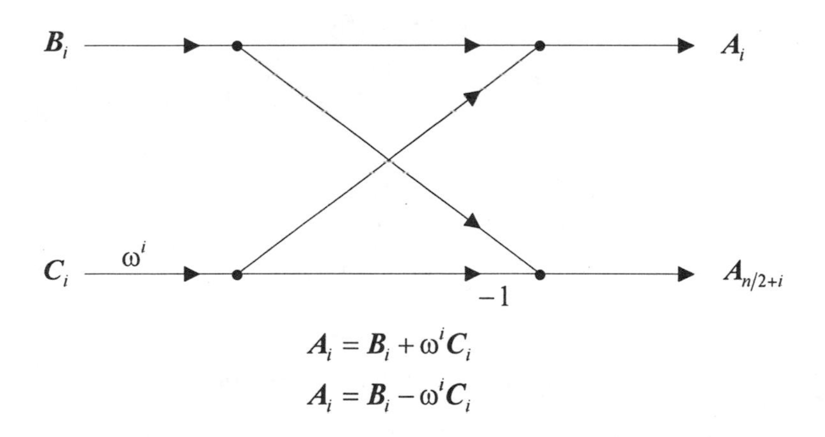

$$A_i = B_i + \omega^i C_i$$
$$A_i = B_i - \omega^i C_i$$

Figure 13.3 Heart of Radix-2 DIT-FFT Algorithm: The Butterfly Operation

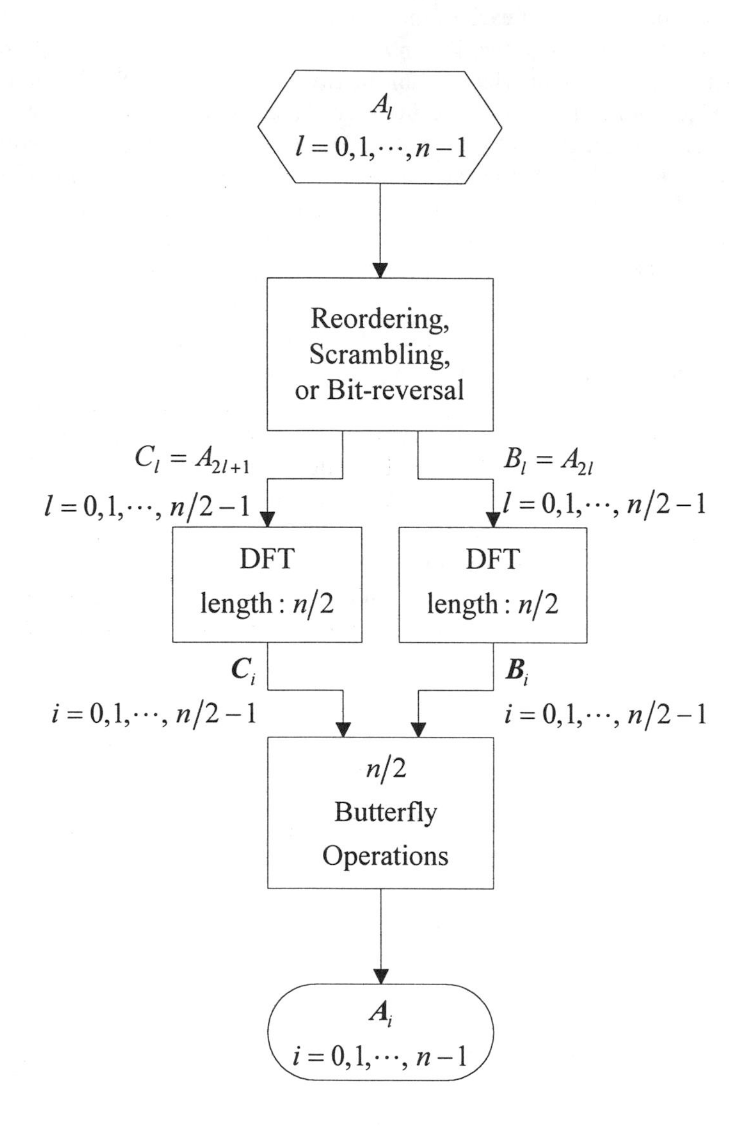

Figure 13.4 Radix-2 DIT-FFT Structure, $n = 2^c$

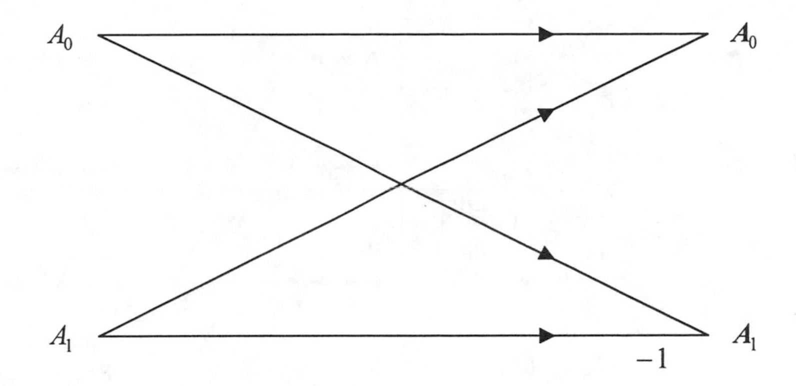

Figure 13.5 Radix-2 DIT-FFT Structure, $n = 2$

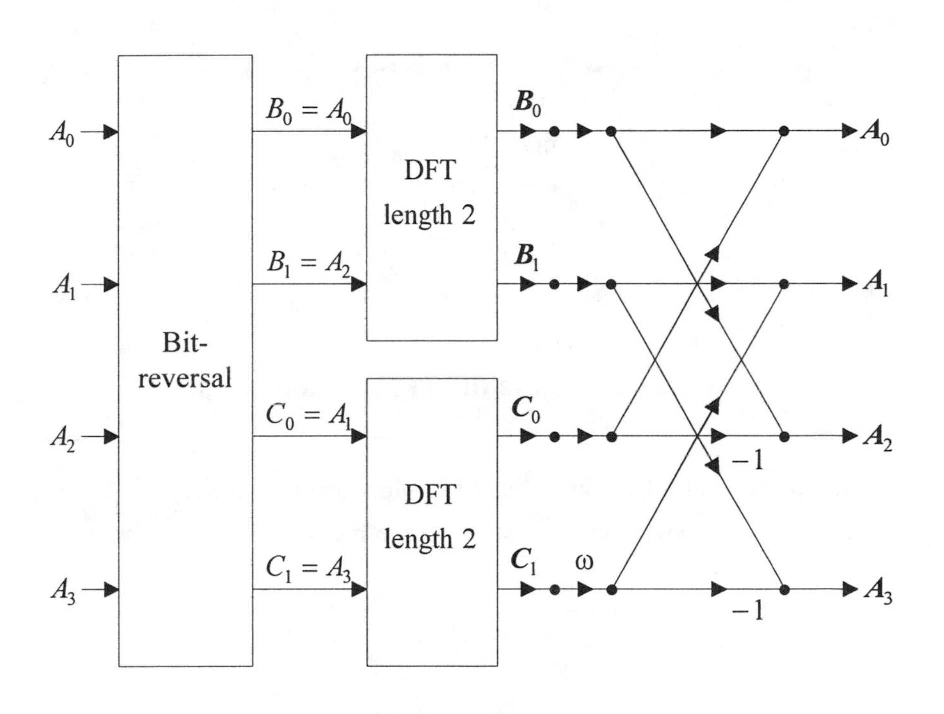

Figure 13.6 Radix-2 DIT-FFT Structure, $n = 4$

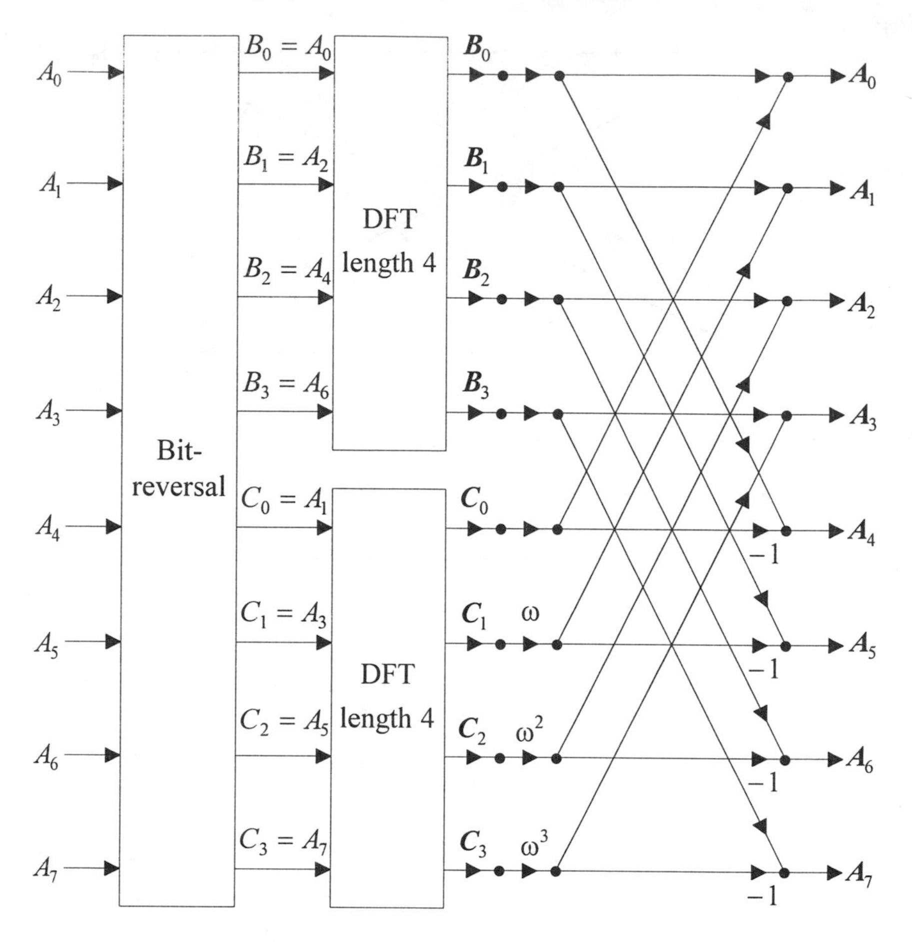

Figure 13.7 Radix-2 DIT-FFT Structure, $n = 8$

It is possible to get another radix-2 FFT algorithm by selecting $a = 2^{c-1}$ and $b = 2$ in (13.4.4). Noting that $\omega^{2^{c-1}} = -1$, it simplifies to

$$
\begin{aligned}
A_{2i_2+i_1} &= \sum_{l_2=0}^{2^{c-1}-1} \omega^{2l_2i_2}\omega^{l_2i_1} \sum_{l_1=0}^{1} A_{l_1,l_2}(-1)^{l_1i_1} \\
&= \sum_{l_2=0}^{2^{c-1}-1} \omega^{2l_2i_2}\omega^{l_2i_1}\left[A_{0,l_2} + (-1)^{i_1} A_{1,l_2}\right], \quad i_1 = 0,\ 1;\ i_2 = 0,\ 1,\ \ldots,\ 2^{c-1} - 1.
\end{aligned}
\tag{13.5.8}
$$

For $i_1 = 0$,

$$A_{2i_2} = \sum_{l_2=0}^{2^{c-1}-1} \omega^{2l_2 i_2} \left[A_{l_2} + A_{2^{c-1}+l_2} \right], \quad i_2 = 0, 1, ..., n/2 - 1. \qquad (13.5.9)$$

Similarly, for $i_1 = 1$,

$$A_{2i_2+1} = \sum_{l_2=0}^{2^{c-1}-1} \omega^{2l_2 i_2} \omega^{l_2} \left[A_{l_2} - A_{2^{c-1}+l_2} \right], \quad i_2 = 0, 1, ..., n/2 - 1. \qquad (13.5.10)$$

In summary, the FFT algorithm based on the factors $a = 2^{c-1}$ and $b = 2$ can be described in the following manner:

Step 1. Partition the input sequence into two sequences B_l and C_l, $l = 0, 1, \ldots, n/2 - 1$, each of length $n/2$. Here,

$$B_l = A_l,$$
$$C_l = A_{n/2+l}, \quad l = 0, 1, \ldots, n/2 - 1. \qquad (13.5.11)$$

Clearly, B_l sequence contains the first half and C_l sequence contains the second half of the input.

Step 2. Given B_l and C_l, form the sequences

$$D_l = B_l + C_l,$$
$$E_l = (B_l - C_l) \cdot \omega^l, \quad l = 0, 1, \ldots, n/2 - 1. \qquad (13.5.12)$$

Step 3. Take the length $n/2$ DFT of each of these sequences to get $\boldsymbol{D}_i$ and $\boldsymbol{E}_i$, $i = 0, 1, \ldots, n/2 - 1$.

Step 4. The even index terms of the original DFT sequence are given by

$$\boldsymbol{A}_{2i} = \boldsymbol{D}_i, \quad i = 0, 1, \ldots, n/2 - 1, \qquad (13.5.13)$$

and the odd index terms are given by

$$\boldsymbol{A}_{2i+1} = \boldsymbol{E}_i, \quad i = 0, 1, \ldots, n/2 - 1. \qquad (13.5.14)$$

The computation of each of the two length $n/2$ DFTs required in Step 2 can be expressed in terms of two length $n/4$ DFTs, and so on. This leads to an FFT algorithm known as the **decimation-in-frequency** FFT (DIF-FFT) algorithm. DIF-FFT algorithm requires that the output sequence be constructed from its even and odd index terms as in (13.5.13) and (13.5.14). Once again, this requires *bit-reversal*. Finally, we observe that DIF-FFT is also an *in-place algorithm* as D_l and E_l can be stored in the same locations as B_l and C_l. Figure 13.8 shows the process of computing D_l and E_l in (13.5.12), once B_l and C_l are known. This is recognized as the *butterfly*. The figure is self-explanatory. Based on the butterfly operation, Figure 13.9 shows the overall configuration of the radix-2 DIF-FFT. Figures 13.10 and 13.11 show the DIF-FFT algorithm structure for $n = 4$ and 8, respectively. DIF-FFT and DIT-FFT are the same for $n = 2$. We note that the two DFTs of length 4 in Figure 13.11 need not be computed by DIF-FFT. Any other FFT is equally valid. In general, the computation of length $n/2$ DFTs $\boldsymbol{D}_i$ and $\boldsymbol{E}_i$, $i = 0, 1, \ldots, n/2 - 1$, as required in the DIF-FFT expressions (13.5.13) and (13.5.14) need not be performed using DIF-FFT.

Let us now turn to the computational complexity of radix-2 FFT algorithms. Both DIT-FFT and DIF-FFT are identical in form as far as computational complexity is concerned and we can analyze any one of them. In the following, we focus on the DIT-FFT. The DIT-FFT computes a length n DFT in terms of

(i) 2 DFTs of length $n/2$,
(ii) (at most) $n/2$ MULTs by twiddle factors ω^i, $i = 0, 1, \ldots, n/2$, and
(iii) n ADDs.

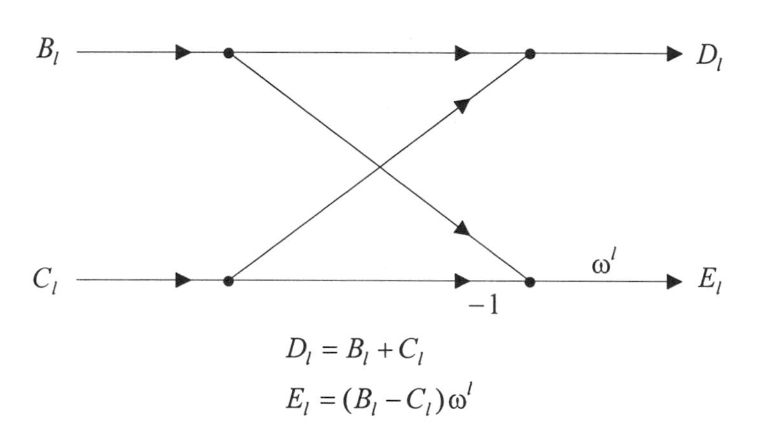

Figure 13.8 Heart of Radix-2 DIF-FFT Algorithm: The Butterfly Operation

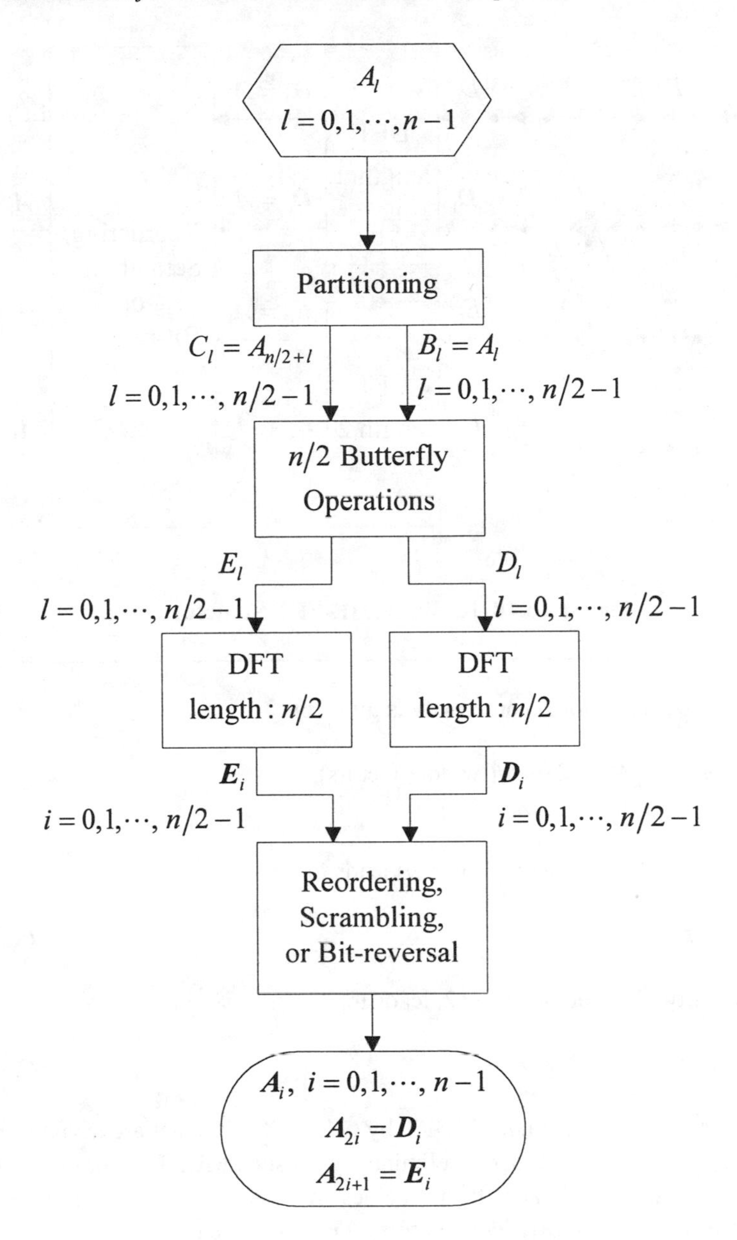

Figure 13.9 Radix-2 DIF-FFT Structure, $n = 2^c$

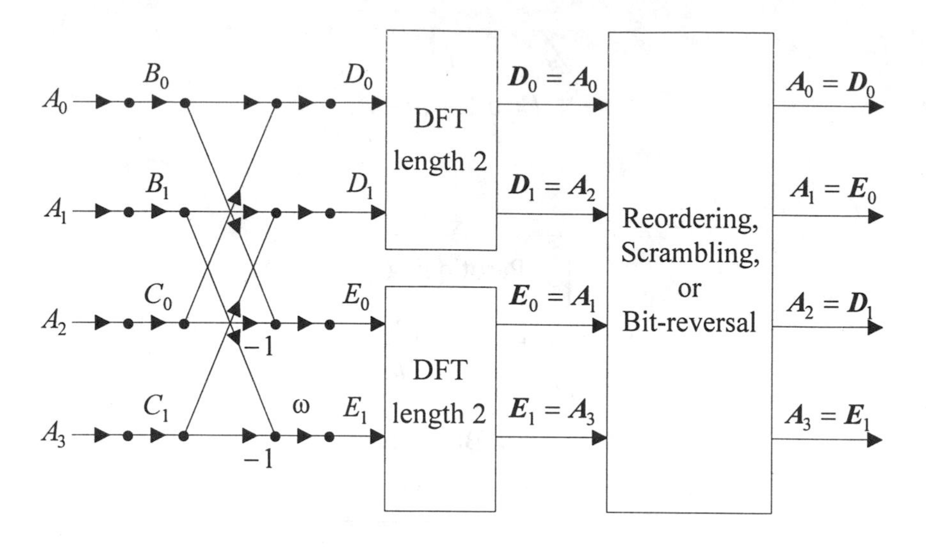

Figure 13.10 Radix-2 DIF-FFT Structure, $n = 4$

Thus, the computational complexity is given by

$$M(n) = 2 \cdot M(n/2) + M(\text{twiddle factors}),$$
$$A(n) = 2 \cdot A(n/2) + n. \tag{13.5.15}$$

A straightforward evaluation for $A(n)$ leads to

$$A(n) = n \cdot \log_2 n. \tag{13.5.16}$$

Setting M(twiddle factors) $= n/2$, leads to

$$M(n) = (n/2) \cdot \log_2 n. \tag{13.5.17}$$

However, this count includes MULT by twiddle factors that are trivial in nature. We may reduce $M(n)$ further by eliminating all such MULTs. For example, ω^i, $i = 0, 1, \ldots, n/2 - 1$ takes trivial values of 1 and $-j$ for $i = 0$ and $i = n/4$, respectively. Setting M(twiddle factors) $= n/2 - 2$, we get

$$M(n) = 2 \cdot M(n/2) + n/2 - 2.$$

Recalling that $M(4) = 0$ (MULT only by ± 1, $\pm j$), we get

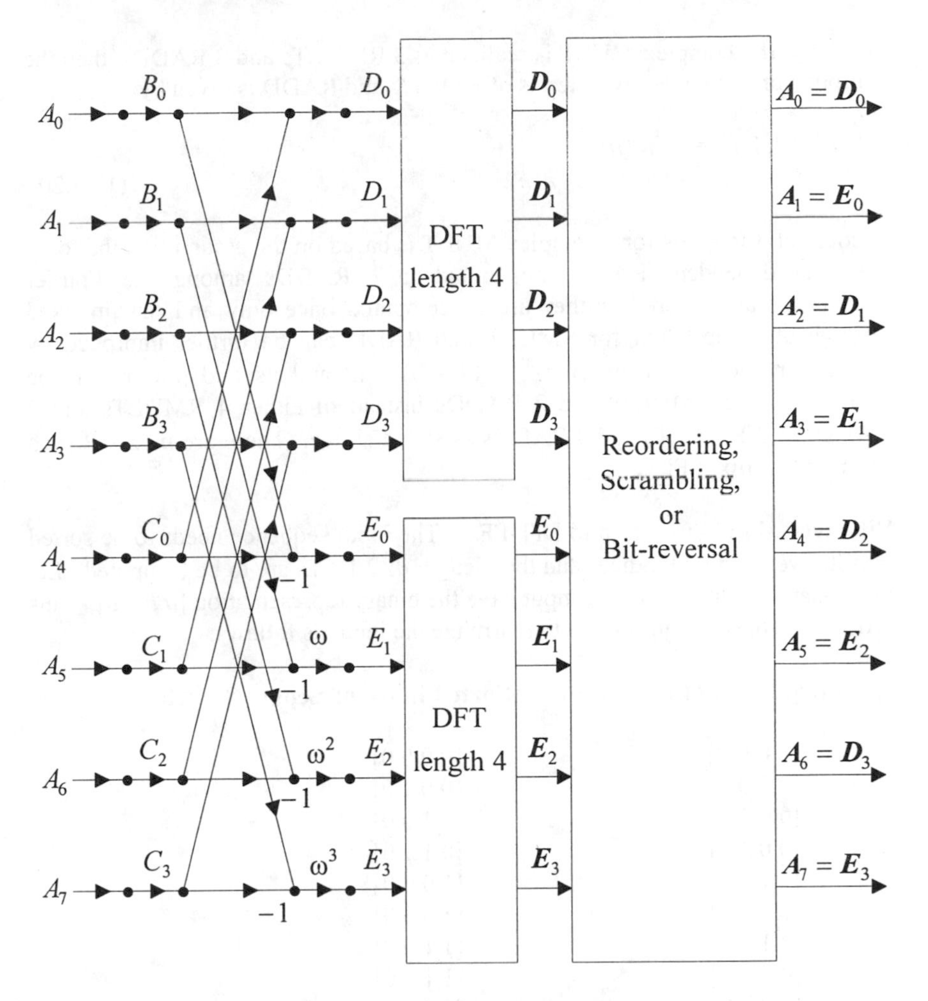

Figure 13.11 Radix-2 DIF-FFT Structure, *n* = 8

$$M(n) = n/2 \cdot (\log_2 n - 3) + 2. \tag{13.5.18}$$

All of these counts are complex for complex-valued data, an assumption we have made up to this point. If a complex MULT is realized as 4 real MULTs (RMULT) and 2 real ADDs (RADD) and a complex ADD is realized as 2 real ADDs, then the computational complexity in terms of RMULT and RADD is given by

$$\begin{aligned} \text{RMULT}(n) &= 4 \cdot M(n), \\ \text{RADD}(n) &= 2 \cdot M(n) + 2 \cdot A(n). \end{aligned} \tag{13.5.19}$$

Similarly, if a complex MULT is realized as 3 RMULTs and 3 RADDs, then the computational complexity in terms of RMULT and RADD is given by

$$\begin{aligned} \text{RMULT}(n) &= 3 \cdot M(n), \\ \text{RADD}(n) &= 3 \cdot M(n) + 2 \cdot A(n). \end{aligned} \tag{13.5.20}$$

A count of 3 RADDs for a complex MULT is based on the assumption that only the data dependent RADD are included; 2 RADDs among the Fourier coefficients are ignored as they are precomputed once only and remain fixed throughout. The count for RMULT and RADD can be further improved by noting that the MULT by $\omega^{(n/8)} = (1 - j)/\sqrt{2}$ and its odd powers can be performed in 2 RMULTs and 2 RADDs instead of either 4 RMULTs and 2 RADDs or 3 RMULTs and 3 RADDs as stated above. There are $n/2 - 2$ such MULTs in radix-2 FFT.

Bit-reversal. Let us return to DIT-FFT. The input sequence needs to be sorted into its even and odd indices and then length $n/2$ DFTs are to be computed. Let $n = 16$, and the indices of the input have the binary representation $[a\ b\ c\ d]$. This process of sorting requires one to rearrange the input as follows:

Input Index of Length 16	**Sorted Index of Length 8 Each**
[0 0 0 0]	[0 0 0 0]
[0 0 0 1]	[0 0 1 0]
[0 0 1 0]	[0 1 0 0]
[0 0 1 1]	[0 1 1 0]
[0 1 0 0]	[1 0 0 0]
[0 1 0 1]	[1 0 1 0]
[0 1 1 0]	[1 1 0 0]
[0 1 1 1]	[1 1 1 0]
[1 0 0 0]	[0 0 0 1]
[1 0 0 1]	[0 0 1 1]
[1 0 1 0]	[0 1 0 1]
[1 0 1 1]	[0 1 1 1]
[1 1 0 0]	[1 0 0 1]
[1 1 0 1]	[1 0 1 1]
[1 1 1 0]	[1 1 0 1]
[1 1 1 1]	[1 1 1 1]

If the length $n/2$ (= 8 here) DFT is once again computed using DIT-FFT, then we need to sort the above sorted sequence again. This process of sorting requires one to rearrange the input as follows:

Sorted Index of Length 8 Each	Sorted Index of Length 4 Each
[0 0 0 0]	[0 0 0 0]
[0 0 1 0]	[0 1 0 0]
[0 1 0 0]	[1 0 0 0]
[0 1 1 0]	[1 1 0 0]
[1 0 0 0]	[0 0 1 0]
[1 0 1 0]	[0 1 1 0]
[1 1 0 0]	[1 0 1 0]
[1 1 1 0]	[1 1 1 0]
[0 0 0 1]	[0 0 0 1]
[0 0 1 1]	[0 1 0 1]
[0 1 0 1]	[1 0 0 1]
[0 1 1 1]	[1 1 0 1]
[1 0 0 1]	[0 0 1 1]
[1 0 1 1]	[0 1 1 1]
[1 1 0 1]	[1 0 1 1]
[1 1 1 1]	[1 1 1 1]

This is continued until the indices are sorted as:

Input Index of Length 16	Sorted Index of Length 2 Each
[0 0 0 0]	[0 0 0 0]
[0 0 0 1]	[1 0 0 0]
[0 0 1 0]	[0 1 0 0]
[0 0 1 1]	[1 1 0 0]
[0 1 0 0]	[0 0 1 0]
[0 1 0 1]	[1 0 1 0]
[0 1 1 0]	[0 1 1 0]
[0 1 1 1]	[1 1 1 0]
[1 0 0 0]	[0 0 0 1]
[1 0 0 1]	[1 0 0 1]
[1 0 1 0]	[0 1 0 1]
[1 0 1 1]	[1 1 0 1]
[1 1 0 0]	[0 0 1 1]
[1 1 0 1]	[1 0 1 1]
[1 1 1 0]	[0 1 1 1]
[1 1 1 1]	[1 1 1 1]

The readers will recognize that the final sorted sequence has indices that are exactly the reverse of the original input sequence indices. This explains the name of this type of sorting. The result holds for all values of n. A number of techniques have been described in the literature for bit-reversal. It is also possible to avoid sorting by rearrangement of the intermediate sequences. Such FFTs are termed as **self-sorting** FFTs.

It is seen that DIT-FFT needs to reorder the input sequence but produces the output in the natural sequence. Similarly, DIF-FFT takes the input in the natural sequence but needs to reorder the output sequence. Advantage of this can be taken if one has to compute both DFT and IDFT simultaneously and the operations in the transform domain can be done on bit-reversed coefficients. Let us say that this is indeed so. One could use the DIF-FFT to obtain the DFT in bit-reversed form (no reordering is performed at the output, that is, Step 4 in DIF-FFT is ignored), perform the operations on these bit-reversed coefficients, and then take the IDFT using the DIT-FFT (no reordering is performed at the input, that is, Step 1 in DIT-FFT is ignored).

13.5.2 Radix-4 FFT

In this section, we explore FFT algorithms for $n = 4^c$. It is obvious that n is also a power of 2 and radix-2 FFT can be used here. The main purpose of studying it separately is to show that if n is an even power of 2, then the computational complexity of the FFT algorithm can be reduced beyond the radix-2 FFT. Once again, the primary motivation for describing them in a separate section is their popularity among the scientific community.

Consider the expression for the FFT in (13.4.4) for $a = 4$ and $b = n/4 = 4^{c-2}$. The resulting FFT algorithm is known as the **radix-4 DIT-FFT**. Noting that $\omega^{n/4} = -j$, (13.4.4) simplifies to

$$A_{(n/4)i_2+i_1} = \sum_{l_2=0}^{3}(-j)^{l_2 i_2}\,\omega^{l_2 i_1}\sum_{l_1=0}^{n/4-1} A_{l_1,l_2}\,\omega^{4 l_1 i_1},$$

$$i_1 = 0, 1, \ldots, n/4-1;\ i_2 = 0, 1, 2, 3. \qquad (13.5.21)$$

Writing it in a matrix form for $i_2 = 0, 1, 2, 3$, we get

$$\begin{bmatrix} A_{i_1} \\ A_{(n/4)+i_1} \\ A_{(n/4)2+i_1} \\ A_{(n/4)3+i_1} \end{bmatrix} = \begin{bmatrix} 1 & 1 & 1 & 1 \\ 1 & -j & -1 & j \\ 1 & -1 & 1 & -1 \\ 1 & j & -1 & -j \end{bmatrix} \begin{bmatrix} \sum_{l_1=0}^{n/4-1} A_{4l_1}\,\omega^{4l_1 i_1} \\ \omega^{i_1}\sum_{l_1=0}^{n/4-1} A_{4l_1+1}\,\omega^{4l_1 i_1} \\ \omega^{2i_1}\sum_{l_1=0}^{n/4-1} A_{4l_1+2}\,\omega^{4l_1 i_1} \\ \omega^{3i_1}\sum_{l_1=0}^{n/4-1} A_{4l_1+3}\,\omega^{4l_1 i_1} \end{bmatrix},$$

$$i_1 = 0, 1, \ldots, n/4-1. \qquad (13.5.22)$$

The matrix-vector of the kind (it corresponds to a length-4 DFT)

$$\begin{bmatrix} a_0 \\ a_1 \\ a_2 \\ a_3 \end{bmatrix} = \begin{bmatrix} 1 & 1 & 1 & 1 \\ 1 & -j & -1 & j \\ 1 & -1 & 1 & -1 \\ 1 & j & -1 & -j \end{bmatrix} \begin{bmatrix} b_0 \\ b_1 \\ b_2 \\ b_3 \end{bmatrix}$$

can be computed in 8 ADDs (and 0 MULT) in the following manner:

$$\begin{aligned} a_0 &= (b_0 + b_2) + (b_1 + b_3), \\ a_1 &= (b_0 - b_2) - j(b_1 - b_3), \\ a_2 &= (b_0 + b_2) - (b_1 + b_3), \\ a_3 &= (b_0 - b_2) + j(b_1 - b_3). \end{aligned} \tag{13.5.23}$$

In summary, the FFT algorithm based on the factors $a = 4$ and $b = n/4$ can be described in the following manner:

Step 1. Partition the input sequence into 4 sequences B_l, C_l, D_l, and E_l, $l = 0, 1, \ldots, n/4 - 1$, each of length $n/4$. Here,

$$\begin{aligned} B_l &= A_{4l}, \\ C_l &= A_{4l+1}, \\ D_l &= A_{4l+2}, \\ E_l &= A_{4l+3}, \quad l = 0, 1, \ldots, n/4 - 1. \end{aligned} \tag{13.5.24}$$

Step 2. Take the length $n/4$ DFT of each of these sequences to get $\boldsymbol{B}_i$, $\boldsymbol{C}_i$, $\boldsymbol{D}_i$, and $\boldsymbol{E}_i$, $i = 0, 1, \ldots, n/4 - 1$.

Step 3. The original DFT sequence is constructed as

$$\begin{aligned} \boldsymbol{A}_i &= (\boldsymbol{B}_i + \omega^{2i} \cdot \boldsymbol{D}_i) + (\omega^i \cdot \boldsymbol{C}_i + \omega^{3i} \cdot \boldsymbol{E}_i), \\ \boldsymbol{A}_{n/4+i} &= (\boldsymbol{B}_i - \omega^{2i} \cdot \boldsymbol{D}_i) - j(\omega^i \cdot \boldsymbol{C}_i - \omega^{3i} \cdot \boldsymbol{E}_i), \\ \boldsymbol{A}_{n/2+i} &= (\boldsymbol{B}_i + \omega^{2i} \cdot \boldsymbol{D}_i) - (\omega^i \cdot \boldsymbol{C}_i + \omega^{3i} \cdot \boldsymbol{E}_i), \\ \boldsymbol{A}_{3n/4+i} &= (\boldsymbol{B}_i - \omega^{2i} \cdot \boldsymbol{D}_i) + j(\omega^i \cdot \boldsymbol{C}_i - \omega^{3i} \cdot \boldsymbol{E}_i), \quad i = 0, 1, \ldots, n/4 - 1. \end{aligned} \tag{13.5.25}$$

The computation of each of the four length $n/4$ DFTs required in Step 2 can be expressed in terms of four length $n/16$ DFTs and so on. This leads to DIT-FFT radix-4 algorithm. It also requires *bit-reversal*. Finally, we observe that DIT-FFT is an *in-place algorithm*. It is possible to get DIF-FFT radix-4 FFT algorithm by selecting $a = n/4$ and $b = 4$ in (13.4.4). Figure 13.12 shows the butterfly for radix-4 DIT-FFT. This figure is based on (13.5.25) and is self-explanatory. Based on the butterfly operation, Figure 13.13 shows the overall configuration of the radix-4 DIT-FFT. Once again, the computation of length $n/4$ DFTs as required in (13.5.25) need not be performed using DIT-FFT. Any other FFT is equally valid.

Let us now turn to the computational complexity of radix-4 FFT algorithms. The DIT-FFT computes a length n DFT in terms of

(i) 4 DFTs of length $n/4$,
(ii) (at most) $3 \cdot n/4$ MULTs by twiddle factors ω^i, ω^{2i}, ω^{3i}, $i = 0, 1, \ldots, n/4$, and
(iii) $8 \cdot (n/4) = 2\,n$ ADDs.

Thus, the computational complexity is given by

$$M(n) = 4 \cdot M(n/4) + M(\text{twiddle factors}),$$
$$A(n) = 4 \cdot A(n/4) + 2\,n. \tag{13.5.26}$$

A straightforward evaluation for $A(n)$ leads to

$$A(n) = 2\,n \cdot c = 2\,n \cdot (0.5 \cdot \log_2 n) = n \cdot \log_2 n. \tag{13.5.27}$$

Setting $M(\text{twiddle factors}) = 3 \cdot n/4$, we get

$$M(n) = (3 \cdot n/8) \cdot \log_2 n - 3 \cdot n/4. \tag{13.5.28}$$

However, this count includes MULT by twiddle factors that are trivial in nature. We may reduce $M(n)$ further by eliminating all such MULTs. For example, for $i = 0, 1, \ldots, n/4 - 1$, ω^i takes trivial values of 1 for $i = 0$, ω^{2i} takes trivial values of 1 and $-j$ for $i = 0$ and $i = n/8$, respectively, and ω^{3i} takes trivial values of 1 for $i = 0$. Setting $M(\text{twiddle factors}) = 3 \cdot n/4 - 4$, we get

$$M(n) = 2 \cdot M(n/2) + 3 \cdot n/4 - 4.$$

Recalling that $M(4) = 0$ (MULT only by ± 1, $\pm j$), we get

$$M(n) = (3 \cdot n/8) \cdot \log_2 n - (13/12) \cdot n + 4/3. \tag{13.5.29}$$

All of these counts are complex for complex-valued data, an assumption we have made up to this point. If a complex MULT is realized as 4 real MULTs (RMULTs) and 2 real ADDs (RADDs) and a complex ADD is realized as 2 real

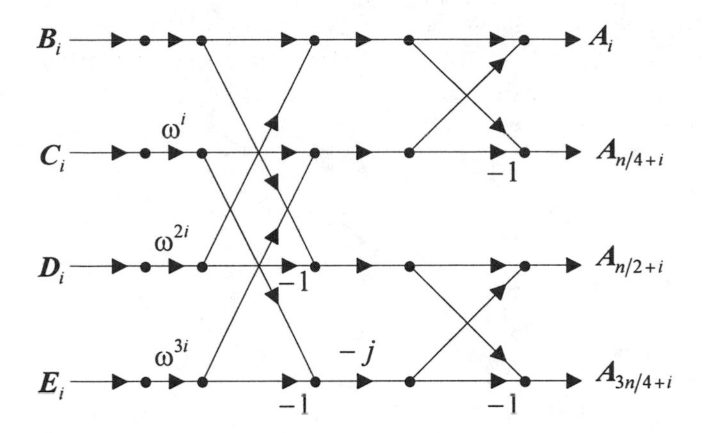

Figure 13.12 Heart of Radix-4 DIT-FFT Algorithm: The Butterfly Operation

---

ADDs, then the computational complexity in terms of RMULT and RADD is obtained by using (13.5.19). Similarly, if a complex MULT is realized as 3 RMULT and 3 RADDs, then the computational complexity in terms of RMULT and RADD is obtained by using (13.5.20). The count for RMULT and RADD can be further improved by noting that the MULT by $\omega^{(n/8)} = (1 - j)/\sqrt{2}$ and its odd powers can be performed in 2 RMULTs and 2 RADDs instead of either 4 RMULTs and 2 RADDs or 3 RMULTs and 3 RADDs as stated above. There are $n/4 - 3$ such MULTs in radix-4 FFT.

A comparison of the computational complexities of radix-2 and radix-4 FFT algorithms is in order here. Comparing (13.5.16) with (13.5.27) shows that both algorithms require the same number of ADDs. However, a comparison of (13.5.18) with (13.5.29) shows that the radix-4 FFT algorithm requires

$$(n/8) \cdot \log_2 n - (5/12) \cdot n + 2/3$$

fewer MULTs as compared to the radix-2 FFT algorithm.

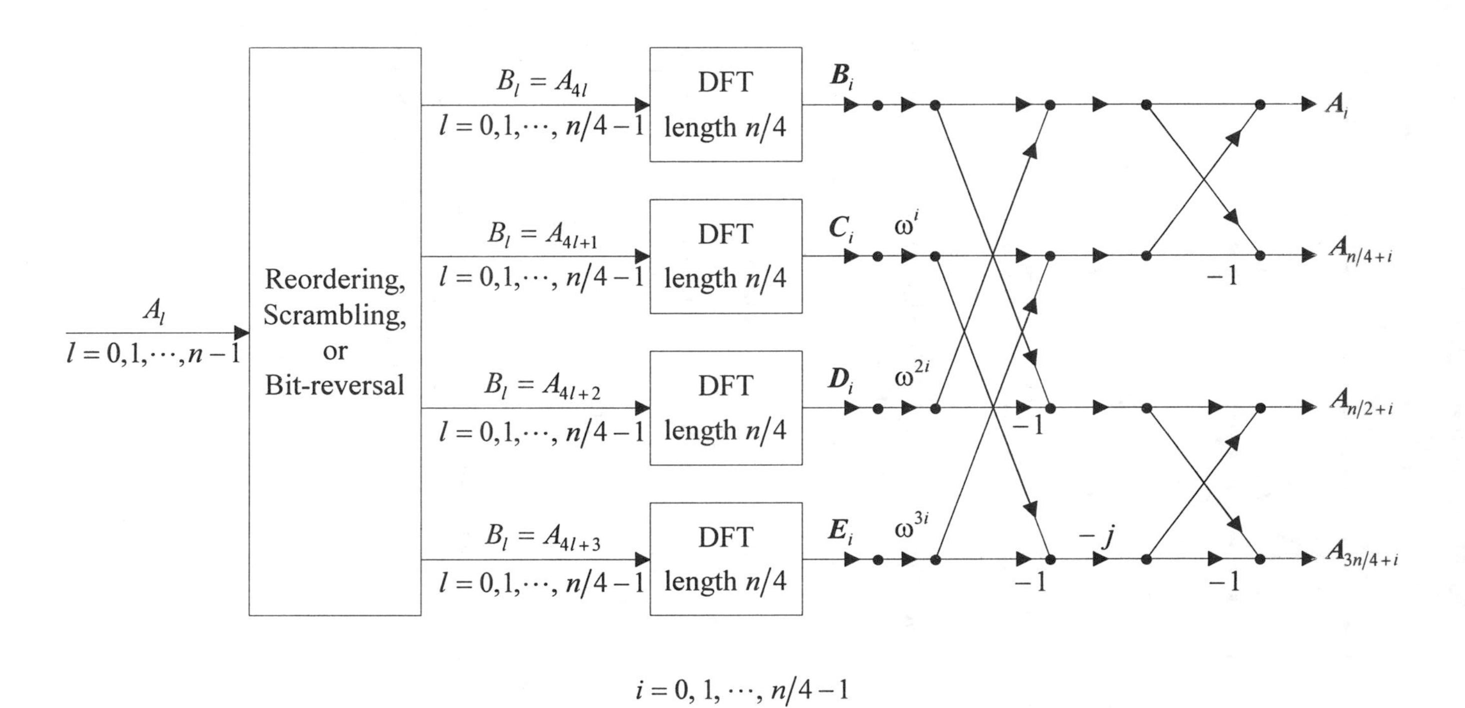

Figure 13.13 Radix-4 DIT-FFT Structure, $n = 4^c$

13.5.3 Split Radix FFT

To describe **split radix FFT**, we consider only the case $n = 2^c$ and the DIF-FFT formulation in (13.5.9) and (13.5.10) as reproduced below:

$$A_{2i_2} = \sum_{l_2=0}^{2^{c-1}-1} \omega^{2l_2 i_2} \left[A_{l_2} + A_{2^{c-1}+l_2}\right], \; i_2 = 0, 1, \ldots, n/2-1. \qquad (13.5.30)$$

and

$$A_{2i_2+1} = \sum_{l_2=0}^{2^{c-1}-1} \omega^{2l_2 i_2} \omega^{l_2} \left[A_{l_2} - A_{2^{c-1}+l_2}\right], \; i_2 = 0, 1, \ldots, n/2-1. \qquad (13.5.31)$$

Splitting (13.5.31) again in terms of its even and odd components, we get

$$A_{4i_2+1} = \sum_{l_2=0}^{2^{c-2}-1} \omega^{4l_2 i_2} \omega^{l_2} \left[\left(A_{l_2} - A_{2^{c-1}+l_2}\right) - j\left(A_{2^{c-2}+l_2} - A_{3\cdot 2^{c-2}+l_2}\right)\right],$$
$$i_2 = 0, 1, \ldots, n/4-1 \qquad (13.5.32)$$

and

$$A_{4i_2+3} = \sum_{l_2=0}^{2^{c-2}-1} \omega^{4l_2 i_2} \omega^{3l_2} \left[\left(A_{l_2} - A_{2^{c-1}+l_2}\right) - j\left(A_{2^{c-2}+l_2} - A_{3\cdot 2^{c-2}+l_2}\right)\right],$$
$$i_2 = 0, 1, \ldots, n/4-1. \qquad (13.5.33)$$

Thus, the split radix DIF-FFT decomposes the DFT of size $n = 2^c$ in terms of one DFT of size $n/2$ as in (13.5.30), and two DFTs of size $n/4$ each as in (13.5.32) and (13.5.33). This procedure is then repeated for these $n/2$ and $n/4$ size DFTs and so on. Figure 13.14 shows the overall configuration of the split-radix DIT-FFT algorithm.

Let us now turn to the computational complexity of split radix DIT-FFT algorithm. It computes a length n DFT in terms of

(i) 1 DFT of length $n/2$,

(ii) 2 DFTs of length $n/4$,

(iii) $n/2 - 4$ complex MULTs by twiddle factors (not including 8th root of unity) and 2 MULTs by 8th root of unity, and

(iv) 1.5 n ADDs.

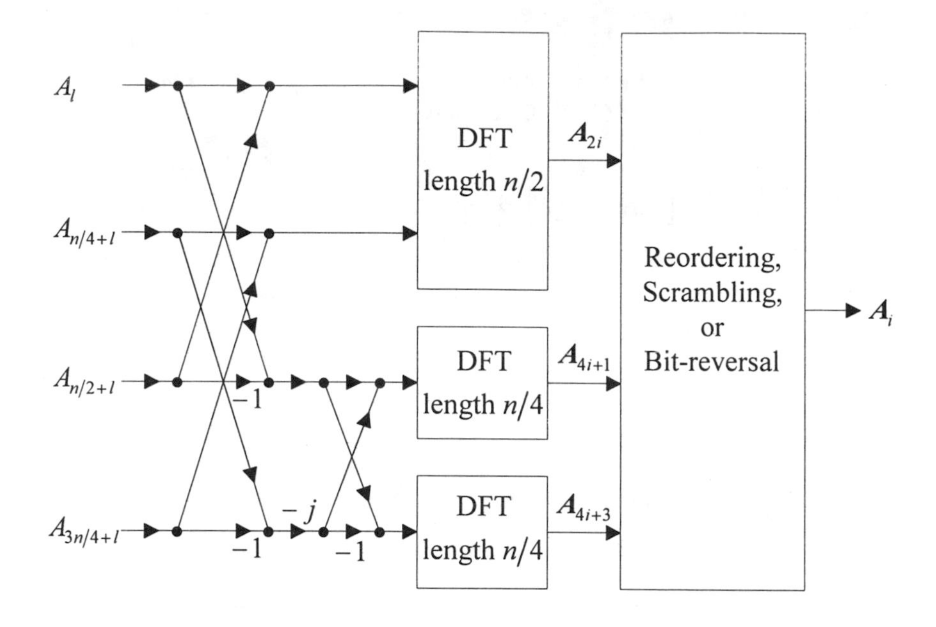

Figure 13.14 Split Radix DIT-FFT Structure, $n = 2^c$

Thus, the computational complexity is given by

$$\begin{aligned} M(n) &= M(n/2) + 2 \cdot M(n/4) + M(\text{twiddle factors}), \\ A(n) &= A(n/2) + 2 \cdot A(n/4) + 1.5\, n. \end{aligned} \tag{13.5.34}$$

$A(n)$ is solved as $A(n) = n \cdot \log_2 n$. If a complex MULT is realized as 3 RMULTs and 3 RADDs, and MULT by 8th root of unity by 2 RMULTs and 2 RDD, then the computational complexity in terms of RMULT and RADD is obtained by

$$\begin{aligned} \text{RMULT}(n) &= \text{RMULT}(n) + 2 \cdot \text{RMULT}(n/4) + 3 \cdot (n/2 - 4) + 4 \\ \text{RADD}(n) &= \text{RMULT}(n) + 2 \cdot A(n). \end{aligned}$$

Finally, we have

$$\text{RMULT}(n) = n \cdot \log_2 n - 3 \cdot n + 4,$$
$$\text{RADD}(n) = 3n \cdot \log_2 n - 3 \cdot n + 4. \tag{13.5.35}$$

It is seen that this algorithm is superior to radix-2 and radix-4 FFTs in terms of computational complexity.

13.5.4 Rader-Brenner Algorithm

The primary motivation behind this algorithm is to convert all the MULTs in the FFT algorithm by purely real (or equivalently, purely imaginary) numbers. We describe this algorithm for the radix-2 DIT-FFT formulation of the DFT in (13.5.6) and (13.5.7) as reproduced below:

$$\boldsymbol{A}_i = \boldsymbol{B}_i + \omega^i \cdot \boldsymbol{C}_i, \; i = 0, 1, \ldots, n/2 - 1 \tag{13.5.36}$$

and

$$\boldsymbol{A}_{n/2+i} = \boldsymbol{B}_i - \omega^i \cdot \boldsymbol{C}_i, \; i = 0, 1, \ldots, n/2 - 1. \tag{13.5.37}$$

It is the 2nd term in (13.5.36) and (13.5.37) that involves MULT by a complex number of the type ω^i. It is this MULT that we seek to simplify in the following analysis. Recall that $\boldsymbol{C}_i$ is defined as

$$\boldsymbol{C}_i = \sum_{l=0}^{n/2-1} A_{2l+1} \omega^{2li}, \; i = 0, 1, \ldots, n/2 - 1, \tag{13.5.38}$$

where ω is the n-th root of unity. Define a sequence D_l, $l = 0, 1, \ldots, n/2 - 1$ as follows:

$$D_l = A_{2l+1} - A_{2l-1} + Q, \; l = 0, 1, \ldots, n/2 - 1, \tag{13.5.39}$$

where

$$Q = \frac{2}{n} \sum_{l=0}^{n/2-1} A_{2l+1}. \tag{13.5.40}$$

The DFT of D_l, $l = 0, 1, \ldots, n/2 - 1$ is $\boldsymbol{D}_i$, $i = 0, 1, \ldots, n/2 - 1$. For $i = 0$,

$$\boldsymbol{D}_0 = Q = \boldsymbol{C}_0. \tag{13.5.41}$$

Q does not appear in any of $\boldsymbol{D}_i$, $i = 1, 2, \ldots, n/2 - 1$. Therefore,

$$\boldsymbol{D}_i = (\boldsymbol{C}_i - \omega^{2i} \cdot \boldsymbol{C}_i) = \boldsymbol{C}_i\,(1 - \omega^{2i}) = \omega^i \cdot \boldsymbol{C}_i(\omega^{-i} - \omega^i)$$

or

$$\omega^i \cdot \boldsymbol{C}_i = -j\,\boldsymbol{D}_i/[2\sin(2\pi\, i/n)]. \tag{13.5.42}$$

It is clear that this division is not defined for $i = 0$, and one needs to use (13.5.41) in that case. The Rader-Brenner algorithm can be described in the following manner:

Step 1. Partition the input sequence into two sequences B_l and C_l, $l = 0, 1, \ldots, n/2 - 1$, each of length $n/2$. Here,

$$B_l = A_{2l},$$
$$C_l = A_{2l+1},\quad l = 0, 1, \ldots, n/2 - 1. \tag{13.5.43}$$

Clearly, B_l sequence contains the even index terms and C_l sequence contains the odd index terms of the input. Next, define a sequence D_l, $l = 0, 1, \ldots, n/2 - 1$ as

$$D_l = C_l - C_{l-1} + Q,\quad l = 0, 1, \ldots, n/2 - 1, \tag{13.5.44}$$

where Q is defined in (13.5.40).

Step 2. Take the length $n/2$ DFT of each of these sequences to get $\boldsymbol{B}_i$ and $\boldsymbol{D}_i$, $i = 0, 1, \ldots, n/2 - 1$.

Step 3. The first half of the original DFT sequence is given by $\boldsymbol{A}_0 = \boldsymbol{B}_0 + \boldsymbol{D}_0$, and

$$\boldsymbol{A}_i = \boldsymbol{B}_i - j\,\boldsymbol{D}_i/[2\sin(2\pi i/n)],\quad i = 1, \ldots, n/2 - 1, \tag{13.5.45}$$

and the second half is given by $\boldsymbol{A}_{n/2} = \boldsymbol{B}_0 - \boldsymbol{D}_0$,

$$\boldsymbol{A}_{n/2+i} = \boldsymbol{B}_i + j\,\boldsymbol{D}_i/[2\sin(2\pi i/n)],\quad i = 0, 1, \ldots, n/2 - 1. \tag{13.5.46}$$

It is clear that all complex MULT are now replaced by purely imaginary numbers. Care must be taken in the use of this algorithm. It involves division by $[2\sin(2\pi i / n)]$ that may become small for certain values of i. A detailed analysis of the computational complexity of Rader-Brenner algorithm is being left as an exercise. We end our discussion by stating that it requires fewer MULT with a slight increase in ADD as compared to radix-2 and radix-4 FFTs.

13.5.5 Brunn's Algorithm

Brunn's algorithm involves only real arithmetic except for the last stage. Once again, let $n = 2^c$. Since ω is the n-th root of unity, we have

$$u^n - 1 = \prod_{i=0}^{n-1} \left(u - \omega^i\right). \tag{13.5.47}$$

If we express $u^n - 1$ as a product of two or more polynomials then the powers of ω^i will distribute themselves accordingly. For example, if $u^n - 1 = (u^{n/2} - 1) \cdot (u^{n/2} + 1)$, then all the even powers of ω are roots of $u^{n/2} - 1$ and all the odd powers are roots of $u^{n/2} + 1$. Therefore, we can write

$$A_i = A(u) \bmod (u - \omega^i) = [A(u) \bmod (u^n - 1)] \bmod (u - \omega^i).$$

For even values of i, we get

$$A_i = [A(u) \bmod (u^{n/2} - 1)] \bmod (u - \omega^i), \tag{13.5.48}$$

and for odd values of i,

$$A_i = [A(u) \bmod (u^{n/2} + 1)] \bmod (u - \omega^i). \tag{13.5.49}$$

This idea when applied recursively to further break down each of $u^{n/2} - 1$ and $u^{n/2} + 1$ leads to the Brunn's algorithm. This process continues until each of modulo polynomials in [.] are reduced to degree 2 polynomials. Finally, u is replaced by the corresponding ω^i in the degree one polynomials. Such a procedure is based on the factorization,

$$u^{4q} + au^{2q} + 1 = \left(u^{2q} + \sqrt{2-a}\, u^q + 1\right)\left(u^{2q} - \sqrt{2-a}\, u^q + 1\right). \tag{13.5.50}$$

The major advantage of Brunn's algorithm is its use of real-valued modulo polynomials throughout the computation. Analysis of its computational complexity is left as an exercise to the readers.

Example13.3 Let $n = 8$. The Brunn's algorithm will split the computations as

Stage 1. Compute $B_1(u) = A(u) \bmod (u^4 - 1)$, $B_2(u) = A(u) \bmod (u^4 + 1)$.
Stage 2. Compute $C_1(u) = B_1(u) \bmod (u^2 - 1)$, $C_2(u) = B_1(u) \bmod (u^2 + 1)$, $C_3(u) = B_2(u) \bmod (u^2 + \sqrt{2}\, u + 1)$, $C_4(u) = B_2(u) \bmod (u^2 - \sqrt{2}\, u + 1)$.
Stage 3. Finally, substitute appropriate values of ω^i in each of the degree one polynomials to obtain the DFT coefficients.

A DFT-filter tree based on the Brunn's algorithm for $n = 8$ is shown in Figure 13.15.

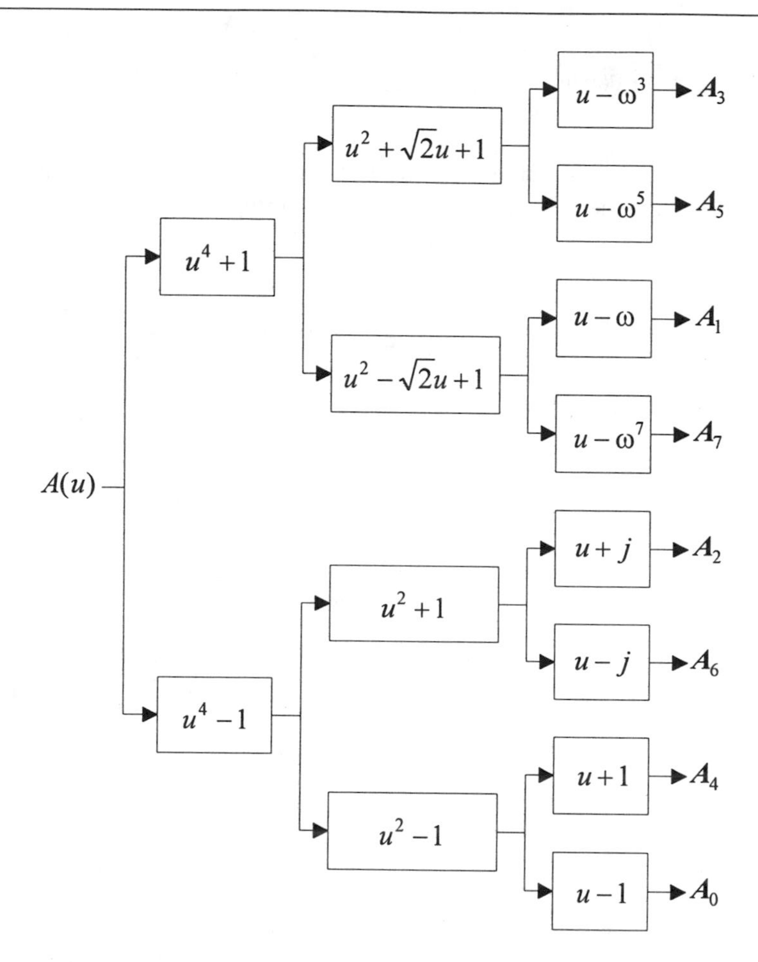

Figure 13.15 DFT-Filter Tree for Brunn's Algorithm, $n = 8$

13.6 The Prime Factor FFT $n = a \cdot b, (a, b) = 1$

This case covers a large number of n except for n a prime number. Collectively, the FFT algorithms of this section are called **prime factor FFT algorithms**. We restrict the factors of n to the case $n = a \cdot b$, $(a, b) = 1$.

The DFT is defined as

$$\mathcal{A}_i = \sum_{l=0}^{n-1} A_l \omega^{li}, \quad i = 0, 1, \ldots, n-1. \tag{13.6.1}$$

Since $n = a \cdot b$, $(a, b) = 1$, we use the CRT-I to express the indices as

$$l = a \cdot t_1 \cdot l_1 + b \cdot t_2 \cdot l_2,$$

$$i = a \cdot i_1 + b \cdot i_2, \; l_1, i_1 = 0, 1, \ldots, b-1; \; l_2, i_2 = 0, 1, \ldots, a-1. \tag{13.6.2}$$

It is clear that

$$\begin{aligned} l_1 &\equiv l \bmod b, \\ l_2 &\equiv l \bmod a, \\ t_1 \cdot a &\equiv 1 \bmod b, \\ t_2 \cdot b &\equiv 1 \bmod a, \\ i_1 &\equiv (a^{-1} \cdot i) \bmod b, \\ i_2 &\equiv (b^{-1} \cdot i) \bmod a. \end{aligned} \tag{13.6.3}$$

Thus, the indices (i_1, i_2) define a permutation on i. We can equivalently define this permutation on the (l_1, l_2) indices. Converting the one-dimensional DFT in (13.6.1) to a two-dimensional DFT using the indices as defined in (13.6.2) and (13.6.3), we get

$$\mathcal{A}_{ai_1+bi_2} = \mathcal{A}_{i_1,i_2} = \sum_{l_1=0}^{b-1} \sum_{l_2=0}^{a-1} A_{at_1l_1+bt_2l_2} \omega^{(at_1l_1+bt_2l_2)(bi_2+ai_1)}. \tag{13.6.4}$$

This expression, when simplified, results in a two-dimensional FFT formulation. Recalling that the powers of ω, ω^a, and ω^b are defined mod n, mod b, and mod a, respectively, (13.6.4) is simplified to

$$\mathcal{A}_{ai_1+bi_2} = \mathcal{A}_{i_1,i_2} = \sum_{l_1=0}^{b-1} \sum_{l_2=0}^{a-1} A_{l_1,l_2} \omega^{al_1i_1+bl_2i_2},$$

$$i_1 = 0, 1, \ldots, b-1; \; i_2 = 0, 1, \ldots, a-1. \tag{13.6.5}$$

This expression leads to some of the prime factor FFT algorithms. Several observations are in order here.

Observation 1. It converts the one-dimensional input vector $\underline{A}$ into a two-dimensional array via the indices as defined in (13.6.3). The sequence A_l, $l = 0, 1, \ldots, a \cdot b - 1$ is arranged as $(b \times a)$ two-dimensional matrix **A** whose (l_1, l_2)-th element is given by $A_{at_1l_1+bt_2l_2}$, $l_1 = 0, 1, \ldots, b-1$, $l_2 = 0, 1, \ldots, a-1$.

Observation 2. It converts the one-dimensional output vector $\underline{A}$ into a two-dimensional array via the indices as defined in (13.6.3). The sequence A_i, $i = 0, 1, \ldots, a \cdot b - 1$ is arranged as $(b \times a)$ two-dimensional matrix **C** whose (i_1, i_2)-th element is given by $A_{ai_1+bi_2}$, $i_1 = 0, 1, \ldots, b-1$, $i_2 = 0, 1, \ldots, a-1$.

Observation 3. ω^a and ω^b are the b-th and a-th roots of unity, respectively.

A comparison of (13.4.4) for the splitting algorithm for FFT and (13.6.5) for the prime factor FFT reveals that the prime factor FFT avoids the need of any twiddle factors. Thus, it is expected to be computationally superior to the splitting algorithm. However, it is more restrictive in terms of its applicability due to the condition $(a, b) = 1$. Furthermore, it requires a more sophisticated arrangement of the input and the output data when one-dimensional sequences are converted to two-dimensional sequences and vice versa. A computational procedure for the FFT algorithm based on (13.6.5) can be described as follows.

A Computational Method for Prime Factor FFT

Step 1. Arrange the input sequence A_l, $l = 0, 1, \ldots, a \cdot b - 1$ into a $(b \times a)$ two-dimensional matrix **A** whose (l_1, l_2)-th element is given by $A_{at_1l_1+bt_2l_2}$, $l_1 = 0, 1, \ldots, b-1$, $l_2 = 0, 1, \ldots, a-1$.

Step 2. Column DFTs. For each value of l_2, $l_2 = 0, 1, \ldots, a-1$, evaluate the sum

$$B_{i_1,l_2} = \sum_{l_1=0}^{b-1} A_{l_1,l_2} \omega^{al_1i_1}, \quad i_1 = 0, 1, \ldots, b-1. \tag{13.6.6}$$

This corresponds to taking a length b DFT of each of the a columns of $\boldsymbol{A}$. Let the resulting $(b \times a)$ matrix be denoted by $\boldsymbol{B}$. The (i_1, l_2)-th element of **B** is $\boldsymbol{B}_{i_1,l_2}$.

Step 3. Row DFTs. For each value of i_1, $i_1 = 0, 1, \ldots, b-1$, evaluate the sum

$$A_{ai_1+bi_2} = \sum_{l_2=0}^{a-1} \omega^{bl_2 i_2} B_{i_1,l_2}, \quad i_2 = 0, 1, \ldots, a-1. \tag{13.6.7}$$

This corresponds to taking a length a DFT of each of the b rows of **C**.

Step 4. Finally, $i = a\,i_1 + b\,i_2$ is used to recover A_i from **C**.

End of Method

It is clear that this FFT algorithm requires computation of (i) a DFTs of length b and (ii) b DFTs of length a. This FFT algorithm is an *in-place algorithm*. Once, the input sequence is arranged as elements of **A**, each of the columns of **A** can be over-written by their DFT to get the matrix **B**. Similarly, each of the rows of **B** can be overwritten by their DFT coefficients. Also, the order of the computation can be reversed by first taking the DFT of the rows.

If $M(n)$ and $A(n)$ denote the multiplicative and additive complexities of a length n FFT, then we have

$$M(n) = b \cdot M(a) + a \cdot M(b) \tag{13.6.8}$$

and

$$A(n) = b \cdot A(a) + a \cdot A(b). \tag{13.6.9}$$

The length a and length b DFTs can also be computed using the same approach if they factor in terms of relatively coprime factors or other FFTs may be used.

Example 13.4 Let $n = 4 \cdot 5 = 20$, $a = 4$, $b = 5$, $t_1 = 4$, $t_2 = 1$. Based on (13.6.3), the two-dimensional indices are

$l_1 \equiv l \bmod 5$,
$l_2 \equiv l \bmod 4$,
$i_1 \equiv 4 \cdot i \bmod 5$,
$i_2 \equiv i \bmod 4$.

The indices are given by

(a) $i \to (i_1, i_2)$ or $0 \to (0, 0)$, $1 \to (4, 1)$, $2 \to (3, 2)$, $3 \to (2, 3)$,
$4 \to (1, 0)$, $5 \to (0, 1)$, $6 \to (4, 2)$, $7 \to (3, 3)$, $8 \to (2, 0)$,
$9 \to (1, 1)$, $10 \to (0, 2)$, $11 \to (4, 3)$, $12 \to (3, 0)$, $13 \to (2, 1)$,
$14 \to (1, 2)$, $15 \to (0, 3)$, $16 \to (4, 0)$, $17 \to (3, 1)$, $18 \to (2, 2)$,
$19 \to (1, 3)$;

(b) $l \rightarrow (l_1, l_2)$ or $0 \rightarrow (0, 0)$, $1 \rightarrow (1, 1)$, $2 \rightarrow (2, 2)$, $3 \rightarrow (3, 3)$, $4 \rightarrow (4, 0)$, $5 \rightarrow (0, 1)$, $6 \rightarrow (1, 2)$, $7 \rightarrow (2, 3)$, $8 \rightarrow (3, 0)$, $9 \rightarrow (4, 1)$, $10 \rightarrow (0, 2)$, $11 \rightarrow (1, 3)$, $12 \rightarrow (2, 0)$, $13 \rightarrow (3, 1)$, $14 \rightarrow (4, 2)$, $15 \rightarrow (0, 3)$, $16 \rightarrow (1, 0)$, $17 \rightarrow (2, 1)$, $18 \rightarrow (3, 2)$, $19 \rightarrow (4, 3)$.

Based on these indices, the length-20 one-dimensional DFT can now be cast as a two-dimensional DFT. Casting (13.6.5) in a matrix-vector form, we have

$$\begin{bmatrix} A_0 & A_5 & A_{10} & A_{15} \\ A_4 & A_9 & A_{14} & A_{19} \\ A_8 & A_{13} & A_{18} & A_3 \\ A_{12} & A_{17} & A_2 & A_7 \\ A_{16} & A_1 & A_6 & A_{11} \end{bmatrix}$$

$$= \begin{bmatrix} 1 & 1 & 1 & 1 & 1 \\ 1 & \omega & \omega^2 & \omega^3 & \omega^4 \\ 1 & \omega^2 & \omega^4 & \omega & \omega^3 \\ 1 & \omega^3 & \omega & \omega^4 & \omega^2 \\ 1 & \omega^4 & \omega^3 & \omega^2 & \omega \end{bmatrix} \begin{bmatrix} A_0 & A_5 & A_{10} & A_{15} \\ A_{16} & A_1 & A_6 & A_{11} \\ A_{12} & A_{17} & A_2 & A_7 \\ A_8 & A_{13} & A_{18} & A_3 \\ A_4 & A_9 & A_{14} & A_{19} \end{bmatrix} \begin{bmatrix} 1 & 1 & 1 & 1 \\ 1 & -j & -1 & j \\ 1 & -1 & 1 & -1 \\ 1 & j & -1 & j \end{bmatrix}.$$

In the *rhs*, the first DFT matrix is for a length-5 DFT and, therefore, ω is the 5-th root of unity. The second DFT matrix is for a length-4 DFT. It is seen that pre-MULT by the length-5 DFT matrix corresponds to taking length-5 DFT of each column of the two-dimensional data matrix. Similarly, post-MULT by the length-4 DFT matrix corresponds to taking length-4 DFT of each row of the two-dimensional data matrix.

13.6.1 Matrix Formulation of Prime Factor FFT

Consider the formulation of the prime factor FFT algorithm in (13.6.5). The one-dimensional data sequence A_l, $l = 0, 1, \ldots, a \cdot b - 1$ is arranged as elements of a $(b \times a)$ two-dimensional matrix $\mathbf{A}$ whose (l_1, l_2)-th element is given by $A_{at_1l_1+bt_2l_2}$, $l_1 = 0, 1, \ldots, b-1$, $l_2 = 0, 1, \ldots, a-1$, using the mapping defined in (13.6.2) and (13.6.3). Similarly, the one-dimensional DFT sequence $\boldsymbol{A}_i$, $i = 0, 1, \ldots, a \cdot b - 1$ is recovered from the $(b \times a)$ two-dimensional matrix $\mathbf{C}$ whose (i_1, i_2)-th element is given by $\boldsymbol{A}_{ai_1+bi_2}$, $i_1 = 0, 1, \ldots, b-1$, $i_2 = 0, 1, \ldots, a-1$. The process of DFT involves the computation of length b DFT of each of the a columns of $\mathbf{A}$ which gives rise to the $\mathbf{B}$ matrix. This is followed by a length a

DFT of each of the rows of **B** to obtain **C**. One can also reverse the order of this computation without changing the final result.

Let Λ_b and Λ_a denote the DFT matrix for length b and length a DFTs, respectively. Then one can write

$$\mathbf{B} = \Lambda_b\,\mathbf{A},$$
$$\mathbf{C} = \mathbf{B}\,\Lambda_a. \tag{13.6.10}$$

Combining, we get

$$\mathbf{C} = \Lambda_b\,\mathbf{A}\,\Lambda_a$$
$$= \Lambda_b\,[\underline{A}_0\,\underline{A}_1\,\dots\,\underline{A}_{a-1}]\,\Lambda_a. \tag{13.6.11}$$

Here $\underline{A}_l$ denotes the l-th column of **A**. Let us arrange the elements of **A** as a single column in the following manner:

$$\underline{A}^P = \begin{bmatrix} \underline{A}_0 \\ \underline{A}_1 \\ \vdots \\ \underline{A}_{a-1} \end{bmatrix}. \tag{13.6.12}$$

It is clear that $\underline{A}^P$ is not the same as the original one-dimensional data sequence. It is obtained by performing the permutations as required by the prime factor FFT algorithm. The superscript P denotes such permutations. The column-by-column length b DFT of each of the columns of **A** can now be expressed as the matrix-vector product

$$\begin{bmatrix} \underline{\boldsymbol{B}}_0 \\ \underline{\boldsymbol{B}}_1 \\ \vdots \\ \underline{\boldsymbol{B}}_{a-1} \end{bmatrix} = \begin{bmatrix} \Lambda_b & \mathbf{0} & \cdots & \mathbf{0} \\ \mathbf{0} & \Lambda_b & \mathbf{0} & \vdots \\ \vdots & \mathbf{0} & \ddots & \mathbf{0} \\ \mathbf{0} & \cdots & \mathbf{0} & \Lambda_b \end{bmatrix} \begin{bmatrix} \underline{A}_0 \\ \underline{A}_1 \\ \vdots \\ \underline{A}_{a-1} \end{bmatrix} \tag{13.6.13}$$

or

$$\underline{\boldsymbol{B}} = [\mathbf{I}_a\,\kappa\,\Lambda_b]\,\underline{A}^P. \tag{13.6.14}$$

Here, $\underline{\boldsymbol{B}}$ denotes the length n vector obtained by arranging the columns of **B**, $\mathbf{I}_a$ denotes the identity matrix of dimension a, and κ denotes the Kronecker product of two matrices. $[\mathbf{I}_a \,\kappa\, \Lambda_b]$ is a block-diagonal matrix having Λ_b on the main diagonal and zeros elsewhere.

The i-th row of $\mathbf{B}$ corresponds to the i-th elements of each of $\boldsymbol{B}_0, \boldsymbol{B}_1, \ldots, \boldsymbol{B}_{a-1}$, $i = 0, 1, \ldots, b-1$. Therefore, the computation of $\mathbf{C}$, that is, the computation of length a DFT of each of the b rows of $\mathbf{B}$ is equivalent to the matrix-vector product

$$\begin{bmatrix} \underline{\boldsymbol{C}}_0 \\ \underline{\boldsymbol{C}}_1 \\ \vdots \\ \underline{\boldsymbol{C}}_{a-1} \end{bmatrix} =$$

$$= \left[\begin{array}{c|c|c|c}
\mathbf{I}_b & \mathbf{I}_b & \cdots & \mathbf{I}_b \\
\hline
\mathbf{I}_b & \begin{matrix} \omega & 0 & \cdots & 0 \\ 0 & \omega & 0 & \vdots \\ \vdots & 0 & \ddots & 0 \\ 0 & \cdots & 0 & \omega \end{matrix} & \cdots & \begin{matrix} \omega^{(a-1)} & 0 & \cdots & 0 \\ 0 & \omega^{(a-1)} & 0 & \vdots \\ \vdots & 0 & \ddots & 0 \\ 0 & \cdots & 0 & \omega^{(a-1)} \end{matrix} \\
\hline
\vdots & \vdots & \ddots & \vdots \\
\hline
\mathbf{I}_b & \begin{matrix} \omega^{a-1} & 0 & \cdots & 0 \\ 0 & \omega^{a-1} & 0 & \vdots \\ \vdots & 0 & \ddots & 0 \\ 0 & \cdots & 0 & \omega^{a-1} \end{matrix} & \cdots & \begin{matrix} \omega^{(a-1)(a-1)} & 0 & \cdots & 0 \\ 0 & \omega^{(a-1)(a-1)} & 0 & \vdots \\ \vdots & 0 & \ddots & 0 \\ 0 & \cdots & 0 & \omega^{(a-1)(a-1)} \end{matrix}
\end{array}\right]$$

$$\cdot \begin{bmatrix} \underline{\boldsymbol{B}}_0 \\ \underline{\boldsymbol{B}}_1 \\ \vdots \\ \underline{\boldsymbol{B}}_{a-1} \end{bmatrix} \tag{13.6.15}$$

or

$$\underline{\boldsymbol{C}} = [\Lambda_a \,\kappa\, \mathbf{I}_b]\, \underline{\boldsymbol{B}}. \tag{13.6.16}$$

In (13.6.15), ω denotes the a-th root of unity. $[\Lambda_a \,\kappa\, \mathbf{I}_b]$ is a block matrix having the (r, s)-th block equal to $\omega^{r \cdot s}\, \mathbf{I}_b$; $r, s = 0, 1, \ldots, a-1$.

Combining (13.6.16) with (13.6.14), we have a compact representation for the prime factor FFT algorithm given by

$$\underline{C} = [\Lambda_a \kappa \mathbf{I}_b]\,[\mathbf{I}_a \kappa \Lambda_b]\,\underline{A}^P. \tag{13.6.17}$$

Recalling the result from matrix theory,

$$(\mathbf{A} \kappa \mathbf{B}) \cdot (\mathbf{C} \kappa \mathbf{D}) = (\mathbf{A} \cdot \mathbf{C}) \kappa (\mathbf{B} \cdot \mathbf{D}), \tag{13.6.18}$$

(13.6.17) simplifies to

$$\underline{C} = [\Lambda_a \kappa \Lambda_b]\,\underline{A}^P. \tag{13.6.19}$$

Finally, we note that the permutation required to obtain $\underline{A}^P$ from $\underline{A}$ can also be written as the matrix-vector product

$$\underline{A}^P = \mathbf{P}\,\underline{A}, \tag{13.6.20}$$

where **P** is a permutation matrix consisting of exactly one 1 in any given row or column with all other elements equal to 0. Expressions in (13.6.19) and (13.6.20) also lead to the following factorization for the original length-n DFT matrix:

$$\Lambda_n = [\Lambda_a \kappa \Lambda_b] \cdot \mathbf{P}. \tag{13.6.21}$$

In a similar manner as above, we can derive the matrix-vector formulation for the Cooley-Tukey algorithm. It is given by

$$\underline{C} = [\Lambda_a \kappa \mathbf{I}_b]\,\mathbf{T}\,[\mathbf{I}_a \kappa \Lambda_b]\,\underline{A}^P \tag{13.6.22}$$

or

$$\Lambda_n = [\Lambda_a \kappa \mathbf{I}_b]\,\mathbf{T}\,[\mathbf{I}_a \kappa \Lambda_b] \cdot \mathbf{P}. \tag{13.6.23}$$

Here **T** is a diagonal matrix and corresponds to the MULT by twiddle factors. Note that (13.6.21) is valid only for $(a, b) = 1$, while (13.6.23) is valid in all situations. It may be possible to compare the prime factor algorithm to the Cooley-Tukey algorithm for the case $(a, b) = 1$. The prime factor algorithm avoids the MULT by twiddle factors by a rearrangement of the input sequence that is more sophisticated than the one required by the Cooley-Tukey algorithm.

13.7 The Winograd FFT Algorithm

The Winograd FFT algorithm, known as the WFTA (**Winograd Fourier transform algorithm**), is based on two key points:

1. A length n DFT can be expressed as a Kronecker product of DFTs of length a and b, respectively, provided $(a, b) = 1$, $n = a \cdot b$. This is based on (13.6.19). Thus, a large length DFT is expressed in terms of smaller length DFTs under suitable conditions.

2. In most instances, a length-s DFT (s small) $\underline{Y} = \Lambda_s\, \underline{y}$ can be computed using a bilinear algorithm of the kind

$$\underline{Y} = \mathbf{P}_s\, \mathbf{Q}_s\, \mathbf{R}_s\, \underline{y}. \tag{13.7.1}$$

Here, $\mathbf{P}_s$ and $\mathbf{R}_s$ are matrices having simple elements such as 0, ±1; they correspond to input and output ADDs required in the length-s FFT algorithm. The matrix $\mathbf{Q}_s$ is a diagonal matrix whose diagonal entries correspond to the MULT required in the length-s FFT algorithm. The dimension of $\mathbf{Q}_s$ is equal to the multiplicative complexity of the length-s FFT algorithm. The diagonal elements of $\mathbf{Q}_s$ are either purely real or imaginary. Such a computation also results in the factorization

$$\Lambda_s = \mathbf{P}_s\, \mathbf{Q}_s\, \mathbf{R}_s. \tag{13.7.2}$$

The second point can be justified in the light of length p^c (p prime) DFT algorithms being derived using the Rader's algorithm by first converting them to a cyclic convolution with the cyclic convolution, in turn, being computed using a bilinear algorithm.

For the lengths a and b, let $\mathbf{P}_a$, $\mathbf{Q}_a$, $\mathbf{R}_a$ and $\mathbf{P}_b$, $\mathbf{Q}_b$, $\mathbf{R}_b$ be the corresponding $\mathbf{P}$, $\mathbf{Q}$, and $\mathbf{R}$ matrices in the factorization of Λ_a and Λ_b, respectively. Then, using (13.6.21), we get

$$\begin{aligned}\Lambda_n &= [\Lambda_a \,\mathrm{K}\, \Lambda_b] \cdot \mathbf{P} = [\{\mathbf{P}_a\, \mathbf{Q}_a\, \mathbf{R}_b\} \,\mathrm{K}\, \{\mathbf{P}_b\, \mathbf{Q}_b\, \mathbf{R}_b\}] \cdot \mathbf{P} \\ &= [\mathbf{P}_n\, \mathbf{Q}_n\, \mathbf{R}_n] \cdot \mathbf{P}.\end{aligned} \tag{13.7.3}$$

Recalling (13.6.18), we get

$$\begin{aligned}\mathbf{P}_n &= \mathbf{P}_a \,\mathrm{K}\, \mathbf{P}_b, \\ \mathbf{Q}_n &= \mathbf{Q}_a \,\mathrm{K}\, \mathbf{Q}_b, \\ \mathbf{R}_n &= \mathbf{R}_a \,\mathrm{K}\, \mathbf{R}_b.\end{aligned} \tag{13.7.4}$$

The permutation matrix $\mathbf{P}$ can also be absorbed in the form of $\mathbf{R}_n$. In WFTA, the steps for the computation are given by

Step 1. Arrange the input vector by performing the necessary permutation as required by **P**.

Step 2. Perform the input ADDs on the input using $\mathbf{R}_n$.

Step 3. Multiply the resulting vector by the diagonal matrix $\mathbf{Q}_n$.

Step 4. Perform the output ADDs using $\mathbf{P}_n$.

Since the elements of $\mathbf{Q}_a$ and $\mathbf{Q}_b$ are either real or purely imaginary, so are the elements of $\mathbf{Q}_n$ in (13.7.4). Following the same analysis for the computational complexity as for the Agarwal-Cooley algorithm for computing one-dimensional cyclic convolution via two-dimensional cyclic convolutions, we get

$$\begin{aligned} M(n) &= M(a) \cdot M(b), \\ A(n) &= a \cdot A(b) + M(b) \cdot A(a). \end{aligned} \tag{13.7.5}$$

Thus, we see that while the multiplicative complexity does not depend on the order of the arrangement of the two-dimensional data, the additive complexity does. Therefore, one may choose the arrangement in a way so as to minimize the additive complexity. WFTA gives some of the best algorithms in terms of multiplicative complexity. However, their additive complexity is higher than those FFTs obtained by using other methods. This is seen by comparing (13.7.5) to (13.6.8) and (13.6.9) for the prime factor FFT. It is seen from (13.7.5) and (13.6.9) that WFTA always requires more ADDs than prime factor FFT as $M(b) > b$, except for some trivial cases.

Example 13.5 Let us continue with the previous example of $n = 20 = 4 \cdot 5 = a \cdot b$. The various matrices involved in the length 4 and 5 FFTs in the form required by WFTA can be constructed from the following:

Length 4 FFT

Step 1. Input ADDs

$$s_1 = A_0 + A_2,\ s_2 = A_0 - A_2,\ s_3 = A_1 + A_3,\ s_4 = A_1 - A_3,\ s_5 = s_1 + s_3,\ s_6 = s_1 - s_3.$$

Step 2. MULTs

$$m_1 = 1 \cdot s_5,\ \ m_2 = 1 \cdot s_6,\ \ m_3 = 1 \cdot s_2,\ \ m_4 = j \cdot s_4.$$

Step 3. Output ADDs

$A_0 = m_1,\ A_1 = m_3 + m_4,\ A_2 = m_2,\ A_3 = m_3 - m_4.$

Length 5 DFT

Let $t = 2\pi/5$.

Step 1. Input ADDs

$$s_1 = A_1 + A_4,\ s_2 = A_1 - A_4,\ s_3 = A_3 + A_2,\ s_4 = A_3 - A_2,\ s_5 = s_1 + s_3,$$
$$s_6 = s_1 - s_3,\ s_7 = s_2 + s_4,\ s_8 = s_5 + A_0.$$

Step 2. MULTs

$$m_0 = 1 \cdot s_8,\ m_1 = [(\cos t + \cos 2t)/2 - 1] \cdot s_5,\ m_2 = [(\cos t - \cos 2t)/2] \cdot s_6,$$
$$m_3 = j(\sin t + \sin 2t) \cdot s_2,\ m_4 = j \sin 2t \cdot s_7,\ m_5 = j(\sin t - \sin 2t) \cdot s_4.$$

Step 3. Output ADDs

$$s_9 = m_0 + m_1,\ s_{10} = s_9 + m_2,\ s_{11} = s_9 - m_2,\ s_{12} = m_3 - m_4,\ s_{13} = m_4 + m_5,$$
$$s_{14} = s_{10} + s_{12},\ s_{15} = s_{10} + s_{12},\ s_{16} = s_{11} + s_{13},\ s_{17} = s_{11} - s_{13}.$$

$$A_0 = m_0,\ A_1 = s_{14},\ A_2 = s_{16},\ A_3 = s_{17},\ A_4 = s_{15}.$$

Notes

It is quite an arduous task to describe all the facets of FFT algorithms in one book, let alone one chapter. We have chosen to restrict our attention to the mathematical aspects of DFT computation and the resulting FFT algorithms. Over the years, DFTs have been studied by researchers from many diverse disciplines. Within the signal processing community, *IEEE Transactions on Signal Processing*, *IEEE Transactions on Circuits and Systems*, and *IEEE Transactions on Computers* have been the prime journals for reporting new research results. DFTs in finite integer fields and rings have also been described. We have used several books and research papers to strengthen our understanding of FFTs. An attempt is made to present the FFT algorithms as a product of the number-theoretic properties of the indices and the data sequences.

Among the books we have used, the books by McClellan and Rader [1979], Nussbaumer [1981], Oppenheim and Schafer [1989], Blahut [1984], Tolimieri, An, and Lu [1989], Myers [1990], and Briggs and Henson [1995] come to mind as having a direct impact on the selection and presentation of the algorithms. The book by Oppenheim and Schafer [1989] had a strong influence on the figures for radix-2 FFT. The book by Briggs and Henson [1995] influenced the matrix representations of the FFTs. The exposition of split radix FFT, Rader-Brenner FFT, and Brunn's FFT algorithms in this chapter follows the original papers by Duhamel [1986], Rader and Brenner [1976], and Brunn [1978]. Historically, the paper by Cooley and Tukey [1965] is considered one of the most important milestones in the developments of FFTs. The length 4 and 5 FFTs described in Example 13.5 are taken from Winograd [1978].

We have chosen to restrict our attention to FFTs in the field of complex numbers. However, FFTs in finite fields and rings exist and find applications in digital signal processing and error control coding.

Bibliography

[13.1] H.J. Nussbaumer, *Fast Fourier Transform and Convolution Algorithms*, Springer-Verlag, 1981.

[13.2] D.G. Myers, *Digital Signal Processing: Efficient Convolution and Fourier Transform Techniques*, Prentice Hall, 1990.

[13.3] J.H. McClellan and C.M. Rader, *Number Theory in Digital Signal Processing*, Prentice Hall, 1979.

[13.4] R.E. Blahut, *Fast Algorithms for Digital Signal Processing*, Addison-Wesley Publishing Co., 1984.

[13.5] M.T. Heideman, *Multiplicative Complexity, Convolution, and the DFT*, Springer-Verlag, 1988.

[13.6] R. Tolimieri, M. An, and C. Lu, *Algorithms for Discrete Fourier Transform and Convolution*, Springer-Verlag, 1989.

[13.7] A.V. Oppenheim and R.W. Schafer, *Discrete-Time Signal Processing*, Prentice Hall, 1989.

[13.8] W.L. Briggs and V.E. Henson, *The DFT: An Owner's Manual for the Discrete Fourier Transforms*, Society for Industrial and Applied Mathematics, Philadelphia, 1995.

[13.9] S. Winograd, "On Computing the Discrete Fourier Transform," *Mathematics of Computation*, Vol. 32, pp. 175-199, Jan. 1978.

[13.10] S. Winograd, *Arithmetic Complexity of Computations*, SIAM Publications, Philadelphia, PA, 1980.

[13.11] J.W. Cooley and J.W. Tukey, "An Algorithm for the Machine Computation of Complex Fourier Series," *Mathematics of Computation*, Vol. 19, pp. 297-301, 1965.

[13.12] C.M. Rader and N.M. Brenner, "A New Principle for Fast Fourier Transform," *IEEE Transactions on Acoustics, Speech, and Signal Processing*, Vol. 24, pp. 264-266, 1976.

[13.13] G. Brunn, "*z*-Transform DFT Filters and FFTs," *IEEE Transactions on Acoustics, Speech, and Signal Processing*, Vol. 26, pp. 56-63, 1978.

[13.14] P. Duhamel, "Implementation of "Split-Radix" FFT Algorithms for Complex, Real, and Real-Symmetric Data," *IEEE Transactions on Acoustics, Speech, and Signal Processing*, Vol. 34, pp. 285-295, 1986.

[13.15] J.-B. Martens, "Recursive Cyclotomic Factorization—A New Algorithm for Calculating the Discrete Fourier Transform," *IEEE Transactions on Acoustics, Speech, and Signal Processing*, Vol. 32, pp. 750-761, 1984.

[13.16] J. Glassman, "A Generalization of the Fast Fourier Transform," *IEEE Transactions on Computers*, Vol. C-19, pp. 105-116, 1970.

[13.17] D.M.W. Evans, "An Improved Digit Reversal Permutation Reversal Algorithm for Fast Fourier and Hartley Transforms," *IEEE Transactions on Acoustics, Speech, and Signal Processing*, Vol. 35, pp. 1120-1125, 1987.

[13.18] J.-B. Martens, "Discrete Fourier Transform Algorithms for Real-Values Sequences," *IEEE Transactions on Acoustics, Speech, and Signal Processing*, Vol. 32, pp. 390-396, 1984.

[13.19] M. Vetterli and P. Duhamel, "Split Radix Algorithms for length p^m DFTs," *IEEE Transactions on Acoustics, Speech, and Signal Processing*, Vol. 34, pp. 57-64, 1989.

[13.20] C.M. Rader, "Discrete Fourier Transforms When the Number of Data Samples is Prime," *Proceedings of the IEEE*, Vol. 56, pp. 1107-1108, 1968.

[13.21] J.M. Pollard, "The Fast Fourier Transform in a Finite Field," *Mathematics of Computation*, Vol. 25, pp. 365-374, 1971.

Problems

13.1 Prove Parseval's theorem.

13.2 Prove properties K, L, and M.

13.3 Describe the method to compute the DFT of two real-valued sequences in one complex DFT.

13.4 Describe a method to compute the DFT of two conjugate symmetric sequences in one complex DFT.

13.5 Describe the method to compute the DFT of real-valued symmetric sequences.

13.6 Construct FFT algorithms for small n, $n = 3, 4, 5, 6, 7, 8, 9, 10$.
Note: Many of these algorithms can be found in Winograd [1978]. They can be derived using Rader's algorithm and splitting methods.

13.7 In Bluestein's algorithm, we need to multiply the input sequence by $\omega^{l^2/2}$ and the output sequence by $\omega^{i^2/2}$.

(a) For $n = p$, an odd prime, find the distinct values taken by $l^2/2$ as l takes values $0, 1, \ldots, n-1$.

(b) Repeat it for $n = p^c$ and, hence, generalize it for the standard factorization of n.
Note: It is a multiplicative function in the same way as the Euler's totient function.

13.8 Describe the Bluestein's algorithm for $n = 32$.

13.9 Design a length $n = 5, 7, 11, 13$ FFT using Rader's algorithm.

13.10 Design a length $n = 9, 25$ FFT using Rader's algorithm.

13.11 Design a length 81 FFT using Rader's algorithm.

13.12 Consider Rader's algorithm for $n = p^c$. It is expressed as a convolution mod $u^{\phi(n)} - 1$ along with some other terms. Show that when the generalized Rader's polynomial dependent on the Fourier coefficients in (13.3.13) is reduced mod $u^{(p-1)/2} - 1$, all the coefficients of the resulting polynomial are real-valued. Similarly, when it is reduced mod $u^{(p-1)/2} + 1$, all the coefficients of the resulting polynomial are purely

imaginary. Finally, when it is reduced modulo $u^{\phi(n)/p} - 1$ ($c > 1$ for this case), all the coefficients are zero.
Note: This property is used to simplify the computational complexity of the cyclic convolution associated with Rader's algorithm.

13.13 Consider the Rader's algorithm for $n = 2p$. Show that it can also be computed using only cyclic convolutions of length $p - 1$ along with some extra arithmetic operations. Can you make the same statement for $n = 4p$?

13.14 Describe Rader's algorithm for $n = 2^c$. In the following, we attempt to develop the various pieces of the algorithm.

(a) Show that A_i, for even i, can be written as a DFT of size $n/2$ at the cost of some ADDs.

(b) For odd i, express A_i as a sum of two terms T1 and T2, T1 for even l and T2 for odd l.

(c) Show that T1 can be computed as a length $n/4$ DFT at the cost of some ADDs and MULTs.

(d) Show that T2 can be written as a $(2 \times n/4)$ two-dimensional cyclic convolution by recalling that all the odd integers in $0, 1, \ldots, n-1$ can be written as $(-1)^a \cdot \alpha^b$, where $a = 0, 1$, and α is an element of order $n/4 = 2^{c-2}$ in $Z(2^c)$. Let this convolution be expressed as a computation $A(u, v) \cdot B(u, v) \bmod [(u^{n/4} - 1), (v^2 - 1)]$, where $A(u, v)$ is data dependent and $B(u, v)$ is a two-dimensional polynomial dependent on Fourier coefficients. Show that (i) $B(u, v) \bmod (v - 1)$ has real coefficients; (ii) $B(u, v) \bmod (v + 1)$ has imaginary coefficients; and (iii) $B(u, v) \bmod (u^{n/8} - 1) = 0$.

(e) Note: This is similar to the Rader's algorithm for $n = p^c$, p an odd prime, with the exception that there is no primitive element in $Z(2^c)$ except for $c = 1, 2$. One possible value for α is 3.

13.15 Design a FFT algorithm for $n = 15$ using the Cooley-Tukey method.

13.16 Design a FFT algorithm for $n = 35$ using the Cooley-Tukey method.

13.17 Cast the radix-2 and radix-4 DIT-FFT and DIF-FFT algorithms as a way to factorize the DFT matrix Λ_n.

13.18 Repeat Problem 13.17 for the radix-2 DIF-FFT.

13.19 Repeat Problem 13.17 for the radix-4 DIT-FFT.

13.20 Describe radix-4 DIF-FFT algorithm.

13.21 Repeat Problem 13.17 for the radix-4 DIF-FFT.

13.22 Write the Cooley-Tukey FFT algorithm for $n = 32$ by letting $a = 4$ and $b = 8$. Split further the length 8 DFT by letting $a = 2$, $b = 4$. Evaluate its computational complexity.

13.23 Write the Cooley-Tukey FFT algorithm for $n = 32$ by letting $a = 2$ and $b = 16$. Further split the length 16 DFT by letting $a = 4$, $b = 4$. Evaluate its computational complexity.

13.24 Compare the computational complexity of the methods in problems 13.22 and 13.23 and radix-2 FFT for $n = 32$.

13.25 Write the Cooley-Tukey FFT algorithm for $n = 64$ by letting $a = b = 8$. Compare its computational complexity with radix-4 FFT.

13.26 Show that radix-4 FFT can be written for n an odd power of 2 by letting $a = 4$ until the last stage which is computed using radix-2. Equivalently, the first stage can be a radix-2 stage followed by radix-4 FFT.

13.27 Derive a generalization of Problem 13.25 for $n = 2^c$, c even, by letting $a = b = 2^{c/2}$ in the Cooley-Tukey algorithm. Modify it appropriately for $n = 2^c$, c odd.

13.28 Describe a radix-3 FFT algorithm for $n = 3^c$.

13.29 Compare the radix-3 FFT derived in Problem 13.28 to the Rader's algorithm for $n = 27$.

13.30 Consider the cyclotomic field $CF(3)$. Describe a radix-3 FFT algorithm in $CF(3)$ for $n = 3^c$.

13.31 Consider the cyclotomic field $CF(p)$, p an odd prime. Describe a radix-p algorithm in $CF(p)$ for $n = p^c$.

13.32 Describe the Cooley-Tukey algorithm for $n = 2^c$ when the data are defined in $CF(8)$, $CF(16)$, $CCF(8)$, $CCF(16)$.

13.33 Design radix-2 and radix-4 algorithms for $n = 16$ that require no bit-reversal but are not in-place.
Note: This can be accomplished by first deriving the signal flow diagram for the in-place algorithm and then alter it suitably.

13.34 Write a computer program for (i) Rader's algorithm, (ii) Cooley-Tukey algorithm, (iii) radix-2 DIT-FFT, (iv) radix-2 DIF-FFT, (v) radix-4 DIT-FFT, (vi) radix-4 DIF-FFT, (vii) split radix FFT, (viii) Rader-Brenner algorithm, (ix) Brunn algorithm, and (x) WFTA algorithms. Compare the execution times for large values of n.
Note: Many such programs are available commercially.

13.35 Describe the split radix DIT-FFT algorithm.

13.36 Describe the radix-4 based split radix FFT algorithm and evaluate its computational complexity. Is it superior to the radix-4 FFT?

13.37 Describe other strategies for split radix approach and obtain their computational complexity.

13.38 Describe split radix algorithm for real-valued data. Show that we do not need to compute A_{4i+3} as it can be obtained from the conjugate of the A_{4i+1} coefficients. Evaluate the computational complexity of this algorithm.

13.39 Describe a radix-2 FFT algorithm that alternates between the DIT and DIF approach and draw its signal flow diagram. Assume $n = 2^4$.

13.40 Describe the Rader-Brenner algorithm for $n = 16$ and obtain its signal flow diagram.

13.41 Evaluate the computational complexity of Rader-Brenner algorithm for real-valued and complex-valued data sequences.

13.42 Describe the DIF Rader-Brenner algorithm.

13.43 Derive FFT algorithms for $n = 12, 15, 20$ using a suitable approach and draw signal flow diagrams. Are these algorithms in-place?

13.44 Describe the length 16 FFT using the Brunn algorithm and draw its filter tree.

13.45 Evaluate the computational complexity of the Brunn's algorithm for real-valued and complex-valued data sequences.

13.46 Cast the FFT for $n = 3, 4, 5, 6, 7, 8$ as a bilinear form suitable for WFTA.

13.47 Describe WFTA for $n = 15, 35$. Evaluate the computational complexity of the algorithms.

13.48 Compare WFTA in Problem 13.47 to the FFT algorithms derived in Problems 13.15 and 13.16. This comparison is in terms of computational cost and other features such as indexing and signal flow diagrams.

13.49 Describe all the possible ways to compute length 105 DFT using WFTA. Identify the best strategy using the computational complexity as a criterion of performance.

13.50 Repeat Problem 13.49 for $n = 120$.

13.51 Prove matrix formulation of Cooley-Tukey algorithm in (13.6.23).

13.52 Describe FFT of length 8, 16 in $GF(17)$.

13.53 Describe radix-2 and radix-4 FFT algorithms in $GF(3^m)$ $GF(5^m)$, if they exist.

13.54 Describe a length 16 FFT in $Z(17^a)$. Let $a = 3, 4$.

13.55 Consider the sequence $A_l = (0.9375)^l$, $l = 0, 1, \ldots, 15$. Compute the DFT using infinite precision arithmetic. Now, let us say that we are using fixed-point representation with 6 bits. Compute the DFT by rounding all computations to 6 bits. Compare the two DFTs in terms of the maximum error and also the sum of the squares of the errors in each DFT coefficient.

13.56 Take the IDFT of the DFT obtained by rounding the computations. Once again, assume that IDFT is computed by rounding all computations to 6 bits. Compare the IDFT sequence to the original sequence.

13.57 Repeat problems 13.55 and 13.56 using 16-bit arithmetic.

13.58 Is it reasonable to say that as the length of the DFT is increased, one needs more bits in the number representation for the same accuracy?

Chapter 14

Fast Fourier Transforms: Multi-Dimensional Data Sequences

In this chapter, we study FFT algorithms for computing DFT of two- and higher-dimensional data sequences. In most cases, algorithms are derived for two-dimensional data sequences and then generalized to higher dimensions. It is our feeling that FFT algorithms for two-dimensional data sequences depart significantly from their one-dimensional counterparts, but extend naturally to higher dimensions. The framework for the various algorithms is developed in a way that the extension of two-dimensional FFTs to higher dimensional FFTs could be obtained in a straightforward manner.

The readers are advised to read this chapter in tandem with Chapter 13. This is important for a number of reasons. One, multidimensional FFTs are used extensively to simplify the computation of one-dimensional DFT. Examples of such cases are the Cooley-Tukey FFT and prime factor FFT algorithms. Two, many multidimensional FFT algorithms consist in reducing the given DFT computation in to a number of one-dimensional DFTs. Examples of such cases are the row-column method and polynomial transform based FFT algorithms. Polynomial transform based FFT algorithms make extensive use of cyclotomic factorization in cyclotomic fields such as $\mathcal{CF}(p)$, $\mathcal{CF}(p^c)$, $\mathcal{CF}(2^n)$, and $\mathcal{CCF}(2^n)$. Thus, the distinction between one- and multidimensional FFTs may be not as clear cut as one may think at the start.

Our presentation in this chapter follows the material in Chapter 13 quite closely. All the quantities in this chapter are defined over the field $\mathcal{C}$. We begin our description by defining multidimensional DFT.

14.1 The Multidimensional DFT: Definition and Properties

Definition 14.1 Multidimensional discrete Fourier transform. Given a d dimensional data sequence of length n_t in the t-th dimension, $A_{l_1,l_2,\ldots,l_d}$, $l_1 = 0, 1, \ldots, n_1 - 1$, $l_2 = 0, 1, \ldots, n_2 - 1$, …, $l_d = 0, 1, \ldots, n_d - 1$, the d-dimensional DFT in $\mathcal{C}$ is defined as the sequence $\mathbf{A}_{i_1,i_2,\ldots,i_d}$, $i_1 = 0, 1, \ldots, n_1 - 1$, $i_2 = 0, 1, \ldots, n_2 - 1$, …, $i_d = 0, 1, \ldots, n_d - 1$, given by

$$\mathbf{A}_{i_1,i_2,\ldots,i_d} = \sum_{l_d=0}^{n_d-1} \cdots \sum_{l_2=0}^{n_2-1} \sum_{l_1=0}^{n_1-1} A_{l_1,l_2,\ldots,l_d}\, \omega_1^{l_1 i_1} \omega_2^{l_2 i_2} \omega_d^{l_d i_d} , \tag{14.1.1}$$

where ω_t is the t-th root of unity.

Definition 14.2 Multidimensional inverse discrete Fourier transform. Given the DFT $\mathbf{A}_{i_1,i_2,\ldots,i_d}$, $i_1 = 0, 1, \ldots, n_1 - 1$, $i_2 = 0, 1, \ldots, n_2 - 1$, …, $i_d = 0, 1, \ldots, n_d - 1$ of a sequence $A_{l_1,l_2,\ldots,l_d}$, $l_1 = 0, 1, \ldots, n_1 - 1$, $l_2 = 0, 1, \ldots, n_2 - 1$, …, $l_d = 0, 1, \ldots, n_d - 1$, the original data sequence $A_{l_1,l_2,\ldots,l_d}$, is said to constitute the *inverse discrete Fourier transform* of the DFT sequence $\mathbf{A}_{i_1,i_2,\ldots,i_d}$.

Theorem 14.1 Given the DFT $\mathbf{A}_{i_1,i_2,\ldots,i_d}$, of a sequence $A_{l_1,l_2,\ldots,l_d}$, the IDFT can be computed as

$$A_{l_1,l_2,\ldots,l_d} = \frac{1}{n_1 n_2 \cdots n_d} \sum_{l_d=0}^{n_d-1} \cdots \sum_{l_2=0}^{n_2-1} \sum_{l_1=0}^{n_1-1} \mathbf{A}_{i_1,i_2,\ldots,i_d}\, \omega_1^{-l_1 i_1} \omega_2^{-l_2 i_2} \omega_d^{-l_d i_d} . \tag{14.1.2}$$

Proof: The proof is similar to the one-dimensional case.

The IDFT computation in (14.1.2) is identical to the DFT computation in (14.1.3). The computation of DFT and IDFT can be performed in an identical manner. Consequently, we will discuss only algorithms for DFT.

As stated earlier, we will dwell on the two-dimensional DFT and then extend the results to higher dimensions. For $d = 2$, the two-dimensional DFT is given by

$$\mathbf{A}_{i_1,i_2} = \sum_{l_2=0}^{n_2-1}\sum_{l_1=0}^{n_1-1} A_{l_1,l_2}\,\omega_1^{\,l_1 i_1}\,\omega_2^{\,l_2 i_2}\,. \tag{14.1.3}$$

This expression can also be written as

$$\mathbf{A}_{i_1,i_2} = \sum_{l_1=0}^{n_1-1} \omega_1^{\,l_1 i_1} \sum_{l_2=0}^{n_2-1} A_{l_1,l_2}\,\omega_2^{\,l_2 i_2} \tag{14.1.4}$$

$$= \sum_{l_2=0}^{n_2-1} \omega_2^{\,l_2 i_2} \sum_{l_1=0}^{n_1-1} A_{l_1,l_2}\,\omega_1^{\,l_1 i_1}\,. \tag{14.1.5}$$

A similar expression for $d > 2$ is easily seen for (14.1.1). Therefore, multi-dimensional DFTs are called **half-separable**. These expressions also suggest a simple method for computing multidimensional DFTs. The method called **row-column method** is as follows for $d = 2$.

Step 0. Arrange the input A_{l_1,l_2} as a $(n_1 \times n_2)$ matrix **A**.

Step 1. Take the one-dimensional DFT of length n_1 of each column of **A**, that is, for different values of l_2 in (14.1.5), compute the inner sum. Let the resulting matrix be **B**.

Step 2. Take the one-dimensional DFT of length n_2 of each row of **B** to get a matrix **C**. The (i_1, i_2)-th element of **C** is the Fourier coefficient $\mathbf{A}_{i_1,i_2}$.

Alternately, we could perform the row DFTs first and then compute the column DFTs as in (14.1.4). In either case, we require two FFT algorithms, one for length n_1 and the other for length n_2. The two-dimensional FFT algorithm computes n_1 DFTs of length n_2 and n_2 DFTs of length n_1. Therefore, the computational complexity of the two-dimensional row-column method is given by

$$M(n_1 \times n_2) = n_1 \cdot M(n_2) + n_2 \cdot M(n_1), \tag{14.1.6}$$
$$A(n_1 \times n_2) = n_1 \cdot A(n_2) + n_2 \cdot A(n_1). \tag{14.1.7}$$

It is evident that the structure of the row-column method is simple. It is an *in-place* algorithm. It also has the nice feature that the algorithm operates only on one vector at any given time. Therefore, it may be a preferred method when there is constraint on the size of the memory.

Consider the d-dimensional DFT expressed as

$$A_{i_1,i_2,\ldots,i_d} = \sum_{l_d=0}^{n_d-1} \omega_d^{\,l_d i_d} \cdots \sum_{l_2=0}^{n_2-1} \omega_2^{\,l_2 i_2} \sum_{l_1=0}^{n_1-1} A_{l_1,l_2,\ldots,l_d}\, \omega_1^{\,l_1 i_1}\,. \tag{14.1.8}$$

The *row-column method* for the above computation is as follows.

Step 0. Arrange the input $A_{l_1,l_2,\ldots,l_d}$ as a $(n_1 \times n_2 \times \cdots \times n_d)$ matrix $\mathbf{A}$.

Step 1. Take the one-dimensional DFT of length n_1 for each distinct value of l_2, l_3, …, l_d. There are $n_2 \times \cdots \times n_d$ such DFTs. Let the resulting matrix be $\mathbf{B}_1$.

Step 2. Take the one-dimensional DFT of length n_2 for each distinct value of i_1, l_3, …, l_d, to get a matrix $\mathbf{B}_2$. There are $n_1 \times n_3 \times \ldots \times n_d$ such DFTs. This process is continued.

Step t. Take the one-dimensional DFT of length n_t for each distinct value of i_1, i_2, …, i_{t-1}, l_{t+1}, …, l_d to get a matrix $\mathbf{B}_t$. There are $n_1 \times n_3 \times \cdots \times n_{t-1} \times n_{t+1} \times \cdots \times n_d$ such DFTs.

$t = 1, 2, \ldots, d$.

These DFTs can be performed in any sequence. In all cases, we require d FFT algorithms, one for each of the lengths n_t. The d-dimensional FFT algorithm computes $n_1 \times n_3 \times \cdots \times n_{t-1} \times n_{t+1} \times \cdots \times n_d$ DFTs of length n_t, $t = 1, 2, \ldots, d$. Therefore, the computational complexity of the d-dimensional row-column method is given by

$$M(n_1 \times n_2 \times \cdots \times n_d) = \sum_{t=1}^{d} \left(\frac{N}{n_t} \right) M(n_t) \tag{14.1.9}$$

and

$$A(n_1 \times n_2 \times \cdots \times n_d) = \sum_{t=1}^{d} \left(\frac{N}{n_t} \right) A(n_t), \tag{14.1.10}$$

where

$$N = (n_1 \cdot n_2 \cdot \cdots \cdot n_d). \tag{14.1.11}$$

Once again, it is also an *in-place* algorithm. In many cases, $n_1 = n_2 = \ldots = n_d = n$, and, in such cases, (14.1.9) and (14.1.10) simplify to

$$M(n \times n \times \cdots \times n) = d \cdot n^{d-1} \cdot M(n) \tag{14.1.12}$$

and

$$A(n \times n \times \cdots \times n) = d \cdot n^{d-1} \cdot A(n). \tag{14.1.13}$$

The expressions in (14.1.9) to (14.1.13) can be used as a benchmark to compare various multidimensional FFTs.

In the following, n denotes any one of $n_1, n_2, \ldots, n_d$.

14.2 FFT for $n = p$, p an Odd Prime

Let $d = 2$, $n_1 = p_1$, $n_2 = p_2$, p_1, p_2 odd primes. The FFT algorithm to be described is a straightforward generalization of the *Rader's algorithm*. In this case, we have

$$\boldsymbol{A}_{0,0} = \sum_{l_2=0}^{p_2-1} \sum_{l_1=0}^{p_1-1} A_{l_1,l_2} \tag{14.2.1}$$

and

$$\boldsymbol{A}_{i_1,0} - \boldsymbol{A}_{0,0} = \sum_{l_1=1}^{p_1-1} \left(\omega_1{}^{l_1 i_1} - 1\right) \sum_{l_2=0}^{p_2-1} A_{l_1,l_2}, \quad i_1 = 1, 2, \ldots, p_1 - 1, \tag{14.2.2}$$

$$\boldsymbol{A}_{0,i_2} - \boldsymbol{A}_{0,0} = \sum_{l_2=1}^{p_2-1} \left(\omega_2{}^{l_2 i_2} - 1\right) \sum_{l_1=0}^{p_1-1} A_{l_1,l_2}, \quad i_2 = 1, 2, \ldots, p_2 - 1, \tag{14.2.3}$$

$$\boldsymbol{B}_{i_1,i_2} = \boldsymbol{A}_{i_1,i_2} - \boldsymbol{A}_{i_1,0} - \boldsymbol{A}_{0,i_2} + \boldsymbol{A}_{0,0} = \sum_{l_2=1}^{p_2-1} \sum_{l_1=1}^{p_1-1} \left(\omega_1{}^{l_1 i_1} - 1\right)\left(\omega_2{}^{l_2 i_2} - 1\right) A_{l_1,l_2},$$
$$i_1 = 1, 2, \ldots, p_1 - 1, \; i_2 = 1, 2, \ldots, p_2 - 1. \tag{14.2.4}$$

Thus the two-dimensional DFT can be expressed as

(a) A sum of $p_1 \times p_2$ terms in (14.2.1),
(b) A one-dimensional cyclic convolution of length $p_1 - 1$ as in (14.2.2),
(c) A one-dimensional cyclic convolution of length $p_2 - 1$ as in (14.2.3), and
(d) A two-dimensional cyclic convolution of length $(p_1 - 1) \times (p_2 - 1)$ as in (14.2.4).

The one-dimensional cyclic convolutions in (14.2.2) and (14.2.3) have the same form as the Rader's algorithm in one-dimension. The two-dimensional cyclic convolution in (14.2.4) can be expressed as

$$
\begin{aligned}
\boldsymbol{B}(u) &= \sum_{g=0}^{p_1-2}\sum_{f=0}^{p_2-2} \boldsymbol{B}_{\alpha^g,\beta^f} u^g v^f = \Delta(u,v)A(u,v) \bmod \left(u^{p_1-1}-1, v^{p_2-1}-1\right) \\
&= \left[\sum_{h=0}^{p_1-2}\sum_{k=0}^{p_2-2}\left(\omega_1^{\alpha^h}-1\right)\left(\omega_2^{\beta^k}-1\right)u^h v^k\right] \\
&\quad \cdot \left[\sum_{h=0}^{p_1-2}\sum_{k=0}^{p_2-2} A_{\alpha^{-h},\beta^{-k}} u^h v^k\right] \bmod \left(u^{p_1-1}-1, u^{p_2-1}-1\right). \qquad (14.2.5)
\end{aligned}
$$

The derivation of (14.2.5) follows the same arguments as the one-dimensional case. Finally, due to the form of $\Delta(u, v)$, it can be simplified to

$$
\Delta(u,v) = \left(\sum_{h=0}^{p_1-2}\left(\omega_1^{\alpha^h}-1\right)u^h\right)\left(\sum_{k=0}^{p_2-2}\left(\omega_2^{\beta^k}-1\right)v^k\right). \qquad (14.2.6)
$$

Fast algorithms for computing two-dimensional cyclic convolution are described in Chapter 10. All of them can be used to compute (14.2.5) in an efficient manner.

A generalization of Rader's algorithm to $n = p^c$ and to higher dimensions $(d > 2)$ is quite straightforward. However, in both instances, it involves a large number of terms that need to be handled separately. Therefore, the usefulness of this approach may be limited to two-dimensions and when the length in each dimension is an odd prime. Even in those cases, it may be useful only when $n_1 = p_1$, $n_2 = p_2$, where $p_1 \neq p_2$. When $p_1 = p_2$, then the polynomial transform approach may be preferred. The polynomial transform method for computing DFT will be described in Section 14.5.

Example 14.1 Let $n_1 = 17$, $n_2 = 5$. Then, the two-dimensional DFT is expressed as

(a) A sum of $17 \cdot 5 = 85$ terms (84 ADDs),
(b) A cyclic convolution of length 16, to be computed using

$$u^{16} - 1 = (u+1)(u-1)(u+j)(u-j)(u^2+1)(u^2-j)(u^4+j)(u^4-j),$$

in 28 MULTs,
(c) A cyclic convolution of length 4, to be computed using

$$v^4 - 1 = (v+1)(v-1)(v+j)(v-j),$$

in 4 MULTs, and

(d) A 16×4 two-dimensional cyclic convolution using the above factorizations in 112 MULTs.

Some minor improvements in multiplicative complexity are possible by noting the properties in Problem 13.12.

14.3 Multidimensional FFT Algorithms for *n* a Power of 2

A number of different approaches for computing multidimensional DFT are a generalization of one-dimensional FFTs. It is possible to design a d-dimensional FFT algorithm using the Cooley-Tukey splitting technique of Section 13.4. Such an approach is quite straightforward. In this section, we describe this approach only in the context of radix-2 FFT algorithms with $d = 2$, $n_1 = 2^c$, $n_2 = 2^b$, $b \geq c$.

14.3.1 Two-Dimensional Radix-2 FFT

The following analysis assumes that radix-2 DIT-FFT is used in both dimensions and follows (13.5.1) to (13.5.7) closely. In this case, the two-dimensional radix-2 FFT algorithm can be described as follows.

Step 1. Partition the input sequence into four sequences B_{l_1,l_2}, C_{l_1,l_2}, D_{l_1,l_2}, E_{l_1,l_2}, $l_1 = 0, 1, \ldots, n_1/2 - 1$, $l_2 = 0, 1, \ldots, n_2/2 - 1$. Here,

$$B_{l_1,l_2} = A_{2l_1,2l_2},$$
$$C_{l_1,l_2} = A_{2l_1+1,2l_2},$$

$$D_{l_1,l_2} = A_{2l_1,2l_2+1},$$
$$E_{l_1,l_2} = A_{2l_1+1,2l_2+1}, \quad l_1 = 0, 1, \ldots, \quad n_1/2-1, \; l_2 = 0, 1, \ldots, n_2/2-1. \quad (14.3.1)$$

Step 2. Take the length $n_1/2 \times n_2/2$ DFT of each of these sequences to get $\boldsymbol{B}_{i_1,i_2}$, $\boldsymbol{C}_{i_1,i_2}$, $\boldsymbol{D}_{i_1,i_2}$, and $\boldsymbol{E}_{i_1,i_2}$, $i_1 = 0, 1, \ldots, n_1/2-1$, $i_2 = 0, 1, \ldots, n_2/2-1$.

Step 3. The DFT of the original sequence is given by

$$\begin{bmatrix} \boldsymbol{A}_{i_1,i_2} \\ \boldsymbol{A}_{i_1,i_2} \\ \boldsymbol{A}_{i_1,i_2} \\ \boldsymbol{A}_{i_1,i_2} \end{bmatrix} = \begin{bmatrix} 1 & 1 & 1 & 1 \\ 1 & -1 & 1 & -1 \\ 1 & 1 & -1 & -1 \\ 1 & -1 & -1 & 1 \end{bmatrix} \begin{bmatrix} \boldsymbol{B}_{i_1,i_2} \\ \omega_1^{i_1} \boldsymbol{C}_{i_1,i_2} \\ \omega_2^{i_2} \boldsymbol{D}_{i_1,i_2} \\ \omega_1^{i_1} \omega_2^{i_2} \boldsymbol{E}_{i_1,i_2} \end{bmatrix},$$
$$i_1 = 0, 1, \ldots, \quad n_1/2-1, \; i_2 = 0, 1, \ldots, n_2/2-1. \quad (14.3.2)$$

Thus, two-dimensional radix-2 FFT consists in writing $n_1 \times n_2$ DFT as 4 $n_1/2 \times n_2/2$ DFTs. Once they are computed as in Step 2 above, computation of the original DFT coefficients in (14.3.2) requires $(3/4) \cdot n_1 \cdot n_2$ MULTs (including the MULT by twiddle factors that take trivial values) and $2n_1n_2$ ADDs. Therefore, the computational complexity is given by

$$M(n_1 \times n_2) = 4 \cdot M(n_1/2 \times n_2/2) + M(\text{twiddle factors}),$$
$$A(n_1 \times n_2) = 4 \cdot A(n_1/2 \times n_2/2) + 2n_1n_2. \quad (14.3.3)$$

Setting $M(\text{twiddle factors}) = (3/4) \cdot n_1 \cdot n_2$ and simplifying, we get

$$M(n_1 \times n_2) = 2^{2c-4} \cdot M(4 \times 2^{b-c+2}) + (\log_2 n_1 - 2) \cdot (3/4) \cdot n_1 \cdot n_2.$$

Noting that $M(4 \times n) = M(n) = (n/2)(\log_2 n - 3) + 2$ from (13.5.18), we get

$$M(n_1 \times n_2) = (1/8) \cdot (n_1n_2)\,[\log_2 n_2 + 5 \log_2 n_1 - 13] + n_1^2/8. \quad (14.3.4)$$

In a similar manner,

$$A(n_1 \times n_2) = (n_1n_2)\,[\log_2 n_2 + \log_2 n_1]. \quad (14.3.5)$$

These values must be compared to the row-column method that gives

$$M(n_1 \times n_2) = 0.5 \cdot (n_1n_2)\,[\log_2 n_2 + \log_2 n_1 - 6] \quad (14.3.6)$$

and

$$A(n_1 \times n_2) = (n_1 n_2)\,[\log_2 n_2 + \log_2 n_1]. \tag{14.3.7}$$

In a similar manner as above, one can design FFT algorithms that are based on

1. Radix-2 DIF-FFT in both dimensions,
2. Radix-2 DIT-FFT in one dimension and DIF-FFT in the other,
3. Radix-4 DIT and/or DIF FFT, and
4. Two-dimensional split radix FFT algorithm based on Section 13.5.3.

Many other combinations are possible. It is straightforward to derive their algorithmic structure and analyze the computational complexity. Some of these are dealt with as exercises at the end of the chapter.

14.4 Matrix Formulation of Multidimensional DFT and Related Algorithms

The material in this section is similar to the matrix formulation of prime factor FFT in Section 13.6.1. Note that here we place no restriction on the lengths unlike Section 13.6.1 where $(n_1, n_2) = 1$. We have kept the details to a minimum as in large parts, matrix formulation does not depend on the nature of lengths.

We begin with the two-dimensional DFT of length $n_1 \times n_2$ defined as

$$A_{i_1,i_2} = \sum_{l_2=0}^{n_2-1}\sum_{l_1=0}^{n_1-1} A_{l_1,l_2}\,\omega_1^{l_1 i_1}\,\omega_2^{l_2 i_2} = \sum_{l_1=0}^{n_1-1}\omega_1^{l_1 i_1}\sum_{l_2=0}^{n_2-1} A_{l_1,l_2}\,\omega_2^{l_2 i_2}\,. \tag{14.4.1}$$

The two-dimensional data sequence A_{l_1,l_2}, $l_1 = 0, 1, \ldots, n_1 - 1$, $l_2 = 0, 1, \ldots, n_2 - 1$ is arranged as elements of a $(n_1 \times n_2)$ two-dimensional matrix **A** whose (l_1, l_2)-th element is given by A_{l_1,l_2}. The two-dimensional DFT sequence $\boldsymbol{A}_{i_1,i_2}$, $i_1 - 0, 1, \ldots, n_1 - 1$, $i_2 - 0, 1, \ldots, n_2 - 1$ is recovered from the $(n_1 \times n_2)$ two-dimensional matrix **C** whose (i_1, i_2)-th element is given by $\boldsymbol{A}_{i_1,i_2}$. The process of DFT involves the computation of length n_1 DFT of each of the n_2 columns of **A** which gives rise to the **B** matrix. This is followed by length n_2 DFT of each of the n_1 rows of **B** to obtain **C**. One can also reverse the order of this computation without changing the final result.

Let Λ_{n_1} and Λ_{n_2} denote the DFT matrix for length n_1 and length n_2 DFTs, respectively. Then one can write

$$\mathbf{B} = \Lambda_{n_1}\,\mathbf{A},$$

$$\mathbf{C} = \mathbf{B}\ \Lambda_{n_2}\,. \tag{14.4.2}$$

Combining, we get

$$\mathbf{C} = \Lambda_{n_1}\ \mathbf{A}\ \Lambda_{n_2} = \Lambda_{n_1}\ [\underline{A}_0\, \underline{A}_1 \cdots \underline{A}_{n_2-1}]\ \Lambda_{n_2}\,. \tag{14.4.3}$$

Here $\underline{A}_l$ denotes the l-th column of **A**. Let us arrange the elements of **A** as a single column in the following manner:

$$\underline{A} = \begin{bmatrix} \underline{A}_0 \\ \underline{A}_1 \\ \vdots \\ \underline{A}_{n_2-1} \end{bmatrix}. \tag{14.4.4}$$

The column-by-column length n_1 DFT of each of the columns of **A** can now be expressed as the matrix-vector product

$$\begin{bmatrix} \underline{\boldsymbol{B}}_0 \\ \underline{\boldsymbol{B}}_1 \\ \vdots \\ \underline{\boldsymbol{B}}_{n_2-1} \end{bmatrix} = \begin{bmatrix} \Lambda_{n_1} & \mathbf{0} & \cdots & \mathbf{0} \\ \mathbf{0} & \Lambda_{n_1} & \mathbf{0} & \vdots \\ \vdots & \mathbf{0} & \ddots & \mathbf{0} \\ \mathbf{0} & \cdots & \mathbf{0} & \Lambda_{n_1} \end{bmatrix} \begin{bmatrix} \underline{A}_0 \\ \underline{A}_1 \\ \vdots \\ \underline{A}_{n_2-1} \end{bmatrix} \tag{14.4.5}$$

or

$$\underline{\boldsymbol{B}} = [\,\mathbf{I}_{n_2}\ \kappa\ \Lambda_{n_1}\,]\,\underline{A}. \tag{14.4.6}$$

Here, $\underline{\boldsymbol{B}}$ denotes the length $n_1 \times n_2$ vector obtained by arranging the columns of **B**, $\mathbf{I}_{n_2}$ denotes the identity matrix of dimension n_2, and κ denotes the Kronecker product of two matrices. $[\,\mathbf{I}_{n_2}\ \kappa\ \Lambda_{n_1}\,]$ is a block-diagonal matrix having Λ_{n_1} on the main diagonal and zeros elsewhere.

The i-th row of **B** corresponds to the i-th elements of each of $\underline{\boldsymbol{B}}_0$, $\underline{\boldsymbol{B}}_1$, …, $\underline{\boldsymbol{B}}_{n_2-1}$, $i = 0, 1, \ldots, n_1 - 1$. Therefore, the computation of **C**, that is, the computation of length n_2 DFT of each of the n_1 rows of **B** is equivalent to the matrix-vector product

$$\begin{bmatrix} \underline{C}_0 \\ \underline{C}_1 \\ \vdots \\ \underline{C}_{n_2-1} \end{bmatrix} = [\Lambda_{n_2} \kappa \mathbf{I}_{n_1}] \begin{bmatrix} \underline{\boldsymbol{B}}_0 \\ \underline{\boldsymbol{B}}_1 \\ \vdots \\ \underline{\boldsymbol{B}}_{n_2-1} \end{bmatrix} \tag{14.4.7}$$

or

$$\underline{C} = [\Lambda_{n_2} \kappa \mathbf{I}_{n_1}] \underline{\boldsymbol{B}}. \tag{14.4.8}$$

Combining (14.4.8) with (14.4.6), we have a compact representation for the two-dimensional FFT algorithm given by

$$\underline{C} = [\Lambda_{n_2} \kappa \mathbf{I}_{n_1}] [\mathbf{I}_{n_2} \kappa \Lambda_{n_1}] \underline{A}. \tag{14.4.9}$$

Recalling the result from matrix theory,

$$(\mathbf{A} \kappa \mathbf{B}) \cdot (\mathbf{C} \kappa \mathbf{D}) = (\mathbf{A} \cdot \mathbf{C}) \kappa (\mathbf{B} \cdot \mathbf{D}), \tag{14.4.10}$$

(14.4.9) simplifies to

$$\underline{C} = [\Lambda_{n_2} \kappa \Lambda_{n_1}] \underline{A}. \tag{14.4.11}$$

We may also arrange the two-dimensional arrays as a $(n_2 \times n_1)$ matrix. In that case, (14.4.11) takes the form

$$\underline{C} = [\Lambda_{n_1} \kappa \Lambda_{n_2}] \underline{A}. \tag{14.4.12}$$

In summary, the d-dimensional DFT can be written as the matrix-vector product

$$\underline{C} = [\Lambda_{n_1} \kappa \Lambda_{n_2} \kappa \cdots \kappa \Lambda_{n_d}] \underline{A}, \tag{14.4.13}$$

where the order in which the lengths n_1, n_2, …, n_d DFTs are arranged can be selected by the user. The expression in (14.4.13) introduces a number of degrees of freedom in the multidimensional FFTs, some of which are described below.

14.4.1 Prime factor Algorithm in a Multidimensional Setting

Consider the two-dimensional DFT in (14.4.12). If n_1 factors as $n_1 = a \cdot b$, $(a, b) = 1$, then we can use the prime factor approach in Section 13.6 to write (14.4.12) as

$$\underline{C} = [\,\Lambda_{n_1}\ \kappa\ \Lambda_{n_2}\,]\,\underline{A} = [\Lambda_a\ \kappa\ \Lambda_b\ \kappa\ \Lambda_{n_2}\,]\,\underline{A}^P. \tag{14.4.14}$$

Here A^P denotes a suitable arrangement of the two-dimensional data. Similarly, if n_2 also factors as $n_2 = e \cdot f$, $(e, f) = 1$, then a similar expression can be written converting the two-dimensional DFT to a four-dimensional DFT. Note that we do not require $(n_1, n_2) = 1$ here. Thus, a two-dimensional DFT of length 15×15 can be converted to a four-dimensional DFT of size $(3 \times 3) \times (5 \times 5)$ DFT and can be computed as such. The possibilities are numerous and one has the freedom to select an algorithm and organize the computations in the best way for a given computer system.

14.4.2 Nesting and Other Algorithms

Once again, consider the two-dimensional DFT in (14.4.12). As in WFTA, we can organize the one-dimensional DFT computation of length-n using the factorization

$$\Lambda = \mathbf{P}\,\mathbf{Q}\,\mathbf{R}. \tag{14.4.15}$$

Here, **P** and **R** are matrices having simple elements such as 0, ±1; they correspond to input and output ADDs required in the length-n FFT algorithm. The matrix **Q** is a diagonal matrix whose diagonal entries correspond to the MULT required in the length-s FFT algorithm. The dimension of **Q** is equal to the multiplicative complexity of the length-n FFT algorithm. The diagonal elements of **Q** are either purely real or imaginary.

Given the factorization for $n = n_1$ and $n = n_2$, we can express the two-dimensional DFT as

$$\underline{C} = [\,\Lambda_{n_1}\ \kappa\ \Lambda_{n_2}\,]\,\underline{A} = [\{\mathbf{P}_1\,\mathbf{Q}_1\,\mathbf{R}_1\}\ \kappa\ \{\mathbf{P}_2\,\mathbf{Q}_2\,\mathbf{R}_2\}]\,\underline{A}. \tag{14.4.16}$$

The matrix identity in (14.4.10) can be used in a variety of ways to give rise to different algorithms and computational organizations. Using it to write

$$\underline{C} = [\mathbf{P}_a\ \kappa\ \mathbf{P}_b] \cdot [\mathbf{Q}_a\ \kappa\ \mathbf{Q}_b] \cdot [\mathbf{R}_a\ \kappa\ \mathbf{R}_b]\,\underline{A} \tag{14.4.17}$$

gives rise to the two-dimensional FFT that is exactly same as the WFTA. There are other possibilities such as

$$\underline{C} = [\mathbf{P}_1\ \kappa\ \mathbf{P}_2] \cdot [(\mathbf{Q}_1 \cdot \mathbf{R}_1)\ \kappa\ (\mathbf{Q}_2 \cdot \mathbf{R}_2)]\,\underline{A} \tag{14.4.18}$$

$$= [(\mathbf{P}_1 \cdot \mathbf{Q}_1)\ \kappa\ (\mathbf{P}_2 \cdot \mathbf{Q}_2)] \cdot [\mathbf{R}_1\ \kappa\ \mathbf{R}_2]\,\underline{A}. \tag{14.4.19}$$

The number of different ways in which the computations can be organized increases even further if any of the pre- and/or post-ADD matrices factor further. For example, if

$$\mathbf{P}_1 = \mathbf{S} \cdot \mathbf{W}, \tag{14.4.20}$$

then one can further exploit it in the FFT algorithm. Such factorizations may appear naturally in the way the pre- and/or post-ADDs are computed.

We now turn to the final aspect of the multidimensional FFT algorithms, namely, the polynomial transform based FFT algorithms. We begin the design of these algorithms by first examining the Rader's algorithm in polynomial form for $n = p$, and $n = p^c$, p an odd prime.

14.5 Polynomial Version of Rader's Algorithm

Consider $n = p$ first. In this case, we can write

$$A(u) = A(u) \bmod (u^p - 1) \tag{14.5.1}$$

as a mechanism to compute DFT. Using CRT-P and cyclotomic factorization

$$(u^p - 1) = (u - 1)\,(u^{p-1} + u^{p-2} + \ldots + u + 1), \tag{14.5.2}$$

we can write $A(u)$ as a direct sum of

$$B_1(u) = A(u) \bmod (u - 1) = A_0 + A_1 + \ldots + A_{p-1} \tag{14.5.3}$$

and

$$\begin{aligned} B_2(u) &= A(u) \bmod (u^{p-1} + u^{p-2} + \ldots + u + 1) \\ &= (A_0 - A_{p-1}) + (A_1 - A_{p-1})u + (A_2 - A_{p-1})u^2 + \ldots + (A_{p-2} - A_{p-1})u^{p-2} \\ &= B_0 + B_1\, u + \ldots + B_{p-2}\, u^{p-2}. \end{aligned} \tag{14.5.4}$$

Noting that (i) evaluating the polynomial at ω^i is equivalent to substituting $u = \omega^i$ and (ii) ω^i, $i = 1, 2, \ldots, p - 1$ are roots of $(u^{p-1} + u^{p-2} + \ldots + u + 1)$, we get

$$A_0 = B_1(1)$$

and

$$A_i = B_2(\omega^i), \quad i = 1, 2, \ldots, p - 1. \tag{14.5.5}$$

Thus, we see that the primary computation in a two-dimensional DFT of length p is the computation

$$A_i = B_0 + \sum_{l=1}^{p-2} B_l \omega^{li}, \quad i = 1, \ldots, p-1. \tag{14.5.6}$$

Using the approach of Section 13.2, the term in the summation sign can once again be converted to a cyclic convolution of length $p - 1$, where the last input point $B_{p-1} = 0$. This is same as the *reduced DFT*. Some simplifications may, therefore, be possible in the algorithm for cyclic convolution in this case.

Consider the case $n = p^c$. Now, the cyclotomic factorization is given by

$$u^{p^c} - 1 = \left(u^{p^{c-1}} - 1\right) C_{p^c}(u), \tag{14.5.7}$$

where

$$C_{p^c}(u) = \sum_{k=0}^{p-1} u^{kp^{c-1}}. \tag{14.4.8}$$

Since ω is the p^c-th root of unity, all the roots ω^i can be partitioned into two parts: (A) i such that $(i, p) = 1$, thereby making ω^i also a p^c-th root of unity, and (B) i such that $(i, p) > 1$. There are $p^{c-1}(p - 1)$ roots in Part (A) and p^{c-1} roots in Part (B).

Using CRT-P, we can write $A(u)$ as a direct sum of

$$B_1(u) = A(u) \bmod \left(u^{p^{c-1}} - 1\right) \tag{14.5.9}$$

and

$$B_2(u) = A(u) \bmod C_{p^c}(u). \tag{14.5.10}$$

Each of $B_1(u)$ and $B_2(u)$ is computed in $p^{c-1}(p - 1)$ ADDs. Based on (14.5.9) and (14.5.10), we can express the length p^c DFT as equivalent to

A. Length p^{c-1} DFT as in (14.5.9), and
B. A reduced DFT given by

$$A_i = B_2\left(\omega^i\right) = B_{2,0} + \sum_{l=1}^{p^c - p^{c-1} - 1} B_{2,l}\,\omega^{li}\ , \ (i, p) = 1. \tag{14.5.11}$$

This is same as the *reduced DFT*, where the last p^{c-1} input points are zero and only those DFT coefficients are computed for which $(i, p) = 1$.

With the polynomial formulation of Rader's algorithm, we now present the polynomial transform based FFT algorithms.

14.6 Polynomial Transform Based FFT Algorithms

We present these algorithms in three categories:

Category 1. $n_1 = n_2 = \ldots = n_d = n = p$, p an odd prime,

Category 2. $n = p^c$, p an odd prime, and

Category 3. $n = 2^c$.

In Categories 2 and 3, n is any one of $n_1, n_2, \ldots, n_d$, and the lengths are not necessarily equal. The polynomial transform based FFT algorithms have their origins in the cyclotomic factorization. Since the form of cyclotomic factorization is different for different values of n, we have divided the analysis to three categories as above.

14.6.1 $n_1 = n_2 = \ldots = n_d = n = p$, p an Odd Prime

Let $d = 2$. Express the two-dimensional DFT as

$$A(u, v) = A(u, v) \bmod (u^p - 1, v^p - 1). \tag{14.6.1}$$

The (i_1, i_2)-th Fourier coefficient is obtained by setting $u = \omega^{i_1}$ and $v = \omega^{i_2}$, where ω is the p-th root of unity. Now use the cyclotomic factorization

$$v^p - 1 = (v - 1)\,(v^{p-1} + v^{p-2} + \ldots + v + 1) \tag{14.6.2}$$

in $\mathbf{Z}$ to express $A(u, v)$ as a direct sum of

$$A_1(u, v) = A(u, v) \bmod (u^p - 1, v - 1) = A(u, 1) \bmod (u^p - 1) \tag{14.6.3}$$

and

$$A_2(u, v) = A(u, v) \bmod (u^p - 1, v^{p-1} + v^{p-2} + \ldots + v + 1). \tag{14.6.4}$$

Computation of $A_1(u, v)$ requires $p(p - 1)$ ADDs. The key to the polynomial transform based FFTs is in further breaking down the computation in (14.6.4) by using the cyclotomic factorization of $u^p - 1$ in $CF(p)$ as

$$u^p - 1 = \prod_{k=0}^{p-1} \left(u - v^k\right). \tag{14.6.5}$$

Using CRT-P, $A_2(u, v)$ can be written as a direct sum of p one-dimensional polynomials

$$B_k(v) = A_2(v^k, v) \bmod (v^{p-1} + v^{p-2} + \ldots + v + 1)$$

$$= A(v^k, v) \bmod (v^{p-1} + v^{p-2} + \ldots + v + 1), \;\; k = 0, 1, \ldots, p - 1. \tag{14.6.6}$$

The computation of two-dimensional DFT has been broken into $(p + 1)$ *one-dimensional* parts given by $A_1(u, 1) = A(u, 1)$ as given in (14.6.3), and $B_k(v)$, $k = 0, 1, \ldots, p - 1$ as given in (14.6.6).

The computation of the DFT proceeds as follows. $A_1(u, 1)$ corresponds to $A(u, \omega^0)$. Setting $u = \omega^i$, we get

$$\boldsymbol{A}_{i,0} = A(\omega^i, \omega^0) = A_1(\omega^i, \omega^0), \;\; i = 0, 1, \ldots, p - 1. \tag{14.6.7}$$

Thus, the p Fourier coefficients $\boldsymbol{A}_{i,0}$ are obtained as a one-dimensional DFT from $A_1(u, 1)$. $B_k(v)$ corresponds to $A(v^k, v) \bmod (v^{p-1} + v^{p-2} + \ldots + v + 1)$. Setting $v = \omega^i$, $i = 1, 2, \ldots, p - 1$, we get,

$$A(\omega^{i \cdot k}, \omega^i) = B_k(\omega^i), \;\; k = 0, 1, \ldots, p - 1, \;\; i = 1, 2, \ldots, p - 1. \tag{14.6.8}$$

This expression is based on the fact that $(v - \omega^i)$, $i = 1, 2, \ldots, p - 1$ is a factor of $(v^{p-1} + v^{p-2} + \ldots + v + 1)$. Therefore,

$$\boldsymbol{A}_{(i \cdot k) \bmod p,\, i} = B_k(\omega^i), \;\; k = 0, 1, \ldots, p - 1, \;\; i = 1, 2, \ldots, p - 1. \tag{14.6.9}$$

Note that the first index is defined mod p, as the powers of ω are defined mod p. For each value of k, computation of $B_k(\omega^i)$, $i = 1, 2, \ldots, p - 1$ is a computation of the type

$$\sum_{l=0}^{p-2} B_l \omega^{li}, \quad i = 1, 2, \ldots, p-1. \tag{14.6.10}$$

This is recognized as an one-dimensional *reduced DFT*, where the first term is not computed and the last input is set equal to 0. We have a total of p reduced DFTs to evaluate. Each of them can be computed using Rader's algorithm as described in previous section.

We now show that the expressions in (14.6.7) and (14.6.9) give all the Fourier coefficients $\boldsymbol{A}_{i_1,i_2}$, $i_1 = 0, 1, \ldots, p-1$, $i_2 = 0, 1, \ldots, p-1$. (14.6.7) gives $\boldsymbol{A}_{i,0}$, $i = 0, 1, \ldots, p-1$. Therefore, we need to show that $\boldsymbol{A}_{i_1,i_2}$, $i_1 = 0, 1, \ldots, p-1$, $i_2 = 1, 2, \ldots, p-1$ can be obtained from $\boldsymbol{A}_{(i \cdot k) \bmod p,\, i}$, $k = 0, 1, \ldots, p-1$, $i = 1, 2, \ldots, p-1$. In other words, we need to show that for any $0 < i < p$, $(i \cdot k \bmod p)$ takes the values $0, 1, \ldots, p-1$, as k goes from $0, 1, \ldots, p-1$. Since p is prime, we have $(i, p) = 1$ and, therefore, $(i \cdot k \bmod p)$ defines a permutation on the values of $k \in \{0, 1, \ldots, p-1\}$. In other words, as $(i, p) = 1$, $k \in \{0, 1, \ldots, p-1\}$ also implies that $(i \cdot k \bmod p) \in \{0, 1, \ldots, p-1\}$.

Example 14.2 Consider $p = 5$. The polynomial transform breaks the computation of 25 Fourier coefficients into 2 parts. The Fourier coefficients $\boldsymbol{A}_{i,0}$ come from the one-dimensional DFT of $A_1(u, 1)$, $i = 0, 1, 2, 3, 4$. The remaining 20 Fourier coefficients come from the reduced one-dimensional DFTs of $B_k(v)$, $k = 0, 1, \ldots, 4$. It is seen that the reduced DFT of

(i) $B_0(v)$ gives $\boldsymbol{A}_{0,1}$, $\boldsymbol{A}_{0,2}$, $\boldsymbol{A}_{0,3}$ and $\boldsymbol{A}_{0,4}$;
(ii) $B_1(v)$ gives $\boldsymbol{A}_{1,1}$, $\boldsymbol{A}_{2,2}$, $\boldsymbol{A}_{3,3}$ and $\boldsymbol{A}_{4,4}$;
(iii) $B_2(v)$ gives $\boldsymbol{A}_{2,1}$, $\boldsymbol{A}_{4,2}$, $\boldsymbol{A}_{1,3}$ and $\boldsymbol{A}_{3,4}$;
(iv) $B_3(v)$ gives $\boldsymbol{A}_{3,1}$, $\boldsymbol{A}_{1,2}$, $\boldsymbol{A}_{4,3}$ and $\boldsymbol{A}_{2,4}$; and
(v) $B_4(v)$ gives $\boldsymbol{A}_{4,1}$, $\boldsymbol{A}_{3,2}$, $\boldsymbol{A}_{2,3}$ and $\boldsymbol{A}_{1,4}$.

The final aspect of this method is the computational complexity of obtaining $B_k(v)$, $k = 0, 1, \ldots, p-1$ in (14.6.6). We perform such a computation as

$$B_k(v) = [A(v^k, v) \bmod (v^p - 1)] \bmod (v^{p-1} + v^{p-2} + \ldots + v + 1), \quad k = 0, 1, \ldots, p-1, \tag{14.6.11}$$

to get p polynomials in the indeterminate v having degree $p-2$. Let us first consider the computation of

$$B_k(v) = [A(v^k, v) \bmod (v^p - 1)]. \tag{14.6.12}$$

The reduction modulo $(v^{p-1} + v^{p-2} + \ldots + v + 1)$ can be performed on each of $B_k(v)$ at a cost of $p - 1$ ADDs. Let

$$A(u,v) = \sum_{l_1=0}^{p-1} \sum_{l_2=0}^{p-1} A_{l_1,l_2} u^{l_1} v^{l_2} . \tag{14.6.13}$$

Setting $u = v^k$, we get

$$B_k(v) = \sum_{l_1=0}^{p-1} \sum_{l_2=0}^{p-1} A_{l_1,l_2} v^{kl_1+l_2} . \tag{14.6.14}$$

For $k = 0$, we get

$$B_0(v) = \sum_{l_2=0}^{p-1} \left[\sum_{l_1=0}^{p-1} A_{l_1,l_2} \right] v^{l_2} . \tag{14.6.15}$$

This is computed in $p(p - 1)$ ADDs. For $k \neq 0$, $k \cdot l_1 \bmod p$ defines a permutation on l_1. Therefore, we can set $k \cdot l_1 \bmod p = j_1$, where j_1 takes values in the set $\{0, 1, \ldots, p - 1\}$ as well. Of course, this permutation depends on k. Substituting in (14.6.14), we get

$$B_k(v) = \sum_{j_1=0}^{p-1} \sum_{l_2=0}^{p-1} A_{(k^{-1} \cdot j_1) \bmod p, l_2} v^{j_1+l_2} . \tag{14.6.16}$$

This represents a circular convolution as the sum $j_1 + l_2$ is defined mod p. There are exactly p pairs of values (j_1, l_2) for which $j_1 + l_2 = l$, $l = 0, 1, \ldots, p - 1$. Therefore, each coefficient of v^l in (14.6.16) is computed in $p - 1$ ADDs. Consequently, $p(p - 1)$ ADDs are required for each of $B_k(u)$, $k = 1, 2, \ldots, p - 1$. Overall, the computation of $B_k(u)$, $k = 0, 1, \ldots, p - 1$ in (14.6.12) requires $p\,[p(p - 1)]$ ADDs. Finally, reducing each of them modulo $(v^{p-1} + v^{p-2} + \ldots + v + 1)$ takes $(p - 1)$ ADDs. This brings the total cost of computing $B_k(u)$, $k = 0, 1, \ldots, p - 1$, as defined in (14.6.6) for the polynomial transforms to $p^2(p - 1) + p(p - 1)$. A further $p(p - 1)$ ADDs are required to compute $A_1(u, v)$ in (14.6.3). Thus, the total cost of two-dimensional polynomial transform of size $p \times p$ for computing two-dimensional DFT is $p^3 + p^2 - 2p$ ADDs.

This method involves computing $(p + 1)$ one-dimensional DFTs. This is the worst-case scenario as p of the DFTs are reduced DFTs. Therefore, it requires at most $(p + 1) \cdot M(p)$ MULTs and $(p + 1) \cdot A(p)$ ADDs beyond the polynomial transform computation. This is to be compared to $2p \cdot M(p)$ MULTs and $2p \cdot A(p)$ ADDs as required by the row-column method for the two-dimensional DFT.

The extension of the method to d-dimensional FFT where $n_1 = n_2 = \ldots = n_d = p$ is quite straightforward. Once again, we use the cyclotomic factorization in Z along one of the dimensions and in $\mathsf{CF}(p)$ along all other dimensions. This results in a $(d - 1)$-dimensional FFT (which is also computed using the polynomial transforms) and p^{d-1} one-dimensional reduced DFTs. The overall cost of this polynomial transform computation is $p^{2d-1} + p^d - 2p^{d-1}$ ADDs. The multiplicative complexity of this method is $[(p^d - 1) / (p - 1)] \cdot M(p)$ in the worst-case scenario. The additive complexity is

$$[(p^d - 1)/(p - 1)] \cdot A(p) + [p^{2d+1} + p^{d+2} - p^{d+1} - 2p^d]/(p^2 - 1).$$

These are to be compared to $d \cdot p^{d-1} \cdot M(p)$ MULTs and $d \cdot p^{d-1} \cdot A(p)$ ADDs as required by the row-column method for the d-dimensional DFT.

14.6.2 $n = p^c$, p an Odd Prime

In the above, n denotes any one of $n_1, n_2, \ldots, n_d$. Since, much of the analysis is similar to the previous case, we limit our discussion to only some salient features. Let $d = 2$, $n_1 = p^c$, $n_2 = p^b$, $b \geq c$. Express the two-dimensional DFT as

$$A(u, v) = A(u, v) \bmod \left(u^{n_1} - 1, v^{n_2} - 1\right). \tag{14.6.17}$$

The (i_1, i_2)-th Fourier coefficient is obtained by setting $u = \omega_1{}^{i_1}$ and $v = \omega_2{}^{i_2}$, where ω_1 and ω_2 are the p^c-th and p^b-th roots of unity, respectively. Now use the cyclotomic factorization in Z given by

$$v^{p^b} - 1 = \left(v^{p^{b-1}} - 1\right)C_{p^b}(v). \tag{14.6.18}$$

Here, the cyclotomic polynomial $C_{p^b}(v)$ is a polynomial of degree $p^b - p^{b-1}$. The CRT-P can be used to express $A(u, v)$ as a direct sum of

$$A_1(u, v) = A(u, v) \bmod \left(u^{p^c} - 1, v^{p^{b-1}} - 1\right) \tag{14.6.19}$$

and

$$A_2(u, v) = A(u, v) \bmod \left(u^{p^c} - 1, C_{p^b}(v)\right). \tag{14.6.20}$$

The key to the polynomial transform based FFTs is in further breaking down the computation in (14.6.20) by using the cyclotomic factorization of $\left(u^{p^c} - 1\right)$ in $CF(p^b)$ as

$$u^{p^c} - 1 = \prod_{k=0}^{p^c-1}\left(u - \left(v^{p^{b-c}}\right)^k\right). \tag{14.6.21}$$

Thus, $A_2(u, v)$ is a direct sum of p^c one-dimensional polynomials

$$\begin{aligned} B_k(v) &= A_2\left(v^{p^{b-c}k}, v\right) \bmod C_{p^b}(v) \\ &= A\left(v^{p^{b-c}k}, v\right) \bmod C_{p^b}(v), \quad k = 0, 1, \ldots, p^c - 1. \end{aligned} \tag{14.6.22}$$

The computation of $p^c \times p^b$ two-dimensional DFT has been broken into a $p^c \times p^{b-1}$ two-dimensional DFT in (14.6.19) and p^c *one-dimensional* parts given in (14.6.22). This process is further continued until $p^c \times p^c$ two-dimensional DFT is obtained along with the corresponding one-dimensional parts. Beyond that, one proceeds by converting it to a $p^c \times p^{c-1}$ two-dimensional DFT which is then converted to a $p^{c-1} \times p^{c-1}$ two-dimensional DFT, and so on.

The computation of the DFT proceeds as follows. $A_1(u, v)$ corresponds to $p^c \times p^{b-1}$ two-dimensional DFT and is computed in a similar manner. Setting $v = \omega_2^i$, $i < p^b$, $(i, p) = 1$, we get

$$A\left(\omega_1^{ik}, \omega_2^i\right) = B_k\left(\omega_2^i\right), \quad k = 0, 1, \ldots, p^c - 1, (i, p) = 1. \tag{14.6.23}$$

This expression is based on the fact that $(v - \omega_2^i)$, $(i, p) = 1$ is a factor of $C_{p^b}(v)$, and $\omega_1 = \omega_2^s$, $s = p^{c-b}$. Therefore,

$$A_{(i \cdot k) \bmod p^c, i} = B_k(\omega_2^i), \quad k = 0, 1, \ldots, p^c - 1, (i, p) = 1. \tag{14.6.24}$$

Note that the first index is defined mod p^c, as the powers of ω_1 are defined mod p^c. For each value of k, computation of $B_k(\omega_2^i)$, $(i, p) = 1$ is a computation of the type

$$\sum_{l=0}^{p^b - p^{b-1} - 1} B_l \omega_2^{\ li}, \ (i, p) = 1. \tag{14.6.25}$$

This is recognized as the computation of one-dimensional *reduced DFT* of length p^b, where only $p^b - p^{b-1}$ terms corresponding to $(i, p) = 1$ are computed and the last p^{b-1} inputs are set equal to 0. Thus, we have a total of p^c reduced DFTs to evaluate. Each of them can be computed using Rader's algorithm as described in the previous section or any other suitable method.

We now show that the expressions in (14.6.19) and (14.6.24) give all the Fourier coefficients $\boldsymbol{A}_{i_1, i_2}$, $i_1 = 0, 1, \ldots, p^c - 1$, $i_2 = 0, 1, \ldots, p^b - 1$. (14.6.19) gives $\boldsymbol{A}_{i,j}$, $i = 0, 1, \ldots, p^c - 1, j = 0, p, \ldots, p^b - p$. Therefore, we need to show that $\boldsymbol{A}_{i_1, i_2}$, $i_1 = 0, 1, \ldots, p^c - 1$, $(i_2, p) = 1$ can be obtained from $\boldsymbol{A}_{(i \cdot k) \bmod p^c, i}$, $k = 0, 1, \ldots, p^c - 1$, $(i, p) = 1$. In other words, we need to show that for any i such that $(i, p) = 1$, $(i \cdot k \bmod p^c)$ takes values $0, 1, \ldots, p^c - 1$, as k goes from $0, 1, \ldots, p^c - 1$. Since $(i, p) = 1$, $(i \cdot k \bmod p^c)$ defines a permutation on the values $\{0, 1, \ldots, p^c - 1\}$. In other words, as $(i, p) = 1$, $k \in \{0, 1, \ldots, p^c - 1\}$ also implies that $(i \cdot k \bmod p^c) \in \{0, 1, \ldots, p^c - 1\}$.

Example 14.3 Consider $p = 3$, $n_1 = n_2 = 9$; $b = c = 2$. The polynomial transform breaks the computation of 81 Fourier coefficients into 2 parts. The Fourier coefficients $\boldsymbol{A}_{i,j}$, $i = 0, 1, \ldots, 8, j = 0, 3, 6$, come from the 9×3 two-dimensional DFT of $A_1(u, v)$. The remaining 54 Fourier coefficients come from the reduced one-dimensional DFTs of $B_k(v)$, $k = 0, 1, \ldots, 8$. It is seen that the reduced DFT of

(i) $B_0(v)$ gives $\boldsymbol{A}_{0,1}, \boldsymbol{A}_{0,2}, \boldsymbol{A}_{0,4}, \boldsymbol{A}_{0,5}, \boldsymbol{A}_{0,7}$ and $\boldsymbol{A}_{0,8}$;
(ii) $B_1(v)$ gives $\boldsymbol{A}_{1,1}, \boldsymbol{A}_{2,2}, \boldsymbol{A}_{4,4}, \boldsymbol{A}_{5,5}, \boldsymbol{A}_{7,7}$ and $\boldsymbol{A}_{8,8}$;
(iii) $B_2(v)$ gives $\boldsymbol{A}_{2,1}, \boldsymbol{A}_{4,2}, \boldsymbol{A}_{8,4}, \boldsymbol{A}_{1,5}, \boldsymbol{A}_{5,7}$ and $\boldsymbol{A}_{7,8}$;
(iv) $B_3(v)$ gives $\boldsymbol{A}_{3,1}, \boldsymbol{A}_{6,2}, \boldsymbol{A}_{3,4}, \boldsymbol{A}_{6,5}, \boldsymbol{A}_{3,7}$ and $\boldsymbol{A}_{6,8}$;
(v) $B_4(v)$ gives $\boldsymbol{A}_{4,1}, \boldsymbol{A}_{8,2}, \boldsymbol{A}_{7,4}, \boldsymbol{A}_{2,5}, \boldsymbol{A}_{1,7}$ and $\boldsymbol{A}_{5,8}$;
(vi) $B_5(v)$ gives $\boldsymbol{A}_{5,1}, \boldsymbol{A}_{1,2}, \boldsymbol{A}_{2,4}, \boldsymbol{A}_{7,5}, \boldsymbol{A}_{8,7}$ and $\boldsymbol{A}_{4,8}$;
(vii) $B_6(v)$ gives $\boldsymbol{A}_{6,1}, \boldsymbol{A}_{3,2}, \boldsymbol{A}_{6,4}, \boldsymbol{A}_{3,5}, \boldsymbol{A}_{6,7}$ and $\boldsymbol{A}_{3,8}$;
(viii) $B_7(v)$ gives $\boldsymbol{A}_{7,1}, \boldsymbol{A}_{5,2}, \boldsymbol{A}_{1,4}, \boldsymbol{A}_{8,5}, \boldsymbol{A}_{4,7}$ and $\boldsymbol{A}_{2,8}$; and
(ix) $B_8(v)$ gives $\boldsymbol{A}_{8,1}, \boldsymbol{A}_{7,2}, \boldsymbol{A}_{5,4}, \boldsymbol{A}_{4,5}, \boldsymbol{A}_{2,7}$ and $\boldsymbol{A}_{1,8}$.

The computational complexity analysis of the polynomial transform based FFT algorithm for $n_1 = p^c$, $n_2 = p^b$, is being left as an exercise. Finally, the extension of the above analysis to d-dimensional situation, $d > 2$, is straightforward.

14.6.3 Polynomial Transform Based FFT Algorithms for $n = 2^c$

In the above, n denotes any one of $n_1, n_2, \ldots, n_d$. Let $d = 2$, $n_1 = 2^c$, $n_2 = 2^b$, $b \geq c$. Express the two-dimensional DFT as

$$A(u, v) = A(u, v) \bmod \left(u^{n_1} - 1, v^{n_2} - 1\right). \tag{14.6.26}$$

The (i_1, i_2)-th Fourier coefficient is obtained by setting $u = \omega_1^{i_1}$ and $v = \omega_2^{i_2}$, where ω_1 and ω_2 are the 2^c-th and 2^b-th roots of unity, respectively. Now we use the cyclotomic factorization in $\mathbf{Z}$ given by

$$v^n - 1\ (v^{n/2} - 1)\ (v^{n/2} + 1), \tag{14.6.27}$$

for n a power of 2. Here, the $v^{n/2} + 1$ is a cyclotomic polynomial having all the odd powers of ω as its roots, ω being the n-th root of unity. Thus, we express $A(u, v)$ as a direct sum of

$$A_1(u, v) = A(u, v) \bmod \left(u^{2^c} - 1, v^{2^{b-1}} - 1\right) \tag{14.6.28}$$

and

$$A_2(u, v) = A(u, v) \bmod \left(u^{2^c} - 1, v^{2^{b-1}} + 1\right). \tag{14.6.29}$$

Express the computation in (14.6.29) by using the cyclotomic factorization of $\left(u^{2^c} - 1\right)$ in $CF(2^b)$ as

$$u^{2^c} - 1 = \prod_{k=0}^{2^c-1} \left(u - \left(v^{2^{b-c}}\right)^k\right). \tag{14.6.30}$$

Thus, $A_2(u, v)$ is a direct sum of 2^c one-dimensional polynomials

$$B_k(v) = A_2\left(v^{2^{b-c}k}, v\right) \bmod \left(v^{2^{b-1}} + 1\right)$$

$$= A\left(v^{2^{b-c}k}, v\right) \bmod \left(v^{2^{b-1}} + 1\right),\ k = 0, 1, \ldots, 2^c - 1. \tag{14.6.31}$$

The computation of $2^c \times 2^b$ two-dimensional DFT has been broken into a $2^c \times 2^{b-1}$ two-dimensional DFT in (14.6.28) and $2^c - 1$ *one-dimensional* parts given in (14.6.31). This process is further continued till $2^c \times 2^c$ two-dimensional DFT is obtained along with the corresponding one-dimensional parts. Beyond that, one proceeds by converting it to a $2^c \times 2^{c-1}$ two-dimensional DFT which then is converted to a $2^{c-1} \times 2^{c-1}$ two-dimensional DFT and so on.

The computation of the DFT proceeds as follows. $A_1(u, v)$ corresponds to $2^c \times 2^{b-1}$ two-dimensional DFT and is computed in a similar manner. Setting $v = \omega_2^i$, $i < 2^b$, i odd, we get

$$A\left(\omega_1^{ik}, \omega_2^{i}\right) = B_k\left(\omega_2^{i}\right), \; k = 0, 1, \ldots, 2^c - 1, \; i: \text{odd}. \tag{14.6.32}$$

This expression is based on the fact that $(v - \omega_2^i)$, i odd is a factor of $\left(v^{2^{b-1}} + 1\right)$, and $\omega_1 = \omega_2^s$, $s = 2^{c-b}$. Therefore,

$$A_{(i \cdot k) \bmod 2^c, i} = B_k(\omega_2^i), \; k = 0, 1, \ldots, 2^c - 1, \; i: \text{odd}. \tag{14.6.33}$$

Note that the first index is defined mod 2^c, as the powers of ω_1 are defined mod 2^c. For each value of k, computation of $B_k(\omega_2^i)$, i odd is a computation of the type

$$\sum_{l=0}^{2^{b-1}-1} B_l \omega_2^{li} = \sum_{l=0}^{2^{b-1}-1} B_l \omega_2^{l(2e+1)} = \sum_{l=0}^{2^{b-1}-1} \left(B_l \omega_2^{l}\right)\left(\omega_2^{2}\right)^{le},$$
$$e = 0, 1, \ldots, 2^{b-1} - 1. \tag{14.6.34}$$

This is recognized as the computation of length 2^{b-1} one-dimensional DFT of the sequence $B_l\omega_2^l$ and is computed as such. Thus, we have a total of 2^c DFTs of length 2^{b-1} to evaluate. Each of them can be computed using any of the one-dimensional FFT algorithms for length n a power of 2. Based on this analysis, the multiplicative complexity of computing $A_{(i \cdot k) \bmod 2^c, i}$, $k = 0, 1, \ldots, 2^c - 1$, i odd is given by $2^c \cdot [2^{b-1} + M(2^{b-1})]$. Using the same arguments in a recursive manner, we have

$$\begin{aligned} M(2^c \times 2^b) &= 2^c[2^{b-1} + 2^{b-2} + \ldots + 2^c] \\ &\quad + 2^c[M(2^{b-1}) + M(2^{b-2}) + \ldots + M(2^c)] + M(2^c \times 2^c) \\ &= 2^c[2^b - 2^c] + 2^c[M(2^{b-1}) + M(2^{b-2}) + \ldots + M(2^c)] + M(2^c \times 2^c). \end{aligned} \tag{14.6.35}$$

In the following, we study the case $n_1 = n_2 = 2^c$ separately.

Algorithms for $n = n_1 = n_2 = 2^c$

Since the above analysis is valid for $b = c$, we have the result that $A(u, v)$ is a direct sum of

$$A_1(u, v) \bmod (u^n - 1, v^{n/2} - 1) \tag{14.6.36}$$

and

$$A_2(u, v) \bmod (u^n - 1, v^{n/2} + 1). \tag{14.6.37}$$

The Fourier coefficients $A_{k,i}$ corresponding to $k = 0, 1, \ldots, 2^c - 1$, i odd are computed from $A_2(u, v)$ via 2^c DFTs of length 2^{c-1} in a manner given by (14.6.34). Let us analyze (14.6.36) now. We express it as

$$A_1(v, u) \bmod (v^{n/2} - 1, u^n - 1), \tag{14.6.38}$$

and split it as a direct sum of

$$A_3(v, u) \bmod (v^{n/2} - 1, u^{n/2} - 1) \tag{14.6.39}$$

and

$$A_4(v, u) \bmod (v^{n/2} - 1, u^{n/2} + 1). \tag{14.6.40}$$

We further use the cyclotomic factorization of $v^{n/2} - 1$ in $\mathsf{CF}(n)$ given by

$$v^{n/2} - 1 = \prod_{k=0}^{n/2-1} \left(v - \left(u^2\right)^k \right) \bmod (u^{n/2} + 1) \tag{14.6.41}$$

to express $A_4(v, u)$ as a direct sum of

$$C_k(u) = A_4(u^{2k}, u), \;\; k = 0, 1, \ldots, n/2 - 1. \tag{14.6.42}$$

Appropriate Fourier coefficients are extracted from $C_k(u)$, $k = 0, 1, \ldots, n/2 - 1$ via $n/2$ DFTs of the same kind as in (14.6.34). The multiplicative complexity of the method can be evaluated by using the recursive formula,

$$M(2^c \times 2^c) = 3 \cdot 4^{c-1} + 3 \cdot 2^{c-1} \cdot M(2^{c-1}) + M(2^{c-1} \times 2^{c-1}). \tag{14.6.43}$$

This leads to

$$M(2^c \times 2^c) \approx 2 \cdot c \cdot 4^{c-1} - (7/6)\, 4^c. \tag{14.6.44}$$

Substituting in (14.6.35), we get

$$M(2^c \times 2^b) = 0.5 \cdot n_1 n_2 \cdot \log_2 n_2 + 3.5 \cdot n_1(n_2 - n_1) - (7/6) \cdot n_1^2. \tag{14.6.45}$$

Here, we have used (13.6.18) for one-dimensional computations. This is to be compared to the row-column method that has a multiplicative complexity given by

$$M(2^c \times 2^b) = 0.5 \cdot n_1 n_2 \cdot (\log_2 n_2 - 3) + 0.5 \cdot n_1 n_2 \cdot (\log_2 n_1 - 3). \tag{14.6.46}$$

The additive complexity analysis of the polynomial transform based FFT algorithm for $n_1 = 2^c$, $n_2 = 2^b$ is being left as an exercise. Finally, the extension of the above analysis to d-dimensional situation, $d > 2$, is straightforward.

We end this chapter by stating that the polynomial transforms required in the multidimensional FFT computation when $n = 2^c$ are all of length 2 and radix-2 and/or radix-4 FFT algorithms can be used for their computation. This is a very interesting result that impacts the computational complexity of the algorithms as well as their structure and organization.

Notes

The story of the FFTs is perhaps the most interesting one of all. We convert one-dimensional DFT into multidimensional DFT and multidimensional DFT into a number of one-dimensional DFT to perform the computations in an efficient manner. Once again, the number-theoretic properties play a very important role in the construction of FFT algorithms.

The matrix description of FFTs brings about a unified representation for the various FFTs. Also, it helps one to derive several related forms and their computational organization on a given computer system. In this regard, the book by Toliemieri, An, and Lu [1989] is an important one. Polynomial transform based FFTs were first described by Nussbaumer and his associates, and are included in Nussbaumer [1981]. Among other books that have influenced this chapter are McClellan and Rader [1979], Oppenheim and Schafer [1989], Blahut [1984], Myers [1990], and Briggs and Henson [1995].

We have chosen to restrict our attention to multidimensional FFTs in the field of complex numbers. In-place multidimensional FFTs are described by Pitas and Strintzis [1986] for real-valued sequences. Some other multi-dimensional FFT algorithms can also be found in the works of Auslander, Feig, and Winograd [1983], and Gertner [1988]. The computational complexity of the FFT algorithms in these papers is quite close to those of polynomial transform based FFT algorithms. Traditionally, it was assumed that MULTs were always more time consuming than ADDs and therefore a great deal of effort was spent on designing algorithms with a view to reducing the multiplicative complexity. Polynomial transforms based FFT algorithms fall into this class as they trade off ADDs in order to reduce MULTs. They also require a more sophisticated computing mechanism than an in-place algorithm such as the row-column method.

Finally, multidimensional FFTs in finite fields and rings exist and can be described in a similar manner.

Bibliography

[14.1] H.J. Nussbaumer, *Fast Fourier Transform and Convolution Algorithms*, Springer-Verlag, 1981.

[14.2] D.G. Myers, *Digital Signal Processing: Efficient Convolution and Fourier Transform Techniques*, Prentice Hall, 1990.

[14.3] J.H. McClellan and C.M. Rader, *Number Theory in Digital Signal Processing*, Prentice Hall, 1979.

[14.4] R.E. Blahut, *Fast Algorithms for Digital Signal Processing*, Addison-Wesley Publishing Co., 1984.

[14.5] M.T. Heideman, *Multiplicative Complexity, Convolution, and the DFT*, Springer-Verlag, 1988.

[14.6] R. Tolimieri, M. An, and C. Lu, *Algorithms for Discrete Fourier Transform and Convolution*, Springer-Verlag, 1989.

[14.7] A.V. Oppenheim and R.W. Schafer, *Discrete-Time Signal Processing*, Prentice Hall, 1989.

[14.8] W.L. Briggs and V.E. Henson, *The DFT: An Owner's Manual for the Discrete Fourier Transforms*, Society for Industrial and Applied Mathematics, Philadelphia, 1995.

[14.9] S. Winograd, "On Computing the Discrete Fourier Transform," *Mathematics of Computation*, Vol. 32, pp. 175-199, Jan. 1978.

[14.10] I. Pitas and M.G. Strintzis, "General In-place Calculation of Discrete Fourier Transforms of Multidimensional Sequences," *IEEE Transactions on Acoustics, Speech, and Signal Processing*, Vol. 34, pp. 565-572, 1986.

[14.11] L. Auslander, E. Feig, and S. Winograd, "New Algorithms for the Multidimensional Discrete Fourier Transform," *IEEE Transactions on Acoustics, Speech, and Signal Processing*, Vol. 31, pp. 388-403, 1983.

[14.12] I. Gertner, "A New Efficient Algorithm to Compute the Two-Dimensional Discrete Fourier Transform," *IEEE Transactions on Acoustics, Speech, and Signal Processing*, Vol. 36, pp. 1036-1050, 1988.

Problems

14.1 Prove Theorem 14.1.

14.2 Show that the row-column method is an in-place algorithm.

14.3 Derive the Rader's algorithm based FFT for $d = 2$, $p_1 = 5$, $p_2 = 13$. Evaluate its computational complexity.

14.4 Derive the Rader's algorithm based FFT for $d = 2$, $p_1 = 17$, $p_2 = 17$. Evaluate its computational complexity for data sequences defined in $\mathbf{R}$ and $\mathbf{C}$.

14.5 Derive the Rader's algorithm based FFT for $d = 3$ for the case $n = n_1 = n_2 = n_3 = p$, p an odd prime.

14.6 Construct two-dimensional FFT algorithms for small n, $n = n_1 = n_2 = 3$, 4, 5, 6, 7, 8, 9, 10.
Note: They can be derived using Rader's algorithm and splitting methods.

14.7 Derive the general two-dimensional Cooley-Tukey algorithm and evaluate its computational complexity.

14.8 Derive the radix-4 FFT algorithm for $d = 2$. Evaluate its computational complexity.

14.9 Derive the split radix FFT algorithm (similar to Section 13.5.3) for $d = 2$. Evaluate its computational complexity and compare it to the radix-2 and radix-4 FFTs.

14.10 Describe all the matrix formulations of two-dimensional FFT algorithm for $n_1 = 15$, $n_2 = 20$ and evaluate their computational complexity.

14.11 Describe all the matrix formulations for one-dimensional FFT of length $n = 35$. Evaluate their computational complexity.

14.12 Repeat Problem 14.11 for $n = 2520$.

14.13 Let us attempt to express a two-dimensional DFT as an one-dimensional DFT. Let $n_1 = 4$, $n_2 = 5$. Show that it can be expressed as a one-dimensional DFT of length 20. Now, think of a one-dimensional FFT algorithm for $n = 20$ and evaluate its computational complexity. Compare the computational complexity of the two-dimensional FFT to the one-dimensional FFT.
Note: In the end, the objective is to compute a two-dimensional DFT.

14.14 Repeat Problem 14.12 for $n_1 = 5$, $n_2 = 7$.

14.15 Derive a two-dimensional Rader's algorithm based FFT for $n_1 = p$, $n_2 = 2p$, p an odd prime.

14.16 Repeat Problem 14.15 for $n_1 = n_2 = 2p$.

14.17 Can we derive two-dimensional Rader's algorithm based FFT when one or both of n_1 or n_2 are of the type $4p$?

14.18 Describe the polynomial transform based FFT for $d = 2$, $n = 5$. Compare it to the Rader's algorithm based FFT for $d = 2$, $n = 5$.

14.19 Repeat Problem 14.18 for $d = 2$, $n = p$, p an odd prime.

14.20 Derive and verify the expressions for the d-dimensional polynomial transform based FFT algorithm for $n = p$.

14.21 In many of the algorithms that have been derived here, we need reduced DFTs. Derive reduced DFT algorithms for length equal to 3, 5, 7, and 9.

14.22 Describe a polynomial transform based FFT algorithm for $n_1 = p$, $n_2 = 2p$, p an odd prime.

14.23 Describe a polynomial transform based FFT algorithm for $n_1 = n_2 = 2p$, p an odd prime.

14.24 Describe a polynomial transform based FFT algorithm for $d = 2$, $n_1 = p$, $n_2 = 4p$, p an odd prime. For this problem, let the original factorization be in the field $\mathcal{CZ}$.

14.25 Describe a polynomial transform based FFT algorithm for $d = 2$, $n_1 = 2p$, $n_2 = 4p$, p an odd prime. For this problem, let the original factorization be in the field $\mathcal{CZ}$.

14.26 Describe a polynomial transform based FFT algorithm for $d = 2$, $n_1 = n_2 = 4p$, p an odd prime. For this problem, let the original factorization be in the field $\mathcal{CZ}$.

14.27 Describe a d-dimensional polynomial transform based FFT algorithm for $n_1 = n_2 = \ldots = n_d = 2p$, p an odd prime. For this problem, let the original factorization be in the field $\mathcal{Z}$.

14.28 Describe a d-dimensional polynomial transform based FFT algorithm for $n_1 = n_2 = \ldots = n_d = 4p$, p an odd prime. For this problem, let the original factorization be in the field $\mathcal{CZ}$.

14.29 Describe a two-dimensional polynomial transform based FFT algorithm for $n_1 = p$, $n_2 = p^2$, p an odd prime.

14.30 For the Problems 14.22 to 14.28, describe the prime factor algorithm based FFT algorithms. Which approach (prime factor vs. polynomial transforms) is superior from a computational point of view.

14.31 Describe a two-dimensional polynomial transform based FFT algorithm for $n_1 = n_2 = 25$.

14.32 Describe a two-dimensional polynomial transform based FFT algorithm for $n_1 = n_2 = 512$. Evaluate the computational complexity of (i) computing the polynomial transforms using radix-2, radix-4, split radix algorithms, and (ii) the FFT algorithm.

14.33 Describe a two-dimensional polynomial transform based FFT algorithm for $n_1 = 2^c$, $n_2 = 2^b$, $b \geq c$. Use the fields $\mathcal{CZ}$ and appropriate $\mathcal{CCF}(n)$ to write the polynomial transforms. Evaluate the computational complexity of (i) computing the polynomial transforms using radix-2, radix-4, split radix algorithms, and (ii) the FFT algorithm.

14.34 Thus far, we have used factorizations in Z and then CF as a way to obtain polynomial transforms. Explore the possibility of using Re and its extensions to obtain polynomial transforms and then the d-dimensional FFTs. Let $d = 2$ and $n = n_1 = \ldots = n_d = 2^c$. For example, can we use the factorization for the Brunn's algorithm to write such a FFT algorithm?

14.35 Explain how polynomial transform based FFT algorithm can be used to compute the DFT of real-valued sequences in an efficient manner.

14.36 One may claim that polynomial transform based FFT algorithm is an in-place algorithm. Is this claim valid? Let $d = 2$, $n_1 = n_2 = p$, p an odd prime, for simplicity.

14.37 Derive radix-p FFT algorithms for computing polynomial transform of length p^c in $\mathsf{CF}(p^b)$, $b \geq c$. Evaluate its computational complexity. Also, evaluate the overall computational complexity of the polynomial transform method for computing two-dimensional DFT of length $p^c \times p^b$, $b \geq c$.

14.38 Repeat Problem 14.37 for $p = 2$.

14.39 Describe fast algorithms for computing the two-dimensional number-theoretic transform in $GF(17)$, $n = n_1 = n_2 = 16$.

14.40 Describe fast algorithms for computing the two-dimensional number-theoretic transform in $Z(17^2)$, $n = n_1 = n_2 = 16$.

14.41 Describe fast algorithms for computing the two-dimensional number-theoretic transform in $CZ(3)$, $n = n_1 = n_2 = 8$.

14.42 Describe fast algorithms for computing the two-dimensional number-theoretic transform in $CZ(3^2)$, $n = n_1 = n_2 = 8$.

14.43 Develop computer programs for computing two-dimensional DFT of lengths $p^c \times p^b$, $b \geq c$, for p an odd prime, and $p = 2$.

PART IV

RECENT RESULTS ON ALGORITHMS IN FINITE INTEGER RINGS

Thoughts on Part IV

It is generally believed that the algorithms for one- and higher-dimensional digital signal processing of sequences defined in finite integer rings $Z(M)$ and $CZ(M)$ are more complex computationally than their counterparts in the infinite fields. This belief manifests itself in what is termed as the **word sequence length constraint** (WSLC). There has been some progress on tackling WSLC but, by and large, it has remained a bottleneck in the finite integer ring based algorithms. In this part, we present four research papers that the author has recently submitted for publication. The first two papers are on two- and one-dimensional cyclic convolution in $Z(M)$ and $CZ(M)$, where it is shown that under the non-restrictive conditions $(N_1, M) = (N_2, M) = 1$ and $(N, M) = 1$, the two- and one-dimensional cyclic convolution are as computationally complex as their counterparts in infinite fields Z and CZ only in the **worst case**.

The third paper deals with an extension of the Euler's theorem in finite integer polynomial rings and its application to compute the cyclotomic factorization in finite integer rings. The fourth paper is on the design of coding techniques for error protection of data defined in finite integer and integer polynomial rings.

The notation adopted in the papers as submitted for publication is slightly different from the rest of the book. We have left it unchanged in the hope that it will not cause any difficulty to the reader. A brief description of the contents of the papers is as follows.

Paper one. H.K. Garg, "A Number Theoretic Approach to Fast Algorithms for Two-Dimensional Digital Signal Processing in Finite Integer Rings," submitted to *IEEE Transactions on Circuits and Systems*.
Abstract. In this work, we present a number theoretic approach to obtaining new polynomial transforms that may be used to compute two-dimensional cyclic convolution of sequences defined in finite integer and complex integer rings. A fundamental result of this work is that under the non-restrictive condition, $(N_1, M) = (N_2, M) = 1$, the polynomial transforms defined in finite integer rings are as intensive computationally as the polynomial transforms defined in rational and complex rational number system only in the worst case. They simplify considerably for many special cases of significance in digital signal processing.

Paper two. H.K. Garg, "On Fast Algorithms for One-Dimensional Digital Signal Processing in Finite Integer and Complex Integer Rings," submitted to *Circuits, Systems, and Signal Processing*.
Abstract. In this work, we present and analyze a number theoretic approach to computing one-dimensional cyclic convolution of sequences defined in finite

integer and complex integer rings. A fundamental result of this work is that under the non-restrictive condition, $(N, M) = 1$, the algorithms defined in finite integer and complex integer rings are as intensive computationally as the corresponding algorithms defined in rational and complex rational number system **only in the worst case**. They simplify considerably for a large number of cases of importance in digital signal processing.

Paper three. H.K. Garg, "Cyclotomic Polynomial Factorization in Finite Integer Rings with Applications to Digital Signal Processing," submitted to *IEEE Transactions on Circuits and Systems.*

Abstract. In this paper, we present results that can be used to obtain all the possible generators for a number theoretic transform (NTT) defined in a finite integer ring and its polynomial extensions. A generalization of the well-known Euler's theorem is derived which can be used to determine all the generators of a given NTT once the generators in the underlying finite field are identified. Based on this extension, we also describe a procedure to compute cyclotomic factorization in these rings. This factorization and Chinese remainder theorem lead to computationally efficient algorithms for computing cyclic convolution of two sequences defined in finite integer and complex integer rings.

Paper four. H.K. Garg, "Error Control Techniques for Data Sequences Defined in Finite Integer Rings," submitted to *Applicable Algebra in Engineering, Communication and Computing.*

Abstract. In this paper, we present results that can be used to design cyclic codes for error control coding of data sequences defined in finite integer and complex integer rings. This follows from our recent work on generalization of the well-known Euler's theorem in finite integer rings and their polynomial extensions. The idea is to describe BCH and Reed-Solomon codes in these rings along with a decoding algorithm. The decoding algorithm in the ring employs the decoder in the finite field in an iterative manner. All the algebraic properties of the resulting codes follow from the underlying finite fields.

The following example is based on the codes presented in Paper four. All items referenced in the example are from that paper.

Example. This example is a continuation of Examples 1 and 3 in Paper four. Let $n = 7$, $p = 2$, and $a = 3$. We have the factorization of $u^7 - 1$ over $Z(2^3)$,

$$u^7 - 1 = (u - 1)(u^3 + 3u^2 + 2u + 7)(u^3 + 6u^2 + 5u + 7). \qquad \text{(E-1)}$$

Based on this factorization, the primitive polynomial over $Z(2^3)$ that we employ in our code construction is

$$Q(\theta) = \theta^3 + 3\theta^2 + 2\theta + 7.$$

All polynomial computations involving the indeterminate θ are simplified using the polynomial $Q(\theta)$ as the modulo polynomial.

Let us construct a single error-correcting BCH code (same as Hamming code) over $Z(2^3)$. Based on the factorization in (E-1), the generator polynomial of such a code is given by

$$G(u) = (u^3 + 3u^2 + 2u + 7). \qquad \text{(E-2)}$$

Let the transmitted code-polynomial $V(u)$ be the all-zero polynomial and the received polynomial $R(u)$ be

$$R(u) = 1 + 2u^3 + 4u^6. \qquad \text{(E-3)}$$

It is clear that Hamming weight of $R(u) = 3 > 1$, the error-correcting capability of the code. We show that correct decoding takes place in this instance as Hamming weight of each of $e_b(u)$, $b = 0, 1, 2$ is either 0 or 1 which is less than or equal to the error-correcting capability of the code. In the following, we give step-by-step working of the decoding algorithm for this case.

Step 1. Given $R(u)$ in (E-3), the syndromes are computed as

$$\begin{aligned} S_1 = R(\theta) &= 1 + 2\theta^3 + 4\theta^6 \\ &= 1 + 2\,(5\theta^2 + 6\theta + 1) + 4\,(\theta^2 + 3\theta + 2) \\ &= 6\theta^2 + 3. \end{aligned} \qquad \text{(E-4)}$$

S_2 can either be computed as

$$S_2 = R(\theta^2)$$

or by using the conjugate symmetry property (this is a BCH code) in (26). We prefer to use the latter in order to establish and exploit all the mathematical properties associated with the code construction. Setting $p = 2$ and $i = 2$ in (26), we get

$$\begin{aligned} S_2 = S_1(\theta^2) &= 6\theta^4 + 3 \\ &= 6\,(7\theta^2 + 7\theta + 5) + 3 \\ &= 2\theta^2 + 2\theta + 1. \end{aligned} \qquad \text{(E-5)}$$

Step 2.

$$s_{1,0} = S_{1,0} = S_1 \bmod 2 = 1,$$

$$s_{2,0} = S_{2,0} = S_2 \bmod 2 = 1. \tag{E-6}$$

Step 3. A (7, 4) binary BCH decoder will output an error (since it is a binary decoder, error magnitude = 1) in the 0-th location. Thus

$$e_0(u) = 1. \tag{E-7}$$

Step 4. $b = 1$.

Step 5.

$$s_{i,1} = \{2^{-1}[S_{i,1} - e_0(\theta^i)]\} \bmod 2, \quad i = 1, 2;$$

$$\begin{aligned} s_{1,1} &= \{2^{-1}[2\theta^2 + 3 - e_0(\theta)]\} \bmod 2 \\ &= \{2^{-1}[2\theta^2 + 3 - 1]\} \bmod 2 \\ &= \theta^2 + 1; \end{aligned}$$

$$\begin{aligned} s_{2,1} &= \{2^{-1}[2\theta^2 + 2\theta + 1 - e_0(\theta^2)]\} \bmod 2 \\ &= \theta^2 + \theta. \end{aligned} \tag{E-8}$$

Step 6. A (7, 4) binary BCH decoder will output an error in the 3rd location. Thus,

$$e_1(u) = u^3. \tag{E-9}$$

Step 7. $b = 2$. Return to Step 5.

Step 5.

$$s_{i,2} = \{2^{-2}[S_{i,2} - e_0(\theta^i) - 2e_1(\theta^i)]\} \bmod 2, \quad i = 1, 2;$$

$$\begin{aligned} s_{1,2} &= \{2^{-2}[6\theta^2 + 3 - 1 - 2\theta^3]\} \bmod 2 \\ &= \theta^2 + \theta; \end{aligned}$$

$$\begin{aligned} s_{2,2} &= \{2^{-2}[2\theta^2 + \theta + 1 - 1 - 2\theta^6]\} \bmod 2 \\ &= \theta + 1. \end{aligned} \tag{E-10}$$

Step 6. A (7, 4) binary BCH decoder will output an error in the 6th location. Thus,

$$e_2(u) = u^6. \tag{E-11}$$

Step 7. $b = 3$. Go to Step 8.

Step 8. Using (E-7), (E-9), and (E-11), we get

$$E(u) = e_0(u) + 2e_1(u) + 4e_2(u) = 1 + 2u^3 + 4u^6.$$

Substituting in (33), we get

$$V(u) = R(u) - E(u) = 1 + 2u^3 + 4u^6 - [1 + 2u^3 + 4u^6] = 0.$$

Thus, correct decoding takes place in this instance. This completes the example.

The abbreviations used in the papers are as follows.

Paper one

CRT-I: Chinese remainder theorem in integer form
CRT-P: Chinese remainder theorem in polynomial form
AICE-CRT: Extensions to CRT-P in integer and integer polynomial rings
CC: Cyclic convolution
PT: Polynomial transforms
DFT: Discrete Fourier Transform
POLY: Polynomial
MOD: Modulo

Paper two

CRT-I: Chinese remainder theorem in integer form
CRT-P: Chinese remainder theorem in polynomial form
AICE-CRT: Extensions to CRT-P in integer and integer polynomial rings
CC: Cyclic convolution
DFT: Discrete Fourier Transform
POLY: Polynomial
MOD: Modulo
DSP: Digital Signal Processing
NTT: Number Theoretic Transform
CNTT: Complex NTT
iff: If and only if
ADD: Addition
MULT: Multiplication
WSLC: Word Sequence Length Constraint

Paper three

NTT: Number Theoretic Transform

CNTT: Complex NTT
POLY: Polynomial
DSP: Digital Signal Processing
DFT: Discrete Fourier Transform
CRT: Chinese Remainder Theorem
BCH: Bose-Chaudhary-Hoquenghem
RS: Reed-Solomon

Paper four

ECC: Error Control Coding
POLY: Polynomial
DSP: Digital Signal Processing
BCH: Bose-Chaudhary-Hoquenghem
RS: Reed-Solomon
NTT: Number Theoretic Transform

The research work that is being reported in these papers is supported by Academic Research Grant No. RP960699, awarded by the National University of Singapore.

Paper one

A Number Theoretic Approach to Fast Algorithms for Two-Dimensional Digital Signal Processing in Finite Integer Rings

Abstract. In this work, we present a number theoretic approach to obtaining new polynomial transforms that may be used to compute two-dimensional cyclic convolution of sequences defined in finite integer and complex integer rings. A fundamental result of this work is that under the non-restrictive condition, $(N_1, M) = (N_2, M) = 1$, the polynomial transforms defined in finite integer rings are as intensive computationally as the polynomial transforms defined in rational and complex rational number systems only in the worst case. They simplify considerably for many special cases of significance in digital signal processing.

I. INTRODUCTION

The main idea behind the computation of convolution has remained unchanged regardless of the number system over which they are defined [1]-[4]. In all instances, a large size problem is expressed as a direct sum of a number of smaller size problems which are then computed in parallel. The Chinese remainder theorem in its integer form (CRT-I), polynomial form (CRT-P), and its polynomial extensions in integer and integer polynomial rings (AICE-CRT) provides the basis for all these direct sums [5]. The nature of the direct sum depends on the number system it is defined over.

In this paper, we study the computation of two-dimensional cyclic convolution (CC) of length $N_1 \times N_2$ of data sequences defined over finite integer and finite complex integer rings, denoted by $Z(M)$ and $CZ(M)$, respectively. The emphasis is on studying the polynomial transforms (PTs) approach which has been primarily developed for processing sequences defined over infinite number systems, such as the rational, real, and complex numbers [1]. It is shown that under the condition that the lengths are relatively coprime to M, that is, $(N_1, M) = (N_2, M) = 1$, the PTs as defined for rational valued sequences can also be used to compute the two-dimensional CC over the finite integer ring of interest. We extend the results further and show that the PTs for rational valued sequences can be further simplified in a large number of instances when the computation is defined over finite integer rings.

Fast algorithms for two-dimensional CC of sequences defined over $Z(M)$ and $CZ(M)$ have been studied extensively. Traditionally, these algorithms were based on the number theoretic transforms [6]-[10]. Recently, a Chinese remainder theorem (CRT) based approach was developed in [11], [12]. Also, it has been shown that the PTs as defined in rational number system, Z, can be used in a finite integer rings provided M is relatively coprime to the lengths [13].

The purpose of this paper is to present a number theoretic framework for PTs in finite integer rings. It generalizes the previous results in many important respects and leads to new algorithms that are among the most computationally efficient algorithms for the task at hand. An interesting result of this work is that under a non-restrictive condition, the PTs defined in finite integer rings are as intensive computationally as the PTs defined in rational number system only in the worst case. They simplify considerably for many special cases of significance in digital signal processing. These cases include (i) $M = 2^a$ and $N = N_1 = N_2 = q^c$, q a Mersenne or a Fermat prime, and (ii) $N = N_1 = N_2 = 2^c$ and M is a Mersenne or a Fermat prime.

The focus is on algorithms for the 2-D CC of length $N_1 \times N_2$ over $Z(M)$ and $CZ(M)$:

$$Z(u, v) = X(u, v)\, Y(u, v) \mod (u^{N_1} - 1, v^{N_2} - 1) \tag{1}$$

$X(u, v)$, $Y(u, v)$ and $Z(u, v)$ being polynomials (POLYs) in u and v with degrees N_1 and N_2, respectively. The finite integer ring, $Z(M)$, and the finite complex integer ring, $CZ(M)$, consist of integers $\{0, 1, ..., M-1\}$ and $\{a + jb,\ a, b \in Z(M), j^2 = -1 \mod M\}$, with the operations of ADD and MULT defined mod M

and mod $(M, j^2 + 1)$, respectively. **A condition of fundamental importance that will be assumed to be satisfied throughout this work is that**

$$(N_1, M) = (N_2, M) = 1, \tag{2}$$

that is, the lengths and M are relatively coprime. This condition ensures that the multiplicative inverses of N_1 and N_2 exist in $Z(M)$ and $CZ(M)$. Five cases that we feel are of most interest to digital signal processing are studied in depth. They are $N = q$, $N = 2 \cdot q$, $N = q^c$, $N = 2 \cdot q^c$ and $N = 2^c$; N is either one of N_1 or N_2 and q is an odd prime. These are also to be treated separately from a number theoretic point of view also as the theory of primitive roots has a different treatment for these five cases.

This paper is organized as follows. In Section II, a brief description of PTs, as defined over the rational number system, Z, is presented. Several special cases of lengths for which the PTs are most useful are also mentioned. Section III is on POLY factorization over $Z(M)$ and $CZ(M)$. In Section IV, results pertaining to cyclotomic factorization are presented that lead to new PTs over $Z(M)$ and $CZ(M)$. In Section V, the special cases of most interest in digital signal processing are analyzed. Section VI is on recursive computation of POLY products required in PTs. Conclusions are presented in Section VII.

II. POLYNOMIAL TRANSFORMS OVER RATIONAL NUMBER SYSTEM

Polynomial transforms have been most widely studied for the cases (a) $N = N_1 = N_2 = q$, q an odd prime, (b) $N = N_1 = N_2 = q^c$, (c) $N_1 = 2 \cdot q$, $N_2 = q$, (d) $N = N_1 = N_2 = 2 \cdot q$, (e) $N_1 = 2 \cdot q^c$, $N_2 = q^c$, and (f) $N = N_1 = N_2 = 2^c$ [1]. In essence, it consists in converting a two-dimensional CC into a number of one-dimensional convolutions.

Consider the cyclotomic factorization

$$u^{N_1} - 1 = \prod_{d_1 | N_1} C_{d_1}(u) = C_{N_1}(u) \prod_{\substack{d_1 | N_1 \\ d_1 < N_1}} C_{d_1}(u) = C_{N_1}(u) \cdot C_1'(u), \tag{3}$$

over Z. In (3), $C_{N_1}(u)$ is a cyclotomic POLY of degree $\phi(N_1)$, $\phi(.)$ being the Euler's totient function of N_1. Thus, $Z(u, v)$ in (2) can be written as a direct sum of $Z_1(u, v)$ and $Z_2(u, v)$ using the CRT corresponding to the factorization in (3), where

$$Z_1(u, v) = X(u, v)\, Y(u, v) \bmod \left(C_1'(u), v^{N_2} - 1\right) \tag{4}$$

and

$$Z_2(u, v) = X(u, v)\, Y(u, v) \bmod \left(C_{N_1}(u), v^{N_2} - 1\right). \tag{5}$$

With respect to the above direct sum the CRT reconstruction requires division by N_1. Similarly, $Z_1(u, v)$ can be written as a direct sum of $Z_3(u, v)$ and $Z_4(u, v)$ using the CRT corresponding to the cyclotomic factorization of $v^{N_2} - 1$, where

$$Z_3(u, v) = X(u, v)\, Y(u, v) \bmod \left(C_1'(u), C_{N_2}(v)\right) \tag{6}$$

and

$$Z_4(u, v) = X(u, v)\, Y(u, v) \bmod \left(C_1'(u), C_2'(v)\right). \tag{7}$$

On many occasions, the computation in (5) and (6) can be performed as a discrete Fourier transform (DFT) in the cyclotomic extension field. The computation in (7) can be expressed in a form similar to the original two-dimensional CC with the lengths reduced by an integral factor, thereby implying that the original factorizations can be used in a recursive manner. These statements hold especially true for the six types of lengths mentioned earlier in this section.

Example 1. Let $N = N_1 = N_2 = q^c$. In this case, we express

$$\left(u^{q^c} - 1\right) = \left(u^{q^{c-1}} - 1\right) C_{q^c}(u). \tag{8}$$

In (8), $C_{q^c}(u)$ is a degree $q^c - q^{c-1}$ cyclotomic POLY (CP). It is crucial to note that $\deg(C_{q^c}(u)) = \phi(q^c)$. Here, $Z(u, v)$ can be written as a direct sum of

$$Z_1(u, v) = X(u, v)\, Y(u, v) \bmod \left(u^{q^{c-1}} - 1, v^{q^c} - 1\right) \tag{9}$$

and

$$Z_2(u, v) = X(u, v)\, Y(u, v) \bmod \left(C_{q^c}(u), v^{q^c} - 1\right). \tag{10}$$

Similarly, $Z_1(u, v)$ can be written as a direct sum of

$$Z_3(u, v) = X(u, v)\, Y(u, v) \mod \left(u^{q^{c-1}} - 1, C_{q^c}(v)\right) \tag{11}$$

and

$$Z_4(u, v) = X(u, v)\, Y(u, v) \bmod \left(u^{q^{c-1}} - 1, v^{q^{c-1}} - 1\right). \tag{12}$$

The computation of $Z_2(u, v)$ in (10) and $Z_3(u, v)$ in (11) are performed using DFT algorithm corresponding to the factorization in the cyclotomic extension field,

$$v^{q^c} - 1 = \prod_{i=0}^{q^c - 1} (v - u^i) \bmod \left(C_{q^c}(u)\right) \tag{13}$$

and

$$u^{q^{c-1}} - 1 = \prod_{i=0}^{q^{c-1} - 1} (u - (v^q)^i) \bmod \left(C_{q^c}(v)\right). \tag{14}$$

The CRT reconstructions corresponding to (9) and (10), (11) and (12), and (13) and (14) require division by q^c and q^{c-1}. Based on (9) to (14), we see that a $q^c \times q^c$ CC is computed in the following manner:

(i) q^c one-dimensional POLY products mod $C_{q^c}(u)$ in (13),

(ii) q^{c-1} one-dimensional POLY products mod $C_{q^c}(v)$ in (14), and

(iii) a two-dimensional CC of length $q^{c-1} \times q^{c-1}$.

The two-dimensional CC of length $q^{c-1} \times q^{c-1}$ can be computed using the above expressions in a recursive manner.

Similar expressions can also be derived for the other cases mentioned earlier. For the case when $N = N_1 = N_2 = 2^c$, each of the one-dimensional POLY products corresponds to a circular correlation due to the form of the cyclotomic POLY,

$$C_{2^c}(u) = u^{2^{c-1}} + 1. \tag{15}$$

Also, it is worthwhile to mention that except when N is a power of 2, $N > 4$, **the cyclotomic POLYs are irreducible over the complex rational number system, *CZ*** [14]. Consequently, one cannot further simplify the PTs for processing complex valued sequences for the first five cases mentioned earlier.

III. BASIC FACTORIZATION AND POLYNOMIAL TRANSFORMS OVER FINITE INTEGER AND COMPLEX INTEGER RINGS

Given the standard factorization of M as

$$M = \prod_{i=1}^{l} p_i^{a_i} = \prod_{i=1}^{l} r_i, \tag{16}$$

the following direct sums are obtained using the CRT:

$$Z(M) = \sum_{\substack{\oplus \\ i=1}}^{l} Z(r_i) \tag{17}$$

$$CZ(M) = \sum_{\substack{\oplus \\ i=1}}^{l} CZ(r_i) \tag{18}$$

A straightforward consequence of the above analysis is that we need to derive algorithms for the convolutions over $Z(r_i)$ and $CZ(r_i)$ and then combine the resulting matrices using the direct sum properties. This approach is adopted throughout this paper. Also the subscript i will be deleted. This will simplify notation without creating any confusion.

Our primary objective is to derive fast algorithms for the following:

$$Z(u, v) = X(u, v)\, Y(u, v) \mod (r, u^{N_1} - 1, v^{N_2} - 1). \tag{19}$$

In $CZ(r)$, all the quantities are degree one POLY in j with the computation defined mod $(j^2 + 1)$. Following PTs over the rational number system, Z, we will pursue PTs over the finite integer and complex integer rings, $Z(r)$ and $CZ(r)$, in the following manner.

Step 1. Seek the *monic* factorization of $u^{N_1} - 1$ in $GF(p)$ and $GF(p^2)$, $r = p^a$;

Step 2. Seek the *monic* factorization of $u^{N_1} - 1$ in $Z(r)$ and $CZ(r)$;

Step 3. Express the two-dimensional CC as the direct sum of one-dimensional POLY products using the cyclotomic factorization in $Z(r)$, $CZ(r)$, and their extension rings.

The basic idea that leads to PTs in all instances is as follows. Given an irreducible cyclotomic POLY $C(u)$ containing the nth root of unity in a field F, let E be an cyclotomic extension field obtained by adjoining $C(u)$ to F. Then, u represents the nth root of unity in E. Consequently, the following factorization holds:

$$v^N - 1 = \prod_{i=0}^{N-1}\left(v - \left(u^{n/N}\right)^i\right) \bmod C(u),$$

for every N that is a factor of n. This leads to a DFT based computation for expressions of the type

$$X(u, v)\ Y(u, v) \bmod (v^N - 1, C(u)),$$

where the quantities $X(u, v)$ and $Y(u, v)$ are defined in F. Let $n = N$ and $\mathsf{F} = Z$, then $C(u) = C_N(u)$ which is a POLY of degree $\phi(N)$. In this case, PTs express $X(u, v)\ Y(u, v) \bmod (v^N - 1, C_N(u))$ as a direct sum of N computations $X(u, u^i)$ $Y(u, u^i) \bmod C(u)$, $i = 0, 1, \ldots, N - 1$, each one being a product of two POLYs of degree $\phi(N) - 1$.

The crucial task now is to study the form of PTs over Z and CZ and establish the corresponding versions over $Z(r)$ and $CZ(r)$. Let $C_N(u)$ over Z factor further over $Z(r)$ or $CZ(r)$ as

$$C_N(u) = A_1(u) \cdot A_2(u) \cdots A_L(u),$$

each of degree m, $m = \phi(N) / L$. Based on this, we can define a PT in $Z(r)$ or $CZ(r)$ which expresses $X(u, v)\ Y(u, v) \bmod (v^N - 1, C_N(u))$ as a direct sum of $N \times m$ computations $X_k(u, u^i)\ Y_k(u, u^i) \bmod A_k(u)$, $i = 0, 1, \ldots, N - 1$, $k = 1, 2, \ldots, L$, each one being a product of two POLYs of degree $m - 1$. The CRT in $Z(r)$, $CZ(r)$ and their POLY extensions forms the basis of these statements [5]. PTs over $Z(r)$ and $CZ(r)$ are computationally simpler to their counterparts over Z and CZ as long as $L > 1$. If $L = 1$, then they are identical in form.

In this regard, all the six cases mentioned in the beginning of Section II will be studied. A basic property of the factorization of $u^N - 1$ in $Z(r)$ and $CZ(r)$ is as follows.

Lemma 1. If $(N, r) = 1$, then the relatively prime monic factors of $u^N - 1$ in Z or CZ are also relatively prime monic factors of $u^N - 1$ in $Z(r)$ or $CZ(r)$, respectively.
Proof: Let $A(u)$ and $B(u)$ be two such factors in Z. Then since $(A(u), B(u)) = 1$, we have

$$u^N - 1 = A(u) \cdot B(u) \cdot C(u). \tag{20}$$

Clearly, this factorization holds when all the terms are defined mod r. Taking formal derivative, multiplying by (u/N) and subtracting (15) from it, we get

$$1 \equiv (u/N) \cdot A'(u) \cdot B(u) \cdot C(u) \bmod (r, A(u))$$
$$1 \equiv (u/N) \cdot A(u) \cdot B'(u) \cdot C(u) \bmod (r, B(u)).$$

Both these equations imply that $A(u)$ and $B(u)$ are relatively coprime in $Z(r)$. Note that $1/N$ denotes the multiplicative inverse of N in $Z(r)$ which exists and is unique as $(N, r) = 1$. A similar set of statements hold for the complex case.

This lemma implies that the cyclotomic POLY factorization used to obtain PTs in Z remains valid in $Z(r)$ and, therefore, can be used to obtain PTs in $Z(r)$. In the following, we will pursue the factorization of $u^N - 1$ directly in $Z(r)$ and obtain considerably simplified forms for same.

A factorization of $u^N - 1$ in terms of monic factors in $Z(r)$ is an expression of the type

$$u^N - 1 = A(u) \cdot B(u) \bmod (r), \tag{21}$$

where $A(u)$ and $B(u)$ are monic. Taking mod p on both sides, we see that it also provides us with the corresponding factorization over $GF(p)$,

$$u^N - 1 = A_0(u) \cdot B_0(u) \bmod (p), \tag{22}$$

where $A_0(u)$ and $B_0(u)$ are equal to $A(u)$ mod p and $B(u)$ mod p, respectively. One can derive the factorization of $u^N - 1$ over $GF(p)$ using finite field algebra [15]. In Appendix A, we describe a Hensel's theorem based procedure to obtain the factors in $Z(r)$ starting from the factors in $GF(p)$ as given in (22).

In a manner similar to (20) to (22), a factorization of $u^N - 1$ in terms of monic factors in $CZ(r)$ can be obtained starting from its monic factorization in $CZ(p, j^2 + 1)$. For $p = 2, j^2 + 1 = (j + 1)^2 \mod 2$. For $p \equiv 3 \mod 4$, $j^2 + 1$ is irreducible over $GF(p)$ and for $p \equiv 1 \mod 4$, $j^2 + 1$ factors in terms of two mutually coprime degree one POLYs over $GF(p)$ as $j^2 + 1 = (j + a_0) \cdot (j + b_0) \mod p$. Therefore, $j^2 + 1$ is irreducible over $Z(r)$, $p \equiv 3 \mod 4$ and factors as $j^2 + 1 = (j + a) \cdot (j + b)$ over $Z(r)$, $p \equiv 1 \mod 4$, $a_0 = a \mod p$ and $b_0 = b \mod p$. Three distinct cases follow from these statements.

Case 1. $p = 2$. In this case, the algorithms for CC over $CZ(r)$ are the same as those over $Z(r)$. Only the input and output sequences are defined over $CZ(r)$.

Case 2. $p \equiv 1 \mod 4$. Based on the CRT, we have the direct sum,

$$CZ(r) = Z(r, j + a) \oplus Z(r, j + b). \tag{23}$$

Therefore, the CC of complex integer-valued sequences in $CZ(r)$, $p \equiv 1 \mod 4$, can be expressed as a direct sum of two integer valued sequences in $Z(r)$.

Case 3. $p \equiv 3 \mod 4$. In this case, $CZ(r)$ will be studied as the p-adic expansion ring of $GF(p^2)$. The analysis for $CZ(r)$ is similar to the analysis for $Z(r)$ in (20) to (22), (A-1) to (A-5).

In the context of designing algorithms over $CZ(r)$, we will deal only with $p \equiv 3 \mod 4$. The analysis for the other two cases as outlined above is complete.

III.1 Factorization of $u^N - 1$ in $GF(p)$ and $Z(r)$, $r = p^a$. The factorization of $u^N - 1$ in $GF(p)$ begins with finding the least integer m such that [15]

$$N \mid (p^m - 1). \tag{24}$$

In other words,

$$p^m \equiv 1 \mod N \tag{25}$$

and, therefore, m is the order of $p \mod N$. Recalling Euler's theorem,

$$p^{\phi(N)} \equiv 1 \mod N, \tag{26}$$

we have

$$m \mid \phi(N). \tag{27}$$

In general, $m = \phi(N)$ if and only if p is a **primitive element** mod N. Primitive elements exist only for those N (besides 2, 4) that are of the form q^c or $2 \cdot q^c$, q an odd prime. In all other cases, the order of p has to be less than $\phi(N)$. Once m is determined, one employs the finite field $GF(p^m)$ to compute the factors of $u^N - 1$ over $GF(p)$. Since in $GF(p^m)$, each element can have at most m conjugate elements, we have the important result that the factors of $u^N - 1$ in $GF(p)$ and hence over $Z(p^a)$ will have degree at most equal to m. The expression in (26) brings out one of the contrasts between the cyclotomic factorization of $u^N - 1$ over Z and $GF(p)$ (and consequently, over $Z(p^a)$). Over Z, the degree of the cyclotomic POLY $C_N(u)$ having the Nth primitive root of unity as a root is exactly $\phi(N)$, while over $GF(p)$ and $Z(p^a)$, it is a factor of $\phi(N)$. Therefore, as long as $(N, p) = 1$, the degrees of the cyclotomic factors of $u^N - 1$ over $GF(p)$ and $Z(p^a)$ are always lower than (or at most equal to) the degrees of the cyclotomic factors of $u^N - 1$ over Z. Once again, equality holds only when p is primitive mod N.

Example 2. Let $N = 17 = q$ and $r = p^a = 2^a$. Here, we seek to find the order of 2 mod 17. The order 2 mod 17 is 8. Therefore, we will use $GF(2^8)$ to factorize $u^{17} - 1$ over $GF(2)$. If β denotes the element of order 17 in $GF(2^8)$, then the order of β^k is 17, $k = 1, 2, \ldots, 16$, since N is prime. The complete factorizations of $u^{17} - 1$ are given by

$$u^{17} - 1 = (u-1)\left(\sum_{i=0}^{16} u^i\right)$$

over Z and

$$\begin{aligned} u^{17} - 1 = (u-1)&\{(u-\beta)(u-\beta^2)(u-\beta^4)(u-\beta^8)(u-\beta^{16})(u-\beta^{15})(u-\beta^{13})(u-\beta^9)\} \\ &\cdot\{(u-\beta^3)(u-\beta^6)(u-\beta^{12})(u-\beta^7)(u-\beta^{14})(u-\beta^{11})(u-\beta^5)(u-\beta^{10})\} \end{aligned}$$

over $GF(2)$. A PT defined over Z for a 17×17 two-dimensional CC would require 18 POLY products mod $C_{17}(u)$, a POLY of degree 16. In contrast, a PT defined over $GF(2)$ and hence over $Z(2^a)$ for a 17×17 two-dimensional CC would require 36 POLY products mod cyclotomic POLYs of degree 8.

III.2 Factorization of $u^N - 1$ in $GF(p^2)$ and $CZ(r)$, $r = p^a$, $p \equiv 3$ mod 4. The factorization of $u^N - 1$ in $GF(p^2)$ begins with finding the least integer s such that [4]

$$N \mid (p^{2s} - 1). \tag{28}$$

In other words,

$$p^{2s} \equiv 1 \bmod N, \tag{29}$$

and, therefore, $2 \cdot s$ is the order of p mod N. Recalling from (25) that m is the order of p mod N, we have

$$s = m/2, \tag{30a}$$

if m is even, or

$$s = m, \tag{30b}$$

if m is odd. Combining (29) with Euler's theorem in (26), we get

$$s \mid \{\phi(N)/2\}, \tag{31}$$

since $\phi(N)$ is always an even number except for the trivial value $N = 2$. In all cases, s is at most one-half of $\phi(N)$. Once s is determined, one employs the finite field $GF(p^{2s})$ to compute the factors of $u^N - 1$ over $GF(p^2)$. Since in $GF(p^{2s})$, each element can have at most s conjugate elements in $GF(p^2)$, we have the important result that the factors of $u^N - 1$ in $GF(p^2)$ and hence over $CZ(p^a)$ will have degree at most equal to s. The expression in (30) brings out further contrast between the cyclotomic factorization of $u^N - 1$ over Z and $GF(p^2)$ (and consequently, over $CZ(p^a)$). Over Z, the degree of the cyclotomic POLY having the N-th primitive root of unity as a root is exactly $\phi(N)$, while over $GF(p^2)$ and $CZ(p^a)$, it is a factor of $\phi(N)/2$. Therefore, as long as $(N, p) = 1$, the degrees of the cyclotomic factors of $u^N - 1$ over $GF(p^2)$ and $CZ(p^a)$ **are at most equal to one-half the degrees** of the cyclotomic factors of $u^N - 1$ over Z and CZ. Also, if p is primitive mod N, then $C_N(u)$ as defined over Z and CZ is also irreducible over $Z(p^a)$, but it will always factor further over $CZ(p^a)$.

Example 3. Let $N = 7 = q$ and $r = p^a = 3^a$. Here, we seek to find the order of $3^2 = 9$ mod 7. The order of 9 mod 7 is 3. Therefore, we will use $GF(9^3)$ to factorize $u^7 - 1$ over $GF(9)$. If β denotes the element of order 7 in $GF(9^3)$, then the order of β^k is 7, $k = 1, 2, ..., 6$, since N is prime. The complete factorizations of $u^7 - 1$ are given by

$$u^7 - 1 = (u-1)\left(\sum_{i=0}^{6} u^i\right)$$

over Z, CZ, and $Z(3^a)$, and

$$u^7 - 1 = (u-1)\{(u-\beta)(u-\beta^2)(u-\beta^4)\}\cdot\{(u-\beta^3)(u-\beta^6)(u-\beta^5)\}$$
$$= (u-1)\cdot\{u^3 + (2+j)u^2 + (1+j)u + 2\}$$
$$\cdot\{u^3 + (2+2j)u^2 + (1+2j)u + 2\} \bmod (3, j^2+1\},$$

over $GF(3^2)$. The corresponding factorization over $CZ(3^2)$ and $CZ(3^3)$ is given by (obtained using the Hensel's theorem and the analysis in Appendix A),

$$u^7 - 1 = (u-1)\cdot\{u^3 + (5+7j)u^2 + (4+7j)u + 8\}$$
$$\cdot\{u^3 + (5+2j)u^2 + (4+2j)u + 8\} \bmod (9, j^2+1\}$$

and

$$u^7 - 1 = (u-1)\cdot\{u^3 + (14+7j)u^2 + (13+7j)u + 26\}$$
$$\cdot\{u^3 + (14+20j)u^2 + (13+20j)u + 26\} \bmod (27, j^2+1\},$$

respectively. A PT defined over Z, CZ, and $Z(3^a)$ for a 7×7 two-dimensional CC would require 4 POLY products mod $C_7(u)$, a POLY of degree 6. In contrast, a PT defined over $GF(3^2)$ and hence over $CZ(3^a)$ for a 7×7 two-dimensional CC would require 8 POLY products mod cyclotomic POLYs of degree 3.

IV. CYCLOTOMIC FACTORIZATION OVER $Z(r)$ AND $CZ(r)$

It is seen from the description of PTs that one seeks the cyclotomic POLY $C_N(u)$ having the N-th primitive root of unity as its root. In $GF(p)$, if β is the element of order N in $GF(p^m)$, m being the order of p mod N, then β has exactly m conjugates. Therefore, $C_N(u)$ as found over Z is either irreducible in $GF(p)$ (and $Z(r)$), which happens if $m = \phi(N)$, or it has $\phi(N)/m$ factors in $GF(p)$ (and $Z(r)$), each of degree m, which happens if m is a factor of $\phi(N)$. Such a factorization is obtained in the following manner.

Procedure for obtaining cyclotomic factors of $C_N(u)$ in $GF(p)$

1. Find m, the order of p mod N.
2. Let β be an element of order N in $GF(p^m)$. Now, $\beta^N = 1$ and $\beta^i \neq 1$, $i < N$, in $GF(p^m)$. Let $S = \{i, i = 1, 2, ..., N; (i, N) = 1\}$. Partition S into subsets S_{i_1}, S_{i_2}, ...; $S_{i_k} = \{i_k, pi_k, p^2 i_k, \cdots\}$, where i_k is the smallest element in S not covered in earlier subsets. The elements in subsets S_{i_1}, S_{i_2}, ... are defined

mod N. There are $\phi(N)/m$ subsets, each subset having m elements. The factorization of $C_N(u)$ is obtained as

$$C_N(u) = \prod_k \left[\prod_{l \in S_{i_k}} \left(u - \beta^l\right) \right] \bmod (p, Q(\alpha)) \tag{32}$$

$$= \prod_k P_k(u) \bmod p, \tag{33}$$

where $Q(\alpha)$ is a primitive POLY in $GF(p^m)$. Each of $P_k(u)$ in (33) is a POLY of degree m having β^{i_k} and its conjugates in $GF(p^m)$ as its roots. Therefore, it is irreducible and has its coefficients in $GF(p)$. Once the factors of $C_N(u)$ in $GF(p)$ are determined, the Hensel's Theorem as described in Appendix A can be used to compute the factors in $Z(r)$.

Similarly, $C_N(u)$ as found over Z is **always reducible** in $CZ(r)$. It factors into $\phi(N)/s$ factors in $GF(p^2)$ (and $CZ(r)$), each of degree s, s being the order of p^2 mod N. Such a factorization is obtained in the following manner.

Procedure for obtaining cyclotomic factors of $C_N(u)$ in $GF(p^2)$

1. Find s, the order of p^2 mod N.
2. Let β be an element of order N in $GF(p^{2s})$. Now, $\beta^N = 1$ and $\beta^i \neq 1$, $i < N$, in $GF(p^{2s})$. Let $S = \{i, i = 1, 2, ..., N; (i, N) = 1\}$. Partition S into subsets S_{i_1}, S_{i_2}, ...; $S_{i_k} = \{i_k, p^2 i_k, p^4 i_k, \cdots\}$, where i_k is the smallest element in S not covered in earlier subsets. The elements in subsets S_{i_1}, S_{i_2}, ... are defined mod N. There are $\phi(N)/s$ subsets, each subset having s elements. The factorization of $C_N(u)$ is obtained as

$$C_N(u) = \prod_k \left[\prod_{l \in S_{i_k}} \left(u - \beta^l\right) \right] \bmod (p, Q(\alpha)) \tag{34}$$

$$= \prod_k P_k(u) \bmod p, \tag{35}$$

where $Q(\alpha)$ is a primitive POLY of degree s in $GF(p^{2s})$ with coefficients in $GF(p^2)$. Each of $P_k(u)$ in (35) is a POLY of degree s having β^{i_k} and its conjugates in $GF(p^{2s})$ as its roots. Therefore, it is irreducible and has its coefficients in $GF(p^2)$. Once the factors of $C_N(u)$ in $GF(p^2)$ are determined, the

Hensel's Theorem as described in Appendix A can be used to compute the factors in $CZ(r)$.

Finally, we note that except for values of 2 and 4, $Z(N)$ has no primitive elements for $N = 2^c$. Therefore, in the case $N = 2^c$, $C_N(u) = u^{2^{c-1}} + 1$ will always factor further over $Z(r)$. Interestingly, each of these factors over $Z(r)$ will always factor further into two POLYs over $CZ(r)$.

Example 4. Let $r = 2^a$ and $N = 2^e - 1$, a Mersenne prime integer. In this case, order of 2 mod $N = m = e$. Therefore, $u^N - 1$ factors as $(u - 1)$ and $C_N(u) = u^{N-1} + u^{N-2} + \ldots + u + 1$ in Z while it factors as $(u - 1)$ and $(N - 1)/e$ primitive POLYs in $Z(r)$, each of degree e, as N is prime.

Example 5. Let $r = 2^a$ and $N = 2^n + 1$, a Fermat prime integer, $n = 2^t$. In this case, order of 2 mod $N = m = 2^{t+1}$. Therefore, $u^N - 1$ factors as $(u - 1)$ and $C_N(u) = u^{N-1} + u^{N-2} + \ldots + u + 1$ in Z, while it factors as $(u - 1)$ and 2^{n-t-1} primitive POLYs in $Z(r)$, each of degree 2^{t+1}, as N is prime.

Example 6. Let $r = 3^a$ and $N = 2^n + 1$, a Fermat prime integer, $n = 2^t$. It is well-known that 3 is a primitive element for Fermat primes [1]. Therefore, in this case, order of 3 mod $N = N - 1$ and $u^N - 1$ factors as $(u - 1)$ and $C_N(u) = u^{N-1} + u^{N-2} + \ldots + u + 1$ in Z and $Z(3^a)$. Over, $CZ(3^a)$, it factors as $(u - 1)$ and two POLYs, each of degree $(N - 1)/2$.

Now, we turn to the six specific cases for the lengths that were mentioned in the beginning of Section III.

V. POLYNOMIAL TRANSFORMS OVER FINITE INTEGER RINGS: SPECIAL CASES

In this section, we analyze the construction of POLY transforms over $Z(p^a)$ and $CZ(p^a)$ for the following six special cases:

Case 1. $N = N_1 = N_2 = q$, q an odd prime,
Case 2. $N_1 = 2 \cdot q$, $N_2 = q$,
Case 3. $N = N_1 = N_2 = 2 \cdot q$,
Case 4. $N = N_1 = N_2 = q^c$,
Case 5. $N_1 = 2 \cdot q^c$, $N_2 = q^c$, and
Case 6. $N = N_1 = N_2 = 2^c$.

Case 1: $N = N_1 = N_2 = q$, q an odd prime. In this case, Euler's theorem simplifies to the Fermat's theorem and we seek the order of p mod q. If p is primitive, then $C_q(u)$ over Z is also irreducible over $Z(p^a)$ and the overall PT over $Z(p^a)$ is identical in form to the PT over Z. If p is not primitive, then $C_q(u)$ over Z factors into m factors over $Z(p^a)$, each of degree $(q-1)/m$. In this case, the PT over $Z(p^a)$ will consist of $(q+1)/m$ POLY products mod cyclotomic POLYs, each of degree m. In Table 1, the properties of some PTs that follow from this analysis are presented.

Example 7. Let $N_1 = 7$ and $N_2 = 7$ and the ring be $Z(4)$. The factorization of $v^7 - 1$ over $Z(4)$ is given by

$$v^7 - 1 = (v-1)(v^3 + 3v^2 + 2v + 3)(v^3 + 2v^2 + v + 3).$$

In this case, we express the computation $X(u, v)\ Y(u, v) \bmod (u^7 - 1, v^7 - 1)$ as a direct sum of

$$X(u, v)\ Y(u, v) \bmod (u^7 - 1, v - 1) = X(u, 1)\ Y(u, 1) \bmod u^7 - 1,$$
$$X(u, v)\ Y(u, v) \bmod (u^7 - 1, v^3 + 3v^2 + 2v + 3),$$

and

$$X(u, v)\ Y(u, v) \bmod (u^7 - 1, v^3 + 2v^2 + v + 3).$$

The first term is a one-dimensional CC of length 7 which is computed the factorization of $u^7 - 1$ as stated above. The second and third terms employ the following factorization over Z(4),

$$u^7 - 1 = \prod_{i=0}^{6}\left(u - v^i\right) \bmod \left(v^3 + 3v^2 + 2v + 3\right)$$

and

$$u^7 - 1 = \prod_{i=0}^{6}\left(u - v^i\right) \bmod \left(v^3 + 2v^2 + v + 3\right),$$

to define a PT. This implies that

$$X(u, v)Y(u, v) \bmod (u^7 - 1, v^3 + 3v^2 + 2v + 3)$$

and

$$X(u, v)Y(u, v) \bmod (u^7 - 1, v^3 + 2v^2 + v + 3)$$

can be written as a direct sum of

$$X(v^i, v)Y(v^i, v) \bmod v^3 + 3v^2 + 2v + 3,\ i = 0, 1, \ldots, 6,$$

and

$$X(v^i, v)Y(v^i, v) \bmod v^3 + 2v^2 + v + 3,\ i = 0, 1, \ldots, 6,$$

respectively. If the product of two POLYs of degree 2 each is computed in 6 MULTs, the two-dimensional CC of length 7×7 over $Z(4)$, and hence over $Z(2^a)$, requires 97 MULTs. It is worthwhile to note here that the corresponding algorithm over Z requires 121 MULTs.

For PT over $CZ(p^a)$, we seek order of p^2 mod q. If p is primitive or if m is even, then $C_q(u)$ over Z factors further over $CZ(p^a)$; otherwise, the PT over $CZ(p^a)$ is identical in form to the PT over $Z(p^a)$. In all cases, $C_q(u)$ over Z factors into s factors over $CZ(p^a)$, each of degree $(q-1)/s$. The PT over $CZ(p^a)$ will consist of $(q+1)/s$ POLY products mod cyclotomic POLYs, each of degree s. In Table 2, the properties of some PTs that follow from this analysis are presented.

Case 2: $N_1 = 2 \cdot q$, $N_2 = q$. Since p is an odd prime (p cannot be even as $(p, N_1) = 1$), we have

$$\text{order of } p \bmod q = m = \text{order of } p \bmod 2 \cdot q. \tag{36}$$

Consequently, the factorization of $u^{2q} - 1$ is obtained as

$$u^{2q} - 1 = (u^q - 1)(u^q + 1). \tag{37}$$

The factorization for $u^q - 1$ has already been discussed. The factorization for $u^q + 1$ can be obtained by replacing u by $-u$ in the factorization for $u^q - 1$. Thus, the two-dimensional convolution $X(u, v)\, Y(u, v) \bmod (u^{2q} - 1, v^q - 1)$ can be written as a direct sum of

$$X(u, v)\, Y(u, v) \bmod (u^q - 1, v^q - 1)$$

and

$$X(u, v)\, Y(u, v) \bmod (u^q + 1, v^q - 1).$$

The second term can be converted to a $q \times q$ two-dimensional CC by making the substitution $u \leftarrow -u$. Thus, the multiplicative complexity of a $2 \cdot q \times q$ two-dimensional CC is twice the multiplicative complexity of a $q \times q$ two-dimensional CC.

Case 3: $N = N_1 = N_2 = 2 \cdot q$. Following the analysis for Case 2 above, the two-dimensional convolution $X(u, v)\, Y(u, v) \bmod (u^{2q} - 1, v^{2q} - 1)$ can be written as a direct sum of

$$X(u, v)\, Y(u, v) \bmod (u^q - 1, v^q - 1),$$

$X(u, v)\, Y(u, v) \bmod (u^q - 1, v^q + 1),$
$X(u, v)\, Y(u, v) \bmod (u^q + 1, v^q - 1),$

and

$X(u, v)\, Y(u, v) \bmod (u^q + 1, v^q + 1).$

The second, third, and the fourth terms can be converted to a $q \times q$ two-dimensional CC by making the substitution $u \leftarrow -u$ and $v \leftarrow -v$ in appropriate places. Thus, the multiplicative complexity of a $2q \times 2q$ two-dimensional CC is four times the multiplicative complexity of a $q \times q$ two-dimensional CC.

Observation. In cases 2 and 3, we have assumed that $p \neq 2$ to ensure that $(p, N_1) = (p, N_2) = 1$. However, if $p = 2$, that is, one wishes to design the algorithm over $Z(2^a)$ or $CZ(2^a)$, then the following approach may be adopted. Convert the $2q \times 2q$ CC to a four-dimensional CC having dimensions $(q \times q) \times (2 \times 2)$ using Agarwal-Cooley algorithm [1]. Now the 2×2 convolution is computed using the direct approach in 9 MULTs and the algorithm for $q \times q$ two-dimensional CC can be designed in the usual manner. The overall algorithm designed in this manner will require 9 times the number of MULTs required for the $(q \times q)$ convolution algorithm. Similarly, the algorithm for $(q \times 2q)$ convolution designed in this manner will require 3 times the number of MULTs required for the $(q \times q)$ convolution algorithm.

Case 4: $N = N_1 = N_2 = q^c$. This is one of the most interesting cases from the standpoint of digital signal processing algorithms and number theory. We are establishing further factorization of $C_M(u)$ (the cyclotomic POLY having all the primitive roots of unity over Z) over $Z(p^a)$ and $CZ(p^a)$. The first step is once again to determine the order of p mod q^c for the factorization over $Z(p^a)$ and the order of p^2 mod q^c for the factorization over $CZ(p^a)$.

Factorization of $C_N(u)$ over $Z(p^a)$. We seek to determine the order of p mod q^c, that is m, for the factorization over $Z(p^a)$. Based on the number theoretic results of Appendix B, a procedure to determine m follows:

Step 1. Determine the order of p mod q. Let it be n. Recall that $n \mid (q - 1)$. If $n = q - 1$, then p is primitive mod q.

Step 2. Determine α such that

$$p^n \equiv 1 \bmod q^i,\ i = 1, 2, \ldots, \alpha \tag{38}$$

but

$$p^n \neq 1 \bmod q^{\alpha+1}. \tag{39}$$

Step 3. The value of m is given by

$$m = n, \tag{40}$$

if $\alpha \leq c$, otherwise

$$m = q^{c-\alpha} \cdot n. \tag{41}$$

If $\alpha = 1$ and $n = q - 1$, then p is primitive mod q^c, for all c. In our work, we have discovered many cases that cover all possibilities. In Table 3, we list prime numbers p and q less than 100 for which $\alpha > 1$.

Example 8. We have the result $3^5 \equiv 1 \bmod 11$ and $3^5 \equiv 1 \bmod 11^2$, but $3^5 \not\equiv 1 \bmod 11^3$. Therefore, $\alpha = 2$, order of 3 mod 11 = order of 3 mod $11^2 = 5$, and order of 3 mod $11^c = 5 \cdot 11^{c-2}$, $c > 2$. This example can be explored further for $p = 47, 113, 487$ and $q = 11$.

Example 9. We have the result $7^4 \equiv 1 \bmod 5$ and $7^4 \equiv 1 \bmod 5^2$, but $7^4 \not\equiv 1 \bmod 5^3$. Therefore, $\alpha = 2$, order of 7 mod 5 = order of 7 mod $5^2 = 4$ and order of 7 mod $5^c = 4 \cdot 5^{c-2}$, $c > 2$. This example can be further explored for $p = 101, 107, 157, 257$, and $q = 5$.

Given m as in (40) or (41), we will use $GF(p^m)$ to factorize $C_N(u)$ over Z further over $GF(p)$ and subsequently use Hensel's theorem to compute the corresponding factors over $Z(p^a)$. If p is primitive mod q^c, that is $m = \phi(q^c) = q^{c-1}(q-1)$, then $C_N(u)$ over Z is irreducible over $Z(p^a)$ as well. In other cases, $C_N(u)$ factors further over $GF(p)$ and $Z(p^a)$ as a product of $\phi(q^c)/m$ factors, each of degree m. If $m = n$, then one can obtain such factors over $GF(p)$ directly by using $GF(p^m)$. In case m is given by (41), then one first factors $C_{q^\alpha}(u)$ over Z in terms of $\phi(q^\alpha)/n$ POLYs over $Z(p)$, each of degree n, using $GF(p^n)$. Next, the corresponding factors over $Z(p^a)$ are computed using Hensel's theorem. Let $A(u)$ be one such factor. The corresponding factor of $C_N(u)$ over $GF(p)$ and $Z(p^a)$ are given by

$$A\left(u^{q^{c-\alpha}}\right) \bmod p \tag{42}$$

and

$$A\left(u^{q^{c-\alpha}}\right) \bmod p^a, \tag{43}$$

respectively.

As a special case, consider $p = 2$ and $N = q^c$, q being a Fermat Prime. The order of 2 mod q is 2^{t+1}. It is seen that $\alpha = 1$. Therefore, the order of 2 mod N is $2^{t+1} \cdot q^{c-1}$. $C_N(u)$ over Z will factor into $(q-1) \cdot 2^{-t-1}$ factors over $Z(2^a)$, each of degree $2^{t+1} \cdot q^{c-1}$. Another special case is $p = 2$ and $N = q^c$, q being a Mersenne prime, $q = 2^e - 1$. The order of 2 mod q is e. It is seen that $\alpha = 1$. Therefore, the order of 2 mod N is $e \cdot q^{c-1}$. $C_N(u)$ over Z will factor into $(q-1)/e$ factors over $Z(2^a)$, each of degree $e \cdot q^{c-1}$. These factorizations also follow the form in (42) and (43).

Example 10. Let us continue Example 8. In this case, $3^5 \equiv 1 \bmod 11^2$, but $3^5 \neq 1 \bmod 11^3$ implies that degree 110 POLY $C_{121}(u)$ over Z will factor into 22 POLYs over $Z(3^a)$, each having degree 5. $GF(3^5)$ and Hensel's theorem are to be used to obtain such factors. A PT over Z for computing 121×121 length convolution over Z requires 132 POLY products mod $C_{121}(u)$ and 12 POLY products mod $C_{11}(u)$. The same convolution over $Z(3^a)$ requires $132 \times 22 + 12 \times 2 = 2928$ POLY products mod cyclotomic POLYs of degree 5 each.

Factorization of $C_N(u)$ over $CZ(p^a)$. In this case, we seek the order of p^2 mod q^c. The results of (30a) and (30b) are valid in this instance, thereby implying that p^2 can never be primitive mod q^c and that $C_N(u)$ always factors further over $CZ(p^a)$. The remainder of the procedure remains the same as described in the context of factorization over $Z(p^a)$ including the expressions in (42) and (43) and the details can be omitted here.

Example 11. Let us continue Example 9. In this case, $7^4 \equiv 1 \bmod 5^2$, but $7^4 \neq 1 \bmod 5^3$ implies that degree 20 POLY $C_{25}(u)$ over Z will factor into 5 POLYs over $Z(7^a)$, each having degree 4 and 10 POLYs over $CZ(7^a)$, each having degree 2. $GF(7^4)$, $GF(49^2)$, and Hensel's theorem are to be used to obtain such factors. A PT over Z for computing 25×25 length convolution over Z requires 31 POLY products mod $C_{25}(u)$ and 6 POLY products mod $C_5(u)$. The same convolution over $Z(7^a)$ requires $31 \times 5 + 6 = 161$ POLY products mod cyclotomic POLYs of degree 4 each. Over $CZ(7^a)$, it requires $(31 \times 5 + 6) \times 2 = 322$ POLY products mod cyclotomic POLYs of degree 2 each.

Case 5: $N_1 = 2 \cdot q^c$, $N_2 = q^c$. This case can be analyzed in the same way as Case 2. The two-dimensional convolution can be written as a direct sum of two $q^c \times q^c$ two-dimensional CCs. Similarly, if $N = N_1 = N_2 = 2 \cdot q^c$, then the two-dimensional convolution can be written as a direct sum of four $q^c \times q^c$ two-dimensional CCs.

Case 6: $N = N_1 = N_2 = 2^c$. In this case, as we seek the order of p mod 2^c, we recall that p can never be primitive. Therefore, degree 2^{c-1} POLY $C_N(u)$ over Z always factors over $Z(p^a)$; each such factor over $Z(p^a)$ in turn always factors further over $CZ(p^a)$. It has been established in Appendix B that the order of p mod 2^c is given by 1 for $c \leq d$ and 2^{c-d} for $c > d$, if p has the form

$$p = 1 + 2^d \cdot B, \tag{44}$$

B being an odd integer and $d \geq 2$ ($p \equiv 1$ mod 4). Similarly, order of p mod 2^c is given by 2 for $c \leq d + 2$ and 2^{c-d-1} for $c > d + 2$, if p has the form

$$p = 1 + 2 + \ldots + 2^d + 2^f \cdot B = 2^{d+1} - 1 + 2^f \cdot B, \tag{45}$$

B being an odd integer and $f \geq d + 2$ ($p \equiv 3$ mod 4), $d > 0$.

These statements establish that if $p \equiv 1$ mod 4, then $C_N(u)$ factors into degree 1 factors for $c \leq d$ and 2^{d-1} factors of degree 2^{c-d} if $c > d$ over $Z(p^a)$. For $c > d$, these 2^{d-1} factors over $Z(p^a)$ are obtained by first obtaining 2^{d-1} factors each of degree 2 for $c = d + 1$. Then the degree 2^{c-d} factors are obtained by replacing u with $u^{2^{c-d-1}}$ in each of the degree 2 POLYs.

Similarly, if $p \equiv 3$ mod 4, then $C_N(u)$ factors into degree 2 factors for $c \leq d + 2$ and 2^d factors of degree 2^{c-d-1}, if $c > d + 2$ over $Z(p^a)$. For $c > d + 2$, these 2^d factors over $Z(p^a)$ are obtained by first obtaining 2^d factors each of degree 4 for $t = d + 3$. Then the degree 2^{c-d-1} factors are obtained by replacing u with $u^{2^{c-d-3}}$ in each of the degree 4 POLYs. All factors over $Z(p^a)$ for $p \equiv 3$ mod 4 factorize further over $CZ(p^a)$ and can be obtained in a similar manner.

Example 12. Let p be a Fermat prime, $p = 2^n + 1$, $n = 2^t$, $p > 3$. Then $p \equiv 1$ mod 4 holds with $d = n$. Therefore, $C_N(u)$ factors into degree 1 factors for $c \leq n$ and 2^{n-1} factors of degree 2^{c-n}, if $c > n$ over $Z(p^a)$.

Example 13. Let p be a Mersenne prime, $p = 2^e - 1$. Then $p \equiv 3$ mod 4 holds with $d + 1 = e$. Therefore, $C_N(u)$ factors into degree 2 factors for $c \leq e + 1$ and 2^{e-1} factors of degree 2^{c-e}, if $c > e + 1$ over $Z(p^a)$. Once again, all of these factors factor further over $CZ(p^a)$. In this analysis, the Fermat prime 3 is to be treated as a Mersenne prime.

Example 14. Let $p = 151 = 1 + 2 + 4 + 16 \cdot 9$ and $c = 9$. Here, (44) holds with $d = 2$ and $C_{512}(u)$ factors into 4 factors of degree 64 over $Z(151^a)$. Over $CZ(151^a)$, it factors into 8 factors of degree 32.

VI. RECURSIVE COMPUTATION IN FINITE INTEGER RINGS

We now return to Cases 4, 5, and 6. In these cases, there are instances where the mod POLY defining the PT has the form $A(u^b)$, $b = q^d$. If $A(u)$ has the q^fth root of unity in $Z(r)$ or $CZ(r)$, then $A(u^b)$ has the q^{d+f}th root of unity in $Z(r)$ or $CZ(r)$. The one-dimensional computation (obtained as a result of PT) is an expression of the type

$$X(u)\ Y(u) \bmod A(u^b). \tag{46}$$

This can be computed using recursive nesting in the following manner. Let $v = u^g$. Now, (46) can be expressed as

$$X(u)\ Y(u) \bmod (u^g - v, A(v^h)) \tag{47}$$

or

$$X(u)\ Y(u) \bmod A(v^h), \tag{48}$$

where $b = g \cdot h$. The ordinary POLY product in u (with coefficients that are POLYs in v defined mod $A(v^h)$) has degree $2g - 2$. It can, therefore, be computed by interpolating u on $2g - 1$ powers of v, that is, $v^0, v^1, ..., v^{2g-2}$, provided that $(2g - 1) \leq h \cdot q^f$. Due to the special form of b, suitable g and h can always be found to satisfy the conditions. This technique may be most useful for $q = 2$. In case $q \neq 2$, interpolating powers of v can also be chosen as $\pm v^0, \pm v^1, ..., \pm v^{g-2}, v^g$, provided that $(g + 1) \leq h \cdot q^f$.

TABLE 1. Properties of Polynomial Transforms Over $Z(p^a)$ for $N = q$, q an Odd Prime, $(p, q) = 1$

N	$p = 2$	$p = 3$	$p = 5$
3	2^*; 1, 2; 4, 2	-	2^*; 1, 2; 4, 2
5	4^*; 1, 4; 6, 4	4^*; 1, 4; 6, 4	-
7	3; 1, 3, 3; 16, 3	6^*; 1, 6; 8, 6	6^*; 1, 6; 8, 6
11	10^*; 1, 10; 12, 10	5; 1, 5, 5; 24, 5	5; 1, 5, 5; 24, 5
13	12^*; 1, 12; 14, 12	3; 1, 3, 3, 3, 3; 42, 3	4; 1, 4, 4, 4; 42, 3
17	8; 1, 8, 8; 36, 8	16^*; 1, 16; 18, 16	16^*; 1, 16; 18, 16
19	18^*; 1, 18; 20, 18	18^*; 1, 18; 20, 18	9; 1, 9, 9; 40, 9
23	11; 1, 11, 11; 48, 11	11; 1, 11, 11; 48, 11	22^*; 1, 22; 24, 22
29	28^*; 1, 28; 30, 28	28^*; 1, 28; 30, 28	14; 1, 14, 14; 60, 14
31	5; 1, 5, 5, 5, 5, 5; 160, 5	30^*; 1, 30; 32, 30	3; 1, 10 polys of degree 3; 320, 3
37	36^*; 1, 36; 38, 36	18; 1, 18, 18; 76, 18	36^*; 1, 36; 38, 36
$2^e - 1$: a Mersenne Prime	e; 1, $(2^e - 2)/e$ polys of deg e; $e \cdot 2^e$, e	-	-
$2^n + 1$, $n = 2^t$: a Fermat prime	2^{t+1}; 1, 2^{n-t-1} polys of degree 2^{t+1}; $2^{t+2}(2^n + 1)$, 2^{t+1}	2^{n*}; 1, 2^n; 2^n + 2, 2^n	-

Note: The entries are marked with a; b_1, ..., b_n; c; d. Here, a denotes the order of p mod q. A star denotes a primitive element. b_1, ..., b_n denote the degrees of the factors of $u^N - 1$ over $Z(p^a)$. c and d denote the number of POLY products and the degree of the mod cyclotomic POLYs for the two-dimensional PT in $Z(p^a)$.

TABLE 2. Properties of Polynomial Transforms Over $CZ(p^a)$ for $N = q$, q an Odd Prime, $(p, q) = 1$, $p \equiv 3 \bmod 4$

N	$p = 3$	$p = 7$	$p = 11$
3	-	1; 1, 1, 1; 9, 1	1; 1, 1, 1; 9, 1
5	2; 1, 2, 2; 12, 2	2; 1, 2, 2; 12, 2	1; 1, 1, 1, 1, 1; 25, 1
7	3; 1, 3, 3; 16, 3	-	3; 1, 3, 3; 16, 3
11	5; 1, 5, 5; 24, 5	5; 1, 5, 5; 24, 5	-
13	3; 1, 3, 3, 3, 3; 42, 3	6; 1, 6, 6; 28, 6	6; 1, 6, 6; 28, 6
17	8; 1, 8, 8; 36, 8	8; 1, 8, 8; 36, 8	4; 1, 4, 4, 4, 4; 72, 4
19	9; 1, 9, 9; 40, 9	3; 1, 3, 3, 3, 3, 3, 3; 120, 3	3; 1, 3, 3, 3, 3, 3, 3; 120, 3
23	11; 1, 11, 11; 48, 11	11; 1, 11, 11; 48, 11	11; 1, 11, 11; 48, 11
29	14; 1, 14, 14; 60, 14	7; 1, 7, 7, 7, 7; 120, 7	14; 1, 14, 14; 60, 14
31	15; 1, 15, 15; 64, 15	15; 1, 15, 15; 64, 15	15; 1, 15, 15; 64, 15
37	9; 1, 9, 9, 9, 9; 152, 9	9; 1, 9, 9, 9, 9; 152, 9	3; 1, 12 polys of deg 3; 456, 3
$2^n + 1$, $n = 2^t$: a Fermat Prime	2^{n-1}; 1, 2^{n-1}, 2^{n-1}; $2(2^n + 2)$, 2^{n-1};	-	-

Note: The entries are marked with a; $b_1, \ldots, b_n$; c; d. Here, a denotes the order of p^2 mod q. There are no primitive elements. $b_1, \ldots, b_n$ denote the degrees of the factors of $u^N - 1$ over $CZ(p^a)$. c and d denote the number of POLY products and the degree of the mod cyclotomic POLYs for the two-dimensional PT in $CZ(p^a)$.

Table 3. Prime pairs p, $q < 100$ for which $\alpha > 1$

q	11	3	3	3	3	3	3	5	5	7	7	7	7
p	3	17	19	37	53	71	89	7	43	19	31	67	79
order of p mod q	5	16	1	1	2	2	2	4	4	6	6	3	3
α	2	2	2	2	3	2	2	2	2	3	2	2	2
q	7	71	13	13	13	43	29	79	47	47	47	59	97
p	97	11	19	23	89	49	41	31	53	67	71	53	53
order of p mod q	2	70	12	6	12	42	4	39	23	46	23	29	48
α	2	2	2	2	2	2	2	2	2	2	2	2	2

VII. CONCLUSIONS

Our efforts have been focused on unifying number theoretic results as they relate to the structure of fast digital signal processing algorithms over different number systems. In this regard, the present work deals with the polynomial transforms in finite integer and complex integer rings. A fundamental result of this work is that under the non-restrictive condition, $(N_1, M) = (N_2, M) = 1$, the polynomial transforms defined in finite integer and complex integer rings are as intensive computationally as the polynomial transforms defined in rational and complex rational number system **only in the worst case**. They simplify considerably for many special cases of significance in digital signal processing.

REFERENCES

[1] H.J. Nussbaumer, *Fast Fourier Transform and Convolution Algorithms*, Springer-Verlag, 1981.

[2] D.G. Myers, *Digital Signal Processing: Efficient Convolution and Fourier Transform Techniques*, Prentice Hall, 1990.

[3] J.H. McClellan and C.M. Rader, *Number Theory in Digital Signal Processing*, Prentice Hall, 1979.

[4] H. Krishna, B. Krishna, K.-Y. Lin, and J.-D. Sun, *Computational Number Theory and Digital Signal Processing*, CRC Press, 1994.

[5] H. Krishna Garg, *Digital Signal Processing Algorithms*, CRC Press, 1997.

[6] B. Arambepola and P.J.W. Rayner, "Discrete Transforms Over Polynomial Rings with Applications in Computing Multidimensional Convolutions," *IEEE Trans. on Acoustics, Speech & Signal Proc.*, Vol. ASSP-28, pp. 407-414, 1980.

[7] R.H.V. Kraats and A.N. Venetsanopoulos, "Hardware for Two-Dimensional Digital Filtering using Fermat Number Transforms," *IEEE Trans. on Acoustics, Speech & Signal Proc.*, Vol. ASSP-30, pp. 155-161, 1982.

[8] J.B. Martens, "Two-Dimensional Convolutions by Means of Number Theoretic Transforms Over Residue Class Polynomial Rings," *IEEE Trans. on Acoustics, Speech and Signal Proc.*, Vol. ASSP-32, pp. 862-871, 1984.

[9] T.K. Troung, I.S. Reed, R.G. Lipes, and C. Wu, "On the Application of a Fast Polynomial Transform and the Chinese Remainder Theorem to Compute a

Two-Dimensional Convolution," *IEEE Trans. on Acoustics, Speech and Signal Processing*, Vol. ASSP-29, pp. 91-97, 1981.

[10] G. Martinelli, "Long Convolutions Using Number Theoretic and Polynomial Transforms," *IEEE Trans. on Acoustics, Speech & Signal Proc.*, Vol. ASSP-32, pp. 1090-1092, 1984.

[11] H. Krishna Garg, "On the Factorization of Polynomials and Direct Sum Properties of Integer Polynomial Rings," *Circuits, Systems, and Signal Processing*, Vol. 15, no. 3, pp. 415-435, 1996.

[12] H. Krishna Garg and C.C. Ko, "Fast Algorithms for Computing One- and Two-Dimensional Convolution in Integer Polynomial Rings," *Circuits, Systems, and Signal Processing*, Vol. 16, no. 1, pp. 121-139, 1997.

[13] H. Krishna Garg, C.C. Ko, and H. Liu, "On Algorithms for Digital Signal Processing of Sequences," *Circuits, Systems, and Signal Processing*, Vol. 15, no. 4, pp. 437-452, 1996.

[14] M.T. HeideMan, *Multiplicative Complexity, Convolution, and the DFT*, Springer-Verlag, 1988.

[15] R.J. McElice, *Finite Fields for Computer Scientists and Engineers*, Kluwer Academic Publishers, 1987.

APPENDIX A
An Algorithm for Computing Factorization Over Finite Integer Rings

Given the factorization of $u^N - 1$ in terms of monic factors in $GF(p)$, $(N, p) = 1$, as

$$u^N - 1 = A_0(u)\, B_0(u) \bmod p, \tag{A-1}$$

we proceed to compute

$$u^N - 1 = A(u)\, B(u) \bmod p^a \tag{A-2}$$

in a recursive manner by computing

$$u^N - 1 = A_k(u)\, B_k(u) \bmod p^k,\ k = 2, 3, \ldots, a. \tag{A-3}$$

This is done using the Hensel's theorem. Given the factorization mod p^k as in (A-3), we express the factorization mod p^{k+1} as

$$u^N - 1 = \{A_k(u) + p^k A^{(k)}(u)\}\ \{B_k(u) + p^k B^{(k)}(u)\} \bmod p^{k+1}. \tag{A-4}$$

Cross-multiplying and simplifying using (A-3), we get

$$\left[p^{-k}\left\{u^N - 1 - A_k(u)B_k(u)\right\}\right] \bmod p = A_0(u)B^{(k)}(u) + B_0(u)A^{(k)}(u) \bmod p. \tag{A-5}$$

Using the formal derivative of (A-1), it is seen that

$$B^{(k)}(u) \equiv \left\{\frac{u}{N} B_0'(u)\left[p^{-k}\left\{u^N - 1 - A_k(u)B_k(u)\right\}\right] \bmod p\right\} \bmod\left(B_0(u), p\right) \tag{A-6}$$

and

$$A^{(k)}(u) \equiv \left\{\frac{u}{N} A_0'(u)\left[p^{-k}\left\{u^N - 1 - A_k(u)B_k(u)\right\}\right] \bmod p\right\} \bmod\left(A_0(u), p\right). \tag{A-7}$$

Similar equations can also be derived when one wishes to obtain factorization over $CZ(r)$ starting from the factorization over $GF(p^2)$. In that case, the mod operation is defined as mod $(p, j^2 + 1)$. Note that the procedure as described here is computationally simpler and completely bypasses the Euclid's algorithm for solving POLY recurrence in (A-5).

APPENDIX B
Some Number Theoretic Results of Interest

Let p be a primitive root of q, q being an odd prime. Here, p does not have to be a prime integer. We seek the order of p mod q^c. In other words, given

$$p^{q-1} \equiv 1 \bmod q, \tag{B-1}$$

we seek the smallest integer m such that

$$p^m \equiv 1 \bmod q^c. \tag{B-2}$$

Recalling Euler's theorem and combining it with (B-1), we find that m must have the form

$$m = q^a(q-1). \tag{B-3}$$

Given (B-1), we now seek the value of α such that

$$p^{q-1} \equiv 1 \bmod q^i,\ i = 1, 2, \ldots, \alpha, \tag{B-4}$$

but

$$p^{q-1} \not\equiv 1 \bmod q^{\alpha+1}. \tag{B-5}$$

Theorem. Given α as defined in (B-4) and (B-5), the value of m is given by

$$m = q^{c-\alpha}(q-1). \tag{B-6}$$

Proof: The relations in (B-4) and (B-5) imply that

$$p^{q-1} = 1 + b \cdot q^{\alpha}, \tag{B-7}$$

such that $(b, q) = 1$. Raising both sides to q-th power, we get

$$p^{q(q-1)} = (1 + b \cdot q^{\alpha})^q = 1 + b \cdot q^{\alpha+1} + b^2 \cdot q^{2\alpha+1} \cdot (q-1)/2.$$

This leads to

$$p^{q(q-1)} \equiv 1 \bmod q^i,\ i = 1, 2, \ldots, \alpha + 1, \tag{B-8}$$

but

$$p^{q(q-1)} \not\equiv 1 \bmod q^{\alpha+2}. \tag{B-9}$$

Comparison of (B-8) and (B-9) with (B-6) and (B-7) reveals that we can apply the same argument in a recursive manner to get

$$p^{q^t(q-1)} \equiv 1 \bmod q^{\alpha+t}.$$

Setting $\alpha + t = c$, we get the final result.

If $\alpha = 1$, then we get the well-known result that if p is primitive mod q^2, then it is also primitive mod q^c. Now, let p be a root of order n mod q, q being an odd prime. Once again, p does not have to be a prime integer. It is clear that n is a factor of $(q-1)$. We seek the order of p mod q^c. In other words, given

$$p^n \equiv 1 \bmod q, \tag{B-10}$$

we seek the smallest integer m such that

$$p^m \equiv 1 \bmod q^c. \tag{B-11}$$

Lemma 1. Given that order of p mod q is n, order of p mod q^c is an integer of the type

$$m = q^k \cdot n. \tag{B-12}$$

Proof: Let $c = 2$. We show that order of p mod q^2 is either n or $q \cdot n$. Given (B-10), we have

$$p^n = 1 + a \cdot q. \tag{B-13}$$

If $(a, q) = q$, order of p mod q^2 is n. Otherwise, raising both sides of (B-13) by b such that $(b, q) = 1$, we get

$$p^{bn} = (1 + a \cdot q)^b = 1 + b \cdot a \cdot q + [b(b-1)/2]\, a^2 \cdot q^2 + \dots .$$

This leads to

$$p^{bn} \equiv 1 + b \cdot a \cdot q \bmod q^2.$$

Since $(a, q) = (b, q) = 1$, we have

$$p^{bn} \neq 1 + b \cdot a \cdot q \bmod q^2.$$

It is clear that order of p mod q^2 is of the type $q \cdot n$ in this case. Thus, order of p mod q^2 is either n or $q \cdot n$. Using the same argument recursively, we can show that the order of p mod q^c is $m = q^k \cdot n$.

Once again, following Theorem 1, we now seek the value of α such that

$$p^n \equiv 1 \bmod q^i,\ i = 1, 2, \dots, \alpha \tag{B-14}$$

but

$$p^n \neq 1 \bmod q^{\alpha+1}. \tag{B-15}$$

Theorem. Given α as defined in (B-14) and (B-15), the value of m is given by

$$m = q^{c-\alpha} \cdot n. \tag{B-16}$$

Proof: The proof of this theorem is based on Lemma 1 and follows the proof of Theorem 1 very closely. Therefore, it is omitted here.

Now, let us turn to $q = 2$. Here, p does not have to be a prime integer; p is any odd integer. We seek the order of p mod 2^c, $c > 2$. There are no primitive elements mod 2^c. Given p, we have two cases, $p \equiv 1$ mod 4 and $p \equiv 3$ mod 4. They have to be analyzed separately.

Case 1. $p \equiv 1$ mod 4. In this case, we can write

$$p = 1 + 2^d \cdot B,$$

B being an odd integer and $d \geq 2$. Squaring both sides, we get

$$p^2 = 1 + 2^{d+1}(B + 2^{d-1}B^2).$$

These expressions lead to $p \equiv 1 \bmod 2^i$, $i = 1, 2, ..., d$, but $p \neq 1 \bmod 2^{d+1}$, $p^2 \equiv 1 \bmod 2^{d+1}$, but $p^2 \neq 1 \bmod 2^{d+2}$. Using the same argument in a recursive manner, we get $p^{2^k} \equiv 1 \bmod 2^{d+k}$, but $p^{2^k} \neq 1 \bmod 2^{d+k+1}$.

Therefore, order of $p = 1$ for $c \leq d$ and order of $p = 2^{c-d}$ for $c > d$.

Case 2. $p \equiv 3 \bmod 4$. In this case, we can write

$$p = 1 + 2 + ... + 2^d + 2^f \cdot B = 2^{d+1} - 1 + 2^f \cdot B,$$

B being an odd integer and $f \geq d + 2$, $d > 0$. $B = 0$ is also to be treated the same way if B is odd. Squaring both sides, we get

$$p^2 = 1 + 2^{d+2}(-1 - B \cdot 2^{f-d-1} + B \cdot 2^f + 2^d + B^2 \cdot 2^{2f-d-2}).$$

The term in the parenthesis is always odd. Therefore, $p^2 \equiv 1 \bmod 2^i$, $i = 1, 2, ..., d + 2$, but $p^2 \neq 1 \bmod 2^{d+3}$. The form for p^2 is the same as in Case 1. Following the same analysis as in Case 1, we get $p^{2^{k+1}} \equiv 1 \bmod 2^{d+k+2}$ but $p^{2^{k+1}} \neq 1 \bmod 2^{d+k+3}$. Therefore, order of $p = 2$ for $c \leq d + 2$ and order of $p = 2^{c-d-1}$ for $c > d + 2$.

Paper two

On Fast Algorithms for One-Dimensional Digital Signal Processing in Finite Integer and Complex Integer Rings

Abstract. In this work, we present and analyze a number theoretic approach to computing one-dimensional cyclic convolution of sequences defined in finite integer and complex integer rings. A fundamental result of this work is that under the non-restrictive condition, $(N, M) = 1$, the algorithms defined in finite integer and complex integer rings are as intensive computationally as the corresponding algorithms defined in rational and complex rational number system **only in the worst case**. They simplify considerably for a large number of cases of importance in digital signal processing.

I. INTRODUCTION

The main idea behind the computation of convolution in digital signal processing (DSP) consists in expressing a large size problem as a direct sum of a number of smaller size problems which are then computed in parallel [1] to [4]. The Chinese remainder theorem in its integer form (CRT-I), polynomial form (CRT-P), and its polynomial extensions in integer and integer polynomial rings (AICE-CRT) provides the basis for all these direct sums [4].

In this paper, we study the computation of one-dimensional cyclic convolution (CC) of length N of data sequences defined in finite integer and finite complex integer rings, denoted by $Z(M)$ and $CZ(M)$, respectively. It is generally accepted that the CC algorithms in these number systems suffer from word sequence length constraint (WSLC) problem due to the non-existence of discrete Fourier transform (DFT) in them except under very restrictive conditions. The DFTs defined in $Z(M)$ and $CZ(M)$ are termed as the number theoretic transform (NTT) and complex NTT (CNTT), respectively. This issue has been addressed in the recent work of the author and AICE-CRT based CC algorithms have been described in $Z(M)$ and $CZ(M)$ [4].

The main purpose of this paper is as follows. We compare the CC algorithms in the field of rational and complex rational numbers (denoted by Z and CZ, respectively), to their counterparts in $Z(M)$ and $CZ(M)$ using a number theoretic approach. A fundamental result of this work is that under the non-restrictive condition, $(N, M) = 1$, the algorithms defined in $Z(M)$ and $CZ(M)$ are as intensive computationally as the corresponding algorithms defined in Z and CZ **only in the worst case**. In fact, they are computationally simpler in a large number of cases of interest in DSP. These cases include (i) $M = 2^a$ and $N = q^c$, q a Mersenne or a Fermat prime, and (ii) $N = 2^c$ and M is a power of a Mersenne or a Fermat prime.

The focus is on algorithms for the 1-D CC of length N in $Z(M)$ and $CZ(M)$:

$$Z(u) = X(u)\,Y(u) \mod (u^N - 1), \tag{1}$$

$X(u)$, $Y(u)$, and $Z(u)$ being polynomials (POLYs) having degree $N - 1$. The finite integer ring, $Z(M)$, and the finite complex integer ring, $CZ(M)$, consist of integers $\{0, 1, ..., M - 1\}$ and $\{a + jb,\ a, b \in Z(M), j^2 = -1 \bmod M\}$, with the operations of ADD and MULT defined mod M and mod $(M, j^2 + 1)$, respectively. **A condition of fundamental importance that will be assumed to be satisfied throughout this work is that**

$$(N, M) = 1, \tag{2}$$

that is, the length N and M are relatively coprime. This condition ensures that the multiplicative inverse of N exists in $Z(M)$ and $CZ(M)$. All those cases that we feel are of most interest to DSP are studied in depth. They are $N = q$, $N = q^c$, $N = 2^c$, $N = 2^{c_0} \cdot q^c$, $N = 2^{c_0} \cdot q_1^{c_1} \cdot q_2^{c_2}$, q, q_1, and q_2 odd primes, and N contains more than two odd prime factors in its standard factorization. These also have to be treated separately from a number-theoretic point of view, as the theory of primitive roots has a different treatment for these cases.

This paper is organized as follows. In Section II, a brief description of CC algorithms in *Z*, *CZ*, *Z(M)*, and *CZ(M)* is presented. Several special cases of lengths which arise in DSP are also mentioned. Section III is on cyclotomic POLY factorization in *Z(M)* and *CZ(M)*. In Section IV, results pertaining to further factorization of cyclotomic POLYs in *Z* and *CZ* in *Z(M)* and *CZ(M)* are presented. In Section V, the special cases of most interest in DSP are analyzed. Section VI is on recursive computation of POLY products required in CC algorithms. Section VII is on the application of the results to CC of length $N = 60$. Conclusions are presented in Section VIII.

II. INTERRELATIONSHIPS AMONG VARIOUS FACTORIZATIONS AND ALGORITHMS

Consider the cyclotomic factorization in *Z*,

$$u^N - 1 = \prod_{i=0}^{N-1}\left(u-\omega^i\right) = \prod_{\substack{(i,d)=1 \\ d|N}}\left(u-\omega^i\right) = \prod_{d|N} C_d(u), \qquad (3)$$

where $\omega = \exp(-2\pi j/N)$. In (3), $C_d(u)$ is a cyclotomic POLY of degree $\phi(d)$, having integer coefficients, $\phi(.)$ being the Euler's totient function. Thus, $Z(u)$ in (1) can be written as a direct sum of $Z_d(u)$ using the CRT-P corresponding to the factorization in (3), where

$$Z_d(u) \equiv Z(u) \equiv X_d(u)\, Y_d(u) \bmod C_d(u). \qquad (4)$$

The CRT-P reconstruction with respect to the above direct sum requires division by *N* [4]. The computational complexity of the CC algorithm is determined by the degrees of the modulo POLYs ($\phi(d)$ in this instance) and in general a large number of factors (each having smaller degree) result in a more efficient algorithm. Those cyclotomic POLYs in (3) for which $4 \mid d$ factor further in *CZ* [7]. The degree of such factors is given by $\phi(d)/2$. The structure of the rest of the CC algorithm remains the same as the one in *Z*. Once again, the CRT reconstruction requires division by *N*.

The cyclotomic factorization in *Z* and *CZ* is valid in *Z(M)* and *CZ(M)*, respectively, under the condition that $(N, M) = 1$. The second aspect of the CC algorithm is the existence of algorithms for acyclic convolution of two degree $\phi(d) - 1$ POLYs in *Z* or two degree $\phi(d)/2$ POLYs for *CZ* (if $4 \mid d$) as shown in (4). If these algorithms are valid in *Z(M)* and *CZ(M)* also, then the computational complexity and structure of the CC algorithm in *Z(M)* and *CZ(M)* is exactly same as those in *Z* and *CZ* [5]. In our study, we have come across a wide range of small degree acyclic convolution algorithms that satisfy such a condition. If d is of the type 2^c, q^c or $2 \cdot q^c$, q an odd prime, then recursive nesting can be used very effectively to derive these acyclic convolution algorithms due to the form of $C_d(u)$ [1]. These algorithms require division by d and their validity in $Z(M)$ and $CZ(M)$ requires that $(d, M) = 1$, a condition trivially satisfied as $(N, M) = 1$ and $d \mid N$. Consequently, recursive nesting can be used for sequences defined in $Z(M)$ and $CZ(M)$ as well.

Example 1. The cyclotomic factorization

$$u^6 - 1 = (u - 1)(u + 1)(u^2 - u + 1)(u^2 + u + 1)$$

is valid in Z, CZ, and those $Z(M)$ and $CZ(M)$ for which $(6, M) = 1$. This will lead to a CC algorithm requiring 8 multiplications (MULTs) in all of these number systems.

Example 2. The cyclotomic factorization

$$u^{12} - 1 = (u - 1)(u + 1)(u^2 - u + 1)(u^2 + u + 1) \cdot (u - j)(u + j)(u^2 - ju - 1)(u^2 + ju - 1)$$

is valid in CZ and those $CZ(M)$ for which $(12, M) = 1$. This will lead to a CC algorithm requiring 16 MULTs in these number systems.

Another approach that is widely used is the Agarwal-Cooley algorithm which consists in converting 1-D CC of length N to t-dimensional CC of length $N_1 \times N_2 \times \cdots \times N_t$ provided that $N = N_1 \cdot N_2 \cdots N_t$, and $(N_i, N_k) = 1$, $i \neq k$. **This approach is valid for sequences defined in all number systems as it is based on number-theoretic properties of indices.**

We now turn to the direct design of CC algorithms in $Z(M)$ and $CZ(M)$ and show that in a vast majority of cases they are computationally simpler than the CC algorithms in Z and CZ. This is accomplished by showing that the degrees of the cyclotomic factors in $Z(r)$ and $CZ(r)$ are lower than the degrees of corresponding cyclotomic factors in Z and CZ.

III. FACTORIZATION IN FINITE INTEGER AND COMPLEX INTEGER RINGS

Given the standard factorization of M as

$$M = \prod_{i=1}^{l} p_i^{a_i} = \prod_{i=1}^{l} r_i, \tag{5}$$

the following direct sums are obtained using the CRT-I:

$$Z(M) = \sum_{\substack{\oplus \\ i=1}}^{l} Z(r_i) \tag{6}$$

$$CZ(M) = \sum_{\substack{\oplus \\ i=1}}^{l} CZ(r_i). \tag{7}$$

A straightforward consequence of the above analysis is that we need to derive algorithms for the convolutions in $Z(r_i)$ and $CZ(r_i)$ and then combine the resulting matrices using the direct sum properties. This approach is adopted throughout this paper. Also, the subscript i will be deleted. This will simplify notation without creating any confusion.

Our primary objective is to derive fast algorithms for the following:

$$Z(u) = X(u)\ Y(u) \mod (r, u^N - 1). \tag{8}$$

In $CZ(r)$, all the quantities are degree one POLY in j with the computation defined mod $(j^2 + 1)$. Following CC algorithms in Z and CZ, we will pursue CC algorithms in $Z(r)$ and $CZ(r)$ in the following manner.

Step 1. Seek the *monic* factorization of $u^N - 1$ in $GF(p)$ and $GF(p^2)$, $r = p^a$;
Step 2. Seek the *monic* factorization of $u^N - 1$ in $Z(r)$ and $CZ(r)$;
Step 3. Express the CC as the direct sum of acyclic convolution using the factorization in $Z(r)$ and $CZ(r)$.

A major effort of this work is in showing that the cyclotomic POLY $C_d(u)$ factors further in $Z(r)$ and $CZ(r)$ in a number of cases. All such cases lead to reduced complexity CC algorithms.

It is well known that $u^N - 1$ factors as the product of N degree one POLYs in $Z(r)$ if and only if (*iff*) N divides $(p - 1)$. Similarly, $u^N - 1$ factors as the product of N degree one POLYs in $CZ(r)$ *iff* N divides $(p^2 - 1)$. Collectively, this constraint is known as the WSLC.

The crucial task now is to study the form of cyclotomic POLYs in Z and CZ and establish the corresponding versions in $Z(r)$ and $CZ(r)$. Let $C_d(u)$ in Z or CZ factor further in $Z(r)$ or $CZ(r)$ as

$$C_d(u) = A_1(u) \cdot A_2(u) \cdots A_L(u),$$

each of degree m, $m = \phi(d) / L$. Based on this, we can express $X(u)\ Y(u)$ mod $C_d(u)$ as a direct sum of L computations $X_k(u)\ Y_k(u)$ mod $A_k(u)$, $k = 1, 2, ..., L$ in $Z(r)$ or $CZ(r)$, each one being a product of two POLYs of degree $m - 1$. The AICE-CRT in $Z(r)$ and $CZ(r)$ forms the basis of these statements [4]. The CC algorithms in $Z(r)$ and $CZ(r)$ are computationally simpler to their counterparts in Z and CZ as long as $L > 1$. If $L = 1$, then they are identical in form. In this regard, all the six cases mentioned in Section I will be studied. In the following, we will pursue the factorization of $u^N - 1$ directly in $Z(r)$ and $CZ(r)$ and obtain considerably simplified forms for same.

A factorization of $u^N - 1$ in terms of monic factors in $Z(r)$ is an expression of the type

$$u^N - 1 = A(u) \cdot B(u) \bmod r, \tag{9}$$

where $A(u)$ and $B(u)$ are monic and relatively coprime. Taking mod p on both sides, we see that it also provides us with the corresponding factorization in $GF(p)$,

$$u^N - 1 = A_0(u) \cdot B_0(u) \bmod p, \tag{10}$$

where $A_0(u)$ and $B_0(u)$ are equal to $A(u)$ mod p and $B(u)$ mod p, respectively. One can derive the factorization of $u^N - 1$ in $GF(p)$ using the finite field algebra [8]. In Appendix A (Paper one), we describe a Hensel's theorem based procedure which can be used to compute the corresponding factors in $Z(r)$ and $CZ(r)$.

In a similar manner, the factorization of $u^N - 1$ in terms of monic factors in $CZ(r)$ can also be obtained. For $p = 2, j^2 + 1 = (j + 1)^2$ mod 2. For $p \equiv 3$ mod 4, $j^2 + 1$ is irreducible in $GF(p)$ and for $p \equiv 1$ mod 4, $j^2 + 1$ factors in terms of two mutually coprime degree one POLYs in $GF(p)$ as $j^2 + 1 = (j + a_0) \cdot (j + b_0)$ mod p. Therefore, $j^2 + 1$ is irreducible in $Z(r)$, $p \equiv 3$ mod 4 and factors as $j^2 + 1 = (j + a) \cdot (j + b)$ in $Z(r)$, $p \equiv 1$ mod 4, $a_0 = a$ mod p and $b_0 = b$ mod p. Three distinct cases follow from these statements.

Case 1. $p = 2$. In this case, the algorithms for CC in $CZ(r)$ are the same as those in $Z(r)$. Only the input and output sequences are defined in $CZ(r)$.

Case 2. $p \equiv 1 \bmod 4$. Based on the CRT, we have the direct sum,

$$CZ(r) = Z(r, j + a) \oplus Z(r, j + b). \tag{11}$$

Therefore, the CC of complex integer-valued sequences in $CZ(r)$, $p \equiv 1 \bmod 4$, can be expressed as a direct sum of two sequences in $Z(r)$.

Case 3. $p \equiv 3 \bmod 4$. In this case, $CZ(r)$ will be studied as the p-adic expansion ring of $GF(p^2)$. The analysis for $CZ(r)$ is similar to the analysis for $Z(r)$ in (9) to (11), (A-1) to (A-9).

In the context of designing algorithms in $CZ(r)$, we will deal only with $p \equiv 3 \bmod 4$. The analysis for the other two cases as outlined above is complete.

III.1 Factorization of $u^N - 1$ in $Z(r)$, $r = p^a$. The factorization of $u^N - 1$ in $Z(r)$ begins with finding the least integer m such that [4]

$$N \mid (p^m - 1). \tag{12}$$

In other words,

$$p^m \equiv 1 \bmod N \tag{13}$$

and, therefore, m is the order of p mod N. Recalling Euler's theorem,

$$p^{\phi(N)} \equiv 1 \bmod N, \tag{14}$$

we have

$$m \mid \phi(N). \tag{15}$$

In general, $m = \phi(N)$ *iff* p is a **primitive element** mod N. Primitive elements exist only for those N (besides 2, 4) that are of the form q^c or $2 \cdot q^c$, q an odd prime. In all other cases, the order of p has to be less than $\phi(N)$.

Let the standard factorization of N be given as

$$N = 2^{c_0} \prod_{i=1}^{k} q_i^{c_i} = 2^{c_0} \prod_{i=1}^{k} s_i = \prod_{i=0}^{k} s_i\,. \tag{16}$$

Note that if N is odd, then $c_0 = 0$ and the index i runs from 1 to k. Define the **universal exponent** $\lambda(N)$ of N by

$$\lambda(N) = lcm\left[\lambda\left(2^{c_0}\right), \phi\left(q_1^{c_1}\right), \phi\left(q_2^{c_2}\right), \ldots, \phi\left(q_k^{c_k}\right)\right], \tag{17}$$

where $\lambda(2) = 1$, $\lambda(2^2) = 2$, and $\lambda(2^k) = 2^{k-2}$ for $k \geq 3$ [6]. It is seen that (i) $\lambda(N) \mid \phi(N)$ and (ii) $\lambda(N) = \phi(N)$, *iff* N is an integer of the type 2, 4, q^c, $2q^c$, q an odd prime. In all other cases, $\lambda(N) \leq \phi(N)/2$. Euler's theorem can be used to show that

$$p^{\lambda(N)} \equiv 1 \bmod N \tag{18}$$

and, thus,

$$m \mid \lambda(N). \tag{19}$$

For example, for $N = 5040 = 2^4 \cdot 3^2 \cdot 5 \cdot 7$, $\lambda(5040) = 12$ and $\phi(5040) = 1152$. Once m is determined, one employs the finite field $GF(p^m)$ to compute the factors of $u^N - 1$ in $Z(r)$. Since in $GF(p^m)$, each element can have at most m conjugate elements, we have the important result that the factors of $u^N - 1$ in $Z(r)$ will have degree at most equal to m. The expression in (19) brings out one of the contrasts between the cyclotomic factorization of $u^N - 1$ in Z and $Z(r)$. In Z, the degree of the cyclotomic POLY $C_N(u)$ having the Nth primitive root of unity as a root is exactly $\phi(N)$ while in $Z(r)$, it is a factor of $\lambda(N)$. Therefore, as long as $(N, p) = 1$, the degrees of the cyclotomic factors of $u^N - 1$ in $Z(p^a)$ are always lower than (or at most equal to) the degrees of the cyclotomic factors of $u^N - 1$ in Z. Once again, equality can hold only when p is primitive mod N which requires that N be an integer of the type 2, 4, q^c, $2q^c$, q an odd prime, as a necessary condition. Even when this condition holds, order of p mod N can take all possible values that satisfy (15).

Example 3. Let $N = 17 = q$ and $r = p^a = 2^a$. Here, we seek the order of 2 mod 17. The order 2 mod 17 is 8. Therefore, we will use $GF(2^8)$ to factorize $u^{17} - 1$ in $Z(2^a)$. If β denotes the element of order 17 in $GF(2^8)$, then the order of β^k is 17, $k = 1, 2, \ldots, 16$, since N is prime. The complete factorizations of $u^{17} - 1$ are given by

$$u^{17} - 1 = (u-1)\left(\sum_{i=0}^{16} u^i\right)$$

in Z and

$$u^{17}-1=(u-1)\{(u-\beta)(u-\beta^2)(u-\beta^4)(u-\beta^8)(u-\beta^{16})(u-\beta^{15})(u-\beta^{13})(u-\beta^9)\}$$
$$\cdot\{(u-\beta^3)(u-\beta^6)(u-\beta^{12})(u-\beta^7)(u-\beta^{14})(u-\beta^{11})(u-\beta^5)(u-\beta^{10})\}$$

in $GF(2)$. A length 17 CC algorithm defined in Z would require 1 POLY products mod $C_{17}(u)$, a POLY of degree 16. In contrast, a length 17 CC algorithm defined in $Z(2^a)$ would require 2 POLY products mod cyclotomic POLYs of degree 8.

Example 4. Let $N = 3 \cdot 5 \cdot 7 = 105$. The cyclotomic POLY $C_{105}(u)$ in Z has degree equal to 48. In this case, $\lambda(105) = \text{lcm}(2, 4, 6) = 12$. Therefore, $C_{105}(u)$ in Z will necessarily factor in $Z(r)$ into at least 4 POLYs having degrees at most equal to 12. Here, r can have any prime factors except for 3, 5, and 7. We have found primes having order that takes all the possible values (factors of $\lambda(105)$). For $p = 13, 43, 83, 97$, $m = 4$, and for $p = 29, 41, 71$, $m = 2$.

III.2 Factorization of $u^N - 1$ in $CZ(r)$, $r = p^a$, $p \equiv 3 \bmod 4$. The factorization of $u^N - 1$ in $CZ(r)$ begins with finding the least integer s such that [4]

$$N \mid (p^{2s} - 1). \tag{20}$$

In other words,

$$p^{2s} \equiv 1 \bmod N \tag{21}$$

and, therefore, $2 \cdot s$ is the order of p mod N. Recalling from (25) that m is the order of p mod N, we have

$$s = m/2, \tag{22a}$$

if m is even, or

$$s = m, \tag{22b}$$

if m is odd. Combining (22) with (19), we get

$$s \mid \{\lambda(N)/2\}, \tag{23}$$

since $\lambda(N)$ is always an even number except for the trivial value $N = 2$. In all cases, s is at most one-half of $\lambda(N)$. Once s is determined, one employs the finite field $GF(p^{2s})$ to compute the factors of $u^N - 1$ in $GF(p^2)$ and $CZ(r)$. Since in $GF(p^{2s})$, each element can have at most s conjugate elements in $CZ(r)$, we have the important result that the factors of $u^N - 1$ in $CZ(r)$ will have degree at most equal to s. The expression in (23) brings out further contrast between the cyclotomic factorization of $u^N - 1$ in Z, CZ and $CZ(r)$. In CZ, the degree of the cyclotomic POLY having the Nth primitive root of unity as a root is exactly $\phi(N)$ or $\phi(N)/2$ (if $4 \mid N$) while in $CZ(r)$, it is a factor of $\lambda(N)/2$. Therefore, as long as $(N, p) = 1$, the degrees of the cyclotomic factors of $u^N - 1$ in $CZ(r)$ **are at most equal to one half the degrees** of the cyclotomic factors of $u^N - 1$ in Z and CZ. Also, if p is primitive mod N, then $C_N(u)$ as defined in Z and CZ is also irreducible in $Z(r)$, but it will always factor further in $CZ(r)$.

Example 5. Let $N = 7 = q$ and $r = p^a = 3^a$. Here, we seek to find the order of $3^2 = 9$ mod 7. The order of 9 mod 7 is 3. Therefore, we will use $GF(9^3)$ to factorize $u^7 - 1$ in $CZ(3^a)$. If β denotes the element of order 7 in $GF(9^3)$, then the order of β^k is 7, $k = 1, 2, ..., 6$, since N is prime. The complete factorizations of $u^7 - 1$ are given by

$$u^7 - 1 = (u-1)\left(\sum_{i=0}^{6} u^i\right)$$

in Z, CZ, and $Z(3^a)$, and

$$\begin{aligned} u^7 - 1 &= (u-1)\left\{(u-\beta)(u-\beta^2)(u-\beta^4)\right\}\cdot\left\{(u-\beta^3)(u-\beta^6)(u-\beta^5)\right\} \\ &= (u-1)\cdot\{u^3 + (2+j)u^2 + (1+j)u + 2\} \\ &\quad \cdot \{u^3 + (2+2j)u^2 + (1+2j)u + 2\} \bmod (3, j^2+1\} \end{aligned}$$

in $CZ(3)$. The corresponding factorization in $CZ(3^2)$ and $CZ(3^3)$ is given by

$$\begin{aligned} u^7 - 1 &= (u-1)\cdot\{u^3 + (5+7j)u^2 + (4+7j)u + 8\} \\ &\quad \cdot \{u^3 + (5+2j)u^2 + (4+2j)u + 8\} \bmod (9, j^2+1\}, \end{aligned}$$

and

$$\begin{aligned} u^7 - 1 &= (u-1)\cdot\{u^3 + (14+7j)u^2 + (13+7j)u + 26\} \\ &\quad \cdot \{u^3 + (14+20j)u^2 + (13+20j)u + 26\} \bmod (27, j^2+1\}, \end{aligned}$$

respectively. A length 7 CC algorithm defined in Z, CZ, and $Z(3^a)$ would require 1 POLY product mod $C_7(u)$, a POLY of degree 6. In contrast, a length 7 CC defined in $CZ(3^a)$ would require 2 POLY products mod cyclotomic POLYs of degree 3.

IV. CYCLOTOMIC FACTORIZATION IN $Z(r)$ AND $CZ(r)$

It is seen from the description of CC algorithm that one seeks the cyclotomic POLY $C_N(u)$ having the N-th primitive root of unity as its root. In $GF(p)$, if β is the element of order N in $GF(p^m)$, m being the order of p mod N, then β has exactly m conjugates. Therefore, $C_N(u)$ as found in Z is either irreducible in $Z(r)$, which happens if $m = \phi(N)$, or it has $\phi(N)/m$ factors in $Z(r)$, each of degree m, which happens if m is a factor of $\lambda(N)$. Such a factorization is obtained in the following manner.

Procedure for obtaining cyclotomic factors of $C_N(u)$ in $Z(r)$.

1. Find m, the order of p mod N.
2. Let β be an element of order N in $GF(p^m)$. Now, $\beta^N = 1$ and $\beta^i \neq 1$, $i < N$ in $GF(p^m)$. Let $S = \{i, i = 1, 2, ..., N; (i, N) = 1\}$. Partition S into subsets S_{i_1}, S_{i_2}, ...; $S_{i_k} = \{i_k, pi_k, p^2 i_k, ...\}$, where i_k is the smallest element in S not covered in earlier subsets. The elements in subsets S_{i_1}, S_{i_2}, ... are defined mod N. There are $\phi(N)/m$ subsets, each subset having m elements. The factorization of $C_N(u)$ is obtained as

$$C_N(u) = \prod_k \left[\prod_{l \in S_{i_k}} \left(u - \beta^l \right) \right] \text{ mod } (p, Q(\alpha)) \tag{24}$$

$$= \prod_k P_k(u) \text{ mod } p, \tag{25}$$

where $Q(\alpha)$ is a primitive POLY in $GF(p^m)$. Each of $P_k(u)$ in (25) is a POLY of degree m or less having β^{i_k} and its conjugates in $GF(p^m)$ as its roots. Therefore, it is irreducible and has its coefficients in $GF(p)$.

Similarly, $C_N(u)$ as found in Z is **always reducible** in $CZ(r)$. It factors into $\phi(N)/s$ factors in $CZ(r)$, each of degree s, s being the order of p^2 mod N. Such a factorization is obtained in the following manner.

Procedure for obtaining cyclotomic factors of $C_N(u)$ in $CZ(r)$

1. Find s, the order of p^2 mod N.
2. Let β be an element of order N in $GF(p^{2s})$. Now, $\beta^N = 1$ and $\beta^i \neq 1$, $i < N$, in $GF(p^{2s})$. Let $S = \{i, i = 1, 2, ..., N; (i, N) = 1\}$. Partition S into subsets S_{i_1}, S_{i_2}, ...; $S_{i_k} = \{i_k, p^2 i_k, p^4 i_k, ...\}$, where i_k is the smallest element in S not covered in earlier subsets. The elements in subsets S_{i_1}, S_{i_2}, ... are defined mod N. There are $\phi(N)/s$ subsets, each subset having s elements. The factorization of $C_N(u)$ is obtained as

$$C_N(u) = \prod_k \left[\prod_{l \in S_{i_k}} \left(u - \beta^l \right) \right] \text{ mod } (p, j^2 + 1, Q(\alpha)) \tag{26}$$

$$=\prod_k P_k(u) \bmod \left(p, j^2+1\right), \tag{27}$$

where $Q(\alpha)$ is a primitive POLY of degree s in $GF(p^{2s})$ with coefficients in $GF(p^2)$. Each of $P_k(u)$ in (27) is a POLY of degree s or less having β^{l_k} and its conjugates in $GF(p^{2s})$ as its roots. Therefore, it is irreducible and has its coefficients in $GF(p^2)$.

Finally, we note that except for trivial values of 2 and 4, $Z(N)$ has no primitive elements for $N = 2^c$. Therefore, in case $N = 2^c$, $C_N(u) = u^{2^{c-1}} + 1$ will always factor further in $Z(r)$. Interestingly, each of these factors in $Z(r)$ will always factor further into two POLYs in $CZ(r)$.

Example 6. Let $N = 64$. In Z, $\deg(C_{64}(u)) = 32$ and $C_{64}(u)$ factors into two POLYs of degree 16 in CZ. As order of 3 mod 64 is 16, $C_{64}(u)$ factors into 2 POLYs of degree 16 in $Z(3^a)$ and 4 POLYs of degree 8 in $CZ(3^a)$.

Example 7. Let $r = 2^a$ and $N = 2^e - 1$, a Mersenne prime integer. In this case, order of 2 mod $N = m = e$. Therefore, $u^N - 1$ factors as $(u - 1)$ and $C_N(u) = u^{N-1} + u^{N-2} + \ldots + u + 1$ in Z and CZ, while it factors as $(u - 1)$ and $(N - 1)/e$ primitive POLYs in $Z(r)$, each of degree e, as N is prime.

Example 8. Let $r = 2^a$ and $N = 2^n + 1$, a Fermat prime integer, $n = 2^t$. In this case, order of 2 mod $N = m = 2^{t+1}$. Therefore, $u^N - 1$ factors as $(u - 1)$ and $C_N(u) = u^{N-1} + u^{N-2} + \ldots + u + 1$ in Z, while it factors as $(u - 1)$ and 2^{n-t-1} primitive POLYs in $Z(r)$, each of degree 2^{t+1}, as N is prime.

Example 9. Let $r = 3^a$ and $N = 2^n + 1$, a Fermat prime integer, $n = 2^t$. It is well known that 3 is a primitive element for Fermat primes [1]. In this case, order of 3 mod $N = N - 1$ and $u^N - 1$ factors as $(u - 1)$ and $C_N(u) = u^{N-1} + u^{N-2} + \ldots + u + 1$ in Z and $Z(3^a)$. In $CZ(3^a)$, it factors as $(u - 1)$ and two POLYs, each of degree $(N - 1)/2$.

Now, we turn to the six specific cases for the lengths that were mentioned in Section I.

V. FACTORIZATION IN FINITE INTEGER RINGS: SPECIAL CASES

In this section, we analyze the construction of CC algorithms in $Z(p^a)$ and $CZ(p^a)$ for the following six special cases:

Case 1. $N = q$, q an odd prime,

Case 2. $N = q^c$,
Case 3. $N = 2^c$,
Case 4. $N = 2^{c_0} \cdot q^c$,
Case 5. $N = 2^{c_0} \cdot q_1^{c_1} \cdot q_2^{c_2}$, and
Case 6. N contains more than two odd prime factors.

Case 1: $N = q$, q an odd prime. In this case, universal exponent and the totient function of N are equal. The Euler's theorem simplifies to Fermat's theorem and we seek the order of p mod q. If p is primitive, then $C_q(u)$ in Z is also irreducible in $Z(r)$ and the CC algorithm in $Z(r)$ is identical in form to the CC algorithm in Z. If p is not primitive, then $C_q(u)$ in Z factors into $(q-1)/m$ factors in $Z(r)$, each of degree m. In this case, the CC algorithm in $Z(r)$ will consist of $(q-1)/m$ POLY products mod cyclotomic POLYs, each of degree m. In Table 1, the properties of some CC algorithms that follow from this analysis are presented.

Example 10. Let $N = 7$ and the ring be $Z(4)$. The factorization of $u^7 - 1$ in $Z(4)$ is given by

$$u^7 - 1 = (u-1)(u^3 + 3u^2 + 2u + 3)(u^3 + 2u^2 + u + 3).$$

In this case, we express the computation $X(u)\,Y(u)$ mod $(u^7 - 1)$ as a direct sum of $X(u)\,Y(u)$ mod $(u-1)$, $X(u)\,Y(u)$ mod $(u^3 + 3u^2 + 2u + 3)$, and $X(u)\,Y(u)$ mod $(u^3 + 2u^2 + u + 3)$. If the product of two POLYs of degree 2 each is computed in 6 MULTs, the length 7 CC algorithm in $Z(4)$ and, hence, in $Z(2^a)$, requires 13 MULTs. It is worthwhile to note here that the corresponding algorithm in Z requires 16 MULTs.

For CC algorithm in $CZ(r)$, we seek order of p^2 mod q. If p is primitive or if m is even, then $C_q(u)$ in $Z(r)$ factors further in $CZ(r)$; otherwise, the CC algorithm in $CZ(r)$ is identical in form to the CC algorithm in $Z(r)$. In all cases, $C_q(u)$ in Z factors into s factors in $CZ(r)$ each of degree $(q-1)/s$. In Table 2, the properties of some CC algorithms that follow from this analysis are presented.

Case 2: $N = q^c$. This is one of the most interesting cases from the standpoint of DSP algorithms and number theory. We are establishing further factorization of $C_N(u)$ (the cyclotomic POLY having all the primitive roots of unity in Z) in $Z(p^a)$ and $CZ(p^a)$. The first step is once again to determine the order of p mod q^c for the factorization in $Z(p^a)$ and the order of p^2 mod q^c for the factorization in $CZ(p^a)$. In this case, universal exponent and the totient function of N are equal as well.

Factorization of $C_N(u)$ in $Z(p^a)$. We seek to determine the order of p mod q^c, that is m, for the factorization in $Z(p^a)$. Based on the number theoretic results of [9], a procedure to determine m is as follows:

Step 1. Determine the order of p mod q. Let it be n. Recall that $n \mid (q-1)$. If $n = q - 1$, then p is primitive mod q.

Step 2. Determine γ such that

$$p^n \equiv 1 \bmod q^i,\ i = 1, 2, \ldots, \gamma \tag{28}$$

but

$$p^n \neq 1 \bmod q^{\gamma+1}. \tag{29}$$

Step 3. The value of m is given by

$$m = n \tag{30}$$

if $\gamma \leq c$, otherwise,

$$m = q^{c-\gamma} \cdot n. \tag{31}$$

If $\gamma = 1$ and $n = q - 1$, then p is primitive mod q^c, for all c. In our work, we have discovered many cases for all possibilities. In Table 3, we list prime numbers p and q less than 100 for which $\gamma > 1$.

Example 11. We have the result $3^5 \equiv 1 \bmod 11$ and $3^5 \equiv 1 \bmod 11^2$, but $3^5 \neq 1 \bmod 11^3$. Therefore, $\gamma = 2$, order of 3 mod 11 = order of 3 mod $11^2 = 5$, and order of 3 mod $11^c = 5 \cdot 11^{c-2}$, $c > 2$. This example can be explored further for $p = 47, 113, 487$ and $q = 11$.

Example 12. We have the result $7^4 \equiv 1 \bmod 5$ and $7^4 \equiv 1 \bmod 5^2$, but $7^4 \neq 1 \bmod 5^3$. Therefore, $\gamma = 2$, order of 7 mod 5 = order of 7 mod $5^2 = 4$, and order of 7 mod $5^c = 4 \cdot 5^{c-2}$, $c > 2$. This example can be further explored for $p = 101$, 107, 157, 257 and $q = 5$.

Given m as in (30) or (31), we will use $GF(p^m)$ to factor $C_N(u)$ in Z further in $Z(p^a)$. If p is primitive mod q^c, that is $m = \phi(q^c) = q^{c-1}(q-1)$, then $C_N(u)$ in Z is irreducible in $Z(p^a)$ as well. In other cases, $C_N(u)$ factors further in $Z(p^a)$ as a product of $\phi(q^c)/m$ factors, each of degree m. If $m = n$, then one can obtain such factors by using $GF(p^m)$. In case m is given by (31), then one first factors $C_{q^\gamma}(u)$ in Z in terms of $\phi(q^\gamma)/n$ POLYs in $Z(r)$, each of degree n, using $GF(p^n)$.

Next, the corresponding factors in $Z(p^a)$ are computed. Let $A(u)$ be one such factor of degree n. The corresponding factor of $C_N(u)$ in $Z(p^a)$ is given by

$$A\left(u^{q^{c-\gamma}}\right) \bmod p^a . \tag{32}$$

As a special case, consider $p = 2$ and $N = q^c$, q being a Fermat Prime. The order of 2 mod q is 2^{t+1}. It is seen that $\gamma = 1$. Therefore, the order of 2 mod N is $2^{t+1} \cdot q^{c-1}$. $C_N(u)$ in Z will factor into $(q - 1) \cdot 2^{-t-1}$ factors in $Z(2^a)$, each of degree $2^{t+1} \cdot q^{c-1}$. Another special case is $p = 2$ and $N = q^c$, q being a Mersenne prime, $q = 2^e - 1$. The order of 2 mod q is e. It is seen that $\gamma = 1$. Therefore, the order of 2 mod N is $e \cdot q^{c-1}$. $C_N(u)$ in Z will factor into $(q - 1)/e$ factors in $Z(2^a)$, each of degree $e \cdot q^{c-1}$. These factorizations also follow the form in (32).

Example 13. Let us continue Example 11. In this case, $3^5 \equiv 1 \bmod 11^2$, but $3^5 \neq 1 \bmod 11^3$ implies that degree 110 POLY $C_{121}(u)$ in Z will factor into 22 POLYs in $Z(3^a)$, each having degree 5. $GF(3^5)$ is to be used to obtain such factors. A length 121 CC algorithm in Z requires POLY product mod $C_{121}(u)$, a degree 110 POLY. The same CC in $Z(3^a)$ requires 22 POLY products mod cyclotomic POLYs of degree 5 each.

In Table 4, further factorization of $C_N(u)$ in $Z(r)$ is shown for selected values of $N = q^c$.

Factorization of $C_N(u)$ in $CZ(p^a)$. In this case, we seek the order of p^2 mod q^c. The results of (22a) and (22b) are valid in this instance, thereby implying that p^2 can never be primitive mod q^c and that $C_N(u)$ always factors further in $CZ(p^a)$. The remainder of the procedure remains the same as described in the context of factorization in $Z(p^a)$ including the expression in (32) and the details can be omitted here.

Example 14. Let us continue Example 12. In this case, $7^4 \equiv 1 \bmod 5^2$, but $7^4 \neq 1 \bmod 5^3$ implies that degree 20 POLY $C_{25}(u)$ in Z will factor into 5 POLYs in $Z(7^a)$, each having degree 4, and 10 POLYs in $CZ(7^a)$, each having degree 2. $GF(7^4)$ and $GF(49^2)$ are used to obtain such factors. A length 25 CC algorithm in Z requires 31 POLY products mod $C_{25}(u)$, a degree 20 POLY. The same convolution in $Z(7^a)$ requires 5 POLY products mod cyclotomic POLYs of degree 4 each. In $CZ(7^a)$, it requires 10 POLY products mod cyclotomic POLYs of degree 2 each.

Case 3: $N = 2^c$. In this case, as we seek the order of p mod 2^c, we recall that p can never be primitive. Therefore, degree 2^{c-1} POLY $C_N(u)$ in Z always factors in $Z(p^a)$; each such factor in $Z(p^a)$ in turn always factors further in $CZ(p^a)$. It has been established in [9] that the order of p mod 2^c is given by 1 for $c \leq d$ and 2^{c-d} for $c > d$, if p has the form

$$p = 1 + 2^d \cdot B, \tag{33}$$

B being an odd integer and $d \geq 2$ ($p \equiv 1$ mod 4). Similarly, order of p mod 2^c is given by 2 for $c \leq d + 2$ and 2^{c-d-1} for $c > d + 2$, if p has the form

$$p = 1 + 2 + \ldots + 2^d + 2^f \cdot B = 2^{d+1} - 1 + 2^f \cdot B, \tag{34}$$

B being an odd integer and $f \geq d + 2$ ($p \equiv 3$ mod 4), $d > 0$.

These statements establish that if $p \equiv 1$ mod 4, then $C_N(u)$ factors into degree 1 factors for $c \leq d$ and 2^{d-1} factors of degree 2^{c-d} if $c > d$ in $Z(p^a)$. For $c > d$, these 2^{d-1} factors in $Z(p^a)$ are obtained by first obtaining 2^{d-1} factors each of degree 2 for $c = d + 1$. Then the degree 2^{c-d} factors are obtained by replacing u with $u^{2^{c-d-1}}$ in each of the degree 2 POLYs.

Similarly, if $p \equiv 3$ mod 4, then $C_N(u)$ factors into degree 2 factors for $c \leq d + 2$ and 2^d factors of degree 2^{c-d-1} if $c > d + 2$ in $Z(p^a)$. For $c > d + 2$, these 2^d factors in $Z(p^a)$ are obtained by first obtaining 2^d factors each of degree 4 for $t = d + 3$. Then the degree 2^{c-d-1} factors are obtained by replacing u with $u^{2^{c-d-3}}$ in each of the degree 4 POLYs. All factors in $Z(p^a)$ for $p \equiv 3$ mod 4 factorize further in $CZ(p^a)$ and can be obtained in a similar manner.

Example 15. Let p be a Fermat prime, $p = 2^n + 1$, $n = 2^t$, $p > 3$. Then $p \equiv 1$ mod 4 holds with $d = n$. Therefore, $C_N(u)$ factors into degree 1 factors for $c \leq n$ and 2^{n-1} factors of degree 2^{c-n} if $c > n$ in $Z(p^a)$.

Example 16. Let p be a Mersenne prime, $p = 2^e - 1$. Then $p \equiv 3$ mod 4 holds with $d + 1 = e$. Therefore, $C_N(u)$ factors into degree 2 factors for $c \leq e + 1$ and 2^{e-1} factors of degree 2^{c-e} if $c > e + 1$ in $Z(p^a)$. Once again, all of these factors factor further in $CZ(p^a)$. In this analysis, the Fermat prime 3 is to be treated as a Mersenne prime.

Example 17. Let $p = 151 = 1 + 2 + 4 + 16 \cdot 9$ and $c = 9$. Here, (34) holds with $d = 2$ and $C_{512}(u)$ factors into 4 factors of degree 64 in $Z(151^a)$. In $CZ(151^a)$, it factors into 8 factors of degree 32.

In Table 5, further factorization of $C_N(u)$ in $Z(r)$ is shown for selected values of $N = 2^c$.

Case 4: $N = 2^{c_0} \cdot q^c$. In this case, we determine the order of p mod 2^{c_0} and mod q^c using the analysis for cases 2 and 3 given above. Recall that p can never be primitive mod $2^{c_0} \cdot q^c$ unless $c_0 = 1$. In all other cases,

$$\text{order of } p \bmod 2^{c_0} \cdot q^c = lcm(\text{order of } p \bmod 2^{c_0}, \text{ order of } p \bmod q^c). \quad (35)$$

Thus, if $c_0 = 2$, then the value for m is a factor of $\phi(N)/2$. If $c_0 > 2$, then the value of m is a factor of $\phi(N)/4$. Consequently, as long as $c_0 > 1$, $C_N(u)$ always factors in $Z(r)$. Similarly, order of p^2 mod N is $\phi(N)/2$ for $c_0 = 1$, $\phi(N)/4$ for $c_0 = 2$ and $\phi(N)/8$ for $c_0 > 2$, thereby implying that $C_M(u)$ always factors in $CZ(r)$.

Consider the situation $N = 2 \cdot q$. Since p is an odd prime (p is odd as $(p, N) = 1$), we have

$$\text{order of } p \bmod q = m = \text{order of } p \bmod 2 \cdot q. \quad (36)$$

Consequently, the factorization of $u^{2q} - 1$ is obtained as

$$u^{2q} - 1 = (u^q - 1)(u^q + 1). \quad (37)$$

The factorization for $u^q - 1$ has been already discussed. The factorization for $u^q + 1$ can be obtained by replacing u by $-u$ in the factorization for $u^q - 1$.

Observation. In case 4, we have assumed that $p \neq 2$ to ensure that $(p, N) = 1$. However, if $N = 2 \cdot q^c$ and $p = 2$, that is, one wishes to design CC algorithm in $Z(2^a)$ or $CZ(2^a)$, then the following approach may be adopted. Convert the $2 \cdot q^c$ CC to a two-dimensional CC having dimension $2 \times q^c$ using Agarwal-Cooley algorithm [1]. Now the length 2 CC is computed using the direct approach in 3 MULTs and the length q^c CC algorithm can be designed in the usual manner. This approach may also be adopted for $N = 4 \cdot q^c$ and $p = 2$.

In Table 6, further factorization of $C_N(u)$ in $Z(r)$ is shown for selected values of $N = 2^{c_0} \cdot q^c$.

Case 5: $N = 2^{c_0} \cdot q_1^{c_1} \cdot q_2^{c_2}$. The value of universal exponent $\lambda(N)$ grows very slowly compared to $\phi(N)$ as the number of odd prime factors of N increases. Therefore, $C_N(u)$ factors into a large number of factors in all such cases. For the present case, we determine the order of p mod N in a manner similar to cases 1 to 4. Recall that p can never be primitive in this case. For $c_0 = 0, 1, 2, > 2$, the value of m is a factor of $\phi(N)/2$, $\phi(N)/2$, $\phi(N)/4$, and $\phi(N)/8$, respectively.

Consequently, $C_N(u)$ always factors in $Z(r)$. In Table 7, further factorization of $C_N(u)$ in $Z(r)$ is shown for selected values of $N = 2^{c_0} \cdot q_1^{c_1} \cdot q_2^{c_2}$.

Case 6: N contains more than two odd prime factors. The smallest value of N for which this case is applicable is $N = 105$. The value of universal exponent $\lambda(N)$ is a factor of $\phi(N)/4$ and grows very slowly in all situations that appear in this case. For example, for $N = 105$, $\phi(105) = 48$ and $\lambda(105) = 12$. The order of $p = 13$, 19, 31, 61, and 97 is 4, 6, 6, 6, and 4, respectively. Similarly, for $N = 165$, $\phi(165) = 80$, $\lambda(165) = 20$, and the order of $p = 2$, 23, 31, 43, and 89 is 20, 4, 5, 4, and 2, respectively. Once again, p can never be primitive in this case and $C_N(u)$ always factors in $Z(r)$.

VI. RECURSIVE NESTING IN FINITE INTEGER RINGS

We now return to cases 2 and 3. In these cases, there are instances where the cyclotomic POLY has the form $A(u^b)$, $b = q^d$ as shown in (32). If $A(u)$ has the q^fth root of unity in $Z(r)$ or $CZ(r)$, then $A(u^b)$ has the q^{d+f}th root of unity in $Z(r)$ or $CZ(r)$. The one-dimensional CC involves computation of the type,

$$X(u)\, Y(u) \bmod A(u^b). \tag{38}$$

This can be computed using recursive nesting in the following manner. Let $v = u^g$. Now, (38) can be expressed as

$$X(u)\, Y(u) \bmod (u^g - v, A(v^h)) \tag{39}$$

or

$$X(u)\, Y(u) \bmod A(v^h), \tag{40}$$

where $b = g \cdot h$. The ordinary POLY product in u (with coefficients that are POLYs in v defined mod $A(v^h)$) has degree $2g - 2$. It can, therefore, be computed by interpolating u on $2g - 1$ powers of v, that is, $v^0, v^1, \ldots, v^{2g-2}$, provided that $(2g - 1) \leq h \cdot q^f$. Due to the special form of b, suitable g and h can always be found to satisfy the conditions. This technique may be most useful for $q = 2$. In case $q \neq 2$, interpolating powers of v can also be chosen as $\pm v^0, \pm v^1, \ldots, \pm v^{g-2}, v^g$, provided that $(g + 1) \leq h \cdot q^f$. This form of recursive nesting is the same in all number systems as long as the factorizations and mapping are defined appropriately.

VII. CYCLIC CONVOLUTION ALGORIHM OF LENGTH $N = 60$

In this section, we bring together the results of this paper and discuss the design of CC algorithm of length $N = 60$ in Z, CZ, $Z(M)$, and $CZ(M)$. The condition $(N, M) = 1$ requires that M can have any primes in its standard factorization except 2, 3, and 5. The factors d of 60 are 1, 2, 3, 4, 5, 6, 10, 12, 15, 20, 30, and 60. The degrees of the cyclotomic POLYs in Z, $\phi(d)$ are 1, 1, 2, 2, 4, 2, 4, 4, 8, 8, 8, and 16. Recalling that $C_d(u)$ in Z factors further in CZ iff $4 \mid d$, the degrees of the cyclotomic factors in CZ are 1, 1, 2, (1, 1), 4, 2, 4, (2, 2), 8, (4, 4), 8 and (8, 8). The degrees put in parentheses correspond to cyclotomic POLYs in CZ that are obtained from further factorization of corresponding cyclotomic POLYs in Z. The degrees of the cyclotomic POLYs in $Z(r)$ and hence in $Z(M)$ are a factor of $\lambda(d)$ and these values are given by 1, 1, 2, 2, 4, 2, 4, (2, 2), (4, 4), (4, 4), (4, 4), and (4, 4, 4, 4). All those POLYs that have even degree factor further in CZ(r) and hence in $CZ(M)$. The degrees of cyclotomic POLYs in $CZ(M)$ are a factor of 1, 1, (1, 1), (1, 1), (2, 2), (1, 1), (2, 2), ((1, 1), (1, 1)), ((2, 2), (2,2)), ((2, 2), (2, 2)), ((2, 2), (2,2)) and ((2, 2), (2, 2), (2, 2), (2, 2)). If the acyclic convolution of two length 2^t sequences is computed in 3^t MULTs, then the multiplicative complexity of the CC algorithm in Z and CZ is 200 MULTs and 160 MULTs, respectively. Based on the upper-bounds on the degrees of the cyclotomic factors in $Z(M)$ and $CZ(M)$, the multiplicative complexity of the CC algorithm in $Z(M)$ and $CZ(M)$ is at most 125 MULTs and 84 MULTs, respectively. The actual values of the degrees of the cyclotomic factors in $Z(83^a)$ and $CZ(83^a)$ is the same as the upperbounds. This holds for $p = 47$ as well. There are other values of p for which the actual multiplicative complexity is lower than the upperbound.

TABLE 1. Properties of CC Algorithms in $Z(r)$ for $N = q$, q an Odd Prime, $(p, q) = 1$

N	$\phi(N) = \lambda(N)$	$p = 2$	$p = 3$	$p = 5$
3	2	2^*; 1, 2	-	2^*; 1, 2
5	4	4^*; 1, 4	4^*; 1, 4	-
7	6	3; 1, 3, 3	6^*; 1, 6	6^*; 1, 6
11	10	10^*; 1, 10	5; 1, 5, 5	5; 1, 5, 5
13	12	12^*; 1, 12	3;1,3,3,3,3	4; 1, 4, 4, 4
17	16	8; 1, 8, 8	16^*; 1, 16	16^*; 1, 16
19	18	18^*; 1, 18	18^*; 1, 18	9; 1, 9, 9
23	22	11; 1, 11, 11	11;1,11,11	22^*; 1, 22
29	28	28^*; 1, 28	28^*; 1, 28	14; 1, 14, 14
31	30	5; 1, 5, 5, 5, 5, 5	30^*; 1, 30	3; 1, 10 POLYs of degree 3
37	36	36^*; 1, 36	18;1,18,18	36^*; 1, 36
$2^e - 1$: Mersenne Prime	$2^e - 2$	e; 1, $(2^e - 2) / e$ POLYs of deg e	-	-
$2^n + 1$, $n = 2^t$: Fermat prime	2^n	2^{t+1}; 1, 2^{n-t-1} POLYs of degree 2^{t+1}	2^{n*}; 1, 2^n	-

Note: The entries are marked with m; $b_1, ..., b_n$. Here, m denotes the order of p mod q. A star denotes a primitive element. $b_1, ..., b_n$ denote the degrees of the factors of $u^N - 1$ in $Z(p^a)$.

TABLE 2. Properties of CC Algorithms in $CZ(r)$ for $N = q$, q an Odd Prime, $(p, q) = 1, p \equiv 3 \bmod 4$

N	$\phi(N)$	$\lambda(N)/2$	$p = 3$	$p = 7$	$p = 11$
3	2	1	-	1; 1, 1, 1	1; 1, 1, 1
5	4	2	2; 1, 2, 2	2; 1, 2, 2	1; 1, 1, 1, 1, 1
7	6	3	3; 1, 3, 3	-	3; 1, 3, 3
11	10	5	5; 1, 5, 5	5; 1, 5, 5	-
13	12	6	3; 1, 3, 3, 3, 3	6; 1, 6, 6	6; 1, 6, 6
17	16	8	8; 1, 8, 8	8; 1, 8, 8	4; 1, 4, 4, 4, 4
19	18	9	9; 1, 9, 9	3; 1, 3, 3, 3, 3, 3, 3	3; 1, 3, 3, 3, 3, 3, 3
23	22	11	11; 1, 11, 11	11; 1, 11, 11	11; 1, 11, 11
29	28	14	14; 1, 14, 14	7; 1, 7, 7, 7, 7	14; 1, 14, 14
31	30	15	15; 1, 15, 15	15; 1, 15, 15	15; 1, 15, 15
37	36	18	9; 1, 9, 9, 9, 9	9; 1, 9, 9, 9, 9	3; 1, 12 POLYs of deg 3
$2^n + 1$, $n = 2^t$: Fermat Prime	2^n	2^{n-1}	2^{n-1}; 1, 2^{n-1}, 2^{n-1}	-	-

Note: The entries are marked with s; b_1, ..., b_n. Here, s denotes the order of p^2 mod q. **There are no primitive elements**. b_1, ..., b_n denote the degrees of the factors of $u^N - 1$ in $CZ(r)$.

TABLE 3. Prime pairs $p, q < 100$ for which $\gamma > 1$

q	11	3	3	3	3	3	3	5	5	7	7	7	7
p	3	17	19	37	53	71	89	7	43	19	31	67	79
order of p mod q	5	16	1	1	2	2	2	4	4	6	6	3	3
γ	2	2	2	2	3	2	2	2	2	3	2	2	2
q	7	71	13	13	13	43	29	79	47	47	47	59	97
p	97	11	19	23	89	49	41	31	53	67	71	53	53
order of p mod q	2	70	12	6	12	42	4	39	23	46	23	29	48
γ	2	2	2	2	2	2	2	2	2	2	2	2	2

TABLE 4. Further Factorization of $C_N(u)$ in $Z(r)$ for $N = q^c$, q an Odd Prime, $(p, q) = 1$

N	$\phi(N) = \lambda(N)$	$p = 17$	$p = 31$	$p = 97$
3^2	6	3, 2	2, 3	2, 3
5^2	20	1, 20*	4, 5	1, 20*
3^3	18	3, 6	2, 9	2, 9
7^2	42	1, 42*	7, 6	21, 2
3^4	54	3, 18	2, 27	2, 27
11^2	110	1, 110*	2, 55	2, 55
5^3	100	1, 100*	4, 25	1, 100*

Note: The entries are marked with a, m. Here, a denotes the number of factors of $C_N(u)$ in $Z(r)$ and m denotes the degree of each factor. A star denotes a primitive element. If m is even, then all the factors in $Z(r)$ factor further in $CZ(r)$ for $p \equiv 3 \bmod 4$.

TABLE 5. Further Factorization of $C_N(u)$ in $Z(r)$ for $N = 2^c$, q an Odd Prime, $(p, q) = 1$

N	$\phi(N)$	$\lambda(N)$	$p = 17$	$p = 47$	$p = 79$	$p = 89$
2^2	2	2	2, 1	1, 2	1, 2	2, 1
2^3	4	2	4, 1	2, 2	2, 2	4, 1
2^4	8	4	8, 1	4, 2	4, 2	4, 2
2^5	16	8	8, 2	8, 2	8, 2	4, 4
2^6	32	16	8, 4	8, 4	8, 4	4, 8
2^8	64	32	8, 8	8, 8	8, 8	4, 16

Note: The entries are marked with a, m. Here, a denotes the number of factors of $C_N(u)$ in $Z(r)$ and m denotes the degree of each factor. There are no primitive elements. If $p \equiv 3 \bmod 4$, then all the factors in $Z(r)$ factor further in $CZ(r)$.

TABLE 6. Further Factorization of $C_N(u)$ in $Z(r)$ for $N = 2^{c_0} \cdot q^c$, q an Odd Prime, $(p, q) = 1$

N	$\phi(N)$	$\lambda(N)$	$p = 17$	$p = 31$	$p = 97$
$2^2 \cdot 3^2$	12	6	6, 2	2, 6	4, 3
$2^3 \cdot 5$	16	4	4, 4	8, 2	4, 4
$2^3 \cdot 7$	24	6	4, 6	4, 6	12, 2
$2^4 \cdot 5$	32	4	8, 4	16, 2	8, 4
$2^2 \cdot 5^2$	40	20	2, 20	4, 10	2, 20
$2^2 \cdot 3^3$	36	18	6, 6	2, 18	4, 9
$2^4 \cdot 7$	48	12	8, 6	8, 6	24, 2

Note: The entries are marked with a, m. Here, a denotes the number of factors of $C_N(u)$ in $Z(r)$ and m denotes the degree of each factor. If m is even, then all the factors in $Z(r)$ factor further in $CZ(r)$ for $p \equiv 3 \bmod 4$.

TABLE 7. Further Factorization of $C_N(u)$ in $Z(r)$ for $N = 2^{c_0} \cdot q_1^{c_1} \cdot q_2^{c_2}$, q_1, q_2 Odd Primes, $(p, q) = 1$

N	$\phi(N)$	$\lambda(N)$	$p = 7$	$p = 41$	$p = 79$
$3 \cdot 5$	8	4	2, 4	4, 2	4, 2
$2 \cdot 3 \cdot 11$	20	10	2, 10	2, 10	2, 10
$3 \cdot 5^2$	40	20	10, 4	4, 10	4, 10
$5 \cdot 17$	64	16	4, 16	4, 16	4, 16
$2^3 \cdot 3 \cdot 5$	32	4	8, 4	16, 2	16, 2
$2 \cdot 5 \cdot 13$	48	12	4, 12	4, 12	24, 2
$2^2 \cdot 3 \cdot 13$	48	12	4, 12	4, 12	24, 2

Note: The entries are marked with a, m. Here, a denotes the number of factors of $C_N(u)$ in $Z(r)$ and m denotes the degree of each factor. If m is even, then all the factors in $Z(r)$ factor further in $CZ(r)$ for $p \equiv 3 \bmod 4$.

VIII. CONCLUSIONS

Our efforts have been focused on unifying number theoretic results as they relate to the structure of fast digital signal processing algorithms in different number systems. In this regard, the present work deals with the one-dimensional cyclic convolution algorithms in finite integer and complex integer rings. A fundamental result of this work is that under the non-restrictive condition, $(N, M) = 1$, the cyclic convolution algorithms defined in finite integer and complex integer rings are as intensive computationally as their counterpart in rational and complex rational number systems **only in the worst case**. They simplify considerably for most cases of significance in digital signal processing.

REFERENCES

[1] H.J. Nussbaumer, *Fast Fourier Transform and Convolution Algorithms*, Springer-Verlag, 1981.

[2] D.G. Myers, *Digital Signal Processing: Efficient Convolution and Fourier Transform Techniques*, Prentice Hall, 1990.

[3] J.H. McClellan and C.M. Rader, *Number Theory in Digital Signal Processing*, Prentice Hall, 1979.

[4] H. Krishna, B. Krishna, K.-Y. Lin, and J.-D. Sun, *Computational Number Theory and Digital Signal Processing*, CRC Press, 1994.

[5] H. Krishna Garg, *Digital Signal Processing Algorithms*, CRC Press, 1998.

[6] D.M. Burton, *Elementary Number Theory*, WCB Publishers, 1994.

[7] M.T. HeideMan, *Multiplicative Complexity, Convolution, and the DFT*, Springer-Verlag, 1988.

[8] R.J. McElice, *Finite Fields for Computer Scientists and Engineers*, Kluwer Academic Publishers, 1987.

[9] H.Krishna Garg, "A Number Theoretic Approach to Fast Algorithms for Two-Dimensional Digital Signal Processing in Finite Integer Rings," *IEEE Trans. on Circuits and Systems*, submitted for publication.

Paper three

Cyclotomic Polynomial Factorization in Finite Integer Rings with Applications to Digital Signal Processing

Abstract. In this paper, we present results that can be used to obtain all the possible generators for a number theoretic transform (NTT) defined in a finite integer ring and its polynomial extensions. A generalization of the well-known Euler's theorem is derived which can be used to determinc all the generators of a given NTT once the generators in the underlying finite field are identified. Based on this extension, we also describe a procedure to compute cyclotomic factorization in these rings. This factorization and Chinese remainder theorem lead to computationally efficient algorithms for computing cyclic convolution of two sequences defined in finite integer and complex integer rings.

I. INTRODUCTION

This paper deals with the algebra of finite integer rings and their polynomial (POLY) extensions with a view to establish certain number theoretic properties that can be of fundamental importance in digital signal processing (DSP) and error control coding. Within the domain of DSP, these properties can be used to obtain all the possible generators for number theoretic transforms (NTTs) and cyclotomic factorization for the computation of one- and two-dimensional cyclic convolution. In the area of error control coding, these results can be used to obtain Bose-Chaudhary-Hoquenghem (BCH) and Reed-Solomon (RS) codes which may find applications in communication and fault tolerant computing systems.

The finite integer rings that we study in this paper are as follows:

(1) $Z(M)$, consisting of integers $\{0, 1, ..., M-1\}$,
(2) $CZ(M)$, consisting of integers $\{a+jb,\ a,\ b \in Z(M), j^2=-1\}$,
(3) $Z(M, Q(\theta))$, consisting of POLYs $\{a_0 + a_1\theta + a_2\theta^2 + ... + a_{m-1}\theta^{m-1} + \theta^m,\ a_i \in Z(M)\}$, and
(4) $CZ(M, Q(\theta))$, consisting of POLYs $\{ a(\theta) + jb(\theta), a(\theta), b(\theta) \in Z(M, Q(\theta)), j^2=-1\}$.

The operations of addition (ADD) and multiplication (MULT) in $Z(M, Q(\theta))$ and $CZ(M, Q(\theta))$ are defined modulo $(M, Q(\theta))$ and $(M, j^2+1, Q(\theta))$, respectively. Here, $Q(\theta)$ is a *monic* degree m POLY with coefficients in $Z(M)$ for $Z(M, Q(\theta))$ and in $CZ(M)$ for $CZ(M, Q(\theta))$. The rings $Z(M)$ and $CZ(M)$ can also be studied as special cases of $Z(M, Q(\theta))$.

In order to define a NTT, we seek the factorization of $u^N - 1$ in terms of N degree one *relatively prime* factors. In other words, we wish to find the generator G that satisfies

$$u^N - 1 = \prod_{k=0}^{N-1}\left(u - G^k\right). \tag{1}$$

Thus, G is an Nth root of unity in the ring. The condition that the factors be relatively prime requires that the multiplicative inverse of $(G^i - G^k)$ exist in the ring for all $i, k = 0, 1, ..., N-1, i \neq k$.

The methodology adopted in this paper consists in developing a unified approach to study a finite integer ring as p-adic expansions of the underlying finite field. All the properties of finite fields are shown to hold for their p-adic expansion integer rings as well. Interestingly, all the properties and quantities in a finite integer ring can be determined solely from the knowledge of the corresponding properties and quantities in the finite field. We begin our analysis by first obtaining a generalization of the Euler's theorem. The results for finite fields, Chinese remainder theorem (CRT), and this generalization are then applied to obtain all the generators of the associated NTTs. The extension also leads to cyclotomic factorization in finite integer rings which are valuable for computing one- and two-dimensional cyclic convolution of sequences.

The organization of this paper is as follows. Section II is on the relevant results on the CRT. In Section III, generalizations of the Euler's theorem are derived. Section IV is on finding the generators for NTTs in $Z(M, Q(\theta))$ and $CZ(M, Q(\theta))$. Section V is on the application of these results for finding the cyclotomic factorization in $Z(q)$ and $CZ(q)$. In Section VI, the conjugate symmetry property of the NTTs is analyzed. A discussion of the results and conclusions are presented in Section VII.

II. RESULTS BASED ON THE CHINESE REMAINDER THEOREM

Given the standard factorization of M as

$$M = \prod_{t=1}^{r} p_t^{a_t} = \prod_{t=1}^{r} q_t ,$$

the CRT can be used to express the ring $Z(M, Q(\theta))$ as the direct sum

$$Z\big(M, Q(\theta)\big) = \sum_{\substack{t=1 \\ \oplus}}^{r} Z\left(p_t^{a_t}, Q^{(t)}(\theta)\right),$$

where $Q^{(t)}(\theta) \equiv Q(\theta) \bmod p_t^{a_t}$. Consequently, the generator for NTT can be expressed as the direct sum

$$G = \sum_{\substack{t=1 \\ \oplus}}^{r} G_t ,$$

where G_t is the generator for the NTT with $M = p_t^{a_t}$. In the following, we will pursue the generator for NTT G_t. Also, the subscript or the superscript t is deleted and results are stated for $M = q = p^a$. It simplifies the form of the various expressions.

If G defines a NTT of length N in a ring, then so does G^s, where s is any integer less than N and relatively coprime to it, that is, $(s, N) = 1$. There are $\phi(N)$ such integers, $\phi(N)$ being the Euler's totient function of N [1]. Consequently, for the rings $Z(q, Q(\theta))$ and $CZ(q, Q(\theta))$, either there does not exist a generator for NTT of length N or there exist exactly $\phi(N)$ generators. For the ring $Z(M, Q(\theta))$

and $CZ(M, Q(\theta))$, either there does not exist a generator for NTT of length N or there exist exactly $(\phi(N))^r$ generators.

II.1 Generators for NTT in $Z(q, Q(\theta))$

Given $Q(\theta)$) as a POLY with coefficients defined in $Z(q)$, we define its p-adic expansion as

$$Q(\theta) = Q_0(\theta) + pQ_1(\theta) + \ldots + p^{a-1}Q_{a-1}(\theta). \tag{2}$$

The coefficients of $Q_d(\theta)$ are defined in $GF(p)$. Since $Q(\theta)$ is a monic degree m POLY, so is $Q_0(\theta)$. All other POLYs in (2) have degree less than m. Given the unique factorization of $Q_0(\theta)$ in $GF(p)$ in terms of its monic irreducible factors as

$$Q_0(\theta) = \prod_{t=1}^{v} \left[P_{0,t}(\theta)\right]^{\gamma_t} = \prod_{t=1}^{v} Q_{0,t}(\theta), \tag{3}$$

the factorization of $Q(\theta)$ in $Z(q)$ in terms of monic irreducible factors is given by $Q(\theta) = \prod_{t=1}^{v} Q_t(\theta)$, where $Q_t(\theta) \equiv Q_0(\theta) \bmod p$. The Hensel's Lemma constitutes a basis for this statement [2].

It has been shown that for the form of $Q(\theta)$ in (3), there exists an extension of the CRT for POLYs which leads to the direct sum [3],

$$Z(q, Q(\theta)) = \sum_{\substack{t=1 \\ \oplus}}^{v} Z(q, Q_t(\theta)).$$

Therefore, we need to pursue only those $Q(\theta)$ POLYs that the corresponding $Q_0(\theta)$ has the form $Q_0(\theta) = [P_0(\theta))]^\gamma$, $P_0(\theta)$ being an irreducible POLY in $GF(p)$. Furthermore, we wish to work with the largest finite field possible for a given $Q(\theta)$. This requires that $\gamma = 1$ or that $Q(\theta)$ be *irreducible* when reduced mod p, an assumption that we make throughout this paper. Such polynomials are well documented in any text on finite field algebra and we assume here that they are known for all values of p and m that are of interest.

III. GENERALIZATION OF EULER'S THEOREM

All expressions in $Z(q)$ reduce to corresponding expressions in $GF(p)$ when taken MOD p. Therefore, we have

$$Z(q, Q(\theta)) = GF(p^m) \bmod p. \tag{4}$$

$GF(p^m)$ is obtained by taking the POLY extension of $GF(p)$ where the operations of MULT and ADD are defined MOD a monic irreducible POLY of degree m. Consider $Z(q, Q(\theta))$, $Q(\theta)$ being an *irreducible* POLY in $GF(p)$. It is clear that the choice of $Q_1(\theta), ..., Q_{a-1}(\theta)$ is completely arbitrary. They can all be set to zero or assigned other values in order to satisfy some desired mathematical feature. All the elements of $Z(q, Q(\theta))$ are POLYs in θ with coefficients in $Z(q)$. Therefore, an arbitrary element $A(\theta)$ can be expressed as the p-adic expansion of the type

$$A(\theta) = A_0(\theta) + pA_1(\theta) + ... + p^{a-1}A_{a-1}(\theta),$$

where $A_i(\theta)$, $i = 0, 1, ..., a-1$ are elements of $GF(p^m)$.

Consider all those elements in $Z(q, Q(\theta))$ for which $A_0(\theta)$ is non-zero, that is, they are non-zero when reduced mod p. It is clear that the set of all such elements is closed under MULT. There are $p^{ma} - p^{m(a-1)} = p^{m(a-1)}\{p^m - 1\}$ such elements. This discussion leads to the following theorem.

Theorem 1. Generalization of Euler's theorem. If $A(\theta)$ is any one of the elements of $Z(q, Q(\theta))$ for which $A_0(\theta)$ is a non-zero element of $GF(p^m)$, then

$$A(\theta)^{p^{m(a-1)}\left(p^m - 1\right)} = 1 \bmod \left(q, Q(\theta)\right). \tag{5}$$

The expression in (5) will play a key role in establishing the results reported herein. The Euler's theorem is a special case of the general expression derived above and is obtained by setting $m = 1$. Similarly, setting $a = 1$ corresponds to the result in finite field theory that every non-zero element in $GF(p^m)$ has order that is a factor of $p^m - 1$.

A more general result can be derived by considering all those elements in $Z(q, Q(\theta))$ for which $A_0(\theta)$ is a non-zero element having order equal to n or a factor of n. There are exactly n such elements and they exist for all values of n that are a factor of $p^m - 1$. It is clear that the set of all such elements is closed under MULT. There are $p^{m(a-1)} \cdot n$ such elements. This discussion leads to the following theorem.

Theorem 2. Generalization of Euler's theorem. If $A(\theta)$ is any one of the elements of $Z(q, Q(\theta))$ for which $A_0(\theta)$ is a non-zero element of $GF(p^m)$ having order equal to n or a factor of n, $n \mid (p^m - 1)$, then

$$A(\theta)^{p^{m(a-1)} \cdot n} = 1 \bmod (q, Q(\theta)). \tag{6}$$

Let us now analyze $CZ(q, Q(\theta))$. In this case, we would like to have the expression

$$CZ(q, Q(\theta)) = GF(p^{2m}) \bmod p. \tag{7}$$

For $p = 2$, $j^2 + 1 = (j + 1)^2 \bmod 2$. For $p \equiv 3 \bmod 4$, $j^2 + 1$ is irreducible in $GF(p)$ and for $p \equiv 1 \bmod 4$, $j^2 + 1$ factors in terms of two mutually coprime degree one POLYs in $GF(p)$ as $j^2 + 1 = (j + a_0) \cdot (j + b_0) \bmod p$. Therefore, $j^2 + 1$ is irreducible in $Z(q)$, $p \equiv 3 \bmod 4$ and factors as $j^2 + 1 = (j + a) \cdot (j + b)$ in $Z(q)$, $p \equiv 1 \bmod 4$, $a_0 = a \bmod p$ and $b_0 = b \bmod p$. Three distinct cases follow from these statements.

Case 1. $p = 2$. In this case, (7) cannot hold. All the factorizations in $CZ(q, Q(\theta))$ are the same as those in $Z(q, Q(\theta))$. Only the input and output quantities are defined in $CZ(q, Q(\theta))$.

Case 2. $p \equiv 1 \bmod 4$. Based on the CRT, we have the direct sum,

$$CZ(q, Q(\theta)) = Z(q, Q(\theta), j + a) \oplus Z(q, Q(\theta), j + b). \tag{8}$$

Therefore, the factorizations in $CZ(q, Q(\theta))$, $p \equiv 1 \bmod 4$, can be expressed as a direct sum of two factorizations in $Z(q, Q(\theta))$.

Case 3. $p \equiv 3 \bmod 4$. In this case, $CZ(q)$ and $CZ(q, Q(\theta))$ will be studied as the p-adic expansion rings of $GF(p^2)$ and $GF(p^{2m})$, respectively, with $Q_0(\theta)$ being a monic, irreducible POLY of degree m in $GF(p^2)$. Consider all those elements in $CZ(q, Q(\theta))$ for whom $A_0(\theta)$ is non-zero, that is, they are non-zero when reduced mod p. It is clear that the set of all such elements is closed under MULT. There are $p^{2ma} - p^{2m(a-1)} = p^{2m(a-1)}\{p^{2m} - 1\}$ such elements. This discussion leads to the generalization of the Euler's theorem in $CZ(q, Q(\theta))$.

Theorem 3. Generalization of Euler's theorem. If $A(\theta)$ is any one of the elements of $CZ(q, Q(\theta))$ for which $A_0(\theta)$ is a non-zero element of $GF(p^{2m})$, then

$$A(\theta)^{p^{2m(a-1)}\left(p^{2m}-1\right)} = 1\,\text{mod}\left(q, Q(\theta)\right). \tag{9}$$

In a manner similar to Theorem 2, we have the following theorem.

Theorem 4. Generalization of Euler's theorem. If $A(\theta)$ is any one of the elements of $CZ(q, Q(\theta))$ for which $A_0(\theta)$ is a non-zero element of $GF(p^{2m})$ having order equal to n or a factor of n, $n \mid (p^{2m} - 1)$, then

$$A(\theta)^{p^{2m(a-1)}\cdot n} = 1\,\text{mod}\left(q, Q(\theta)\right). \tag{10}$$

In our further study of $CZ(q, Q(\theta))$, we will deal only with $p \equiv 3 \bmod 4$. The analysis for the other two cases as outlined above is complete.

IV. GENERATORS FOR NUMBER THEORETIC TRANSFORMS

If $A_0(\theta)$ is a primitive element in $GF(p^m)$, then

$$A_0(\theta)^{p^{m(a-1)}k} \neq 1\,\text{mod}\left(q, Q(\theta)\right),\ i = 1, 2, \ldots, p^m - 2.$$

This leads to the unique factorization of $u^N - 1$ in the finite integer ring $Z(q, Q(\theta))$ once it is known in the underlying finite field $Z(p, Q_0(\theta)) = GF(p^m)$.

Given $Q(\theta)$, let n be the smallest integer for which $Q_0(\theta)$ divides $\theta^n - 1$ in $GF(p)$. It is known that n is a factor of $p^m - 1$. If $n = p^m - 1$, then $Q_0(\theta)$ is a *primitive* POLY in $GF(p)$. In this case, we have the factorization

$$u^N - 1 = \prod_{k=0}^{N-1}\left(u - \theta^{(n/N)k}\right)\text{mod}\left(p, Q_0(\theta)\right) \tag{11}$$

for every N that divides n. The above expression is based on the facts that (1) $Z(p, Q_0(\theta))$ is same as $GF(p^m)$ whose elements can be expressed as POLYs in θ of degree up to $m - 1$ in $GF(p)$; and (2) θ is an element of order n in this field.

If $Q_0(\theta)$ is a primitive POLY, then θ is a primitive element in $GF(p^m)$ and we have

$$u^N - 1 = \prod_{k=0}^{N-1}\left(u - \theta^{\left(\left(p^m-1\right)/N\right)k}\right)\text{mod}\left(p, Q_0(\theta)\right) \tag{12}$$

for every N that divides $p^m - 1$. It is clear from these expressions that a length N NTT in $Z(p, Q_0(\theta))$ exists for all those values of N that divide n, n being the order of θ in $GF(p^m) = Z(p, Q_0(\theta))$, the generator being $G = \theta^{s(n/N)}$, $(s, N) = 1$.

If $Q_0(\theta)$ is not primitive, then one can still define a NTT of length $p^m - 1$ (and all its factors), but the form of the generator is not as simple as $G = \theta$ for values of N other than n and its factors. The primitive element is a POLY in θ, say $\alpha(\theta)$. In such a case, we have

$$u^N - 1 = \prod_{k=0}^{N-1}\left(u - \alpha(\theta)^{\left(\left(p^m-1\right)/N\right)k}\right) \mathrm{mod}\left(p, Q_0(\theta)\right) \tag{13}$$

for every N that divides $p^m - 1$, and the generator is given by $G = \alpha(\theta)^{s\left(p^m-1\right)/N}$, $(s, N) = 1$. We prefer the form of expression in (11) over that in (13) to define a length N NTT for N a factor of n and $Q_0(\theta)$ a non-primitive POLY.

Let us now turn to the integer ring $Z(q, Q(\theta))$. In the following, we first assume that $Q_0(\theta)$ is irreducible in $GF(p)$ while $Q_i(\theta)$ are arbitrary. This will be the case when we know only the properties of $GF(p^m)$ and not much else as far as the construction of $Z(q, Q(\theta))$ is concerned.

Case 1. $Q_0(\theta)$ irreducible and $Q_i(\theta)$ arbitrary. Let $Q_0(\theta)$ divide $\theta^n - 1$ in $GF(p)$, n a factor of $p^m - 1$, and N be a factor of n. Based on (6), we get

$$u^N - 1 = \prod_{k=0}^{N-1}\left(u - \theta^{p^{m(a-1)}(n/N)k}\right) \mathrm{mod}\left(q, Q(\theta)\right). \tag{14}$$

This is shown by using (6) to verify that $\theta^{p^{m(a-1)}(n/N)}$ satisfies all the properties of the generator of a NTT. Thus, all the generators of a length N NTT are given by $\theta^{p^{m(a-1)}(n/N)s}$, $(s, N) = 1$.

Case 2. $Q_0(\theta)$ primitive and $Q_i(\theta)$ arbitrary. Let $Q_0(\theta)$ divide $\theta^n - 1$ in $GF(p)$, $n = p^m - 1$, and N be a factor of n. Based on (5), we get

$$u^N - 1 = \prod_{k=0}^{N-1}\left(u - \theta^{\left(p^{m(a-1)}\left(p^m-1\right)/N\right)k}\right) \mathrm{mod}\left(q, Q(\theta)\right). \tag{15}$$

Thus, all the generators of a length N NTT are given by $\theta^{\left(p^{m(a-1)}\left(p^m-1\right)/N\right)s}$, $(s, N) = 1$.

Case 3. $Q_0(\theta)$ irreducible and $Q_i(\theta)$ arbitrary. Let $Q(\theta)$ divide $\theta^n - 1$, n a factor of $p^m - 1$, and N not be a factor of n. Here, we use the expression in (13) as a starting point. Based on (5), the factorization for the NTT is given by

$$u^N - 1 = \prod_{k=0}^{N-1}\left(u - \alpha(\theta)^{\left(p^{m(a-1)}\left(p^m-1\right)/N\right)k}\right) \bmod\left(q, Q(\theta)\right) \tag{16}$$

for every N that divides $p^m - 1$, and the generators are given by $G = \alpha(\theta)^{sp^{m(a-1)}\left(p^m-1\right)/N}$, where $\alpha(\theta)$ is a primitive element in $GF(p^m)$ and $(s, N) = 1$.

We now seek to simplify the form of factorization for the NTTs based on appropriate selection of $Q_i(\theta)$, $i = 1, 2, ..., a - 1$ in the p-adic expansion of $Q(\theta)$ in (2). We know that the irreducible POLY $Q_0(\theta)$ divides $\theta^n - 1$ in $GF(p)$ for some n that is a factor of $p^m - 1$. It is shown in the subsequent section that it is possible to select $Q_i(\theta)$, $i = 1, 2, ..., a - 1$ in a way that $Q(\theta)$ divides $\theta^n - 1$ in $Z(q)$. The following simplifications to the form of NTT are based on the assumption that $Q(\theta)$ divides $\theta^n - 1$.

Case 1, simplified. $Q_0(\theta)$ irreducible. $Q(\theta)$ divides $\theta^n - 1$, n is a factor of $p^m - 1$, and N a factor of n:

$$u^N - 1 = \prod_{k=0}^{N-1}\left(u - \theta^{(n/N)k}\right) \bmod\left(q, Q(\theta)\right). \tag{17}$$

This expression is based on the facts that (1) $Z(p, Q_0(\theta)) = GF(p^m)$, whose elements can be expressed as POLYs in θ of degree up to $m - 1$ in $GF(p)$; and (2) θ is an element of order n in this field. It also leads to $G = \theta^{s(n/N)}$, $(s, N) = 1$.

Case 2, simplified. $Q_0(\theta)$ primitive. $Q(\theta)$ divides $\theta^n - 1$, $n = p^m - 1$, N a factor of n:

$$u^N - 1 = \prod_{k=0}^{N-1}\left(u - \theta^{\left(\left(p^m-1\right)/N\right)k}\right) \bmod\left(q, Q(\theta)\right). \tag{18}$$

This expression is based on the facts that (1) $Z(p, Q_0(\theta))$ is same as the finite field $GF(p^m)$ whose elements can be expressed as POLYs in θ of degree up to

$m - 1$ in $GF(p)$; and (2) θ is an element of order $p^m - 1$ in this field. It also leads to $G = \theta^{s(p^m-1)/N}$, $(s, N) = 1$.

No further simplifications are possible in Case 3.

IV.1 Generators for NTT in $CZ(q, Q(\theta))$

The analysis for $CZ(q, Q(\theta))$ is similar to the analysis for $Z(q, Q(\theta))$. It is summarized in the following for the various cases that arise.

Case 1. $Q_0(\theta)$ irreducible and $Q_i(\theta)$ arbitrary. Let $Q_0(\theta)$ divide $\theta^n - 1$ in $GF(p^2)$, n be a factor of $p^{2m} - 1$, and N a factor of n. Here,

$$u^N - 1 = \prod_{k=0}^{N-1}\left(u - \theta^{p^{2m(a-1)}(n/N)k}\right) \bmod\left(q, Q(\theta)\right). \tag{19}$$

All the generators of a length N NTT are given by $\theta^{p^{2m(a-1)}(n/N)s}$, $(s, N) = 1$.

Case 2. $Q_0(\theta)$ primitive and $Q_i(\theta)$ arbitrary. Let $Q_0(\theta)$ divide $\theta^n - 1$ in $GF(p)$, $n = p^{2m} - 1$, and N be a factor of n. Here,

$$u^N - 1 = \prod_{k=0}^{N-1}\left(u - \theta^{\left(p^{2m(a-1)}\left(p^{2m}-1\right)/N\right)k}\right) \bmod\left(q, Q(\theta)\right). \tag{20}$$

Thus, all the generators of a length N NTT are given by $\theta^{\left(p^{2m(a-1)}\left(p^{2m}-1\right)/N\right)s}$, $(s, N) = 1$.

Case 3. $Q_0(\theta)$ irreducible and $Q_i(\theta)$ arbitrary. $Q(\theta)$ divides $\theta^n - 1$, n is a factor of $p^{2m} - 1$, N not a factor of n: Here,

$$u^N - 1 = \prod_{k=0}^{N-1}\left(u - \alpha(\theta)^{\left(p^{2m(a-1)}\left(p^{2m}-1\right)/N\right)k}\right) \bmod\left(q, Q(\theta)\right). \tag{21}$$

for every N that divides $p^m - 1$, and the generators are given by $G = \alpha(\theta)^{sp^{2m(a-1)}\left(p^{2m}-1\right)/N}$, where $\alpha(\theta)$ is a primitive element in $GF(p^{2m})$ and $(s, N) = 1$.

Once again, the following simplifications to the form of NTT are based on the assumption that $Q(\theta)$ divides $\theta^n - 1$.

Case 1, simplified. $Q_0(\theta)$ irreducible. $Q(\theta)$ divides $\theta^n - 1$, n is a factor of $p^{2m} - 1$, and N a factor of n:

$$u^N - 1 = \prod_{k=0}^{N-1}\left(u - \theta^{(n/N)k}\right) \bmod (q, Q(\theta)). \tag{22}$$

It leads to $G = \theta^{s(n/N)}$, $(s, N) = 1$.

Case 2, simplified. $Q_0(\theta)$ primitive. $Q(\theta)$ divides $\theta^n - 1$, $n = p^{2m} - 1$, N a factor of n:

$$u^N - 1 = \prod_{k=0}^{N-1}\left(u - \theta^{\left(\left(p^{2m}-1\right)/N\right)k}\right) \bmod (q, Q(\theta)). \tag{23}$$

It leads to $G = \theta^{s\left(p^{2m}-1\right)/N}$, $(s, N) = 1$.

No further simplifications are possible in Case 3.

We end this section by stating that the generalizations of the Euler's theorem presented in this section are also valid for $m = 1$. For $m = 1$, one can use them to identify the generators of NTT in $Z(q)$ and $CZ(q)$. Thus, they include the previously reported results for the generators of NTT in $Z(q)$ and $CZ(q)$ as a special case [1], [4].

V. CYCLOTOMIC FACTORIZATION IN FINITE INTEGER RINGS

The factorization of $u^N - 1$ in $GF(p)$ is obtained by first writing it as a product of N degree one factors in $GF(p^m)$ and then combining these factors selectively to obtain factors in $GF(p)$. We will follow the same approach in the following to obtain the factors in $Z(q)$ (and $CZ(q)$) starting from the degree one factors in $Z(q, Q(\theta))$ (and $CZ(q, Q(\theta))$).

All expressions in $Z(q)$ reduce to corresponding expressions in $GF(p)$ when taken MOD p. $GF(p^m)$ is obtained by taking POLY extension of $GF(p)$, where the operations of MULT and ADD are defined MOD $Q(\theta)$, a monic, primitive POLY of degree m. Therefore, in this section, we require that $Z(q, Q(\theta))$ be a POLY extension of $Z(q)$ where the operations of MULT and ADD are defined MOD $Q(\theta)$, such that $Q(\theta)$ is a monic, primitive POLY. In this instance, a primitive POLY $Q(\theta)$ in $Z(q)$ is defined as a POLY such that $Q_0(\theta) \equiv Q(\theta)$ mod p is primitive in $GF(p)$.

In the following, we employ Theorem 1. If $A_0(\theta)$ is a primitive element in $GF(p^m)$, then $A(\theta)^{p^{m(a-1)}k} \neq 1 \bmod (q, Q(\theta))$, $k = 1, 2, \ldots, p^m - 2$. Since $Q(\theta)$ is a primitive POLY, $A(\theta) = \theta$ is a primitive element and Theorem 1 can be expressed as

$$\theta^{p^{m(a-1)}(p^m-1)} = 1 \bmod (q, Q(\theta)). \tag{24}$$

It is possible to let $Q(\theta) = Q_0(\theta)$ for simplicity in the form of $Q(\theta)$. Given (24), a complete factorization of $u^n - 1$ in $Z(q, Q(\theta))$, $n = p^m - 1$, in terms of degree one factors is given by

$$u^{p^m-1} - 1 = \prod_{i=1}^{p^m-1} \left(u - \theta^{p^{m(a-1)} i} \right) \bmod (q, Q(\theta)). \tag{25}$$

There is a one-to-one correspondence between the factorization in $GF(p)$ and $Z(q)$, and $GF(p^m)$ and $Z(q, Q(\theta))$. These factorizations are unique in all instances as the POLYs are monic [3]. This property leads to the following procedure to obtain the factors in $Z(q)$ starting from the degree one factors in (25).

Partition the set $\Omega = \{1, 2, \ldots, p^m - 1\}$ into subsets $\Omega_{j_1}, \Omega_{j_2}, \ldots$. A cyclotomic set Ω_j begins with j, where j is the smallest power not included in the preceding subsets. Other elements in the subset Ω_j are obtained as

$$\Omega_j = \{j, jp, jp^2, jp^3, \ldots\}. \tag{26}$$

The powers are defined mod $p^m - 1$. Also, $jp^m \equiv j \bmod (p^m - 1)$ implies that there are at most m elements in each Ω_j. It is easy to verify that no elements in two different cyclotomic sets are equal. Let Δ be the set of indices $j_1, j_2, \ldots$. Based on this partitioning and (25), we write the factorization of $u^n - 1$, $n = p^m - 1$, as

$$u^{p^m-1} - 1 = \prod_{j \in \Delta} \left[\left\{ \prod_{k \in \Omega_j} \left(u - \theta^{p^{m(a-1)} \cdot k} \right) \right\} \bmod (q, Q(\theta)) \right] = \prod_{j \in \Delta} R_j(u). \tag{27}$$

The POLY $R_j(u)$ is defined as

$$R_j(u) = \left(u - \theta^{p^{m(a-1)} \cdot j}\right)\left(u - \theta^{p^{m(a-1)} \cdot jp}\right) \cdots \left(u - \theta^{p^{m(a-1)} \cdot jp^{l-1}}\right) \bmod (q, Q(\theta)), \tag{28}$$

such that

$$jp^l \equiv j \bmod (p^m - 1). \tag{29}$$

We now show that each of $R_j(u)$ has coefficients in $Z(q)$ and the factorization is complete.

Theorem 5. Each of $R_j(u)$ as defined in (28) has coefficients in $Z(q)$.
Proof: We use mathematical induction on a to prove this result. For $a = 1$, $Z(q, Q(\theta)) = GF(p^m)$, $Z(q) = GF(q)$, $Q(\theta) = Q_0(\theta)$ and (28) becomes

$$R_j(u) = \left(u - \theta^j\right)\left(u - \theta^{jp}\right) \cdots \left(u - \theta^{jp^{l-1}}\right) \bmod (p, Q_0(\theta)). \tag{30}$$

This is a well-known expression in finite field theory. Since $R_j(u)$ has θ^j and all its conjugates in $GF(p^m)$ as roots, it has coefficients in $GF(p)$. Hence the statement of the theorem is valid for $a = 1$. Now let us assume that it is valid for $a = K$. Also, to show the recursive nature of the expressions, we label the quantities in (28) with the superscript (K) to write

$$R_j^{(K)}(u) = \left(u - \theta^{p^{m(K-1)} \cdot j}\right)\left(u - \theta^{p^{m(K-1)} \cdot jp}\right) \cdots \left(u - \theta^{p^{m(K-1)} \cdot jp^{l-1}}\right) \bmod (p^K, Q^{(K)}(\theta)), \tag{31}$$

where

$$Q^{(K)}(\theta) = Q_0(\theta) + p\, Q_1(\theta) + \ldots + p^{K-1}\, Q_{K-1}(\theta). \tag{32}$$

As per the assumption, $R_j^{(K)}(u)$ in (31) has coefficients in $Z(p^K)$. For $a = K + 1$, $R_j^{(K+1)}(u)$ is given by

$$R_j^{(K+1)}(u) = \left(u - \theta^{p^{mK} \cdot j}\right)\left(u - \theta^{p^{mK} \cdot jp}\right) \cdots \left(u - \theta^{p^{mK} \cdot jp^{l-1}}\right) \bmod (p^{K+1}, Q^{(K+1)}(\theta)), \tag{33}$$

where

$$Q^{(K+1)}(\theta) = Q_0(\theta) + p\, Q_1(\theta) + \ldots + p^{K-1}\, Q_{K-1}(\theta) + p^K\, Q_K(\theta)$$

$$= Q^{(K)}(\theta) + p^K Q_K(\theta) \tag{34}$$

Step 1. Compare $R_j^{(K+1)}(u)$ in (33) to $R_j^{(K)}(u)$ in (31). Writing Theorem 1 for $A(\theta) = \theta$ and $a = K$, we have

$$\theta^{p^{mK}} = \theta^{p^{m(K-1)}} 1 \bmod\left(p^K, Q^{(K)}(\theta)\right). \tag{35}$$

Also,

$$Q^{(K+1)}(\theta) \equiv Q^{(K)}(\theta) \bmod p^K. \tag{36}$$

Therefore, taking mod p^K on both sides of (33), we get

$$R_j^{(K+1)}(u) \equiv R_j^{(K)}(u) \bmod\left(p^K, Q^{(K)}(\theta)\right). \tag{37}$$

This expression is important in two respects. It indicates that the distribution of the roots of $u^n - 1$, $n = p^m - 1$, in $Z(p^K, Q^{(K)}(\theta))$ has a one-to-one correspondence with the distribution of the roots in $Z(p^{K+1}, Q^{(K+1)}(\theta))$. Second, it leads to the following form for $R_j^{(K+1)}(u)$:

$$R_j^{(K+1)}(u) = R_j^{(K)}(u) + p^K S_j(u) \bmod\left(p^{K+1}, Q^{(K+1)}(\theta)\right), \tag{38}$$

$S_j(u)$ being a POLY with coefficients in $GF(p^m)$. In order to complete the proof of the theorem, we need to show that $S_j(u)$ has coefficients in $GF(p)$. Substituting for $R_j^{(K+1)}(u)$ in (27), we get

$$u^{p^m-1} - 1 = \prod_{j \in \Delta} R_j^{(K+1)}(u) \bmod\left(p^{K+1}, Q^{(K+1)}(\theta)\right)$$

$$= \prod_{j \in \Delta} \left\{R_j^{(K)}(u) + p^K S_j(u)\right\} \bmod\left(p^{K+1}, Q^{(K)}(\theta) + p^K Q_K(\theta)\right). \tag{39}$$

Recalling that $R_j^{(K)}(u)$ has coefficients in $Z(p^K)$, (39) can be written as

$$u^{p^m-1} - 1 - \left\{\prod_{j \in \Delta} R_j^{(K)}(u) \bmod p^{K+1}\right\}$$

$$= \sum_{J \in \Delta} p^K S_J(u) \prod_{\substack{j \in \Delta \\ j \neq J}} R_j^{(K)}(u) \mod\left(p^{K+1}, Q^{(K)}(\theta) + p^K Q_K(\theta)\right). \qquad (40)$$

Since p^K is a common factor throughout, dividing by p^K, we get

$$\left[p^{-K}\left(u^{p^m-1} - 1 - \left\{\prod_{j \in \Delta} R_j^{(K)}(u) \bmod p^{K+1}\right\}\right)\right] \bmod p$$
$$= \sum_{J \in \Delta} S_J(u) \prod_{\substack{j \in \Delta \\ j \neq J}} R_j^{(1)}(u) \mod\left(p, Q^{(1)}(\theta)\right). \qquad (41)$$

Note that $R_j^{(1)}(u)$ is the same as $R_j(u)$ as expressed in (30) and $Q^{(1)}(\theta) = Q_0(\theta)$. Thus, $S_J(u)$ can be obtained by solving the following POLY congruence in $GF(p^m)$:

$$\left[p^{-K}\left(u^{p^m-1} - 1 - \left\{\prod_{j \in \Delta} R_j^{(K)}(u) \bmod p^{K+1}\right\}\right)\right] \mod\left(p, R_J(u)\right)$$
$$\equiv S_J(u) \prod_{\substack{j \in \Delta \\ j \neq J}} R_j^{(1)}(u) \mod\left(p, Q^{(1)}(\theta), R_J(u)\right). \qquad (42)$$

We now observe that this congruence is of the form $A(u) \equiv B(u)\, C(u) \bmod D(u)$ with $A(u)$, $C(u)$ and $D(u)$ having coefficients in $GF(p)$ and $(C(u), D(u)) = 1$. Therefore, a unique solution to the unknown POLY $B(u)$ with coefficients in $GF(p)$ exists. In our case, that unknown POLY with coefficients in $GF(p)$ is $S_J(u)$. This proves that $R_j^{(K+1)}(u)$ in (38) has coefficients in $Z(p^{K+1})$. This proves the theorem for $a = K + 1$, thereby establishing the theorem for all a by mathematical induction. Furthermore, the uniqueness of $S_J(u)$ establishes the uniqueness of $R_j^{(K)}(u)$ for all values of K.

The cyclotomic factorization of $u^N - 1$ in $Z(q)$, N a factor of $p^m - 1$, can be obtained by partitioning the set $\{1, 2, ..., N\}$ into cyclotomic sets with the powers defined mod N. In this case,

$$u^N - 1 = \prod_{j\in\Delta}\left[\left\{\prod_{k\in\Omega_j}\left(u - \theta^{p^{m(a-1)}\{(p^m-1)/N\}\cdot k}\right)\right\}\bmod\left(q, Q(\theta)\right)\right] = \prod_{j\in\Delta} R_j(u). \tag{43}$$

Alternatively, one may obtain these factors from the original factorization of $u^n - 1$, $n = p^m - 1$.

Similar equations can also be derived when one wishes to obtain factorization in $CZ(q)$ starting from the factorization in $CZ(q, Q(\theta))$ with $Q_0(\theta)$ being a primitive POLY in $GF(p^2)$. In this case, the mod q operation is replaced by mod $(q, j^2 + 1)$. All the expressions and statements remain valid with p replaced by p^2 in (24) to (43). The powers in the cyclotomic sets are defined mod $p^{2m} - 1$. Once again, there are at most m elements in each cyclotomic set.

It is to be noted that the procedure as established above is of vital importance if one wishes to obtain a POLY $Q(\theta)$ that divides $u^n - 1$ in $Z(q)$, starting from the primitive POLY $Q_0(\theta)$ that divides $u^n - 1$ in $GF(p)$, $n = p^m - 1$. Tables of primitive POLYs in finite fields are available, but no such information is available for the primitive POLYs in finite integer rings. From the description of the cyclotomic POLY construction, such $Q(\theta)$ in $Z(q)$ are obtained as

$$Q_0(u) = R_1^{(1)}(u) \bmod p \tag{44}$$

and

$$Q(u) = R_1^{(a)}(u) \bmod p^a. \tag{45}$$

From (45), $Q(\theta)$ is obtained by replacing u by θ. These statements are valid in $CZ(q)$ as well. In the remainder of this paper, we assume that $Q(\theta)$ is selected in a way that (i) $Q_0(\theta)$ is primitive in $GF(p)$ for factorization in $Z(q)$ and in $GF(p^2)$ for factorization in $CZ(q)$, and (ii) (45) is satisfied. In such a case, the expression for $R_j^{(K)}(u)$ in (31) simplifies to

$$R_j^{(K)}(u) = \left(u - \theta^j\right)\left(u - \theta^{jp}\right)\cdots\left(u - \theta^{jp^{l-1}}\right) \bmod (p^K, Q^{(K)}(\theta)), \tag{46}$$

and the factorization of $u^N - 1$ is as given in (18). Under this assumption, the NTTs in $Z(q, Q(\theta))$ and $CZ(q, Q(\theta)$ correspond to the simplified version of Case 2.

VI. CONJUGATE SYMMETRY PROPERTY OF NTT

In this section, we deal only with the simplified version of Case 2 for $Z(q, Q(\theta))$ and $CZ(q, Q(\theta))$. Given a sequence x_i, $i = 0, 1, ..., N-1$, N a factor of $p^m - 1$, its NTT is defined as the sequence X_k, $k = 0, 1, ..., N-1$, where

$$X_k = \sum_{i=0}^{N-1} x_i \left[\theta^{\{p^m-1\}/N}\right]^{i \cdot k}. \tag{47}$$

Each of X_k, in general, is a POLY in θ of degree up to $m - 1$. In many DSP applications, the data sequence x_i may be defined in $Z(q)$ even though one needs to employ $Z(q, Q(\theta))$ in order to describe a NTT. This is similar to the use of complex number system in order to compute the discrete Fourier transform (DFT) of a real-valued data sequence. In such a case, the NTT X_k satisfies additional properties that may be used to simplify the computational burden. In the following, we first assume that $N = p^m - 1$.

Define the generating function of input sequence as

$$x(u) = x_0 + x_1 u + ... + x_{N-1} u^{N-1}. \tag{48}$$

The computation of NTT can be written as a two-step process:

Step 1. Reduce $x(u)$ mod each of $R_j(u)$ in the cyclotomic factorization of $u^N - 1$. Let it be $x_j(u)$. The POLYs $x_j(u)$ have coefficients in $Z(q)$.

Step 2. Obtain

$$X_k = x_j(\theta^k),\ k = j, jp, jp^2, ..., jp^{l-1}, \tag{49}$$

where $jp^l \equiv 1 \bmod \{p^m - 1\}$. Since the POLY $x_j(u)$ is independent of θ, we have the conjugate symmetry property satisfied by X_k:

$$X_{kp} = X_k \Big|_{\theta=\theta^p}. \tag{50}$$

In (50), X_k is a POLY in θ. Once X_k is determined for $k = j$, (50) can be used to determine X_k for $k = jp, jp^2, ..., jp^{l-1}$. Therefore, it is sufficient to determine the NTT only for those indices that belong to the set Δ. The NTT for all other values can be determined using the conjugate symmetry property. We also note that the conjugate symmetry property in $Z(q)$ as expressed in (50) is different from the conjugate symmetry property in $GF(p)$ given by $X_{jp} = (X_j)^p$.

If N is a factor of $p^m - 1$, then the cyclotomic sets are defined mod N. In that case, the conjugate symmetry property remains the same as in (50) with the difference that θ is replaced by the generator of NTT of length N and the cyclotomic sets are constructed accordingly. In such a case, it may be more convenient to employ $\beta = \theta^{\{p^m-1\}/N}$ directly as a generator of length N NTT with the entire computation defined in $Z(q, Q(\beta))$, $Q(\beta)$ being the irreducible POLY in $Z(q)$ having β and all its conjugates as its roots in $Z(q, Q(\beta))$. The analysis for $CZ(q, Q(\theta))$ is very similar to the analysis for $Z(q, Q(\theta))$.

Example 1. Consider $p = 2$, $q = 2^a$, $N = 2^m - 1$, N being a Mersenne prime. Order of p mod $N = m$. Thus, we work in $Z(2^a, Q(\theta))$, $Q(\theta)$ being a degree m primitive POLY in $Z(2^a)$. Since N is prime, order of θ^k is N, $k = 1, 2, ..., N - 1$. Each element has exactly m conjugates (except 1) and the degree of each of the cyclotomic factor is m. All cyclotomic factors are primitive in nature. (In general, there are $\phi(p^m - 1)/m$ primitive POLYs of degree m in $Z(q)$, $\phi(.)$ being the Euler's totient function.) Let $m = 5$. The cyclotomic sets are (1, 2, 4, 8, 16), (3, 6, 12, 24, 17), (5, 10, 20, 9, 18), (7, 14, 28, 25, 19), (11, 22, 13, 26, 21), (15, 30, 29, 27, 23), (0), and $\Delta = \{0, 1, 3, 5, 7, 11, 15\}$. For a length 31 NTT of a sequence in $Z(2^a)$, it is sufficient to determine its values for $k = 0, 1, 3, 5, 6, 11$, and 15. The NTT at all other points can be determined by using the conjugate symmetry property.

Example 2. Consider $p = 2$, $q = 2^a$, $N = 2^e + 1$, N being a Mersenne prime. Order of p mod $N = m = 2\,e$. Thus, we work in $Z(2^a, Q(\theta))$, $Q(\theta)$ being a degree m primitive POLY in $Z(2^a)$. Since N is prime, order of θ^k is N, $k = 1, 2, ..., N - 1$. Each element has exactly m conjugates (except 1) and the degree of each of the cyclotomic factor is m. Let $e = 4$. The cyclotomic sets are (1, 2, 4, 8, 16, 15, 13, 9), (3, 6, 12, 7, 14, 11, 5, 10), (0), and $\Delta = \{0, 1, 3\}$. For a length 17 NTT of a sequence in $Z(2^a)$, it is sufficient to determine its values for $k = 0$, 1, and 3. The NTT at all other points can be determined using the conjugate symmetry property.

We end this section by stating that the conjugate symmetry property in finite integer rings (and not the finite complex integer rings) was described in [5] using a far more abstract and lengthy approach. The theoretical background in [5] was incomplete and one had to rely on a computer program to generate the results.

VII. DISCUSSION AND CONCLUSIONS

In this paper, we have studied number theoretic transforms in a finite integer ring in terms of their counterparts defined in the underlying finite field. This leads to closed form expressions for the generators of the various transforms in terms of the primitive elements of the finite fields. The results can be used to identify all the possible generators for a length N NTT and their suitability can be analyzed.

The analysis also brings out two approaches for computing the cyclic convolution of two sequences defined in either $Z(q)$ or $CZ(q)$; one is based on expressing it as a NTT in an extension ring and computing it using the conjugate symmetry property. The second approach is based on using the cyclotomic factorization of $u^N - 1$ in either $Z(q)$ or $CZ(q)$ and the recently described forms of the Chinese remainder theorem in them [6]. These approaches may have their own unique features from a hardware implementation point of view.

A very interesting aspect of the work is that only finite field algebra is used to derive all the results in the finite integer rings. To our knowledge, this is the first time that an extension to the Euler's theorem is being described which makes all this analysis possible. The application of the number-theoretic results to designing Bose-Chaudhary-Hoquenghem (BCH) and Reed-Solomon (RS) codes will be described in a future paper.

REFERENCES

[1] H.J Nussbaumer, *Fast Fourier Transform and Convolution Algorithms*, Springer-Verlag, New York, 1982.

[2] L.K. Hua, *Introduction to Number Theory*, Springer-Verlag, New York, 1982.

[3] H. Krishna, B. Krishna, K.Y. Lin, and J.-D. Sun, *Computational Number Theory and Digital Signal Processing*, CRC Press, 1995.

[4] M.C. Vanwormhoudt, "Structural Properties of Complex Residue Rings Applied to Number Theoretic Fourier Transforms," *IEEE Transactions on Acoustics, Speech, and Signal Processing*, Vol. ASSP-26, pp. 99-104, 1978.

[5] J.-B. Martens and M.C. Vanwormhoudt, "Convolution Using a Conjugate Symmetry Property for Number Theoretic Transforms Over Rings of Regular Integers," *IEEE Transactions on Acoustics, Speech, and Signal Processing*, Vol. ASSP-31, pp. 1121-1125, 1983.

[6] H. Krishna Garg, *Digital Signal Processing Algorithms*, CRC Press, 1998.

Paper four

Error Control Techniques for Data Sequences Defined in Finite Integer Rings

Abstract. In this paper, we present results that can be used to design cyclic codes for error control coding of data sequences defined in finite integer and complex integer rings. This follows from our recent work on generalization of the well-known Euler's theorem in finite integer rings and their polynomial extensions. The idea is to describe BCH and Reed-Solomon codes in these rings along with a decoding algorithm. The decoding algorithm in the ring employs the decoder in the finite field in an iterative manner. All the algebraic properties of the resulting codes follow from the underlying finite fields.

I. INTRODUCTION

This paper deals with the algebra of finite integer rings and their polynomial (POLY) extensions with a view to establish their mathematical properties that can be of fundamental importance in error control coding (ECC). These results can be used to obtain Bose-Chaudhary-Hoquenghem (BCH) and Reed-Solomon (RS) codes that may find applications in communication and fault tolerant computing systems. Within the domain of digital signal processing (DSP), these properties can be used to obtain all the possible number theoretic transforms (NTTs) and cyclotomic factorization for the computation of one- and two-dimensional cyclic convolution [1].

The finite integer rings that we study in this paper are as follows:

(1) $Z(M)$, consisting of integers $\{0, 1, ..., M-1\}$,
(2) $CZ(M)$, consisting of integers $\{a + j\,b,\ a, b \in Z(M), j^2 = -1\}$,
(3) $Z(M, Q(\theta))$, consisting of POLYs $\{a_0 + a_1\,\theta + a_2\,\theta^2 + ... + a_{m-1}\,\theta^{m-1} + \theta^m,\ a_i \in Z(M)\}$, and
(4) $CZ(M, Q(\theta))$, consisting of POLYs $\{\ a(\theta) + j\,b(\theta),\ a(\theta), b(\theta) \in Z(M, Q(\theta)),\ j^2 = -1\}$.

The operations of addition (ADD) and multiplication (MULT) in $Z(M, Q(\theta))$ and $CZ(M, Q(\theta))$ are defined modulo $(M, Q(\theta))$ and $(M, j^2 + 1, Q(\theta))$, respectively. Here, $Q(\theta)$ is a *monic* degree m POLY with coefficients in $Z(M)$ for $Z(M, Q(\theta))$ and in $CZ(M)$ for $CZ(M, Q(\theta))$. The rings $Z(M)$ and $CZ(M)$ can also be studied as special cases of $Z(M, Q(\theta))$.

In order to construct an (n, k) code for data sequences defined in $Z(M)$ and $CZ(M)$, we notice the following salient features of cyclic code construction that are valid for data sequences defined in all number systems:

1. Cyclic codes are specified in terms of the generator POLY $G(u)$, which is a *monic* POLY of degree $n - k$.
2. $G(u)$ divides $u^n - 1$ in the number system.
3. The message POLY $M(u)$ and the code POLY $V(u)$ are related as $V(u) = M(u) \cdot G(u)$.

The procedure to compute the factorization of $u^n - 1$ in finite fields is well known and can be found in any book on coding theory. We will describe the corresponding results for $Z(M)$ and $CZ(M)$. Also, the decoding algorithm for all such cyclic codes will be described in terms of the decoding algorithm for the cyclic code in the underlying finite field.

The methodology adopted in this paper consists in developing a unified approach to study a finite integer ring as p-adic expansion of the underlying finite fields. Interestingly, all the properties and quantities in a finite integer ring can be determined solely from the knowledge of the corresponding properties and quantities in the finite field. We begin our analysis by describing the algebraic construction of finite integer rings. The results for finite fields, Chinese remainder theorem (CRT), and a recently described generalization of the Euler's theorem are then applied to obtain all the cyclic codes in the finite integer rings.

The organization of this paper is as follows. Section II is on basic mathematical results on cyclic code construction in finite integer rings. In

Section III, results on cyclotomic factorization in finite integer rings are described. Section IV is on the construction of BCH and RS codes in finite integer rings. In Section V, a decoding algorithm for the BCH and RS codes is presented. A discussion of the results and conclusions are presented in Section VI.

II. BASIC RESULTS ON CYCLIC CODE CONSTRUCTION IN FINITE INTEGER RINGS

Given the standard factorization of M as

$$M = \prod_{l=1}^{r} p_l^{a_l} = \prod_{l=1}^{r} q_l , \tag{1}$$

the CRT can be used to express the ring $Z(M, Q(\theta))$ as the direct sum,

$$Z(M, Q(\theta)) = \sum_{\substack{l=1 \\ \oplus}}^{r} Z\left(p_l^{a_l}, Q^{(l)}(\theta)\right), \tag{2}$$

where $Q^{(l)}(\theta) \equiv Q(\theta) \bmod p_l^{a_l}$. Consequently, the generator matrix for a cyclic code can be expressed as the direct sum

$$\mathbf{G} = \sum_{\substack{l=1 \\ \oplus}}^{r} \mathbf{G}_l , \tag{3}$$

where $\mathbf{G}_l$ is the generator matrix for the code with $M = p_l^{a_l}$. A necessary and sufficient condition for a (n, k, d) code in $Z(M)$, $CZ(M)$, $Z(M, Q(\theta))$, $CZ(M, Q(\theta))$ to exist is that there also exists (n, k, d_l) codes in $Z(p_l^{a_l})$, $CZ(p_l^{a_l})$, $Z(p_l^{a_l}, Q^{(l)}(\theta))$, $CZ(p_l^{a_l}, Q^{(l)}(\theta))$, respectively, $d_l \geq d$, $l = 1, 2, \ldots, r$ [2]. Consequently, we will pursue only the generator matrix (or, equivalently, the generator POLY) $\mathbf{G}_l$ in subsequent analysis. Also, the subscript or the superscript l is deleted and results are stated for $M = q = p^a$. It simplifies the form of the various expressions.

II.1 BCH and RS Codes in Finite Fields

Given θ as the primitive element of order $p^m - 1$ in $GF(p^m)$, the generator POLY of the BCH and RS code has $\theta, \theta^2, \ldots, \theta^{d-1}$ as roots for a designed minimum distance equal to d. Thus, for a (n, k) BCH code defined in $GF(p)$,

$$\begin{aligned} n &= p^m - 1, \\ n - k &\leq m\,t \\ d_{min} &\geq 2\,t + 1. \end{aligned} \tag{4}$$

The generator POLY for a BCH code contains $\theta, \theta^2, \ldots, \theta^{d-1}$ and all their conjugates as roots. For a RS code defined in $GF(p^m)$, the parameters are

$$\begin{aligned} n &= p^m - 1, \\ n - k &= d + 1 \\ d_{min} &= d. \end{aligned} \tag{5}$$

Note that the parameters mentioned in (4) correspond to the BCH codes in the narrow sense. It is possible to describe more general forms of BCH codes [4]. However, for the sake of the analysis to be presented here, it is sufficient to describe only the codes included in (4) and (5). They can be easily generalized to other forms of BCH and RS codes. These codes have been studied extensively from the standpoint of their algebraic properties and methods for encoding and decoding [3] to [6].

III. ALGEBRA OF FINITE INTEGER RINGS AND CYCLOTOMIC FACTORIZATION

III.1 Algebraic Construction of $Z(q, Q(\theta))$ and $CZ(q, Q(\theta))$

The key idea in this section is to describe a finite integer ring as p-adic expansion of the underlying finite field. Given a POLY $Q(\theta)$ with coefficients defined in $Z(q)$, we define its p-adic expansion as

$$Q(\theta) = Q_0(\theta) + pQ_1(\theta) + \ldots + p^{a-1}Q_{a-1}(\theta). \tag{6}$$

The coefficients of $Q_b(\theta)$, $b = 0, 1, \ldots, a - 1$ are defined in $GF(p)$. In that regard, the elements of $Z(q, Q(\theta))$ are POLYs in θ of degree up to $m - 1$ with the arithmetic operations of MULT and ADD defined a primitive POLY $Q(\theta)$ in $Z(q)$. Here, a **primitive POLY** in $Z(q)$ is defined as a monic POLY $Q(\theta)$ of degree m such that $Q_0(\theta) \equiv Q(\theta) \bmod p$ is a primitive, degree m POLY in $GF(p)$. Such $Q_0(\theta)$ POLYs are well documented in any text on finite field algebra and we assume here that they are known for all values of p and m that are of interest.

All expressions in $Z(q)$ reduce to corresponding expressions in $GF(p)$ when taken MOD p. Therefore, we have

$$Z(q, Q(\theta)) = GF(p^m) \bmod p. \tag{7}$$

It is clear that the choice of $Q_1(\theta), \ldots, Q_{a-1}(\theta)$ is completely arbitrary. They can all be set to zero or assigned other values in order to satisfy some desired mathematical feature.

Let us now analyze $CZ(q, Q(\theta))$. In this case, we would like to have the expression

$$CZ(q, Q(\theta)) = GF(p^{2m}) \bmod p. \tag{8}$$

For $p = 2, j^2 + 1 = (j + 1)^2 \bmod 2$. For $p \equiv 3 \bmod 4, j^2 + 1$ is irreducible in $GF(p)$ and for $p \equiv 1 \bmod 4, j^2 + 1$ factors in terms of two mutually co-prime degree one POLYs in $GF(p)$ as $j^2 + 1 = (j + a_0) \cdot (j + b_0) \bmod p$. Therefore, $j^2 + 1$ is irreducible in $Z(q)$, $p \equiv 3 \bmod 4$ and factors as $j^2 + 1 = (j + a) \cdot (j + b)$ in $Z(q)$, $p \equiv 1 \bmod 4$, $a_0 = a \bmod p$, and $b_0 = b \bmod p$. Three distinct cases follow from these statements.

Case 1. $p = 2$. In this case, (8) cannot hold. All the cyclic codes in $CZ(q, Q(\theta))$ are the same as those in $Z(q, Q(\theta))$. Only the input and output quantities are defined in $CZ(q, Q(\theta))$.

Case 2. $p \equiv 1 \bmod 4$. Based on the CRT, we have the direct sum,

$$CZ(q, Q(\theta)) = Z(q, Q(\theta), j + a) \oplus Z(q, Q(\theta), j + b). \tag{9}$$

Therefore, the codes in $CZ(q, Q(\theta))$, $p \equiv 1 \bmod 4$, can be expressed as a direct sum of two codes in $Z(q, Q(\theta))$.

Case 3. $p \equiv 3 \bmod 4$. In this case, $CZ(q)$ and $CZ(q, Q(\theta))$ will be studied as the p-adic expansion rings of $GF(p^2)$ and $GF(p^{2m})$, respectively, with $Q_0(\theta)$ being a primitive POLY of degree m in $GF(p^2)$. The analysis for the other two cases as outlined above is complete.

III.2 Cyclotomic Factorization in $Z(q, Q(\theta))$ and $CZ(q, Q(\theta))$

In a recent paper, it was established that the following generalizations to the Euler's theorem hold [1].

Theorem 1. Generalization of Euler's Theorem. The following result holds in $Z(q, Q(\Theta))$:

$$\theta^{p^{m(a-1)}\left(p^m-1\right)} = 1 \bmod\left(q, Q(\theta)\right). \tag{10}$$

Theorem 2. Generalization of Euler's theorem. The following result holds in $CZ(q, Q(\Theta))$:

$$\theta^{p^{2m(a-1)}\left(p^{2m}-1\right)} = 1 \bmod\left(q, Q(\theta)\right). \tag{11}$$

There are several generalizations to the Euler's theorem reported in [1]; only the ones to be employed in our analysis are reproduced.

Based on (10), we have the following cyclotomic factorization in $Z(q, Q(\Theta))$:

$$u^{p^m-1} - 1 = \prod_{i=0}^{p^m-2}\left(u - \theta^{p^{m(a-1)}i}\right) \bmod\left(q, Q(\theta)\right). \tag{12}$$

Similarly, based on (11), we have the following cyclotomic factorization in $CZ(q, Q(\Theta))$:

$$u^{p^{2m}-1} - 1 = \prod_{i=0}^{p^{2m}-2}\left(u - \theta^{p^{2m(a-1)}i}\right) \bmod\left(q, j^2+1, Q(\theta)\right). \tag{13}$$

There is a one-to-one correspondence between the factorization in $GF(p)$ and $Z(q)$, and $GF(p^m)$ and $Z(q, Q(\Theta))$. These factorizations are unique in all instances as the POLYs are monic [7]. This property leads to the following procedure to obtain the factors in $Z(q)$ starting from the degree one factors in (12).

Partition the set $\Omega = \{1, 2, ..., p^m - 1\}$ into subsets $\Omega_{j_1}, \Omega_{j_2}, \ldots$. A cyclotomic set Ω_j begins with j, where j is the smallest power not included in the preceding subsets. Other elements in the subset Ω_j are obtained as

$$\Omega_j = \{j, jp, jp^2, jp^3, \ldots\}. \tag{14}$$

The powers are defined mod $p^m - 1$. Also, $jp^m \equiv j \bmod (p^m - 1)$ implies that there are at most m elements in each Ω_j. It is easy to verify that no elements in two different cyclotomic sets are equal. Let Δ be the set of indices $j_1, j_2, \ldots$.

Based on this partitioning and (12), we write the factorization of $u^n - 1$, $n = p^m - 1$ as

$$u^{p^m-1} - 1 = \prod_{j\in\Delta}\left[\left\{\prod_{i\in\Omega_j}\left(u - \theta^{p^{m(a-1)}\cdot i}\right)\right\}\mathrm{mod}\left(q, Q(\theta)\right)\right] = \prod_{j\in\Delta} R_j(u). \qquad (15)$$

The POLY $R_j(u)$ is defined as

$$R_j(u) = \left(u - \theta^{p^{m(a-1)}\cdot j}\right)\left(u - \theta^{p^{m(a-1)}\cdot jp}\right)\cdots\left(u - \theta^{p^{m(a-1)}\cdot jp^{l-1}}\right) \bmod (q, Q(\theta)), \qquad (16)$$

such that

$$jp^l \equiv j \bmod (p^m - 1). \qquad (17)$$

Each of $R_j(u)$ has coefficients in $Z(q)$ and the factorization is complete [1].

Similar equations can be derived when one wishes to obtain factorization in $CZ(q)$ starting from the factorization in $CZ(q, Q(\theta))$ with $Q_0(\theta)$ being a primitive POLY in $GF(p^2)$. In this case, mod q operation is replaced by mod $(q, j^2 + 1)$. All the expressions and statements remain valid with p replaced by p^2 in (14) to (17). The powers in the cyclotomic sets are defined mod $p^{2m} - 1$. Once again, there are at most m elements in each cyclotomic set.

It is to be noted that the procedure as established above is of vital importance if one wishes to obtain a POLY $Q(\theta)$ that divides $u^n - 1$ in $Z(q)$, starting from the primitive POLY $Q_0(\theta)$ that divides $u^n - 1$ in $GF(p)$, $n = p^m - 1$. Tables of primitive POLYs in finite fields are available, but no such information is available for the primitive POLYs in finite integer rings. From the description of the cyclotomic POLY construction, such $Q(\theta)$ in $Z(q)$ are obtained as

$$Q(u) = R_1(u) \bmod p^a. \qquad (18)$$

From (18), $Q(\theta)$ is obtained by replacing u by θ. These statements are valid in $CZ(q)$ as well. In the remainder of this paper, we assume that $Q(\theta)$ is selected in a way that $Q_0(\theta)$ is primitive in $GF(p)$ for factorization in $Z(q)$ and in $GF(p^2)$ for factorization in $CZ(q)$ and that (18) is satisfied. Further, in those cases when (18) is satisfied, the expression for $R_j(u)$ in (16) simplifies to

$$R_j(u) = \left(u - \theta^j\right)\left(u - \theta^{jp}\right)\cdots\left(u - \theta^{jp^{l-1}}\right) \bmod (q, Q(\theta)) \qquad (19)$$

and the factorization of $u^n - 1$, $n = p^m - 1$, is as given below:

$$u^{p^m - 1} - 1 = \prod_{i=0}^{p^m - 2} \left(u - \theta^i\right) \bmod\left(q, Q(\theta)\right). \tag{20}$$

In summary, the elements of $Z(q, Q(\theta))$ ($CZ(q, Q(\theta))$) are POLYs in θ of degree up to $m - 1$ with coefficients in $Z(q)$ ($CZ(q)$) with the arithmetic operations of MULT and ADD defined mod $Q(\theta)$, $Q(\theta)$ being a primitive POLY in $Z(q)$ ($CZ(q)$) such that $Q(\theta)$ divides $\theta^n - 1$, $n = p^m - 1$ ($n = p^{2m} - 1$).

Example 1. Consider the factorization in $Z(2^2)$:

$$u^7 - 1 = (u - 1) \cdot (u^3 + 3u^2 + 2u + 3) \cdot (u^3 + 2u^2 + u + 3).$$

It is obtained by noting that $\theta^3 + \theta^2 + 1$ is a primitive POLY in $GF(2)$ and then following the steps as described above. It is seen that either one of $(\theta^3 + 3\theta^2 + 2\theta + 3)$ or $(\theta^3 + 2\theta^2 + \theta + 3)$ can be used as the primitive POLY to generate $Z(2^2)$ that satisfies all the properties.

Example 2. Consider the factorization in $CZ(3^3)$:

$$\begin{aligned} u^7 - 1 = (u - 1) &\cdot \{u^3 + (14 + 7j)u^2 + (13 + 7j)u + 26\} \\ &\cdot \{u^3 + (14 + 20j)u^2 + (13 + 20j)u + 26\} \bmod (27, j^2 + 1). \end{aligned}$$

It is seen that either $\{\theta^3 + (14 + 7j)\,\theta^2 + (13 + 7j)\,\theta + 26\}$ or $\{\theta^3 + (14 + 20j)\,\theta^2 + (13 + 20j)\,\theta + 26\}$ can be used as the irreducible (not primitive) POLY to generate $CZ(3^3)$.

IV. BCH AND RS CODES IN FINITE INTEGER RINGS

Given the cyclotomic factorization of $u^n - 1$, the BCH and RS codes of length n ($n = p^m - 1$ for $Z(q)$ and $n = p^{2m} - 1$ for $CZ(q)$) in $Z(q)$ and $CZ(q)$ are obtained by letting the generator POLY $G(u)$ to have $\theta, \theta^2, \ldots, \theta^{d-1}$, as roots. For the BCH codes, all the conjugates of these elements are roots of $G(u)$ as well. These codes are cyclic as $G(u)$ is a monic POLY that divides $u^n - 1$ in the stated number system. The overall code parameters of the BCH codes in $Z(q)$ and $CZ(q)$ remain the same as the code parameters of the BCH codes in $GF(p)$ and $GF(p^2)$, respectively.

Theorem 3. The minimum distance of the codes in $Z(q)$ ($CZ(q)$) generated by $G(u)$ is the same as the minimum distance of the code in $GF(p)$ ($GF(p^2)$).

Proof: The code-POLYs are obtained as $V(u) = M(u) \cdot G(u)$. Taking the p-adic expansion of the quantities, we get

$$v_0(u) + p\, v_1(u) + \ldots + p^{a-1}\, v_{a-1}(u) = \{m_0(u) + p\, m_1(u) + \ldots + p^{a-1}\, m_{a-1}(u)\} \cdot \{g_0(u) + p\, g_1(u) + \ldots + p^{a-1}\, g_{a-1}(u)\} \bmod p^a.$$

It is clear that $g_0(u) \equiv G(u) \bmod p$ is the generator POLY of a BCH code having $\theta, \theta^2, \ldots, \theta^{d-1}$ as its roots in $GF(p)$ $(GF(p^2))$. Let $M(u)$ be a non-zero message POLY and b be the smallest integer such that $m_b(u) \neq 0$, $b < a$. Such a b always exists or else $M(u) = 0$. Now, $v_b(u) = m_b(u) \cdot g_0(u) \bmod p$, that is, $v_b(u)$ is a code-POLY in the BCH or RS code defined in the finite field. Thus,

$$\text{Hamming weight } V(u) \geq \text{Hamming weight } v_b(u) \geq d.$$

Let $m(u)$ be a message POLY in the finite field such that the corresponding code-POLY $v(u)$ has Hamming weight equal to d. Then, for the message $M(u) = p^{a-1}\, m(u)$, $V(u)$ $(= p^{a-1}\, v(u))$ has Hamming weight equal to d. This proves the statement of the theorem.

It is also worthwhile to mention that if one is not interested in a **cyclic** code in $Z(q)$ or $CZ(q)$, but merely an **equivalent** error correcting code, then one can use the generator POLY of the minimum distance d BCH code in $GF(p)$ or $GF(p^2)$ to obtain a minimum distance code d in $Z(q)$ or $CZ(q)$, respectively.

Example 3. Let us continue Example 1 with $\theta^3 + 3\theta^2 + 2\theta + 3$ as the primitive POLY in $Z(2^2)$. A (7, 4) minimum distance 3 cyclic BCH code in $Z(2^2)$ is generated by $G(u) = u^3 + 3u^2 + 2u + 3$. Similarly, a minimum distance 4 BCH code is generated by $G(u) = (u + 1) \cdot (u^3 + 3u^2 + 2u + 3)$.

Example 4. Let us continue Example 2. Based on the factorization of $u^7 - 1$ in $CZ(3^3)$ and the description of the BCH code, we see either $\{u^3 + (14 + 7j)u^2 + (13 + 7j)u + 26\}$ or $\{u^3 + (14 + 20j)u^2 + (13 + 20j)u + 26\}$ can be used to generate a (7, 4) minimum distance 3 BCH code in $CZ(3^3) \bmod (27, j^2 + 1)$. This is an example of a non-primitive BCH code.

For the RS codes in $Z(q, Q(\theta))$ having length $n = p^m - 1$, the generator POLY $G(u)$ has $\theta, \theta^2, \ldots, \theta^{d-1}$ as roots and, in general, has coefficients in $Z(q, Q(\theta))$. A similar statement holds for RS codes in $CZ(q, Q(\theta))$ having length $n = p^{2m} - 1$. The overall code parameters of the RS codes in $Z(q, Q(\theta))$ and $CZ(q, Q(\theta))$ remain the same as the code parameters of the RS codes in $GF(p^m)$ and $GF(p^{2m})$, respectively.

V. A DECODING ALGORITHM FOR BCH AND RS CODES IN FINITE INTEGER RINGS

Given the received POLY $r(u)$ in $GF(p)$, the decoding algorithm for the BCH codes in $GF(p)$ is a five-step procedure given by

Step 1. Compute the syndromes s_i, $i = 1, 2, \ldots, d-1$,

$$s_i = r(\theta^i) = e(\theta^i). \tag{21}$$

Use is made of the conjugate symmetry property $s_{jp} = (s_j)^p$ to simplify the computation.

Step 2. Determine the number of errors and the error locator POLY $\sigma(u)$ by solving the key equation.

Step 3. Determine the error locations as the roots of $\sigma(u)$.

Step 4. Determine the error values by solving a Vandermonde system of linear equations.

Step 5. Recover the $v(u)$ as

$$v(u) = r(u) - e(u). \tag{22}$$

Let us say that Steps 2 to 4 are performed by a decoder Đ based on the relationship in (21). Such a decoder operates in $GF(p)$ with s_i, $i = 1, 2, \ldots, d-1$ as the input and $e(u)$ as the output. This covers a very wide range of possibilities for the realization of Đ.

We now show that Đ can be used in an order-recursive manner to perform decoding in $Z(q)$. Given

$$R(u) = V(u) + E(u) \bmod p^a, \tag{23}$$

the first step of the decoding procedure is to compute the syndromes in $Z(q)$ given by

$$S_i = R(\theta^i),\ i = 1, 2, \ldots, d-1. \tag{24}$$

$R(u)$ is a code vector if and only if

$$S_i = 0,\ i = 1, 2, \ldots, d-1. \tag{25}$$

The computation of the syndrome can be simplified by noting the conjugate symmetry property in $Z(q)$ [1]. It is given by

$$S_{ip} = S_i\Big|_{\theta=\theta^p}, \; i = jp, jp^2, \ldots, jp^{l-1}, \tag{26}$$

where $jp^l \equiv 1 \bmod \{p^m - 1\}$. Once S_i is determined for $i = j$, (26) can be used to determine S_{ip}.

All the polynomials in (23) can be expressed in their p-adic form as:

$$\begin{aligned} R(u) &= \sum_{h=0}^{a-1} r_h(u)p^h, \\ V(u) &= \sum_{h=0}^{a-1} v_h(u)p^h, \\ E(u) &= \sum_{h=0}^{a-1} E_h(u)p^h, \end{aligned} \tag{27}$$

where the polynomials $r_h(u)$, $v_h(u)$, and $e_h(u)$ are polynomials in $GF(p)$. The approach is to make use of Đ over $GF(p)$ recursively $a - 1$ times to compute $E(u)$ using its p-adic form. Over $Z(q_{b+1})$, $q_{b+1} = p^{b+1}$,

$$\begin{aligned} V_b(u) &= V(u) \bmod p^{b+1}, \\ R_b(u) &= R(u) \bmod p^{b+1}, \\ &= V_b(u) + E_b(u) \bmod p^{b+1}, \\ E_b(u) &= E(u) \bmod p^{b+1}, \; b = 1, 2, \ldots, a - 1. \end{aligned} \tag{28}$$

Based on the syndromes in (24), the following partial syndromes can be computed:

$$\begin{aligned} S_{i,b} \equiv S_i \bmod p^{b+1} &= R(\theta^i) \bmod p^{b+1} \\ &= E(\theta^i) \bmod p^{b+1}, \; b = 0, 1, \ldots, a - 1; \; i = 1, 2, \ldots, d - 1. \end{aligned} \tag{29}$$

Let

$$s_{i,0} = S_{i,0} = e_0(\theta^i) \bmod p. \tag{30}$$

Given $s_{i,0}$ in (30), $e_0(u)$ can be found using Đ. The next step is to find $e_1(u)$.

Consider $S_{i,1}$.

$$S_{i,1} \equiv R_1(\theta^i) \bmod p^2$$
$$\equiv E_1(\theta^i) \bmod p^2 \equiv e_0(\theta^i) + p\, e_1(\theta^i) \bmod p^2. \quad (31)$$

Rearranging the terms,

$$s_{i,1} = \left\{ p^{-1}\left[S_{i,1} - e_0(\theta^i) \right] \right\} \bmod p \equiv e_1(\theta^i) \bmod p.$$

One must interpret the above expression carefully. The quantities in [.] must be computed mod $(Q_0(\theta) + p\, Q_1(\theta), p^2)$. Again, $s_{i,1}$ is used to find $e_1(u)$ using the same Đ. In general,

$$s_{i,b} = \left\{ p^{-b}\left[S_{i,b} - \sum_{j=0}^{b-1} e_j(\theta^i) p^j \right] \right\} \bmod p \equiv e_b(\theta^i) \bmod p. \quad (32)$$

Again, the quantities in [.] must be computed mod $(Q_0(\theta) + p\, Q_1(\theta) + \ldots + p^{b-1} Q_{b-1}(\theta), p^b)$. Given $s_{i,b}$, Đ can be employed to find $e_b(u)$, $b = 1, 2, \ldots, a - 1$ (or equivalently, $E(u)$). At the end of this recursive procedure, the decoded code POLY $V(u)$ is computed as

$$V(u) = R(u) - E(u) \bmod p^a. \quad (33)$$

In summary, the decoding algorithm in $Z(p^a)$ is given as follows:

Step 1. Given $R(u)$, compute the syndromes S_i, $i = 1, 2, \ldots, d - 1$.
Step 2. Let $s_{i,0} = S_{i,0} = e_0(\theta^i) \bmod p$.
Step 3. Employ Đ to determine $e_0(u)$.
Step 4. Let $b \leftarrow 1$.
Step 5. Compute $s_{i,b} \equiv e_b(\theta^i) \bmod p$ using (32).
Step 6. Employ Đ to determine $e_b(u)$.
Step 7. $b \leftarrow b + 1$. If $b < a$, go to Step 5.
Step 8. Decode $R(u)$ to $V(u)$ using (33).

A flowchart for this algorithm is shown in Figure 1.

Such a decoding algorithm has some attractive features. One, it computes the syndromes based on $R(u)$ (which can be a POLY of large degree) only once. Two, it leads to correct decoding as long as the Hamming weight of each of $e_b(u)$

is less than t even though the Hamming weight of $E(u)$ may exceed it, t being the error correcting capability of the code. The structure of the decoding algorithm in $CZ(q)$ is very similar to its structure in $Z(q)$. The decoder Đ now operates in $GF(p^2)$. Finally, the decoder for the RS codes in $Z(q)$ and $CZ(q)$ is similar to the decoder for the BCH codes with the exception that the conjugate property does not hold any longer. In this case, the received POLY has coefficients in $Z(q, Q(\theta))$ and the coefficients of the various POLYs in the p-adic expansion are elements of $GF(p^m)$.

VI. DISCUSSION AND CONCLUSIONS

In this paper, we have studied algebraic construction of cyclic codes, specifically the BCH and RS codes, in a finite integer ring in terms of their counterparts defined in the underlying finite field. This leads to closed form expressions for the generator polynomials of the various codes in terms of the primitive element of the rings. A complete step-by-step decoding algorithm is also described for all the codes. Interestingly, it employs the decoder of the underlying finite field code as a building block. An important aspect of the work is that only the finite field algebra is used to derive all the results in the finite integer rings.

REFERENCES

[1] H. Krishna Garg, "Cyclotomic Factorization in Finite Integer Rings with Applications to Digital Signal Processing," *IEEE Transactions on Circuits and Systems*, submitted for publication.

[2] H. Krishna, "A Mathematical Framework for Algorithm-Based Fault-Tolerant Computing Over a Ring of Integers," *Circuits, Systems, and Signal Processing*, Vol. 13, no. 5, pp. 625-653, 1995.

[3] S. Lin and D.J. Costello, Jr., *Error Control Coding: Fundamentals and Applications*, Prentice Hall, 1983.

[4] W.W. Peterson and E.J. Weldon, Jr., *Error-Correcting Codes*, The MIT Press, 1978.

[5] F.J. MacWilliams and N.J.A. Sloane, *The Theory of Error Correcting Codes*, North Holland Publishing Company, 1977.

[6] G.C. Clark and J.B. Cain, *Error-Correction Coding for Digital Communication*, Plenum Press, 1981.

[7] H. Krishna Garg, *Digital Signal Processing Algorithms*, CRC Press, 1998.

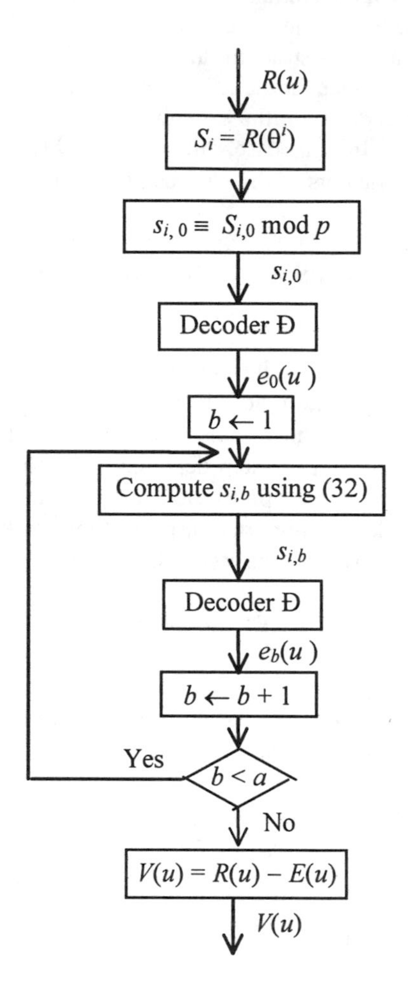

Figure 1. A flowchart for decoding BCH and RS codes over finite Integer Rings

Appendix A

Small Length Acyclic Convolution Algorithms

In this appendix, we present acyclic convolution algorithms for small degree polynomials. These algorithms are valid over all fields. In this appendix, we use the indeterminate input sequences $\{x_0, x_1, ...\}$ and $\{y_0, y_1, ...\}$ and the output sequence $\{z_0, z_1, ...\}$. MULTs are designated as $m_0, m_1, ...$.

Degree 0

$$z_0 = x_0 \cdot y_0.$$

Complexity: 1 MULT.

Degree 1

$$\left(\sum_{i=0}^{1} x_i u^i\right)\left(\sum_{i=0}^{1} y_i u^i\right) = \sum_{i=0}^{2} z_i u^i .$$

Let

$$\begin{aligned} m_0 &= x_0 \cdot y_0, \\ m_1 &= x_1 \cdot y_1, \\ m_2 &= (x_0 + x_1) \cdot (y_0 + y_1); \end{aligned}$$

then

$$\begin{aligned} z_0 &= m_0, \\ z_1 &= -m_0 - m_1 + m_2, \\ z_2 &= m_1. \end{aligned}$$

Complexity: 3 MULTs, 4 ADDs.

Degree 2

$$\left(\sum_{i=0}^{2} x_i u^i\right)\left(\sum_{i=0}^{2} y_i u^i\right) = \sum_{i=0}^{4} z_i u^i .$$

Let

$$m_0 = x_0 \cdot y_0,$$
$$m_1 = x_1 \cdot y_1,$$
$$m_2 = x_2 \cdot y_2,$$
$$m_3 = (x_0 + x_1) \cdot (y_0 + y_1),$$
$$m_4 = (x_1 + x_2) \cdot (y_1 + y_2),$$
$$m_5 = (x_0 + x_2) \cdot (y_0 + y_2);$$

then

$$z_0 = m_0,$$
$$z_1 = -m_0 - m_1 + m_3,$$
$$z_2 = -m_0 + m_1 - m_2 + m_5,$$
$$z_3 = -m_1 - m_2 + m_4,$$
$$z_4 = m_2.$$

Complexity: 6 MULTs, 13 ADDs.

Note: $Z(u) \bmod u^3$ requires 5 MULTs; delete m_5, z_3, z_4 in the above and replace m_5 with m_4 in z_2.

Degree 3

$$\left(\sum_{i=0}^{3} x_i u^i\right)\left(\sum_{i=0}^{3} y_i u^i\right) = \sum_{i=0}^{6} z_i u^i .$$

Let

$$m_0 = x_0 \cdot y_0,$$
$$m_1 = x_1 \cdot y_1,$$
$$m_2 = (x_0 + x_1) \cdot (y_0 + y_1),$$
$$m_3 = x_2 \cdot y_2,$$
$$m_4 = x_3 \cdot y_3,$$
$$m_5 = (x_2 + x_3) \cdot (y_2 + y_3),$$
$$m_6 = (x_0 + x_2) \cdot (y_0 + y_2),$$
$$m_7 = (x_1 + x_3) \cdot (y_1 + y_3),$$
$$m_8 = (x_0 + x_1 + x_2 + x_3) \cdot (y_0 + y_1 + y_2 + y_3);$$

then

$$z_0 = m_0,$$
$$z_1 = -m_0 - m_1 + m_2,$$

$$
\begin{aligned}
z_2 &= -m_0 + m_1 - m_3 + m_6, \\
z_3 &= m_0 + m_1 - m_2 + m_3 + m_4 - m_5 - m_6 - m_7 + m_8, \\
z_4 &= -m_1 + m_3 - m_4 + m_7, \\
z_5 &= -m_3 - m_4 + m_5, \\
z_6 &= m_4.
\end{aligned}
$$

Complexity: 9 MULTs, 25 ADDs.

Note: $Z(u) \bmod u^4$ requires 8 MULTs; delete m_8, z_4, z_5 and z_6 in the above; replace m_2 with m_4 in z_1, $m_1 - m_3 + m_6$ with $-m_2 + m_5$ in z_2; and delete $m_4 - m_5$ in z_3.

Can you derive other algorithms for degrees 3 and 4 that requires fewer MULTs at the expense of increase in ADDs and the scalars taking values other than 0, 1, and −1 (these values can now be fractions)?

Appendix B

Classification of Cyclotomic Polynomials

In this appendix, we present results on the classification of the cyclotomic polynomial $C_n(u)$, where n is the product of three distinct odd primes p, q, and r, that is, $n = p \cdot q \cdot r$. We present the classification of the coefficients for $C_n(u)$, $P(u)$ and the remainders $u^i \bmod C_n(u)$, $i = 1, 2, \ldots, \phi(n) - 1$. The cases considered cover most values of length n that are of interest from the point of view of cyclic convolution algorithms. The notation for the classification is as follows:

Type 0 ... $\{0, \pm 1\}$
Type 1 ... $\{0, \pm 1, 2\}$
Type 2 ... $\{0, \pm 1, -2\}$
Type 21 ... Types 1 and 2 but not including Type 32
Type 32 ... $\{0, \pm 1, \pm 2\}$
Type 43 ... $\{0, \pm 1, \pm 2, \pm 3\}$
Type 44 ... $\{0, \pm 1, \pm 2, \pm 3, \pm 4\}$
Type 45 ... $\{0, \pm 1, \pm 2, \pm 3, \pm 4, \pm 5\}$
Type 49 ... others

A total of 354 cases are considered for $p \cdot q \cdot r < 5{,}000$. Let p be the smallest of the three primes. There are 224 cases for $p = 3$, 86 cases for $p = 5$, 35 cases for $p = 7$, 8 cases for $p = 11$, and 1 case for $p = 13$. The entries in this table are generated by exhaustively tabulating all the polynomials. The smallest value of n which is not covered is $n = 1{,}155$, the product of four odd prime numbers.

p	q	r	$p \cdot q \cdot r$	$C(u)$	$P(u)$	Remainder Polynomials
3	5	7	105	2	0	21
3	5	11	165	1	0	21
3	5	13	195	2	0	21
3	5	17	255	1	0	21
3	5	19	285	2	0	21
3	5	23	345	1	0	32
3	5	29	435	0	0	0
3	5	31	465	0	0	0
3	5	37	555	2	0	32
3	5	41	615	1	0	21
3	5	43	645	2	0	21
3	5	47	705	1	0	21

3	5	53	795	1	0	32
3	5	59	885	0	0	0
3	5	61	915	0	0	0
3	5	67	1,005	2	0	32
3	5	71	1,065	1	0	21
3	5	73	1,095	2	0	21
3	5	79	1,185	2	0	21
3	5	83	1,245	1	0	32
3	5	89	1,335	0	0	0
3	5	97	1,455	2	0	32
3	5	101	1,515	1	0	21
3	5	103	1,545	2	0	21
3	5	107	1,605	1	0	21
3	5	109	1,635	2	0	21
3	5	113	1,695	1	0	32
3	5	127	1,905	2	0	32
3	5	131	1,965	1	0	21
3	5	137	2,055	1	0	21
3	5	139	2,085	2	0	21
3	5	149	2,235	0	0	0
3	5	151	2,265	0	0	0
3	5	157	2,355	2	0	32
3	5	163	2,445	2	0	21
3	5	167	2,505	1	0	21
3	5	173	2,595	1	0	32
3	5	179	2,685	0	0	0
3	5	181	2,715	0	0	0
3	5	191	2,865	1	0	21
3	5	193	2,895	2	0	21
3	5	197	2,955	1	0	21
3	5	199	2,985	2	0	21
3	5	211	3,165	0	0	0
3	5	223	3,345	2	0	21
3	5	227	3,405	1	0	21
3	5	229	3,435	2	0	21
3	5	233	3,495	1	0	32
3	5	239	3,585	0	0	0
3	5	241	3,615	0	0	0
3	5	251	3,765	1	0	21
3	5	257	3,855	1	0	21
3	5	263	3,945	1	0	32
3	5	269	4,035	0	0	0
3	5	271	4,065	0	0	0
3	5	277	4,155	2	0	32

3	5	281	4,215	1	0	21
3	5	283	4,245	2	0	21
3	5	293	4,395	1	0	32
3	5	307	4,605	2	0	32
3	5	311	4,665	1	0	21
3	5	313	4,695	2	0	21
3	5	317	4,755	1	0	21
3	7	11	231	0	0	32
3	7	13	273	1	0	32
3	7	17	357	2	0	32
3	7	19	399	0	0	32
3	7	23	483	0	0	32
3	7	29	609	2	0	32
3	7	31	651	0	0	32
3	7	37	777	1	0	21
3	7	41	861	0	0	21
3	7	43	903	0	0	21
3	7	47	987	2	0	21
3	7	53	1,113	0	0	32
3	7	59	1,239	2	0	32
3	7	61	1,281	0	0	32
3	7	67	1,407	1	0	32
3	7	71	1,491	2	0	32
3	7	73	1,533	0	0	32
3	7	79	1,659	1	0	21
3	7	83	1,743	0	0	21
3	7	89	1,869	2	0	21
3	7	97	2,037	1	0	32
3	7	101	2,121	2	0	32
3	7	103	2,163	0	0	32
3	7	107	2,247	0	0	32
3	7	109	2,289	1	0	32
3	7	113	2,373	2	0	32
3	7	127	2,667	0	0	21
3	7	131	2,751	2	0	21
3	7	137	2,877	0	0	32
3	7	139	2,919	1	0	32
3	7	149	3,129	0	0	32
3	7	151	3,171	1	0	32
3	7	157	3,297	0	0	32
3	7	163	3,423	1	0	21
3	7	167	3,507	0	0	21
3	7	173	3,633	2	0	21

3	7	179	3,759	0	0	32
3	7	181	3,801	1	0	32
3	7	191	4,011	0	0	32
3	7	193	4,053	1	0	32
3	7	197	4,137	2	0	32
3	7	199	4,179	0	0	32
3	7	211	4,431	0	0	21
3	7	223	4,683	1	0	32
3	7	227	4,767	2	0	32
3	7	229	4,809	0	0	32
3	7	233	4,893	0	0	32
3	11	13	429	2	0	21
3	11	17	561	1	32	32
3	11	19	627	2	0	32
3	11	23	759	1	0	21
3	11	29	957	1	0	32
3	11	31	1,023	2	0	32
3	11	37	1,221	2	0	32
3	11	41	1,353	0	0	21
3	11	43	1,419	2	0	21
3	11	47	1,551	1	0	32
3	11	53	1,749	1	0	32
3	11	59	1,947	0	0	32
3	11	61	2,013	2	0	32
3	11	67	2,211	0	0	0
3	11	71	2,343	1	0	32
3	11	73	2,409	0	0	32
3	11	79	2,607	2	0	32
3	11	83	2,739	1	0	32
3	11	89	2,937	1	0	21
3	11	97	3,201	2	0	32
3	11	101	3,333	1	0	32
3	11	103	3,399	2	0	32
3	11	107	3,531	0	0	21
3	11	109	3,597	2	0	21
3	11	113	3,729	1	0	32
3	11	127	4,191	2	0	32
3	11	131	4,323	0	0	0
3	11	137	4,521	1	0	32
3	11	139	4,587	0	0	32
3	11	149	4,917	1	0	32
3	11	151	4,983	2	0	32

3	13	17	663	0	0	21
3	13	19	741	0	32	32
3	13	23	897	2	0	32
3	13	29	1,131	2	0	21
3	13	31	1,209	0	0	32
3	13	37	1,443	0	0	32
3	13	41	1,599	0	0	32
3	13	43	1,677	1	0	32
3	13	47	1,833	0	0	32
3	13	53	2,067	2	0	32
3	13	59	2,301	0	0	32
3	13	61	2,379	0	0	32
3	13	67	2,613	1	0	21
3	13	71	2,769	2	0	32
3	13	73	2,847	1	0	32
3	13	79	3,081	0	0	21
3	13	83	3,237	2	0	32
3	13	89	3,471	2	0	21
3	13	97	3,783	0	0	32
3	13	101	3,939	2	0	32
3	13	103	4,017	1	0	32
3	13	107	4,173	2	0	21
3	13	109	4,251	0	0	32
3	13	113	4,407	2	0	32
3	13	127	4,953	1	0	21

3	17	19	969	2	0	21
3	17	23	1,173	0	32	32
3	17	29	1,479	0	32	32
3	17	31	1,581	0	0	32
3	17	37	1,887	2	0	32
3	17	41	2,091	1	0	32
3	17	43	2,193	2	0	32
3	17	47	2,397	1	0	32
3	17	53	2,703	1	0	32
3	17	59	3,009	1	0	32
3	17	61	3,111	2	0	32
3	17	67	3,417	2	0	21
3	17	71	3,621	0	0	32
3	17	73	3,723	0	0	32
3	17	79	4,029	0	0	32
3	17	83	4,233	1	0	32
3	17	89	4,539	0	0	21
3	17	97	4,947	2	0	32

3	19	23	1,311	0	0	32
3	19	29	1,653	0	0	32
3	19	31	1,767	1	32	43
3	19	37	2,109	1	0	32
3	19	41	2,337	0	0	32
3	19	43	2,451	1	0	21
3	19	47	2,679	2	0	32
3	19	53	3,021	2	0	32
3	19	59	3,363	0	0	32
3	19	61	3,477	1	0	32
3	19	67	3,819	1	0	32
3	19	71	4,047	2	0	21
3	19	73	4,161	0	0	32
3	19	79	4,503	1	0	32
3	19	83	4,731	2	0	32
3	23	29	2,001	1	32	43
3	23	31	2,139	2	0	32
3	23	37	2,553	2	0	32
3	23	41	2,829	0	32	32
3	23	43	2,967	2	0	32
3	23	47	3,243	1	0	21
3	23	53	3,657	1	0	32
3	23	59	4,071	0	0	32
3	23	61	4,209	0	0	32
3	23	67	4,623	2	0	32
3	23	71	4,899	1	0	32
3	29	31	2,697	2	0	21
3	29	37	3,219	2	0	32
3	29	41	3,567	1	32	32
3	29	43	3,741	2	0	21
3	29	47	4,089	1	32	43
3	29	53	4,611	1	32	32
3	31	37	3,441	0	32	32
3	31	41	3,813	0	0	21
3	31	43	3,999	1	32	32
3	31	47	4,371	0	0	32
3	31	53	4,929	2	0	32
3	37	41	4,551	0	0	21
3	37	43	4,773	1	32	32

5	7	11	385	43	0	43
5	7	13	455	32	0	32
5	7	17	595	43	32	44
5	7	19	665	43	32	44
5	7	23	805	32	0	32
5	7	29	1,015	1	0	32
5	7	31	1,085	32	0	32
5	7	37	1,295	1	0	32
5	7	41	1,435	2	0	32
5	7	43	1,505	32	0	43
5	7	47	1,645	32	0	32
5	7	53	1,855	43	0	43
5	7	59	2,065	43	0	43
5	7	61	2,135	32	0	32
5	7	67	2,345	32	0	32
5	7	71	2,485	0	0	32
5	7	73	2,555	32	0	32
5	7	79	2,765	32	0	32
5	7	83	2,905	32	0	32
5	7	89	3,115	43	0	43
5	7	97	3,395	32	0	43
5	7	101	3,535	32	0	32
5	7	103	3,605	2	0	32
5	7	107	3,745	1	0	32
5	7	109	3,815	32	0	32
5	7	113	3,955	32	0	43
5	7	127	4,445	32	0	32
5	7	131	4,585	32	0	32
5	7	137	4,795	32	0	32
5	7	139	4,865	0	0	32
5	11	13	715	32	0	43
5	11	17	935	43	32	45
5	11	19	1,045	2	0	43
5	11	23	1,265	2	0	32
5	11	29	1,595	2	0	43
5	11	31	1,705	1	0	43
5	11	37	2,035	32	0	43
5	11	41	2,255	1	0	32
5	11	43	2,365	1	0	43
5	11	47	2,585	32	0	32
5	11	53	2,915	0	0	32

5	11	59	3,245	2	0	32
5	11	61	3,355	1	0	43
5	11	67	3,685	2	0	43
5	11	71	3,905	1	0	43
5	11	73	4,015	32	0	43
5	11	79	4,345	2	0	32
5	11	83	4,565	1	0	32
5	11	89	4,895	2	0	32
5	13	17	1,105	32	32	43
5	13	19	1,235	43	0	43
5	13	23	1,495	43	43	49
5	13	29	1,885	2	0	43
5	13	31	2,015	32	32	43
5	13	37	2,405	32	0	32
5	13	41	2,665	43	32	44
5	13	43	2,795	43	32	43
5	13	47	3,055	32	0	43
5	13	53	3,445	32	0	43
5	13	59	3,835	32	0	32
5	13	61	3,965	2	0	43
5	13	67	4,355	1	0	32
5	13	71	4,615	32	0	32
5	13	73	4,745	43	0	43
5	17	19	1,615	32	32	21
5	17	23	1,955	43	32	32
5	17	29	2,465	43	32	32
5	17	31	2,635	2	0	32
5	17	37	3,145	43	32	32
5	17	41	3,485	2	0	32
5	17	43	3,655	43	0	32
5	17	47	3,995	32	32	32
5	17	53	4,505	32	0	32
5	19	23	2,185	2	32	43
5	19	29	2,755	1	32	45
5	19	31	2,945	2	0	43
5	19	37	3,515	32	0	32
5	19	41	3,895	2	0	43
5	19	43	4,085	2	32	44
5	19	47	4,465	32	32	43
5	23	29	3,335	1	0	43

5	23	31	3,565	32	32	43
5	23	37	4,255	32	32	44
5	23	41	4,715	1	32	43
5	23	43	4,945	43	43	45
5	29	31	4,495	2	0	43
7	11	13	1,001	2	32	43
7	11	17	1,309	43	32	44
7	11	19	1,463	43	32	45
7	11	23	1,771	32	0	43
7	11	29	2,233	32	43	45
7	11	31	2,387	32	32	44
7	11	37	2,849	44	32	45
7	11	41	3,157	32	0	43
7	11	43	3,311	1	0	43
7	11	47	3,619	43	32	45
7	11	53	4,081	44	32	45
7	11	59	4,543	44	0	44
7	11	61	4,696	43	0	43
7	13	17	1,547	32	0	43
7	13	19	1,729	43	32	49
7	13	23	2,093	32	32	43
7	13	29	2,639	2	0	44
7	13	31	2,821	43	0	43
7	13	37	3,367	32	0	32
7	13	41	3,731	1	0	43
7	13	43	3,913	2	0	44
7	13	47	4,277	43	0	44
7	13	53	4,823	32	0	44
7	17	19	2,261	43	32	44
7	17	23	2,737	44	32	45
7	17	29	3,451	43	32	45
7	17	31	3,689	44	43	49
7	17	37	4,403	43	32	44
7	17	41	4,879	43	43	49
7	19	23	3,059	43	32	45
7	19	29	3,857	43	0	44
7	19	31	4,123	43	32	49
7	19	37	4,921	32	32	44

7	23	29	4,669	43	0	45
7	23	31	4,991	44	32	49
11	13	17	2,431	44	32	49
11	13	19	2,717	45	32	49
11	13	23	3,289	43	32	49
11	13	29	4,147	44	32	45
11	13	31	4,433	43	32	49
11	17	19	3,553	44	32	49
11	17	23	4,301	32	32	49
11	19	23	4,807	43	32	49
13	17	19	4,199	44	43	49

Index